Dan Dascălu
Department of Electronics
and Telecommunications
Polytechnical Institute of Bucharest

Electronic processes in unipolar solid-state devices

Editura Academiei
București
România

Abacus Press
Tunbridge Wells, Kent
England
1977

First published in 1977
from the author's original English language manuscript
by
EDITURA ACADEMIEI
București, Calea Victoriei 125, Romania
and
ABACUS PRESS
Abacus House, Speldhurst Road, Tunbridge Wells, Kent, England

British Library Cataloguing in Publication Data
Dascălu, Dan

Electronic processes in unipolar solid-state devices.

Bibl.—Index.
ISBN 0—85626—025—8

1. Title   2. Hammel, John
621.3815′2   TK7871.95
Metal oxide semiconductors

Copyright © Abacus Press, 1977
All rights reserved. No part of this publication may be reproduced, stored in a retrieval system, or transmitted in any form or by any means, electronic, mechanical, photocopying, recording or otherwise, without the prior permission of the publishers.

Printed in Romania in December 1976

Electronic processes
in unipolar solid-state devices

Contents

Preface 9
List of Main Symbols 11
Introduction 19

Part 1
Basic Phenomena

1. Electronic Conduction in Semiconductor Crystals 23

1.1. Introduction 23
1.2. Crystal Structure 23
1.3. Energy Bands 29
1.4. Energy Distribution of Electrons in Semiconductors 34
1.5. Electron Transport Phenomena 42
1.6. Carrier Mobility 47
1.7. High-field Effects: I. Hot-carriers in Semiconductors 56
1.8. High-field Effects: II. Negative Mobility in Semiconductors 62
1.9. High-field Effects: III. Impact Ionization and Avalanche Breakdown 71
1.10. Carrier Transport in Non-homogeneous Semiconductors 75
1.11. Particle Continuity Equations 79
1.12. Basic Semiconductor Equations 81
References 82

2. Carrier Injection in Solids 86

2.1. Ohmic Contacts and Injecting Contacts 86
2.2. Potential Barrier at a Metal-semiconductor Contact 87
2.3. The Surface Layer at Thermal Equilibrium 93
2.4. Image Force Lowering of the Barrier (Schottky Effect) 98
2.5. Effect of Other Contact Phenomena upon the Barrier Height 101
2.6. Carrier Injection at a Metal-semiconductor Contact 108
2.7. Semiconductor Homojunctions 117
2.8. Heterojunctions 123
2.9. Practical Problems of Injecting Contacts 124
References 126
Problems 128

Part 2
Bulk- and Injection-controlled Devices

3. Schottky Diode 131

3.1. Introduction 131
3.2. Steady-state Characteristic 133

3.3.	Minority-carrier Injection	143
3.4.	Edge and Surface Effects. Various Practical Constructions	152
3.5.	Noise in Schottky barrier Diodes	162
3.6.	Schottky-barrier Capacitance	165
	References	166
	Problems	168

4. Space-charge-limited Current Solid-state Devices — 169

4.1.	Introduction	169
4.2.	Fundamental Law of Space-charge-limited Current	171
4.3.	Effect of Diffusion Current	173
4.4.	Insulator SCLC Diode	177
4.5.	Semiconductor SCLC Diode	184
4.6.	Transit-time Effects at Very High Frequencies	188
4.7.	Noise in SCLC Diodes	193
4.8.	SCLC Triodes (Transistors)	199
4.9.	Conclusion	201
	References	203
	Problems	206

5. Punch-through Semiconductor Diodes. Microwave BARITT Diodes — 208

5.1.	Space-charge-limited Current in a Punch-through p^+np^+ (or n^+pn^+) Structure	208
5.2.	High-frequency Negative Resistance Computed for an SCLC p^+np^+ Structure	214
5.3.	Small-signal Behaviour of Punch-through BARITT Diodes	216
5.4.	Noise Properties of Punch-through BARITT Diodes	224
5.5.	Large-signal Behaviour of the Punch-through BARITT Diode	227
5.6.	Current Transport in Punched-through Metal-semiconductor-metal (MSM) Structures	228
5.7.	Microwave Impedance of Punched-through MSM Diodes	236
5.8.	Construction and Utilization of BARITT Diodes	242
	References	244
	Problems	246

6. Impact-avalanche Transit-time (IMPATT) Diodes — 247

6.1.	Principle of Operation	247
6.2.	Static Analysis	251
6.3.	Small-signal High-frequency Impedance	258
6.4.	Large-signal Theory of a Read-type Structure	266
6.5.	Computer Analysis of Large-signal IMPATT Mode	277
6.6.	Intrinsic Noise of IMPATT Diodes	293
6.7.	Experimental	298
	References	300
	Problems	303

7. Transferred-electron Diodes for Microwave Amplification — 305

7.1.	Introduction	305
7.2.	Steady-state Characteristic and Low-frequency Negative Resistance	307
7.3.	Small-signal Space-charge Waves	314
7.4.	Subcritically Doped Diode	323
7.5.	Negative-resistance Diodes with Injection-limiting Cathodes	329
7.6.	Stabilization Mechanisms in Supercritically-doped Transferred-electron Amplifiers	330
7.7.	Space-charge Waves in NDM Semiconductor Layers	334
7.8.	Field Profile in a GaAs Layer Biased above Threshold	338
7.9.	GaAs Travelling Wave Amplifiers	341
7.10.	Noise in Transferred-electron Amplifiers (TEA)	346
	References	348
	Problems	352

8. Transferred-electron Devices used as Oscillators and as Logic Elements 353

- 8.1. Introduction 353
- 8.2. Formation of a High-field Domain 354
- 8.3. Stable High-field Domains 355
- 8.4. Effect of Bias Field upon Stable High-field Domains 362
- 8.5. Quasi-stationary (QS) Characteristic and Bistable Switching . . . 365
- 8.6. Equivalent Circuit for a Gunn Diode Operating in Domain Mode . . . 366
- 8.7. High-field Domain Nucleation by a Doping Notch 371
- 8.8. Effect of Electrode Boundary Conditions upon Behaviour of NDM Samples . 377
- 8.9. Stable Space-charge Layers Propagating in an NDM Semiconductor . . . 384
- 8.10. Gunn Diodes in Resonant Circuits. Domain Modes 387
- 8.11. Space-charge Controlled Modes 394
- 8.12. Various Phenomena in Practical Gunn Diodes 403
- 8.13. Functional Devices 413
- 8.14. Gunn-effect Digital Circuits 416
- References 420
- Problems 424

9. Junction-gate Field-effect Transistor 425

- 9.1. Physical Principles 425
- 9.2. Ideal Field-effect Transistor: a Generalized Steady-state Model 429
- 9.3. Elementary Theory of Junction-gate Field-effect Transistor 433
- 9.4. Effect of Mobility-field Dependence on Steady-state Characteristics of Junction gate FET 436
- 9.5. Numerical Analysis of the Junction-gate Field-effect Transistor 440
- 9.6. High-frequency Behaviour 447
- 9.7. Schottky-barrier Gate Field-effect Transistor 455
- 9.8. Noise of Junction-gate FET 459
- References 464
- Problems 466

Part 3

Surface-controlled Devices

10. Metal-insulator-semiconductor Capacitor 469

- 10.1. Introduction 469
- 10.2. Electric Charge in the Semiconductor Surface Layer 472
- 10.3. A Simplified Model of the Surface Layer 475
- 10.4. Ideal MIS (or MOS) Structure 477
- 10.5. Real MOS (MIS) Structures 485
- 10.6. The Silicon-dioxide Silicon System 490
- 10.7. Surface Quantization and Surface Mobility in Inversion Layers 497
- 10.8. Effect of Semiconductor Doping upon Characteristics of MOS Capacitor . 504
- 10.9. Effect of Interface States 517
- 10.10. Determination of Minority Carrier Lifetime 527
- 10.11. Surface Space-charge Region in Non-equilibrium Conditions 532
- 10.12. Gate-controlled Diode Structure 534
- 10.13. Surface Breakdown and Avalanche Injection into the Oxide 538
- 10.14. MOS (MIS) Varactor 543
- References 543
- Problems 547

11. Metal-oxide-semiconductor (MOS) Transistor 548

- 11.1. Introduction 548
- 11.2. Simplified Theory (Substrate Effect Neglected) 551

11.3. Theory of MOSFET Including Substrate Effect	553
11.4. MOSFET Theory Including Diffusion: Current Saturation and Operation in Weak Inversion	556
11.5. Analysis of MOSFET Saturation Region	563
11.6. Drain Voltage Limitations	569
11.7. Effect of Substrate Bias on Threshold Voltage	573
11.8. Dynamic Behaviour of a MOSFET	578
11.9. Noise Sources in MOSFET's	585
References	588
Problems	590
12. Charge-coupled Devices and other MIS Structures used for Memory Applications	**592**
12.1. MOS (MIS) Memories	592
12.2. Charge-coupled Semiconductor Devices	594
12.3. The Potential in Charge-coupled Devices	596
12.4. Charge Transfer Mechanisms in Charge-coupled Devices	600
12.5. Noise in Charge-transfer Devices	605
12.6. Various Charge-transfer Devices	607
12.7. 'Memory' Transistors	612
References	614
Subject Index	617

Preface

Two decades after the invention of the transistor the well-known book "Physics and technology of semiconductors devices" by A. S. Grove of University of California, Berkeley, described the basic principles of planar devices. The silicon planar technology (including the MOS process) maintains its dominant place in the semiconductor industry, although for certain devices used in power, microwave, display applications, other technologies are preferred.

The variety of solid-state devices makes difficult any attempt to cover this topic exhaustively. Not only the technology but also the operating principle, the basic physical phenomena and the device modelling are quite different. The last few years have indicated a tendency to consolidate the important achievements of previous decades, through refined technologies (for example ion implantation) and deeper understanding of device physics and its behaviour in circuit (microscopic analysis and device-circuit interaction studied through computer simulation). This prepares the ground for new discoveries and better exploitation of the solid-state in electronic applications through functional devices, far from any tube-like and transistor-like circuit concepts (for example, transferred electron functional devices and integrated circuits based on direct charge transfer).

The present book represents a monograph of basic electronic phenomena in unipolar (one-carrier) solid-state devices.

The theory of device operation is accompanied by certain results of computer simulation and by experimental data. Technological and circuit details are beyond the scope of this book. The background of this text is given by papers published in specialized journals. The references used are acknowledged at the end of each chapter and in figure captions. The bibliography is, of course, far from being exhaustive. A historical presentation was not attempted. Problems are suggested at the end of most chapters. The high-frequency behaviour of unipolar devices was the subject of our monograph entitled "Transit-time effects in unipolar solid-state devices" published in 1974 by the same firms. These two books are somewhat complementary, and the superposition between them is minor. The earlier one concentrates on the mathematical approach of carrier dynamics in rapidly varying electric fields.

We have first discussed the topic of unipolar devices in another book published in Romanian*. *However, the present book is far from being a translation or even a revised version of the above Romanian book. Both the content and the scope of this new monograph are considerably different. We have introduced Schottky diodes, BARITT and IMPATT devices, charge-coupled devices, etc. in our presentation, whereas the thin film transistor, for example, is no longer discussed.*

On the other hand, the aim of this book is to expound the variety of electronic phenomena encountered in unipolar devices, whereas the previous one underlines the similarity of behaviour and especially the space-charge and transit-time effects.

This book is intended for electronic engineers and physicists involved in solid-state device research and for postgraduate students as well.

The reader should have a general background of semiconductor electron devices and circuits (see for example the first chapters in "Electronic principles. Physics, models and circuits", by P. E. Gray and C. L. Searle, 1969, John Wiley).

The author would like to express his gratitude to some of his colleagues and students for reading part of the manuscript of this book. The presentation of the first chapter was somehow influenced by a book published in Romanian, "Solid-state electronics", by M. Drăgănescu (Bucharest, 1972). The author wishes to thank the Editors for their kind and efficient co-operation in the preparation of this book for publication.

January, 1976

---

* *"Unipolar injection in semiconductor electron devices", Ed. Academiei, Bucharest, 1972.*

List of Main Symbols *

Symbol	Explanation	First occurrence in (defined by)
A	area of injecting contact in a device with plane-parallel electrodes	
A^*	Richardson constant for thermionic emission at a metal-semiconductor contact	Equation (2.6.5)
A^{**}	effective (field-dependent) constant of thermionic emission	Figure (2.6.3)
a	lattice constant	Chapter 1
a	half thickness of semiconductor layer	Figure 7.7.1
B	magnetic induction	Equation (1.5.3)
C	differential capacitance per unit of electrode area, e.g. capacitance of the MOS structure	Equation (3.6.3)
C_a	avalanche region capacitance (per unit of electrode area)	Equation (6.3.22)
C_D	domain capacitance (per unit of electrode area)	Equation (8.6.10)
C_{D0}	capacitance of a triangular domain (per unit area)	Equation (8.6.3)
C_d	depletion layer capacitance	Figure 10.8.11
C_{FB}	flat-band capacitance (per unit area)	Equation (10.4.16)
C_{gc}	gate-to-channel capacitance	Figure 9.6.2
C_{gd}	gate-to-drain capacitance	Equation (11.8.4)
C_{gs}	gate-to-source capacitance	Equation (11.8.1)
C_j	junction depletion capacitance	Equation (10.8.11)
C_m	minimum low-frequency MOS capacitance (per unit area)	Figure 10.4.2
C_{min}	minimum high-frequency MOS capacitance (per unit area)	Equation (10.4.14)
C_{ox}	oxide capacitance (per unit area)	Equation (10.4.9)
C_s	semiconductor capacitance (per unit area)	Equation (10.4.6)
C_{st}	quasi-stationary capacitance (per unit area)	Equation (4.6.22)
C_{ss}	surface (interface)-state capacitance (per unit area)	Equation (10.9.1)
C_0	geometrical capacitance of the active (drift) space (per unit area)	Table 4.2.1
$\mathscr{D}$	displacement vector	Equation (1.12.1)
$D(D_n, D_p)$	diffusion coefficient (for electrons, holes)	Equations (1.10.2) and (1.10.3)
D'	derivative of diffusion coefficient	Figure 7.3.1
D_s	surface state density (per unit of surface area and unit-energy)	Figure 2.5.4
d	depletion layer thickness in a high-field domain	Equation (8.13.14)

* See also "Physical constants" on inside cover.

Symbol	Explanation	First occurrence in (defined by)
d_s	source depth	Figure 11.6.1
E	electron energy	Equation (1.3.1)
E_A	acceptor level	Figure 1.4.7
E_c	bottom of the conduction band	Equation (1.4.8)
E_D	donor level	Figure 1.4.7
E_F	Fermi level (electrochemical potential)	Equation (1.4.6)
E_g	band gap width	Section 1.3
E_i	intrinsic Fermi level	Equation (1.4.13)
E_t	trapping level	Figure 7.2.2
E_v	top of the valence band	Equation (1.4.9)
$\mathscr{E}(\mathscr{E})$	electric field vector (intensity)	Equation (1.5.3)
$\mathscr{E}_a$	anode field	Equation (7.2.6)
$\mathscr{E}_b$	bias field	Equation (8.11.1)
$\mathscr{E}_c$	critical field for mobility field-dependence	Equation (1.7.3), etc.
$\mathscr{E}_c$	cathode field	Equation (7.2.6)
$\mathscr{E}_{cr}$	critical field for avalanche breakdown	Equation (2.7.8)
$\mathscr{E}_D$	maximum domain field	Figure 8.3.1
$\mathscr{E}_M$	threshold field for negative mobility	Figure 1.7.1
$\mathscr{E}_m$	valley field	Figure 1.7.1
$\mathscr{E}_N$	outside field	Equation (8.2.3)
$\mathscr{E}_{N\min}$	minimum outside field	Figure 8.3.2
$\mathscr{E}_{ox}$	oxide field	Equation (10.4.7)
$\mathscr{E}_s$	surface field in semiconductor	Figure 2.5.4
$\mathscr{E}_\times$	crossover field	Figure 7.2.1
f	frequency	Equation (4.7.2)
$f(E)$	Fermi-Dirac distribution	Equation (1.4.6)
f_T	transit-time frequency	Equation (8.4.2)
G	conductance	Equation (7.4.1)
G_b	bulk generation rate	Equation (10.10.5)
G_i	differential (incremental) conductance	Equation (3.5.4)
G_N	intrinsic channel conductance	Equation (9.3.3)
$G_n(G_p)$	electron (hole) generation rate	Equations (1.11.3) and (1.11.4)
G_0	gate-controlled channel conductance	Equation (9.2.9)
G_s	surface generation-recombination rate	Equation (10.10.7)
g	channel conductance	Equation (10.4.19)
g_d	drain conductance	Equation (9.2.8)
g_{d0}	zero current drain conductance	Equation (9.2.9)
$g_{d\,\text{sat}}$	drain conductance in saturation	Equation (9.2.17)
g_m	transconductance	Equation (9.2.16)
$g_{m\,\text{sat}}$	transconductance in saturation	Equation (9.2.18)
H	channel thickness	Equation (9.3.2)
I	total current	Equation (3.2.24)
I_D	drain current	Figure 9.1.2
I_{DD}		Equation (8.6.5)
$I_{Ds\,\text{at}}$	drain current in saturation	Equation (9.2.12)
I_{DSS}		Equation (9.3.12)
$\overline{i_d^2}$	mean-square drain noise current	Equation (11.9.1)
I_{eq}	equivalent noise current	Equation (3.5.2)
I_G	gate current	Section 9.1
I_s	saturation current	Equation (3.5.2)
I_R	reverse current	Section 10.12
i	$\sqrt{-1}$ (imaginary unit)	
$\overline{i_n^2}$ $(\overline{\Delta i^2})$	mean-square value of the noise current	Equations (3.5.4), (4.7.1), etc.

Symbol	Explanation	First occurrence in (defined by)
$J(J_n, J_p)$	total current density (electron, hole current density)	Equations (1.5.20), (1.5.24)
$\bar{J}$	normalized current density	Equation (7.4.2)
J_M	threshold current (neutral sample)	Figure 7.4.5
J_{MS}	current density of electron flow from metal to semiconductor	Section 2.6
$J_{n\,\text{sat}}$	saturation current of an n-type Schottky diode	Equation (3.3.3)
J_s	saturation current density	Equations (3.2.16) and (6.4.6)
J_{SM}	current density of electron flow from semiconductor to metal	Equation (2.6.1)
$J_\times$	crossover current	Figure 7.2.1
j	particle current density	Equation (6.3.7)
j_c	control characteristic	Equation (7.2.8)
$j_i(t)$	induced current	Equation (6.1.4)
j_n	neutral characteristic	Equation (7.2.9)
$\mathbf{k}$	wave vector	Equation (1.2.5)
$\mathbf{k}_0$	wave vector for an extremum of an energy band	Equation (1.3.7)
k	complex wave number	Equation (7.3.8)
L	sample length, length of the active region, channel length	Equation (4.1.3), Figure 3.3.1, etc.
L_a	length of the avalanche region	Section 6.3
L_D	Debye length	Equation (2.3.14)
L_{DE}	extrinsic Debye length	Equation (2.3.18)
L_g	gate length	Figure 9.5.1
L_n, L_p	diffusion length for electrons, holes	Equation (1.12.4), Section 3.3
$\mathscr{L}_a$	avalanche inductance	Equation (6.3.29)
l	effective channel length	Equation (11.5.1)
l'	length of the drain (depleted) region	Equation (11.5.3)
M	noise measure	Equation (5.4.5)
M_c	number of equivalent minima in the conduction band	Equation (1.4.4)
M_n, M_p	electron, hole current multiplication ratio	Equations (1.9.6) and (1.9.9)
m^*	effective electron mass	Equation (1.3.9)
m_l^*, m_t^*	longitudinal, transversal effective mass	Section 1.3
m_n^*, m_h^*	electron, hole effective mass	
N	effective impurity concentration	Equation (2.6.13)
N_A	acceptor concentration	Section 1.4
N_c	effective density of states in the conduction band	Equation (1.4.8)
N_D	donor concentration	Section 1.4
N_I	total impurity concentration	Equation (1.6.13)
N_t	trap density	Equation (4.4.2)
N_v	effective density of states in the valence band	Equation (1.4.9)
$N_\square$	implant dose	Equation (10.8.4)
N_{ss}	surface (interface) state density	Section 10.9
$\bar{N}_{ss}$	interface state density per unit of surface and of energy	Equation (10.9.3)
n	free electron density	Equation (1.4.8)
n	ideality factor	Equation (3.2.23)
n_i	intrinsic carrier concentration	Equation (1.4.13)
n_m		Equation (3.2.13)

Symbol	Explanation	First occurrence in (defined by)
n_n	electron density in n-type semiconductor	
$n_p(n_{p0})$	electron density in p-type semiconductor (the same at thermal equilibrium)	
n_s	surface electron concentration	Equation (2.3.9)
n_t	density of trapped electrons	Equation (4.4.2)
n_0		Equation (3.2.12)
p	hole density	
$p_n(p_{n0})$	hole density in n-type semiconductor (the same at thermal equilibrium)	
p_p	hole density in p-type semiconductor	
p_s	surface hole concentration	Equation (2.3.10)
Q_B	bulk fixed space-charge (per unit area)	Equation (10.3.4)
Q_d	fixed charge in depletion layer	Equation (3.6.2)
Q_G	charge on metallic gate (per unit area)	Figure 10.1.2
Q_{mob}	mobile electron charge (per unit of electrode area)	Equation (4.1.3)
Q_n	mobile electron charge in semiconductor, per unit of gate area	Equation (10.2.9)
Q_p	mobile hole charge in semiconductor, per unit of gate area	Equation (10.2.10)
Q_s	total semiconductor charge, per unit of gate area	Figure 10.1.2
Q_{ss}	surface density of charge in interface states	Equations (2.5.9), (10.5.6)
R_c	contact resistance	Equation (2.6.16)
R_i	incremental resistance	Equation (4.6.18)
R_{i0}, R_0	low-field incremental resistance	Figure 7.4.4, Equation (8.6.4)
R_p	effective implantation depth	Equation (8.10.4)
R_{sc}	space charge resistance (for a unit area device)	Equation (5.3.13)
R_{ss}	surface (interface) state resistance	Section 10.9
r_d, r_s	drain, source series resistance	Figure 9.3.3
$S_i(f)$	spectral intensity of short-circuit noise current	Equation (4.7.2)
$S_v(f)$	spectral intensity of open-circuit noise voltage	Equation (4.7.4)
s_0	surface generation velocity	Equation (10.10.7)
T	temperature	Equation (1.4.6)
T_a	transit time through avalanche region	Equation (6.3.32)
T_{eq}	equivalent noise temperature	Equation (3.5.4)
T_L	carrier transit time	Equation (4.1.2)
T_n	electron temperature	Section 1.7
T_x	transit time across the distance x	Equation (4.5.5)
t	time	
t_{ep}	thickness of epitaxial layer	Figure 7.8.1
t_{ox}	oxide thickness	Equation (10.4.9)
U_n, U_p	electron, hole recombination rates	Equations (1.11.3) and (1.11.4)
u	electric potential normalized to kT/e	Equation (2.3.4)
u_F	potential Φ_F normalized to kT/e	Equation (2.3.4)
u_s	surface potential normalized to kT/e	Equation (2.3.8)
V	applied potential difference (bias voltage)	Equation (2.6.15)
$V(y)$	channel potential with respect to source	Equation (9.2.1)
$\bar{V}$	normalized bias voltage	Equation (7.4.2)
V_B	breakdown voltage	Figure 2.7.6, Equation (10.4.22)
V_{bi}	built-in voltage	Figure 2.7.1, Equation (2.7.4)
$V_{\mathcal{D}}$	domain excess voltage	Equation (8.2.6)
V_D	drain to source voltage	Figure 9.1.1

Symbol	Explanation	First occurrence in (defined by)
$V_{D(B)}$	drain voltage for avalanche breakdown	Figure 4.6.5
$V_{D\,\text{sat}}$	saturation drain voltage	Figure 9.11.7
V_{FB}	flat-band voltage	Figure 3.4.3
V_{FR}	field reversal point voltage drop	Section 11.5
V_G	gate to substrate voltage	Figure 3.4.2
V_G	gate to source voltage	Figure 9.1.1
V_G'	effective gate voltage	Equation (10.5.2)
V_J	bias applied to a pn junction	
V_M	threshold voltage $\mathcal{E}_M L$	Figure 7.4.6
V_m	minimum voltage sustaining dipole domains	Figure 8.10.1
V_{ox}	oxide voltage drop	Equation (10.4.1)
V_{PT}	punch-through voltage	Equation (5.1.9)
V_R	reverse bias	Equation (10.12.1)
V_{RT}	reach-through voltage	Figure 5.1.1
V_{sub}	substrate to source potential difference	Figure 11.7.1
V_T	threshold voltage	Figure 9.1.4
V_T	threshold voltage for strong inversion	Equation (10.4.13)
V_T'	threshold voltage for the onset of inversion	Equation (10.6.1)
V_{TFL}	trap-filled limit voltage	Equation (4.4.12)
V_{th}	threshold voltage for instabilities	Section 8.8
$\mathbf{v}$	average quantum velocity	Equation (1.5.3)
v	drift velocity	
$\bar{v}$	average thermal velocity	Equation (1.6.9)
v_c	collection velocity	Equation (3.2.11)
v_{cp}	hole collection velocity	Equation (3.3.10)
v_D	Debye diffusion velocity	Equation (3.2.21)
v_D	domain velocity	Equation (8.3.1)
v_d	effective diffusion velocity	Equation (3.2.15)
v_d	diffusion induced velocity	Equation (7.3.5)
v_M	maximum (peak) velocity	Figure 7.1.1
v_m	valley velocity	Figure 7.1.1
v_N	outside velocity	Equation (8.3.6)
$v_{N\min}$	minimum outside velocity	Figure 8.3.2
v_n, v_p	electron, hole drift velocity	Equations (1.11.1), (1.11.5)
v_p	phase velocity	Equation (7.3.19)
v_R	recombination velocity	Equation (3.3.8)
v_{sat}	high-field saturated drift velocity of charge carriers	Equation (1.7.3)
v_{th}	electron thermal velocity	Equation (1.11.6)
v_{0s}	injection velocity	Equation (5.3.2)
W	electron potential energy	Equations (1.3.1), (2.4.2)
W	gate width	Figure 9.2.1
x	distance (perpendicular to semiconductor surface, electrode)	
x_d	depletion layer thickness	Equation (2.3.22)
$x_{d\max}$	depletion layer thickness with strong inverted surface	Equation (10.3.3)
x_i	inversion layer depth	Equation (10.4.19)
x_m	position of potential minimum	Figures 2.4.1 and 5.6.2
Y	diode admittance (per unit area)	Equation (4.6.20)
y	distance along the surface channel	
Z	diode impedance, for unit area device	Equation (4.6.16)
$Z(E)$	distribution in energy of electron states	Section 1.4
$\bar{Z}$	impedance normalized to incremental resistance	

Symbol	Explanation	First occurrence in (defined by)
$\alpha(\alpha_n, \alpha_p)$	ionization coefficient (for electrons, holes)	Equation (1.9.1)
α_1	field derivative of ionization coefficient	Equation (6.3.15)
β		Equation (7.3.12)
Γ	Euler function	Equation (1.5.19)
Γ	injection parameter	Figure 5.3.4 and Equation (5.7.4)
γ	minority carrier injection ratio	Equation (3.3.4)
Δf	frequency band	Equation (3.5.1)
ΔR_p	standard deviation for a Gaussian implanted profile	Equation (8.10.4)
$(\Delta \Phi)_{es}$	barrier lowering due to electrostatic screening	Equation (2.5.1)
$(\Delta \Phi)_i$	barrier lowering due to interfacial layer	Equation (2.5.6) and Figure 2.5.2
$(\Delta \Phi)_s$	barrier lowering due to Schottky effect	Equation (2.4.4)
δ	thickness of interfacial insulating layer at a metal semiconductor contact	Figure 2.2.2
δ	trapping factor	Equation (4.4.7)
ε	semiconductor permittivity	Equation (1.12.3)
ε_i	insulator permittivity	Figure 2.5.3
ε_{opt}	optical dielectric constant	Section 2.4
ζ_n, ζ_p	electron, hole quasi-Fermi level	Figures 2.6.1 and 2.7.3
θ	transit angle	Equation (4.6.17)
θ_r	transit-time normalized to dielectric relaxation time	Equation (4.5.10)
θ_r'		Equation (5.1.8)
χ	electron affinity, activation energy	Figure 2.2.1 and Equation (2.2.1)
λ	mean free path	Equation (1.6.5)
λ	wavelength	Equation (7.3.23)
λ_D	modified Debye length	Equation (7.3.13)
μ_d	differential mobility	Equation (7.1.3)
μ_s	surface mobility	Equation (10.7.3)
ν	normalized doping-length product	Equation (7.4.3)
ξ	channel potential normalized to kT/e	Equation (10.11.4)
σ	conductivity	Equation (1.5.15)
σ_c	collision cross-section	Equation (1.6.4)
σ_n, σ_p	electron, hole capture cross-section for recombination centers	Section 1.11
σ_t	capture cross-section for empty traps	Equation (4.4.15)
$\sigma_1(0)$	differential conductivity of the source (cathode) plane	Equation (5.3.3)
$\tau, \tau(\mathbf{k})$	momentum relaxation time	Equation (1.5.1)
$\tau = \tau(E)$	transmission probability	Equation (2.6.10)
$\tau(\tau_n, \tau_p)$	minority carrier lifetime (electron, hole)	Sections 1.11, 1.12, 10.10
$\tilde{\tau}$	equivalent momentum relaxation time	Equation (1.5.22)
τ_d	differential relaxation time	Equation (7.1.3)
τ_e	energy relaxation time	Section 1.7
τ_f	lifetime of free carriers	Equation (4.4.15)
τ_g	minority carrier generation lifetime	Equation (10.10.5)
τ_n	equivalent relaxation time	Equation (1.10.6)
τ_s	minority carrier storage time	Equation (3.3.9)
τ_{ss}	surface-state trapping time constant	Equation (10.9.1)
τ_1	intrinsic response time	Equation (6.6.5)
φ	electric potential	Equation (2.3.3)

Symbol	Explanation	First occurrence in (defined by)
$\phi_B(\phi_{B_n}, \phi_{B_p})$	metal-semiconductor potential barrier (electron, holes), also ϕ_n, ϕ_p	Figure 2.3.1
ϕ_F		Equation (2.3.4), Figure 2.3.1
ϕ_g	potential difference E_g/e	Figure 2.2.2
ϕ_M	extraction potential from metal	Figure 2.2.1
ϕ_{MS}	metal-semiconductor work function difference	Equation (2.2.1) and Figure 2.6.1
ϕ_s	surface potential (relative to the neutral bulk)	Figure 2.2.1
ϕ_{s0}	surface potential at thermal equilibrium (no drain current)	Equation (10.11.1)
ϕ_0		Figures 2.2.2, 2.2.4, 2.5.4
ψ	wave function	Equations (1.3.1) and (3.2.8)
ψ_k	Bloch function	Equation (1.3.3)
ω	angular frequency of oscillation	
ω_a	avalanche frequency	Equation (6.3.22)
ω_c	cutoff frequency	Equation (6.3.31)
$\omega_D, \omega_r, \omega_r'$	differential dielectric relaxation frequency	Equations (7.3.3), (4.5.6), (5.1.8)

◇

Boldface symbols stand for vector quantities.
Boldface underlined symbols denote phasors.
Subscript 0 (zero) indicates a zero-order of magnitude (or direct current, steady-state) quantity.
Subscript 1 denotes a first-order (or small-signal) quantity.
$\mathcal{Re}$ and $\mathcal{Jm}$ denote the real and, respectively, imaginary part.
d.c. means direct current, a.c. — alternating current, r.f. — radio frequency, c.w. — continuous wave, etc.

Introduction

Most of today's solid-state devices are charge-controlled devices: the electric current passing through the external electrodes is determined by an internal transport of charge carriers. These devices can be divided into two broad categories: unipolar (one-carrier) devices and bipolar (two-carrier) devices.

We assume that the reader is familiar with the junction-type bipolar devices: the *pn* junction diode and the *pnp* or *npn* transistors are typical examples. This is the only kind of bipolar device to be considered below. In a simplified analysis the semiconductor is divided into high-resistivity space-charge regions with negligible mobile carriers which surround the metallurgical junctions and also neutral regions. The minority carriers diffuse into neutral regions where they disappear gradually by recombination. The injection of these carriers into the neutral region is controlled by the potential barrier developed across the space-charge region at the *pn* junction. Note the fact that the major part of the external voltage bias drops on the high-resistivity space-charge regions and therefore the voltage controls the minority carrier injection. However, the magnitude of the current is determined by diffusion and recombination of *excess* minority carriers into the neutral regions. This excess concentration is a deviation from thermal equilibrium. A. S. Grove pointed out that the operation of these devices is governed by their tendency to return to equilibrium: the typical process is carrier recombination described by parameters as minority carrier lifetime, diffusion length and surface recombination velocity. Two *pn* junctions can be coupled electrically by making the distance between them sufficiently small, such that the minority carrier flow injected by the first junction reaches the second one with a negligible recombination loss: this leads to the transistor effect.

The operation of unipolar devices is quite different. Only one type of charge carrier is present and recombination is practically impossible. The active device regions are, as a rule, far from neutrality, the electric field is quite high and diffusion plays a minor rôle: the current is carried by drift. The magnitude of this current can be influenced in several ways: by controlling the potential barrier at the electrode which injects charge carriers, or the electric field at the semiconductor surface; bulk processes are also important: the field distortion by the mobile space charge in transit between electrodes, the dielectric relaxation mechanism. Although the operation mechanisms of various unipolar devices are quite diverse, important similarities exist and this was the basic idea of writing this book.

Two introductory chapters are devoted to the basic processes: electronic conduction in semiconductors and means of injecting free charge carriers in solids.

This is an elementary treatment and represents a background for the device physics which is presented, in general, at a phenomenological level in Chapters 3—12.

The second part of this book (Chapters 3—9) describes bulk and injection-controlled devices. The Schottky diode (Chapter 3) is a typical injection-controlled (or barrier-controlled) device. The magnitude of the majority carriers current is determined by the potential barrier occurring at the metal-semiconductor contact.

The space-charge-limited current (SCLC) diode is a bulk-controlled device. The injecting contact is ohmic, the injecting mobile charge sets up an electric field which opposes to injection and limits the magnitude of the current. We note the important fact that a semiconductor resistor becomes a symmetrical SCLC diode if the carrier transit time from one electrode to another is made very short compared with the dielectric relaxation time. Transit-time effects are essential in high-frequency operation.

Both barrier-injection and space-charge effects are important in BARITT diodes: the transit time effects determine power generation at microwave frequencies (Chapter 5).

The IMPATT diode (Chapter 6) is not a true unipolar device because both electrons and holes are involved in device operation. However, a single-drift IMPATT diode is somewhat similar to a BARITT diode, except that the injecting barrier-type contact is replaced by a quite wide avalanche region. The TRAPATT mode is not presented in this book.

Transferred-electron devices (TED) (or Gunn devices) are also transit-time structures utilized at microwave frequencies. Chapter 7 presents small-signal operation of TED's used for microwave amplification. Chapter 8 describes domain-type operation (Gunn oscillators, logic and functional devices). The high-field domain is a stable propagating dipole layer formed due to the negative-mobility (transferred-electron effect). The cyclic nucleation and propagation of these domains lead to Gunn type oscillations of the external current. In amplifier TED's, the domains are prevented to form and the high-frequency negative resistance is the result of combined transit-time and negative-mobility phenomena.

The junction-gate field-effect transistor is simply a semiconductor resistor whose cross-sectional area is controlled by the extension of the space-charge region of a *pn* junction which acts as a gate (Chapter 9). This unipolar transistor has the advantage of low temperature and radiation effects and high operating frequencies (when a Schottky contact replaces the *pn* junction gate).

The third part of this book describes the basic phenomena in surface-controlled devices of metal-oxide-semiconductor (MOS) type. Chapter 10 presents the physics of the basic structure — the MOS capacitor, with particular emphasis to the $Si-SiO_2$ system. The usefulness of the MOS capacitor as an experimental tool for studying semiconductor, oxide (insulator) and interface properties is widely discussed.

Chapter 11 presents the operation of MOS-type transistor (MOSFET), whereas Chapter 12 describes the principle of charge-coupled devices.

Part 1
Basic Phenomena

1
Electronic Conduction in Semiconductor Crystals

1.1 Introduction

The analysis of unipolar solid-state devices is preceded by a summary of semiconductor physics. This summary is specifically oriented towards the most important processes occurring in these devices. For a more complete understanding of semiconductor physics and solid-state physics in general the reader will consult basic references such as [1] — [8].

This chapter starts with a brief discussion on crystal structure. The basic concepts of energy bands of crystalline solids and electron statistics are reviewed. The Boltzmann transport equation is presented and the simplest results on electronic conduction are derived on this basis. Carrier mobility is discussed at a phenomenological level. Three sections of this chapter are devoted to high-field effects: hot carrier effects, electron transfer in multivalley semiconductors and impact ionization.

The basic equations of semiconductor device analysis are presented, with particular emphasis on carrier transport equations in non-homogeneous semiconductors subjected to intense electric fields.

Magnetic and optical effects are not considered. The generation-recombination (trapping) phenomena, less important for unipolar devices, are only briefly mentioned. Surface effects and interface phenomena are discussed in other chapters of this book.

1.2 Crystal Structure

A crystalline solid is characterized by the periodicity of the spatial arrangement of the lattice atoms. The entire crystal can be obtained by translating a cell. A *primitive* cell has a minimum volume. The crystalline structure is obtained by attaching the same primitive cell with atoms (a base) to a lattice of geometrical points, periodically arranged in space. Consider for example the two-dimensional lattice of Figure 1.2.1. The shaded area is a primitive cell. $\mathbf{a}$ and $\mathbf{b}$ are primitive vectors. All lattice sites can be obtained by a translation with $n_1\mathbf{a} + n_2\mathbf{b}$ where n_1 and n_2 are integers. The point defined by $\mathbf{r} = \mathbf{r}' + n_1\mathbf{a} + n_2\mathbf{b}$ has exactly the same environment as the point $\mathbf{r}'$.

The primitive vectors for a three-dimensional lattice are denoted by $\mathbf{a}, \mathbf{b}, \mathbf{c}$. The parallelepiped constructed on these vectors is a primitive cell *. Its volume is $|\mathbf{a} \cdot \mathbf{b} \times \mathbf{c}|$.

---

* A primitive cell can always be constructed as follows. We draw the straight segments connecting one node and all its neighbours in the lattice. The smallest continuous space determined by the bisector planes of the above segments is the *Wigner-Seitz cell*. The whole lattice can be constructed by using this cell as a building block. It contains only one node and is a primitive cell.

Miller indices are used to indicate families of parallel planes or parallel directions in a lattice. Usually we consider crystalline planes and directions, which contain lattice atoms. The primitive vectors **a**, **b**, **c** or any other basic vectors of an elementary cell may be taken as reference. The Miller indices of a plane are denoted by $(h\,k\,l)$ where h, k, l are integers (minus sign is indicated

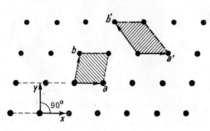

Figure 1.2.1. A two-dimensional lattice. Shown are the lattice sites; in the simplest case one atom is placed in each site. The vectors **a**, **b** as well as **a**′, **b**′ are primitive. Both shaded parallelograms are primitive cells (their area is the same). The lattice reproduces itself by a translation with the fundamental vector $n_1\mathbf{a} + n_2\mathbf{b} = \mathbf{T}$, where n_1 and n_2 are integers. The primitive vectors (here **a** and **b**) are often used as the versors of crystallographic axes. The parallelogram constructed on these versors has an atom in each corner. However, other axes may be used if this is more convenient. Our figure shows, for example, rectangular axes (**x**, **y**) which allow us to define the position of each atom in a cell.

by a bar above the number). This plane, by definition, intercepts the **a**, **b**, **c** axes in three points determined respectively by $\dfrac{a}{h}, \dfrac{b}{k}, \dfrac{c}{l}$. Figure 1.2.2 illustrates this definition. Figure 1.2.3 shows several planes for a cubic lattice.

The Miller indices of a direction in crystal are denoted by $[h\,k\,l]$. h, k and l are the smallest integers proportional to the components of a vector colinear with this direction. In a cubic lattice the direction $[h\,k\,l]$ is perpendicular to the plane $(h\,k\,l)$, but this is not true in general for other lattices.

There exist fourteen types of spatial lattices (Bravais lattices). Three of them are shown in Figure 1.2.4: simple cubic, body-centred cubic and face-centred cubic. Also shown are the corresponding primitive cells. Note the fact that the primitive cells contain only one node per cell because each node belongs to eight cells.

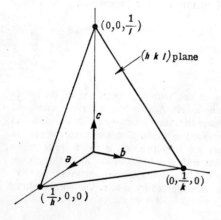

Figure 1.2.2. The plane shown intercepts the **a**, **b**, **c** axes in three points having the coordinates $\left(\dfrac{1}{h}, 0, 0\right)$, $\left(0, \dfrac{1}{k}, 0\right)$ and $\left(0, 0, \dfrac{1}{l}\right)$, respectively, where h, k, l are integers. $(h\,k\,l)$ are Miller indices of the family of planes parallel to that indicated in our figure.

The most important semiconductors have a crystalline structure corresponding to the face-centred cubic lattice, with a base of two atoms associated to each node of the spatial lattice. The sites of atoms in an elementary cell are shown in Figure 1.2.5. There are four internal atoms and an average of four boundary atoms per cell (each atom on a cube face belongs to two cells, etc.)

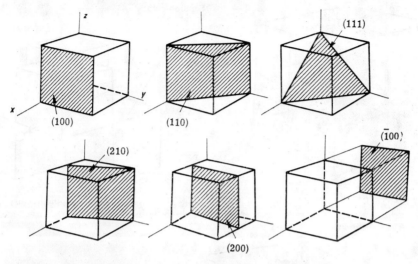

Figure 1.2.3. Shown are several important planes for a cubic lattice and their Miller indices ($hk\,l$). Planes (200) and (100) are parallel but not equivalent from a crystallographic point of view (they do not contain the same arrangement of atoms). However, all faces of the cube (100), (010), etc. are equivalent by crystallographic symmetry and are denoted by {100}. The full set of equivalent crystalline directions, such as [100], [010], etc. in a cubic lattice are denoted by <100>.

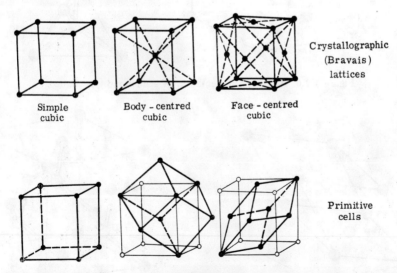

Figure 1.2.4. The figure shows three examples of three-dimensional Bravais lattices (these spatial lattices are defined by symmetries with respect to a plane or a point, see for example [1]). Also indicated are the primitive cells of these lattices. The body-centred cubic lattice for example, contains an average of two atoms per crystallographic cell (one of them is internal, each of the other eight atoms is shared between eight cells). The primitive cell contains only one atom.

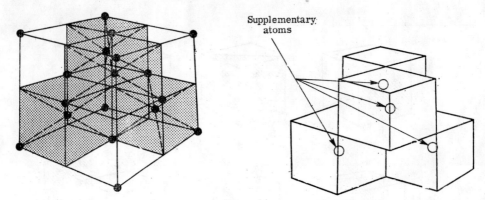

Figure 1.2.5. Crystallographic cell of a diamond-type lattice, with four supplementary atoms inside a face-centred cubic lattice.

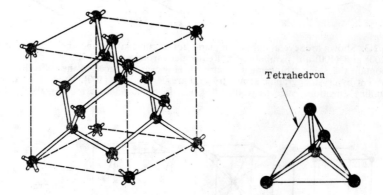

Figure 1.2.6. Another representation of a diamond-type lattice.

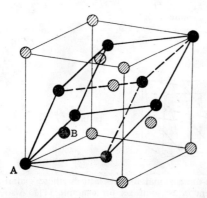

Figure 1.2.7. The primitive cell of a diamond-type lattice, containing two atoms, A and B.

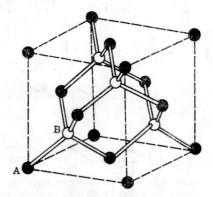

Figure 1.2.8. Zincblende-type lattice structure, with two different atoms in a primitive cell.

1. Electronic Conduction in Semiconductors

therefore eight atoms per cell. Each atom has four equidistant nearest neighbours which lie at the corners of a tetrahedron. The tetrahedral arrangement inside a cell is indicated in Figure 1.2.6 by bars connecting these neighbours. The primitive cell is indicated in Figure 1.2.7 and contains two atoms, for example A (external) and B (internal). There is no possibility to choose the primitive cell such that to contain one atom only.

The two atoms, A and B, may be identical or not. In the first case one obtains the diamond-type structure. Silicon and germanium crystals belong to this type.

If the atoms A and B are different, the crystal structure is a zincblende-type lattice structure. Each atom is surrounded by four atoms of opposite type, as shown conventionally in Figure 1.2.8. GaAs InP, CdS and other semiconductors crystallize in this structure. Table 1.2.1 shows the value of the lattice constant, a (referring to the basic cubic cell) for several materials of interest.

Table 1.2.1

Lattice Constant at 300 K [1], [8]

	C	Si	Ge	CdS	GaAs	InP	SiC
a (Å)	3.56	5.43	5.65	5.82	5.65	5.87	4.36
Structure	diamond-type			zincblende-type			

The diamond-type lattice is less compact than other lattices. This is related to the type of chemical bound. Silicon and germanium are covalent crystals. Each atom shares its four valence electrons with their nearest neighbours. A covalent bond between two atoms consists of a pair of electrons (with antiparallel spins), one from each atom. Each atom has therefore four covalent bonds and will have four neighbours placed in the corners of a tetrahedron, whereas the atom considered is placed in the centre. This is just the diamond-type lattice.

In a zincblende-type crystalline structure the atoms are partly ionized: the chemical bond is a combination of covalent and ionic. There is no clear boundary between covalent crystals and ionic crystals. Table 1.2.2 shows a semi-empirical degree of ionization of the chemical bond for several crystals [1].

Table 1.2.2

Degree of Ionization of the Chemical Bond in Binary Crystals [1]

Crystal	Si, Ge	SiC	GaAs	InP	CdS	NaCl
Ionization degree	0	0.18	0.32	0.44	0.69	0.94

GaAs is covalent to a large extent whereas NaCl is practically ionic (and crystallizes in a centred-face cubic lattice).

The reciprocal lattice is an important concept in crystal physics. The lattice associated to the actual crystal structure is called the direct lattice. If the versors of the direct lattice coordinate system associated with a primitive cell are denoted by $\mathbf{a}$, $\mathbf{b}$ and $\mathbf{c}$, then the reciprocal lattice is defined by the versors $\mathbf{a}^*$, $\mathbf{b}^*$ and $\mathbf{c}^*$:

$$\mathbf{a}^* = 2\pi \frac{\mathbf{b} \times \mathbf{c}}{\mathbf{a} \cdot \mathbf{b} \times \mathbf{c}}, \quad \mathbf{b}^* = 2\pi \frac{\mathbf{c} \times \mathbf{a}}{\mathbf{a} \cdot \mathbf{b} \times \mathbf{c}}, \quad \mathbf{c}^* = 2\pi \frac{\mathbf{a} \times \mathbf{b}}{\mathbf{a} \cdot \mathbf{b} \times \mathbf{c}}, \quad (1.2.1)$$

where $\mathbf{a} \cdot \mathbf{b} \times \mathbf{c} = \mathbf{b} \cdot \mathbf{c} \times \mathbf{a} = \mathbf{c} \cdot \mathbf{a} \times \mathbf{b} = \Omega$ is the volume of the basic cell in the direct lattice. If

$$\mathbf{T} = u\mathbf{a} + v\mathbf{b} + w\mathbf{c} \quad (u, v, w = \text{integers}) \quad (1.2.2)$$

is a vector of a node in the direct lattice, and

$$\mathbf{U} = m\mathbf{a}^* + n\mathbf{b}^* + l\mathbf{c}^* \quad (1.2.3)$$

is a similar arbitrary vector for the reciprocal lattice, then [1], [2]

$$\frac{\mathbf{T}\cdot\mathbf{U}}{2\pi} = \text{an integer, and } \exp(i\,\mathbf{T}\cdot\mathbf{U}) = 1. \tag{1.2.4}$$

The Brillouin zone is a Wigner-Seitz cell (see above) of the reciprocal lattice. For illustration we consider first the two-dimensional reciprocal lattice shown in Figure 1.2.9. The Wigner-Seitz cell is the central zone (square), which is the smallest area around the origin limited by the bisector lines of the vectors of the reciprocal lattice, **U**. The so-called first Brillouin zone coincides with this cell. The following zones are also defined with respect to the bisector lines (or bisector planes in the three-dimensional case). Brillouin zones were first introduced in the band theory of solids. However, they are also useful for diffraction studies. For example, the wave vectors **k** satisfying the condition for Bragg reflection, are those vectors in the space of reciprocal lattice which start in the origin and end on a bisector line in Figure 1.2.9. The diffraction condition* for the wave vector **k** is [1].

$$2\mathbf{k}\cdot\mathbf{U} = \mathbf{U}^2. \tag{1.2.5}$$

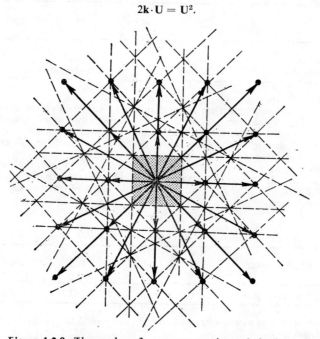

Figure 1.2.9. The nodes of a square reciprocal lattice are indicated by points, the lattice vectors by arrows and the bisector lines of these vectors are shown dotted. The central square (shaded) is the Wigner-Seitz cell of the reciprocal lattice or the first Brillouin zone.

It may be also written $\mathbf{k}\cdot\left(\frac{1}{2}\mathbf{U}\right) = \left(\frac{1}{2}\mathbf{U}\right)^2$, has the geometrical interpretation of **Figure 1.2.10** and proves the above statement.

---

* This is equivalent to the well-known Bragg condition $2d \sin \theta = n\lambda$ where d is the distance between two crystallographic planes, θ is the angle between the wave direction and these planes, λ is the wavelength and n an integer.

Consider now a three-dimensional lattice. The bisector planes of the reciprocal lattice vectors are parallel to the crystallographic planes. A wave will be diffracted by the crystal lattice if, and only if, the wave vector **k** in the reciprocal lattice space represents a point of the above bisector

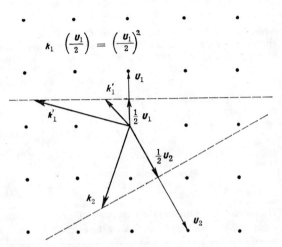

Figure 1.2.10. This figure shows a square reciprocal lattice and bisector lines of arbitrary lattice vectors. Consider the bisector line corresponding to the lattice vector U_1. Any vector k_1 (or k_1' satisfies the diffraction condition $k_1 \cdot \left(\dfrac{1}{2} U_1\right) = \left(\dfrac{1}{2} U_1\right)^2$.

planes. The images obtained by X-ray diffraction should yield a map of the reciprocal lattice of the crystal studied.

Figure 1.2.4 shows the face-centred cubic lattice and its primitive cell, defined by versors **a′**, **b′**, **c′**. The reciprocal lattice constructed according to equation (1.2.1), with **a′**, **b′** and **c′** as versors of the direct lattice, yields a volume centred cubic lattice. The corresponding Wigner-Seitz cell (the first Brillouin zone) is shown in Figure 1.2.11.

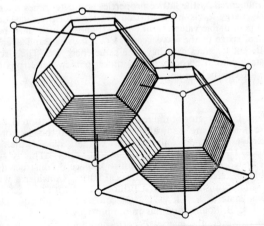

Figure 1.2.11. Wigner-Seitz cells for a face-centred cubic crystalline structure (the reciprocal lattice is volume-centered cubic).

1.3 Energy Bands

The electron energy in an atom is quantized. In a crystalline solid each atom has its own electronic shell but the outer electrons, situated on higher energy levels, are less bound and, in this sense, belong to the entire crystal. In other words the interaction between lattice atoms determines the splitting of the higher discrete levels of an isolated atom in energy bands of the entire solid, bands containing discrete levels (one per atom) which form a quasi-continuum.

The band structure can be obtained approximately by solving the Schrödinger equation for one electron

$$\frac{\hbar^2}{2m}\nabla^2\psi + [E - W_p(\mathbf{r})]\psi = 0, \tag{1.3.1}$$

where ψ is the electron wave function, E is the electron total energy and $W_p(\mathbf{r})$ is the potential energy in the electric field of lattice atoms (which are ionized if the less-bound electrons are considered separately). $W_p(\mathbf{r})$ has the periodicity of the crystalline lattice

$$W_p(\mathbf{r} + \mathbf{T}) = W_p(\mathbf{r}), \tag{1.3.2}$$

where $\mathbf{T}$ is an arbitrary vector (1.2.2) of this lattice. The solutions of equation (1.3.1) are of the form

$$\psi_\mathbf{k}(\mathbf{r}) = u_\mathbf{k}(\mathbf{r}) \exp i\mathbf{k}\mathbf{r} = \text{Bloch function} \tag{1.3.3}$$

and describe plane waves (a factor $\exp - i\omega t$ is omitted in equations (1.3.3)) 'spatially modulated' by the periodic function $u_\mathbf{k}(\mathbf{r})$

$$u_\mathbf{k}(\mathbf{r} + \mathbf{T}) = u_\mathbf{k}(\mathbf{r}). \tag{1.3.4}$$

The probability of finding the $\mathbf{k}$ electron in the volume element $d\mathbf{r}$ is $|\psi_\mathbf{k}|^2 d\mathbf{r}$. The electrons are individualized by their wave vector $\mathbf{k}$ (if the electron is free, $\hbar \mathbf{k}$ will be its momentum, and $\mathbf{k}$ the wave vector of a true plane wave associated to this particle) and not by their position which is uncertain. The wave function will be normalized by integrating the position probability $|\psi_\mathbf{k}|^2 d\mathbf{r}$ for the crystal considered. The result must be unity.

The equation (1.3.1) has a solution of the form (1.3.3) only for certain values of the electron energy, E. The dependence

$$E = E(\mathbf{k}) \tag{1.3.5}$$

is influenced by the lattice properties and determines the energy-band diagram for a certain solid *. The energy levels are quantized but they form a quasi-continuum. Some E values are forbidden: the permissible values form energy 'bands' which must be imagined as regions in the three-dimensional space of the wave vector, $\mathbf{k}\,(k_x, k_y, k_z)$. Band overlapping is possible because electrons with different $\mathbf{k}$ values may have the same energy, E. The electron state is described by $\mathbf{k}$. A quantum state $\mathbf{k}$ cannot be occupied by more than two electrons (with antiparallel spins).

It can be shown that

$$\psi_\mathbf{k}(\mathbf{r}) = \psi_{\mathbf{k}'}(\mathbf{r}), \quad \mathbf{k}' = \mathbf{k} + \mathbf{U}, \tag{1.3.6}$$

where $\mathbf{U}$ is a vector of the reciprocal lattice. Equation (1.2.4) was used in this demonstration. Therefore $\mathbf{k}$ and $\mathbf{k}'$ are equivalent with respect to the electron wave function. From this point of view we have to consider only that part of the $\mathbf{k}$ space (which is identical to the space of the reciprocal lattice, $\mathbf{U}$) which contains independent vectors. This is just the Wigner-Seitz cell in the origin (or the first Brillouin zone). The entire $\mathbf{k}$ space is constructed by translating this basic cell with an arbitrary vector, $\mathbf{U}$, of the reciprocal lattice. Consider a wave vector $\mathbf{k}'$ in the Wigner-Seitz cell obtained by translating the first Brillouin zone with the vector $\mathbf{U}$. The vector $\mathbf{k}'$ describes the same electronic state as $\mathbf{k} = \mathbf{k}' - \mathbf{U}$. Of course, from all equivalent wave vectors, that corresponding to the first Brillouin zone has minimum length.

However, the energy E corresponding to two equivalent wave vectors, $\mathbf{k}$ and $\mathbf{k} + \mathbf{U}$ is not the same. We have to compute $E = E(\mathbf{k})$ for the entire $\mathbf{k}$ space and then reduce this space to the first Brillouin zone such that for each $\mathbf{k}$ in this zone will correspond a set of energy values. The $\mathbf{k}$

---

* For a free electron $E = \dfrac{(\hbar \mathbf{k})^2}{2m} = \dfrac{\hbar^2}{2m} k^2 \, (W_p = 0)$. If the electron is quasi-free (constrained to remain in a potential well) the k values are quantized.

space will be divided into Brillouin zones, each Brillouin zone containing only one vector equivalent to a vector **k** in the first Brillouin zone (or reduced zone). An energy band corresponds to each Brillouin zone and therefore to a wave vector in the reduced zone will correspond one different energy in each energy band.

The successive Brillouin zones are constructed by using the perpendicular bisector planes for the **U** vectors of the reciprocal lattice. Figure 1.3.1 shows the first three zones for a square reciprocal

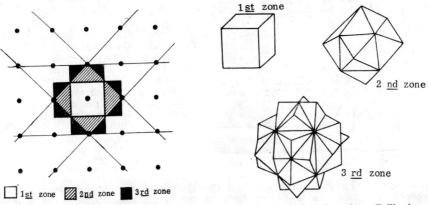

Figure 1.3.1. The first three Brillouin zones in a square reciprocal lattice. The area of all Brillouin zones is the same.

Figure 1.3.2. The first three Brillouin zones (a, b and c, respectively) for a simple cubic lattice.

lattice. For an arbitrary three-dimensional lattice the construction is similar. Each zone is formed from 'pieces' which wrap the preceding zone, as may be seen in Figure 1.3.2 for a simple cubic lattice. The volume of all zones is the same.

The first Brillouin zone for face-centred cubic lattice was shown in Figure 1.2.11: the diamond-type and the zincblende-type crystalline structures have the same Brillouin zone as the face-centred cubic lattice.

Consider a one-dimensional lattice with identical atoms separated by the distance a (lattice constant). The k-space divided in Brillouin zones is the abscissa of the $E = E(k)$ dependence in Figure 1.3.3. The total length of each zone is $2\pi/a$. The $E = E(k)$ diagram corresponding to the

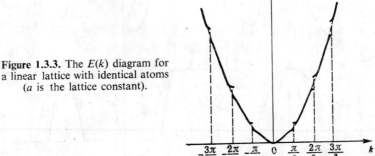

Figure 1.3.3. The $E(k)$ diagram for a linear lattice with identical atoms (a is the lattice constant).

free electron is shown dashed ($E \propto k^2$). For a one-dimensional lattice the $E = E(k)$ curve is discontinuous, thus giving rise to forbidden energy gaps, as shown in Figure 1.3.4. This discontinuity corresponds to Bragg reflection of the electron wave. The electron energy may be written $E = \hbar\omega$ and $\partial\omega/\partial k$ is the group velocity which stands for particle velocity in the quantum-mechanical theory.

The Bragg reflection (Section 1.2.2) on the periodical lattice structure occurs for wave vectors corresponding to the boundaries of the Brillouin zones constructed as shown above. This reflection imposes $\partial \omega / \partial k = 0$ at these boundaries and therefore the discontinuities in $E = E(k)$.

In a three-dimensional k space the energy bands, represented as families of constant-energy surfaces, can, however, overlap. Bands of forbidden energy will however, remain. At 0 K the

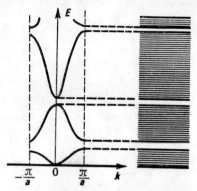

Figure 1.3.4. The energy bands for the one-dimensional lattice, reduced to the first Brillouin zone: detailed structure in the **k** space (left); global structure, with the energy bands 'filled' with discrete levels one per lattice atom in each band (right).

electrons are situated on the lowest energy levels (two electrons with antiparallel spins in each state). The upper band occupied (at least in part) with electrons is the valence band. In *semiconductors* (or insulators) this band is fully occupied and is separated from the next band (conduction band) by band gap of width E_g. The minimum energy E_g must be communicated to an electron in order to bring it into the conduction band. Therefore the number of conduction electrons in semiconductors increases (exponentially) with the temperature and is higher for narrower band gap semiconductors. An insulator is a wide-gap (several eV) semiconductor, with very few free carriers in normal conditions.

The computation of constant-energy surfaces for an actual crystal is based on physical approximations and requires numerical techniques. The result can be represented in graphical form along various directions in the **k** space. Figure 1.3.5 shows results of such computations for

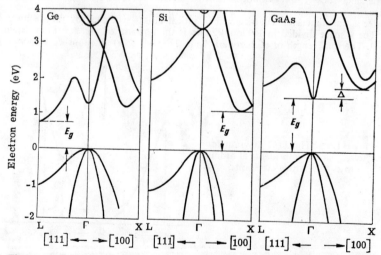

Figure 1.3.5. Energy band structure calculated for Ge, Si and GaAs, reduced to the first Brillouin zone, (the directions in the **k** space are indicated in the next figure) (after Sze, (8)). The experimental verification of bands structures is performed by optical transitions and cyclotron resonance (1).

germanium, silicon and gallium arsenide. The axes in the **k** space and the corresponding directions in the crystal are indicated for convenience in Figure 1.3.6. The valence band consists of four sub-bands when the electron spin is not taken into account in the Schroedinger equation (as above), and each band is doubled when spin is taken into account. Three of these sub-bands are degenerate at $\mathbf{k} = 0$ (Γ point in Figure 1.3.6) and form the upper edge of the band whereas the fourth

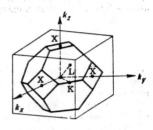

Figure 1.3.6. Important directions and points in the first Brillouin zone of a face-centred cubic lattice, diamond lattice or zincblende lattice (see Figure 1.5.11.). For all, a is the lattice constant of the lattice constant of the elementary cubic lattice, as shown in Figures 1.2.4–1.2.8. One point Γ, co-ordinates $(0, 0, 0)$, zone centre. Six X points, co-ordinates $\left(\dfrac{2\pi}{a}, 0, 0\right)$, etc., situated at the zone edge, along the $\{100\}$ axes (perpendicular to the $\{100\}$ planes in the direct lattice). Eight L points, $\left(\dfrac{\pi}{a}, \dfrac{\pi}{a}, \dfrac{\pi}{a}\right)$, zone edge, along the $\langle 111 \rangle$ axes. Twelve K points, $\left(\dfrac{3\pi}{2a}, \dfrac{3\pi}{2a}, 0\right)$, zone edge, along the $\langle 110 \rangle$ axes (after Sze, [8]).

one forms the bottom. The spin-orbit interaction causes a splitting of the band at $\mathbf{k} = 0$, which is not shown in Figure 1.3.5; see, for example, Chapter 11 in reference [1].

The conduction band also consists of a number of sub-bands. The most important are those which give the lower energy minimum (the bottom of the conduction band). This minimum occurs along the $\langle 111 \rangle$-axes for germanium, $\langle 100 \rangle$-axes for silicon and at $k = 0$ for GaAs. The shapes of the constant energy surface approximated near the bottom of the conduction band are shown in Figure 1.3.7. For Ge there are eight ellipsoids of revolution along the $\langle 111 \rangle$-axes. However, only half of each ellipsoid is situated in the first Brillouin zone. For Si there are six ellipsoids along the $\langle 100 \rangle$-axes, whereas for GaAs the constant energy surface is a sphere at the zone centre [8], [9].

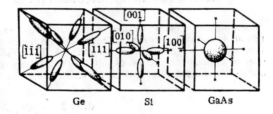

Figure 1.3.7. Shapes of constant energy surfaces in Ge, Si and GaAs relative to the lowest energy minimum. The eight ellipsoids of revolution for Ge are centred in the L points (Figure 1.3.6). For Si there are six, centred on the $\langle 100 \rangle$ axes inside the Brillouin zone (Figure 1.3.5). For GaAs the constant energy surface is a sphere centred in the Γ point (zone centre): however the higher energy surfaces are distorted and there exist also constant energy surfaces centred on L points as well as along the $\langle 100 \rangle$ directions (Figure 1.3.5).

The electron energy in the vicinity of an extremum of $E = E(\mathbf{k})$ situated at $\mathbf{k}_0$ may be written

$$E(\mathbf{k}) = E(\mathbf{k}_0) + \frac{1}{2} \sum_{\alpha,\beta} \left(\frac{\partial^2 E}{\partial k_\alpha \partial k_\beta}\right)_{\mathbf{k}_0} (k_\alpha - k_{\alpha 0})(k_\beta - k_{\beta 0}), \qquad (1.3.7)$$

where $\alpha, \beta = x, y, z$ and $k_{\alpha 0}(\alpha = x, y, z)$ are the coordinates of $\mathbf{k}_0$ in the x, y, z system. The nine coefficients $\left(\dfrac{\partial^2 E}{\partial k_\alpha \partial k_\beta}\right)_{\mathbf{k}_0}$ are the components of a second-order symmetrical tensor $\left(\dfrac{\partial^2 E}{\partial k_\alpha \partial k_\beta} = \dfrac{\partial^2 E}{\partial k_\beta \partial k_\alpha}\right)$. The constant-energy surfaces are ellipsoidal. Given an extremum $\mathbf{k}_0$, the axes can be always changed to coincide with the principal axes of this ellipsoid, and

$$E(\mathbf{k}) = E(\mathbf{k}_0) + \frac{1}{2} \sum_\alpha \left(\frac{\partial^2 E}{\partial k_\alpha^2}\right)_{\mathbf{k}_0} (k_\alpha - k_{\alpha 0})^2, \qquad (1.3.8)$$

where $\alpha = u, v, w =$ the new coordinates. The notation

$$\frac{1}{m_\alpha^*} = \frac{1}{\hbar^2} \left(\frac{\partial^2 E}{\partial k_\alpha^2}\right)_{\mathbf{k}_0}; \quad \alpha = u, v, w \tag{1.3.9}$$

introduces the components of the inverse *effective mass* tensor. If the constant energy surfaces are spherical in the vicinity of $\mathbf{k}_0$ then the above tensor will become a scalar, $m^* = m_\alpha(\alpha = u, v, w)$ and the 'excess' of electron energy is ($\mathbf{k}_{\alpha_0} = 0; \alpha = u, v, w$)

$$E(\mathbf{k}) - E(\mathbf{k}_0) = \frac{\hbar^2}{2m^*} k^2 \tag{1.3.10}$$

and is of the same form as the energy of a free electron with the effective electron mass m^* substituted for the free electron mass. It should be stressed, however, that $\hbar \mathbf{k}$ is no longer the electron momentum.

If the constant energy surfaces are ellipsoids of revolution (Figure 1.3.7), the effective mass will have two components: the longitudinal effective mass m_l^* and the transversal mass m_t^*.

The effective mass tensor describes the constant energy surfaces around a certain extremum (energy maximum or minimum) of a definite energy band corresponding to a particular crystal. This tensor reflects the crystal anisotropy through the details of the band structure. The components of the effective mass calculated for the maximum of the valence band are negative. If this maximum is degenerated, the calculation of the constant energy surfaces follows a different way [8], [9]. The above surfaces have a complicated shape and are approximated in general by spherical surfaces, thus defining a scalar effective mass for each maximum. This mass will be considered positive and denoted by m_{lh}^* (light holes, the lower branch in Figure 1.3.6) and m_{hh}^* (heavy holes) for reasons which will be clarified below.

1.4 Energy Distribution of Electrons in Semiconductors

According to the Pauli principle one electron state defined by the wave vector $\mathbf{k}$ can be occupied by only two electrons with antiparallel spins. If the electron spin is included in the state definition, then only one electron per state is accepted. The total number, dv, of available states, including spin, in an infinitesimal cube, $dk_x\, dk_y\, dk_z$ of the $\mathbf{k}$ space is proportional to the crystal volume, V, and does not depend upon its particular structure *

$$dv = 2\frac{V}{(2\pi)^3} dk_x\, dk_y\, dk_z. \tag{1.4.1}$$

The energy distribution of electron states $Z(E)$ for a given crystal can be calculated if the band structure $E = E(\mathbf{k})$ is known. Consider, for simplicity, a minimum of the conduction band energy, E_0, with spherical energy surfaces in its vicinity:

$$\Delta E = E - E_0 = \frac{\hbar^2}{2m^*} k^2 = \text{small}. \tag{1.4.2}$$

For ellipsoidal surfaces (such as those for Si and Ge, Figure 1.3.7), an equivalent density-of-state effective mass for electrons can be introduced

$$m_{de} = [m_l^*(m_t^*)^2]^{1/3}. \tag{1.4.3}$$

---

* This finite number of states is due to quantization occurring by solving the Schroedinger equation (1.3.1) with proper boundary conditions (the so-called 'cyclic' conditions which refer to a definite volume without interrupting crystal periodicity).

The density of states will be multiplied by M_c, the number of equivalent minima in the conduction band. The number of states in the energy interval dE becomes

$$Z(E)\,dE = 2VM_c \frac{(2m_{de})^{3/2}}{4\pi^2\hbar^3}\sqrt{E-E_0}\,dE. \qquad (1.4.4)$$

In many cases this spherical band approximation is satisfactory because only the bottom of the conduction band is occupied by electrons.

The occupancy of the available states in governed by the Fermi-Dirac statistics. The total number of electrons with energies between $E(0)$ and E is

$$N = \int_{E(0)}^{E} Z(E)\,f(E)\,dE = N(E), \qquad (1.4.5)$$

where $f(E)$ is the Fermi-Dirac distribution function

$$f(E) = \frac{1}{1 + \exp\dfrac{E - E_F}{kT}}, \qquad (1.4.6)$$

E_F being the electrochemical potential, or the Fermi level [1]–[6]. Figure 1.4.1a shows $f(E)$ at 0 K (when all states are occupied below E_F and empty above E_F) and two other temperatures. Note the fact that for $E - E_F \gg kT$ the distribution is approximately Maxwellian (or non degenerate) $f(E) \approx \exp\dfrac{E_F - E}{kT}$.

Figure 1.4.1b shows the electron distribution in energy for a parabolic conduction band ($m^* =$ = const. in equation (1.4.2) and $Z(E)$ given by equation (1.4.4)). The area under $N(E)$ curve is the total number of electrons in that conduction band. For $V =$ unity in equation (1.4.4), one obtains the electron density.

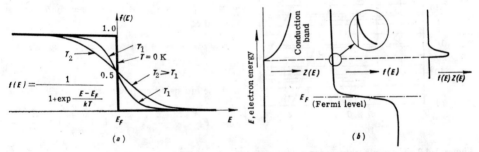

Figure 1.4.1. (a) Fermi-Dirac distribution, $f(E)$, at $T = 0$ K and also T_1 and $T_2 > T_1$; (b) electron distribution in energy in a parabolic conduction band at 0 K and $T_1 > 0$ K.

If E_F is several kT below E_c, the electron density is

$$n = \int_{E_c}^{\infty} Z(E)\bigg|_{V=\text{unity}} \exp\left(\frac{E_F - E}{kT}\right) dE \qquad (1.4.7)$$

and may be written

$$n = N_c \exp\left(-\frac{E_c - E_F}{kT}\right),\quad N_c = 2\left(\frac{2\pi m_{de}kT}{h^2}\right)^{3/2} M_c, \qquad (1.4.8)$$

where N_c is the effective density of states in the conduction band, and E_c is the bottom energy level of this band.

In a semiconductor at 0 K the valence band is fully occupied whereas the conduction band is empty. At a certain crystal temperature, $T > 0$ K, the average electron energy increases and a few electrons occupy levels at the bottom of the conduction band (the free electron density is n) leaving empty states in the valence band. The distribution of these empty states, or 'holes' (the hole concept will be discussed below), is also important. The probability of 'occupancy' of a level E by a 'hole' is $1 - f(E)$. A similar calculation for the valence band with similar approximations yields the hole density

$$p = N_v \exp\left(-\frac{E_F - E_v}{kT}\right), \quad (1.4.9)$$

where E_F is a few kT above the top of the valence band (E_v), N_v is the effective density of states in the valence band

$$N_v = 2\left(\frac{2\pi m_{dh} kT}{h^2}\right)^{3/2} \quad (1.4.10)$$

and m_{dh} is the density-of-state effective mass of the valence band

$$m_{dh} = [(m_{lh}^*)^{3/2} + (m_{hh}^*)^{3/2}]^{2/3} \quad (1.4.11)$$

(see the end of Section 1.3 and Figure 1.3.6).

For a pure (or intrinsic) semiconductor at thermal-equilibrium

$$n = p = n_i = \text{intrinsic carrier concentration} \quad (1.4.12)$$

and the intrinsic Fermi level E_i

$$E_F = E_i = \frac{E_c + E_v}{2} \times \frac{3kT}{4} \ln\left(\frac{m_{dh}}{m_{de}}\right) \approx \frac{E_c + E_v}{2} \quad (1.4.13)$$

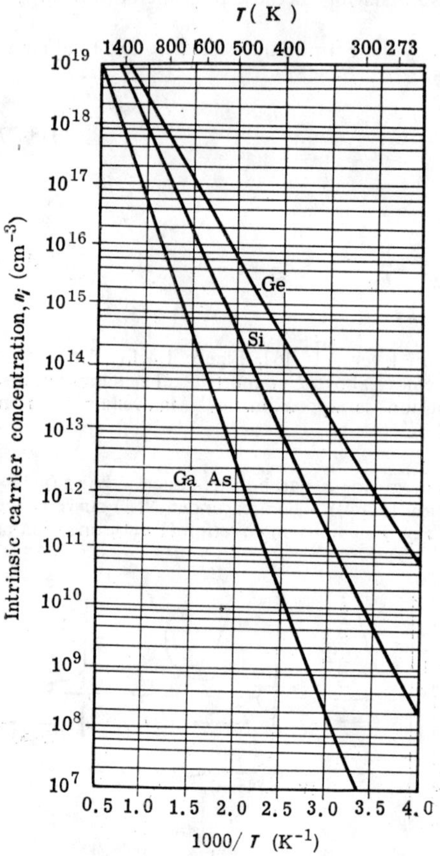

Figure 1.4.2. The temperature dependence of intrinsic carrier concentration, n_i, in Ge, Si and GaAs. The temperature dependence of E_g, is also taken into account (*after Sze*, [8]).

lies very close to the middle of the band gap. The intrinsic carrier concentration becomes

$$n_i = \sqrt{np} = \sqrt{N_c N_v} \exp\left(-\frac{E_g}{2kT}\right), \quad E_g = E_c - E_v \quad (1.4.14)$$

$$n_i = 4.9 \times 10^{15} \left(\frac{m_{de} m_{dh}}{m_0^2}\right)^{3/4} T^{3/2} \exp\left(-\frac{E_g}{2kT}\right) \quad (1.4.15)$$

(m_0 is the free electron mass). The variation of n_i with temperature is shown in Figure 1.4.2 for Ge, Si and GaAs.

The room temperature intrinsic concentration is extremely low as compared to the density of the lattice atoms. Therefore, even a small number of impurity atoms may have an important effect upon the semiconductor electric conductivity. The most important impurities in Si and Ge are the atoms belonging to the group 5 and group 3 of the periodic table of elements.

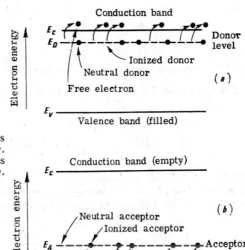

Figure 1.4.3. Donor and acceptor levels in the band gap, shown filled or empty. E_D and E_A are the ionization energies for donor and acceptor atoms, respectively.

A group 5 atom (donor) has a valence electron which does not participate in the covalent bond. The energy levels corresponding to this electron are discrete, which means that they are spatially localized because there is no interaction between the low-density donor atoms. The lowest energy level is usually the only one situated in the band gap and is close to the bottom of the conduction band (the energy $E_c - E_D$ is quite small, see Figure 1.4.3). The position of a number of impurity levels in silicon band gap can be found in Figure 1.4.4 [8]–[10].

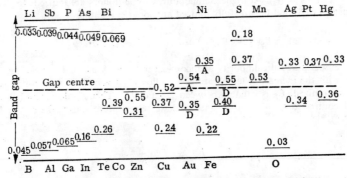

Figure 1.4.4. Impurity levels in silicon. The ionization energies indicated are measured from the top of the valence band for energy levels in the lower half of the band gap (and correspond to acceptors unless indicated by D for donor level), and from the bottom of the conduction band in the upper half of the band gap (donors, unless indicated by A). The band gap is $E_g = 1.12$ at 300 K (*after* Sze [8]).

The impurity level can accommodate only one electron, irrespective of its spin. Therefore the level cannot be regarded as doubly degenerated. The Fermi-Dirac statistics gives the following value for the probability of occupancy of the level E_D

$$f(E_D) = \frac{1}{1 + \frac{1}{2} \exp \frac{E_D - E_F}{kT}} \qquad (1.4.16)$$

$\left(\frac{1}{g} \text{ will be substituted for } \frac{1}{2} \text{ if more than one level, the fundamental one, must be considered}\right)$. The density of neutral donors (occupied levels) is therefore

$$n_D = \frac{N_D}{1 + \frac{1}{2} \exp \frac{E_D - E_F}{kT}}, \qquad (1.4.17)$$

where N_D is the density of donors.

The energy level, E_A, corresponding to acceptor atoms (group 3, one covalent bond unsatisfied), density N_A, is situated close to the valence band (Figure 1.4.3). The density of neutral acceptor is

$$n_A = \frac{N_A}{1 + \frac{1}{2} \exp - \frac{E_A - E_F}{kT}}. \qquad (1.4.18)$$

The density of free electrons in the conduction band and holes in the valence band will be calculated with the same formulae as for an intrinsic semiconductor. The neutrality condition

$$n + (N_A - n_A) = p + (N_D - n_D) \qquad (1.4.19)$$

should yield, in principle, the Fermi level E_F.

Consider, for example, an n-type semiconductor containing N_D donors $\times$ cm^{-3} ($N_A = 0$). At 0 K, $n = p = 0$ and $n_D = N_D$. At higher temperatures some donors are ionized. In many practical cases the impurities in Si and Ge are fully ionized at room temperature

$$n_D \simeq 0, \quad n \geqslant N_D, \quad p = n_i^2/n \simeq n_i^2/N_D. \qquad (1.4.20)$$

At very high temperatures the impurity semiconductor behaves as an intrinsic one. The Fermi level E_F varies with the increasing temperature, as follows. At very low temperatures, when only a few donors are ionized, E_F is situated between E_D and E_c. Then E_F decreases below E_D as n_A becomes small compared to N_A. In the intermediate temperature range where $n \approx N_D$ we have (equation (1.4.8))

$$E_c - E_F \approx kT \ln \frac{N_c}{N_D}. \qquad (1.4.21)$$

The dependence of $E_c - E_F$ upon temperature and impurity concentration is confirmed by the results of more exact computations, Figure 1.4.5. This figure also shows that at higher temperatures $E_F \to E_i$, and the semiconductor becomes practically intrinsic. We recall the fact that E_F was assumed to be a few kT/e within the band gap, the statistics of electrons in the conduction band and holes (vacant levels) in the valence band is non-degenerate, equations (1.4.8), (1.4.9) and (1.4.14) are valid and the semiconductor itself is called non-degenerate. In a more general case, the degenerate statistics formula (1.4.6) should be used in equation (1.4.5) and equation (1.4.8) will be replaced by

$$n = N_c \frac{2}{\sqrt{\pi}} F_{1/2}\left(\frac{E_F - E_c}{kT}\right), \quad (1.4.22)$$

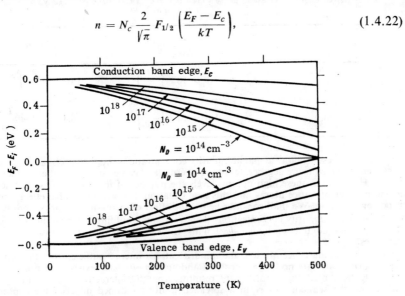

Figure 1.4.5. Fermi level in silicon as a function of temperature, for several impurity concentrations. The impurities are assumed fully ionized and non-degenerate statistics is used. (*after Grove* [7]).

where $F_{1/2}(\eta)$ is defined and plotted in Figure 1.4.6. If the temperature is sufficiently high, almost all donors will be ionized and $n \simeq N_D$ in the above equation. It can be shown that if $N_D > 0.765 \, N_c$ the Fermi level will be situated in the conduction band ($E_F > E_c$ in equation (1.4.22)). The

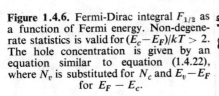
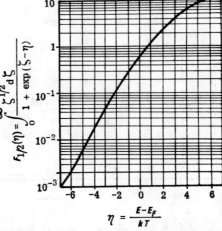

Figure 1.4.6. Fermi-Dirac integral $F_{1/2}$ as a function of Fermi energy. Non-degenerate statistics is valid for $(E_c - E_F)/kT > 2$. The hole concentration is given by an equation similar to equation (1.4.22), where N_v is substituted for N_c and $E_v - E_F$ for $E_F - E_c$.

effective density of states N_c and N_v, defined by equations (1.4.8) and (1.4.10) are given in Table 1.4.1 for Si and Ge.

Table 1.4.1
Effective Density of States in the Conduction and Valence Bands, N_c and N_v for Si and Ge
(Temperature in K)

	$N_c(\text{cm}^{-3})$	$N_v(\text{cm}^{-3})$
Silicon	$2.8 \times 10^{19} \left(\dfrac{T}{300}\right)^{3/2}$	$1.02 \times 10^{19} \left(\dfrac{T}{300}\right)^{3/2}$
Germanium	$1.4 \times 10^{19} \left(\dfrac{T}{300}\right)^{3/2}$	$6.1 \times 10^{18} \left(\dfrac{T}{300}\right)^{3/2}$

The above results indicate that at impurity concentrations (doping levels) above $10^{18}-10^{19}$ cm^{-3} these semiconductors are degenerate and the Fermi level can be situated in the conduction or valence bands. However, at these high doping levels the number of impurity atoms becomes comparable to the number of lattice atoms and this *new* situation has a considerable effect upon the energy band configuration. We shall consider below the effect of increasing impurity density [11].

At *moderate* impurity concentrations there is a small but finite probability of transfer between impurity states (conduction by *hopping*). This transfer mechanism is observable at very low temperatures.

At *large* doping concentration, the impurity atoms interact sufficiently with each other so that the impurity states are no longer localized and an *impurity band* exists. Because the impurities are randomly distributed, the impurity band resulting from this interaction will vary in width (in energy) throughout the semiconductor. In a so-called uniformly-doped material the concentration, spatially averaged on small domains, is uniform and the impurity band width should be also regarded as uniform but the spatial averaging of this width will result in an impurity band with an edge which is not sharp but 'tails off'. If part of the impurity levels are free, the electrons can change their places and an impurity-band electronic conduction occurs.

Another important effect at large dopings is the distortion of the conduction band edge which also forms tails. At *very large* dopings the conduction band and the impurity band overlap, as

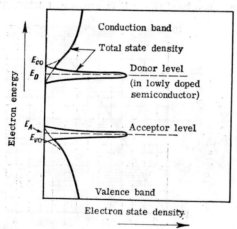

Figure 1.4.7. Density of states *versus* energy in a heavily doped semiconductor. The same diagram for low doping is shown dashed. The total density of states is assumed equal to the envelope of the conduction band and the impurity band, according to reference [11]. Note the fact that the impurity atoms take the places of silicon atoms in the crystal.

shown in Figure 1.4.7, and the ionization energy becomes zero. It was indeed shown experimentally that both the ionization energy and the semiconductor band gap depend upon the impurity concentration. If the semiconductor doping is non-uniform the band structure will depend upon position.

Figure 1.4.8 shows the impurity-concentration dependence of the Fermi level in silicon calculated by van Overstraeten et al. [12]. Both donors and acceptors are considered and $N_D > N_A$. At room temperature the acceptor levels are filled with electrons and the effective donor density is $N_D - N_A$ (abscissa in Figure 1.4.6). This is a compensated n-type semiconductor. The dependence of the

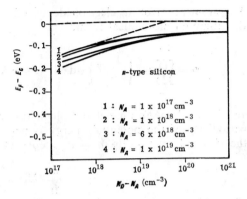

Figure 1.4.8. Fermi level versus $N_D - N_A$ with N_A as parameter for n-type silicon. For $N_A < 10^{17}$ cm^{-3}, the curve $N_A = 10^{17}$ cm^{-3} should be used. The Fermi level is calculated from the neutrality condition, using $m_n^* = 1.11 \ m_0$, $m_p^* = 0.81 \ m_0$, $\varepsilon = 11.8 \ \varepsilon_0$ (silicon) and $E_c - E_D = 0.044$ eV (phosphorus as donor level). (after van Overstraeten et al., [12]).

Fermi level upon the acceptor density is, clearly, an effect of the impurity dependence of energy levels. The product $np = n_{ie}^2$, where n_{ie} is the so-called effective intrinsic concentration, was also calculated (Figure 1.4.9). Note the fact that n_{ie} is always greater than $n_i (np > n_i^2)$, whereas the classical band theory gives $np \leqslant n_i^2$, the equality corresponding to the non-degenerate statistics (equation (1.4.22) approximated by equation (1.4.8)). The difference between $np = n_{ie}^2$ and n_i^2 increases with increasing effective donor concentration and with increasing compensation. The above theo-

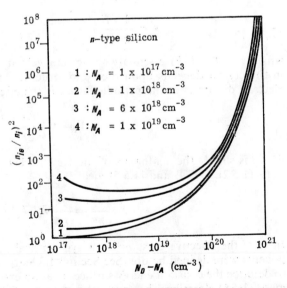

Figure 1.4.9. $(n_{ie}/n_i)^2$ versus $N_D - N_A$ with N_A as parameter for n-type silicon. For $N_A > 10^{17}$ cm^{-3} the curve $N_A = 10^{17}$ cm^{-3} will be used (after van Overstraeten et al., [12]).

retical results are confirmed by experiment. The calculations were repeated [13] for high temperatures (900–1200 °C) corresponding to phosphorus diffusion in silicon, in an attempt to clarify the impurity diffusion mechanism itself (for example the calculated values of the Fermi level are used to find the impurity concentration dependence of the total number of vacancies generated).

1.5 Electron Transport Phenomena

1.5.1 Electron Dynamics in a Periodical Lattice

The momentum of an electron described by the Bloch function (1.3.3) is given by the quantum-mechanical formula:

$$\mathbf{p} = \iiint_{-\infty}^{+\infty} \psi^* \left(\frac{\hbar}{i} \nabla\right) \psi \, dx \, dy \, dz = \hbar \mathbf{k} + \frac{\hbar}{i} \iiint_{-\infty}^{+\infty} u_\mathbf{k}^*(\mathbf{r}) \nabla u_\mathbf{k}(\mathbf{r}) \, dx \, dy \, dz \qquad (1.5.1)$$

and is equal to $\hbar \mathbf{k}$ only if the Bloch function has a constant amplitude $u_\mathbf{k}(\mathbf{r})$ (the periodicity of the potential energy is neglected). It can be shown, however, that the classical relation $d\mathbf{p}/dt = \mathbf{F}$ is replaced here by

$$\frac{d}{dt}(\hbar \mathbf{k}) = \mathbf{F}, \qquad (1.5.2)$$

where $\mathbf{F}$ is the 'external' force acting upon the electron with the wave vector $\mathbf{k}$. In an electromagnetic field *

$$\mathbf{F} = -e(\mathscr{E} + \mathbf{v} \times \mathbf{B}), \qquad (1.5.3)$$

where $\mathbf{v}$ is the average quantum velocity

$$\mathbf{v} = \frac{1}{\hbar} \frac{\partial E}{\partial \mathbf{k}} = \frac{1}{\hbar} \nabla_\mathbf{k} E, \qquad E = E(\mathbf{k}), \qquad (1.5.4)$$

which is ($\omega = E/\hbar$) essentialy the group velocity. Therefore, the electron motion in the crystal depends upon the band structure. The parameter characterizing the electron dynamics in the periodic lattice is the effective mass m^* defined by

$$\mathbf{F} = m^* \frac{d\mathbf{v}}{dt}, \qquad (1.5.5)$$

which is simply the analogous of the Newton law in the classical mechanics. Equations (1.5.2), (1.5.4) and (1.5.5) yield

$$\frac{1}{m^*} = \frac{1}{\hbar^2} \nabla_\mathbf{k}(\nabla_\mathbf{k} E). \qquad (1.5.6)$$

In fact, the effective mass m^* is a symmetrical second order tensor and its components were defined before (See Section 1.3 starting with equation (1.3.7)) in order to describe the constant energy surfaces near an extremum of $E = E(\mathbf{k})$. The effective mass (1.5.6) describes, however, the motion of an electron in an arbitrary state

---

* This field is extremely low as compared to the internal periodic field of the crystal lattice and is practically uniform over distances much longer than the lattice constant.

k and not only in the vicinity of an extremum. As the electron wave vector, **k**, changes under the action of an external force, the effective mass (1.5.6) also changes. Note the fact that the electron acceleration, $d\mathbf{v}/dt$, has not the same direction, in general, as the force acting upon this electron. This fact occurs due to crystal anisotropy.

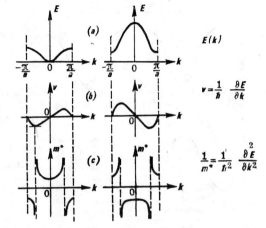

Figure 1.5.1. The energy (*a*), velocity (*b*) and effective mass (*c*) shown in the *k* space, for two energy bands of a one-dimensional lattice.

Figure 1.5.1 shows the electron energy, velocity and effective mass for two bands of a one-dimensional lattice (see also Figures 1.3.3 and 1.3.4). Here, m^* is simply a scalar. The most interesting feature of Figure 1.5.1 is the negative value taken by the effective mass. Consider for example the band exhibiting a minimum (a conduction-band type). The negative mass near the band edge denotes electron braking instead of electron acceleration. At the band edge (the boundary of the Brillouin zone) the electron suffers a Bragg reflection (its velocity becomes zero). This reflection takes place without energy loss, the velocity changes its sign, the wave vector changes from $-\pi/a$ to $+\pi/a$ and the electron appears at the opposite boundary of the Brillouin zone (see also Section 1.2).

The electronic conduction is possible only if the electron can change its state in the external electric field $\mathscr{E}$, according to equation (1.5.2) where $\mathbf{F} = -e\mathscr{E}$ is the electric force. The electron states are quantized but the energy levels are so dense that the acceleration process within one energy band may be regarded as continuous. The electron acceleration is, however, interrupted by 'collisions'. It will be shown below that the electron motion in a real crystal is dominated by various 'collision' mechanisms and can be described by the 'mobility' concept.

1.5.2 Electrons and Holes

Assume a few electrons in the conduction band of a semiconductor. They are considered 'free' because plenty of energy levels are available for their acceleration in the electric field. The valence band may also contribute to the current because the empty levels (states) can be occupied by other electrons. The density of free electrons in the conduction band and of empty levels in the valence band in the **k** space are shown in Figure 1.5.2*a* for a one-dimensional $E = E(k)$ diagram. The electron energy distribution is symmetrical around $\mathbf{k} = 0$ and given by a non-degenerate (Boltzmann) statistics. The application of an external electric field determines a displacement of this distribution, for example in the positive direction (Figure 1.5.2b). At a given $\mathscr{E} = $ const., this distribution is 'stabilized' by collisions which reduce successively

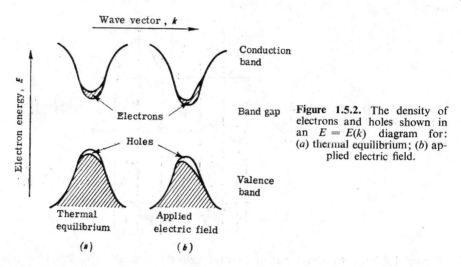

Figure 1.5.2. The density of electrons and holes shown in an $E = E(k)$ diagram for: (a) thermal equilibrium; (b) applied electric field.

the electron energy and prevents a continuous increase of k and $E(k)$. Note the fact that the electrons near the top of the valence band have a negative mass (Figure 1.5.1) and they move in the opposite direction (opposite sign of the wave vector k) as compared to the conduction-band electrons (Figure 1.5.2). This anomalous situation is better characterized by the hole concept [1], [2].

The hole is a *positively* charged 'particle' (charge $+e$) with positive (effective) mass. The motion of the hole gives an electric current equivalent to that carried by valence-band electrons with negative mass. The electric force $\mathbf{F}$ is therefore $+e\mathscr{E}$. The absolute value of the effective mass is introduced in equations (1.5.5) and (1.5.6). The hole density (1.4.9) was already calculated by using the probability of occupancy $1 - f(E)$ of the energy level $E\ldots$ by a hole, where $f(E)$ is the Fermi-Dirac function (1.4.6). In a p-type semiconductor the conduction is provided mainly by holes in the valence band, these holes occur by the ionization of the acceptor impurities, namely energetic electrons from the valence band which occupy the acceptor levels thus leaving free energy levels in this band.

Note the fact that the hole motion is in fact a process of rearrangement of an assembly of electrons. However, the concept of a definite positive particle with properties symmetrical to those of an electron is satisfactory in almost all situations encountered in device analysis.

1.5.3 Boltzmann Equation

The one-electron distribution function $f = f(\mathbf{k}, \mathbf{r}, t)$ defines the probability of finding at time t and position $\mathbf{r}$ an electron with the wave vector $\mathbf{k}$. The available number of states in the infinitesimal element $dk_x\, dk_y\, dk_z$ is $2/(2\pi)^3$ for a crystal with unity volume (equation (1.4.1)). Therefore the particle density is

$$n(\mathbf{r}, t) = \frac{1}{4\pi^3} \int f(\mathbf{k}, \mathbf{r}, t)\, dk_x dk_y dk_z \qquad (1.5.7)$$

whereas the electron current density is

$$\mathbf{J}(\mathbf{r}, t) = \frac{-e}{4\pi^3} \int \mathbf{v} f(\mathbf{k}, \mathbf{r}, t)\, dk_x dk_y dk_z. \qquad (1.5.8)$$

In the above formulas f also depends upon the energy band considered, $E = E(\mathbf{k})$. The integrations will be carried out on the entire Brillouin zone. Similar expressions should be written for all bands and the total current density will be the sum of components (1.5.8). In the semiconductor valence band, however, the computations may be done for holes, with electric charge $+e$ and probability $1-f$.

The distribution function f will be found by solving the Boltzmann continuity equation which is simply a statement about the conservation of particle number as they move in space $\mathbf{r}$ and change their quantic state $\mathbf{k}$ in time. The total derivative of f is

$$\frac{df}{dt} = \frac{\partial f}{\partial t} + \frac{\partial f}{\partial \mathbf{k}} \frac{d\mathbf{k}}{dt} + \frac{\partial f}{\partial \mathbf{r}} \frac{d\mathbf{r}}{dt} = \frac{\partial f}{\partial t} + \frac{\mathbf{F}}{\hbar} \nabla_\mathbf{k} f + \mathbf{v} \nabla_\mathbf{r} f. \qquad (1.5.9)$$

However, the particle distribution also changes due to various collision events such that the particle conservation requires [2]

$$\frac{\partial f}{\partial t} + \frac{\mathbf{F}}{\hbar} \nabla_\mathbf{k} f + \mathbf{v} \nabla_\mathbf{r} f = \left(\frac{\partial f}{\partial t}\right)_c, \qquad (1.5.10)$$

where $\left(\frac{\partial f}{\partial t}\right)_c$ is a notation for the collision term. A widely used approximation is

$$\left(\frac{\partial f}{\partial t}\right)_c = -\frac{f - f_0}{\tau}, \qquad (1.5.11)$$

which states that the perturbed distribution f relaxes to the thermal equilibrium distribution f_0 with some time constant (relaxation time). The above formulation will be correct only if the energy band is spherical (spherical constant-energy surfaces around an extremum of $E(\mathbf{k})$) and the collisions are elastic (without energy loss): hence τ is a momentum relaxation time

$$\tau = \tau(\mathbf{k}) = \tau(E). \qquad (1.5.12)$$

Consider the stationary behaviour ($\partial f/\partial t = 0$) in a homogeneous semiconductor at constant temperature ($\nabla_\mathbf{r} f_0 = 0$) under an applied electric field ($\mathbf{F} = -e\mathscr{E}$ for electrons). It is assumed that this field is low and produces only a small perturbation, f_1, of the distribution function

$$f = f_0 + f_1, \; f_1 \ll f_0. \tag{1.5.13}$$

The Boltzmann equation (1.5.10) becomes

$$\frac{e\mathscr{E}}{\hbar} \nabla_\mathbf{k} f_0 = \frac{f_1}{\tau(E)}. \tag{1.5.14}$$

The current density (1.5.8) becomes

$$\mathbf{J} = \sigma \mathscr{E}, \tag{1.5.15}$$

where the conductivity σ is a scalar for spherical energy bands and a tensor in the general case (due to crystal anisotropy). For constant relaxation time (τ independent of E)

$$\sigma = \frac{e^2 \tau}{m^*} n, \tag{1.5.16}$$

where n is the electron density (uniform) and m^* is the effective mass ($E = (\hbar k)^2/2m^*$ for spherical energy bands). If $\tau = \tau(E)$, the conductivity will be

$$\sigma = \frac{e^2 n}{m^*} \langle \tau \rangle + \frac{e^2 n}{3\hbar^2} \left\langle k^2 \frac{\partial \tau}{\partial E} \right\rangle, \tag{1.5.17}$$

where $\langle \mathscr{F} \rangle$ denotes statistical average of function $\langle \mathscr{F} \rangle$:

$$\langle \mathscr{F} \rangle = \frac{\int f_0 \mathscr{F} dk_x dk_y dk_z}{\int f_0 \, dk_x dk_y dk_z}. \tag{1.5.18}$$

The conductivity σ may be formally written $\sigma = e^2 \tilde{\tau} n/m^*$, where the equivalent relaxation time $\tilde{\tau}$ is not equal to the statistical average $\langle \tau \rangle$. It can be shown [5] that for a non-degenerate spherical band where τ varies with energy as $\tau = \tau_0 E^q$, one obtains

$$\tilde{\tau} = \tau_0 (kT)^q \frac{\Gamma(q + 5/2)}{\Gamma(5/2)}, \tag{1.5.19}$$

where $\Gamma(z)$ is the Euler function satisfying $\Gamma(z + 1) = z \Gamma(z)$, $\Gamma(1) = 1$, $\Gamma(1/2) = \sqrt{\pi}$.

The average drift velocity of the electronic charge (density $\rho = -en$), $\mathbf{v}_n$ is defined by

$$\mathbf{J} \Rightarrow \mathbf{J}_n = \rho \mathbf{v}_n = -en\mathbf{v}_n \tag{1.5.20}$$

and therefore

$$\mathbf{v}_n = -\mu_n \mathscr{E}, \tag{1.5.21}$$

where μ_n is the drift mobility* given by

$$\mu_n = \frac{\sigma}{en} = \frac{e\tilde{\tau}}{m^*}. \tag{1.5.22}$$

In a similar way can be defined the hole drift velocity, v_p, and the hole drift mobility, μ_p,

$$\mathbf{v}_p = \mu_p \mathscr{E}, \tag{1.5.23}$$

such that in a homogeneous semiconductor the electron and hole current densities, $\mathbf{J}_n$ and $\mathbf{J}_p$ add to give the total current density

$$\mathbf{J} = \mathbf{J}_n + \mathbf{J}_p = -en\mathbf{v}_n + ep\mathbf{v}_p = e(n\mu_n + p\mu_p)\mathscr{E}. \tag{1.5.24}$$

1.6 Carrier Mobility

1.6.1 Mean Free Time

We shall consider below the mean free time of an electron between successive collisions and assume for the moment that this time is identical for all electrons (in fact it depends upon the electron velocity). Consider a number of electrons which have *not* collided until the time t, namely $n(t)$. In a short time interval dt the number of collisions is proportional to $n(t)$ and dt, such that [5]

$$\frac{dn(t)}{dt} = -\frac{n(t)}{\tau}, \quad \tau = \text{const.} \tag{1.6.1}$$

and

$$n(t) = n(0) \exp\left(-\frac{t}{\tau}\right). \tag{1.6.2}$$

The probability that an electron has *not* made a collision until the time t is therefore $n(t)/n(0) = \exp(-t/\tau)$. It can be shown that the probability of such a collision between t and $t + dt$ is $(t/\tau) \exp(-t/\tau)$, such that the average time between collisions (average free time) is

$$\int_0^\infty \frac{t}{\tau} \exp\left(-\frac{t}{\tau}\right) dt = \tau. \tag{1.6.3}$$

Assume that the electron has zero diameter and collides with obstacles of cross section σ_c and density N. The average time between collisions may be written

$$\tau = \frac{1}{N\sigma_c v}, \tag{1.6.4}$$

---

* The particle mobility is a tensor in the general case. Its components are proportional to the components of the conductivity tensor (multiplied by $1/en$).

where v is the electron velocity. In fact the *collision cross-section* σ_c is a property of particle *and* obstacle. We see that τ should indeed depend upon electron velocity, and therefore, upon electron energy.

The mean free path λ is defined by

$$\lambda = \tau v = \frac{1}{N\sigma_c}. \qquad (1.6.5)$$

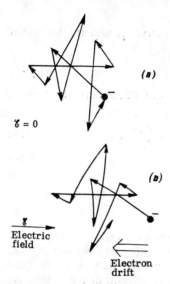

Figure 1.6.1. The electron motion in a crystal: (*a*) at thermal equilibrium; (*b*) with applied electric field.

Consider now the simplest case of electron collision. This is the case of isotropic scattering: the particle loses all knowledge of its previous motion. Scattering by the lattice thermal vibrations (collisions with 'phonons', see below), is isotropic. For many other types of scattering, and in particular for the 'collisions' with ionized impurity centers, a small angle deviation occurs (this is the deviation in the Coulomb field of a charged particle). In the absence of electric field, the electron is subjected to a random motion with a uniform velocity between two successive collisions (see Figure 1.6.1*a*). We shall treat electrons as classical particles, with m^* mass, and m^*v momentum (this is in agreement with the free electron model in crystals: the electrons are described by Bloch functions of constant amplitude). If an external electric field, $\mathscr{E}$, is applied, the electron velocity between collisions will be non-uniform. With $\mathscr{E}$ along the x direction, the only velocity component which changes is v_x, and

$$v_x = v_{x0} - \mathscr{E}\, et/m^*, \qquad (1.6.6)$$

where v_{x0} is the value of v_x at $t = 0$, just after the electron has made a collision. The averaging of equation (1.6.6) over all values of t, by using the collision probability at t (see above), is

$$\bar{v}_x = \bar{v}_{x0} - (\mathscr{E}e/m^*)\int_0^\infty \frac{t}{\tau}\exp\left(-\frac{t}{\tau}\right)dt = \bar{v}_{x0} - \mathscr{E}e\tau/m^* = -\frac{e\mathscr{E}\tau}{m^*} = -\mu_n\mathscr{E}, \qquad (1.6.7)$$

where $\bar{v}_{x0} = 0$ (because, in average, collisions destroy any previous velocity) and μ_n is the drift mobility. The electron, still subjected to random collisions, moves gradually in the direction of the electric field (see Figure 1.6.1*b*) with an average drift velocity $\mu_n\mathscr{E}$. Note that the electron mobility has an expression identical to (1.5.21), where the relaxation time $\tilde{\tau}$ is replaced by the average time between collisions (compare equations (1.5.11) and (1.6.1)).

1. Electronic Conduction in Semiconductors

Let us now consider the order of magnitude of some of the above quantities. The electron mobility in Ge is 3 900 cm² V⁻¹ s⁻¹. With $m^* = 0.3\,m_0$ (m_0 is the free electron mass), the mean free time is $\tau = 6 \times 10^{-13}$ s, much shorter than any time constant usually involved in the device operation. For this reason the electron velocity is almost always assumed to be an instantaneous function of $\mathscr{E}$. The electron velocity in random motion can be calculated by considering that the electron gas in a non-degenerate semiconductor (Boltzmann statistics) is characterized by an average energy per particle

$$\frac{1}{2} m^* \overline{v^2} = \frac{3}{2} kT. \tag{1.6.8}$$

In the above relation, T is the lattice temperature, and the electron energy is due to collisions with thermal vibrations of the lattice (see below). Therefore, the average velocity of this 'thermal' motion is

$$v_{\text{th}} \approx (\overline{v^2})^{1/2} = \left(\frac{3kT}{m^*}\right)^{1/2} \tag{1.6.9}$$

and $T = 300$ K, $v_{\text{th}} \simeq 2.5 \times 10^7$ cm s⁻¹ ($m^* = 0.3\,m_0$). The drift velocity at $\mathscr{E} = 1\,V$ cm⁻¹ is only 3.9×10^3 cm s⁻¹ ($\mu_n = 3\,900$ cm²V⁻¹s⁻¹), which is smaller with about four orders of magnitude. Note, however, that at very high applied fields (a few kV cm⁻¹ for example), the drift velocity will become comparable to the thermal velocity v_{th}. The energy gained by an electron between two collisions cannot be entirely communicated to the lattice and the electron becomes 'hot'. The hot electron problem will be discussed separately [14], [15].

The mean free path is $\lambda = \overline{v}\tau = (2.5 \times 10^7\text{ cm s}^{-1}) \times (6 \times 10^{-13}\text{ s}) = 1.5 \times 10^{-5}$ cm $= 0.15\,\mu$m $= 1500$ Å and extends over several hundreds of interatomic distances. Note the fact that in high-mobility semiconductors and/or at low temperatures (μ_n and τ increase with decreasing lattice temperature) the mean free path may become comparable to certain very short distances which are characteristic of device operation and even to constructive dimensions.

1.6.2 Lattice Vibrations [1], [3], [5[, [15]

An electron travels freely (without collisions) in a perfectly periodic crystal. If it is accelerated by an external electric field, its wave vector **k** will increase and at the boundary of the Brillouin zone a Bragg reflection will occur. The concept of collision describes the electron interaction with the imperfections of a real crystal. A major cause of imperfection is the crystal vibration associated to the thermal energy of lattice atoms. These vibrations are quantized and described by the 'phonon' concept. The electron-phonon 'collision' will be discussed first. Other collisions occur with dislocations, impurities and even between the electrons.

Each atom in the crystal is attracted to its neighbouring atoms by elastic forces (the nature of the chemical bond does not concern us here). At 0 K these atoms are almost fixed in positions corresponding to an ideal periodic lattice (the residual energy is very small). At a temperature $T > 0$ K the crystal atoms oscillate (thermal agitation) around their equilibrium positions. The displacement of any atom influences the motion of the other atoms. The assembly of these coupled oscillations are described by vibration waves corresponding to the entire crystal. These are travelling waves of the form $A \exp i(\mathbf{k}\mathbf{r} - \omega t)$, where $\mathbf{k}$ takes only N discrete values in each Brillouin zone of the reciprocal lattice space (N is the number of cells in the crystal, each containing one or more atoms). The energy bands $E = E(\mathbf{k})$ for electron waves are replaced by frequency bands $\omega = \omega(\mathbf{k})$ for lattice vibration waves.

Consider, for example, a linear lattice with two atoms in a primitive cell: the atoms may be different or identical. The chemical bond may be predominantly ionic (NaCl crystal) or predominantly covalent (GaAs, see Table 1.2.2). The 'frequency bands' $\omega = \omega(k)$ are the acoustic branch and the optical branch, represented in Figure 1.6.2. The group and phase velocity of the lowest branch at low frequencies (linear region in Figure 1.6.2) is just the velocity of sound in the crystal, c. For $a = 5$Å lattice constant and $c = 10^5 - 10^6$ cm s^{-1} the maximum (extrapolated) frequency corresponds to $f = \omega/2\pi = 10^{12} - 10^{13}$ Hz, and the maximum acoustical frequencies are somewhat lower. These are true 'acoustic' waves only below 10^5 Hz. The upper branch (optical) (Figure 1.6.2) contains frequencies around 10^{13} Hz and corresponds to the oscillations of the two atoms in a cell about the centre of the mass. In a polar crystal the optical vibrations which displace the two ionized atoms relative to each other and lead to polar scattering of electrons, thus increasing considerably the electrical resistance. The acoustic vibrations correspond to the displacement of the centre of the mass of these atoms in the cell, lead to deformation of regions in the crystal and to a deformation of the electrons potential energy that reflects or scatters the electron waves (deformation potential scattering). The change in the lattice potential is called deformation potential. Note the fact that the electron thermal velocity (of the order of 10^7 cm s^{-1}) is much larger than the propagation velocity of the acoustical vibrations. In other words, the potential from which the electron is scattered moves slowly with respect to the electron wave. The scattering will determine a small Doppler shift in frequency and a corresponding small change in energy [5]. This electron-lattice interaction corresponds to generation or anni-

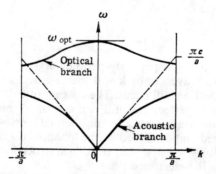

Figure 1.6.2. Dispersion diagram for lattice waves in a linear lattice with the two atoms per cell. The group velocity $\partial\omega/\partial k$ and phase velocity ω/k for the acoustical branch at low frequencies are identical to the velocity of sound, c.

hilation of a quanta of lattice vibration energy, called the *phonon* (see below).

The above model is valid for a linear chain of atoms with longitudinal displacements. Transverse displacements could also be considered. In a practical crystal there exist longitudinal (L) and transverse (T), optical (O) and acoustic (A) modes of vibration, denoted by LA, LO, TA and TO. The vibrations corresponding to TO and TA branches in a two-atom (polar) crystal are shown in Figure 1.6.3. The total number of acoustic branches is three: two correspond to transversal polarized waves and one to longitudinal waves. The total number of optical branches is

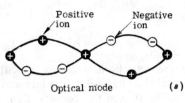

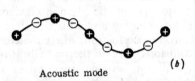

Figure 1.6.3. Transversal optical waves (a) and transversal acoustic waves (b), both with the same wavelength, for linear biatomic lattice of a polar crystal (the atoms are shown ionized) (*after Kittel*, [1]).

$3(s-1)$ where s is the number of atoms in a primitive cell. Note the fact that for diamond, germanium and silicon $s = 2$ because the primitive cell has two atoms. Each branch must be imagined as a frequency band $\omega = \omega(\mathbf{k})$ and will be represented for certain crystallographic directions, as shown in Figure 1.6.4. If these

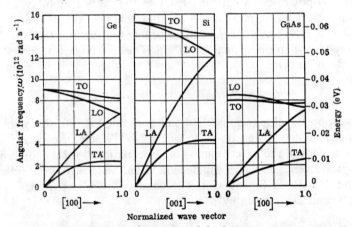

Figure 1.6.4. Experimental spectra of lattice vibrations in Ge, Si and GaAs, reduced to the first Brillouin zone and shown for a certain (crystallographic) direction (after Sze, [8]). The optical branches are determined in principle by measuring the transmission and reflection coefficients of light of various wavelength. The acoustic branches can be determined by neutron scattering [1].

crystallographic directions are axes of symmetry the two transverse branches are identical (degenerate).

We shall define now the phonon (compare with the photon concept). A phonon is a quanta of lattice thermal energy, $\hbar\omega$, associated to the frequency $\omega^q{}_j$ from the vibration spectrum, which corresponds to the wave vector **k** in the branch j. The phonon distribution in energy (in frequency) is governed by the Bose-Einstein statistics (quantum particles without spin). The probability function is

$$f(\omega) = \frac{1}{\exp\dfrac{\hbar\omega_{kj}}{kT} - 1}. \qquad (1.6.10)$$

The number of phonons increases with the lattice temperature. Phonons may be also generated or annihilated in interaction with electrons. This interaction is considered a 'collision' and the conservation of energy and (quasi) momentum $\hbar\mathbf{k}$ should take place. From this point of view the phonon may be regarded as a particle.

A major use of the phonon concept is in studying the thermal properties of a solid (heat capacity and heat transfer) [1] — [5]. Another important application is the electronic conduction, where the electron motion is influenced by phonon scattering ('collisions' due to thermal vibrations of the lattice) [5], [15]. It is demonstrated that a collision between an electron and an acoustical phonon is almost elastic (the variation of electron energy is negligible). It can be understood that the scattering is isotropic because the vibration waves travel in all directions and there is an equal probability of finding the necessary phonons for interaction.

The mobility determined by electron-phonon-collisions is called *the lattice mobility*. In non-polar semiconductors (Si, Ge) the lattice mobility is determined by scattering with acoustic phonons. Assuming spherical constant-energy surfaces, it can be shown that $\tau \propto T^{-1}E^{-1/2}$ and according to equations (1.5.19) and (1.5.22) the lattice mobility is proportional to $T^{-3/2}$. The mobility also depends upon the effective electron mass m_n^*

$$\mu_n \propto (m_n^*)^{-5/2} T^{-3/2} \qquad \text{(lattice mobility)} \qquad (1.6.11)$$

because τ is proportional to $(m_n^*)^{-3/2}$. The same relation holds for holes and therefore (μ_n/μ_p) is approximately proportional to $(m_h^*/m_n^*)^{5/2}$. When the constant energy surfaces are not spherical, the effective mass m^* in equation (1.6.11) should be replaced by an appropriate average value. τ will depend upon the density of state effective mass (see equation (1.4.3) for Si and Ge). The conductivity effective mass in equation (1.5.22) for Si and Ge is given by*

$$\frac{1}{m^*} = \frac{1}{3}\left(\frac{2}{m_t^*} + \frac{1}{m_l^*}\right), \qquad (1.6.12)$$

where m_t^* and m_l^* are the transversal and longitudinal components of the effective mass tensor (1.5.6) for Si and Ge (see Figure 1.3.7). The above results are calculated for scattering with longitudinal modes. Similar temperature dependence is expected for collisions with 'transversal' acoustic phonons. However, the experi-

---

* The anisotropy of conductivity will be discussed in the next section.

mental temperature dependence of carrier mobilities in Si and Ge (Table 1.6.1) deviates considerably from the $T^{-3/2}$ law, as indicated by Table 1.6.1.

Table 1.6.1

Temperature Dependence of Lattice Mobility [9], in cm² V⁻¹s⁻¹

Semi-conductor	Charge carriers	300 K mobility	Temperature dependence (T in K)
Ge	electrons	3900	$4.9 \times 10^7 T^{-1.66}$ (100–300 K)
	holes	1900	$1.05 \times 10^9 T^{-2.33}$ (125–300 K)
Si	electrons	1350	$2.1 \times 10^9 T^{-2.5}$ (160–400 K)
	holes	480	$2.3 \times 10^9 \times T^{-2.7}$ (150–400K)

In polar semiconductors (such as GaAs) electron scattering by optical phonons is dominant* and the theoretical mobility-temperature dependence should be different. The experimental results near room temperature for pure n-type and p-type GaAs is $\mu_n \propto T^{-1}$ and $\mu_p \propto T^{-2.1}$ [8].

1.6.3 Impurity Scattering

The scattering of an electron in the Coulomb field of an ionized impurity is highly anisotropic, small angles of scattering (angle between the wave vectors of incident and scattered wave) being favoured. The approximate result for the theoretical impurity mobility (mobility determined by collisions with impurities) is

$$\mu \propto (m^*)^{-1/2} N_I^{-1} T^{3/2}, \qquad (1.6.13)$$

where N_I is the density of ionized impurities. The experimental dependence of mobilities upon the impurity concentration is shown in Figure 1.6.5 for Ge and Si. Scattering by *neutral* impurities also leads to a mobility inversely proportional to the impurity concentration but it is independent of temperature [5]. Neutral impurity scattering should be important especially at very low temperature (where most of the impurities are neutral) and in compensated semiconductors.

---

* The coupling between an electron and optical modes of lattice vibration is likely to be much stronger in a polar crystal because of the strong dipole moment set in by these oscillations. The optical phonons are highly energetic and the electron-phonon collision will be highly inelastic. Interaction with optical phonons may be responsible for intervalley scattering, in semiconductors with more than a single minimum in the conduction band (see Figure 1.3.5 for GaAs).

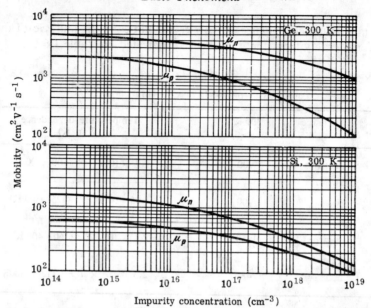

Figure 1.5.5. Drift mobilities in Ge and Si, at 300 K, *versus* impurity concentration (*after Sze*, [8]).

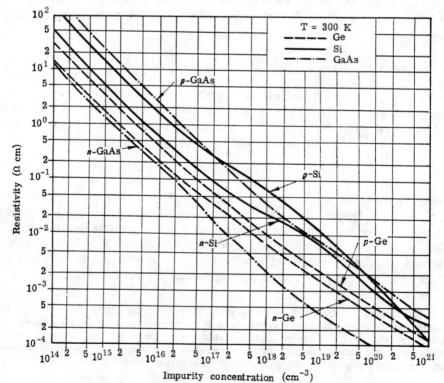

Figure 1.6.6. Measured resistivity *versus* impurity concentration for Ge, Si and GaAs, at 300 K (*after Sze*, [8]).

1. Electronic Conduction in Semiconductors

When both lattice and impurity scattering are taken into account, the mobility μ is given by

$$\frac{1}{\mu} = \frac{1}{\mu_{imp}} + \frac{1}{\mu_{lattice}}, \qquad (1.6.14)$$

because the probability of these two scattering mechanisms is simply added. The impurity mobility dominates at high carrier concentrations and also at low temperatures.

The semiconductor resistivity may be written (n-type) $\rho_n = (e\mu_n n)^{-1}$. If $n \simeq N_D =$ (effective) donor concentration, then ρ_n is not inversely proportional to N_D due to the dependence $\mu_n = \mu_n(N_D)$. Figure 1.6.6 shows, for example, the resistivity *versus* impurity concentration for Ge, Si and GaAs, at 300 K [8], [16]. Of course, the impurity levels are not necessarily ionized at all temperatures. Happily, the free carrier concentration can be determined directly from Hall measurements [5], [8], and thus the drift mobility will be determined. Figure 1.6.7 shows the temperature and impurity concentration dependence for electron and hole mobility in silicon.

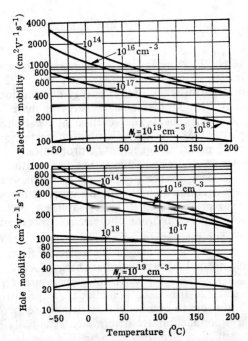

Figure 1.6.7. Electron and hole drift mobilities in silicon as a function of temperature and impurity concentration (*after Sze*, [8]).

1.6.4 Other Scattering Mechanisms

Electron scattering by lattice defects (dislocations) will also occur but this is unlikely to be important except at very low temperatures or in a heavily strained material [5].

The electron-electron (or interelectronic) scattering is due to the mutual repulsion of electric charges. This scattering of mobile electrons does not significantly contribute to the electrical resistance. It merely redistributes the energy and momentum among the carriers. The redistribution of energy is much more effective in these collisions because of the similarity of masses of the colliding particles. This scattering helps to randomize the motion and plays an important rôle in establishing a steady-state under high-field conditions.

1.7 High-field Effects: I. Hot-carriers in Semiconductors

The electron gas in a non-degenerate semiconductor at thermal equilibrium has a Maxwellian distribution function*

$$f(\mathbf{r}, \mathbf{k}) = n\hbar^3 (2\pi m^* kT)^{-3/2} \exp\left[-\frac{(\hbar\mathbf{k})^2}{2m^* kT}\right], \qquad (1.7.1)$$

where T is the lattice temperature. The electrons are in thermal equilibrium with the lattice and their average energy is $3kT/2$ (see equation (1.6.9)). An applied electric field perturbs the carrier energy distribution since the carriers gain energy from the field. This supplementary energy is shared with the lattice through collisions. For low electric fields the solution of the Boltzmann equation consists of a symmetric part (f_0, Maxwellian distribution corresponding to the lattice temperature) and a small perturbation term (f_1, proportional to the electric field). The carriers are still considered to be at the lattice temperature and the only effect of the electric field is to determine a drift velocity, much smaller than the random thermal velocity (see equation (1.6.9)). In the simplest model the carrier drift mobility was shown to be proportional to the relaxation time, which completely describes the effect of collision events upon the electron transport.

The mechanism of energy transfer from electrons to the lattice by collisions is fairly inefficient and at high electric fields the carrier energy exceeds considerably the normal average thermal energy $3kT/2$. The electron energy may be still written $3kT_n/2$, where the electron temperature T_n, corresponding to a hypothetical Maxwellian distribution, is higher than the lattice temperature T. Note the fact that it is possible to obtain 'hot carriers' in semiconductors because the heat capacity and heat conduction are determined essentially by the lattice vibration (and not by the conduction electrons such as in a metal).

The increase of electron temperature should increase the collision probability and thus decrease the mobility. This is an intuitive explanation for the experimental decrease of lattice mobility with increasing field intensity [17].

A more accurate description, even at a phenomenological level, of hot carrier effects is however necessary. Consider, for example, the acoustic phonon scattering. The carrier momentum is practically lost after such a collision, whereas the energy transfer to the lattice is practically negligible. The momentum and energy loss rates are not necessarily related. The collisions most effective for momentum change (those that determine the drift mobility) are not necessarily those most effective for energy transfer. The energy loss rate for the assembly of conduction electrons (or holes) should be described by an appropriate parameter [15].

If the scattering is completely isotropic the *momentum relaxation time* τ_m (which enters into the expression of drift mobility) is equal to the average time between collisions, $\tau_m = \tau$. In general, however, τ_m is longer than τ. An *energy relaxation*

---

* For parabolic bands m^* is constant, the electron momentum is $\hbar\mathbf{k}$ and the electron kinetic energy $(\hbar\mathbf{k})^2/2m^*$.

time, τ_e, can be also introduced. It describes the rate of decay of the supplementary carrier energy (gained from the field) when the field is removed and the average carrier energy tends towards the thermal equilibrium energy. It will be difficult to define τ_e mathematically if the energy lost per collision is an appreciable fraction of this supplementary energy. However, it is still useful to retain the concept of energy relaxation time, τ_e. This time is, in general, energy-dependent.

It was shown in Section 1.6 that the collision term in the Boltzmann equation is approximated by assuming that the perturbed distribution function, f, relaxes towards the thermal equilibrium distribution with a time constant, τ. This is justifiable for low carrier energies (low applied fields), where τ is essentially determined by the momentum relaxation time. Sometimes a similar collision term is postulated in studying the hot carrier effects. However, the single relaxation time introduced in this way cannot account, in general, for both momentum and energy relaxation processes [15].

A central assumption in the hot-carrier theory is that of a quasi-Maxwellian distribution function. The unknown parameters of such a distribution are determined from the Boltzmann equation. This assumption can be accepted only if there is a scattering mechanism which randomizes the electron distribution. This mechanism is electron-electron (or inter-carrier) scattering. The magnitude of this term is proportional to the carrier concentration, whereas the others are independent of concentration [18], [19].

At *low carrier concentrations* no special assumptions can be made and the Boltzmann equation should be solved directly. Consider for example only acoustic phonon scattering. The distribution function is non-Maxwellian and no carrier temperature may be defined. The mobility decreases with the increasing electric field. The higher the mobility, the lower the fields where this decreasing occurs. The dependence upon the crystallographic axes and the velocity saturation are confirmed by experiment and will be discussed later.

At *intermediate carrier concentrations* ($10^{16} - 10^{17}$ cm^{-3}) the distribution is assumed Maxwellian with the electron temperature, replacing the crystal temperature. The energy exchange between carriers is much faster than that between carriers and lattice, the carriers share their energy and attain a common temperature, T_e. This temperature is determined by using the energy balance equation (the energy gained from the field is equal to the energy lost through collisions).

At *very large carrier concentration* the solution is assumed to be the Maxwellian function displaced in the momentum space. The carriers share between them both the energy and momentum, and are characterized by a common temperature, as well as a common momentum. The interelectronic scattering randomizes the charge carriers momentum about its mean drift velocity. This is possible because in this case the scattering centres move with the particle stream. The displaced Maxwellian distribution function is

$$f(\mathbf{r}, \mathbf{k}) = n\hbar^3(2\pi m^* k T_n)^{-3/2} \exp\left[-\frac{\hbar^2(\mathbf{k} - \mathbf{k}_0)^2}{2m^* k T_n}\right]. \qquad (1.7.2)$$

The parameters T_n and $\mathbf{k}_0$ will be determined by using the energy and momentum balance equations.

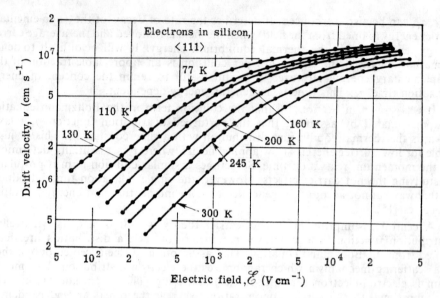

Figure 1.7.1. Electron drift velocity in silicon, as a function of the electric field parallel to the $\langle 111 \rangle$ direction, measured at several temperatures (*after Canali et al.*, [23]).

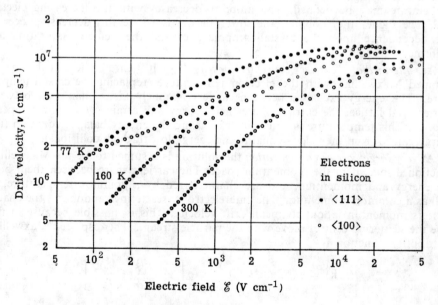

Figure 1.7.2. Electron drift velocities in silicon parallel to the $\langle 111 \rangle$ and $\langle 100 \rangle$ directions, measured at several temperatures (*after Canali et al.*, [23]).

It was shown [20], [21] that in certain circumstances the electrons may be 'cooled' rather than heated by applying an electric field. Such a result was obtained by Blötekjaer [21] for polar semiconductors by assuming a displaced Maxwellian distribution function and dominant scattering by optical phonons. This carrier 'cooling' occurs because when the distribution function is displaced due to an electric field the energy transferred to the lattice exceeds the energy supplied by the field which shifts the distribution. This is a pure quantum effect deduced from momentum and energy balance equations. However, the effect does not occur if scattering mechanisms which are not due to optical phonons are present. High-purity InSb at 100 K should be suitable for an experimental verification.

The Boltzmann equation can be also solved by using the Monte Carlo technique, briefly described in the next section. Such a calculation was performed by McCombs, Jr., and Milnes [22] for hot carriers in silicon.

Canali et al. [23] recently published results of extensive measurements of carrier drift velocity in silicon (77—300 K, 0.1—50 kV cm^{-1}). These measurements were performed by a time-of-flight technique: the time required by a charge carrier for travelling a sample of known length, L, is experimentally found thus yielding $v = v(\mathscr{E})$. Figure 1.7.1 shows $v = v(\mathscr{E})$ for electrons at a few temperatures. Figure 1.7.2 illustrates the conductivity anisotropy. When the electric field $\mathscr{E}$ is applied parallel to a $\langle 111 \rangle$ direction, all the six $\langle 100 \rangle$ valleys are oriented alike with respect to $\mathscr{E}$, and give the same contribution to the transport properties of the sample. The multivalley nature of the silicon conduction band is apparent only in determining the relaxation of electron distribution by the intervalley phonon scattering (Figure 1.7.3) [24].

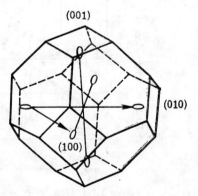

Figure 1.7.3. First Brillouin zone for Si, showing the position of conduction band valleys and (indicated by arrows) the two possible types of intervalley scattering mechanism (*after Duh and Moll*, [24]).

When $\mathscr{E}$ is parallel to the $\langle 100 \rangle$ direction, the two valleys $\langle 100 \rangle$ and $\langle \overline{1}00 \rangle$ have their longitudinal axes parallel to the electric field (Figure 1.7.3) so that they have states populated predominantly along these axes, characterized by a larger effective mass (0.9 m_0). The effective mass of electrons in the remaining four valleys, whose longitudinal axes are oriented perpendicularly to the field, is only 0.2 m_0 and the mobility of these 'light' electrons is higher than that of 'heavy' electrons in the first two valleys. At low temperatures and relatively low electric field intensities the (average) drift velocity is lower when the field is applied parallel to the $\langle 100 \rangle$ direction (Figure 1.7.2). This is due to the presence of intervalley scattering which determines the repopulation of these valleys with a net shift towards the heavy-electron valleys, because the light electrons are more easily accelerated and thus heated up by the field [24]. This repopulation, and thus the conduction anisotropy, is reduced when the electron energy increases and becomes comparable to the intervalley phonon energy such that the population tends to be equalized in all valleys. Therefore the anisotropy is reduced at higher lattice temperatures and very intense electric fields, as Figure 1.7.2 does indeed show.

The drift velocity saturation was first explained by electron interaction with optical phonons [5], [17], [18], with energy $\hbar\omega \simeq$ const. (Figure 1.6.2). At very intense fields the electron will be accelerated and gain energy $\hbar\omega_{opt}$ in such a short time that it is not scattered by acoustical modes. The maximum velocity attained is given by $m^* v_{max}^2/2 = \hbar\omega$ and the average velocity is $v_{sat} = v_{max}/2 = (\hbar\omega_{opt}/2m^*)^{1/2} =$ saturation drift velocity. This simple computation gives order of magnitude in agreement with the experimental value for germanium [18]. The saturation of electron velocity in silicon was also explained [25] by the field-dependence of the effective mass: the energy band is non-parabolic (experimental evidence) and the mass of free electrons heated by the electric field changes. The saturation velocity decreases slowly with increasing lattice temperature. A phenomenological theory explaining this decrease was given by Duh and Moll [24], who concluded that the principal loss mechanism is through the generation of optical phonons (the effective mass was assumed constant).

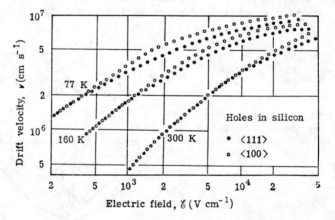

Figure 1.7.4. Hole drift velocities in silicon along the $\langle 111 \rangle$ and $\langle 100 \rangle$ directions, measured for three temperatures (*after Canali et al.*, [23]).

Figure 1.7.4 shows the field and temperature dependence of hole velocity in silicon, for $\langle 111 \rangle$ and $\langle 100 \rangle$ directions [23]. The higher $\langle 100 \rangle$ drift velocity may be associated to the lower effective mass (higher $\partial E/\partial k$) in this direction (Figure 1.3.5). At the highest applied field (50 kV cm^{-1}) and lower temperature (77 K) the hole drift velocity does not reach a limiting value [23]. The same paper presents a comparison of the available experimental data in the literature [23].

The theory of electron devices requires a simple analytical relationship $v = v(\mathscr{E})$ such that the empirical formula given below

$$v = v_{\text{sat}} \frac{\mathscr{E}/\mathscr{E}_c}{[1 + (\mathscr{E}/\mathscr{E}_c)^\beta]^{1/\beta}}, \quad (1.7.3)$$

with the following data indicated by Caughey and Thomas [26] for electrons and holes in silicon at 300 K

Table 1.7.1

Carrier velocity (1.7.3) in Silicon, at 300 K [26]

	v_{sat} (cm s^{-1})	$\mathscr{E}_c$ (V cm^{-1})	β
Holes	9.5×10^6	1.95×10^4	1
Electrons	1.1×10^7	8×10^3	2

The above approximation was evaluated for a certain range of field intensity and does not imply that the carrier velocity indeed saturates at v_{sat} for very intense fields.

All the results discussed in this section are for negligible impurity scattering. At very low temperatures this scattering mechanism could become important even at very low impurity concentrations.

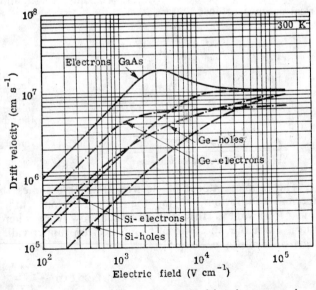

Figure 1.7.5. Measured carrier velocity *versus* electric field for high-purity Ge, Si and GaAs. For highly-doped samples the velocity is, in general, lower than indicated. However, at higher fields the velocity is essentially independent of doping (*after Sze*, [8]).

Figure 1.7.5 compares the electron and hole drift velocities in germanium, silicon and GaAs. The case of GaAs (negative differential mobility $dv/d\mathscr{E}$) will be discussed in the next section.

1.8 High-field Effects: II. Negative Mobility in Semiconductors

The velocity-field $v = v(\mathscr{E})$ curve measured for GaAs, InP and other semiconductors exhibits a range with negative slope: the differential mobility $dv/d\mathscr{E}$ is negative. This property is extremely important for construction of negative-resistance devices with applications in power generation and amplification. The negative-mobility effect was explained by considering the detailed band structure. A two-valley model for the conduction band of GaAs (and other semiconductors) is sufficient for a qualitative explanation. Consider the lower valley with a lower effective mass (higher mobility, μ_1) and the upper valley with a higher effective mass (lower mobility, μ_2). Only the lower valley is populated by electrons at low electric field intensities. At higher fields the electrons gain sufficient energy to be transferred in the upper valley and therefore the average carrier velocity may decrease with increasing field, as shown in Figure 1.8.1.

The conduction band of GaAs exhibits a minimum at $k = 0$ (Figure 1.3.5) with spherical energy surfaces and an effective mass $0.067\,m_0$ at the bottom of the band. However, the band is not parabolic, i.e. the scalar effective mass depends on energy. Photoemission studies indicated subsidiary minima at about 0.35, 0.45 and 0.95 eV above the central ($k = 0$) minimum. These were identified as being at the X_1, L_1, and X_3 points of the Brillouin zone (see Figure 1.3.6) [27]. The most important for negative-mobility are the six $\langle 100 \rangle$ minima located at the edge of the Brillouin zone and situated 0.36 eV above the $k = 0$ minimum. The integrated conductivity mass for $\langle 100 \rangle$ valleys is $0.35\,m_0$: these minima are represented by spherical energy surfaces [27].

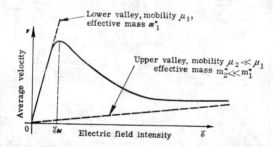

Figure 1.8.1. Average electron velocity in a two-valley semiconductor. The electron mass and mobility are m_1, μ_1 respectively in the lower valley and m_2, μ_2 in the upper valley (see also Figure 1.3.5). $\mathscr{E}_M$ is the threshold field for negative differential mobility.

The simplest model of electron transfer assumes that all the electrons are in thermal equilibrium, having the same electron temperature, T_n, in both $k = 0$ and $\langle 100 \rangle$ valleys (this is indeed so at low fields where $T_n =$ the lattice temperature). Therefore, the ratio of heavy electron (upper valley) density, n_2, and the light electron density, n_1, is [15]

$$\frac{n_2}{n_1} = \text{const.} \times \exp -\frac{\Delta}{kT_n}, \qquad (1.8.1)$$

where Δ is the energy distance between valleys ($\Delta = 0.36$ eV for GaAs). The electron temperature increases with increasing electric field ($T_n - T_{\text{lattice}} \propto$ energy) thus increasing the population of the upper valley. The average electron velocity is $(\mu_2 \to 0)$

$$v = \frac{n_1\mu_1 + n_2\mu_2}{n_1 + n_2}\mathscr{E} \simeq \frac{\mu_1\mathscr{E}}{1 + n_2/n_1} \qquad (1.8.2)$$

and decreases with increasing $\mathscr{E}$ only if n_2/n_1 increases sufficiently fast with $\mathscr{E}$. The negative mobility does not occur if the lattice temperature is too high or the energy difference Δ is too small. The density of states in the upper valley should be high thus increasing the probability of capture in this valley and reducing the threshold field (Figure 1.8.1).

However, the electron temperature in the two valleys cannot be the same. Whereas the electrons in the lower valley are heated by the field (high mobility), the upper valley electrons remain practically at the lattice temperature: they attain lower velocities due to much lower mobilities and also the electrons just transferred in this valley are no longer energetic.

A more exact treatment involves an approximate computer solution of two coupled Boltzmann equations (one for each valley). The basic simplification is the assumption of a particular form of the distribution function, for example a displaced Maxwellian distribution [28], [29].

The electron velocity-field dependence in GaAs was obtained by Fawcett et al. [27] by using the Monte Carlo method. The motion of an electron in the momentum space with a large number of random scattering processes is simulated by a computer. The time spent in each element of momentum space during this motion is proportional to the value of the distribution function in these elements. The time averaged velocity is just the drift velocity (average for an assembly of particles). The computer generates random numbers with equal probabilities and converts these numbers in complex probability distributions used to describe the time of flight between two collisions, the type of scattering and the state after scattering. Some results of these calculations [27], [30] are given below.

Figure 1.8.2 shows the spherically symmetric part of the distribution function. The distribution is close to Maxwellian at low field intensities. The carrier temperature is inversely proportional to the slope of the Maxwellian distribution function (an exponential is represented by a straight line). Below $\Delta = 0.36$ eV, the 'temperature' of the distribution (which in fact is not Maxwellian) increases rapidly with increasing field strength but above this energy (in the $k = 0$ valley) the temperature is close to that of the lattice, the distribution being almost parallel to the zero-field distribution function. These distributions are, of course, related to the scattering mechanisms involved. Below $\Delta = 0.36$ eV the dominant mechanism is optical polar scattering*. The rate of this process, however, decreases above 0.1 eV and this determines a large number of hot electrons. The rapid fall-off of the distribution function above $\Delta = 0.36$ eV is due to the intervalley scattering. A population inversion occurs at 15 and 25 kV cm^{-2} and can be explained [27] by the combined action of these scattering processes. The influence of impurity scattering was taken into account by Ruch and Fawcett [30], using the same Monte Carlo method and the

---

* The longitudinal optic phonon frequency at the centre of the zone ω_{opt}, is 5.37×10^{13} rad s^{-1}, and the static and high frequency dielectric constants are 12.53 and 10.82, respectively [27]. For theory and experimental data on frequency-dependent permittivity, see, for example, references [31], [32].

same parameters. It was shown that only the low energy part of the distribution function is altered by impurity scattering. The calculated drift mobility and drift velocity for various temperatures and carrier concentration are given below.

Figure 1.8.3 shows the low-field mobility versus reciprocal temperature for various impurity concentrations. At room temperature the mobility is 9, 8.6, 7.2

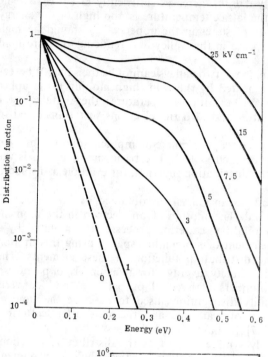

Figure 1.8.2. Energy dependence of the spherically symmetric part of the distribution function in the $k = 0$ valley for an intervalley deformation potential (see Section 1.6) of 1×10^9 eV cm^{-1} and for the field strengths indicated. The broken line is the zero-field Maxwellian distribution (after Fawcett et al., [27]).

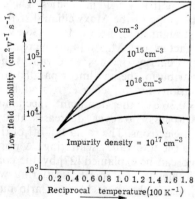

Figure 1.8.3. Calculated low-field mobility of electrons in GaAs as a function of temperature and impurity concentration (after Ruch and Fawcett, [30]).

and 5.3×10^3 cm^{2}V^{-1}s^{-1} for doping densities of 0, 10^{15}, 10^{16} and 10^{17}cm^{-3}, respectively [30] (at room temperature and with no impurity scattering, the upper valley mobility is about 325 cm^{2}V^{-1}s^{-1} [27]).

Figure 1.8.4 shows the calculated $v = v(\mathscr{E})$ dependence at six temperatures between 77 K and 500 K. At 77 K and low fields (a few hundred V cm^{-1}) the electron mobility decreases abruptly because of the onset of polar scattering which occurs only after the carriers were heated by the field. This abrupt change disappears when the impurity scattering is high (Figure 1.8.5b).

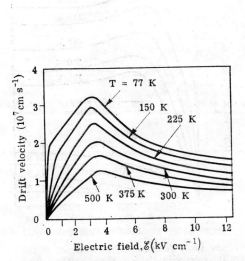

Figure 1.8.4. Calculated temperature dependence of the velocity-field characteristic in intrinsic GaAs. These results are in overall agreement with the results obtained by Clarke [33] by using a two-temperature model (*after Ruch and Fawcett*, [30]).

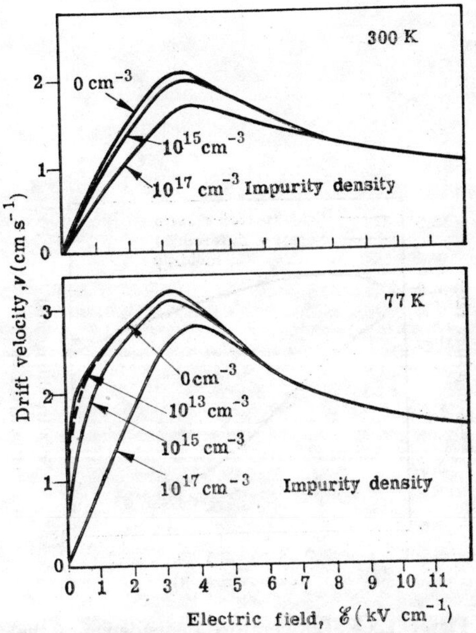

Figure 1.8.5. Influence of impurity scattering on the velocity-field characteristic for impurity concentrations of 10^{15} and 10^{17} cm^{-3} at 300 K (a) and 10^{13}, 10^{15} and 10^{17} cm^{-3} at 77 K (b) (*after Ruch and Fawcett*, [30]).

The threshold field for the onset of negative mobility is, surprisingly, almost unaffected by temperature (Figure 1.8.4) and impurity scattering at 300 K (Figure 1.8.5a). The ratio of the peak-velocity (at the threshold field) to the minimum (valley) velocity, and the magnitude of negative differential mobility are changed by the variation of the crystal temperature as shown in Figure 1.8.6 and this is important in designing microwave devices with GaAs. Another essential feature of the characteristic is the tendency of velocity saturation at higher electric fields [15], [27], [30], [33]. This is satisfactorily explained by a two-temperature model [34] corrected for the field dependence of the upper valley mobility, μ_2. However, the analysis by the Monte-Carlo method gives a somewhat different picture [30]. Figure 1.8.7 shows the field and temperature dependence of drift velocity in $\mathbf{k} = 0$ and $\langle 100 \rangle$ valleys. Note the negative mobility range for the $\mathbf{k} = 0$ valley velocity-field curves which gives a significant contribution to the overall negative differential mobility. Both $\mathbf{k} = 0$ and $\langle 100 \rangle$ valley velocities are only slightly field dependent in the high-field range. In the same field-range the population of the two valleys are comparable (equal at about 8 kV cm^{-1}, room temperature) although the fraction of electrons in the $\mathbf{k} = 0$ valley decreases somewhat with increasing field, as shown in Figure 1.8.8. Therefore, the high-field region of the room-temperature $v(\mathscr{E})$ characteristic shown in Figure 1.8.4 is essentially determined by the lower valley velocity.

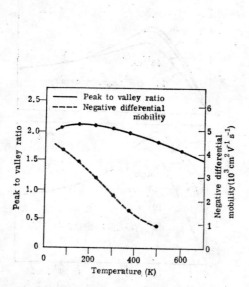

Figure 1.8.6. Temperature dependence of the peak-to-valley ratio and maximum negative differential mobility computed for GaAs. (after Ruch and Fawcett, [30]).

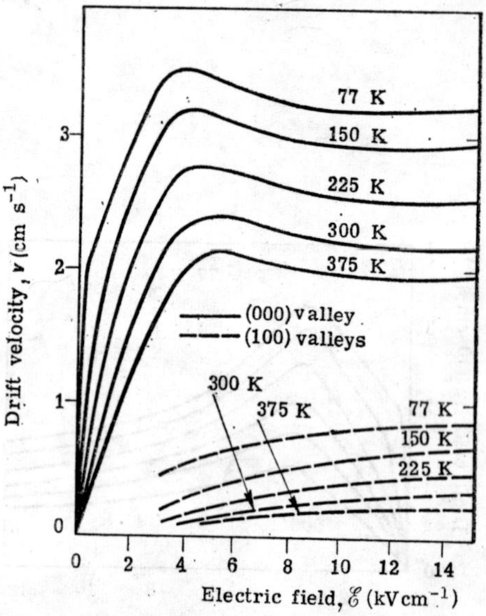

Figure 1.8.7. Temperature dependence of the drift velocity in the $k = 0$ and $\langle 100 \rangle$ valleys as functions of field intensity (after Ruch and Fawcett, [30]).

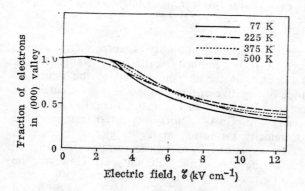

Figure 1.8.8. Fraction of electrons in the (000) valley as a function of field strength at several temperatures. (*after Ruch and Fawcett,* [30]).

The fact should be stressed that the above calculations depend upon details of the band structure and parameters of microscopic interactions which are not accurately known or directly accessible to experiment. For example, the good agreement obtained between the results of Monte Carlo calculations and the experimental results, as shown in Figure 1.8.9, is somewhat fortuitous because it was obtained for an intervalley deformation potential of 10^9 eV cm^{-1}, this potential being used as a parameter. Other data included in calculations are also uncertain such that it is not possible to conclude that this potential is indeed 10^9 eV cm^{-1}.

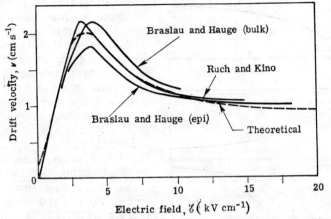

Figure 1.8.9. Experimental velocity-field characteristic determined by Ruch and Kino [35] for insulating GaAs and by Braslau and Hauge [36] for 1.5 Ω cm bulk and epitaxial material. The broken curve is calculated by the Monte Carlo method for an intervalley deformation potential of 1×10^9 eV cm^{-1}. (after Fawcett et al., [27]).

The experimental determination of $v = v(\mathscr{E})$ in negative-mobility semiconductors is complicated by the charge readjustment which takes place in the material bulk due to the negative differential mobility and negative (differential) relaxation time associated with this (see Chapter 7). These difficulties were avoided by Ruch and Kino [35] who used insulating GaAs (10^8 Ω cm) and barrier contacts (see the next chapter) to prevent electron injection by the metallic contacts when the electric potential is applied. Therefore the field in the sample is uniform. Electrons are injected by the beam of an electron gun and enter the material through the transparent metallic electrode. The electrons are injected in bunches and collected at the opposite end of the sample. The current pulse obtained in the external circuit has the width T_L and the electron velocity is $v = L/T_L$ where L is the sample length.

The velocity-field characteristic can be determined by measuring the microwave mobility (the microwave frequency is so high that no charge readjustment can take place) [32] — [36]. However, it is tacitly assumed that the quasi steady-state equilibrium between the various scattering processes (including the intervalley scattering) is attained in a time which is short compared to the oscillation period, which is not really the case [32], [37], [38]. For the same reason, the 'steady-state' $v = v(\mathscr{E})$ characteristic cannot be used when the electric field changes rapidly in time.

It is obvious that the Monte Carlo method cannot be applied in a non steady-state situation. The time response of the high-field electron distribution in GaAs was studied by Rees [37] who suggested a numerical iterative method. The steady-state distribution function $f = f(\mathbf{k})$ was generated by an iterative process with a large number of steps (the initial function used is arbitrary). It was shown that incorporation of a certain fictitious scattering process in these calculations allows us to consider time-dependent problems; each iteration becomes approximately equivalent to a time step. The small-signal response (in frequency) of the electrons is found by first calculating the response to a small step of electric field. The large-signal behaviour is studied by resetting the value of the electric field for each iteration. Figure 1.8.10 shows the response of the valley velocities and the drift velocity to a step in the electric field (from 6 to 5 kV cm^{-1} at $t = 0$). The overall response speed is limited by scattering processes within the $k = 0$ valley. This is due to the weak scattering of electrons in this valley for energies between 0.1 and 0.35 (0.36) eV, as previously explained (Figure 1.8.2). The response indicated in Figure 1.8.10 approximates the small-signal behaviour for the (mean) field of 5.5 kV cm^{-1}. The Fourier transformation of drift velocity step response yields the frequency response of small-signal mobility shown in Figure 1.8.11. Note the fact that the mobility is a complex quantity: the real part becomes positive above 80 GHz, the imaginary part shows that the free electrons act capacitively. This modifies the dielectric constant as shown in the same figure.

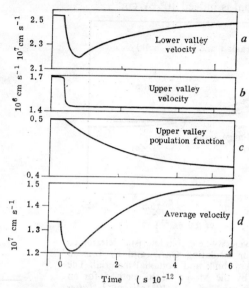

Figure 1.8.10. Response of the lower valley velocity (a), upper valley velocity (b), upper valley population fraction (c) and the average velocity (d) to a field step from 6 to 5 kV cm^{-1} at time $t = 0$. (*after Rees*, [37]).

The negative mobility depends upon both bias field and frequency. Figure 1.8.12 shows that the maximum negative mobility decreases appreciably with increasing frequency and occurs at higher fields; the threshold field also increases. At 50 GHz the maximum mobility decreases to about 35% of its low frequency value, whereas the threshold field increases from 3 to 4.5 kV cm^{-1}. The free-carrier dielectric constant is also bias- and frequency-dependent and may have an important influence upon the behaviour of negative-mobility devices.

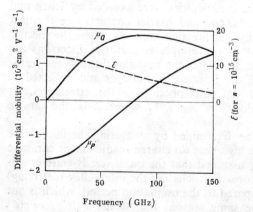

Figure 1.8.11. Frequency dependence of differential mobility for a bias field of 5.5 kV cm^{-1}: μ_P and μ_Q are the components, respectively, in phase and in quadrature with the alternating field and ε is the free-carrier dielectric constant derived from μ_Q. (*after Rees*, [37]).

Figure 1.8.13 shows large-signal results, from which a dynamic $v = v(\mathscr{E})$ curve can be obtained. Note the fact that at high fields the electron distribution follows closely the quasi steady-state variation which would be obtained if the signal fre-

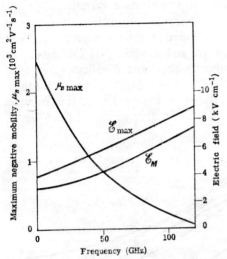

Figure 1.8.12. Frequency dependence of the threshold field $\mathscr{E}_M$, the field for maximum negative mobility ($\mathscr{E}_{max}$) and μ_{nmax} (maximum negative mobility). (*after Rees*, [37]).

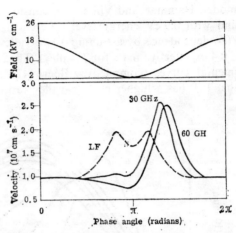

Figure 1.8.13. Large-signal response of the electron drift velocity to a 8 kV cm^{-1} alternating field with 10 kV cm^{-1} bias at low frequency (LF), 30 and 60 GHz. The upper curve shows the electric field and the lower curves show the velocity variation. (*after Rees*, [37]).

quency were low. At lower fields the electron response becomes slower as the field decreases and the dynamic distribution is completely different from the static one. When the field increases again the electron distribution tends to adjust itself and increases but with a delay which depends on frequency.

It was suggested by Hilsum and Rees [39] that *indium phosphide* (InP) is a better negative-mobility material due to the conduction mechanism involving three valleys. These valleys are: (1) the $\mathbf{k} = 0$ central valley, or $\varGamma$ valley (see Figure 1.3.6); (2) the L valley (four equivalent minima corresponding to the $\langle 111 \rangle$ directions); (3) the X valley (three equivalent minima for the $\langle 100 \rangle$ directions). These valleys are schematically shown in Figure 1.8.14, with the corresponding energy levels and effective masses. The parameters of the X and L valleys are less known but this is not too important because the mobilities are very low and the current is carried predominantly by $\varGamma$-valley electrons even when there is an important transfer to

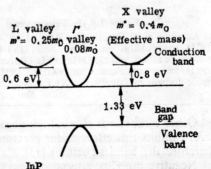

Figure 1.8.14. The $\varGamma$, X, L valleys in InP (with data from [40]).

X and L valleys. It was assumed, in analogy with Ge and GaAs, that the coupling between Γ and L minima should be weak while the $\Gamma - X$ and $L - X$ couplings are strong [40], and therefore that the field-heated electrons should be concentrated in the L valley, that the $v(\mathscr{E})$ characteristic should have a larger peak-to-valley ratio, larger negative mobility, lower sensitivity to temperature variations.

The experimental results were recently explained by considering a two-level model. Hammar and Vinter [40] considered the Γ and X valleys strongly coupled and with 0.9 eV energy level difference. Fawcett and Herbert [41] calculated the $v(\mathscr{E})$ dependence by assuming a two-valley model with Γ and L valleys (they used a 0.4 m_0 effective mass for L valley). The above results [40], [41] together with experimental curves (compiled by Boers [42]) are shown in Figure 1.8.15. It is obvious that InP has a higher peak-to-valley ratio and a higher threshold-field than GaAs.

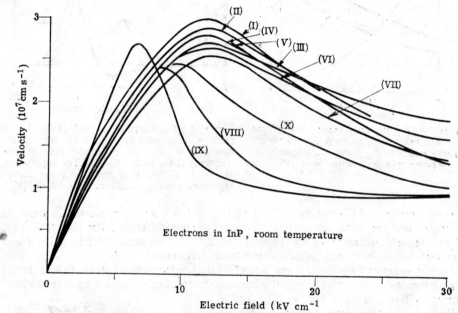

Figure 1.8.15. Velocity-field characteristic of electrons in InP.
(I) James et al. (1970), theory; (II) 'I Lam and Acket (1971), microwave experiment; (III) Boers (1971), space charge wave experiment; (IV) Hammar and Vitner [40], theory; (V) Glover (1972), microwave experiment (33 GHz); (VI) Glover (1972), microwave experiment (9 GHz); (VII) Nielsen (1972), microwave experiment (9 GHz); (VII) Rees and Hilsum (39), theory ; (IX) Prew (1972), dipole-domain measurement; (X) Fawcett and Herbert [41], theory (averaged). *after Boers* [42] *and Fawcett and Herbert* [41].

Other compounds also exhibit negative mobility but the threshold field is higher. InAs exhibits this effect under pressure (which modifies the interatomic spaces and therefore the band structure) [15].

Negative mobility effects also occur in Ge at low temperatures (below 130 K) [43] — [45].

1.9 High-field Effects: III. Impact Ionization and Avalanche Breakdown

At very high electric fields an electron can acquire sufficient energy such as to ionize a lattice atom in a collision. This high field impact ionization generates an electron-hole pair in the semiconductor. The electron and the hole move in opposite directions in the applied electric field and both of them can generate another electron-hole pair. The number of carriers increases rapidly during such an avalanche process which is similar to avalanche breakdown in gases [7], [8], [15].

The electron *ionization coefficient (rate)*, α_n, is defined as the probability of ionization when the electron travels the unit length and therefore has the dimensions (length)$^{-1}$. The similar coefficient for holes is denoted by α_p. The ionization coefficients are strongly field dependent. A simple theory of impact ionization shows that α depends exponentially upon the electric field [8], [15], [46]. Many experimental data for $\alpha = \alpha(\mathscr{E})$ are quite well approximated by [8]

$$\alpha = \alpha_0 \exp\left[-\left(\frac{b}{\mathscr{E}}\right)^m\right], \tag{1.9.1}$$

where $m \approx 1$ for Si and Ge and $m \approx 2$ for GaAs.

The experimental determination of α_n and α_p is extremely difficult. We note two major problems. First, the electric field required for considerable avalanche multiplication is so high and the avalanche current increases so rapidly that it is convenient to use a non-uniform sample where the avalanche occurs in a very thin region. Such a situation occurs in a reverse biased $p - n$ junction (Chapter 2). Because $\mathscr{E}$ is non-uniform, α will be also non-uniform. Secondly, it is practically impossible to separate the effect of ionizing electrons and ionizing holes such that both processes should be considered together. Therefore both $\alpha_n = \alpha_n(\mathscr{E})$ and $\alpha_p = \alpha_p(\mathscr{E})$ have to be determined concomitantly by measuring the current multiplication at various bias voltages.

Consider a high-field (avalanche region) bounded by $x = 0$ and $x = L$ planes. The electric field $\mathscr{E}$ is directed towards negative x, such that the electrons move in the positive x direction and holes in the opposite direction. Let J_n and J_p be the electron and hole current density, respectively at the x plane. At $x + \mathrm{d}x$ the electron current increases by $\mathrm{d}J_n$ and this increase is initiated by electrons ($\alpha_n J_n \mathrm{d}x$) and by holes ($\alpha_p J_p \mathrm{d}x$) such that in the steady-state

$$\frac{\mathrm{d}J_n}{\mathrm{d}x} = \alpha_n J_n + \alpha_p J_p. \tag{1.9.2}$$

The electron stream builds up in the positive x direction, and the hole current in the opposite sense

$$-\frac{\mathrm{d}J_p}{\mathrm{d}x} = \alpha_n J_n + \alpha_p J_p. \tag{1.9.3}$$

(of course $J_n + J_p = J = $ constant, independent of x). Assume that the avalanche is initiated by a pure electron current J_{no} at $x = 0$, whereas the hole current entering at the opposite boundary is zero. We have

$$x = 0, \quad J_n = J_{no}, \quad J_p = J - J_{no} \tag{1.9.4}$$

$$x = L, \quad J_n = J, \quad J_p = 0. \tag{1.9.5}$$

By integrating equations (1.9.2) and (1.9.3) with the above boundary condition we get [15], [47] – [49] the following 'ionization integral'

$$1 - \frac{1}{M_n} = \int_0^L \alpha_n \left[\exp - \int_0^x (\alpha_n - \alpha_p) \, dx' \right] dx, \tag{1.9.6}$$

where

$$M_n = \frac{J}{J_{no}} = \text{electron multiplication ratio} \tag{1.9.7}$$

and α_n, α_p are functions of position through the electric field $\mathscr{E} = \mathscr{E}(x)$. In the special case $\alpha_n = \alpha_p = \alpha$ one obtains

$$1 - \frac{1}{M_n} = \int_0^L \alpha \, dx. \tag{1.9.8}$$

If the avalanche is initiated by a pure hole current at $x = L$, then the hole multiplication ratio M_p will be given by

$$1 - \frac{1}{M_p} = \int_0^L \alpha_p \left[\exp \int_x^L (\alpha_n - \alpha_p) \, dx' \right] dx. \tag{1.9.9}$$

The avalanche breakdown is attained when the multiplication factor is infinite (the ionization integral is unity).

The above equations were approximately solved for $\alpha_n(\mathscr{E})$ and $\alpha_p(\mathscr{E})$, either by assuming a simple field profile [50] or introducing * α as given by equation (1.9.1) with $m = 1$ and α_0, b unknown [47]. A more exact calculation was performed recently by Grant [48]. In this way the ionization rates were derived from measurements of multiplication ratios on p–n junctions. Pure electron and hole currents were generated by exposing the sample to light of suitable wavelength.

Figure 1.9.1 shows electron and hole ionization rates measured for silicon [47], [48], [50] *versus* the reciprocal field. The results obtained by Grant [48] are well fitted by exponentials, as follows

$$\alpha_n = 6.2 \times 10^5 \exp(-1.08 \times 10^6/\mathscr{E}) \; \text{cm}^{-1} \tag{1.9.10}$$

---

* Another approach is to assume $\alpha_n/\alpha_p = \gamma = $ const. [47] and calculate an effective ionization rate $\alpha_{\text{eff}} = \dfrac{\gamma - 1}{\ln \gamma} \alpha_n(\mathscr{E})$ for the avalanche breakdown condition $\int_0^L \alpha_{\text{eff}}(\mathscr{E}) \, dx = 1$.

for
$$2.4 \times 10^5 \, \text{Vcm}^{-1} < \mathscr{E} < 5.3 \times 10^5 \, \text{Vcm}^{-1}$$

$$\alpha_p = 2.10 \times 10^6 \exp(-1.97 \times 10^6/\mathscr{E}) \, \text{cm}^{-1} \qquad (1.9.11)$$

for
$$2.0 \times 10^5 \, \text{Vcm}^{-1} < \mathscr{E} < 5.3 \times 10^5 \, \text{Vcm}^{-1}.$$

At higher fields, $\mathscr{E} > 5.3 \times 10^5 \, \text{V cm}^{-1}$

$$\alpha_n = 5.0 \times 10^5 \exp(-0.99 \times 10^6/\mathscr{E}) \, \text{cm}^{-1} \qquad (1.9.12)$$

$$\alpha_p = 5.6 \times 10^5 \exp(-1.32 \times 10^6/\mathscr{E}) \, \text{cm}^{-1}. \qquad (1.9.13)$$

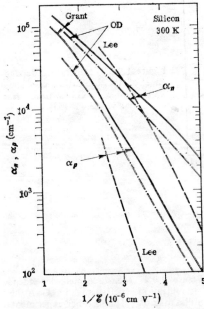

Figure 1.9.1. Room temperature measured ionization rates in silicon as a function of reciprocal field. Data of Lee et al., [50], van Overstraeten and de Man [47] (OD) and Grant [48]). (*after Grant*, [48]).

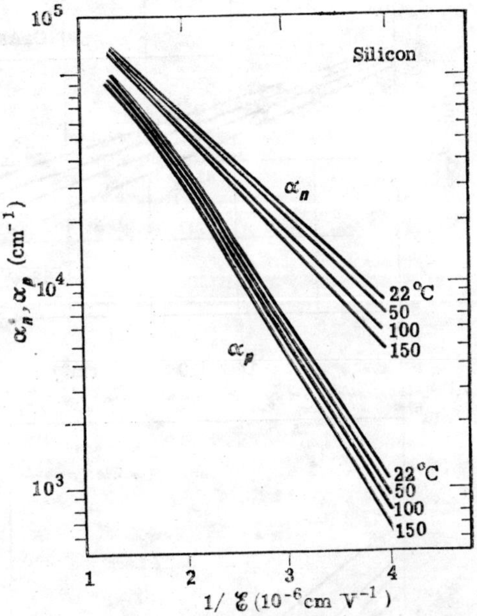

Figure 1.9.2. Experimental field and temperature dependence of ionization rates in silicon. (*after Grant*, [48]).

Figure 1.9.2 indicates the measured temperature dependence for α_n and α_p in silicon (22–150 °C). It was found that α_n and α_p are of the form (1.9.1) where $\alpha_0 \simeq$ const. whereas b changes with temperature. Equations (1.9.10) and (1.9.11) become $(2.4 \times 10^5 \, \text{Vcm}^{-1} < \mathscr{E} < 5.3 \times 10^5 \, \text{Vcm}^{-1})$ [48]

$$\alpha_n \approx 6.2 \times 10^5 \exp - [(1.05 \times 10^6 + 1.3 \times 10^3 T)/\mathscr{E}] \qquad (1.9.14)$$

$$\alpha_p \approx 2.0 \times 10^6 \exp - [(1.95 \times 10^6 + 1.1 \times 10^3 T)/\mathscr{E}] \qquad (1.9.15)$$

where T is in degrees centigrade ($T \geqslant 22$ °C).

The ionization rates in GaAs were usually determined by assuming $\alpha_n = \alpha_p = \alpha$ and $\alpha(\mathscr{E})$ obeying a law of the form 1.9.1. The majority of experimental results indicate $m = 2$ (see reference [51] for a review of α measurements in GaAs). Hall and Leck [52], for example, found (at 300 K) with $\alpha_n = \alpha_p = \alpha_0 \times \exp[-(b/\mathscr{E})^2]$ and a simplified calculation of $\mathscr{E}$ at a $p-n$ junction

$$\alpha_0 = 2.0 \times 10^5 \text{ cm}^{-1}, \quad b = 5.5 \times 10^5 \text{ Vcm}^{-1}, \quad \mathscr{E} = (2.5 - 5) \times 10^5 \text{ Vcm}^{-1}. \quad (1.9.16)$$

The temperature dependence (-20 °C to 80 °C) of α_0 and b in GaAs was also determined by Hall and Leck [53] (see Figure 1.9.3). The temperature coefficients are: $7 \pm 3 \times 10^{-4}$ K^{-1} for α_0 and $12 \pm 4 \times 10^{-4}$ K^{-1} (both increase with increasing temperature).

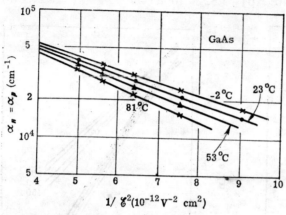

Figure 1.9.3. Temperature dependence of the ionization coefficient $\alpha_p = \alpha_n = \alpha$ in GaAs. (*after Hall and Leck*, [53]).

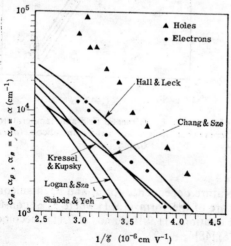

Figure 1.9.4. Measured field-dependence of ionization coefficients in GaAs by Stillman *et al.* [51] and other experimental results obtained by Hall and Leck [52], [53], Kressel and Kupsky (1966), Chang and Sze (1969), Logan and Sze (1966) and Shabde and Yeh (1970). (*after Stillman et al.*, [51]).

Stillman *et al.* [51] have shown recently that the ionization rates in GaAs are not equal (and $\alpha_p > \alpha_n$). The measurements were made on Schottky barrier avalanche photodiodes: the multiplication ratio *versus* bias voltage yielded α_n and α_p according to a mathematical procedure described in reference [54]. These results are plotted in Figure 1.9.4 together with data obtained by other authors which assume *à priori* $\alpha_n = \alpha_p$.

1.10 Carrier Transport in Non-homogeneous Semiconductors

When a uniform stationary electric field is applied to a homogeneous semiconductor, the current is carried by drift with a drift mobility, which will depend upon the electric field intensity if the carrier average kinetic energy increases considerably (hot carriers). Therefore, for an isotropic material (scalar drift mobilities)

$$\mathbf{J} = \mathbf{J}_n + \mathbf{J}_p = en\mu_n(\mathscr{E})\mathscr{E} + ep\mu_p(\mathscr{E})\mathscr{E}. \qquad (1.10.1)$$

In a non-homogeneous semiconductor material the carrier concentration is non-uniform. Non-uniform carrier concentration also arises due to carrier injection or non-uniform semiconductor heating. A diffusion current, proportional to the concentration gradient is calculated for the electronic gas such that the total electron current is written

$$\mathbf{J}_n = en\mu_n\mathscr{E} + eD_n\nabla n, \qquad (1.10.2)$$

where D_n is the diffusion constant for electrons. A similar relation holds for hole transport

$$\mathbf{J}_p = ep\mu_p\mathscr{E} - eD_p\nabla p. \qquad (1.10.3)$$

At high fields $\mu_n = \mu_n(\mathscr{E})$ and it is assumed* that D_n also depends on $\mathscr{E}$ and because $\mathscr{E}$ is non-uniform the diffusion constant is frequently introduced under the gradient sign, such that equation (1.10.3) is written

$$\mathbf{J}_n = en(x)\mu_n(\mathscr{E})\mathscr{E} + e\nabla[D_n(\mathscr{E})n(x)], \qquad (1.10.4)$$

where $\mathscr{E} = \mathscr{E}(x)$. There was some controversy as to whether equation (1.10.2) or equation (1.10.4) yield the 'true' value of the diffusion current when $D_n = D_n(\mathscr{E})$.

However, neither (1.10.3) nor (1.10.4) are correct, except in very limited circumstances. In fact, even the concept of field-dependent mobility and field-dependent diffusion constant is misleading [55] — [60]. When a non-uniform field exists (in Schottky barriers, p—n junctions and other devices) the change in average kinetic energy of charge carriers depends on the local product of electric field and current and can be positive (carrier heating) as well as negative (carrier cooling) [55], [57]. The mobility and the diffusion constant depend upon the distribution function (they arise from the Boltzmann equation (1.5.10) as shown below) and this distribution function depends on electron and hole currents (which are both uniform** in the steady state), on carrier densities (non-uniform) and electric field (non-uniform). Therefore the mobility and the diffusion constant are not uniquely determined by the local field strength but also depend on the current [55], [57], [60].

---

* In a certain special case (see below) the diffusion constant is proportional to the mobility, $D = \dfrac{\mu kT}{e}$ (Einstein relation).

** We further consider a one-dimensional model only.

The Boltzmann equation (1.5.10) written for a non-homogeneous steady distribution is

$$\frac{\mathbf{F}}{\hbar}\nabla_{\mathbf{k}}f + \mathbf{v}\nabla_{\mathbf{r}}f = \left(\frac{\partial f}{\partial t}\right)_c, \qquad (1.10.5)$$

where $\mathbf{F} = -e\mathscr{E}$ is the electric force and the collision terms is approximated as in equation (1.5.11), with $\tau = \tau_n$ as an equivalent relaxation time depending upon energy and position. In a one-dimensional coordinate system, with $\mathscr{E}$ directed along the x axis, equation (1.10.5) becomes

$$f(x, \mathbf{k}) = f_0(x, \mathbf{k}) + \tau_n\left(\frac{e\mathscr{E}}{\hbar}\nabla_{\mathbf{k}}f_0 - v_x\frac{\partial f_0}{\partial x}\right), \qquad (1.10.6)$$

where f_0 is the symmetrical part of the distribution function and f_0 was substituted for f in the left-hand side of equation (1.10.5) because the perturbation of f was assumed to be small (see Section 1.5) as in Section 1.6. The electron momentum is $m_n^* v = \hbar k$ and the electron energy measured with respect to the bottom of the band is equal to its average kinetic energy

$$E = \frac{m_n^* v_x^2}{2}. \qquad (1.10.7)$$

Equation (1.10.6) becomes [56], [57]

$$f = f_0 + \tau_n\left(e\mathscr{E}v_x\frac{\partial f_0}{\partial x} - v_x\frac{\partial f_0}{\partial x}\right) \qquad (1.10.8)$$

and this distribution function will be replaced in the expression (1.5.8) of the electric current. The first term (f_0) gives, by definition, no contribution to this current. The second term in equation (1.10.8) yields the drift current (proportional to $\mathscr{E}$), whereas the last one yields the diffusion current. The drift current is

$$(J_n)_{\text{drift}} = en\mu_n\mathscr{E}, \qquad (1.10.9)$$

where

$$\mu_n = \frac{e}{m_n^*}\left\langle\tau_n\left[1 + \frac{2}{3}\frac{d(\ln \tau_n)}{d(\ln E)}\right]\right\rangle \qquad (1.10.10)$$

with $\langle\ \rangle$ denoting the average (1.5.18) over the distribution function. In our case (parabolic bands) this average may be written

$$\langle\mathscr{F}\rangle = \frac{\int f_0 E^{1/2} \mathscr{F}(E)\,dE}{\int f_0 E^{1/2}\,dE}. \qquad (1.10.11)$$

If τ_n is independent of energy, we reobtain the familiar expression $\mu_n = e\tau_n/m_n^*$ (Section 1.5).

1. Electronic Conduction in Semiconductors

We shall now write the diffusion current. We assume first

$$\partial \tau_n / \partial x = 0, \qquad (1.10.12)$$

which occurs if (a) the temperature gradient is absent, and (b) either the density of ionized impurities is independent of x or the impurity scattering is negligible [57]. In this case [56], [57]

$$(J_n)_{\text{diff}} = e \frac{d}{dx}(D_n n), \quad D_n = \frac{2}{3m_n^*} \langle \tau_n E \rangle \qquad (1.10.13)$$

and the electron current will be

$$J_n = e n \mu_n \mathscr{E} + e \frac{d}{dx}(D_n n), \qquad (1.10.14)$$

whereas, under the similar restrictive conditions, we have

$$J_p = e p \mu_p \mathscr{E} - e \frac{d}{dx}(D_p p). \qquad (1.10.15)$$

We stress the fact that, in contrast to equation (1.10.4), the mobilities and diffusion constants are not dependent of the electric field alone and cannot be replaced by empirical relations $\mu = \mu(\mathscr{E})$ and $D = D(\mathscr{E})$ derived from experiment.

Effective calculation of μ and D implies the knowledge of the distribution function. But this function can be determined only by using the following set of fundamental equations: particle conservation equations, momentum conservation equations (which in this model of quasi-free electrons is equivalent to the current continuity equations [61]), energy conservation equations (all the above written for both electrons and holes), Poisson equation. At this time we recall the four major assumptions already made, namely:

(a) The asymmetric part $f - f_0$ of the distribution function f is small compared to the symmetrical part, f_0;

(b) the collision term is proportional to $f - f_0$, i.e. the collisions are described by a single relaxation time. This is possible [57] for elastic scattering (acoustic mode scattering and ionized impurity scattering) and for velocity-randomizing collisions (non-polar optical mode and intervalley scattering). A relaxation time can sometimes be defined for optical polar scattering. Carrier-carrier collisions cannot be accounted for by a relaxation time. The probability of various collisions mechanisms can be added and thus an equivalent relaxation time can be introduced [57];

(c) scattering relaxation time independent of x (equation (1.10.13));

(d) free-electron approximation (constant scalar effective mass).

To simplify somewhat the picture of current transport we shall assume that f_0 is known and has the Maxwellian form with the lattice temperature replaced by the electron temperature, T_n (see Section 1.6 and equation (1.10.7))

$$f_0 = \frac{1}{2} n (2\pi \hbar)^3 (2\pi m_n^* k T_n)^{-3/2} \exp\left(-\frac{E}{kT_n}\right). \qquad (1.10.16)$$

Here n is given by equation (1.5.7) with $f = f_0$ and T_n depends, in general, on x and should be determined in principle from the energy balance equation. By introducing f_0 in equations (1.10.10) and (1.10.13), one obtains the Einstein relationship

$$D_n = \frac{kT_n}{e} \mu_n. \tag{1.10.17}$$

A similar relation holds for holes. Equation (1.10.17) is known, in general, for the case $T_n = T$ = lattice temperature: the symmetrical part of the distribution function is identical to the thermal equilibrium distribution function. When the electrons are in thermal equilibrium with the lattice but the distribution is degenerate (see equation (1.4.6)), the Einstein relation $D_n = \frac{kT}{e} \mu_n$ will be replaced by a more complicated dependence [56], [62].

The electron current (1.10.14) becomes

$$J_n = en\mu_n \mathscr{E} + kT_n\mu_n \frac{dn}{dx} + en \frac{dD_n}{dx}, \tag{1.10.18}$$

where

$$\frac{dD_n}{dx} = \frac{dD_n}{dT_n} \frac{dT_n}{dx} = \frac{d}{dT_n}\left(\frac{kT_n}{e}\mu_n\right) \frac{dT_n}{dx} = \frac{k}{e}\left(\mu_n + T_n \frac{d\mu_n}{dT_n}\right) \frac{dT_n}{dx} \tag{1.10.19}$$

and $\mu_n = \mu_n(T_n)$ is given by equations (1.10.10), (1.10.11) and (1.10.16) and depends upon $\tau_n = \tau_n(E)$. The spatial dependence of carrier temperature results from energy conservation considerations [58]

$$\mathscr{E} J_n = \frac{1}{\tau_e}\left[\frac{3}{2} nk(T_n - T) + \frac{dS}{dx}\right] \tag{1.10.20}$$

(electron current only, τ_e is the energy relaxation time $\tau_e = \tau_e(T_n)$ and S is the flow of heat transported by carriers) and therefore T_n depends upon field *and* current and may increase or decrease with x. The evaluation of the dS/dx term is extremely complicated: Berz [58] had done this for a $p-n$ junction at very small currents. The third component of the current (1.10.18) may add to or subtract from the total current: its existence is related to the non-uniform carrier temperature (or to the non-uniform field and charge).

If the assumption (1.10.12) is removed, another component should be added* to the total electron current. Stratton [56] shows that

$$J_n = en\mu_n \mathscr{E} + eD_n \frac{dn}{dx} + en\left(\frac{dD_n}{dx} - U\right) \tag{1.10.21}$$

where D_n is given by equation (1.10.13) and U is defined by

$$U = \frac{2}{3m_n^*}\left\langle \frac{\partial \tau}{\partial x} E \right\rangle \tag{1.10.22}$$

---

* van Overstraeten et al. [12] have shown that in heavily doped semiconductors an additional *drift* component occurs due to the position dependent band structure (Section 1.4).

and has velocity dimensions. The added term, proportional to U, is associated to thermal diffusion if τ depends upon the non-uniform lattice temperature. If the carriers are in thermal equilibrium with the lattice ($T_n \simeq T$) then (note that $dD_n/dx = 0$)

$$J_n = en\mu_n \mathscr{E} + eD_n \frac{dn}{dx} + eD_n^T n \frac{dT}{dx}, \qquad (1.10.23)$$

where D_n^T is a constant for thermal diffusion [56].

The current equations for many-valley semiconductors (such that GaAs or InP) are much more complicated. The correct start is to solve two or more coupled Boltzmann equations (one for each valley). Blötekjaer [60] has indicated that as far as diffusion is neglected the average velocity may be considered a function of the local field for slow variations in space and time. When diffusion is taken into account it is impossible to use an approximate single-gas model. It was shown [60] that for GaAs it is convenient to consider that the electron mobility is a function of the average velocity of carriers in the lower valley instead of the electric field (we have seen that the upper valley electrons are practically at the lattice temperature). Even this simplified analysis is too complicated for analytical study of device behaviour.

1.11 Particle Continuity Equations

It was shown that the Boltzmann transport equation is in fact a conservation particle equation written with respect to the distribution function $f_0 = f_z(x, y, z, k_x, k_y, k_z, t)$ which describes statistically the particle ensemble. This equation can be conveniently integrated in the momentum space so as to give [15]

$$\frac{\partial n}{\partial t} + \nabla(n\mathbf{v}_n) = 0, \qquad (1.11.1)$$

which is the *continuity equation* for electrons. Such a relation is fundamental in hydrodynamics (fluid flow). It can be also written by using the total electron current $\mathbf{J}_n = -en\mathbf{v}$, such that for electrons and holes, respectively

$$\frac{\partial n}{\partial t} = \frac{1}{e}\nabla\mathbf{J}_n, \quad \frac{\partial p}{\partial t} = -\frac{1}{e}\nabla\mathbf{J}_p. \qquad (1.11.2)$$

Clearly, in the steady state the divergence of both electron and hole current is zero.

The carrier continuity equations (1.11.2) should be modified to include carrier generation-recombination processes (in a semiconductor a free charge carrier can be created or annihilated):

$$\frac{\partial n}{\partial t} = G_n - U_n + \frac{1}{e}\nabla\mathbf{J}_n, \qquad (1.11.3)$$

$$\frac{\partial p}{\partial t} = G_p - U_p - \frac{1}{e}\nabla\mathbf{J}_p. \qquad (1.11.4)$$

G_n and G_p are the electron and hole generation rates due to the external influences such as optical excitation* or impact ionization (Section 1.9). The generation rate for impact ionization is

$$G_n = G_p = \alpha_n n v_n + \alpha_p p v_p. \qquad (1.11.5)$$

U_n and U_p are the recombination rates for electrons and holes, respectively, and describe thermal electron-hole recombination (and generation) or trapping in defect centers. We briefly discuss the recombination processes (for trapping see Chapter 4).

Consider first a semiconductor at thermal equilibrium. The processes of electron-hole pair generation are exactly balanced by the opposite processes of electron-hole recombination. The simplest case is that of direct band-to-band recombination accompanied by the emission of a photon (radiative recombination) or by transfer of the electron energy to another carrier (Auger process, which is the inverse of impact ionization) [64]. Other recombination processes involve one or multiple energy levels in the band gap and take place sequentially. For example, in the first step the electron is trapped by the recombination centre. In the next step the same electron falls in the valence band which is equivalent to hole trapping by the same centre (and therefore annihilation of the electron-hole pair). This capture process can be described in terms of capture probability and capture cross-section, just like the collision process (Section 1.6) [5] — [8], [64].

Consider an extrinsic semiconductor out of thermal equilibrium, where $\Delta n = \Delta p$ excess carriers occur by optical excitation, injection, etc. It is only the recombination of minority carriers which really matters in continuity equations. Consider for example an n-type semiconductor where $n_n = n_{n0} + \Delta n$ and $p_n = p_{n0} + \Delta p$, where $\Delta n, \Delta p \ll n_{n0}$ (low injection conditions). The minority carrier density at thermal equilibrium is very small, $p_{n0} = n_i^2/n_{n0} \ll n_{n0}$. The number of excess holes is important for hole continuity equation but it is insignificant for electron continuity. The recombination of *excess* holes is described by the rate

$$U_p = \sigma_p v_{\text{th}} N_r (p_n - p_{n0}) = \frac{p_n - p_{n0}}{\tau_p}, \qquad (1.11.6)$$

where σ_p is the hole capture cross-section, v_{th} the thermal velocity $\sqrt{3kT/m^*}$, N_r the density of recombination centres and τ_p the minority hole lifetime. The recombination probability of excess holes is $1/\tau_p$. If $p_n < p_{n0}$, U_p becomes negative and the thermal generation overcomes the recombination in order to restore the equilibrium. A similar relation is valid for minority electrons.

The recombination processes are essential for the bipolar devices [7], [8], but less important and in many cases negligible in unipolar (one-carrier) devices.

---

* A sufficiently energetic photon may ionize at atom and thus generate an electron or create an electron-hole pair (when a valence electron is brought into the conduction band).
Sah and Lindholm [63] introduced a supplementary term directly in the Boltzmann transport equation, to acount for the recombination-generation-trapping-tunneling events, when the frequency of these events becomes comparable to the collision frequency (minority carrier transport in the thin base region of microwave transistors, etc.).

1.12 Basic Semiconductor Equations

The basic semiconductor equations are the Maxwell equations, the current transport equations and the continuity equations.

The Maxwell's equations for a homogeneous and isotropic semiconductor in an electric field are

$$\mathbf{J}(t) = \mathbf{J}_{\text{particle}}(\mathbf{r}, t) + \frac{\partial \mathcal{D}(\mathbf{r}, t)}{\partial t} \tag{1.12.1}$$

$$\nabla \cdot \mathcal{D} = \rho(\mathbf{r}) \quad \text{(Poisson equation)} \tag{1.12.2}$$

$$\mathcal{D} = \varepsilon \mathcal{E} \quad (\varepsilon = \text{const.}) \tag{1.12.3}$$

$$\nabla \times \mathcal{E} = 0 \quad (\mathcal{E} = -\nabla \Phi) \tag{1.12.4}$$

(see List of Symbols).

The transport equations are presented below. The total current sums up the contribution of all energy bands where conduction occurs. In the most familiar case we have

$$\mathbf{J}_{\text{particle}}(\mathbf{r}, t) = \mathbf{J}_n(\mathbf{r}, t) + \mathbf{J}_p(\mathbf{r}, t), \tag{1.12.5}$$

where $\mathbf{J}_n$ is the electron current and $\mathbf{J}_p$ the hole current. These currents depend upon the applied field, $\mathcal{E}$. If the semiconductor is homogeneous and isotropic the energy bands are parabolic and the charge carriers are almost in thermal equilibrium with the lattice at the uniform temperature T and have a non-degenerate (Maxwellian) distribution in energy, then the electron and hole currents will be

$$\mathbf{J}_n = en\mu_n \mathcal{E} + kT\mu_n \nabla n. \tag{1.12.6}$$

$$\mathbf{J}_p = ep\mu_p \mathcal{E} - kT\mu_p \nabla p. \tag{1.12.7}$$

In the above equations μ_n and μ_p are material constants (temperature dependent) and can be found experimentally. In a more general case equations (1.12.6) and (1.12.7) should be replaced by more complicated expressions (Section 1.11) which contain parameters hardly accessible to experiment. If the diffusive contribution of any kind are neglected, the current will have only drift components. At high field intensities the charge carriers are no longer in thermal equilibrium with the lattice and the carrier mobilities μ_n, μ_p cannot be regarded as constants. Assume a pure electron current carried by drift

$$\mathbf{J}_{\text{particle}} = \mathbf{J}_n = en(\mathbf{v}_n)_{\text{drift}} = en\mathbf{v}_n. \tag{1.12.8}$$

The drift velocity depends on the excess carrier energy and is proportional to the local $\mathcal{E}\mathbf{J}$ product, i.e. $\mathcal{E}(en\mathbf{v}_n)$, if $\dfrac{dS}{dx}$ can be neglected in equation (1.10.21). The excess energy per one charge carrier is proportional to $\mathcal{E}v_n$ and thus v_n depends upon $\mathcal{E}v_n$ and can be regarded, formally, as a local function of electric field. Therefore, if diffusion is neglected

$$J_n = env_n(\mathcal{E}), \tag{1.12.9}$$

where the electric field can be non-uniform and $v_n = v_n(\mathscr{E})$ can be found experimentally in a uniform sample of appreciable length and at given doping and temperature and represents a characteristic of a certain semiconductor. However, this is true if the heat transported by the charge carriers in the particular device studied (dS/dx) can, indeed, be neglected; this is not always the case [58].

The particle continuity equations were already given in equations (1.11.3) and (1.11.4). For the special case of minority carrier low-level injection we have

$$\frac{\partial n_p}{\partial t} = G_n - \frac{n_p - n_{p0}}{\tau_n} + \frac{1}{e}\nabla \mathbf{J}_n, \qquad (1.12.10)$$

$$\frac{\partial p_n}{\partial t} = G_p - \frac{p_n - p_{n0}}{\tau_p} - \frac{1}{e}\nabla \mathbf{J}_p. \qquad (1.12.11)$$

For the special case of steady-state diffusion with zero electric field (neutral region) and zero external generation rate, we obtain from equations (1.12.6) and (1.12.10) (one dimensional case)

$$-\frac{n_p - n_{p0}}{\tau_n} + D_n \frac{\partial^2 n_p}{\partial x^2} = 0. \qquad (1.12.12)$$

The carrier concentration $n_p(x)$ will be determined with $n_p(0) = $ const. and $n_p(\infty) = n_{p0}$ (far from $x = 0$ the semiconductor is neutral). Therefore, we have

$$n_p(x) = n_{p0} + [n_p(0) - n_{p0}]\exp\left(-\frac{x}{L_n}\right), \qquad (1.12.13)$$

where

$$L_n = \sqrt{D_n \tau_n} = \text{the diffusion length} \qquad (1.12.14)$$

is a characteristic length for the decay of the minority carrier concentration in excess of thermal equilibrium value.

Grove [7] shows that the operation of many semiconductor devices is determined by their tendency to return to equilibrium. The minority carrier lifetime and diffusion length are quantities which characterize this tendency. However, the operation of unipolar devices in general is considerably different, because of the absence of recombination and neutrality.

References

1. C. KITTEL, 'Introduction to solid state physics', John Wiley, New York, 1971.
2. J. M. ZIMAN, 'Principles of the theory of solids', Cambridge Univ. Press, London, 1964.
3. A. J. DEKKER, 'Solid state physics', Macmillan, London, 1965.
4. T. S. HUTCHINSON and D. C. BAIRD, 'The physics of engineering solids', John Wiley, New York, 1963.
5. R. A. SMITH, 'Semiconductors', Cambridge Univ. Press, London, 1964.
6. J. L. MOLL, 'Physics of semiconductors', McGraw-Hill, New York, 1964.
7. A. S. GROVE, 'Physics and technology of semiconductor devices', John Wiley, New York, 1967.

8. S. M. SZE, 'Physics of semiconductor devices', John Wiley, New York, 1969.
9. E. M. CONWELL, 'Properties of silicon and germanium', Part II, *Proc. I.R.E.*, **46**, 1281–1300 (1958).
10. E. SCHIBLI and A. G. MILNES, 'Deep impurities in silicon', *Mater. Sci. Eng.*, **2**, 173–180 (1967).
11. D. DALE KLEPPINGER and F. A. LINDHOLM, 'Impurity concentration dependence of the density of states in semiconductors', *Solid-St. Electron.*, **14**, 199–206 (1971).
12. R. J. VAN OVERSTRAETEN, H. J. DEMAN and R. P. MERTENS, 'Transport equations in heavy dopes silicon', *IEEE Trans. Electron Dev.*, **ED–20**, 290–298 (1973).
13. R. K. JAIN and R. J. VAN OVERSTRAETEN, 'Theoretical calculations of the Fermi level and of other parameters in phosphorus doped silicon at diffusion temperatures', *IEEE Trans. Electron Dev.*, **ED–21**, 155–165 (1974).
14. E. M. CONWELL, 'High field transport in semiconductors', in *Solid State Physics*, Supplement 9, Academic Press, New York, 1967.
15. J. E. CARROLL, 'Hot electron microwave generators', Edward Arnold, London, 1970.
16. S. M. SZE and J. C. IRVIN, 'Resistivity, mobility and impurity levels in GaAs, Ge and Si at 300 K', *Solid-St. Electron.*, **11**, 599–602 (1968).
17. W. SHOCKLEY, 'Hot electrons in germanium and Ohm's law', *Bell. Syst. Techn. J.*, **30**, 990–1034 (1951).
18. B. R. NAG, 'Hot-carrier d.c. conduction in elemental semiconductors', *Solid-St. Electron.*, **10**, 385–400 (1967).
19. M. COSTATO and L. REGGIANI, 'The displaced distribution approach to the solution of transport problems in semiconductors', *Atti. Sem. Mat. Fis. Modena*, **18**, 144–169 (1969).
20. V. V. PARNAJAPE and E. DE ALBA, 'Decrease of electron temperature by electric fields', *Proc. Phys. Soc.*, **85**, 945 (1965).
21. K. BLÖTEKJAER, 'Mobility and temperature of electrons in polar semiconductors', *Arkiv för Fysik*, **33**, 105–120 (1967); 'Cooled electrons in polar semiconductors', *Phys. Lett.*, **24 A**, 15–17 (1967).
22. A. E. MCCOMBS, Jr and A. G. MILNES, 'Calculation of drift velocity in silicon at high electric fields', *Int. J. Electron.*, **24**, 573–578 (1968).
23. C. CANALI, G. OTTAVIANI and A. ALBERIGI QUARANTA, 'Drift velocity of electrons and holes and associated anisotropic effects in silicon', *J. Phys. Chem. Solids*, **32**, 1707–1720 (1971).
24. C. Y. DUH and J. L. MOLL, 'Temperature dependence of hot electron drift velocity in silicon at high electric field', *Solid-St. Electron.*, **11**, 917–932 (1968).
25. M. COSTATO and S. SCAVO, 'Hot-electron variable effective mass in silicon', *Il Nuovo Cimento*, **56B**, 343–348 (1968).
26. D. M. CAUGHEY and R. E. THOMAS, 'Carrier mobilities in silicon empirically related to doping and field', *Proc. IEEE*, **55**, 2192–2193 (1967).
27. W. FAWCETT, A. D. BOARDMAN and S. SWAIN, 'Monte Carlo determination of electron transport properties in gallium arsenide', *J. Phys. Chem. Solids*, **31**, 1963–1990 (1970).
28. P. N. BUTCHER and W. FAWCETT, 'The intervalley transfer mechanisms of negative resistivity in bulk semiconductors', *Proc. Phys. Soc.*, **86**, 1205–1219 (1965); 'Calculation of the velocity-field characteristic for gallium arsenide', *Phys. Lett.*, **21**, 489–490 (1966).
29. W. HEINLE, 'Influence of nonparabolicity on the Gunn-effect characteristic in GaAs in the displaced Maxwellian approximation', *Phys. Lett.*, **27A**, 629–630 (1968).
30. J. G. RUCH and W. FAWCETT, 'Temperature dependence of the transport properties of gallium arsenide determined by a Monte Carlo method', *J. Appl. Phys.*, **41**, 3843–3849 (1970).
31. S. M. WU, E. BRIDGES and K. C. KAO, 'Microwave complex permittivities of Si and GaAs semiconductors in the presence of high steady electric fields', *Int. J. Electron.*, **31**, 233–241 (1971).
32. S. KANEDA and M. ABE, 'Microwave anisotropy and frequency dependence of hot electron in n-type GaAs', *Japan J. Appl. Phys.*, **10**, 1396–1904 (1971).

33. P. P. BOHN, 'A nonsaturating velocity-field approximation for improved invariant domain analysis', *Proc. IEEE*, **58**, 1397–1398 (1970); J. POKORNY and F. JELINEK, 'Experimental nonsaturating velocity-field characteristic of GaAs', *Proc. IEEE*, **60**, 457–458 (1972).
34. R. J. CLARKE, 'The effect of variations in temperature and in material parameters on the velocity-field characteristic of gallium arsenide', *Radio Electron. Eng.*, **43**, 389–393 (1973).
35. J. S. RUCH and G. S. KINO, 'Measurement of the velocity-field characteristic of gallium arsenide', *Appl. Phys. Lett.*, **10**, 40–43 (1967); 'Transport properties of GaAs', *Phys. Rev.*, **174**, 921–931 (1968).
36. N. BRASLAU and P. S. HAUGE, 'Microwave measurement of the velocity-field characteristic of GaAs', *IEEE Trans. Electron Dev.*, **ED–17**, 616–622 (1970).
37. H. D. REES, 'Time response of the high-field electron distribution function in GaAs', *IBM J. Res. Dev.*, **13**, 537–542 (1969).
38. W. R. CURTICE and J. J. PURCELL, 'Analysis of the LSA mode including effects of space charge and intervalley transfer time', *IEEE Trans. Electron Dev.*, **ED–17**, 1048–1060 (1970).
39. C. HILSUM and H. D. REES, 'Three-level oscillator: a new form of transferred-electron device', *Electron. Lett.*, **6**, 277–278 (1970).
40. C. HAMMAR and B. VINTER, 'Calculation of the velocity-field characteristic of n-InP', *Solid-St. Commun.*, **11**, 751–754 (1972).
41. W. FAWCETT and D. C. HERBERT, 'High-field transport in indium phosphide', *Electron. Lett.*, **9**, 308–309 (1973).
42. P. M. BOERS, 'Comment on determination of the velocity field characteristic for n-type indium phosphide from dipole-domain measurement', *Electron. Lett.*, **9**, 134–135 (1973).
43. W. FAWCETT and E. G. S. PAIGE, 'Negative differential resistance in n-type germanium', *Electron. Lett.*, **3**, 505–507 (1967).
44. R. S. DE BIASI, S. S. YEE, 'Bulk negative differential conductance in n-type germanium at low electric fields', *Proc. IEEE*, **58**, 256–257 (1970); 'Bulk negative differential conductance in high-purity n-type germanium', *J. Appl. Phys.*, **41**, 3863–3869 (1970).
45. R. S. DE BIASI and S. S. YEE, 'Current oscillations in n-type Ge at low temperatures', *J. Appl. Phys.*, **43**, 609–614 (1972).
46. G. A. BARAFF, 'Distribution junctions and ionization rates for hot electrons in semiconductors', *Phys. Rev.*, **128**, 2507–2517 (1962).
47. R. VAN OVERSTRAETEN and H. DE MAN, 'Measurement of the ionization rates in diffused silicon $p-n$ junctions', *Solid-St. Electron.*, **12**, 583–608 (1969).
48. W. N. GRANT, 'Electron and hole ionization rates in epitaxial silicon at high electric fields', *Solid-St. Electron.*, **16**, 1189–1203 (1973).
49. C. D. BULUCEA and D. C. PRISECARU, 'The calculation of the avalanche multiplication factor in silicon $p-n$ junctions taking into account the carrier generation (thermal or optical) in the space-charge region', *IEEE Trans. Electron Dev.*, **ED–20**, 692–701 (1973).
50. C. A. LEE, R. A. LOGAN, R. I. BATDORF, J. J. KLEIMACK and W. WIEGMAN, 'Ionization rates of holes and electrons in silicon', *Phys. Rev.*, **134**, A761–A773 (1964).
51. G. E. STILIMAN, C. M. WOLFE, J. A. ROSSI and A. G. FOYT, 'Unequal electron and hole ionization coefficients in GaAs', *Appl. Phys. Lett.*, **24**, 471–474 (1974).
52. R. HALL and J. H. LECK, 'Avalanche breakdown of gallium arsenide $p-n$ junctions', *Int. J. Electron.*, **25**, 529–537 (1968).
53. R. HALL and J. H. LECK, 'Temperature dependence of avalanche breakdown in gallium arsenide $p-n$ junctions', *Int. J. Electron.*, **25**, 539–546 (1968).
54. M. H. WOODS, W. C. JOHNSON and M. A. LAMPERT, 'Use of Schottky barrier diodes for measurements of impact ionization coefficients', *Solid-St. Electron.*, **16**, 381–394 (1973).

55. R. STRATTON, 'Carrier heating or cooling in a strong built-in electric field', *J. Appl. Phys.*, **40**, 4582–4583 (1969).
56. R. STRATTON, 'Semiconductor current-flow equations (diffusion and degeneracy)', *IEEE Trans. Electron Dev.*, **ED–19**, 1288–1292 (1972).
57. M. SÁNCHEZ, 'Carrier heating or cooling in semiconductor devices', *Solid-St. Electron.*, **16**, 549–557 (1973).
58. F. BERZ, 'Carrier heating effects in junctions at very low currents', *Solid-St. Electron.*, **16**, 1067–1071 (1973).
59. K. BLÖTEKJAER, 'Waves in semiconductors with nonconstant mobility', *Electron. Lett.*, **4**, 357–358 (1968).
60. K. BLÖTEKJAER, 'Transport equations for electrons in two-valley semiconductors', *IEEE Trans. Electron Dev.*, **ED–17**, 38–47 (1970).
61. C. GOLDBERG, 'Electric current in semiconductor space-charge region', *J. Appl. Phys.*, **40**, 4612–4614 (1969).
62. S. S. LI and F. A. LINDHOLM, 'Alternative formulation of generalized Einstein relation for degenerate semiconductors', *Proc. IEEE*, **56**, 1256–1257 (1968).
63. C. T. SAH and F. A. LINDHOLM, 'Transport in semiconductors with low scattering rate and at high frequencies', *Solid-St. Electron.*, **16**, 1447–1449 (1973).
64. I. S. BLACKMORE, 'Semiconductor statistics', Pergamon Press, London, 1962.

2
Carrier Injection in Solids

2.1 Ohmic Contacts and Injecting Contacts

All solid-state devices presented in this book require electric contacts to provide the path of the electric current flowing through the device. The electric current inside the device is sustained, at least in part, by charge transport. Therefore, charge carriers must be injected in and extracted from solid.

Carrier injection can be provided by a metal-semiconductor (insulator) contact. The pn junctions, high-low (p^+p or n^+n) junctions and heterojunctions can also act as injecting contacts. Charge carriers can be generated in a certain region of the crystal by optical radiation or by avalanche multiplication (at high electric fields).

We discuss first the metal-semiconductor (MS) injecting contact. The *injecting contact* encountered in unipolar devices must be distinguished from the *ohmic contact*: The injecting contact may be either ohmic or non-ohmic. The properties of ohmic and injecting contacts are compared in Table 2.1.1.

The term ohmic contact was often used for injecting contacts to insulators, having the meaning of an infinite reservoir of charge carriers. We shall utilize here

Table 2.1.1
Properties of Ohmic and Injecting Contacts

Properties \ Contacts	Ohmic	Injecting (Ohmic or non-ohmic)
Contact region (semiconductor in the vicinity of the contact)	Neutral	Depleted, accumulated or inverted semiconductor
Carrier concentration in semiconductor at contact	Equilibrium carrier concentration (infinite recombination velocity)	Large deviations from equilibrium concentrations are possible
Type of carrier injected	Majority carriers	Either majority or minority carriers
Contact resistance	Linear, very low	May be non-linear and relatively large
The effect upon the device characteristic	The $J-V$ characteristic does not explicitly depend upon contact properties	May depend or not upon contact properties

2. Carrier Injection in Solids

the term *ideal injecting contact* for a contact which provides infinite one-carrier injection in a semiconductor or insulator region.

The charge carrier transfer at the metal-semiconductor interface is limited by the potential barrier which forms at this contact.

The magnitude and shape of the barrier is determined by various interface processes, as shown below.

2.2 Potential Barrier at a Metal-semiconductor Contact

The potential barrier occurring at the interface between a metal and a semiconductor is influenced by phenomena such as the existence of an interfacial layer, surface trapping centres, etc.

The potential barrier at a 'clean' and 'ideal' metal-semiconductor (MS) contact should be determined by the work function difference. This idealized situation is depicted in Figure 2.2.1 for an n-type semiconductor. In Figure 2.2.1a the semiconductor and the metal are isolated and each of them represents a separate system at thermal equilibrium. The energy diagram does not indicate the details at the surface, within a few interatomic distances. The minimum energy necessary for an electron in metal to escape into vacuum is $e\phi_M$ (Figure 2.2.1a), the work function. The semiconductor work function is also defined as the difference between the energy of the free space and the energy of the Fermi level.

If the semiconductor is electrically connected to the metal (Figure 2.2.1b) then

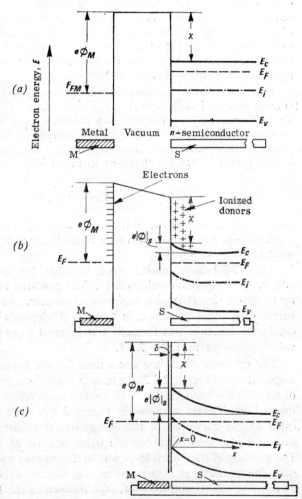

Figure 2.2.1. Effect of work function difference upon the metal-semiconductor contact: (a) energy diagram for isolated semiconductor and metal; (b) metal and semiconductor still separated by vacuum but constituting a system at thermal equilibrium, established by charge carrier transfer through the conductive wire connecting the semiconductor and the metal; (c) metal and semiconductor brought into intimate contact.

the charge carrier transfer will be possible, the MS system will eventually reach equilibrium and the Fermi levels will be aligned at the same energy. If the work function difference

$$e\phi_{MS} = e\phi_M - (\chi + E_c - E_F) \qquad (2.2.1)$$

is positive, then electrons from the *n*-type semiconductor will pass through the conductive wire into the metal, until the equilibrium is reached. The work function difference determines a potential distribution, an electric field and a double layer of electric charges positive (uncompensated) ionized donors in semiconductor and excess electrons on the metal surface. The space-charge layer at the semiconductor surface will be relatively thick because the donor concentration is, generally, much less than the density of metal lattice atoms.

If the metal and semiconductor are brought into intimate contact, then the contact potential difference ϕ_{MS} will occur practically on the semiconductor surface layer (Figure 2.2.1c). The distance δ between the semiconductor and the metal is very small, of the order of interatomic distances (a few angstroms), and is practically transparent for electrons tunnelling through the barrier. Therefore, the effective potential barrier * seen by electrons in metal is

$$e\phi_{Bn} = e\phi_M - \chi \qquad (2.2.2)$$

and may be much lower than the metal work function. This means that free electron injection into the semiconductor is possible at much lower temperatures than thermionic emission in vacuum.

Table 2.2.1 indicates the work function for several metals and the electron affinity for a few semiconductors [1]. The potential barrier $e\phi_{Bn}$ calculated as shown by equation (2.2.2) is also indicated. However, the actual height of the potential barrier (measured as shown in Chapter 3) depends upon the way the metal-semiconductor junction was formed, and in most cases departs considerably from the values indicated in Table 2.2.1.

The presence of *surface states* modifies the barrier height [2]. We consider first a special case (Figure 2.2.2), namely a single acceptor level in the forbidden band of an *n*-type semiconductor. The energy diagram of the isolated semiconductor, at thermal equilibrium is shown in Figure 2.2.2a. A surface space-charge layer (indicated by the curvature of the energy-levels) occurs and, here, the semiconductor surface is depleted of mobile electrons because of trapping in the acceptor level. The position of the Fermi level will be determined by imposing the following condition: the total charge of uncompensated ionized donors in the surface layer should be equal to the magnitude of charge trapped in the surface states.

---

* Depending upon the work-function difference, the semiconductor surface layer may be accumulated, depleted or inverted. There is not a clear boundary between depletion or inversion (see also Chapter 3). We shall consider *here* that a depletion region is characterized by a negligible mobile charge density compared to the ionized impurity density. The potential-barrier at a depleted MS contact is named the Schottky barrier.

Figure 2.2.2b shows the energy diagram of the MS system at thermal equilibrium: the semiconductor surface is still not in contact with the metal, but an electric contact between the semiconductor bulk and the metal was provided, as indicated in Figure 2.2.1b. Because the semiconductor *surface* work function is lower than the

Table 2.2.1

Theoretical Barrier Heights ($e\Phi_{Bn} = e\Phi_M - \chi$), in eV

Metal work function (eV)	Electron affinity (eV)	Silicon 4.05	Ge GaP 4.0	GaAs 4.07	CdS 4.8
Au	4.77	0.72	0.77	0.70	−0.03
Al	4.25	0.20	0.25	0.18	−0.55
Ag	4.30	0.25	0.30	0.23	−0.50
Cu	4.40	0.35	0.40	0.33	−0.40
W	4.50	0.45	0.50	0.43	−0.30
Mo	4.27	0.22	0.27	0.20	−0.53
Ti	3.92	−0.13	−0.08	−0.15	−0.88
Pt	5.30	1.25	1.30	1.23	0.50
Pd	4.78	0.73	0.78	0.71	−0.02
Hg	4.50	0.45	0.50	0.43	−0.30
In	3.77	−0.28	−0.23	−0.30	−1.03
Cr	4.60	0.55	0.60	0.53	−0.20
Ni	4.50	0.45	0.50	0.43	−0.30

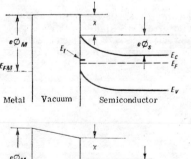

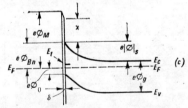

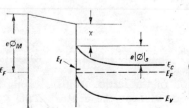

Figure 2.2.2. Metal - n-type semiconductor contact with high-density surface states (acceptor level located at E_t): (a) energy diagram of semiconductor isolated from the metal; (b) metal and semiconductors separated by vacuum but connected through a conductive wire (thermal equilibrium established); (c) intimate metal-semiconductor contact at thermal equilibrium.

metal work function, part of electrons trapped in surface states will leave the semiconductor and form a sheet of negative charge on the metal surface. A small supplementary curvature of the energy bands may correspond to a large amount of charge transferred from the surface states.

If the metal is brought very close to the semiconductor surface (δ is of the order of the interatomic distances in Figure 2.2.2 c) the electron transfer is possible just at the interface. Even in this case the position of the Fermi level with respect to the energy bands in semiconductor will change very little provided that the *density of surface states* is sufficiently large. Therefore, the effective potential barrier $e\phi_{Bn}$ seen by electrons in the metal will be almost independent of the metal work function (the Fermi level is 'pinned' by the surface states at a certain determined value).

The surface states occur due to local lattice imperfections, absorbed and adsorbed impurity atoms, oxide films, etc. Figure 2.2.3 shows the potential barriers measured for several metal contacts on *clean* semiconductor surfaces. The results are plotted *versus* metal electronegativity, a number which differs by a constant from the metal work function, in eV. The plot for a certain semiconductor would approach a straight line of slope unity if the surface states were absent. Figure 2.2.3 confirms this for ZnS but not for GaAs [3]. Therefore *some* clean semiconductors surfaces do exhibit surface states and others do not. It is well known that there exist surface states which are an unavoidable result of the termination of the lattice (Tamm-Shockley states). Mead [3] points out that there are two classes of semiconductors characterized by a different position of the surface levels in the forbidden band. For semiconductors characterized by a low interaction between atoms (ionic bound is a typical example) the energy of the surface states is close to the valence and conduction band edges. The 'upper' surface states are unoccupied and the 'lower' surface states are fully occupied. The surface Fermi level may move within large limits without affecting

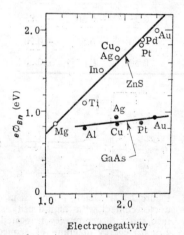

Figure 2.2.3. Barrier energies $e\phi_{Bn}$ measured for various metals on ZnS and GaAs *versus* the metal electronegativity. (*after Mead*, [3]).

the occupancy of surface states. For such semiconductors the energy barrier is determined by work function difference.

In semiconductors where the interaction of the lattice atoms is strong (such as covalent crystals) the surface states are located near the centre of the forbidden band and almost determine the barrier energy, as shown by Figure 2.2.3 for contacts to GaAs.

Theoretical calculations of the Tamm-Shockley states for certain particular semiconductors are almost non-existent [3], [4]. Experimental data [5] plotted in Figure 2.2.4 show that for gold contacts to certain semiconductors the Fermi level at the

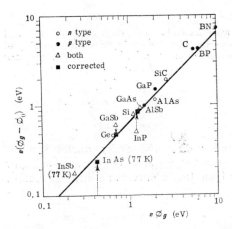

Figure 2.2.4. The location of Fermi level at surface (Figure 2.2.2c) for gold contacts on various surface-controlled materials as a function of band gap $e\phi_g$. The line corresponds to a barrier height equal to $\frac{2}{3} e\phi_g$. The full squares correspond to a theoretical correction. (*after Mead and Spitzer*, [5]).

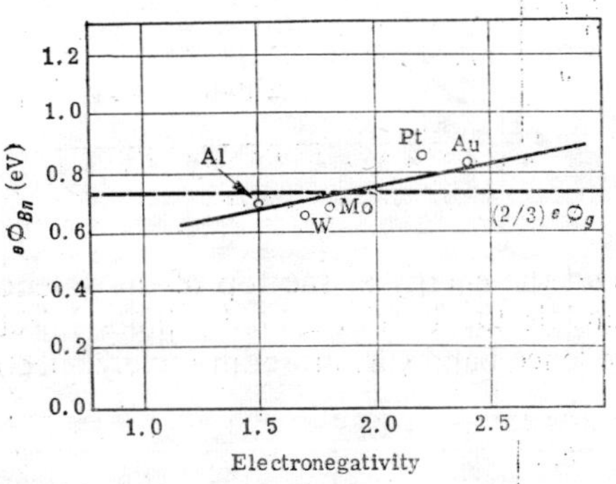

Figure 2.2.5. Barrier height for electrons at various metal-silicon contacts, plotted as a function of metal electronegativity. The dotted line corresponds to the Fermi level 'pinned' two-thirds of band gap below the conduction band. (*after Yu*, [6]).

surface is 'pinned' to about two-thirds of the forbidden gap below the conduction band (solid line in the above figure). Many other semiconductors (InAs, InP, GaSb, CdTe, CdSe, etc.) do not obey the above rule.

Figure 2.2.5, reproduced after Yu [6], shows that the energy barrier $e\phi_{Bn}$ for several metal contacts to silicon is almost independent of metal electronegativity. These data were compiled from papers of various authors. A systematic determination of $e\phi_{Bn}$ for various metal contacts on n-type silicon was reported by Turner and Rhoderick [7]. Their results are shown in Figure 2.2.6 and indicate that $e\phi_{Bn}$ depends upon the surface treatment before the metal deposition. The barrier measured for 'vacuum cleaved' silicon surface are essentially independent of the metal work function, thus indicating a possible high surface state density (and this is what we expect from a 'clean' surface). However [7], the barrier heights of the same metal-silicon contacts evaporated on chemically prepared surfaces do depend upon the metal used and are close to the values given by equation (2.2.2) which ignores the surface states. The authors [6] indicated a possible effect of an interfacial (oxide) layer at contacts on chemically prepared surfaces, a situation to which we refer later*. Smith and Rhoderick [9] also measured the Schottky barriers formed by several metals on p-type silicon chemically prepared surface. The Schottky barrier height, $e\phi_{Bp}$ is now equal to the difference between the energy of the Fermi level

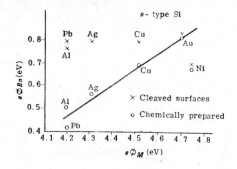

Figure 2.2.6. Barrier height $e\phi_{Bn}$ versus metal work function, measured for various metals and silicon surface treatments. (*after Turner and Rhoderick*, [7]).

and the energy at the top of the valence band, both at the semiconductor surface, and is the barrier seen by holes injected from the metal into the semiconductor valence band (or, more intuitively, electrons from the valence band into the metal).

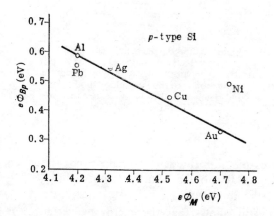

Figure 2.2.7. Barrier height $e\phi_{Bp}$ versus metal work function, measured for Schottky barriers on p-type silicon. (*after Smith and Rhoderick*, [9]).

The sum of the two barrier heights, for electrons and holes, respectively, should equal the band-gap. This is what the experimental results, Figure 2.2.7 [9], do indeed show, thus indicating that the surface state parameters for chemically prepared surfaces were almost identical in both experiments [7], [9].

A number of experimental data on MS barriers, $e\phi_B$, are compiled in Table 2.2.2. The fact should be stressed, however, that the measured $e\phi_B$ depends not only upon the method of contact realization (as shown above for silicon) but also upon the experimental technique used to determine the barrier height (see Chapter 3).

---

* The surface states due to lattice interruption seem to be annihilated due to the fact that the valence bonds of the surface Si atoms are now satisfied. Similar results were obtained by Mönch [8] covering the silicon surface with a monolayer of Cs (although the adsorbed Cs atoms determine the occurrence of other surface states, thus modifying the work function).

2. Carrier Injection in Solids

Table 2.2.2

Measured Schottky Barrier Heights, at 300 K (compiled by Sze, reference [1])

Semiconductor and surface preparation	Metal	Barrier-height measurement (eV)		
		from current-voltage characteristic	from capacitance-voltage characteristic	photo-electric measurement
n–Si (Chem.)	Au	0.79	0.80	0.78
p–Si (Chem.)	Au	0.25		
n–Si (Chem.)	Mo	0.59	0.57	0.56
n–Si (Back sputtering)	PtSi	0.85	0.86	0.85
p–Si (Back sputtering)	PtSi	0.20		
n–Si (Chem.) (see also the text)	W	0.67	0.65	0.65
n–Ge (Vacuum cleave)	Au		0.45	
	Al		0.48	
n–GaAs (Vacuum cleave)	Au		0.95	0.90
	Pt		0.94	0.86
	Ag		0.93	0.88
	Cu		0.87	0.82
	Al		0.80	0.80
p–GaAs (Vacuum cleave)	Au		0.48	0.42
	Al		0.63	0.50
n–InP (Vacuum cleave)	Au		0.49	0.52
	Ag		0.54	0.57
p–InP (Vacuum cleave)	Au		0.76	
n–CdS (Chem.)	Pt	1.2	1.2	1.1
	Au	0.68	0.66	0.68
	Pd	0.61	0.59	0.62
	Cu	0.47	0.41	0.50

2.3 The Surface Layer at Thermal Equilibrium

The magnitude of the energy barrier at the metal semiconductor contact determines whether, for a given semiconductor (n- or p-type), the surface layer at the contact will be an accumulation, a depletion or an inversion layer. Figure 2.3.1 represents these situations for contacts on an n-type semiconductor, at thermal equilibrium. The electron and hole density in the bulk are given by

$$n|_{x \to \infty} = n_n = N_c \exp \frac{E_F - E_c}{kT} = n_i \exp \frac{E_F - E_i}{kT} \qquad (2.3.1)$$

$$p|_{x \to \infty} = p_n = N_v \exp \frac{E_v - E_F}{kT} = n_i \exp \frac{E_i - E_F}{kT} \qquad (2.3.2)$$

(non-degenerate statistics, E_F at least a few kT within the band-gap) where n_i, E_i are the intrinsic concentration and intrinsic Fermi level, respectively. Similar relations are valid for the surface layer at thermal equilibrium but $E_c = E_c(x)$ and

$E_v = E_v(x)$ (Figure 2.3.1). E_c, for example, is the potential energy of an electron at the bottom of the conduction band and is related to the electrostatic potential ϕ by

$$E_c = -e\phi + \text{const.}, \qquad (2.3.3)$$

therefore the band curvature is directly related to the potential distribution in the surface layer. We assume here $\phi = 0$ for $x \to \infty$ (in the semiconductor bulk) and define the following normalized quantities

$$u = \frac{e\phi}{kT}, \quad u_F = \frac{e\phi_F}{kT} = \frac{(E_i)_{\text{bulk}} - E_F}{kT}, \qquad (2.3.4)$$

such that the mobile carrier densities can be written [10]

$$n = n_i \exp(u - u_F) \qquad (2.3.5)$$

$$p = n_i \exp(u_F - u) = \frac{n_i^2}{n}, \qquad (2.3.6)$$

and depend upon the distance x from the semiconductor surface (Figure 2.3.1a) through $u = u(x)$ ($\phi = \phi(x)$). The intrinsic concentration, equation (1.4.14)

$$n_i = (N_c N_v)^{1/2} \exp\left(\frac{-e\phi_g}{kT}\right), \quad e\phi_g = E_c - E_v \qquad (2.3.7)$$

depends upon the semiconductor and temperature.

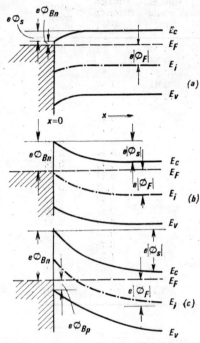

Figure 2.3.1. Surface layer at a contact between a metal and an n-type semiconductor: (*a*) accumulation layer; (*b*) depletion layer (Schottky barrier); (*c*) inversion layer.

We denote by ϕ_s the surface potential, positive, negative or zero ($e|\phi_s|$ is the total band bending) and introduce

$$u_s = \frac{e\phi_s}{kT}. \qquad (2.3.8)$$

The surface concentrations ($x = 0$) are:

$$n_s = n(0) = n_i \exp(u_s - u_F) \qquad (2.3.9)$$

$$p_s = p(0) = n_i \exp(u_F - u_s). \qquad (2.3.10)$$

The accumulation, depletion and inversion layers are defined with respect to surface carrier concentrations and surface potential, for both n-type and p-type

2. Carrier Injection in Solids

semiconductors, as shown in Table 2.3.1[*]. These results are valid for thermal equilibrium and non-degenerate statistics. The surface properties are specified through the normalized surface potential equation (2.3.8). In a special case, the surface conditions are determined by an MS contact. Table 2.3.1 also shows the relation between the potential barrier and the semiconductor properties for accumulation, depletion and inversion layers. Figure 2.3.1 is an illustration for an n-type semiconductor. Note that the whole potential barrier (space-charge layer) occurs in a semiconductor. The relations for an MS contact in Table 2.3.1 will not be valid, for instance, if an interface layer which is not transparent for electrons (comparable to or thicker than the mean free path) is present.

Table 2.3.1
Semiconductor Surface Layers

		Surface layer		
		Accumulation	Depletion	Inversion
n-type semiconductor ($u_F < 0$)	Surface carrier concentrations n_s, p_s	$n_s > n_n$	$n_n > n_s > p_s$	$p_s > n_s$
	Normalized surface potential u_s	$u_s > 0$	$u_s < 0$, $\|u_s\| < \|u_F\|$	$u_s < 0$, $\|u_s\| > \|u_F\|$
	Metal semiconductor barrier ϕ_{Bn}	$e\phi_{Bn} < E_{c(bulk)} - E_F$	$e\phi_{Bn} > E_{c(bulk)} - E_F$, $e\phi_{Bn} < E_c - E_i$	$e\phi_{Bn} > E_{c(bulk)} - E_F$, $e\phi_{Bn} > E_c - E_i$
p-type semiconductor ($u_F > 0$)	Surface carrier concentrations n_s, p_s	$p_s > n_s$	$p_p > p_s > n_s$	$n_s > p_s$
	Normalized surface potential u_s	$u_s < 0$	$u_s > 0$, $u_s < u_F$	$u_s > 0$, $u_s > u_F$
	Metal semiconductor barrier ϕ_{Bp}	$e\phi_{Bp} < E_F - E_{v(bulk)}$	$e\phi_{Bp} > E_F - E_{v(bulk)}$, $e\phi_{Bp} < E_i - E_v$	$e\phi_{Bp} > E_F - E_{v(bulk)}$, $e\phi_{Bp} > E_i - E_v$

The potential distribution in the surface layer should be determined from Poisson equation

$$\frac{d^2\phi}{dx^2} = -\frac{e}{\varepsilon}(p - n + N_D - N_A), \qquad (2.3.11)$$

where $N_D - N_A > 0$ for an n-type and $N_D - N_A < 0$ for a p-type semiconductor. The doping is assumed uniform ($N_D - N_A$ = constant). By using equations (2.3.4) — (2.3.6), Poisson equation becomes

$$\frac{d^2 u}{dx^2} = L_D^{-2}[\sinh(u - u_F) + \sinh u_F], \qquad (2.3.12)$$

---

[*] Note that the definition of the inversion layer differs from that just used in the preceding Section (see also Chapter 3).

where u_F, defined by equation (2.3.4), has the value

$$u_F = \operatorname{arg\,sinh}\left(-\frac{N_D - N_A}{2n_i}\right), \qquad (2.3.13)$$

and

$$L_D = (kT\varepsilon/2e^2 n_i)^{1/2} \qquad (2.3.14)$$

is the so-called *Debye length*. The boundary conditions are

$$x \to +\infty;\ u \to 0,\ \frac{du}{dx} \to 0. \qquad (2.3.15)$$

$$x = 0;\ u = u_s. \qquad (2.3.15')$$

An analytical solution of equation (2.3.12) does not exist. This equation will be again discussed in the next chapter. Here we consider two extreme situations.

(A) Very small deviations from neutrality. If u is small compared to unity, $\exp u$ will be approximated by $1 + u$ and equation (2.3.12) becomes

$$\frac{d^2 u}{dx^2} \approx \frac{u}{(L_D')^2}, \qquad (2.3.16)$$

where

$$L_D' = L_D(\cosh u_F)^{-1/2} = \{kT\varepsilon/e^2[(N_D - N_A)^2 + 4n_i^2]^{1/2}\}^{1/2}. \qquad (2.3.17)$$

By assuming $|N_D - N_A| \gg n_i$ (a true extrinsic semiconductor) we obtain a characteristic length

$$L_D' \simeq L_{DE} = [kT\varepsilon/e^2|N_D - N_A|]^{1/2}, \qquad (2.3.18)$$

known as the *extrinsic Debye length*. The solution of (2.3.16) is

$$u = u_s \exp\left(-\frac{x}{L_{DE}}\right),\ |u_s| \ll \frac{kT}{e}, \qquad (2.3.18')$$

and the effective thickness of the surface layer is of the order of L_{DE}, a length which decreases with increasing doping. However, the above result is valid only for weak accumulation and weak inversion.

(B) If the mobile carrier concentration is assumed negligible in the surface layer, then the Poisson equation (2.3.11) becomes

$$\frac{d^2\phi}{dx^2} \simeq -\frac{e}{\varepsilon}(N_D - N_A) \qquad (2.3.19)$$

and has a solution of the form

$$\phi(x) = Ax^2 + Bx + C. \qquad (2.3.20)$$

Assume that the entire semiconductor is neutral except a totally depleted surface layer of thickness x_d. The constants A, B and C will be determined by introducing $\phi(x)$ in equation (2.3.19), by requiring $d\phi/dx = 0$, $\phi = 0$ for $x = x_d$ and also $\phi = \phi_s$ for $x = 0$. The potential distribution is parabolic ($0 < x < x_d$)

$$\phi = |\phi_s|\left(1 - \frac{x}{x_d}\right)^2, \quad |\phi_s| = e|N_D - N_A|x_d^2/2\varepsilon, \quad (2.3.21)$$

the layer thickness is

$$x_d = \left(\frac{2\varepsilon|\phi_s|}{e|N_D - N_A|}\right)^{1/2}, \quad (2.3.22)$$

and also decreases as the doping increases. For large $|\phi_s|$ inversion occurs at the surface and this result becomes incorrect.

Table 2.3.2 indicates orientative values for the magnitude of the depletion length x_d, by using $|\phi_s| = 1V$ and $\varepsilon = 10\varepsilon_0$ in equation (2.3.22) (ϕ_s at thermal equilibrium is sometimes called *the built-in voltage*, $e|\phi_s|$ is the total band bending).

Table 2.3.2

Depletion-Layer Thickness ($|\phi_s| = 1V$, $\varepsilon = 10\varepsilon_0$)

$|N_D - N_A|$ (cm^{-3})	10^{15}	10^{17}	10^{19}
x_d(cm)	10^{-4}	10^{-5}	10^{-6}
(μm)	1	0.1	0.01
(Å)	10^4	10^3	10^2

Table 2.3.3 gives numerical values for the extrinsic Debye length in silicon ($\varepsilon = 11.7\,\varepsilon_0$). At room temperature L_{DE} is about one order of magnitude shorter than x_d indicated in Table 2.3.2, for the same doping.

Table 2.3.3

Extrinsic Debye Length in Silicon ($\varepsilon = 11.7\,\varepsilon_0$)

$|N_D - N_A|$ (cm^{-3})		10^{13}	10^{15}	10^{17}	10^{19}
$L_{DE}(\mu m)$	77 K	0.66	0.066	0.0066	0.00066
	300 K	1.3	0.13	0.013	0.0013
	450 K	—	$\approx$0.16	0.016	0.0016

Note the fact that if an external voltage is applied and a current flows at the MS junction then equation (2.3.12) is no longer valid because we cannot define a Fermi level and use equations (2.3.5) and (2.3.6). Simple expressions are, however, available for a totally depleted surface layer, as will be shown in Chapter 3.

2.4 Image Force Lowering of the Barrier (Schottky Effect)

The Schottky effect may be understood if one considers a 'microscopic' view of the potential barrier at the metal surface. The electron potential energy at the metal surface should vary gradually, and not abruptly, as indicated in Figure 2.2.1. This potential energy in the vicinity of the metal surface may be associated to the so-called image force. When an electron is brought out of the metal at the distance x, a positive charge will be induced on the metal surface. The attraction force between the electron and the induced positive charge is equal to the electrostatic force which would exist between the electron and an equally positive charge $(+e)$ located at $(-x)$ (hence the term 'image'). The 'image' force is

$$F = - \frac{e^2}{16\pi\varepsilon x^2}, \qquad (2.4.1)$$

where ε is the permittivity of the medium in contact with the metal (see also below). The electric potential associated with this force is $e/16\pi\varepsilon x$. The corresponding potential energy is indicated in Figure 2.4.1. The reference was taken to be the top of the energy barrier which must be surpassed by electrons which leave definitively the metal. Assume now that an electric field $\mathscr{E}$ is applied at the metal

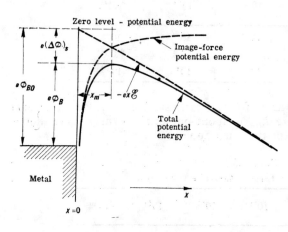

Figure 2.4.1. Schottky effect (image force lowering of the barrier) at a metal-semiconductor contact.

surface (constant and directed towards the metal, $\mathscr{E}$ is the absolute value). The total potential energy will be now

$$W = -\frac{e^2}{16\pi\varepsilon x} - e\mathscr{E}x, \qquad (2.4.2)$$

where the second term is due to the external field. W has a maximum at $x = x_m$ ($dW/dx = 0$)

$$x_m = (e/16\pi\varepsilon\mathscr{E})^{1/2} \qquad (2.4.3)$$

2. Carrier Injection in Solids

and the barrier $e\phi_B$ is lowered by an amount $e(\Delta\phi)_s$ where

$$(\Delta\phi)_s = \left(\frac{e\mathscr{E}}{4\pi\varepsilon}\right)^{1/2}, \quad \varepsilon \to \varepsilon_{opt}. \tag{2.4.4}$$

This is the Schottky effect (or image force lowering of the potential barrier).

It is worth stressing that the permittivity ε in the above relations should be the optical or ultra-high frequency permittivity (ε_{opt}) which is generally lower than the static value. This will occur [1] if during the emission process, the electron transit-time from the metal surface to the barrier maximum located at x_m is shorter than the dielectric relaxation time and therefore the semiconductor will not have enough time to be polarized. However, for Si, Ge and GaAs, the optical dielectric constant is essentially constant up to a frequency of about 3×10^{14} s^{-1} (wavelength of 1 μm) [1]. If we consider $x_m = 10 - 50$ Å (see below) and an electron velocity of 10^7 cm s^{-1}, the carrier transit-time is $10^{-14} - 5 \cdot 10^{-14}$ s, being longer than the oscillation period. Therefore for these materials the static value of ε may be used.

Figure 2.4.2 shows $(\Delta\phi)_s$ and x_m as a function of the electric field intensity for several values of $\varepsilon/\varepsilon_0$ (relative dielectric constant). Consider the image force lowering effect upon the Schottky barrier (depletion layer at a metal-semiconductor surface). For moderate and low dopings, the depletion length is much longer (Table 2.3.2) than x_m, therefore the external electric field may be considered almost constant in the vicinity of the barrier and $\mathscr{E}$ in equations (2.4.3) and (2.4.4) will be replaced by the maximum field in the depletion length, calculated at surface ($x = 0$).

If the semiconductor doping is relatively high, the effect of the image force will be evaluated by considering the parabolic potential distribution, equation (2.3.21), combined with the image-force potential for an electron. If the bottom of the conduction band in the bulk is taken arbitrarily as zero level for the potential energy *of an electron* injected in the depletion region, W, then we have

$$W = -\frac{e^2}{16\pi\varepsilon_{opt}x} + e|\phi_s|\left(1 - \frac{x}{x_d}\right)^2, \quad 0 < x < x_d, \tag{2.4.5}$$

where x_d is given by equation (2.3.22), with ε as static permittivity. This may be written in normalized form

$$\frac{W}{e|\phi_s|} = \left(1 - \frac{x}{x_d}\right)^2 - \frac{1}{8\pi\left(\frac{|\phi_s|}{\phi_1}\right)^{3/2}\left(\frac{x}{x_d}\right)}, \tag{2.4.6}$$

where

$$\phi_1 = \frac{e}{2}\left(\frac{|N_D - N_A|}{\varepsilon\,\varepsilon_{opt}^2}\right)^{1/3} \tag{2.4.7}$$

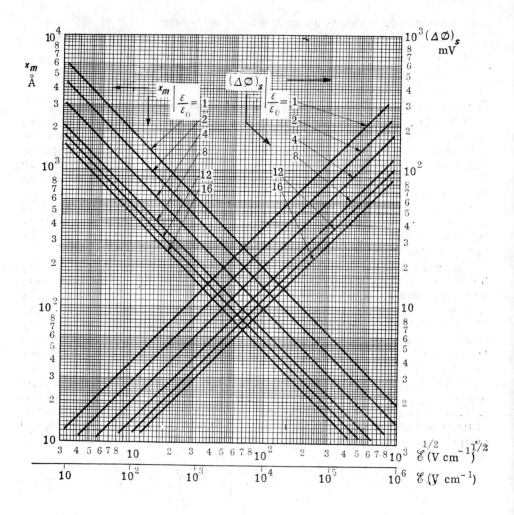

Figure 2.4.2. The Schottky barrier lowering, $(\Delta\phi_s)$, and the position of the potential minimum, x_m, *versus* electric field intensity for several values of relative semiconductor permittivity (ε is the optical dielectric constant).

is a characteristic potential. Figure 2.4.3 shows the shape of the potential barrier for various values of the parameter $(|\phi_s|/\phi_1)^{3/2}$, which decreases with increasing doping for a given band bending $e|\phi_s|$. This figure shows that at relatively high

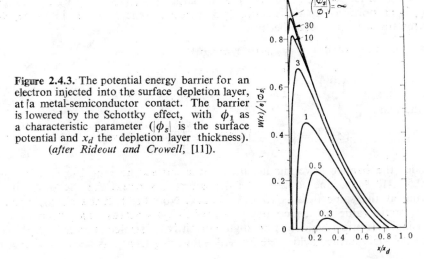

Figure 2.4.3. The potential energy barrier for an electron injected into the surface depletion layer, at a metal-semiconductor contact. The barrier is lowered by the Schottky effect, with ϕ_1 as a characteristic parameter ($|\phi_s|$ is the surface potential and x_d the depletion layer thickness). *(after Rideout and Crowell, [11]).*

dopings x_m becomes indeed comparable to x_d. Rideout and Crowell [11] have shown that the relative magnitude of the barrier lowering $(\Delta\phi)_s/|\phi_s|$ depends only upon the normalized parameter $(|\phi_s|/\phi_1)$ and has almost the same value as calculated within the uniform field approximation (the electric field in equation (2.4.4) is calculated for $x=0$ using the parabolic potential distribution). However, when examining the transport properties at the metal-semiconductor interface not only the height, but also the shape of the barrier is also important (electrons can penetrate through the barrier by a quantum-mechanical effect).

2.5 Effect of Other Contact Phenomena upon the Barrier Height

Other contact phenomena affecting the height of the potential barrier at an MS contact are: the electrostatic screening in the metal, the interfacial layer and the charge density associated with it, the dipole layer formed at an intimate MS contact (a quantum-mechanical effect).

Consider first the effect of electrostatic screening (field penetration) into the metal [12], [13]. This effect consists of the following: free charges which are induced capacitively at the metal surface cannot be supported without an adjustment of the electrostatic potential inside the metal. The total variation of potential inside the metal is $(\Delta\phi)_{es}$, which is just the barrier lowering due to electrostatic screening.

The potential variation inside the metal follows an exponential law and may be written [13], [14].

$$\phi(x) = (\Delta\phi)_{es} \exp x/\lambda, \quad x < 0, \tag{2.5.1}$$

where the coordinate x and the potential reference are defined on Figure 2.5.1 and λ is of the order of 0.5×10^{-8} cm [13] or larger [14], [15]. The barrier lowering $(\Delta\phi)_{es}$ is proportional to the electric field in semiconductor at the interface, $\mathscr{E}(0^+)$ (directed towards the metal) because

$$\varepsilon\mathscr{E}(0^+) = \varepsilon_{metal}\mathscr{E}(0^-) \approx \varepsilon_0 \mathscr{E}(0^-) \tag{2.5.2}$$

$$\mathscr{E}(0^-) = \left.\frac{d\phi}{dx}\right|_{x=0} = \left.\frac{\phi(x)}{\lambda}\right|_{x=0} = \frac{(\Delta\phi)_{es}}{\lambda}, \tag{2.5.3}$$

and therefore

$$(\Delta\phi)_{es} = \left(\frac{\varepsilon}{\varepsilon_{metal}}\right)\lambda\mathscr{E}(0^+) \approx \left(\frac{\varepsilon}{\varepsilon_0}\right)\lambda\mathscr{E}(0^+). \tag{2.5.4}$$

Consider the barrier lowering in silicon at a surface field of 10^5 V cm^{-1}. Using $\lambda = 0.5 \times 10^{-8}$ cm one obtains $(\Delta\phi)_{es} \approx 6$ mV, whereas the Schottky barrier lowering at the same field is $(\Delta\phi)_s \simeq 35$ mV. Note the fact that $(\Delta\phi)_{es}$ increases with ε (static value of semiconductor permittivity) whereas $(\Delta\phi)_s$ decreases with increasing ε (optical value). For high permittivity semiconductors, $(\Delta\phi)_{es}$ may exceed (at high surface fields) the Schottky lowering $(\Delta\phi)_s$, as shown by Perlman

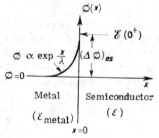

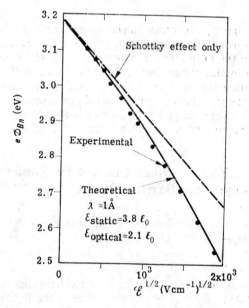

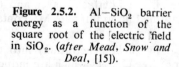

Figure 2.5.1. Potential distribution inside the metal at a metal-semiconductor contact (λ is the Thomas-Fermi screening distance in the metal).

Figure 2.5.2. Al–SiO$_2$ barrier energy as a function of the square root of the electric field in SiO$_2$. (*after Mead, Snow and Deal*, [15]).

[13]. It is interesting that Mead *et al.* reported [15] evidence of field penetration at the Al–SiO$_2$ interface (lower permittivities but higher field intensities) by fitting the experimental results with $\lambda = 10^{-8}$ cm $= 1$ Å (Figure 2.5.2). However, the total

barrier lowering was considered to be $(\Delta\phi)_s + (\Delta\phi)_{es}$, where $(\Delta\phi)_s$ is given by equation (2.4.4), which is not strictly correct because the field penetration into the metal will also have an effect upon the image charges in the metal and hence upon the image potential [15]. On the other hand, any interfacial layer will also produce a barrier lowering which is proportional to the surface field (see below), and this may be responsible for an experimental $\lambda = 1 \text{ Å}$ [15], thicker than the calculated value $\lambda = 0.5 \text{ Å}$ [13].

We shall neglect now the electrostatic screening and image force lowering of the barrier and consider the effect of a very thin interfacial (insulating) layer of thickness δ and permittivity ε_i at the metal-semiconductor contact. No charge is assumed to exist in this layer or at interfaces. The displacement vector is continuous at the insulator-semiconductor interface and (see the notations in Figure 2.5.3)

$$\varepsilon_i \mathscr{E}(\delta^-) = \varepsilon \mathscr{E}(\delta^+). \tag{2.5.5}$$

The electric field is constant in the insulating layer and the total potential drop on this layer, denoted by $(\Delta\phi)_i$, is $\delta\mathscr{E}(\delta^-)$. If δ is very small, the interfacial layer will be transparent for electrons. However, $(\Delta\phi)_i = \delta\mathscr{E}(\delta^-)$ will be an actual barrier lowering due to this layer. If $\mathscr{E} = \mathscr{E}(\delta^+)$ is the field in the semiconductor at the interface, then

$$(\Delta\phi)_i = \delta \frac{\varepsilon}{\varepsilon_i} \mathscr{E}(\delta^+) = \delta \frac{\varepsilon}{\varepsilon_i} \mathscr{E}. \tag{2.5.6}$$

The combined effect of field penetration into the metal and of an interfacial layer may be evaluated by using only one parameter, α'. The total barrier lowering is

$$(\Delta\phi)' = \frac{\varepsilon}{\varepsilon_{\text{metal}}} \lambda\mathscr{E} + \delta \frac{\varepsilon}{\varepsilon_i} \mathscr{E} = \alpha'\mathscr{E}, \tag{2.5.7}$$

where ε is the semiconductor static permittivity. We stress again that the Schottky effect (image force lowering) cannot be taken into account by simply adding $(\Delta\phi)_s$ calculated with no interfacial layer and field penetration (see equation (2.4.4)) to the

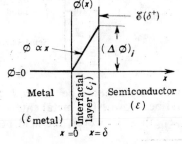

Figure 2.5.3. Potential distribution inside an interfacial layer (containing no electric charge), at an MS contact.

above $(\Delta\phi)$. Note, for example, that if the layer thickness, δ, is comparable to x_m given by equation (2.4.3) (and this situation does indeed occur in practice), then a large error will be made by neglecting the interfacial layer in calculating the image-force potential.

The model of the metal-semiconductor contact should be completed by including the interface charges associated with the interfacial layer. We shall consider that the electric charge is located just at the semiconductor surface. If this charge is due to ionized atoms on semiconductor surface or to electrons trapped in *deep* [*] surface states, it may be considered constant. A fixed negative charge will decrease the magnitude of the lowering of the barrier for electron injection in semiconductor, or even will increase the barrier (see Problem 2.8).

However, in general, the total surface charge density depends upon the parameters of the (metal)-(interfacial insulating layer)-(semiconductor) system. One may discuss the effect of a single trap level [16] (see Problem 2.10) or of a multiple set of levels [17]. Following Cowley and Sze [1], [18] and other authors, we consider below a continuous and uniform distribution of electron traps (acceptor-type states) in energy, having the density D_s (measured in states cm^{-2} eV^{-1}). In fact the demonstration below assumes that D_s is constant only over a limited range in the energy gap.

Figure 2.5.4 shows the energy diagram of the contact at thermal equilibrium. The semiconductor is characterized by χ, ϕ_g and ϕ_0 ($e\phi_0$ specifies the level above the valence band below which all surface states must be filled for an electrically neutral semiconductor surface, or the Fermi level for neutrality, sometimes called the 'neutral' level) and also by its static permittivity ε and by the interface field

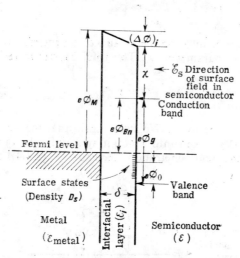

Figure 2.5.4. Energy levels at the MS contact (the energy barrier for electrons is $e\phi_{Bn}$, the Schottky effect is not considered).

in semiconductor, denoted below by $\mathscr{E}_s$.[†] The insulator thickness δ and permittivity ε_i should be known. The metal enters througout its work function $e\phi_M$ and the screening length λ. However, the field penetration into the metal is not considered

---

[*] These are trap levels situated near the valence band, being occupied irrespective of the actual conditions at the contact.

[†] If the surface layer is a Schottky barrier, then $\mathscr{E}_s$ will depend upon the above quantities and also upon the semiconductor doping.

here explicitly; it may be taken into account by simply using an effective length of the interfacial layer, δ'. According to equation (2.5.7) [14]

$$\delta' = \delta + \frac{\varepsilon_i}{\varepsilon_{metal}} \lambda. \tag{2.5.8}$$

We shall calculate the effect of interface states and of the surface field upon the barrier height $e\phi_B$. The net electric charge density at the semiconductor surface is (Figure 2.5.4)

$$Q_{ss} = -e^2 D_s \left[\phi_g - \phi_0 - \phi_M + (\Delta\phi)_i + \frac{\chi}{e} \right]. \tag{2.5.9}$$

Because of this charge, the displacement vector will show a discontinuity at the semiconductor surface and equation (2.5.6) must be replaced by

$$(\Delta\phi)_i = \frac{\delta}{\varepsilon_i} [\varepsilon \mathscr{E}_s + Q_{ss}], \tag{2.5.10}$$

where the electric field at the semiconductor surface is directed towards the interface layer and has the absolute value $\mathscr{E}_s$.

The barrier height is (Figure 2.5.4)

$$e\phi_{Bn} = e\phi_M - \chi - e(\Delta\phi)_i. \tag{2.5.11}$$

By eliminating $(\Delta\phi)_i$ and Q_{ss} between equations (2.5.9)–(2.5.11), one obtains

$$\phi_{Bn} = \gamma \left(\phi_M - \frac{\chi}{e} \right) + (1-\gamma)(\phi_g - \phi_0) - \delta \frac{\varepsilon}{\varepsilon_i} \gamma \mathscr{E}_s, \tag{2.5.12}$$

where

$$\gamma = \frac{\varepsilon_i}{\varepsilon_i + e^2 \delta D_s} \tag{2.5.13}$$

is a parameter characterizing the interface. Two limit cases correspond, respectively, to:

(a) $D_s \to 0$, $\gamma \to 1$ and the barrier is field-lowered by a quantity $(\Delta\phi)_i$ given by equation (2.5.6);

(b) $D_s \to \infty$, $\gamma \to 0$ and

$$\phi_{Bn} \simeq \phi_g - \phi_0, \tag{2.5.14}$$

i.e. the Fermi level is 'pinned' $e\phi_0$ above the valence band (see also Section 2.2) and the barrier is independent of the work function of the metal, of the interfacial layer and of the surface field.

Note the fact that the field-lowering term in equation (2.5.12) is linear in the electric field (the barrier decreases linearly with the increasing surface field $\mathscr{E}_s$).

If the field-lowering is neglected [1], [7], [9], [19], then for a given semiconductor and given surface conditions, the barrier will depend linearly upon the metal work function

$$\phi_{Bn} \simeq \gamma \left(\phi_M - \frac{\chi}{e} \right) + (1 - \gamma)(\phi_g - \phi_0) = \gamma \phi_M + A \qquad (2.5.15)$$

(A = const.), the slope determining the density D_s and the constant A, the location ϕ_0 of surface states [1] (see for example Figure 2.2.3). However, almost all the remaining parameters intervening in equations (2.5.13) and (2.5.15) are uncertain. The thickness of the interfacial layer is considered to be of the order of 10 Å (silicon oxidized naturally in atmosphere), whereas the layer permittivity ε_i is taken sometimes equal to the free space value ε_0, because the interfacial film consists at most of a few atomic layers. Turner and Rhoderick [7] pointed out that ϕ_M determined for bulk material is hardly relevant for film contacts deposited under various conditions, whereas the electron affinity χ of the silicon depends upon surface roughness (about 4 eV for smooth cleavage and as low as 3.75 eV for a rougher cleavage). Therefore, detailed qualitative verification of the above theory is practically impossible at the present time and a more detailed model for interface trapping seems hardly justifiable.

The MS contact leads usually to a Schottky barrier (depletion layer) at the semiconductor surface. The energy band model for such a contact is shown in Figure 2.5.5, indicating the electrostatic screening into the metal and, formally, the image force lowering $(\Delta \phi)_s$. Both $(\Delta \phi)_s$ and $\mathscr{E}_s$ (see Problem 2.6) depend upon the semiconductor doping and therefore the effective barrier height will change with doping. Eimers and Stevens [14] calculated such a dependence for an Au—Si Schottky barrier. They used the model indicated in Figure 2.5.5 and equation (2.5.12) where $\mathscr{E}_s$ was calculated for a Schottky barrier (note that $\mathscr{E}_s = \mathscr{E}_s(\phi_{Bn})$) and the Schottky lowering $(\Delta \phi)_s$ was determined as in the absence of any other interfacial effects (see Section 2.4), which is not strictly correct, and then subtracted from ϕ_{Bn} given by equation (2.5.12). The results were indicated as solid lines in Figure 2.5.6. Note the fact that the 'effective' thickness δ' of the interfacial layer (including the effect of field penetration) has little effect upon the barrier lowering. The dashed curves shown on the same figure are calculated by neglecting the last term in equation (2.5.12) (which is the usual procedure, as already shown) and considering only the image-force field-lowering of the barrier. The above authors [14] concluded that the experimentally observed deviations of the barrier height are probably due to variations of semiconductor doping rather than to variations of oxide layer thickness. Anyway, the correction for ϕ_{Bn} indicated by solid curves in Figure 2.5.6 [14] depends upon doping through the last term in equation (2.5.12), which arises due to interface effects (surface trapping).

An *intimate metal-semiconductor contact*, such as those involving a thin layer of metallic silicide formed at the silicon surface [19]—[24] does not exhibit an interfacial layer or interface states. However, the experimentally observed barrier lowering for these contacts is substantially larger than that predicted by the image-force theory. The mechanism explaining the additional barrier lowering was discussed by Parker et al [25], Andrews et al [19], [20], [24], Broom [22], Kircher [21]. This mechanism consists in formation of a dipole layer at the metal-semiconductor

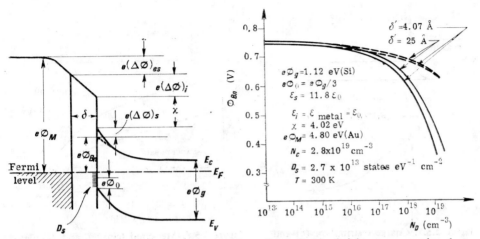

Figure 2.5.5. Energy diagram for a Schottky barrier, indicating barrier lowering due to: field penetration into the metal, $e(\Delta\phi)_{es}$; a thin interfacial layer and interface states, $e(\Delta\phi)_i$; Schottky effect, $e(\Delta\phi)_s$.

Figure 2.5.6. Barrier height *versus* semiconductor doping calculated for Au—Si contact (model of Figure 2.5.5). Dashed curves are calculated by neglecting the field-dependent lowering of the barrier except the image-force lowering, which is included. *(after Eimers and Stevens, [14]).*

interface and is based upon the quantum-mechanical calculations of Heine [26], who showed that metal wave-functions penetrate into the forbidden semiconductor band and these supplementary states are occupied by electrons which induce a positive charge in the metal. The density of charge trapped in these states decays exponentially with distance measured in the semiconductor from the interface (the characteristic length x_0 is of a few Å [26]). The charge distribution at a Schottky-barrier contact is as shown in Figure 2.5.7a. The potential distribution arising from this charge exhibits a minimum and the band bending at the contact is as indicated in Figure 2.5.7 b. The barrier height is $e\phi_{Bn}$ and is located at the distance x_{0m} from the metal. Broom [22] showed that the barrier *lowering* due to this effect depends linearly upon the surface field $\mathscr{E}_s$

$$(\Delta\phi)_0 = x_{0m}\mathscr{E}_s + \text{const.} \approx x_{0m}\mathscr{E}_s \tag{2.5.16}$$

(both the maximum field at the interface $\mathscr{E}_s$ and the depletion region thickness $(x_d \gg x_{0m})$ are calculated by neglecting the negative charge at the semiconductor surface). Results of experimental measurements on Pd-silicide Schottky diodes were compared to the above theory, as shown in Figure 2.5.8 [22]. The solid line is the curve giving the best fit for experiment and yielding a surface-state density of electronic states $N_s^- = 10^{21} \text{cm}^{-3}$ and a decay length $x_0 = 6$ Å. These results were obtained by simply adding to (2.5.16) the barrier lowering due to Schottky effect in the absence of (2.5.16). A correct evaluation of $(\Delta\phi)$ [22] which adds the image-force potential to the potential corresponding to Figure 2.5.7b and determines x_{om} as a unique result of both effects, leads to important modifications (thus $x_{om} = 54$ Å ($x_0 = 8$ Å), instead of 36 Å, at 4.10^{15} cm^{-3} doping).

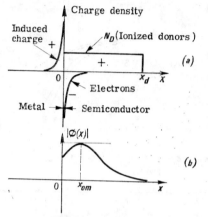

Figure 2.5.7. Charge distribution (a) and potential distribution (b) at an intimate MS contact. The exponentially varying negative charge at semiconductor surface is due to electrons trapped in surface states. (after Broom, [22]).

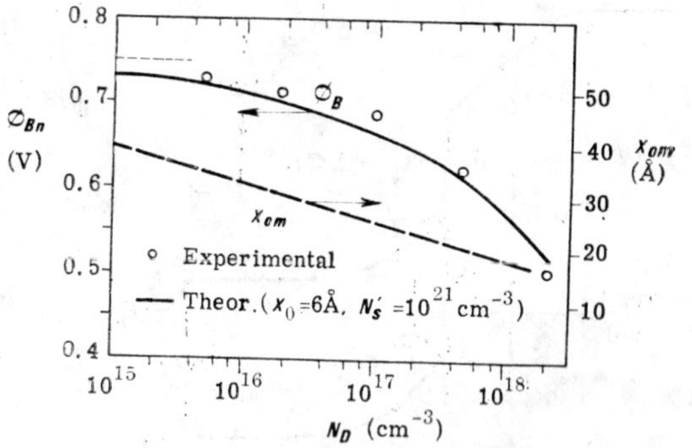

Figure 2.5.8. Measured (circles) and calculated (solid line) potential barrier for electrons at a Pd-silicide Schottky-barrier contact, *versus* semiconductor doping. The position of potential extremum is also indicated. (after Broom, [22]).

Finally we note that all physical phenomena discussed in this Section determine a linear field-lowering of the barrier. Therefore, measuring the barrier height field-dependence is of little value in distinguishing between these mechanisms of barrier lowering.

2.6 Carrier Injection at a Metal-semiconductor Contact

We shall consider below the electron transport at the metal-semiconductor contact, with particular emphasis to *electron* injection in semiconductor (hole injection may be similarly discussed). The metal-semiconductor diode as a device by itself will be studied in a separate chapter entitled 'Schottky diode'.

Any metal-semiconductor contact is characterized by a potential barrier. This means that the electron must acquire a supplementary potential energy passing from the metal into the semiconductor or from the semiconductor bulk into the metal, over the barrier (quantum-mechanical tunnelling is ignored for the moment). At a depletion-type (Schottky) MS contact this barrier is asymmetric (Figure 2.6.1a). The electron energy at the top of the barrier is $e\phi_{Bn}$ above the Fermi level in the metal. Usually $e\phi_{Bn} \gg kT$ and the electrons in the metal have a Maxwellian distribution in energy above the energy corresponding to the top of the barrier (the corresponding 'tail' of electron distribution in energy is indicated in Figure 2.6.1a). A thermionic emission current J_{MS}, per unit area, will flow into the semiconductor*, over the potential barrier ϕ_{Bn}. At thermal equilibrium an equal electron

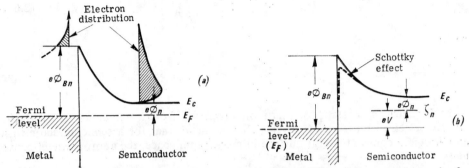

Figure 2.6.1. Schottky-barrier MS contact: (a) at thermal equilibrium, the potential barrier for electrons in metal is ϕ_{Bn}, whereas the barrier for electrons in semiconductor is $\phi_{Bn} - \phi_n$ the energy distribution of electrons passing over the barrier (in both directions) is Maxwellian; (b) forward biased (external voltage V, metal positive) Schottky contact, the energy barrier for electron in semiconductor is reduced by $+eV$, the image-force rounding of the barrier (Schottky effect) may be also taken into account, ζ_n is quasi Fermi level for electrons in the semiconductor bulk.

current density J_{SM} will flow into the metal, such as the total current density is zero. By using the Maxwellian distribution for the electron 'gas' in semiconductor

$$J_{SM}^r = \frac{en(m^*)^{3/2}}{(2\pi kT)^{3/2}} \int_{-\infty}^{+\infty} dv_y \int_{-\infty}^{+\infty} dv_z \int_{v_{0x}}^{\infty} v_x \exp\left[-\frac{m^*(v_x^2 + v_y^2 + v_z^2)}{2kT}\right] dv_x =$$

$$= en \left(\frac{m^*}{2\pi kT}\right)^{1/2} \int_{v_{0x}}^{\infty} v_x \exp\left(-\frac{m^* v_x^2}{2kT}\right) dv_x =$$

$$= en \left(\frac{kT}{2\pi m^*}\right)^{1/2} \exp\left(-\frac{m^* v_{0x}^2}{2kT}\right), \qquad (2.6.1)$$

where n is electron concentration in the semiconductor bulk ($E_c - E_F = e\phi_n$, where E_F is the Fermi level in the bulk)

$$n = N_c \exp\left(-\frac{e\phi_n}{kT}\right) = 2\left(\frac{2\pi m^* kT}{h^2}\right)^{3/2} \exp -\frac{\phi_n}{kT} \qquad (2.6.2)$$

and v_{ox} is the velocity in the x direction corresponding to the minimum kinetic energy required to surmount the barrier (Figure 2.6.1)

$$\frac{1}{2} m^* v_{0x}^2 = e(\phi_{Bn} - \phi_n). \qquad (2.6.3)$$

---

* The conventional positive direction for the electron current going into semiconductor is however, from semiconductor to the metal.

We have assumed that the barrier is thin in comparison to the mean free path [1]. The result of the above three equations is

$$J_{SM} = A^*T^2 \exp\left(-\frac{e\phi_{Bn}}{kT}\right), \qquad (2.6.4)$$

where

$$A^* = \frac{4\pi e m^* k^2}{h^3} \qquad (2.6.5)$$

is the Richardson constant for thermionic emission from a semiconductor with the effective electron mass m^*. Crowell [27] has shown that for a semiconductor with an ellipsoidal constant-energy surface in momentum space, the appropriate effective mass for thermionic emission is

$$m^* = (l^2 m_y^* m_z^* + m^2 m_z^* m_x^* + n^2 m_x^* m_y^*)^{1/2}, \qquad (2.6.6)$$

where l, m and n are the direction cosines of the direction of emitted current relative to the principal axes of the constant energy ellipsoid and m_x^*, m_y^*, m_z^* the corresponding components of the effective mass tensor. If the conduction band has more than one potential minimum in wave vector space, the corresponding effective masses will be summed to yield the effective mass entering into the expression of the Richardson constant. The ratio m^*/m_0 (where m_0 is the free electron mass) for Si, Ge and GaAs, is given in Table 2.6.1.

Table 2.6.1

Values of m^*/m_0 [1], [27]

Semiconductor	Ge	Si	GaAs
p-type	0.34	0.66	0.02
n-type ⟨111⟩	1.11	2.2	0.068 (low field)
n-type ⟨100⟩	1.19	2.1	1.2 (high field)

The electron current injected from metal into the semiconductor should be equal to J_{SM} (thermal equilibrium, zero net current). The Richardson constant entering in the expression of the thermionically emitted current from the metal into the semiconductor is that of the semiconductor such that the thermionic emission process takes place within the semiconductor. We note the fact that the combined action of charge carrier image force and the depletion layer field displaces the potential energy maximum into the semiconductor (although it remains very close to the metal surface).

A more detailed account of the emission process should consider scattering by optical phonons and quantum-mechanical reflection and tunnelling at the potential barrier (Figure 2.6.2). The tunnelling process depends upon the *shape* of the potential barrier, which in turn depends upon the semiconductor doping*. All these processes may be accounted for by introducing an effective Richardson constant A^{**} [19].

If the potential energy diagram is modified due to an applied external bias, then the thermionic emission current from metal to semiconductor J_{MS} will have the same value as at thermal equilibrium provided that the potential barrier remains the same, ϕ_{Bn} (Figure 2.6.1b). However, this current is slightly modified by the applied bias since both the barrier height and the effective Richardson constant A^{**} depend upon the electric field at the contact. Figure 2.6.3 shows the field dependence of A^{**} calculated for *electrons* and *holes* in silicon at a given semiconductor doping [1], [19]. One may approximate $A^{**} \simeq 112$ A cm^{-2} K^{-2} for electrons and $A^{**} \simeq 32$ A cm^{-2}K^{-2} for holes in a wide range of field intensities. We stress the fact that the above results are calculated for thermionic emission over a Schottky barrier. At larger

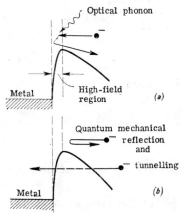

Figure 2.6.2. Schematic representation of kinetic processes at a metal-semiconductor contact, which modify the Richardson constant for thermionic emission of electrons from semiconductor into the metal: (a) scattering by optical phonons into the high field interface region; (b) quantum-mechanical reflection of electrons with energies above the top of the barrier and tunnelling through the potential barrier.

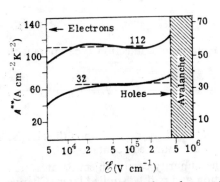

Figure 2.6.3. Electric field dependence of effective Richardson constants calculated for electrons and holes in silicon (the average values of these constants and the field limit due to avalanche breakdown). (after Andrews and Lepselter, [19]).

doping densities the field emission (tunnelling through the barrier) prevails as will be shown later.

The emission process at a metal-semiconductor contact of an electronic device should determine *a boundary condition* for the semiconductor equations which

---

* The drift and diffusion of charge carriers in the depletion region also modifies the emission current (because this region may not be short as compared to the mean free path as assumed above, see Chapter 3).

describe the device behaviour. Consider a unipolar device and electron injection into the semiconductors. Two distinct situations will be discussed:

(a) The device current is determined by bulk processes and its magnitude $J = J_{MS} - J_{SM}$ is smaller than the emission current J_{MS}. If the net current density J is much smaller than J_{MS}, then the electron density at the contact ($x = 0$) is approximately equal to its thermal equilibrium value [28], [29]

$$n(0) = N_c \exp\left(\frac{-e\phi_{Bn}}{kT}\right) \qquad (2.6.7)$$

(for the device analysis the potential barrier may be considered as located just at the contact). The barrier height $e\phi_{Bn}$ may depend upon the electric field.

This is a 'reservoir-type' contact (Lampert and Mark [29]).

(b) The device current is limited by the injection process and the *particle* current injected at the boundary is

$$J = J_{MS} = J_{\text{sat}}. \qquad (2.6.8)$$

For a thermionic-injection type contact the boundary condition is

$$J = J_{\text{sat}} = A^{**} \exp\left(-\frac{e\phi_{Bn}}{kT}\right), \qquad (2.6.9)$$

where A^{**} and ϕ_{Bn} depend upon the electric field in the semiconductor, at the contact. This is a 'saturated' contact [30].

The minority carrier injection at a metal-semiconductor contact deserves special attention and will be studied later (Chapter 3).

We shall consider below *carrier injection into highly doped semiconductors*. Table 2.3.2 shows that the depletion layer thickness at a metal-semiconductor contact will become *very thin* if the semiconductor doping is high (above 10^{18}cm^{-3}, for example). Therefore, carrier tunnelling through the potential barrier (field emission) may become the dominant injection process.

We consider first a simple model for carrier tunnelling. The potential barrier is assumed parabolic and the transmission probability for the barrier is defined as [31]

$$\tau(E) = \exp{-\frac{4\pi}{h} \int_0^{x_1} \{2m^{**}[e\phi(x) - E]\}^{1/2} \, dx}, \quad E < e|\phi_s| \qquad (2.6.10)$$

$$\tau(E) = 1, \quad E > e|\phi_s| \quad \text{(thermionic emission)}. \qquad (2.6.11)$$

The principal quantities in the above relations are defined in Figure 2.6.4. The effective mass (assumed independent of energy) is that of semiconductor because the tunnelling process (in both directions) takes place inside the semiconductor. More rigorously, the tunnelling should also depend upon the band structure in the metal but this effect is small [31]. Crowell [32] has shown that the tunnelling effective mass is the mass component in the direction of current flow

$$m^{**} = \left(\frac{l^2}{m_x^*} + \frac{m^2}{m_y^*} + \frac{n^2}{m_z^*} \right)^{-1} \qquad (2.6.12)$$

(see equation 2.6.6 for notations) and in general differs from that in the Richardson constant (see also references [33]–[36]). Crowell [32] indicates an effective tunnelling

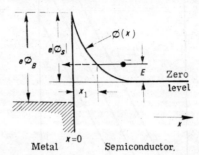

Figure 2.6.4. Electron tunnelling into the metal at a Schottky-barrier contact. E is the kinetic energy of an electron in the semiconductor conduction band.

mass $m^{**} = 0.259\, m_0$ for silicon ($\langle 111 \rangle$ direction of current). The results published by Chang, Fang and Sze [36] confirm this value for very low temperatures and silicon doping below 5×10^{17}. Otherwise, the effective tunnelling mass increases with increasing temperature and doping [36].

By using equations (2.6.10) and (2.6.11), Crowell and Rideout [31] found that the character of the emission process can be appreciated by using the parameter

$$E_{00} = \frac{eh}{4\pi} \left(\frac{N}{\varepsilon m^{**}} \right)^{1/2}, \qquad (2.6.13)$$

where ε is the static dielectric constant and m^{**} the effective semiconductor tunnelling mass. Equation (2.6.13) is plotted in Figure 2.6.5. If E_{00} is small compared to kT (high temperature, low semiconductor doping) the thermionic emission process will be dominant. If $E_{00}/kT \gg 1$, the carrier transport is practically due to field emission. An approximate value for the electron flow injected from semiconductor into the metal by field emission only ($kT/E_{00} \to 0$) is [31]

$$J_{SM} = A^{**}T^2 \exp\left(-\frac{e\phi_s}{E_{00}} \right) \exp \frac{e\phi_n}{kT}, \qquad (2.6.14)$$

where A^{**} is an effective Richardson constant for field emission.

The above theory was improved by including the effect of image force 'rounding' of the barrier and also incorporating the quantum-mechanical reflection of carriers above and near the top of the barrier in the transmission probability [11].

A more accurate analysis, reported by Chang and Sze [34], uses the quantum mechanical transmission coefficient calculated by solving the Schroedinger equation for the potential profile of the metal-semiconductor system (parabolic barrier corrected by the image-force effect, see equation (2.4.5) and Figure 2.4.3). This coefficient is plotted in Figure 2.6.6 versus the electric field at the interface for energies above and below the top of the barrier. Due to quantum-mechanical reflection, the transmission coefficient for energies slightly above the barrier is relatively low and decreases

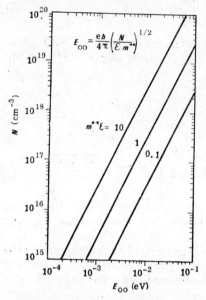

Figure 2.6.5. Semiconductor doping N versus energy E_{00}, defined by equation (2.6.13), where ε is the static dielectric constant and m^{**} the effective tunnelling mass in the direction of current flow. (after Crowell and Rideout, [31]).

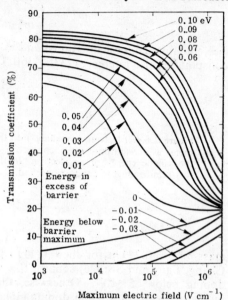

Figure 2.6.6. Transmission coefficient versus electric field in semiconductor, at contact, with electron energy as parameter. The probability of thermionic emission decreases and the probability of field-emission increases as the barrier becomes sharper (electric field increases). (after Chang and Sze, [34]).

with the increasing electric field. Figure 2.6.7 shows the theoretical electron current injected in semiconductor at thermal equilibrium, J_{MS}, at the Au—Si and Pt—Si barrier, as a function of doping (n-type silicon) at several temperatures [34]. Note the existence of a minimum value of J_{MS} as a function of doping, which is the result of increased transmission probability for thermionic emission at lower dopings (lower surface fields), according to Figure 2.6.6 [34]. The dashed curves shown in Figure 2.6.7 are calculated by taking into account the statistical nature of the impurity distribution in the sample [34]. A Gaussian distribution was assumed with Δ as the percentage deviation. The current increases due to this statistical distribution and the increase is higher for higher Δ and at higher dopings.

A Schottky-barrier contact on very low resistivity silicon has a low resistance and may be considered to be ohmic. *The contact resistance R_c for unit area is defined as the differential resistance $(dJ/dV)^{-1}$, at zero bias $(V = 0)$* where J is the net current density as a function on the voltage drop on the barrier, V. The $J - V$ characteristic for a barrier contact to a lowly doped semiconductor (thermionic emission) will be

$$J = J_{SM} - J_{MS} = J_{MS}\left(\exp\frac{eV}{kT} - 1\right), \quad (2.6.15)$$

2. Carrier Injection in Solids 115

if J_{MS} is assumed independent of field (J_{MS} depends exponentially upon the barrier seen by electrons injected from semiconductor into the metal, Figure 2.6.1b). Therefore, the specific contact resistance for high-barrier contacts to lowly doped semi-

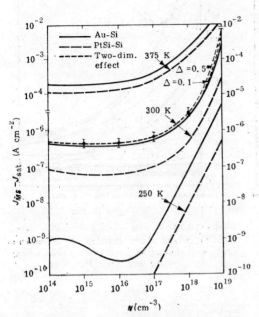

Figure 2.6.7. Metal-semiconductor (saturation) current density *versus* doping, calculated with temperature as parameter. Dotted lines were obtained for statistical variation of impurity concentration, with Δ as percentage deviation in a Gaussian distribution with respect to the device area. (after Chang and Sze, [34]).

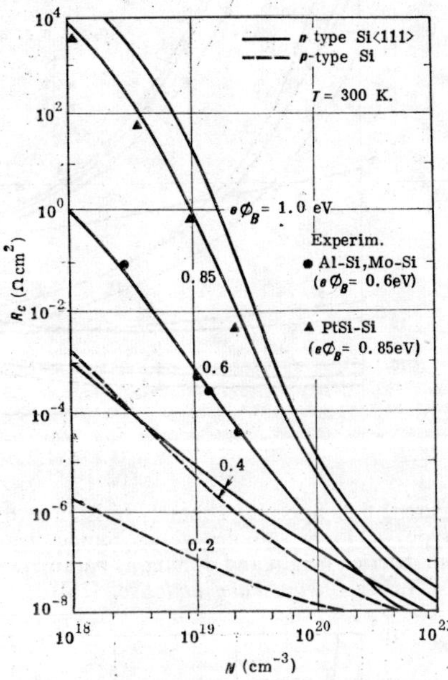

Figure 2.6.8. Theoretical specific contact resistance at 300 K for n-type, $\langle 111 \rangle$ oriented (solid lines) and p-type (dashed lines) silicon. The solid circles are experimental results for Al–Si and Mo–Si barriers ($e\phi_{Bn} = 0.6$ eV), and triangles for PtSi–Si barriers ($e\phi_{Bn} = 0.85$ eV). (after Chang, Fang and Sze, [36]).

conductors

$$R_c = \left.\frac{\partial J}{\partial V}\right|_{V=0} = \frac{kT}{eJ_{MS}} \qquad (2.6.16)$$

is high because J_{MS} is low.

The contact resistance for field emission is of the form [36]

$$R_c \propto \frac{1}{E_{00}} \exp \frac{e|\phi_s|}{E_{00}}, \qquad (2.6.17)$$

where E_{00} is given by equation (2.6.13). Therefore R_c depends upon the tunnelling effective mass and decreases rapidly with increasing doping.

Chang, Fang and Sze [36] reported results of numerical calculations for R_c, based essentially upon the model of Chang and Sze [34]. The results are shown in Figures 2.6.8 to 2.6.12. The contact resistance of highly doped samples is almost independent upon temperature (field-emission dominant), whereas R_c for lowly

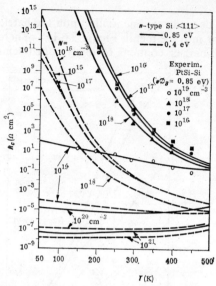

Figure 2.6.9. Specific contact resistance for n-type $\langle 111 \rangle$ Si samples *versus* temperature, with barrier height and doping as parameters. (*after Chang, Fang and Sze*, [36]).

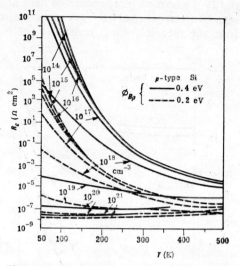

Figure 2.6.10. Specific contact resistance calculated for p-type silicon samples, *versus* temperature, with doping as parameter. (*after Chang, Fang and Sze*, [36]).

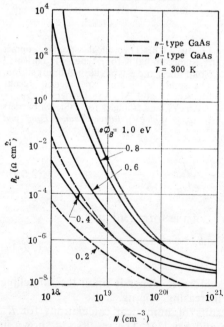

Figure 2.6.11. Specific contact resistance at 300 K *versus* impurity concentration calculated for n-type and p-type GaAs. (*after Chang, Fang and Sze*, [36]).

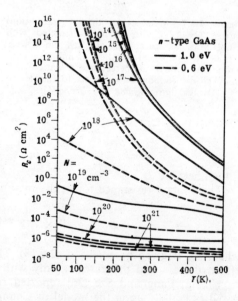

Figure 2.6.12. Specific contact resistance calculated for n-type GaAs, with doping and barrier height as parameters. (*after Chang, Fang and Sze*, [36]).

doped samples depends only slightly upon semiconductor doping (thermionic emission dominant). R_c decreases with decreasing barrier height and the effect is larger for higher impurity concentrations.

2.7 Semiconductor Homojunctions

The most effective method to inject charge carriers into a semiconductor is to form very highly doped regions in that semiconductor. If *electron* injection is desired, an n^+ region will be used, thus forming n^+p (asymmetric *pn* junction) or n^+n (high-low junction) contacts to semiconductor.

The *pn* junction will be considered first. Figure 2.7.1 shows a *pn* junction at thermal equilibrium. A space-charge region develops on both sides of the metallurgical junction ($x = 0$) because of the mobile carrier diffusion determined by a very large concentration gradient. The *n*-side will be positively and the *p*-side-negatively charged, respectively. Mobile holes diffusing to the left introduce a positive charge in the *n*-side and leave uncompensated ionized acceptors (negative charge) in the *p*-side (Figure 2.7.1a), etc. The band-energy diagram (Figure 2.7.1b) and the potential distribution are also shown (Figure 2.7.1c). V_{bi} is the 'built-in' potential or diffusion potential. Note the fact that the electric field created by the 'built-in' potential difference opposes to carrier diffusion such as the total electron current and the total hole current are both zero at thermal equilibrium.

The built-in potential V_{bi} is (Figure 2.7.1b)

$$V_{bi} = \phi_{Fp} + |\phi_{Fn}|, \qquad (2.7.1)$$

where (non-degenerate statistics)

$$|\phi_{Fn}| = \left(\frac{E_F - E_i}{e}\right)_{\substack{\text{neutral} \\ n \text{ region}}} = \frac{kT}{e} \ln \frac{n}{n_i} \simeq \frac{kT}{e} \ln \frac{N_D}{n_i} \qquad (2.7.2)$$

$$\phi_{Fp} = \left(\frac{E_i - E_F}{e}\right)_{\substack{\text{neutral} \\ p \text{ region}}} = \frac{kT}{e} \ln \frac{p}{n_i} \simeq \frac{kT}{e} \ln \frac{N_A}{n_i} \qquad (2.7.3)$$

and thus

$$V_{bi} \simeq \frac{kT}{e} \ln \frac{N_D N_A}{n_i^2}. \qquad (2.7.4)$$

where N_D and N_A are the effective donor and acceptor concentration, respectively.

The potential distribution at an *abrupt* junction ($N = $ const. on both sides) at thermal equilibrium $\phi = \phi(x)$ should be determined by using the Poisson equation (2.3.11). For the *p*-side, equation (2.3.12) is valid ($u_F = u_{Fp} > 0$, $u_{Fp} = e\phi_{Fp}/kT$, $N_D - N_A < 0$ in equation (2.3.13)). The potential reference is chosen in the neutral *p* region ($x \to +\infty$). A similar equation is valid for the *n* region ($x < 0$). These equations will be solved simultaneously by imposing continuity at $x = 0$ and zero electric field in the neutral regions ($x \to \pm \infty$). An exact* analytical solution does not exist. However, the space charge region may be assumed totally depleted of mobile carriers (the *depletion approximation*). This approximation is based upon the fact that very large variations of mobile charge concentrations occur over dis-

---

* Equation (2.3.12) is derived by assuming non-degenerate carriers statistics and is no longer correct for very highly doped regions.

tances comparable to the extrinsic Debye length L_{DE} (see Section 2.3) and therefore the transition between neutral regions and fully depleted regions requires a few L_{DE}. The depletion approximation will be acceptable if the depletion length is much longer than L_{DE}.

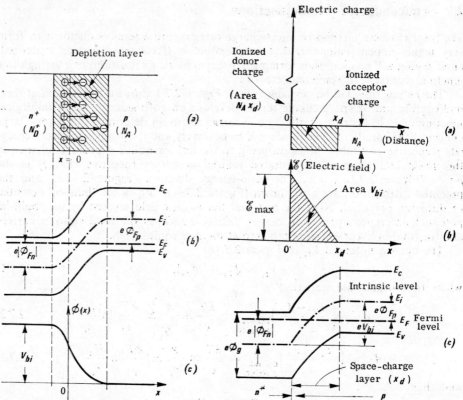

Figure 2.7.1. p-n homojunction: (a) scheme of the depletion layer indicating the uncompensated ionized donors; (b) energy band diagram at thermal equilibrium; (c) internal potential distribution at thermal equilibrium (V_{bi} is the built-in potential).

Figure 2.7.2. Abrupt asymmetrical pn^+ junction at thermal equilibrium: (a) charge distribution; (b) field distribution; (c) energy band diagram (the potential distribution inside the depletion layer is parabolic).

Only the case of asymmetrical (one-sided) pn junctions is of interest here. Assume an abrupt n^+p junction (n-side much highly doped). Because the electric charge on both sides of a metallurgical junction should be equal, the depletion region will extend in practice only into the p-region. By integrating a constant space charge of density eN_A over the length x_d (Figure 2.7.2a) one obtains the maximum field (at $x = 0$, Figure 2.7.2b)

$$\mathscr{E}_{max} = \frac{eN_A x_d}{\varepsilon}, \qquad (2.7.5)$$

where

$$x_d \simeq \left(\frac{2\varepsilon}{e}\frac{V_{bi}}{N_A}\right)^{1/2}. \tag{2.7.6}$$

The potential distribution follows the parabolic law (Figure 2.7.2c)

$$\phi(x) = V_{bi}\left(1 - \frac{x}{x_d}\right)^2. \tag{2.7.7}$$

Note the fact that for a one-sided abrupt junction in a given semiconductor the only quantity of interest is the impurity concentration in the lowly doped side. The formulae are identical to those derived in Section 2.3 for a Schottky barrier ($|\phi_s|$ is replaced by V_{bi}). For silicon $p-n$ junctions V_{bi} is usually much larger than kT/e and the depletion length (2.7.6) is much longer than the Debye length (2.3.18), therefore the depletion approximation is acceptable at thermal equilibrium.

An external bias V_J modifies the potential barrier such that V_{bi} should be replaced by $V_{bi} - V_J$ where $V_J > 0$ means p-side positively biased with respect to n-side [1], [37]. An increasing positive V_J determines a rapid increase of the current (which is mainly due to majority electrons diffusing over the potential barrier $V_{bi} - V_J$ into the p region). The depletion length decreases and may become comparable to L_{DE} thus invalidating the depletion approximation, which is indeed confirmed by results of exact computer calculations [38]. The depletion approximation, however, is well satisfied for reverse biased $p-n$ junctions (Figure 2.7.3) where V_{bi} should be replaced by $V_{bi} + |V_J|$ in equation (2.7.6). The depletion length *versus* reverse bias for several impurity densities (silicon, 300 K) is plotted in Figure 2.7.4, after Grove [37]. Because the ratio maximum field $\mathscr{E}_{max}$ to doping N_A is proportional to the depletion length, the right-hand scale was simply converted into proper units to show $\mathscr{E}_{max}$ per doping.

Most junctions in modern semiconductor technology are made by diffusion. The diffused junction may be approximated abrupt only if the diffusion length is much shorter than the depletion length. Otherwise, the total depletion length will also depend upon the surface concentration and the diffusion length* [37]. Only abrupt and plane junctions will be considered in this book (except where otherwise stated, in computer calculations).

An n^+p junction may be used as an electron injecting contact in unipolar devices. Because the device current is carried by one type of carriers (electrons) the injection efficiency is theoretically equal to unity. A positively biased n^+p junction may inject electrons into a depleted or inverted p-region. The first case corresponds to a punched-through n^+pn^+ diode (floating base transistor) whereas the second occurs for carrier injection from the source into the channel (inversion layer) of a metal-oxide-semiconductor (MOS) field-effect transistor (FET). In a punched-through diode, the current is determined by the electron flow thermionically emitted over the small potential barrier at the forward biased junction †. The injection

---

* The diffusion length = $2\sqrt{\text{diffusion constant} \times \text{diffusion time}}$ (see reference [37], chapter 3).
† The height of this potential barrier depends upon the applied voltage. The electron flow is carried by both diffusion (over the potential barrier) and drift.

may be limited, at large currents to the maximum current which can be extracted from a semiconductor region of finite carrier concentration [39] (see Chapter 5). In an MOS transistor the current flow emitted by the source region is controlled by the electric field induced in the channel.

A reverse biased *pn* junction may also act as an injecting electrode. The reverse current of a *pn* junction is due to electron-hole pairs generated somewhere in the

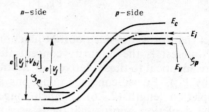

Figure 2.7.3. Energy band diagram of a reverse biased *p-n* junction (bias $|V_J|$). ζ_n and ζ_p are electron and hole quasi Fermi levels, respectively.

semiconductor [37]. The reverse current component due to charge carriers thermally generated in the neutral region (within a diffusion length from the depleted region) is very small, and independent upon the reverse bias provided that $|V_J| \gg kT/e$. For an n^+p junction it mainly consists of electrons transported by the electric field

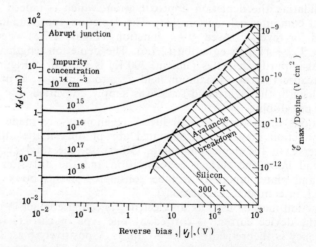

Figure 2.7.4. Depletion width for an abrupt asymmetrical *p-n* junction (silicon, 300 K) *versus* the reverse bias voltage, with background impurity concentration as parameter (*after* Grove, [37]). The figure was completed by indicating the maximum field in the depletion region (the right-hand scale) and the boundary for avalanche breakdown (according to Figure 2.7.5).

into the n^+ region. The reverse current component due to carrier generation inside the space-charge region does however depend upon the applied bias, because both the extent of this space-charge region and the electric field within it depend upon the reverse bias. At large reverse voltages the electric field at junction $\mathscr{E}_{max}$

reaches the critical value $\mathscr{E}_{cr}$ for avalanche breakdown, therefore, according to (2.7.5) and (2.7.6)

$$\mathscr{E}_{max} = \left[\frac{2e}{\varepsilon N_A}(V_{bi} + |V_J|)\right]^{1/2} \simeq \left(\frac{2e|V_J|N_A}{\varepsilon}\right)^{1/2} = \mathscr{E}_{cr} \qquad (2.7.8)$$

and the breakdown voltage is

$$V_B = \frac{\varepsilon \mathscr{E}_{cr}^2}{2eN_A}, \qquad (2.7.9)$$

where N_A is the effective doping of the lowly doped side of the abrupt asymmetrical junction. Therefore, the critical field $\mathscr{E}_{cr}$ can be determined from the experimentally measured breakdown voltages. $\mathscr{E}_{cr}$ increases with increasing doping, as shown in Figure 2.7.5 for silicon [37]. The critical field for Zener breakdown (direct band to band-tunnelling) is also indicated. At impurity concentrations below $10^{18} cm^{-3}$ the breakdown mechanism is by avalanche multiplication (impact ionization).

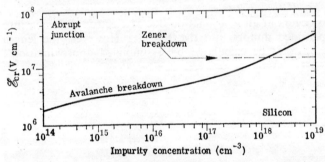

Figure 2.7.5. Critical field for avalanche breakdown *versus* background impurity concentration (abrupt asymmetrical silicon junctions). These data were obtained by using equation (2.7.8) with experimentally measured V_B. The critical field for Zener breakdown is also indicated. (*after Grove*, [37]).

However, the critical field for avalanche breakdown is simply a concept. The correct expression of V_B should be calculated by using the avalanche integral and the electron and hole ionization rates. The avalanche multiplication will take place

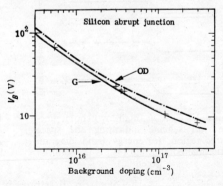

Figure 2.7.6. Breakdown voltage *versus* background impurity concentration for abrupt asymmetrical silicon *pn* junctions, calculated by using the experimental ionization coefficients obtained by Grant[40] (curve G) and by van Overstraeten and de Man [41] (curve OD). Several experimental points were also indicated. (*after Grant*, [40]).

in a certain avalanche region of finite thickness and the breakdown voltage will depend upon the exact doping profile. Recent calculations based upon experimentally determined ionization rates yield for the V_B-doping dependence in an abrupt silicon junction the full curve shown in Figure 2.7.6 [40], [41].

An n^+p junction biased to avalanche breakdown will inject *holes* into the *p*-region (negative with respect to the n^+-side). The 'minority' carrier injection by such a junction is used in avalanche transit-time diodes.

Reverse biased *pn* junctions which do not inject charge carriers are also used in unipolar semiconductor devices as collecting electrodes or as control electrodes. In a junction-gate FET, for example, the channel width is modulated by the depletion region of a reverse biased *pn* junction. The depletion approximation (abrupt transition from neutral to totally depleted semiconductor regions) is valid provided that: (a) the device dimensions are large compared to the Debye length, and (b) carrier injection into the depletion layer is negligibly small.

Electron injection into a semiconductor can be also provided by an n^+n (high-low) junction (Figure 2.7.1a). The internal potential difference (built in voltage) V_{bi} for such a junction (Figure 2.7.1b) can be calculated as shown above for a *pn* junction. By using the non-degenerate statistics

$$V_{bi} = \frac{kT}{e} \ln \frac{N_D^+}{N_D} \quad (n^+n \text{ junction}). \tag{2.7.10}$$

A space charge region also occurs at an n^+n junction due to electron diffusion into the *n*-side and to uncompensated donors into the n^+-side. However, the length of the space-charge region cannot be calculated by a simplified approximation*.

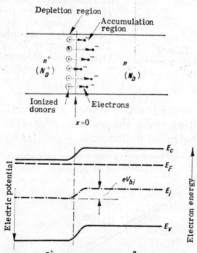

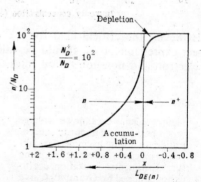

Figure 2.7.7. n^+n (high-low) homojunction: (*a*) scheme indicating the space charge region; (*b*) energy band diagram at thermal equilibrium (V_{bi} is the built-in potential).

Figure 2.7.8. Electron distribution at the nn^+ interface, calculated for a doping ratio $N_D^+/N_D = 10^2$. $L_{DE(n)}$ is the extrinsic Debye length for the lowly doped region. (*after Hauser and Littlejohn*, [42]).

Figure 2.7.8 shows the electron distribution at an n^+n junction, after Hauser and Littlejohn [42]. The thickness of the accumulation layer is of the order of the extrin-

---

* The voltage drop on the n^+ region is smaller than kT/e and the depletion approximation cannot be used [42].

sic Debye length, for n region, as expected. The depleted region inside the n^+ contact is thinner. Therefore, a satisfactory boundary condition for an n^+n contact is

$$n = N_D^+ \text{ (at the } n^+n \text{ interface)}. \tag{2.7.11}$$

2.8 Heterojunctions

A heterojunction is a junction formed between two different semiconductors. An ideal heterojunction requires a perfect match of lattice constants and thermal expansion coefficients, which is not normally possible. Lattice defects located at the heterojunction interface may have considerable effect upon its electrical properties and reduce its usefulness.

We shall consider below the model suggested by Anderson [43] for an ideal abrupt heterojunction, without interface states. Figure 2.8.1a shows the energy-band diagram for isolated semiconductors, namely a narrow-gap n-type semiconductor (all corresponding quantities are denoted by index 1) and a wide-gap p-type semiconductor (index 2). Figure 2.8.1b indicates the energy-band diagram of the np heterojunction at thermal equilibrium. ΔE_c and ΔE_v are the discontinuities in conduction-band and valence-band edges, respectively. The total built-in potential V_{bi} is equal to the sum of the partial built-in voltages, V_{b1} and V_{b2}, as shown in Figure 2.8.1b. A depletion layer is formed at the interface and extends in both sides of the np heterojunction. The potential distribution in this layer may be calculated by using the depletion approximation.

Figure 2.8.2 shows the energy-band diagram for an nn heterojunction. The band bending is opposite to that indicated for an np junction (Figure 2.8.1b) because the work function of the wide-gap semiconductor is smaller. The interface layer of semiconductor 1 is accumulated and the mobile electron concentration should be taken into account in calculations. This heterojunction resembles somewhat a Schottky barrier metal-semiconductor contact, the narrow-gap semiconductor replacing the metal. If the wide-gap semiconductor is negatively biased, the barrier for thermionic electron injection into the narrow-gap semiconductor will decrease and the current will increase almost exponentially with applied voltage. If the applied bias is reversed, the current will be carried by electrons injected into the wide-gap semiconductor. The barrier for this thermionic injection process is not, however, constant (as for a metal-semiconductor contact) but instead decreases with increasing voltage being smaller than $\Delta E_c - eV_{b1}$, at thermal equilibrium (Figure 2.8.2). Therefore, the 'reverse' current never saturates but increases with voltage (see Chapter 3 in reference [1]). Kumar [44] pointed out, however, that the current injected into the wide-gap semiconductor cannot exceed the maximum value determined by finite doping and finite velocity of electrons in semiconductor 1. The same author suggests a generalized diffusion theory (carrier collisions into the barrier region are taken into account, in contrast to the emission theories like that presented in Section 2.6 for a Schottky contact) of isotype heterojunctions.

The injection mechanism at a heterojunction depends upon the band discontinuities (if, for example, the barrier for holes is much higher than that for electrons, the heterojunction current will be an electron current), upon the density of interface states (if this density is high, the dominant current will be a generation-recombination current) and upon the semiconductor doping (tunnelling of the barrier is possible at very high doping concentrations).

There are many possible applications of heterojunctions. Heterojunctions may be used for example as an emitter junction with very high injection efficiency in a heterojunction transistor [45] (the wide-gap emitter introduces a very large barrier

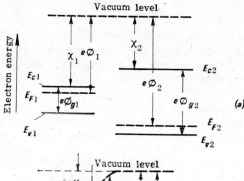

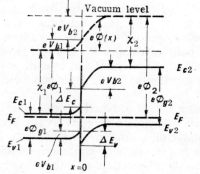

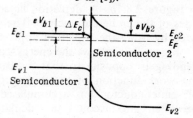

Figure 2.8.1. (a) Energy band diagrams of two isolated semiconductors, having different energy-gap ($e\phi_g$), electron affinity (χ) and work function (ϕ_1 and ϕ_2, respectively), semiconductor 1 is n-type and semiconductor 2 p-type; (b) pn heterojunction formed by the above two semiconductors. (after Sze, chapter 3 in [1]).

Figure 2.8.2. Energy-band diagram of an n-n heterojunction (isotype heterojunction). V_{b1} and V_{b2} are the partial built-in voltages. (after Sze, chapter 3 in [1]).

for carrier injection from base into emitter). Another application is the heterojunction phototransistor. This is a *pnp* or *npn* structure with a wide-gap emitter operating with floating base and the collector junction reverse biased. The transistor is illuminated on the emitter side and the light is absorbed near the emitter junction in the narrow energy-gap side (base) (the photon energy is lower than the emitter energy gap) [46].

Heterojunctions are not particularly useful as injecting contacts in unipolar devices. An example of unipolar transistors using a heterojunction is given in Chapter 4.

2.9 Practical Problems of Injecting Contacts

The problem of providing a suitable injecting contact to a unipolar device must be formulated by considering the type of device and the properties of the semiconductor used (its energy-gap and doping are especially important). We are interested in making ohmic contacts for transferred electron oscillators and junction-gate field-effect transistors, injecting contacts for wide-gap semiconductors and insulators, barrier-type contacts for Schottky diodes and metal-semiconductor-metal microwave diodes. pn junctions are also used as injecting contacts in MOS transistors, IMPATT diodes, punch-through diodes.

The two fundamental types of contacts are: semiconductor junctions and metal-semiconductor contacts. The desirable properties of semiconductor junctions can be obtained by modern technological processes: diffusion, epitaxial growth, ion implantation.

The metal-semiconductor contact is simpler in principle (because it also provides the device electrode which will be connected to the external circuit) but in fact is difficult to be realized with the desired properties.

Metals on highly doped semiconductors form good ohmic contacts (for practical purposes we shall consider a contact to be ohmic even if the contact itself does not obey Ohm's law, provided that this contact can supply the required current with a sufficiently small voltage drop, i.e. its resistance is negligibly small compared to the semiconductor resistance). However, a number of wide-band-gap semiconductors present contacting problems [47] because they cannot be doped heavily enough*. This may occur due to the high temperature required for doping process (the semiconductor decomposes or evaporates), due to limited solubility of dopants or due to self-compensation by native defect centres.

A method used widely to achieve a high doping density is the technique of alloy regrowth. The Al—Si contact, for example, is formed by depositing the metal onto the silicon surface, by heating the system above its eutectic temperature (576 °C) and then cooling it slowly to allow a suitable regrowth of the molten surface layer. The recrystallized silicon (0.5 μm thick) contains about 5×10^{18} cm^{-3} Al atoms [48] and is a highly doped p-type. Such a contact is ohmic on p-type and highly doped n-type silicon. Of course, such an alloyed contact cannot be formed for any metal-semiconductor pair. We note here the difficulties encountered in realizing a stable ohmic contact to GaAs devices [49], [50]. It was shown that a thin high-resistance layer forms underneath the alloy-semiconductor interface in Gunn-effect diodes with alloyed contacts. Cox and Hasty [49] indicated that the above high resistivity layer is probably due to the formation of a high density of lattice defects during the rapid freeze portion of the alloying process (Ag—In—Ge evaporated contact alloyed at 630 °C for 30 s)†. Therefore, good and reliable ohmic contacts should be made by alloying contacts on epitaxially grown n^+ GaAs layers, instead of doing this directly on the n-type GaAs, thus obtaining a metal—n^+—n—n^+—metal structure [51].

Good ohmic contacts to silicon are now made by using metal silicides. The contact is formed by alloying silicon to a metal silicide. Let us consider for example the eutectic system PtSi (platinum silicide) — Si [52]. This silicide can be formed by solid-solid reaction at 300 °C, far below the melting point of the metal-silicon eutectic (980 °C). This contact is superior because the formation of a liquid phase during the process (such as in alloying the Al—Si contact) leads to surface tension

---

* Another difficulty may be a very high work function (wide energy gap) which prevents the formation of a hole-injecting contact.

† The performance of such a device may continue to deteriorate during periods of operation and this was indeed observed and explained as due to the increasing number of defects in the damaged area at high operating temperatures (200—300 °C) [49].

effects which produce 'balling-up' and, finally, a non-uniform penetration. A uniform contact is the result of a solid-solid reaction.

Palladium silicide (Pd_2Si) contacts to silicon are widely studied [20] — [23], [53]. The contact may be formed by annealing the silicon wafer with evaporated Pd layer at 260 °C for 10 minutes [53]. Buckley and Moss have shown [53] that Pd_2Si matches to the $\langle 111 \rangle$ surface so well that it may be considered as epitaxially grown. These contacts are unusually uniform and reproducible [53].

The direct contact between a metal and a lowly-doped semiconductor (moderate or wide band gap) is, as a rule, a Schottky-barrier contact. The barrier height depends slightly upon the metal work function and is determined by the interface phenomena, as previously shown. The barrier height depends upon the conditions under which the contact was prepared. The physical situation at the barrier can barely be envisaged because the surface parameters can be only measured indirectly. However, reproducible Schottky contacts can still be fabricated and used as Schottky diodes or as injecting contacts for other semiconductor devices (MSM structures, IMPATT diodes, transferred electron diodes).

References

1. S. M. SZE, 'Physics of semiconductor devices', John Wiley, New York, etc., 1969; chapter 8.
2. A. MANY, Y. GOLDSTEIN and N. B. GROVER, 'Semiconductor surfaces', North Holland Publish. Comp., Amsterdam, 1965.
3. C. A. MEAD, 'Metal-semiconductor surface barriers', *Solid-St. Electron.*, **9**, 1023–1033 (1966).
4. C. A. MEAD, 'Physics of interfaces', in 'Ohmic contacts to semiconductors', Electrochem. Soc., New York, 1969.
5. C. A. MEAD and W. G. SPITZER, 'Fermi level position at metal-semiconductor interfaces', *Phys. Rev.*, **134**, A713–A716 (1964).
6. A. Y. C. YU, 'The metal-semiconductor contact: an old device with a new future', *IEEE Spectrum*, **7**, No. 3, 83–89 (1970).
7. M. J. TURNER and E. H. RHODERICK, 'Metal-silicon Schottky barriers', *Solid-St. Electron.*, **11**, 291–300 (1968).
8. W. MÖNCH, 'On metal-semiconductor surface barriers', *Surf. Sci.*, **21**, 443–446 (1970); 'Surface states on clean and on cesium-covered cleaved silicon surfaces', *Phys. Stat. Sol.*, **40**, 257–265 (1970).
9. B. L. SMITH and E. H. RHODERICK, 'Schottky barriers on *p*-type silicon', *Solid-St. Electron.*, **14**, 71–75 (1971).
10. J. S. BLAKEMORE, 'Semiconductor statistics', Pergamon Press, London, 1962.
11. V. L. RIDEOUT and C. R. CROWELL, 'Effects of image force and tunnelling on current transport in metal-semiconductor (Schottky barrier) contacts', *Solid-St. Electron.*, **13**, 993–1009 (1970).
12. H. Y. FAN, 'Contacts between metals and between a metal and a semiconductor', *Phys. Rev.*, **62**, 388–394 (1942).
13. S. S. PERLMAN, 'Barrier height diminution in Schottky diodes due to electrostatic screening', *IEEE Trans. Electron Dev.*, **16**, 450–454 (1969).
14. G. W. EIMERS and E. H. STEVENS, 'A composite model for Schottky diode barrier height', *IEEE Trans. Electron Dev.*, **12**, 1185–1186 (1971).
15. C. A. MEAD, E. H. SNOW and B. E. DEAL, 'Barrier lowering and field penetration at metal-dielectric interfaces', *Appl. Phys. Lett.*, **9**, 53–55 (1966).
16. M. A. NICOLET, V. RODRIGUEZ and D. STOLFA, 'Unipolar interface-charge-limited current', *Surf. Sci.*, **10**, 146–164 (1968).
17. P. H. LADBROOKE, 'Reverse current characteristics of some imperfect Schottky barriers', *Solid. St. Electron.*, **16**, 743–749 (1973).
18. A. M. COWLEY and S. M. SZE, 'Surface states and barrier height of metal-semiconductor systems', *J. Appl. Phys.*, **36**, 3212 (1965).

19. J. M. ANDREWS and M. P. LEPSELTER, 'Reverse current-voltage characteristics of metal-silicide Schottky diodes', *Solid. St. Electron.*, **13**, 1011–1023 (1970).
20. J. M. ANDREWS and F. B. KOCH, 'Formation of NiSi and current transport across the NiSi–Si interface', *Solid-St. Electron.*, **14**, 901–908 (1971).
21. C. J. KIRCHER, 'Metallurgical properties and electrical characteristics of palladium silicide-silicon contacts', *Solid-St. Electron.*, **14**, 507–513 (1971).
22. R. F. BROOM, 'Doping dependence of the barrier height of palladium-silicide Schottky-diodes', *Solid-St. Electron.*, **14**, 1087–1092 (1971).
23. A. SHEPELA, 'The specific contact resistance of Pd_2Si contacts on n- and p-Si', *Solid-St. Electron.*, **16**, 477–481 (1973).
24. M. P. LEPSELTER and J. M. ANDREWS, 'Ohmic contacts to silicon' in 'Ohmic contacts to semiconductors', Electrochem. Soc., New York, 1969.
25. G. H. PARKER, T. C. MCGILL, C. A. MEAD and D. HOFFMANN, 'Electric field dependence of GaAs Schottky barriers', *Solid-St. Electron.*, **11**, 201–204 (1968).
26. V. HEINE, *Phys. Rev.*, **138**, A1689 (1965).
27. C. R. CROWELL, 'The Richardson constant for thermionic emission in Schottky barrier diodes', *Solid-St. Electron.*, **8**, 395–399 (1965).
28. D. J. PAGE, 'Some computed and measured characteristics of CdS space-charge-limited diodes', *Solid-St. Electron.*, **9**, 255–264 (1966).
29. M. A. LAMPERT and P. MARK, 'Current injection in solids', Academic Press, New York and London, 1970.
30. D. DASCALU, 'Transit time effects in unipolar solid-state devices', Ed. Acad., București, Romania, and Abacus Press, Kent, England, 1974.
31. C. R. CROWELL and V. L. RIDEOUT, 'Normalized thermionic-field emission in metal-semiconductor (Schottky) barriers', *Solid-St. Electron.*, **12**, 89–105 (1969).
32. C. R. CROWELL, 'Richardson constant and tunnelling effective mass for thermionic and thermionic-field emission in Schottky barrier diodes', *Solid-St. Electron.*, **12**, 55–59 (1969).
33. S. J. FONASH, 'Current transport in metal-semiconductor contacts – a unified approach', *Solid-St. Electron.*, **15**, 783–787 (1972).
34. C. Y. CHANG and S. M. SZE, 'Carrier transport across metal-semiconductor barriers', *Solid-St. Electron.*, **13**, 727–740 (1970).
35. A. Y. C. YU, 'Electron tunnelling and contact resistance of metal-silicon contact barriers', *Solid-St. Electron.*, **13**, 239–247 (1970).
36. C. Y. CHANG, Y. K. FANG and S. M. SZE, 'Specific contact resistance of metal-semiconductor barriers', *Solid-St. Electron.*, **14**, 541–550 (1971).
37. A. S. GROVE, 'Physics and technology of semiconductor devices', John Wiley & Sons, New York, 1967, chapter 6.
38. A. DE MARI, 'An accurate numerical steady-state one-dimensional solution of the p-n junction', *Solid-St. Electron.*, **11**, 33–58 (1968).
39. G. PERSKY, 'Thermionic saturation of diffusion currents in transistors', *Solid-St. Electron.*, **15**, 1345–1351 (1972).
40. W. N. GRANT, 'Electron and hole ionization rates in epitaxial silicon at high electric fields', *Solid-St. Electron.*, **16**, 1189–1203 (1973).
41. R. VAN OVERSTRAETEN and H. DE MAN, 'Measurement of the ionization rates in diffused silicon p–n junctions', *Solid-St. Electron.*, **13**, 583–609 (1969).
42. J. R. HAUSER and M. A. LITTLEJOHN, 'Approximations for accumulation and inversion space-charge layers in semiconductors', *Solid-St. Electron.*, **11**, 667–674 (1968).
43. R. L. ANDERSON, 'Experiments on Ge–GaAs heterojunctions', *Solid-St. Electron.*, **5**, 341–351 (1962).
44. R. C. KUMAR, 'Current transport in isotype heterojunctions', *Internat. J. Electron.*, **25**, 239–247 (1968).
45. H. J. HOVEL and A. G. MILNES, 'ZnSe–Ge heterojunction transistors', *IEEE Trans. Electron Dev.*, **ED–16**, 766–774 (1969).
46. T. MORIIZUMI and K. TAKAHASHI, 'Theoretical analysis of heterojunction phototransistors', *IEEE Trans. Electron Dev.*, **ED–19**, 152–159 (1972).
47. M. AVEN and R. K. SWANK, 'Ohmic contacts to wide-band-gap semiconductors', in 'Ohmic contacts to semiconductors', Electrochem. Soc., New York, 1969.

48. R. M. WARNER (Editor), 'Integrated circuits, design principles and fabrication', McGraw-Hill, New York, 1965, p. 308.
49. R. H. COX and T. E. HASTY, 'Metallurgy of alloyed ohmic contacts for the Gunn oscillator', in 'Ohmic contacts to semiconductors', Electrochem. Soc., New York, 1969.
50. W. S. C. GURNEY, 'Contact effects in Gunn diodes', *Electron. Lett.*, 7, 711–713 (1971).
51. S. KNIGHT and C. PAOLA, 'Ohmic contacts for gallium arsenide bulk effect devices', in 'Ohmic contacts to semiconductors', Electrochem. Soc., New York, 1969.
52. M. P. LEPSELTER and J. M. ANDREWS, 'Ohmic contacts to silicon', in 'Ohmic contacts to semiconductors', Electrochem. Soc., New York, 1969.
53. W. D. BUCKLEY and S. C. Moss, 'Structure and electrical characteristics of epitaxial palladium silicide contacts on single crystal silicon and diffused $p-n$ diodes', *Solid-St Electron.*, 15, 1331–1337 (1972).

Problems

2.1. Figure 2.4.3 shows that for large ϕ_1 the barrier lowering equals the total band bending $|\phi_S|$. Explain the disappearance of the barrier at a given finite band bending.

2.2. How can the statistical distribution of impurities affect the results presented in Figure 2.4.3 ?.
Hint: If $\varepsilon_{opt} \simeq \varepsilon$, then $(|\phi_S|/\phi_1)^{3/2} \simeq |N_D - N_A| x_d^3$.

2.3. Show that the transition from the neutral semiconductor bulk to the totally depleted surface layer occurs within a distance comparable to the extrinsic Debye length L_{DE} and determine the condition required for this distance to be negligible as compared to the thickness of the depletion layer (at thermal equilibrium).

2.4. The condition $\phi = 0$ imposed for $x = x_d$ is not strictly correct because the transition between the totally depleted layer $(0 < x < x_d)$ and the neutral semiconductor bulk $(\phi = 0)$ takes place within a layer of finite thickness which supports a finite voltage drop. Show that due to the presence of the mobile carriers in this transition layer, the depletion length should be, approximately, written [1]

$$x_d \approx \left[2\varepsilon \left(|\phi_S| - \frac{kT}{e} \right) \middle/ e |N_D - N_A| \right]^{1/2}. \qquad \text{(P.2.4.1)}$$

2.5. Calculate L_{DE} for n-type silicon, $120\,\Omega$ cm resistivity at 200 °C.

2.6. Find the maximum electric field intensity in a Schottky barrier occurring at a gold contact on an n-type silicon sample $0.1\,\Omega$ cm resistivity, at thermal equilibrium and room temperature.

2.7. How does the field penetration into the metal depend upon the semiconductor doping, at a metal-semiconductor contact. Found $(\Delta\phi)_{es}$ for silicon at usual doping levels ($10^{15}-10^{18}$cm^{-3}) by neglecting any other contribution to the barrier lowering.

2.8. Show that if a constant charge is trapped at the MS surface in an n-type semiconductor — insulator layer (transparent for electrons)-metal system, then the barrier may increase with the increasing thickness of the interfacial layer.

2.9. Determine the conditions required for the barrier (2.5.12) (Figure 2.5.4) to increase with increasing thickness of the interfacial layer.

2.10. Consider a single acceptor level at the surface of an n-type semiconductor (surface state density is given). Calculate the mobile electron density at surface as a function of surface field in semiconductor.

2.11. Find the built-in voltage of a p-n junction without neglecting the intrinsic carrier concentration with respect to semiconductor doping.

2.12. Calculate the depletion length x_d and the maximum electric field $\mathscr{E}_{max}$ for a linearly graded junction (the gradient of impurity concentration is given).

2.13. Find the length of depletion region at the p-n heterojunction shown in Figure 2.8.1b.

Part 2

Bulk- and Injection-controlled Devices

3
Schottky Diode

3.1 Introduction

The Schottky diode is a majority-carrier device, essentially identical to a Schottky-barrier metal-semiconductor contact (see Chapter 2). Consider such a contact to an n-type semiconductor. Only electron transfer at the interface is considered to a first approximation. The thermal equilibrium energy-band diagram is shown in Figure 3.1.1a. A reverse applied bias (minus on metal) increases the potential barrier seen by electrons going from the semiconductor into the metal (Figure 3.1.1c) and the reverse current becomes practically equal to the current flow of electrons injected from the metal into the semiconductor. A forward bias (plus on metal) decreases the barrier for the electrons into the semiconductor conduction

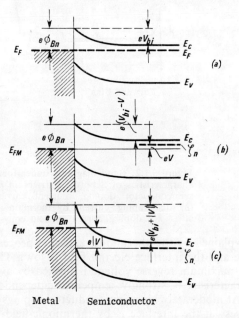

Figure 3.1.1. Energy band diagram for a Schottky contact on an n-type semiconductor: (a) at thermal equilibrium; (b) forward bias; (c) reverse bias. V_{bi} is the built-in voltage drop into the semiconductor, ϕ_{Bn} the potential barrier at the metal-semiconductor interface and ζ_n the quasi Fermi level for electrons (shown in the semiconductor bulk).

band (Figure 3.1.1b) and determines a rapid increase of the forward current which consists of majority carriers injected from the semiconductor into the metal.

The electron transfer at the metal-semiconductor interface takes place by thermionic emission, thermionic-field emission or field emission (Section 2.6). The transfer mechanism depends upon the intensity of the electric field at the metal-semiconductor interface which is directly related to the surface space-charge layer and thus to semiconductor resistivity. The steady-state characteristic of a Schottky-barrier on a lowly doped semiconductor looks as shown in Figure 3.1.2a and

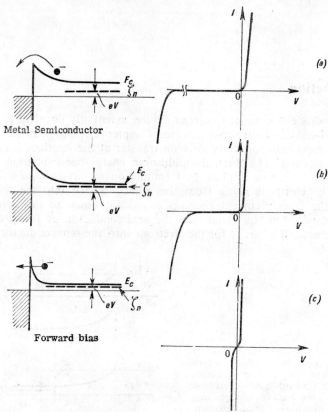

Figure 3.1.2. Band diagrams under forward bias and current-voltage characteristics of Schottky barriers on n-type semiconductor: (a) low semiconductor doping, current carried by thermionic emission; (b) high doping, current by thermionic-field emission; (c) very high doping, current by field emission. (after Yu, [1]).

is explained by the thermionic emission process (the space-charge layer is relatively wide and the interface electric field is low). The reverse current is almost constant. The maximum reverse voltage is limited by avalanche breakdown. The entire $J-V$ characteristic is strongly temperature-dependent.

At moderately high semiconductor dopings (in the 10^{18}cm^{-3} range) the transport process at the interface is by thermionic-field emission (Figure 3.1.2b). The maximum reverse voltage is also determined by thermionic-field emission and is small (a few volts).

If the semiconductor doping is very heavy (for example, $N_D > 10^{19}$cm^{-3} for n-type silicon) the depletion width will be very thin and the potential barrier will be 'penetrated' by carrier tunnelling. The field-emission type characteristic is shown in Figure 3.1.2c. The metal-semiconductor contact behaviour is practically ohmic. The contact resistance is doping dependent but temperature insensitive*.

The Schottky diode is a rectifyer-type device and should be constructed by using a high-resistivity semiconductor (Figure 3.1.2 a). The higher the maximum reverse operating voltage, the lower the semiconductor doping. The ohmic drop at high forward current densities becomes, therefore, very important and the semiconductor thickness should be kept to a minimum (just equal to the depletion width at the maximum reverse voltage) which is of the order of a few μm. A practical solution is to use an n-type epitaxial layer grown on a n^+-substrate. This substrate provides a good ohmic back-contact to the Schottky diode and also facilitates the device handling through the fabrication process.

The energy band diagrams for a practical Schottky diode (metal-n^+-n-metal structure) are shown in Figure 3.1.3. For simplicity, the conduction band in the n^+ region (degenerated) is levelled to the Fermi level into the metal (back contact) and the very thin barrier (Figure 3.1.2 c) is not shown. Both metal contacts are assumed to have the same work-function. Note the existence of the built-in potential at the n^+n junction. Under reverse bias (Figure 3.1.3b) the predominating transport mechanism is thermionic emission into the semiconductor. Figure 3.1.3c shows the energy band diagrams for forward bias. The majority carriers (electrons) diffuse towards the right at the n^+n interface and then are thermionically emitted over the potential barrier, whose height is reduced by the forward voltage bias. Note, however, that part of this voltage bias drops on the high-resistivity epitaxial layer.

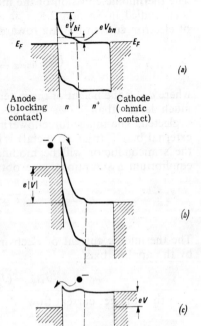

Figure 3.1.3. Band diagrams for a metal-n-n^+ Schottky diode: (a) thermal equilibrium; (b) reverse bias; (c) forward bias.

3.2 Steady-state Characteristic

The current voltage characteristic of the Schottky diode will be now computed by taking into account the transport of majority carriers. There are a number of approaches to this problem, the first to be examined being the thermionic emission theory [2].

---

* If the barrier is low (< 0.2 eV, for example PtSi on p—Si, $\phi_{Bp} \approx 0.2$ eV, hole emission) the thermionic emission current at room temperature can be so large that the current-voltage characteristic may appear linear over a large current range. However, this characteristic is strongly temperature dependent and can be easily distinguished from tunnelling contacts [1].

(a) Thermionic Emission Theory

The thermionic emission of the metal-semiconductor contact at thermal equilibrium was studied in Section 2.6. For a metal-n-type semiconductor contact the density of electron current flowing towards the metal is

$$(J_{SM})_{\text{therm.eq.}} = A^* T^2 \exp\left(-\frac{e\phi_{Bn}}{kT}\right), \tag{3.2.1}$$

where ϕ_{Bn} is the height of the potential barrier at the interface (Figure 2.6.1) assumed much larger than kT/e. The electron collisions within the depletion region are neglected. The image-force-lowering of the barrier is not taken into account. If an external bias V (plus on metal) is applied, the potential barrier seen by electrons in the semiconductor will be modified accordingly and, by maintaining the thermal equilibrium assumptions, one obtains

$$J_{SM} = A^* T^2 \exp\left(-\frac{e\phi_{Bn}}{kT}\right) \exp\left(\frac{eV}{kT}\right). \tag{3.2.2}$$

The thermionic current of electrons injected into the semiconductor is not affected by the applied bias

$$J_{MS} = (J_{MS})_{\text{therm.eq.}} = (J_{SM})_{\text{therm.eq.}}, \tag{3.2.3}$$

such that the net current flow is*

$$J = J_{SM} - J_{MS} = A^* T^2 \exp\left(-\frac{e\phi_{Bn}}{kT}\right)\left[\exp\left(\frac{eV}{kT}\right) - 1\right] \tag{3.2.4}$$

$$J = J_{\text{sat}}\left[\exp\left(\frac{eV}{kT}\right) - 1\right], \quad J_{\text{sat}} = A^* T^2 \exp\left(-\frac{e\phi_{Bn}}{kT}\right). \tag{3.2.5}$$

Equation (3.2.5) is similar to the Shockley equation for pn junction. However, the expression of J_{sat} and the physical processes involved are quite different [2].

(b) Diffusion Theory

The diffusion theory removes the hypothesis of negligible effect of carrier collisions within the space-charge layer. Numerical computations show, indeed, that for most Schottky diodes the width of the space-charge layer is much larger than the mean free path of charge carriers and their collisions cannot be neglected [3]. The current through the depletion layer is computed as

$$J = J_n = en(x)\,\mu_n \mathscr{E}(x) + eD_n \frac{dn}{dx} = \text{const.} \tag{3.2.6}$$

---

* The conventional positive sense of the electric current is from the metal towards the semiconductor. The net electron flux is directed in the opposite sense of J.

3. Schottky Diode

The current density is assumed to be sufficiently small such that the mobile carrier concentration is negligible with respect to ionized impurities, the potential distribution is parabolic and the electric field varies linearly. Two boundary conditions for the electron concentration are also necessary and will be written by assuming that they have their thermal equilibrium values, given by a non-degenerate statistics. The result is [2]

$$J = J_n \simeq \left\{ \frac{e^2 D_n N_c}{kT} \left[\frac{e(V_{bi} - V) 8\pi N_D}{\varepsilon} \right]^{1/2} \exp\left(-\frac{e\phi_{Bn}}{kT}\right) \right\} \times$$

$$\times \left[\exp\left(\frac{eV}{kT}\right) - 1 \right]; \qquad (3.2.7)$$

The $J - V$ characteristic is of the same form as (3.2.5) but the 'saturation' current J_{sat} increases with increasing reverse bias $|V|$. The diffusion theory predicts, however, currents considerably above the value observed experimentally. It was shown that the difference occurs as the result of the improper boundary condition which states the electron concentration at the interface at its thermal equilibrium value (the continuity of the electron quasi-Fermi level is assumed at the interface), independent of the applied voltage. This is based upon the assumption that electron density injected into the metal, even under the forward bias, is too low to disturb the electron distribution into the metal. This fact is not obvious, because the injected electrons have higher energies than the electrons in the metal at thermal equilibrium. A more reasonable assumption is that the electrons are injected into the metal with their thermal velocity. Detailed 'microscopic' calculation based upon the Boltzmann equation and an electron temperature model were presented in references [3], [4]. A phenomenological approach is indicated below.

(c) Thermionic-diffusion Theory (5), (6)

Figure 3.2.1 shows the energy-band diagram of a Schottky barrier metal contact to an n-type semiconductor. $e\psi(x)$ is electron potential energy measured with respect to the metal Fermi level. Within most of the space-charge layer ($x_m \lesssim x < x_d$)

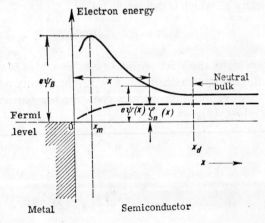

Figure 3.2.1. Electron energy diagram at a metal-semiconductor barrier. ζ_n is the electron quasi-Fermi level, by definition replacing the Fermi level in the expression of electron density, $n(x)$ (see Chapter 1). Here this level, as well as the electron potential energy (the bottom of the conduction band), are measured with respect to the metal Fermi level. The contact space-charge region extends in the semiconductor to the depth x_d.

this energy is directly related to the macroscopic electric potential ϕ ($\psi = -\phi +$ const) and to the electric field

$$\mathcal{E} = -\frac{d\phi}{dx} = \frac{d\psi}{dx}. \tag{3.2.8}$$

The rounded form of the energy barrier $e\psi(x)$ near $x = x_m$ arises from the combined effects of the electric field of ionized donors and the image force.

The Boltzmann statistics is assumed valid for $x_m < x < x_d$

$$n(x) = N_c \exp\{[-e\psi(x) + \zeta_n(x)]/kT\}, \tag{3.2.9}$$

where T is the electron temperature, assumed constant in the above region, and equal to the lattice temperature, and ζ_n is the electron quasi-Fermi level (imref). Note the fact that ζ_n in the semiconductor bulk is eV above the metal Fermi level, where V is the voltage bias applied to the Schottky diode ($V > 0$ means plus on metal).

By using equations (3.2.8), (3.2.9) and the Einstein relation $\left(\mu_n = D_n \dfrac{e}{kT}\right)$, the transport equation (3.2.6) becomes

$$J_n = en\mu_n \frac{d\zeta_n}{dx}. \tag{3.2.10}$$

The very thin region $0 < x < x_m$, in the vicinity of the interface, cannot be described by equations (3.2.9) and (3.2.10) because here the potential energy changes rapidly in distances comparable to the electron mean free path. In the same region the electron distribution cannot be described by an imref ζ_n, nor be associated with an effective density of surface states [2], [5]. The charge carriers should diffuse across this region, being collected by the metal. This process is characterized by a collection velocity v_c, which is defined by [6]

$$J_n = e(n_m - n_0)v_c, \tag{3.2.11}$$

where n_m is the electron density at $x = x_m$ when current is flowing, and n_0 is the quasi-equilibrium density at $x = x_m$, the density which would occur if it were possible to reach equilibrium without changing the barrier occurring at the applied bias V. Therefore (Figure 3.2.1)

$$n_0 = n(x_m)\Big|_{\substack{V=0 \\ \psi = \psi_m(V)}} = N_c \exp(-e\psi_B/kT) \tag{3.2.12}$$

$$n_m = n(x_m)\Big|_{V=V} = N_c \exp\frac{\zeta_n(x_m) - e\psi_B}{kT}. \tag{3.2.13}$$

3. Schottky Diode

The value of the collection velocity* remains to be calculated by considering the microscopic processes at the metal-semiconductor contact. The total current becomes †

$$J = \frac{eN_c v_c}{1 + v_c/v_d} \exp\left(-\frac{e\psi_B}{kT}\right) \times \left[\exp\left(\frac{eV}{kT}\right) - 1\right], \quad (3.2.14)$$

where

$$v_d = \frac{\mu_n kT}{e} \left\{ \int_{x_m}^{x_d} \exp[-e\psi_B + e\psi(x)]\,dx \right\}^{-1} \quad (3.2.15)$$

is an effective diffusion velocity associated with the transport of electrons through the depletion layer towards the metal contact.

The $J-V$ characteristic (3.2.14) is of the form

$$J = J_s \left[\exp\left(\frac{eV}{kT}\right) - 1\right], \quad J_s = J_s(V), \quad (3.2.16)$$

the factor J_s being, however, bias dependent. The result of the diffusion theory, equation (3.2.7) can be readily obtained by introducing $v_c \to \infty$ in equation (3.2.15), by using the parabolic potential distribution for the Schottky barrier and also the Einstein relation $\left(\mu_n = \frac{eD_n}{kT}\right)$.

It is interesting to note that an infinite collection velocity ($v_c \to \infty$) in equation (3.2.11) implies $n_m \to n_0$ and therefore $\zeta_n(x_m) = 0$, such that the quasi-Fermi level for electrons at the potential barrier should coincide with the Fermi level in the metal. Crowell and Beguwala [6] note that this means that it is impossible to inject hot electrons into the metal from the semiconductor and this result cannot be accepted, except as an asymptotic limit.

The thermionic emission can be also obtained as a special case of equation (3.2.14), namely for a very large diffusion velocity as compared to the collection velocity, $v_d \gg v_c$, such that equation (3.2.4) is found, where $\phi_B = \phi_{Bn}$ and the collection velocity should be [6]

$$v_c \to v_{c0} = \frac{A^* T^2}{eN_c}, \quad (3.2.17)$$

where A^* is the Richardson constant and v_{c0} is the collection velocity for a thermionic approximation with negligible phonon scattering and quantum-mechanical reflections (Figure 2.6.2).

Tunnelling and phonon scattering reduce the collection velocity by factors f_q and f_p, respectively,

$$v_c = f_q f_p v_{c0}. \quad (3.2.18)$$

---

* Or the effective recombination velocity [2], [5].
† $\zeta_n(x)$ obtained from equations (3.2.9) and (3.2.10) is integrated between x_m and x_d and then equations (3.2.11)–(3.2.13) are used.

The factor f_q is obtained by averaging the tunnelling probability over the Maxwellian carrier distribution (Section 2.6) incident to the barrier. For thermionic emission this factor is less than unity. The image-force modifies the potential barrier such that at $x = x_m$ we have a maximum (Figure 3.2.1). Electrons that cross the barrier from the semiconductor towards the metal are scattered between $x = x_m$ and $x = 0$ due to optical and acoustic phonons. This process was already indicated schematically in Figure 2.6.2a. A number of electrons will return to the semiconductor bulk and this determines an apparent reduction of the collection velocity by a factor f_p. This factor is derived by averaging the probability of phonon emission over a Maxwellian distribution. Details of these calculations can be found in [2], [5], and were used to compute the effective Richardson constant A^{**} plotted in Figure 2.6.3. This effective constant A^{**} is defined by (3.2.14) where J_s should be identified with the saturation current (2.6.4) for thermionic emission, such that

$$A^* T^2 = \frac{eN_c v_c}{1 + v_c/v_d} \qquad (3.2.19)$$

and equations (3.2.17) – (3.2.19) yield

$$A^{**} = \frac{f_p f_q A^*}{1 + f_p f_q v_{co}/v_d}, \qquad (3.2.20)$$

which is field-dependent through f_p, f_q and v_d. The diffusion velocity v_d was calculated [2], [5] by using the potential distribution corrected by the image-force.

Crowell and Beguwala [6] discussed recently the effect of diffusion upon the $J-V$ characteristic of a Schottky barrier. They used a parabolic potential distribution (depletion approximation, negligible image-force effect) for evaluating the diffusion velocity and expressed it as a function of

$$v_D = \frac{\mu_n kT}{(2eL_{DE})^{1/2}}, \quad L_{DE} = \left(\frac{kT\varepsilon}{e^2 N_D} \right)^{1/2} \qquad (3.2.21)$$

(which has the dimensions of velocity and is called below the Debye diffusion velocity) and the band bending normalized to kT, which is $e|\phi_S|/kT = |u_s|$. The diffusion velocity v_d increases with the band bending as shown by Figure 3.2.2.

The collection velocity was also calculated from equations (3.2.17) and (3.2.18). By approximating $f_p f_q \approx 1$ and using the appropriate material constants [2], [6], the ratio v_D/v_c was obtained as a function of semiconductor doping, as shown in Figure 3.2.3. Note the fact that, whereas v_c is a constant for a certain semiconductor, at a given doping level and temperature, the diffusion velocity v_d also depends upon the metal-semiconductor barrier $e\phi_B$ and the applied bias.

The factor J_s in equation (3.2.16) may be expressed as a function of the thermionic-emission saturation current J_{sat}, as follows (see equations (2.6.4), (3.2.14), (3.2.16), (3.2.17), (3.2.18), with $f_p f_q = 1$)

$$J_s = \frac{J_{sat}}{1 + v_c/v_d} = \frac{J_{sat}}{1 + \left(\frac{v_D}{v_d}\right)\left(\frac{v_D}{v_c}\right)^{-1}}. \qquad (3.2.22)$$

The ratio J_s/J_{sat} was plotted in Figure 3.2.4 as a function of normalized band bending $|u_s|$, with v_D/v_c as a parameter. Note that the numerical values chosen for v_D/v_c

cover most practical cases (see Figure 3.2.3). For a band bending $>10\ kT$, J_s is close to J_{sat} and almost constant if the semiconductor resistivity is low (thin barrier and negligible diffusion effects). Figure 3.2.3 shows that for n-type silicon, the donor concentration should be at least $10^{16} cm^{-3}$ ($v_c \approx v_D$), but even at $10^{18} cm^{-3}$ J_s is a few percent below J_{sat}*.

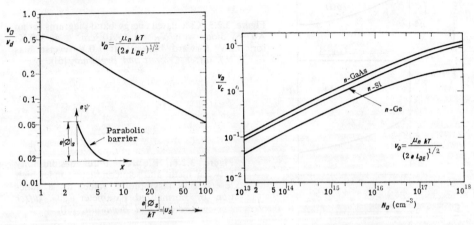

Figure 3.2.2. The diffusion velocity, v_d, as a function of the normalized band bending. (after Crowell and Beguwala, [6]).

Figure 3.2.3. The collection velocity, v_c, as a function of semiconductor doping at a Schottky barrier to an n-type semiconductor. (after Crowell and Beguwala, [6]).

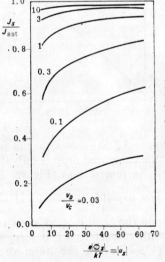

Figure 3.2.4. Ratio of the apparent saturation current to thermionic saturation current, J_s/J_{sat} versus the normalized band bending for several values of v_D/v_c (see Figure 3.2.3). (after Crowell and Beguwala, [6]).

---

* At high doping concentrations the thermionic-field emission mechanism should be taken into account. Crowell and Beguwala [6] estimated a low diffusion effect upon the thermionic-field characteristic.

The electron quasi-Fermi level ζ_n can be computed by integrating equation (3.2.10). This was done by Crowell and Beguwala [6] under the same simplifying assumptions. Two particular results are shown in Figure 3.2.5, for an Au—Si diode,

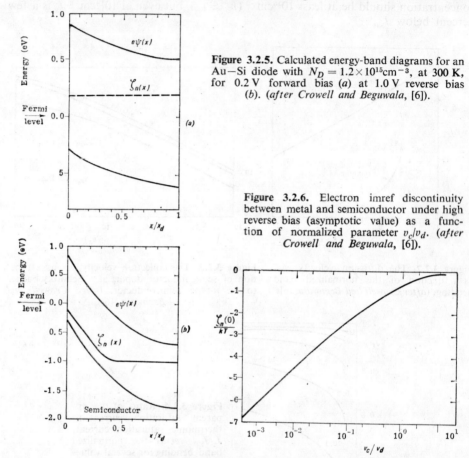

Figure 3.2.5. Calculated energy-band diagrams for an Au—Si diode with $N_D = 1.2 \times 10^{15} \text{cm}^{-3}$, at 300 K, for 0.2 V forward bias (a) at 1.0 V reverse bias (b). (after Crowell and Beguwala, [6]).

Figure 3.2.6. Electron imref discontinuity between metal and semiconductor under high reverse bias (asymptotic value) as a function of normalized parameter v_c/v_d. (after Crowell and Beguwala, [6]).

with $N_D = 1.2 \times 10^{15} \text{cm}^{-3}$, at 300 K. Figure 3.2.5a shows that in forward bias the electron imref ζ_n is practically constant (the computed drop is only 0.3 kT, at higher doping levels this drop is even smaller)*. In reverse bias (Figure 3.2.5b) ζ_n bends towards the Fermi level in the metal. The asymptotic value $\left(|V| \gg \dfrac{kT}{e}\right)$ of the discontinuity at the interface depends upon the ratio v_c/v_d as shown in Figure 3.2.6. This discontinuity becomes negligible with respect to kT if v_c/v_d is large. Figure 3.2.6 also shows the asymptotic value $(V \gg kT/e)$ of the imref drop in semicon-

---

* The continuity of imref at the interface assumed in the diffusion theory requires a very large imref gradient $\dfrac{d\zeta_n}{dx}$, at the interface and predicts a larger value of the electron current. This is not confirmed by experiment [3].

ductor in forward bias as a function of v_c/v_d. This drop increases with v_c/v_d but becomes comparable to kT only if $v_c > v_d$.

The imref in both forward and reverse direction is discontinuous at the metal-semiconductor interface because of the finite collection velocity v_c [6]. In forward bias the metal collection velocity is small compared to the diffusion velocity of electrons and therefore the free carriers in semiconductor tend to be in equilibrium with the semiconductor bulk rather than the metal [6] ($\zeta_n(x) \simeq$ constant). In reverse bias ζ_n at the interface is lower than the Fermi level in the metal (Figure 3.2.5b) and this corresponds to the fact that electrons can diffuse into the semiconductor more rapidly than the metal can emit. The electron imref will become, however, continuous at the interface if the collection velocity v_c becomes infinite (Figure 3.2.6).

(d) Experimental Results

The majority of authors compare the experimental data with the thermionic-emission theory. An empirical relation for the $J-V$ dependence is used, namely

$$J = J_s \left[\exp\left(\frac{eV}{nkT}\right) - 1 \right], \qquad (3.2.23)$$

where n is an ideality factor, determined experimentally. n should equal unity for an ideal behaviour (see equation (3.2.4)). In practice n is somewhat higher than unity and is determined from the $\ln J - V$ plot of the forward current at $V > 2kT/e$, which should be a straight line. Deviations from $n = 1$ may occur due to a variety of causes such that diffusion effects*, tunnelling, the presence of an interfacial layer, recombination current in the depletion region [7].

J_s in equation (3.2.23) is a saturation current. Ideally, J_s is identical with the thermionic-emission current from the metal into the semiconductor $J_{sat} = A^{**}T^2 \exp(-e\phi_B/kT)$ and is used to determine the barrier, ϕ_B.

A typical experimental *forward* characteristic is shown in Figure 3.2.7 (Al-nSi diode). The $J-V$ dependence is exponential over about six decades of current, with n close to unity. The slight departure from the straight line in the semilogarithmic plot at higher currents is due to the voltage drop across the epitaxial layer and substrate (the effect of a series resistance). The intercept of the straight line (which is $J = J_s \exp eV/nkT$) with $V = 0$ yields $I = 1.5 \times 10^{-10}$ A for an area 0.85×10^{-5} cm^{-2} and hence J_s. With $A^{**} = A = 120$ A cm^{-2} K^{-2} (acceptable for electrons in silicon) [2], the barrier $e\phi_B = e\phi_{Bn} = 0.7$ eV is obtained.

The barrier can be also determined † from the so-called *activation energy plot*. The forward current for a given voltage in excess of $2kT/e$ obeys a relationship of the form

$$\frac{I}{T^2} = SA^{**} \exp\left(\frac{eV - e\phi_B}{kT}\right), \qquad (3.2.24)$$

---

* Crowell and Beguwala [6] have shown that thermionic-diffusion theory (see above) predicts in most practical cases for GaAs, Ge and Si an n-value less than 1.01.

† The capacitance voltage dependence for the Schottky barrier is also used for ϕ_B determination. However, the most reliable method of determining the barrier height is by photo-emission experiments (the metal semiconductor light is excited with monochromatic light and the resulting photocurrent is measured, the extrapolated, minimum photon energy required for optical excitation of electrons in the metal is equal to the energy barrier) [2], [3], [7].

such that $\ln I/T^2$ depends linearly upon $1/T$ and ϕ_B can be determined from the slope, as shown in Figure 3.2.8 (Al-nSi diode) [7].

For diodes of similar area Schottky barriers have much larger currents than a p-n junction at the same forward bias; in other words, the forward voltage will

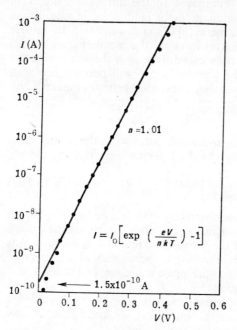

Figure 3.2.7. Experimental forward $I - V$ characteristic of Al-nSi Schottky diode. Area of the diode is 0.85×10^{-5} cm^2, resistivity and thickness of the epitaxial layer are $0.4\,\Omega$ cm and 2 μm, respectively. (*after Yu and Mead*, [7]).

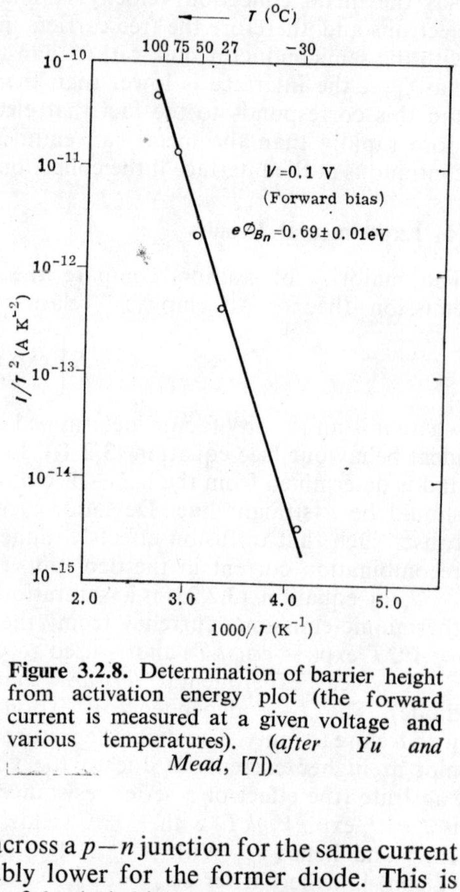

Figure 3.2.8. Determination of barrier height from activation energy plot (the forward current is measured at a given voltage and various temperatures). (*after Yu and Mead*, [7]).

be much less across a Schottky diode than across a $p-n$ junction for the same current and the turn-on voltage will be considerably lower for the former diode. This is illustrated by Figure 3.2.9. For a current of 1 μA (which may define the turn-on voltage), the voltage drop is more than 0.5 V for the p-n junction and only 0.24 V for a Schottky barrier with an area of about 1/30 of that of the $p-n$ diode.

Figure 3.2.10 shows the temperature coefficient dV/dT (forward bias) as a function of current density for a silicon $p-n$ junction and a Schottky barrier on n silicon. The differences are important in the design of circuits utilizing both types of diodes (or transistors and Schottky diodes) [7], [8].

The near-ideal characteristics obtained for Al-n-Si Schottky diodes [7] are explained* by the intimate Al—Si contact (absence of the SiO$_2$ layer) which is obtained by a brief low-temperature temperature (400—500 °C) treatment. Chino

---

* The diode is constructed in planar form with metal overlap of the oxide, thus avoiding the corner breakdown, as discussed in Section 3.4.

[9] studied the effect of heat treatment on Al—Si Schottky barriers. The samples were heated in the 400—650 °C range, in hydrogen atmosphere, for 15 min to 60 min. The height of the barrier was found to increase as the heating temperature increases from 450 to 650 °C (about 0.1 eV change). This effect was attributed to possible

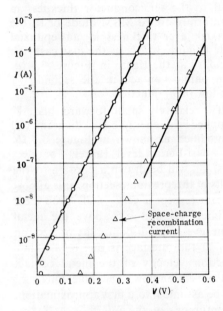

Figure 3.2.9. Forward current-voltage characteristics of an aluminium-n-Si Schottky barrier (circles): area of the diode is $0.85 \times 10^{-5} \mathrm{cm}^{-2}$, $N_D = 10^{16} \mathrm{cm}^{-3}$; and that of a p^+n junction (triangles): area of the diode is $2.5 \times 10^{-4} \mathrm{cm}^2$, $N_D = 10^{16} \mathrm{cm}^{-3}$. Straight lines represent $I \propto \exp(eV/kT)$. (*after Yu*, [1]).

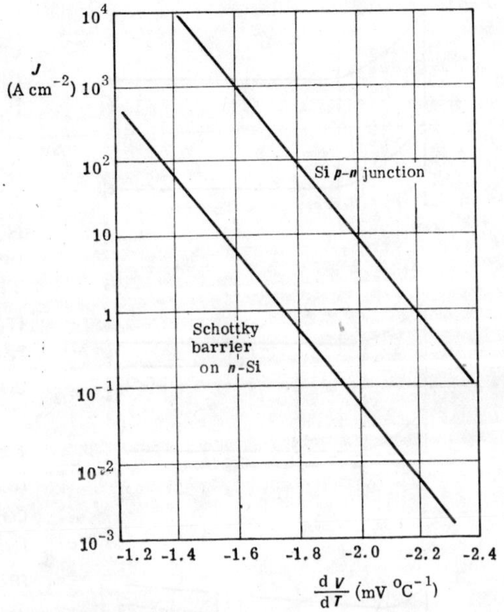

Figure 3.2.10. Plot of forward current density versus dV/dT for n-Si Schottky barrier and pn junction. (*after Yu*, [1]).

diffusion of Si into Al, and silicide formation. Samples heated above 580 °C (eutectic temperature) exhibit anomalous characteristics explainable by the formation of a very thin p-type layer (Al-rich silicon layer).

Saltich and Terry [10] also show that the barrier height is affected by the presence of an intermediate oxide layer. They studied the effect of surface treatment before metal evaporation and also the change in the barrier height with a post-evaporation heat treatment, for Pt, Ti, Mo, Cr and Ni on silicon.

3.3 Minority-carrier Injection

Although Schottky barrier diodes are basically majority carrier devices, the minority carrier injection at the Schottky contact can, however, affect the diode behaviour. A Schottky diode with an n-type semiconductor is considered in Figure 3.3.1.

Figure 3.3.1a shows the energy-band diagram at thermal equilibrium. The space-charge layer width is denoted by x_d and decreases with increasing forward bias. The neutral semiconductor length is L, and because in general $L \gg x_d$, L is assumed constant and approximately equal to the semiconductor thickness in a uniformly doped MSM diode or to the epitaxial layer thickness in an epitaxial diode (Figure 3.1.3). In the first case the boundary at the right, in Figure 3.3.1a, is an ohmic contact, in the second — a high-low (n^+n) junction.

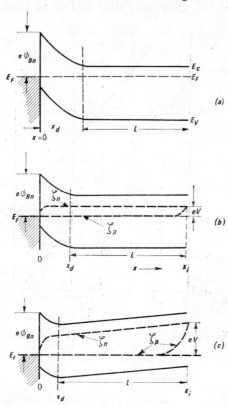

Figure 3.3.1. Energy diagram for a Schottky diode on n-type semiconductor: (a) thermal equilibrium; (b) small forward current; (c) high forward current.

If a relatively small forward bias, V, is applied, the energy-band diagram will be modified as shown in Figure 3.3.1b. The quasi-Fermi level (imref) for electrons, ζ_n, will be drawn according to the results of the previous section: it is almost constant (except a very small drop near the interface) and eV above the metal Fermi level. The discontinuity of electron imref at the interface is due to the finite collection velocity of the metal contact for the electron flow. The hole imref, ζ_p, may be assumed, to a first approximation, as continuous at the interface as far as the hole current is very small and the deviation from a thermal quasi-equilibrium of valence electrons is negligible. Note the fact that ζ_n and ζ_p will coincide at the right boundary (Figure 3.3.1b) if an ohmic contact provides here thermal equilibrium concentrations.

Provided that the whole applied voltage V drops on the depletion layer the hole concentration at the boundary between this layer and the neutral region ($x = x_d$) is

$$p(x_d) = p_{n0} \exp \frac{eV}{kT}. \qquad (3.3.1)$$

The hole current, J_p, is equal to the *diffusion* current through the neutral region, of length L. If $L \ll L_p$ then

$$J_p = \frac{eD_p p_{n0}}{L} \left[\exp \left(\frac{eV}{kT} \right) - 1 \right]. \qquad (3.3.2)$$

The electron current, J_n, will be considered as due to thermionic emission, such that

$$J_n = J_{n\,sat}\left[\exp\left(\frac{eV}{kT}\right) - 1\right]. \tag{3.3.3}$$

Because J_n is usually much greater than J_p, the *minority carrier injection ratio*, γ, becomes [2], [11]

$$\gamma = \frac{J_p}{J_n + J_p} \simeq \frac{J_p}{J_n} = \frac{eD_p p_{n0}}{J_{n\,sat} L} = \frac{eD_p n_i^2}{N_D L J_{n\,sat}} \tag{3.3.4}$$

and is independent of the bias current $J \simeq J_n^*$. The above relationship was qualitatively verified by Yu and Snow [12], who reported data for Schottky barriers ranging from 0.65 to 0.85 eV on n-type silicon with doping densities from 10^{14} to 6×10^{16} cm^{-3} (γ values between 10^{-5} and 10^{-2} are usual). The injection ratio was measured in a metal-emitter transistor structure (metal-n-p), the base contact to n region extracting the majority carrier current and the p-collector driving the minority carrier (hole) current, such that is the ratio of the collector current to the emitter current. These experiments also indicated an increase of γ at higher forward currents, as predicted by Scharfetter [11].

At high bias currents, part of the applied voltage will drop across the neutral region, because of the finite semiconductor conductivity. The corresponding energy-band diagram is shown in Figure 3.3.1c. The increased electric field within the neutral semiconductor will add a substantial drift component to the minority hole flow. Hence the hole current add the injection ratio γ will increase with the total current. At very large currents the almost entire hole current is a drift current and

$$\gamma \simeq \frac{J_p}{J_n} = \frac{p(x_d)}{N_D} \frac{\mu_p}{\mu_n}. \tag{3.3.5}$$

If V_d is the voltage drop on the depletion region, then

$$p(x_d) = p_{n0} \exp\frac{eV_d}{kT} \simeq p_{n0} \frac{J_n}{J_{n\,sat}} \simeq \frac{p_{n0} J}{J_{n\,sat}} = \frac{n_i^2 J}{N_D J_{n\,sat}} \tag{3.3.6}$$

and the injection ratio

$$\gamma = \frac{n_i^2}{N_D^2}\left(\frac{\mu_p}{\mu_n}\right)\frac{J}{J_{n\,sat}} \tag{3.3.7}$$

will be proportional to the injected current. For example [11] a gold-n-type silicon diode with $N_D = 10^{15}$cm^{-3} and $J_{nsat} = 5 \times 10^{-7}$A cm^{-2} would have an injection

---

* Even if $J_{n\,sat}$ is slightly voltage-dependent, according to the diffusion (or thermionic-diffusion theory), the injection ratio will be almost constant because the forward bias current increases very rapidly with V.

ratio of 5% at a current density of about 350 A cm^{-2}, whereas for currents less than about 5 A cm^{-2}, the injection ratio would be constant at about 0.1%.

At very high current densities the injection ratio approaches unity and the diode characteristic is considerably modified. Jäger and Kosak [13] published recently

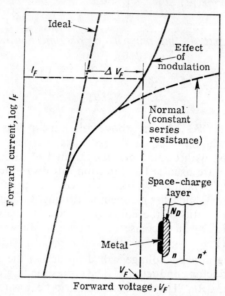

Figure 3.3.2. Schematic forward characteristic of a Schottky barrier diode exhibiting the modulation effect. (*after Jäger and Kosak*, [13]).

experimental data indicating the effect of intense minority hole injection in silicon Schottky diodes. The principle of their experimental technique is explained in Figure 3.3.2. The 'ideal' forward $J-V$ characteristic plotted as log J versus V should be a straight line, because for $V \gg kT/e$, $J = J_n = J_{n\,\text{sat}} \exp(eV/kT)$, or $J \propto \exp eV/nkT$, where $n \geqslant 1$ is a constant. A 'normal' $J-V$ characteristic would show like that indicated in Figure 3.2.2: its shape is determined by the series resistance of the epitaxial layer, resistance assumed constant. However, if the minority carrier injection increases considerably, the conductivity of the epitaxial layer increases and the series resistance is reduced, resulting in an experimental characteristic steeper than the normal one (Figure 3.3.2). Therefore three regions occur on the $J-V$ characteristic: the 'ideal' low-current region, the intermediate-current region where the epitaxial layer resistance is important but the hole conduction is still negligible, and, finally, the high-current region, where the hole injection decreases appreciably the series resistance. The latest effect is called 'modulation' [13]. The higher the hole density (3.3.6), the higher the hole injection in the neutral zone, and the greater the modulation effect. According to equation (3.3.6) the modulation effect should be important for higher barriers (lower $J_{n\,\text{sat}}$) and lower doping densities. This was proved by Jäger and Kosak [13] by plotting the diode current, versus ΔV, where ΔV is the difference between the extrapolated ideal (low-current) curve and the actual characteristic at the diode current J. The ratio $I/\Delta V$ is just the series resistance.

Figure 3.3.3 shows results obtained for planar diodes obtained from epitaxial nn^+ silicon with resistivity 0.65 Ω cm and thickness of the epitaxial layer 1 μm. The Schottky contact is circular (45 μm diameter) and made using five metals with different barrier heights. The series (steady-state) resistance resulting from Figure

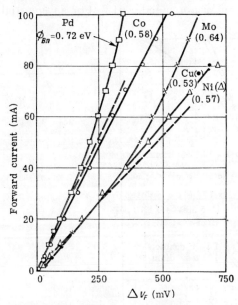

Figure 3.3.3. Forward characteristics of the neutral zone series resistance of several metal n-n^+Si diodes, as obtained by the $\Delta V_F - I_F$ analysis (Figure 3.3.2). (*after Jäger and Kosak,* [13]).

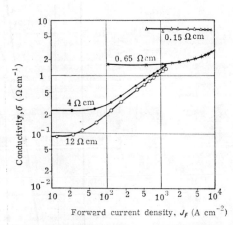

Figure 3.3.4. Conductivity increase in the neutral zone of Pd–Si Schottky barrier diodes exhibiting the modulation effect. The epitaxial layer resistivity is shown for each curve. (*after Jäger and Kosak,* [13]).

3.3.3 changes (decreases) at high current levels, only for high-barrier diodes, as predicted above. The effect of donor concentration upon the 'modulation' effect is indicated by Figure 3.3.4, where the conductivity $\sigma = (I/\Delta V)(L/A)$ (A is the diode area) of the epitaxial layer is plotted *versus* I for palladium silicon Schottky barrier diodes with different resistivities of the n-type epitaxial layer. The effect (increase of σ) occurs at higher currents for larger doping densities [13].

A very suggestive proof of the hole injection effect is given by Figure 3.3.5, which shows the effect of electron irradiation upon the steady-state characteristic. This irradiation increases the series resistance by decreasing the 'modulation' of the neutral layer, induced by minority hole injection. The high-energy irradiation produces recombination centres, thus reducing the minority carrier lifetime and the corresponding diffusion length, L_p. However, the 'modulation' effect occurs as described above only as far as the diffusion length is greater than the semiconductor thickness: an appreciable reduction of L_p will decrease this hole-induced conductivity. For the same reason, the gold doping (introducing recombination centres) increases the series resistance, having an adverse effect upon device operation [13].

The above discussion was based upon the assumption of infinite recombination velocity at the right-hand interface, $x = x_i$ (Figure 3.3.1). This is acceptable when this interface is either an ohmic contact, or a p-n junction [12]. For an epitaxial

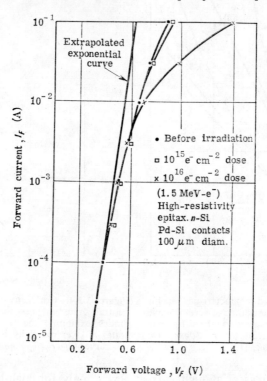

Figure 3.3.5. The effect of electron irradiation on the forward characteristic of an epitaxial n-n^+Si/Pd Schottky barrier diode. (*after Jäger and Kosak*, [13]).

nn^+ diode, the recombination velocity, defined by (Figure 3.3.1)

$$J_p(x_i) = ev_R[p_n(x_i) - p_{n0}] \tag{3.3.8}$$

is finite and sometimes approximated by zero, such as an accumulation of minority carriers should occur at this interface. The hole continuity equation was solved by Scharfetter [11] by using the boundary conditions (3.3.6) and (3.3.8). The computed injection ratio, γ, is plotted in Figure 3.3.6a. Note the fact that the low-current injection ratio decreases considerably due to the finite recombination velocity, v_R. The same effect as v_R, has the ratio $L/D_p (L \to \infty$ is equivalent to $v_R \to \infty$). However, the high-current γ becomes independent of v_R and follows the law (3.3.7). Scharfetter [11] also computed the storage time τ_S defined by (Figure 3.3.1)

$$\tau_S = \frac{\int_{x_d}^{x_i} ep(x)dx}{J}. \tag{3.3.9}$$

Figure 3.3.6b shows that the high-current τ_S is constant and therefore the minority hole charge should be proportional to the total current J, as the hole concentration $p(x_d)$ does (see equation (3.3.6)).

Green and Schewchun [14] performed a numerical investigation of two-carrier Schottky diode equations and found that the minority carrier injection ratio does, in general, exhibit a maximum at a certain forward current. The fact that $\gamma = \gamma(J)$ deviates from Scharfetter's linear law (3.3.7) is due to high-injection level effects in the neutral region or to contact limitations for hole injection*. The model uses exact equations for electron and hole transport, equation (3.2.11) as boundary condition for electron current and a similar condition for hole injection at the Schottky contact ($x = 0$)

$$J_p|_{x=0} = -ev_{cp}[p(0) - p_0(0)], \tag{3.3.10}$$

where $p_0(0)$ is the thermal-equilibrium concentration at the contact and v_{cp} is a hole 'collection' velocity. For positive J_p (hole injection in semiconductors, $p(0) < p_0(0)$ and the hole imref ζ_p is above the Fermi level in the metal. The limit

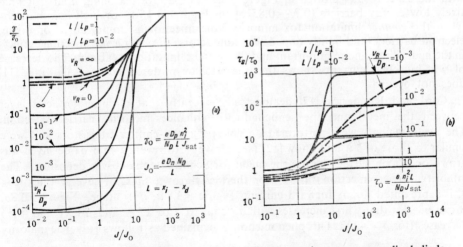

Figure 3.3.6. (a) Normalized minority carrier injection ratio *versus* normalized diode current density; (b) normalized minority carrier storage time *versus* normalized current density. (*after Scharfetter* [11], *reproduced by Sze in* [2]).

---

* Clearly, by definition, γ cannot increase above unity. Equation (3.3.7) is valid only if $J_p \ll J \simeq J_n$, otherwise $\gamma = \dfrac{J_p}{J_n + J_p}$. Scharfetter's theory predicts γ saturation at unity at very high currents. Green and Schewchun [14] show that γ is limited to a considerably lower value.

case $v_{cp} \to \infty$ corresponds to the diffusion theory for injected holes (see above). The opposite case, small v_{cp} (see Section 3.2) corresponds to thermionic injection of holes into the semiconductor (in fact valence electrons in semiconductor are injected into the metal, their energy must be sufficiently high to occupy the free levels situated above the Fermi level). The thermionic minority-carrier emission hypothesis was used by Sze et al. [15] in their theory on current transport in metal-semiconductor-metal structures, and later discussed by Calzolari and Graffi [16]. The numerical approach by Green and Schewchun [14] corresponds to a two-carrier thermionic-diffusion theory (neither ζ_n nor ζ_p is forced to coincide to the metal Fermi level). We note that the back contact was modelled as an ohmic contact such that their results [14] should be considered as a correction to Scharfetter's curves for $v_R \to \infty$ (Figures 3.3.5 and 3.3.6).

Figure 3.3.7 shows the current dependence of γ with electron barrier and doping as parameters. The absolute limitation to γ can be written [14]

$$\gamma_{max} = \frac{\mu_p}{\mu_n + \mu_p p_0(0)/N_D} \frac{p_0(0)}{N_D}, \qquad (3.3.11)$$

where the drift hole current was roughly evaluated by using the thermal equilibrium hole density at $x = 0$, $p_0(0) > p(0)$ and the actual hole density in the semiconductor bulk is always lower than $p(0)$. γ_{max} increases with increasing electron barrier height ($p_0(0)$ increases) and decreases with increasing doping, qualitatively in agreement with the exact results plotted in Figure 3.3.7. The above reasoning holds for relatively low energy barriers (0.7 — 0.85 eV in Figure 3.3.7a), or high *hole* barriers, when the *contact* limitation for minority hole injection is important. For higher electron barriers (and lightly doped semiconductors) the high level injection* effects in the neutral region are important and the contact limitation is absent. The decrease of γ at high bias currents shows that the mechanism described by Scharfetter [11] does not occur †.

Card and Rhoderick [17] demonstrated recently that a thin interfacial layer between the metal and the semiconductor enhances minority carrier injection. The measurements were performed on epilayer metal-emitter transistors and were similar to those of Yu and Snow [12] (see above). However, a thin oxide layer was grown on silicon surface before the metal for Schottky contact was deposited. The minority carrier injection ratio γ for thermally-grown SiO_2 of several thickness, δ, is represented *versus* forward current, Figure 3.3.8 *a*, or voltage, Figure 3.3.8*b*. It can be seen that γ first increases with δ (from 10 Å to 17 Å and 28 Å) and then decreases (for $\delta = 40$Å). This phenomenon is explained as follows [17]. The majority

---

* By definition, the high-level injection occurs when the minority carrier density in a neutral region is comparable to the concentration of majority carriers normally present here (see for example Pritchard [18], p. 77).

† At high-level injection an excess majority carrier concentration will be injected to preserve the neutrality. The boundary condition (3.3.6) becomes incorrect. The electric field in the neutral region is not longer $J_n/e\mu_n N_D$. Clearly, Scharfetter's approach is invalid and exact two-carrier analysis is necessary. Intuitively, when these phenomena occur γ is comparable to unity (see Figure 3.3.7) and this is due to intense hole injection at very small hole barrier.

electron current J_n is limited by the possibility of electron penetration into the metal: this will take place by tunnelling if an oxide is present. Clearly, at a given bias, J_n is smaller for larger oxide thickness. The hole current is, however, limited by diffusion and drift in the semiconductor bulk and not by emission through the

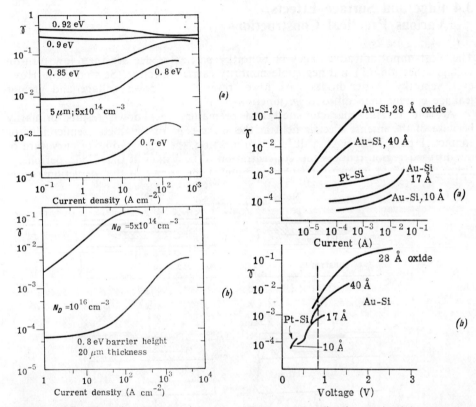

Figure 3.3.7. The bias dependence of the minority carrier injection ratio. The parameter is the barrier height. The distance between contacts is $L = 5$ μm, the donor density is $N_D = 5 \times 10^{14}$ cm^{-3} and the lifetime parameters in the Schottky-Read-Hall recombination formula are $\tau_{no} = \tau_{po} = 5$ μs. (after Green and Schewchun [14]).

Figure 3.3.8. Minority carrier injection ratio of Au–Si junctions with thermally-grown oxide layers at the interface: (a) dependence on the total current; (b) dependence of junction voltage. (after Card and Rhoderick, [17]).

oxide as far as this oxide is not too thick. This explains the increase of γ with δ. If the oxide is relatively thick, both electron and hole currents will be limited by tunnelling through the oxide barriers. In the above experiments [17], J_n and J_p were measured independently and were found to decrease with increasing oxide thickness. Card and Rhoderick [17] pointed out that there exists an asymmetry in the barriers for electrons and holes. This asymmetry favours the electron current and determines the decrease of γ with δ, as shown by Figure 3.3.8.

There exists, therefore, an optimum oxide thickness for maximum minority carrier injection ratio. This phenomenon can be used to enhance light emitting properties of metal contacts on wide gap semiconductors (CdS or GaP)*.

3.4 Edge and Surface Effects. Various Practical Constructions

The most important advantages of Schottky barrier diodes are their low turn-on voltages (Section 3.2) and negligible minority carrier injection (Section 3.3). However, Schottky barrier diodes have lower reverse breakdown voltage and greater leakage current than diffused *pn* junctions.

A 'soft' reverse characteristic † and premature breakdown results primarily because of the intense electric field at the sharp-edge of the metal semiconductor contact (Figure 3.4.1a). At a diffused *pn* junction, the finite radius of curvature of the diffused region reduces the concentration of field lines at the diode periphery, although the field intensity is still maximum at the corner of the depletion region

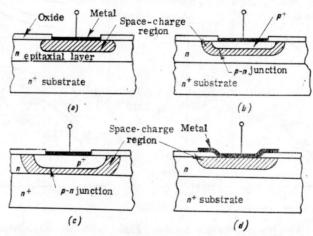

Figure 3.4.1. Schottky diodes and *pn* diodes made by using the planar technology: (*a*) normal Schottky diode; (*b*) *pn* junction; (*c*) 'deep' *pn* junction; (*d*) Schottky diode with metal overlap of the oxide.

closely following the shape of the junction (Figure 3.4.2b). A deep diffusion may reduce the 'corner' field below the bulk field (Figure 3.4.3c) and thus increase the breakdown voltage V_B towards its bulk value ††.

---

* Sputtered oxide layers have similar properties and can be deposited on any semiconductor, including those in which a *pn* junction cannot be made and a metal-semiconductor barrier should be used for a luminiscent diode [19].

† A 'soft' reverse characteristic exhibits a gradual increase of the diode current towards high values.

†† For V_B of diffused junctions see references [2], [20].

The edge effect occurring at a Schottky diode made by planar technology (Figure 3.4.1a) can be reduced by allowing the metal to overlap the oxide (Figure 3.4.1d) and act as a field plate which extends the depletion region outside the periphery of the metal-semiconductor contact [1]. The effect will be important if the oxide is sufficiently thin (1000 Å). A two-step oxide profile may be used (with a thin oxide ring surrounding the contact). The diode capacitance increases, of course, above the value corresponding to the effective diode area. A careful tailoring of the metal profile can increase considerably the breakdown voltage.

Not only the breakdown voltage but the entire characteristics can be modified by the nonuniform electric field at the contact. The reverse 'saturation' current (below breakdown) will increase due to image-force lowering of the barrier or tunnelling at the contact periphery (this high-field region is very narrow but the current-field dependence is exponential) [21].

Yu and Snow [21] demonstrated that large departures from the ideal characteristics occur due to *surface effects*. They used for experiments the gate-controlled Schottky diode structure shown in Figure 3.4.2 which is similar to be gate-controlled *p-n* junction (see Grove [22] and also Chapter 10 of this book). The space-charge region beneath the semiconductor surface is indicated in Figure 3.4.3,

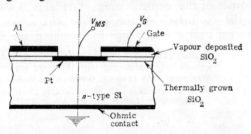

Figure 3.4.2. The gate-controlled Schottky diode structure used by Yu and Snow [21].

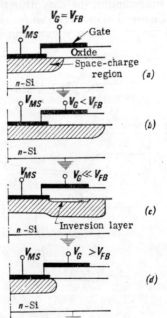

Figure 3.4.3. Space-charge (cross-hatched area) of the gate-controlled metal-semiconductor diode: (a) surface at flat-band condition; (b) surface depletion; (c) surface inversion; (d) surface accumulation. (*after Yu and Snow*, [21]).

for various values of the gate bias relative to the semiconductor substrate, V_G. Figure 3.4.3a corresponds to the application of the so-called flat-band voltage, V_{FB}, which compensates the effect of the work function difference and interface charges such that the energy bands are 'flat' at the surface (the energy band diagram is drawn for a direction perpendicular to the surface). In this case the potential barrier at the edge of the metal contact looks the same going along the surface as it does going into the bulk [21]. A more negative V_G will deplete the surface below the gate (Figure 3.4.3b). A still more negative gate voltage will still further widen the depletion layer and form an inversion layer under the gate (Figure 3.4.3c). If, however, $V_G > V_{FB}$ an accumulation layer will be formed at the surface and the shape of the depletion region will be modified as shown in Figure 3.4.3d.

The above-mentioned effects of the gate voltage were verified by capacitance-voltage and current-voltage measurements [21]. Figure 3.4.4 shows, for example, the forward and reverse current *versus* V_G with the metal-semiconductor diode bias, V_{MS}, as parameter. The currents increase abruptly to a higher level after the surface is inverted ($V_G \leqslant -45$ V). This increase is due to generation-recombination currents associated with the field-induced junction (p-type inversion layer at the surface and the n-type semiconductor). A similar effect occurs for the gate-controlled p-n junction and is discussed in Chapter 10.

The forward and reverse currents increase rapidly when the surface is strongly accumulated ($V_G > 0$). This increase is associated with the increase of the field near the contact periphery (Figure 3.4.3d) by the field-enhancing mechanisms mentioned above, in this section. This explanation is proved by the fact that the above increase is higher for more negative reverse voltages (Figure 3.4.4b) which increases the tunnelling probability near the corner of the depletion region (Figure 3.4.3d).

Figure 3.4.5 clearly shows that the "softness" of the reverse characteristics and the breakdown voltage can be varied over a wide range by changing the gate

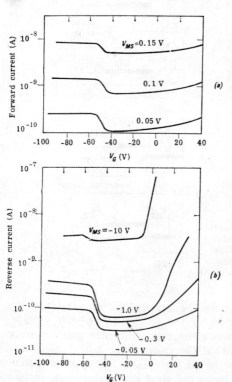

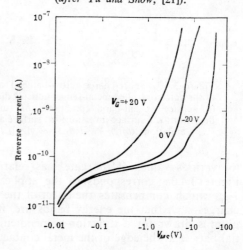

Figure 3.4.4. (a) Forward current *versus* gate voltage, V_G, measured with forward bias as parameter; (b) reverse current as a function of gate voltage, V_G, with reverse bias as parameter. (*after Yu and Snow*, [21]).

Figure 3.4.5. Reverse characteristics of the gate-controlled metal-semiconductor diode, with the gate voltage, V_G, as parameter. (*after Yu and Snow*, [21]).

voltage [21]. The reverse characteristics of Schottky barrier diodes are therefore very sensitive to surface properties of the semiconductor. The semiconductor surface can be protected from the ambient by using a passivating layer, such as the silicon dioxide in silicon planar technology. Unfortunately, the positive oxide

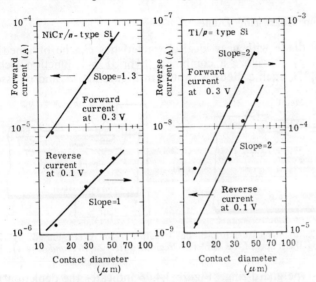

Figure 3.4.6. Forward and reverse current measured at a given voltage bias *versus* diode diameter for: (a) Ni—Cr/n-type silicon acide-passivated diodes; (b) Ti/p-type silicon oxide-passivated diodes. *(after Cowley, [23])*.

charge (Chapter 10) in a SiO_2—Si system creates an accumulation layer at the surface of n-type silicon and deteriorates considerably the characteristics of planar Schottky diodes made with *any metal* on n-type silicon*. The effect of the positive oxide charge is identical to that of a positive gate bias which accumulates the surface beneath the gate: the reverse characteristic is modified as shown in Figure 3.4.5 (curve $V_G = +20$ V).

Cowley [23] has shown, for example, that while p-type silicon Schottky diodes can be fabricated using standard oxide passivation techniques, n-type diodes characteristics are severely degraded by the presence of silicon dioxide at the periphery of the diode. The importance of edge effects is shown by Figure 3.4.6a, where the reverse current of NiCr/n-type silicon oxide-passivated diodes increases almost linearly with the diode diameter. This phenomenon is thought to be due to accumulation of the n-type silicon surface by the positive charge in the oxide. In contrast, the forward and reverse current of a p-type oxide-passivated diode is proportional to the square of the diode diameter (Figure 3.4.6b), thus indicating negligible edge effects. The surface of the silicon around the p-type diodes should be depleted for

---

* The same physical situation in the SiO_2—Si system prevents the construction of inversion-channel n-type MOS transistors which are non-conducting with zero gate bias, because the positive oxide charge inverts the surface of the p-type substrate (see Chapters 10 and 11).

resistivities below 1 Ωcm [23] and the field at the priphery should be lower than in the centre of the contact.

n-type silicon diodes could be fabricated by using modified planar structure such as that of the guard-ring diode, examined below.

(a) p-n Junction Guard-ring Diode

A Schottky diode with near ideal characteristics can be obtained by using the planar technology to realize a combined (hybrid) *p-n* junction Schottky-barrier structure [24], [25]. Such a diode is shown in Figure 3.4.7a and its essential feature

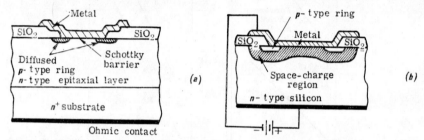

Figure 3.4.7. (a) Typical hybrid-diode structure; (b) reverse-biased hybrid diode, showing the shape of the depletion region. *(after Zettler and Cowley, [25]).*

is a diffused *p*-type guard-ring. Figure 3.4.7b indicates the depletion region of the reverse-biased structure. The maximum reverse voltage is determined by the impact avalanche occurring first either at the *p-n* junction or at the metal-semiconductor junction, depending upon the depth and profile of the diffused junction (see Figure 3.4.1). For an abrupt diffused junction of small radius of curvature, the highest fields will be at the periphery of the junction and the breakdown voltage, V_B, will be lower than that of a plane parallel junction. If the junction were linearly graded, V_B would be higher than for an abrupt plane parallel junction and the reverse voltage would be probably limited by the breakdown at the metal-semiconductor junction [25]. If was indeed demonstrated that the avalanche breakdown voltage of a guard-ring diode can be controlled by tailoring the profile of the diffused junction [24].

We shall first assume that the metal forms an ohmic contact to the *p*-type ring (Figure 3.4.7). Then the total structure can be considered as a parallel combination of a *p-n* junction and an *n*-type Schottky barrier diode. With the metal positive relative to the n^+ substrate, both diodes are forward biased. However, the turn-on voltage of the Schottky diode is much lower than that of the *pn* junction and hence the low-current region of the forward characteristic is essentially that of the metal-semiconductor contact. At higher current levels (larger bias voltages) the *p-n* junction injects an appreciable minority carrier charge and modifies the device characteristic as shown qualitatively in Figure 3.4.8. The minority carrier storage becomes important thus limiting the switching speed and deteriorating the rectification performances of the Schottky diode.

If the guard-ring is a p^+ diffusion region the metal contact will indeed be ohmic because the carrier transport takes place by tunnelling (Chapter 2). A lower surface concentration (say below 10^{18}cm^{-3} for silicon [25]) for the diffused p region will prevent tunnelling and determine a rectifying contact, provided that the potential barrier for holes is not too low and the saturation current which can be supplied by the reverse-biased metal-p-type semiconductor barrier is low (see Figure 3.4.9). The molybdenum contact, with a large hole barrier ϕ_{Bp} (approximately 0.42 eV for electron-beam evaporated material [25]), will be quite good for such a rectifying contact, whereas the gold and platinum silicide contacts act essentially as an ohmic contact because their hole barriers are low (0.30 eV for Au, 0.25 eV for PtSi) and the saturation current for a p-type barrier is relatively high (120 mA

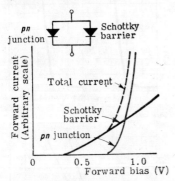

Figure 3.4.8. Qualitative $I-V$ characteristic expected from a hybrid diode (equivalent circuit shown above the characteristics). (*after* Zettler and Cowley, [25]).

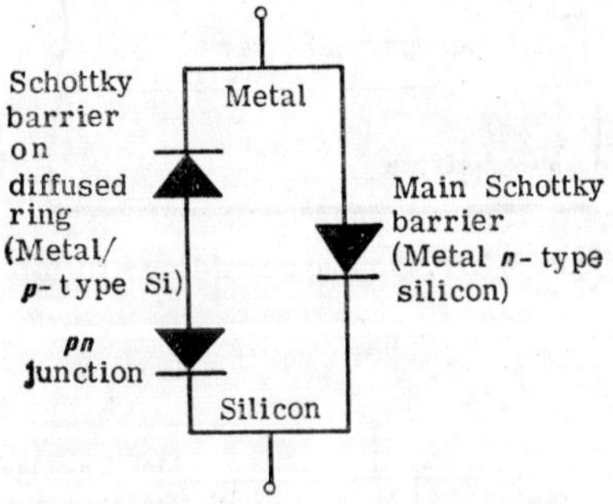

Figure 3.4.9. Equivalent circuit of hybrid diode, assuming rectifying contact between metal and diffused ring.

for Au-contact [25]). A rectifying contact to the diffused ring prevents minority carrier injection by the p-n junction and therefore allows the characteristics of the hybrid diode to be almost ideal up to large current densities. Such a contact requires, however, a selection of the metal used (Mo, Ti, Cr, Ni) and a relatively low surface concentration for the diffused ring. On the other hand the p-type barrier (metal on the diffused ring) suffers from the same problems as a planar Schottky barrier without a guard ring, in particular their reverse voltage for breakdown are quite low, of a few volts [25] and their leakage currents enhance minority carrier injection by the p-n junction.

(b) Double Diffused Guard-ring Schottky Barrier Diodes (26)

The double guard-ring structure shown in Figure 3.4.10 is more effective than a p-type guard ring diode in suppressing the minority carrier injection at high current levels. The n-type ring diffused in the p-type ring suppresses the minority carrier injection to higher current levels because: (a) the saturation current of the top np junction in Figure 3.4.11 is extremely small; (b) guarding of the metal-p region Schottky diode reduces edge effects; (c) the area of the metal p-region contact area is smaller.

Figure 3.4.12 compares the $I-V$ characteristics of two Schottky diodes: (A) with a double diffused guard ring, and (B) with a single guard ring. The parameters of these diodes are identical (6—6.5 μm thick epitaxial layers with (111) orientation and 0.7—1.0 Ωcm resistivity, Mo or Cr contacts, area of each diode 0.46×10^{-4} cm², area of guard ring 0.805×10^{-4} cm²).

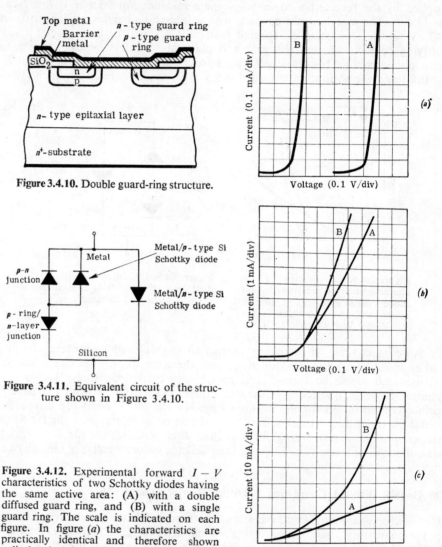

Figure 3.4.10. Double guard-ring structure.

Figure 3.4.11. Equivalent circuit of the structure shown in Figure 3.4.10.

Figure 3.4.12. Experimental forward $I - V$ characteristics of two Schottky diodes having the same active area: (A) with a double diffused guard ring, and (B) with a single guard ring. The scale is indicated on each figure. In figure (a) the characteristics are practically identical and therefore shown displaced. (after Saltich and Clark, [26]).

Figure 3.4.12a indicates that the low-current characteristics of diodes A and B are indeed identical. At moderate currents a difference occurs (Figure 3.4.12b) due to a slightly less series resistance of the diode B. This may be due for example to a lower resistivity of the epitaxial layer, an assumption which is consistent with

3. Schottky Diode

the slightly lower breakdown voltage, indicated by Figure 3.4.13. At higher current levels, the *pn* junction of the diode B injects minority carriers and the characteristics experience a rapid increase (Figure 3.4.12c). This increase is due to the

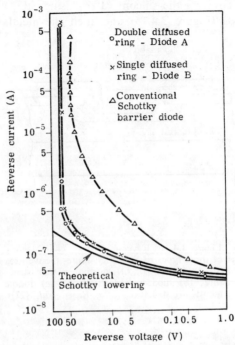

Figure 3.4.13. Reverse $I-V$ characteristics measured for conventional, single and double guard-ring diodes. Note the 'soft' breakdown of the normal planar diode. *(after Saltich and Clark*, [26]).

Figure 3.4.14. Experimental waveforms recorded for diodes A and B. The effective minority carrier lifetime, τ, is determined from the amplitude of the negative spike (whereas the tilt observed in the baseline of the trace is due to the diode capacitance). *(after Saltich and Clark*, [26]).

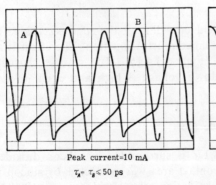

Peak current=10 mA

$\tau_A = \tau_B \leqslant 50$ ps

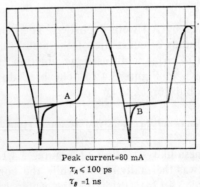

Peak current=80 mA

$\tau_A \leqslant 100$ ps

$\tau_B = 1$ ns

current of the *p-n* junction which adds to the Schottky barrier current and due to the conductivity modulation of the *n*-epitaxial layer as a result of the minority carrier injection. The oscilloscope traces shown in Figure 3.4.14 indicate the fast recovery of the diode A. The effective minority carrier lifetime (indicated for both diodes) was determined indirectly by using the current displayed by the diode driven by a sinusoidal voltage.

(c) Moat-etched Schottky Barrier Diode [27]

The moat-etched diode is fabricated as follows. Circular windows are cut in the SiO$_2$ covering the silicon wafer. Using SiO$_2$ as a mask, moat-etched regions were formed (Figure 3.4.15) and then the metals (Cr followed by Ag—Au and Mo

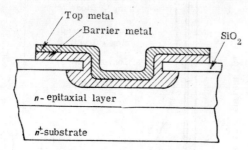

Figure 3.4.15. Moat-etched Schottky diode [27].

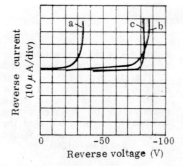

Figure 3.4.16. Reverse $I-V$ characteristic of three types of Schottky diodes (active area 0.245×10^{-4} cm^2): (a) normal planar structure; (b) moat-etched structure; (c) double guard-ring structure. (*after Rhee et al.,* [27]).

followed by Al) were deposited. The field crowding at the edge is avoided by the finite radius of curvature of the metal-semiconductor junction. The effect is similar to that seen at a diffused *pn* junction (Figure 3.4.1*b*). The reverse characteristic is shown in Figure 3.4.16. The breakdown voltage is high and equal to 0.8—0.9 of the bulk breakdown voltage.

(d) Planar Mesa Schottky Barrier Diode [28]

The mesa-like structure of Figure 3.4.17*c* is surrounded by silicon dioxide. This oxide was thermally grown while the contact area was protected by silicon nitride (Figure 3.4.17*b*). Prior to the wafer oxidation, the nitride mask (1000 Å) was formed by etching the deposited silicon nitride layer (1000 Å) everywhere except below the pyrolitic oxide dot (3000 Å) defined by photolithographic techniques (Figure 3.4.17a). The entire process is compatible with the planar technology (see Chapter 10) [22].

Molybdenum-silicon and PtSi—Si Schottky diodes have been fabricated by using the above technique. The breakdown voltage obtained was comparable or higher to that of equivalent p^+ guard-ring diodes, whereas the leakage current was significantly lower [28].

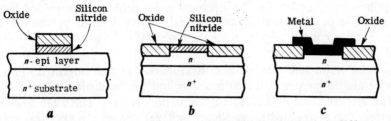

Figure 3.4.17. Processing steps for planar mesa Schottky diode [28].

(e) Mesa Schottky Barrier Diode

Coleman et al. [29] reported ideal avalanche breakdown characteristics from a conventional mesa diode having a truncated cone shape with an overhanging metal electrode, as shown by the inset in Figure 3.4.18. This is a Pt—GaAs diode with 200 μm diameter active contact. The forward and reverse characteristics measured at 460 K were interpreted by using the thermionic-emission theory. The saturation current density of 1.1×10^{-4} A cm^{-2} determined from the forward characteristic (Section 3.2) yields a barrier of 0.9 eV for an effective Richardson constant of 8.2 A cm^{-2} K^{-2}. The reverse characteristic compares well with the theory based on the image force lowering of the barrier (dashed curve in Figure 3.4.18). At about 120 V there is a sharp increase of reverse current due to avalanche multiplication. The experimental V_B and its temperature dependence is very close to the theoretical value calculated for an abrupt asymmetrical junction ($N_D = 6 \times 10^{15}$ cm^{-3}).

The constructions (c)—(e) discussed above have the advantage of low capacitance and no minority carrier storage upon the guard-ring structures.

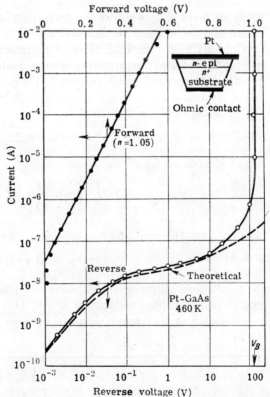

Figure 3.4.18. Forward and reverse current-voltage characteristics of the Pt—GaAs Schottky diode at 460 K. The mesa diode structure is shown in the inset. (after Coleman et al., [29]).

3.5 Noise in Schottky Barrier Diodes

An excellent introduction for the electrical noise problem is provided by reference [30]. The noise in solid-state devices was extensively studied by van der Ziel [31], [32]. In this book we briefly consider the electrical noise of each kind of device discussed.

The term noise refers to random fluctuations of the current flowing through the device and originates in the corpuscular and probabilistic nature of the current flow. The average value of the current fluctuation is zero. The mean square fluctuation current (or the related mean square voltage) defines the electrical noise.

The *shot noise* is due to completely random events and is independent of frequency (white spectrum) at low and intermediate frequencies.

The shot noise current associated to the current flow I is

$$\overline{i_n^2} = 2eI\Delta f, \tag{3.5.1}$$

where Δf is the bandwidth of the measuring instrument. The shot noise is a major source of noise in many semiconductor devices (see also Section 4.7).

Consider below the shot noise of a *Schottky diode*. Under *forward bias* there is a net flow of electrons from the semiconductor to the metal (n-type diode) determining the net steady-state current I. However, equal and opposite components of the saturation current, I_{sat}, also flow in the barrier and produce shot noise [33]. The resulting shot noise current is

$$\overline{i_n^2} = 2e(I + 2I_{sat})\Delta f \tag{3.5.2}$$

for an ideal Schottky diode, where

$$I = I_{sat}\left(\exp\frac{eV}{kT} - 1\right), \quad V > 0 \tag{3.5.3}$$

(Section 4.2). The equivalent noise temperature T_{eq} is defined by

$$\overline{i_n^2} = 4kT_{eq}G_i \tag{3.5.4}$$

(Nyquist's formula for 'thermal' noise, see [30] and also Section 4.7) where G_i is the differential device conductance dI/dV. The noise temperature ratio $t_B = T_{eq}/T$ (noise related to the barrier, hence t_B) can be found by using equations (3.5.2)–(3.5.4)

$$t_B = \frac{T_{eq}}{T} = \frac{1}{2}\left(1 + \frac{I_{sat}}{I + I_{sat}}\right). \tag{3.5.5}$$

The noise temperature ratio will decrease from unity at very low currents ($I \ll I_{sat}$) to one half at large currents ($I \gg I_{sat}$). At very large currents the effect of diode series resistance R_S becomes important. It is expected [33] that this resistance will exhibit thermal noise and that this noise will be uncorrelated with the barrier noise and the noise temperature ratio t of the device will be (Problem 3.8)

$$t = \frac{R_B t_B + R_S}{R_B + R_S}, \qquad (3.5.6)$$

where R_B is the dynamic resistance of the barrier. At very large currents ($R_B \ll R_S$) t approaches unity.

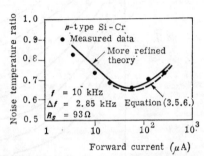

Figure 3.5.1. Noise temperature ratio for a forward-biased p-type nichrome solicon diode. The characteristic is approximated by $I = I_{s0}\left(\exp\dfrac{eV}{nkT} - 1\right)$ (see Section 3.2 and Problem 3.8). (*after Cowley and Zettler,* [33]).

The shot noise of the *reverse-biased* diode will be given by equation (3.5.1), where I is the reverse current. It is assumed that all of the current crosses the barrier as an emission current (the leakage current associated with the edges and the generation current are negligible).

The above simple theory was verified experimentally by Cowley and Zettler [33]. Figure 3.5.1 shows the measured and calculated noise temperature ratio as a function of the forward bias current. The measured noise temperature at currents above 100 μA is higher than expected, probably because of the excess noise [34]. This excess noise (see below) occurs at low frequencies and increases with current. Typical noise spectra for almost ideal Al-nSi Schottky diode (Figure 3.2.7) are shown in Figure 3.5.2. Note the fact that at high frequencies the diode noise is practically equal to the shot noise of the forward bias current.

The low-frequency excess noise was evaluated by Hsu [35]. The generation-recombination and trapping noise current in the space-charge region of the forward-biased Schottky barrier diodes was computed as the result of the barrier height fluctuation due to fluctuation of the charge density by generation-recombination and trapping processes. The equivalent current for noise in excess of the shot noise is shown in Figure 3.5.3. The dashed lines show that the measured spectrum can be decomposed into $1/f$ frequency dependent noise and a supplementary noise which exhibits a low-frequency plateau and then a rapid decrease with frequency. The latter component is due to generation and recombination. The low-frequency $1/f$ noise may arise due to trapping in the space-charge region as well as due to surface effects. Hsu [35] indicated that the random fluctuation of the charge density in the surface states at the diode periphery determines electrical field intensity fluctuations at the metal-semiconductor contact and therefore modulates the diode current due to image-force lowering of the barrier. The surface effect was experimentally demonstrated [35] by using the gate-controlled Schottky diode of Figure 3.4.2 [21]. The best low-frequency noise performances can be obtained when the

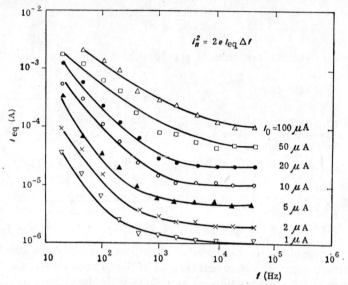

Figure 3.5.2. Noise spectra $I_{eq}=I_{eq}(f)$, where $\overline{i_n^2}=2eI_{eq}\Delta f$ for an Al-nSi Schottky diode. *(after Yu and Mead, [7]).*

surface potential corresponds to the flat-band condition [35]. The device noise increases rapidly when the surface is accumulated. Since positive oxide charge is always associated with thermally grown SiO_2, n-type silicon surfaces are all accumulated. The better low-frequency noise behaviour of p-type silicon diodes was indeed

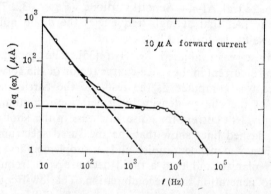

Figure 3.5.3. Excess noise current $I_{eq(ex)}$ of an Au-doped Al-nSi Schottky barrier diode (with oxide overlap, Figure 3.4.10) biased at 10 μA forward current. *(after Hsu [35]).*

demonstrated experimentally [23]. Good noise performance can be obtained by controlling the surface-state charge and using a metal contact overlapping the oxide (see reference [7] and Figures 3.4.1d, 3.5.2). The p-n junction guard ring also improves noise performances [25].

3.6 Schottky-barrier Capacitance

A reverse biased Schottky diode behaves as a capacitor whose capacitance is, to a first approximation, equal to the depletion layer capacitance. Consider a Schottky contact on a uniformly doped n-layer. The depletion layer width, x_d, can be calculated as for a p-n junction (Section 2.7) and the result is (Problem 2.4)

$$x_d = \sqrt{\frac{2\varepsilon}{eN_D}\left(V_{bi} - V - \frac{kT}{e}\right)}, \qquad (3.6.1)$$

where V is the applied (forward) bias (plus on metal). The total fixed charge per unit of contact area is

$$C = eN_D x_d = \sqrt{2e\varepsilon N_D \left(V_{bi} - V - \frac{kT}{q}\right)} \qquad (3.6.2)$$

and the depletion-layer capacitance (per unit area) is

$$C = \frac{\partial Q_d}{\partial V} = \sqrt{\frac{2\varepsilon N_D}{2\left(V_{bi} - V - \frac{kT}{e}\right)}} = \frac{\varepsilon}{x_d}. \qquad (3.6.3)$$

Therefore, the $1/C^2$ versus V plot should be a straight line. The intercept with the vertical axis determines the 'internal' potential difference, V_{bi}, and therefore the barrier height, whereas the slope of this line is related to the semiconductor doping. The Schottky-barrier capacitance method is frequently used to determine the doping in a surface layer. If the doping is non-uniform the $1/C^2$ versus V plot will no longer be a straight line, but the doping $N_D = N_D(x_d)$ can still be found from the slope of the $(1/C^2 - V)$ characteristic [2], [22]

$$N(x_d) = \frac{2}{e\varepsilon} \frac{1}{d(1/C^2)d(-V)}, \quad x_d = x_d(V). \qquad (3.6.4)$$

C is the differential capacitance measured by superimposing a small alternating voltage on the steady-state bias. An increase of the applied voltage by ΔV widens the depletion layer by Δx_d and increases the charge corresponding to the barrier capacitance by an amount ΔQ_d. The differential capacitance is $\Delta Q_d/\Delta V$. Equation (3.6.4) relates the slope of the $1/C^2 - V$ plot, namely $\Delta V/\Delta(C^{-2})$, to the local concentration at the distance x_d from the Schottky contact. A similar relation is valid for an abrupt pn junction [22].

However, anomalous profiles were sometimes obtained by using the $C-V$ technique. Deep impurity levels appear to be responsible for a number of such anomalies. Their effect was thoroughly investigated and it was found that the capacitance measured at various frequencies, temperatures and applied voltages may provide information about the energy level and capture cross-section of these impurities [36]–[40]. The deep levels are electrically inactive in the bulk, where the Fermi level lies several kT/e above them. However, near the metal contact these levels may emerge above the Fermi level and thus contribute to the total electric charge and the depletion-layer capacitance [36]. Figure 3.6.1 illustrates the case of a Schottky barrier with one deep level and one shallow level. The local change of charge density with the surface potential is also shown: this change peaks at the point where the electron quasi-Fermi level crosses the deep level and is due to ionization and deionization of deep donors. The second peak, at the boundary of the depletion region is due to free electrons. At very low frequencies all the charge responds to the voltage change.

At higher frequencies only the charge beyond x (the solid line in Figure 3.6.1b) responds, because a finite time is necessary for charge readjustment. At very high frequencies most of the ions in the depletion layer do not have time enough to be neutralized and only the free electron charge

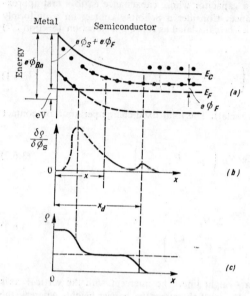

Figure 3.6.1. (a) Electron energy band diagram for Schottky barrier with deep and shallow level impurities; (b) spatial distribution of the charge derivative with respect to surface potential for d.c. voltage measurements (solid plus dashed line) and a.c. measurement at a certain frequency (solid line); (c) total charge distribution in the barrier region. (after Crowell and Nakano [37]).

located at the boundary of the depletion layer responds (the dielectric relaxation time is still assumed negligible) [37]. The $C^{-2} - V$ plots measured at various frequencies [36]−[39] were used for deep level analysis. This analysis is, however, complicated by the complex frequency dependence. Zohta [40] indicated that the impurity profile, as well as the energy level and the electron emission rate of deep impurities, can be determined by measuring $\Delta V/\Delta(C^{-2})$. This quantity depends not only on the signal frequency of the capacitance measurement, but also on the frequency of the incremental bias voltage ΔV, which is used to obtain $\Delta V/\Delta(C^{-2})$. Analytic formulas valid for impurity profiles of shallow and deep impurities were derived by Zohta [40] for a single deep level. Bleicher and Lange [41] and Glover [42] used transient measurements of Schottky barrier capacitance for deep-level determination.

References

1. A. Y. C. Yu, 'The metal-semiconductor contact: an old device with a new future', *IEEE Spectrum*, **7**, 83−89 (1970).
2. S. M. Sze, 'Physics of semiconductor devices', John Wiley, New York, 1969.
3. T. Sugano, F. Koshiga, A. Morino, H. K. Chou, M. Yoshida, K. Mishima, T. Nishi, T. Obunai and S. Matsuda, 'Metal to silicon contact. Conduction theory, fabrication technologies, characteristics and applications', *J. Fac. Eng., Univ. Tokyo (B)*, **31**, No. 1, 125−198 (1971).
4. T. Y. Stokoe and J. E. Parrot, 'Inclusion of carrier temperature effects in thermionic-diffusion theory of the Schottky barrier', *Solid-St. Electron.*, **17**, 477−484 (1974).
5. C. R. Crowell and S. M. Sze, 'Current transport in metal-semiconductor barriers', *Solid-St. Electron.*, **9**, 1035−1048 (1966).
6. C. R. Crowell and M. Beguwala, 'Recombination velocity effects on current diffusion and imref in Schottky barriers'. *Solid-St. Electron.*, **14**, 1149−1157 (1971).

7. A. Y. C. Yu and C. A. Mead, 'Characteristics of aluminium-silicon Schottky barrier diode', *Solid-St. Electron.*, **13**, 97—104 (1970).
8. R. N. Noyce, R. E. Bohn and H. T. Chua, 'Schottky diodes make IC scene', *Electronics*, **42**, 74—80, July 21, 1969.
9. K. Chino, 'Behaviour of Al—Si Schottky barrier diodes under heat treatment', *Solid-St. Electron.*, **16**, 119—121 (1973).
10. J. L. Saltich and L. E. Terry, 'Effects of pre- and post-annealing treatments on silicon Schottky barrier diodes', *Proc. IEEE*, **58**, 492—494 (1970).
11. D. L. Scharfetter, 'Minority carrier injection and charge storage in epitaxial Schottky barrier diodes', *Solid-St. Electron.*, **8**, 299—311 (1965).
12. A. Y. C. Yu and E. H. Snow, 'Minority carrier injection of metal-silicon contacts', *Solid-St. Electron.*, **12**, 155—160 (1969).
13. H. Jäger and W. Kosak, 'Modulation effect by intense hole injection in epitaxial silicon Schottky-barrier-diodes', *Solid-St. Electron.*, **16**, 357—364 (1973).
14. M. A. Green and J. Shewchun, 'Minority carrier effects upon the small signal and steady-state properties of Schottky diodes', *Solid-St. Electron.*, **16**, 1141—1150 (1973).
15. S. M. Sze, D. J. Coleman, Jr. and A. Loya, 'Current transport in metal-semiconductor-metal (MSM) structures', *Solid-St. Electron.*, **14**, 1209—1218 (1971).
16. P. U. Calzolari and S. Graffi, 'Minority carrier transport in metal-semiconductor junctions', *Solid-St. Electron.*, **16**, 1501—1503 (1973).
17. H. C. Card and E. H. Rhoderick, 'The effect of an interfacial layer on minority carrier injection in forward-biased silicon Schottky diodes', *Solid-St. Electron.*, **16**, 365—374 (1973).
18. R. L. Pritchard, 'Electrical characteristics of transistors', McGraw-Hill, New York, 1967.
19. H. C. Card and B. L. Smith, 'Green injection luminescence from forward-biased Au—GaP Schottky barriers', *J. Appl. Phys.*, **42**, 5863—5865 (1971).
20. R. M. Warner, Jr., 'Avalanche breakdown in silicon diffused junctions', *Solid-St. Electron.*, **15**, 1303—1318 (1972); P. R. Wilson, 'Avalanche breakdown voltage of diffused junctions in silicon', *Solid-St. Electron.*, **16**, 991—998 (1973).
21. A. Y. C. Yu and E. H. Snow, 'Surface effects on metal-silicon contacts', *J. Appl. Phys.*, **39**, 3008—3016 (1968).
22. A. S. Grove, 'Physics and technology of semiconductor devices', John Wiley, New York, 1967.
23. A. M. Cowley, 'Titanium-silicon Schottky barrier diodes', *Solid-St. Electron.*, **13**, 403—414 (1970).
24. M. P. Lepselter and S. M. Sze, 'Silicon Schottky barrier diode with near ideal $I-V$ characteristics', *Bell-Syst. Techn. J.*, **47**, 195—208 (1968).
25. R. A. Zettler and A. M. Cowley, 'p-n junction — Schottky barrier hybrid diode', *IEEE Trans-Electron Dev.*, **16**, 58—63 (1969).
26. J. L. Saltich and L. E. Clark, 'Use of a double diffused guard ring to obtain near ideal $J-V$ characteristics in Schottky barrier diodes', *Solid-St. Electron.*, **13**, 857—863 (1970).
27. C. Rhee, J. Saltich and R. Zwernemann, 'Moat-etched Schottky barrier diode displaying near ideal $I-V$ characteristics', *Solid-St. Electron.*, **15**, 1181—1186 (1972).
28. N. G. Anantha and K. G. Ashar, 'Planar mesa Schottky barrier diode', *IBM J. Res. Dev.*, **15**, 442—445 (1971).
29. D. J. Coleman, Jr., J. C. Irvin and S. M. Sze, 'GaAs Schottky diodes with near-ideal characteristics', *Proc. IEEE*, **59**, 1121—1122 (1971).
30. W. R. Bennett, 'Electrical noise', McGraw-Hill, New York, 1960.
31. A. Van Der Ziel, 'Fluctuation phenomena in semiconductors', Butterworth, London, 1959.
32. A. Van Der Ziel, 'Noise in solid-state devices and lasers', *Proc. IEEE*, **58**, 1178—1206 (1970).
33. A. M. Cowley and R. A. Zettler, 'Shot noise in silicon Schottky barrier diodes', *IEEE Trans. Electron Dev.*, **ED-15**, 761—769 (1968).
34. G. W. Neudeck, 'High frequency shot noise in Schottky barrier diodes', *Solid-St. Electron.*, **13**, 1249—1256 (1970).
35. S. T. Hsu, 'Low frequency excess noise in metal-silicon Schottky barrier diodes', *IEEE Trans. Electron Dev.*, **ED-17**, 496—506 (1970).

36. G. L. ROBERTS and C. R. CROWELL, 'Capacitance energy level spectroscopy of deep-lying semiconductor impurities using Schottky barriers', *J. Appl. Phys.*, **41**, 1767–1776 (1970).
37. C. R. CROWELL and K. NAKANO, 'Deep level impurity effects of the frequency dependence of Schottky barrier capacitance', *Solid-St. Electron.*, **15**, 605–610 (1972).
38. G. I. ROBERTS and C. R. CROWELL, 'Capacitive effects of Au and Cu impurity levels in Pt-*n*-type Si Schottky barriers', *Solid-St. Electron.*, **16**, 29–38 (1973).
39. K. HESSE and H. STRACK, 'On the frequency dependence of GaAs Schottky barrier capacitances', *Solid-St. Electron.*, **15**, 767–774 (1972).
40. Y. ZOHTA, 'Frequency dependence of C and $\Delta V/\Delta(C^{-2})$ of Schottky barriers containing deep impurities', *Solid-St. Electron.*, **16**, 1029–1035 (1973).
41. M. BLEICHER and E. LANGE, 'Schottky-barrier capacitance measurements for deep level impurity determination', *Solid-St. Electron.*, **16**, 375–380 (1973).
42. G. H. GLOVER, 'Determination of deep levels in semiconductors from $C-V$ measurements', *IEEE Trans. Electron. Dev.*, ED–19, 138–143 (1972).

Problems

3.1. Find the variation with temperature of the forward voltage for a Schottky diode on *n*-type silicon.

Hint: Assume that $dE_g/dT \simeq -0.2\,mV/°C$, that the Richardson constant is 120 A cm^{-2} K^{-2} and see also Figure 3.2.10.

3.2. Evaluate the carrier concentrations inside the space-charge region of a Schottky-barrier diode at thermal equilibrium and discuss the relative importance of carrier transport mechanisms inside this region (zero applied bias). Consider the case of an ideal Al–Si Schottky contact with $N_D = 10^{16}$cm^{-3}.

Hint: Use the depletion approximation.

3.3. Derive the expression of the minority hole current, equation (3.3.2).

3.4. By using the critical field for avalanche breakdown derived for abrupt asymmetrical *p-n* junction, compute the maximum voltage $V_{\text{inv max}}$ and maximum depletion layer width for an ideal metal-silicon Schottky contact with doping concentrations ranging from 5×10^{14}cm^{-3} to 5×10^{17} cm^{-3}. Repeat the calculations for epitaxial diodes with epilayer thickness of 0.5, 1, 2, 5, 10, 20 μm. Plot on the same graph $V_{\text{inv max}}$ and the series resistance of the epilayer (forward conduction), per unit of surface area, as a function of doping, with layer thickness as parameter.

3.5. Compute the internal potential distribution for an Au–nSi–n^+Si–Al ideal diode with 10^{16}cm^{-3}, 1 μm and 10^{19}cm^{-3}, 200 μm doping and thickness of n and $n+$ regions respectively, at thermal equilibrium and at 50 V applied reverse bias.

3.6. Consider a Schottky contact made on a semiconductor whose surface layer was doped by ion implantation and the active impurities follow a gaussian distribution (see for example Chapter 10). Derive the $C(V)$ characteristic of the reverse-biased contact and show how it can be used to derive the parameters of this distribution.

3.7. Discuss the similarities and differences between an (say abrupt, asymmetrical) *pn* junction and a Schottky-barrier diode.

Hint: Consider the internal potential distribution, built-in voltage, the $I-V$ characteristic, the role of minority carriers, etc.

3.8. Calculate the effect of a series resistance, R_s, upon the noise temperature ratio t (verify equation (3.5.6)).

4
Space-charge-limited Current Solid-state Devices

4.1 Introduction

Consider the simplest model of a unipolar solid-state device, namely a crystal bounded by two plane-parallel electrodes: the cathode ($x = 0$ plane) and the anode ($x = L$). We ignore for the moment the nature of these electrodes and the detailed electronic properties of the crystal itself. Assume that electrons are injected at $x = 0$ and collected at $x = L$ (the anode is positively biased with respect to cathode, as shown in Figure 4.1.1). The solid-state region is totally depleted of mobile holes and no electrons are generated inside the conduction space. Let us neglect the contribution of diffusion to the total current flow; therefore the current density under steady-state conditions is

$$J = env = \text{const.} \qquad (4.1.1)$$

(everywhere below electrons are treated as positively charged particles) where v is the electron drift velocity. By integrating $Jv^{-1} = en$ with respect to x from $x = 0$ to $x = L$, one obtains

$$J = \frac{Q_\text{mob}}{T_L}, \qquad (4.1.2)$$

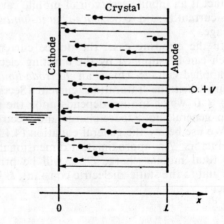

Figure 4.1.1. Plane-parallel solid-state diode. The crystal is bounded by two electrodes denoted conventionally cathode and anode. The field lines starting at the anode end on mobile electrons injected by the cathode and on the cathode itself.

where

$$Q_\text{mob} = \int_0^L en \, dx \qquad (4.1.3)$$

is the total mobile charge (per unit of electrode area) contained in the conduction space, and

$$T_L = \int_0^L \frac{dx}{v} \qquad (4.1.4)$$

is the electron transit-time from $x=0$ to $x=L$. Although highly suggestive and extremely simple, equation (4.1.1) by itself does not allow us to calculate the device current. The Poisson equation, the velocity-field dependence and proper boundary conditions should be also used. We exclude from our discussion the simple case when the crystal is a neutral semiconductor with ohmic contacts and the diode is simply a resistor*. Therefore, the electric field inside the device is non-uniform. The analysis may be simplified by assuming large injection of mobile charge and negligible density of fixed charge (ionized impurities or trapped carriers). The field lines starting at the anode (Figure 4.1.1) end either on the mobile space charge, or on the cathode. The mobile carrier concentration will be also non-uniform. In the cathode vicinity, the charge concentration would be, probably, high, because the electric field intensity is low and the velocity is small. The carrier velocity will increase and the carrier density will decrease with increasing x.

Two limit cases are important. Assume, first, that the cathode is a reservoir-type contact (Section 2.6) which provides an infinite carrier density at $x=0$ (ideal injecting contact). The electric field at such a contact will be zero

$$\mathscr{E}(0) = 0 \qquad (4.1.5)$$

and all field lines originating at the anode will end on mobile electrons. Equation (4.1.5) will be used as a boundary condition to determine the current. Note the fact that the cathode field $\mathscr{E}(0)$ vanishes due to the injected charge accumulated in the cathode vicinity. The device current is determined by the bulk space charge and does not explicitly depend upon the contact properties (provided that this contact is an infinite reservoir of mobile carriers and can supply an infinite current). This current is said to be *space-charge-limited* (SCL) and is determined by the applied voltage.

In the opposite case, the device current is exactly equal to the finite current which can be supplied by the emitting electrode ($x=0$) and is independent upon the applied voltage. This is an *electrode-limited* current. A slightly-modified situation occurs when the 'saturation' current (Section 2.6) depends upon the cathode field $\mathscr{E}(0) \neq 0$, which in turn depends upon the injected space charge. The diode current is, in general, both space-charge and electrode limited.

We use below the general equation (4.1.2) to describe the SCL currents in semiconductors. An approximate ('dimensional') analysis yields the following results. The total mobile charge Q_{mob} will be proportional to the 'average' electric field V/L and to the static dielectric constant, ε. We may also define an average transport velocity

$$v_{\text{average}} = \frac{L}{T_L} \qquad (4.1.6)$$

---

* The $J-V$ characteristic of such a 'resistor' may be, however, non-linear due to non-linear velocity-field dependence.

such that

$$J \propto \frac{\varepsilon V}{L^2} v_{\text{average}} \qquad (4.1.7)$$

At low applied fields, the carrier mobility μ is constant and equation (4.1.7) yields

$$J \propto \varepsilon \mu L^{-3} V^2. \qquad (4.1.8)$$

At very high electric fields the carrier velocity is saturated, $v_{\text{average}} \simeq v_{\text{sat}}$ over the almost entire conduction space, and the current is linear with respect to bias voltage

$$J \propto \varepsilon v_{\text{sat}} L^{-2} V. \qquad (4.1.9)$$

Equation (4.1.7) is also valid for the vacuum diode with plane-parallel electrodes, where $v_{\text{average}} \propto V^{1/2}$ and $\varepsilon = \varepsilon_0$, such that

$$J \propto \varepsilon_0 L^{-2} V^{3/2} \qquad (4.1.10)$$

and we reobtain Child's law ($J \propto V^{3/2}$).

4.2 Fundamental Law of Space-charge-limited Current

A more exact treatment will prove the square-law characteristic, equation (4.1.8). We consider below negligible diffusion current, negligible fixed charge, constant mobility and infinite injection at the emitting contact. The equations are

$$J = en\mu\mathscr{E} = \text{const.} \quad \text{(current continuity equation)} \qquad (4.2.1)$$

$$\frac{d\mathscr{E}}{dx} = \frac{en}{\varepsilon} \quad \text{(Poisson equation).} \qquad (4.2.2)$$

The carrier density will be eliminated and the electric field will be found by integrating the differential equation

$$\frac{d\mathscr{E}}{dx} = \frac{J}{\mu\varepsilon\mathscr{E}} \qquad (4.2.3)$$

with $\mathscr{E}(0) = 0$ as boundary condition (J is considered a parameter). Finally, the applied voltage is

$$V = \int_0^L \mathscr{E}\, dx, \quad \mathscr{E} = \mathscr{E}(x; J) \qquad (4.2.4)$$

and the $J - V$ characteristic is given by

$$J = \frac{9}{8} \varepsilon\mu \frac{V^2}{L^3}, \qquad (4.2.5)$$

which is the fundamental $J - V$ characteristic of SCL current in solids [1] — [3]. The SCL solid-state diode is compared to the SCL vacuum diode in Table 4.2.1.

Table 4.2.1

Comparison between SCL Diodes with Plane-parallel Electrodes

	Vacuum diode	Square-law solid-state diode
'Motion' equation	$m \dfrac{dv}{dt} = e\mathscr{E}$, m = free electron mass	$v = \mu \mathscr{E}$ (μ = const.)
Boundary condition	$v(0) = 0, \left. \dfrac{dv}{dt} \right|_{x=0} = 0$	$\mathscr{E}(0) = 0$
Approximations (see below)	Negligible emission velocities	Negligible diffusion-current
Electric field distribution	$\mathscr{E}(x) \propto \left(\dfrac{x}{L}\right)^{1/3}$	$\mathscr{E}(x) \propto \left(\dfrac{x}{L}\right)^{1/2}$
Carrier distribution	$n(x) \propto \left(\dfrac{x}{L}\right)^{-2/3}$	$n(x) \propto \left(\dfrac{x}{L}\right)^{-1/2}$
Velocity distribution	$v(x) \propto \left(\dfrac{x}{L}\right)^{2/3}$	$v(x) \propto \left(\dfrac{x}{L}\right)^{1/2}$
$J - V$ characteristic	$J = \dfrac{4}{9} \sqrt{\dfrac{2e}{m}} \, \varepsilon_0 \dfrac{V^{3/2}}{L^2}$	$J = \dfrac{9}{8} \varepsilon \mu \dfrac{V^2}{L^3}$
Carrier transit-time	$T_L = \sqrt{\dfrac{9m}{2eV}} \, L$	$T_L = \dfrac{4}{3} \dfrac{L^2}{\mu V}$
Incremental resistance	$R_i = \dfrac{T_L}{2 C_0}, \quad C_0 = \dfrac{\varepsilon}{L}$	$R_i = \dfrac{T_L}{3 C_0}, \quad C_0 = \dfrac{\varepsilon}{L}$

The following important properties of the square-law SCLC diode should be emphasized.

(a) The $J - V$ characteristic depends very little upon the temperature. In general, at a given V, J decreases slowly with increasing temperature due to the mobility-temperature dependence.

(b) The SCLC current is independent upon the contact properties. Its magnitude is determined by the bulk space-charge.

(c) The complete $J - V$ characteristic may be symmetrical or unsymmetrical. In the latter case one contact will be made rectifying. Both kind of devices may be useful in electronic applications [2], [4]—[11], the $J \propto V^2$ dependence being especially attractive.

(d) It will be shown that the only important time-constant for time varying device operation is the carrier transit-time T_L. If this time is made short then the device parameters will be constant up to very high frequencies [11].

We shall consider below a more realistic model for insulator and semiconductor SCLC diodes. The shape of the $J-V$ characteristic and the above properties in general may, consequently, experience important alterations.

4.3 Effect of Diffusion Current

The fundamental parabolic law of SCL current in solids was indeed confirmed by some experimental measurements, but only for a limited range of applied voltages. Some important modifications of the above simple theory should be discussed. The behaviour of SCL currents in solids is by far more complex and diverse than in vacuum.

The basic equation (4.2.5) does not contain any parameter related to the injecting and collecting contacts. The cathode contact was explicitly idealized by assuming infinite mobile carrier density at this contact, in the absence of diffusion. However, this infinite carrier concentration should determine an infinite concentration gradient and infinite diffusion current just at the cathode. Therefore, neglecting diffusion, at least in the cathode vicinity, is an inconsistent approximation. The diffusion current should be included into calculations in order to clarify the nature of the current flow and establish the concept of SCL current in solids.

By assuming positively charged $(+e)$ electrons, constant carrier mobility and using Einstein's relation $eD = \mu kT$ (D is the diffusion constant), the total current is

$$J = en\mu\mathscr{E} - \mu kT(dn/dx) = \text{const.} \tag{4.3.1}$$

If J is considered a parameter and n is replaced from equation (4.2.2), then we will have

$$\frac{d^2\mathscr{E}}{dx^2} - \frac{\mathscr{E}}{kT/e}\frac{d\mathscr{E}}{dx} + \frac{J}{\varepsilon\mu kT/e} = 0, \tag{4.3.2}$$

which is a second-order non-linear differential equation in $\mathscr{E} = \mathscr{E}(x)$. Two boundary conditions are now necessary.

The above equation may be written in normalized form

$$\frac{d^2\bar{\mathscr{E}}}{d\bar{x}^2} - \frac{1}{2}\frac{d\bar{\mathscr{E}}^2}{d\bar{x}} + \bar{J} = 0 \tag{4.3.3}$$

$$\bar{\mathscr{E}} = \frac{\mathscr{E} \cdot L}{kT/e}, \quad \bar{x} = \frac{x}{L}, \quad \bar{J} = \frac{JL^3}{\varepsilon\mu(kT/e)^2} \tag{4.3.4}$$

and its general solution may be expressed in terms of Bessel [12]–[14] or Airy [15] functions. We shall reproduce below a few results of the analytical treatment given by Wright [13]. He considered explicitly a metal-dielectric-metal structure and used as boundary conditions the value of carrier concentration at the cathode ($x = 0$) and the anode ($x = L$). These are reservoir-type contacts (Section 2.6), but the carrier concentration at the anode is extremely small (rectifying contact).

174   Bulk- and Injection-controlled Devices

The integration constants cannot be expressed explicitly in terms of these boundary conditions. Approximate analytic expressions have been developed for certain special cases. Wright [13] found that for small positive voltages (anode — positive, forward bias) the current depends approximately in an exponential fashion upon the applied bias. This is a diffusion current, flowing against the electric field (Figure 4.3.1 a) which arises from the metal work function difference. At very large

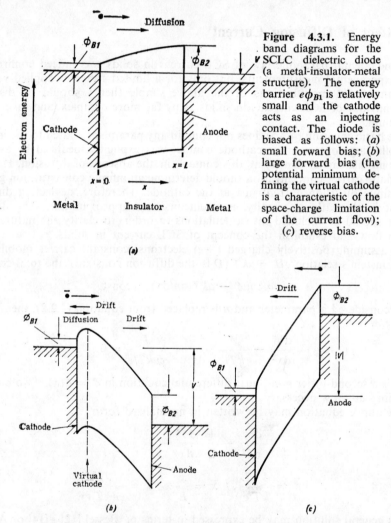

Figure 4.3.1. Energy band diagrams for the SCLC dielectric diode (a metal-insulator-metal structure). The energy barrier $e\phi_{B1}$ is relatively small and the cathode acts as an injecting contact. The diode is biased as follows: (a) small forward bias; (b) large forward bias (the potential minimum defining the virtual cathode is a characteristic of the space-charge limitation of the current flow); (c) reverse bias.

forward voltages the basic law, equation (4.2.5), is reobtained but V should be replaced by the external applied voltage minus a threshold voltage which is almost equal to the work-function difference of the metal electrodes, but less than this value due to the effect of diffusion. Diffusion current is important in the cathode region, between the virtual cathode, defined as the plane of potential minimum (zero electric field), and the cathode proper ($x = 0$). Beyond the virtual cathode, over most of the crystal, carrier drift is the predominant transport mechanism (Figure 4.3.1b). The reverse current is due to electrons emitted by the anode contact and is very small: the electric field inside the crystal is essentially constant (Figure 4.3.1c) and the current is carried by drift and is injection limited (it is proportional to carrier concentration at the anode and to the applied voltage).

Page [16] reported numerical calculations confirming the above results. These were performed for an In—CdS—Au diode with metal-semiconductor barriers determined by the work-function difference *. The boundary conditions were: the electron concentration and the electric field at the cathode. The electron concentration was given by $N_c \exp(-e\phi_{B_1}/kT)$, with ϕ_{B_1} as the cathode barrier †. The electric field was computed iteratively by using the following value for the anode electron density [3], [16]

$$n(L) = \frac{J(6\pi)^{1/2}}{ev_{\text{th}}}, \qquad (4.3.5)$$

because J, the forward current density, is generally larger than the random thermal current of the anode contact, at thermal equilibrium (i.e. than $en_{\text{th eq}}(L)v_{\text{th}}/(6\pi)^{1/2}$, where $n_{\text{th eq}}(L)$ is the thermal equilibrium carrier concentration at the anode, and v_{th} is the rms thermal velocity). Field-enhanced emission was neglected. However, it was found that the computed solutions were insensitive to $n(L)$ and $n(L)$ was taken equal to $n_{\text{th eq}}(L)$ [16] as Wright [13] did in his analytical model.

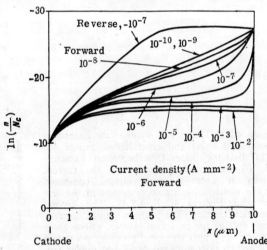

Figure 4.3.2. The electron concentration distribution normalized to the effective density of states in the conduction band, computed numerically for a CdS dielectric diode (details in the text). Note that at high current densities the carrier density is almost uniform beyond the potential minimum (compare to Figures 4.3.3 and 4.3.4). (after Page, [16]).

Figure 4.3.2 shows computed electron distributions within the crystal. The electron concentration is almost uniform for very large positive or negative applied voltages (see also the $J-V$ characteristic below). In both situations the contribution

---

* The following data were used $(e\phi_{Bn})_{\text{cathode}} = e\phi_{B_1} = 0.1$ eV, $J_{\text{sat cathode}} = 35$ A mm^{-2} (thermionic injection), $(e\phi_{Bn})_{\text{anode}} = e\phi_{B_2} = 0.7$ eV, $J_{\text{sat cathode}} = 10^{-7}$ A mm^{-2}, $\varepsilon = 10^{-10}$ F m^{-1}, $\mu = 210$ cm^2 V^{-1}s^{-1}, $T = 300$ K, $N_c = 2.51 \times 10^{19}$ cm^{-3}, $L = 10$ μm [16].

† The current density is much below the saturation current of the cathode contact and the electron concentration is very close to its thermal equilibrium value.

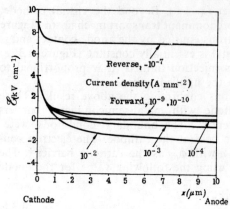

Figure 4.3.3. The electric field intensity inside a SCLC dielectric diode (details in the text). (after Page, [16]).

Figure 4.3.4. The potential distribution inside a SCLC dielectric diode. (after Page, [16]).

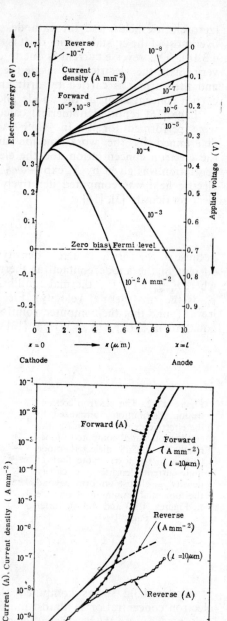

Figure 4.3.5. Theoretical steady-state characteristics (solid lines) and experimental points for SCLC dielectric diodes. One theoretical characteristic (forward and reverse) gives the current density (in A mm^{-2}) versus voltage, computed for the numerical example presented in the text (see also Figures 4.3.2—4.3.4). The second theoretical characteristic is the computed current versus voltage for an actual sample, whose characteristic was also measured point by point and indicated by dots (forward bias) and circles (reverse bias) on the same figure. (after Page, [16]).

of the diffusion current is small. For large forward currents the parabolic distribution given in Table 4.2.1 is approached. Figure 4.3.3 shows the computed electric field intensity *versus* distance. The negative electric field accelerates electrons towards the anode. However, field opposes to drift in the cathode region and the current is carried by diffusion. Figure 4.3.4 indicates the potential distribution. The potential minimum is characteristic of the SCL current because the net current flow is determined by the bulk space-charge rather by potential barriers or emission properties of contacts. The plane of this potential minimum defines the virtual cathode. As the applied voltage increases, the virtual cathode moves closer to the cathode ($x = 0$) and the 'height' of the potential minimum decreases. Thus the elementary theory which assumes zero cathode field is approached, in principle, for very high applied voltages. The computed $J-V$ characteristic, shown in Figure 4.3.5 does indeed confirm this behaviour (square-law dependence $J \propto V^2$, for large V, if the work function difference is not taken into account). The elementary theory seems to be approximately satisfied even at current densities where the cathode region (defined on Figure 4.3.1b) occupies a significant fraction of the crystal (Figure 4.3.4): the mechanism of carrier transport in the cathode vicinity does not seem to have a considerable influence upon the total steady-state current.

4.4 Insulator SCLC Diode

A distinct separation between crystalline insulators and semiconductors does not exist. Conventionally, the energy gap is 'wide' for the former category and 'low' for the latter. For the purpose of our discussion we define these categories more completely, although somewhat arbitrarily, as shown in Table 4.4.1.

Table 4.4.1
SCL Currents in Insulators and Semiconductors

	Insulator	Semiconductor
Examples	CdS, CdSe, ZnS, ZnTe, SiC, anthracene, etc. [3]	silicon, germanium, gallium arsenide
Energy gap	Wide, say above 2 eV	below 2 eV
Ohmic conductivity	Very low	Moderate or low
Trapping effects	Important in crystals Very important in polycrystalline films and amorphous materials	Negligible, except at low temperatures
Hot-carrier effects	Not reported	Very important
Usual injecting contacts	Metal-insulator contact Light excitation	n^+p or n^+n junctions * (for electron injection)

* A *pn* junction injects minority carriers. A minority carrier flow may be also space-charge limited (see the chapter on punched-through diodes).

The possibility of electronic conduction in insulators results from the quantum-mechanical theory of solids, as first indicated by Mott and Gurney [1]. They made the observation that it should be possible to inject electrons into the empty conduction band of an insulator from a suitable metallic contact, even at room temperature (such as a heated cathode injects electrons into vacuum). The most characteristic feature of SCL currents in insulators is the effect of traps. The theory of SCL currents in the presence of trapping was founded by Rose [17] and by Lampert [18]. We first consider, following Lampert [18], [19], the effect of a single trapping level in an insulating crystal. The energy levels in the bulk and at the metal-insulator interface are indicated in Figure 4.4.1a for the thermal equilibrium case. The free electron density is (non-degenerate statistics)

$$n = N_c \exp\left(-\frac{E_c - E_F}{kT}\right), \tag{4.4.1}$$

where $E_c = E_c(x)$. The density of trapped electrons is

$$n_t = \frac{N_t}{1 + \frac{1}{g}\exp\frac{E_t - E_F}{kT}}, \tag{4.4.2}$$

where N_t is the trap density (assumed uniform) and g is a degeneracy factor (for example $g = 2$ [19]). The above formula describes an equilibrium trap occupancy, resulting from the balance between capture and re-emission of electrons.

According to the hypothesis of *quasi-thermal equilibrium*, the presence of a moderate electric field will not affect directly the microscopic processes of electron trapping and release. Therefore, at low and moderate field intensities, the trap occupancy will be modified only through the change of free electron concentration accompanying injection [18]. Equations (4.4.1) and (4.4.2) will be still valid provided that the Fermi level E_F is replaced by a quasi-Fermi level for electrons, ζ_n (Figure 4.4.1b). By eliminating ζ_n between these equations one obtains

$$n_t(x) = \frac{N_t n(x)}{n(x) + N_c'}, \tag{4.4.3}$$

where

$$N_c' = \frac{N_c}{g}\exp\frac{E_t - E_c}{kT} = \text{independent of } x. \tag{4.4.4}$$

Equation (4.4.3) is a state equation describing the quasi-thermal equilibrium between traps and the free carriers.

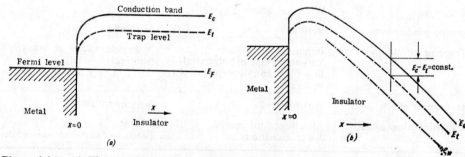

Figure 4.4.1. (a) The metal-insulator contact at thermal equilibrium. E_t indicates a single trapping level; (b) energy levels in insulator under applied bias (ζ_n is the electron quasi-Fermi level).

4. SCLC Solid-state Devices

Two limit cases can be readily examined. Assume first that ζ_n lies several kT below the trap level everywhere in the crystal; the traps are named *shallow*, and are almost unoccupied ($n_t \ll N_t$). Equation (4.4.3) becomes

$$n_t(x) \simeq \frac{N_t}{N_c'} n(x) \qquad (4.4.5)$$

and the ratio of free charge density to the total charge density is

$$\delta = \frac{n(x)}{n(x) + n_t(x)} = \frac{1}{1 + g \dfrac{N_t}{N_c} \exp \dfrac{E_c - E_t}{kT}} \qquad (4.4.6)$$

and does not depend upon the position x. These traps reduce the value of the injected current to the fraction δ of the current in the same device but without traps. Equation (4.2.5) becomes

$$J = \frac{9}{8} \varepsilon \mu \delta \frac{V^2}{L^3} \qquad (4.4.7)$$

(negligible diffusion, constant electron mobility, negligible ohmic conductivity).

The trap level will be named *deep*, if ζ_n is several kT above and practically all trapping states are occupied

$$n_t(x) \simeq N_t. \qquad (4.4.8)$$

This equation should be combined with

$$J = en(x) \mu \mathscr{E}(x) \qquad (4.4.9)$$

$$\frac{d\mathscr{E}(x)}{dx} = \frac{e[n(x) + n_t(x)]}{\varepsilon} \simeq \frac{e[n(x) + N_t]}{\varepsilon} \qquad (4.4.10)$$

and solved with

$$\mathscr{E}(0) = 0, \quad \int_0^L \mathscr{E}(x) \, dx = V. \qquad (4.4.11)$$

The complete analytical solution to (4.4.8)–(4.4.11) will be given later. Here we discuss the $J - V$ characteristic qualitatively. At low applied voltages almost all injected electrons are trapped and the current is negligibly small. A steep increase of the J versus V characteristic should be expected as V increases above the voltage required to fill completely all trap sites in the crystal, the so-called trap-filled limit voltage

$$V_{TFL} = \frac{eN_t L^2}{2\varepsilon}. \qquad (4.4.12)$$

V_{TFL} was calculated by integrating the Poisson equation with $n_t(x) = N_t$ and $n(x) = 0$. At voltages exceeding $2 \ldots 3 V_{TFL}$, the mobile injected charge is considerably greater than the trapped charge and the latter may be neglected such that the $J - V$ characteristic will merge with the ideal trap-free characteristic (4.2.5).

The 'shallow' and 'deep' character of a trap level is relative, depending upon the injection level. As the injection level increases, the quasi-Fermi level ζ_n moves closer to the conduction band. When ζ_n crosses the trap level, this level will be filled. Therefore, a 'shallow' level becomes 'deep'. The $J-V$ characteristic should consist of at least three regions, a low current $J \propto \delta V^2$ region (shallow traps, characterized by δ of equation (4.4.6)), a second region characterized by an abrupt increase of J around V_{TFL} (trap filling), and a high-voltage (trap-free) square-law region. This is indeed indicated by the results of more exact calculations reported by Lampert [18] and reproduced in Figure 4.4.2. The dashed curves are three ideal limit curves:

TRAP-FREE — the ideal square-law (4.2.5)
TRAP-FILLING — the characteristic derived from equations (4.4.8)–(4.4.11)
OHMIC — the ohmic law corresponding to a certain finite crystal resistivity.

The characteristics appearing in Figure 4.4.2 were computed for a certain CdS diode (diffusion neglected, constant mobility). However, the location of the trap level ($E_c - E_t$) was taken as a parameter whereas N_t (and thus V_{TFL}) are fixed. All computed curves lie within the triangle formed by the above limit characteristics.

If more than one discrete level exists, the $J - V$ characteristics should exhibit a 'step-by-step' increase, each abrupt current rise indicating that the quasi-Fermi level crosses a trap level.

A continuous distribution of traps in energy was anticipated for polycrystalline or amorphous insulators*. A reasonable approximation for a set of traps derived from one level is a Gaussian

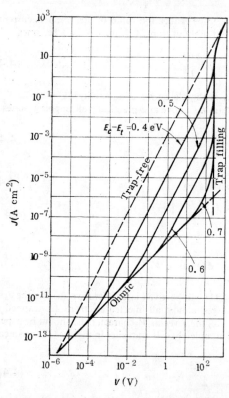

Figure 4.4.2. The current-voltage characteristics computed for an SCLC insulator diode with a single trapping level, located $E_c - E_t$ below the conduction band. The dashed curves are the three limiting characteristics discussed in the text. The insulator is CdS with $\varepsilon = 11\ \varepsilon_0$ and $\mu_n = 200$ cm² V⁻¹ s⁻¹, at T = 300 K. For all curves $N_t = 10^{14}$ cm⁻³, $L = 50\ \mu m$ and $E_c - E_F$ (at thermal equilibrium, in the bulk) is 0.75 eV corresponding to a resistivity of $\approx 3 \times 10^{10}\ \Omega$ cm (free carrier density at thermal equilibrium equal to $\approx 10^6$ cm⁻³); $E_c - E_t$ is taken as a parameter. (after Lampert, [18])

distribution of energy [20] and an exponential distribution may be used for a superposition of many different sets of traps, all smeared out in energy. By assuming [20]

$$N_t(E) = N_0 \exp \frac{E - E_c}{kT_t}, \qquad (4.4.13)$$

---

* 'Owing to the large amount of structural disorder a given kind of electron trap, either of chemical or structural origin, does not have a uniquely defined environment. These will be differences from one trap site to another, in nearest neighbour, next nearest neighbour configurations, and so on. The net result will be a broad smearing out of the level' (quoted from Lampert [20]).

where T_t is a characteristic temperature, an approximate calculation [20] yields ($T < T_t$)

$$J \simeq e\mu N_c \left(\frac{\varepsilon}{eN_0 k T_t}\right)^{T_t/T} \times \frac{V^{(T_t/T+1)}}{L^{(2T_t/T+1)}} \qquad (4.4.14)$$

i.e. the $J - V$ curve is steeper than $J \propto V^2$. This result was first derived by Rose [17].

It was therefore shown that trapping modifies profoundly the $J - V$ characteristic. Some principal advantages of the SCLC diode are compromised. Nor is its parabolic shape distorted but even the square-law regions are less useful. For example, the current given by the shallow trap equation (4.4.7), $\delta \ll 1$, will increase very rapidly with increasing temperature because the trapping coefficient δ increases. Dynamic effects of trapping are also important (see below).

Some SCLC single-crystal diodes did still exhibit a trap-free behaviour. Such were the CdS devices experimented by Wright and his co-workers [21]–[23]. The experimental results reported by Page [16] and reproduced here in Figure 4.3.5 do indeed show such an ideal behaviour. Because the CdS crystals used were grown from *impure* starting material, it was suggested that some form of automatic defect compensation must be taking place during crystal growth [16].

SCLC in insulators were, however, used as an experimental tool for studying defect levels in insulators. Consider as an example the single-level case. The trap density may be determined by measuring V_{TFL} (equation (4.4.12))*. The location of the energy level E_t can be determined from the trapping factor δ (equation (4.4.6)).

Figure 4.4.3 shows $J - V$ characteristics measured by Trodden [25] for two single crystal CdS diodes. Figure 4.4.3 a corresponds to the single-level theory given by Lampert [18] and yields $N_t = 2 \cdot 1 \times 10^{14}$ cm^{-3}, $E_c - E_t = 0.55$ eV. The trap-free $J \propto V^2$ region was not reached because the current was limited by power dissipation. The trap density should be two orders of magnitude lower to obtain, for the same diode, an ideal square-law behaviour above 1 V. Figure 4.4.3b was considered [25] as an indication about the presence of (at least) two discrete trapping levels above the thermal equilibrium Fermi level, levels characterized by 3.6×10^{14} cm^{-3}, 0.51 eV and 3.6×10^{14} cm^{-3}, 0.46 eV, respectively. A comprehensive review of results obtained from SCLC measurements was given recently by Lampert and Mark [3].

The SCL current is very sensitive to extremely small defect state densities (10^{12}cm^{-3} or lower). In this respect, the technique of SCLC measurements is complementary to the usual chemical, optical, etc. procedures which are increasingly difficult to employ as the defect concentration decreases [20].

However, there are a number of difficulties arising in connection with this SCLC technique:

(A) The results interpreted by using the unipolar SCL current model may be due, in fact, to other physical phenomena. For example, the steep rise of the $J - V$ characteristic may be due to electrical breakdown, or, another example, the $J \propto V^2$ region may be explained by a double injection model [3].

(B) The results were almost invariably interpreted in terms of theories assuming uniform spatial trap distribution. Nicolet [26] has shown that a non-uniform spatial distribution of traps is plausible and, if present, may lead to considerable errors in interpreting the results of SCLC measurements. This is especially valid for the case of shallow traps, which are characterized by a single parameter, the trapping factor δ, irrespective of the trap distribution in space and energy. A uniform layer of traps within the device is more efficient in reducing the SCL current when located in the immediate vicinity of the cathode. The effect of traps located just at the cathode interface was also theoretically studied [27]. A generalized approach for insulators with traps non-uniformly distributed in space and energy was given recently [28].

(C) In measuring the $J-V$ characteristics of high-resistivity wide-band materials, the quasi-thermal equilibrium conditions cannot be, in general, established within a reasonable time [29]. This is due to very long time constants associated to trapping and releasing processes. Some computations [3] indicate that the thermal-release time at room temperature may be of the order of minutes, for trap depth $E_c - E_t > 0.75$ eV and weeks for $E_c - E_t > 1.0$ eV. Experiments involving crystals with such trapping levels require special means to approach the equilibrium (by heating

---

* Lampert and Edelman [24] calculated numerically the 'trap-filling' characteristic by taking diffusion into account and found that the correction factor for V_{TFL} does not exceed 2.

the sample or by shining light of suitable wavelength on the sample). These precautions were not, perhaps, observed by numerous experimentators and this may be the cause of exceedingly low trap densities (not confirmed by other tests) reported for some insulators, as well as the reason for poor reproducibility of many room-temperature $I-V$ characteristics [29].

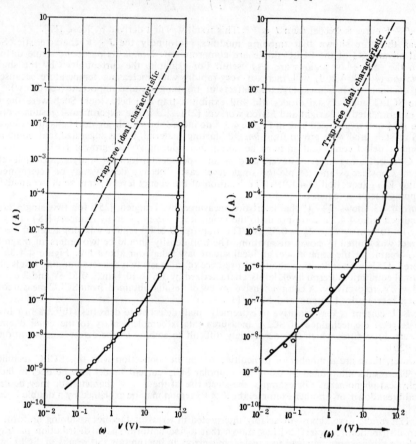

Figure 4.4.3. Experimental results (full curves) measured for CdS single-crystal dielectric diodes. The crystal thickness is 19 μm (a) and 10 μm (b). The dashed curves represent the ideal square law computed for $\varepsilon = 10\, \varepsilon_0$ and $\mu_n = 250$ cm^2 V^{-1}s^{-1}. The measured characteristics indicate the filling of one (a) and two (b) trapping levels. (*after Trodden*, [25]).

Measurements of *transient SCL currents* also represent an experimental tool used to determine certain properties of solids. The basic theory of transient was put forward by Many and Rakavy [30]. This is presented in two recent monographs [3], [31] and will not be repeated here. The solid curve in Figure 4.4.4 represents the theoretical current transient for a step voltage V_0 applied at $t=0$ to an ideal square-law SCLC diode. The current maximum occurs for $t_1 \simeq 0.8\, L^2/\mu V_0$ and indicates that the front of the injected charge reaches the anode. This feature is well confirmed by experiments and permits the determination of carrier mobility μ. The method has the advantages of time-of-flight techniques and is extremely valuable for certain low-mobility materials. In such experiments the ohmic contact for carrier injection is often provided by steady-state illumination.

The current transients can also yield information about trapping parameters [31]. Figure 4.4.5 shows for example, that the experimental points (hole injection in iodine) measured for two voltage steps are well below the theoretical curve calculated with no trapping. The experiment indicated that quasi-stationary conditions corresponding to a free to trapped carrier density ratio of about 0.3

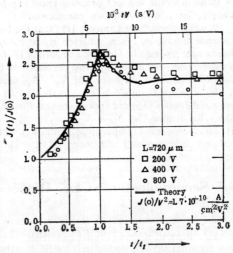

Figure 4.4.4. Theoretical (solid line) and experimental (squares, triangles and circles) transient current (hole injection in an iodine single crystal), when a voltage step V_0 is applied at $t = 0$ to the terminals of an SCLC insulator diode. $J(0)$ is the initial current and t_1 the transit-time of the leading front of charge carriers injected in the insulator at $t = 0$. The theoretical curve is computed by assuming an infinite carrier lifetime (no trapping). The experimental results at a much larger time scale, i.e. seconds (t in this figure is equal to 32,16 and 8 μs for an applied voltage of 300, 400 and 800 V, respectively), indicate a considerable decay of the current due to a slow trapping process. (after Many, Weisz and Simhony, [32]).

(see equations (4.4.5) and (4.4.6)) are reached after 10—20 transit-times. Using this quasi-stationary value, two theoretical curves were calculated, for $\tau_f/t_1 = 5$ and 2.5, respectively, where t_1 is the transit-time of the first front of carriers and t_f is the trapping time (or the lifetime of free carriers)

$$\tau_f \simeq \frac{1}{v_{th}\sigma_t N_t}. \qquad (4.4.15)$$

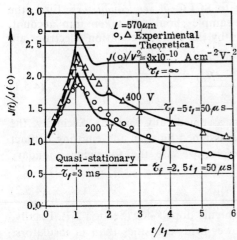

Figure 4.4.5. Transient SCLC indicating fast trapping processes. The solid line $\tau_f = \infty$ is the theoretical response with no trapping. The experimental points (triangles and circles) indicate a hole lifetime τ_f comparable to 'initial' carrier transit-time t_1, which is 20 μs (for 200 V) and 10 μs (for 400 V). The solid lines were both computed for $\tau_f = 50$ μs (see the text) and provide a good fit to experimental results (hole injection in iodine). (after Many, Weisz and Simhony, [32]).

All calculations are made for shallow traps and therefore the density of empty states is taken equal to the trap density N_t (almost all are unoccupied). v_{th} is the thermal velocity of the mobile carrier and σ_t is the capture cross-section of an empty trap. The experimental results in Figure 4.4.5 [32] are well fitted by the two theoretical curves calculated for the same trapping time, $\tau_f = 50\ \mu s$. By using the above results combined with steady-state measurements at various temperatures, $E_c - E_t$, N_t and σ_t may be obtained. Practical determination of these magnitudes are complicated by the fact that little is known about the band structure of insulators (for example the effective density of states in iodine was not precisely known by Many et al. [32]), or due to the presence of many trapping levels with unknown parameters.

It was suggested by Lampert [18] that the best experimental proof of the SCLC model in the presence of trapping can be achieved in *high-purity semiconductors at low temperatures*. Such experiments were reported by Gregory and Jordan [33] for p^+pp^+ silicon samples at 4.2 K (the p^+ regions are degenerate and act as injecting contacts). At this temperature a p-type (compensated) semiconductor contains no free holes but $N_A - N_D$ neutral acceptor centres which could act as trapping levels for *injected* mobile holes. The steady-state and transient measurements, performed by Gregory and Jordan [33] did indeed confirm the filling of deep traps and the space-charge limitation. The value obtained for $N_A - N_D$ from the trap-filled limit voltage V_{TFL} was close to that given by Hall measurements of impurity concentration. The capture cross section for holes was also determined and compared well with theory [33].

4.5 Semiconductor SCLC Diode

An SCLC semiconductor diode is usually made as an n^+nn^+ (or p^+pp^+) structure. This is a symmetrical SCLC resistor. The central region must have a sufficiently low conductivity in order that an SCL current be observed. The highly doped regions act as injecting contacts. A rectifying SCLC diode can be also constructed using different metallic contacts for cathode and anode, but it is difficult to avoid double injection, due to the relatively small energy-gap. Büget and Wright [34] experienced such a difficulty for silicon diodes. For our analysis, however, the nature of the collecting contact is unimportant. We only assume electron injection at the cathode contact and exclude hole injection.

There are a number of effects specific to SCLC *in semiconductors*. First, the ohmic conductivity of the semiconductor sample should be taken into account. The deviations from neutrality are indicated by the Poisson equation (electrons are again considered positively charged)

$$\frac{d\mathscr{E}}{dx} = \frac{e(n - N_D)}{\varepsilon} \tag{4.5.1}$$

(N_D is the effective concentration of donors assumed fully ionized). Secondly, the carrier concentration at the cathode will be assumed finite, if the emitting contact takes the form of an n^+n junction. Therefore, we use the following boundary condition

$$n(0) = N_D^+, \tag{4.5.2}$$

where N_D^+ is the donor concentration on the highly doped side (Section 2.7). Finally, the charge carriers become hot at much lower field intensities than in insulators, and the current (diffusion neglected) will be written

$$J = env(\mathscr{E}) \tag{4.5.3}$$

4. SCLC Solid-state Devices

We shall first discuss the first two effects. By introducing $v = \mu\mathscr{E}$ in equation (4.5.3) and also using equation (4.5.1), one obtains:

$$\varepsilon \frac{d\mathscr{E}}{dx} = \frac{J}{\mu\varepsilon} - eN_D. \tag{4.5.4}$$

A useful substitution is

$$dx = v\,dT_x = \mu\mathscr{E}\,dT_x, \tag{4.5.5}$$

where T_x is the carrier transit time since it was emitted and until it reaches the plane x. Equation (4.5.4) becomes

$$\frac{d\mathscr{E}}{dT_x} + \omega_r\mathscr{E} = \frac{J}{\varepsilon}, \quad \omega_r = \frac{e\mu N_D}{\varepsilon}, \tag{4.5.6}$$

with ω_r as the reciprocal of the dielectric relaxation time. This equation will be integrated by using

$$\mathscr{E}\big|_{x=0} = \mathscr{E}\big|_{T_x=0} = \frac{J}{e\mu n(0)} = \frac{J}{e\mu N_D^+} \tag{4.5.7}$$

and the J–V characteristics will be then obtained in normalized parametric form [31]

$$\bar{J} = \frac{J}{eN_D L\omega_r} = \frac{1}{\theta_r + \left(\frac{\varDelta-1}{\varDelta}\right)[\exp(-\theta_r) - 1]} \tag{4.5.8}$$

$$\bar{V} = \frac{V}{eN_D L^2/\varepsilon} = \left\{\theta_r + 2\left(\frac{\varDelta-1}{\varDelta}\right)[\exp(-\theta_r) - 1] - \right.$$
$$\left. - \frac{1}{2}\left(\frac{\varDelta-1}{\varDelta}\right)^2[\exp(-2\theta_r) - 1]\right\}\left\{\theta_r + \left(\frac{\varDelta-1}{\varDelta}\right)[\exp(-\theta_r) - 1]\right\}^{-2}, \tag{4.5.9}$$

where

$$\theta_r = \omega_r T_L, \quad \varDelta = N_D^+/N_D. \tag{4.5.10}$$

The special case of infinite injection may be obtained for $\varDelta \to \infty$ (this is equivalent to zero electric field at $x = 0$). The corresponding J–V characteristic is plotted in Figure 4.5.1 and shows a gradual transition from the low-voltage ohmic region (large T_L due to slow drift, large θ_r, and very small dielectric relaxation time as compared to the transit time of the injected mobile charge, therefore the sample is almost neutral) to a square-law dependence at large voltage bias (the transit-time of charge carriers becomes shorter than ω_r^{-1} and the dielectric relaxation mechanism has little influence upon the $J-V$ characteristic), where the current is true space-charge limited. A similar $J-V$ characteristic was previously computed by Lampert [18] and van der Ziel [35], and verified experimentally [6], [36]—[38], although it had to be corrected for hot carrier effects [36], [37] (see below).

However, the shape of the $J-V$ characteristic experiences a considerable modification if the doping ratio $\varDelta = N_D^+/N_D$ is taken finite, as shown by dashed curves in Figure 4.5.1. The current is, in fact, emission limited and at large voltages it approaches asymptotically

$$J = eN_D^+\mu V/L. \qquad (4.5.11)$$

Even if the carrier concentration at $x = 0$ is sufficiently high, the experimental $J-V$ characteristic may not behave according to the parabolic law of SCLC in solids. This may be explained by the mobility field-dependence. An analytical theory incorporating high-field effects should, however, neglect the sample conductivity ($N_D = 0$ in equation (4.5.1)). Space-charge-limited emission with $\mathscr{E}(0) = 0$ will be also considered. By assuming (v_{sat} is the saturation velocity and $v_{\text{sat}}/\mathscr{E}_c$ the low field mobility)

$$v = v_{\text{sat}} \frac{\mathscr{E}/\mathscr{E}_c}{1 + \mathscr{E}/\mathscr{E}_c}, \qquad (4.5.12)$$

which is appropriate for holes in silicon, one obtains [39], [40]

$$\bar{J} = \frac{J}{\varepsilon \mathscr{E}_c v_{\text{sat}} L^{-1}} = m - \ln(1 + m) \qquad (4.5.13)$$

$$\bar{V} = \frac{V}{\mathscr{E}_c L} = \left[\frac{m^2}{2} - m + \ln(1 + m)\right][m - \ln(1 + m)]^{-1}, \qquad (4.5.14)$$

where $m = \mathscr{E}(L)/\mathscr{E}_c$ (the ratio between the anode field and the characteristic field). The $J-V$ characteristic is plotted in Figure 4.5.2. The low-current region is parabolic and corresponds to equation (4.2.5). The asymptotic dependence for $V \to 0$ is linear and occurs due to velocity saturation*. The theoretical characteristic is compared to experiment in Figure 4.5.3 [40].

The $J-V$ characteristic of SCL current can be used to determine the velocity-field dependence for electrons or holes in semiconductors [33], [36], [38], [39], [41], [42]. However, the $J-V$ characteristic is not very sensitive to the detailed shape of the $v - \mathscr{E}$ characteristic [43], [44] because the effect of $v(\mathscr{E})$ at a given current occurs 'integrated' over the entire device length and from zero field to the maximum field $\mathscr{E}(L)$. Another difficulty occurs because, in general, the sample conductivity

---

* The finite carrier concentration at the cathode should lead to the same linear behaviour at high voltages ($\varDelta$ = finite in Figure 4.5.1). However, for usual doping levels in an n^+nn^+ structure only the hot-carrier effect is important.

cannot be neglected at bias fields where the mobility-field dependence becomes important [36], [37]. The $J-V$ characteristics tend to show a less pronounced deviation from linearity and experimental data become even less significant. Low temperature measurements are more difficult because the charge carriers become hot at lower field intensities.

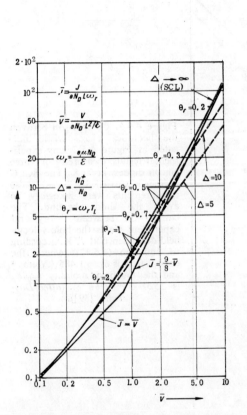

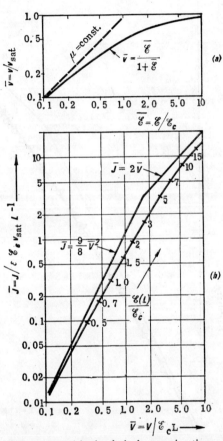

Figure 4.5.1. Steady-state normalized $J-V$ characteristics calculated for an n^+nn^+ resistor. (*after Dascalu*, [31] and [45]).

Figure 4.5.2. (*a*) Analytical aproximation of carrier velocity field-dependence; (*b*) normalized $J-V$ characteristics calculated for SCL injection, negligible conductivity, and velocity field-dependence as indicated in (*a*). The low-field and high-field asymptotes are also shown (*after Dascalu*, [39]).

Due to the residual conductivity and the mobility-field dependence the parabolic law (4.2.5) was not often observed in semiconductors. It will be shown in Section 4.9 that for a given semiconductor, the doping-length product should be sufficiently small in order to obtain a square-law region of the $J-V$ characteristic [31], [45]. At this time this would be possible due to the recently available thinning procedures ($5-10$ μm crystals can be obtained).

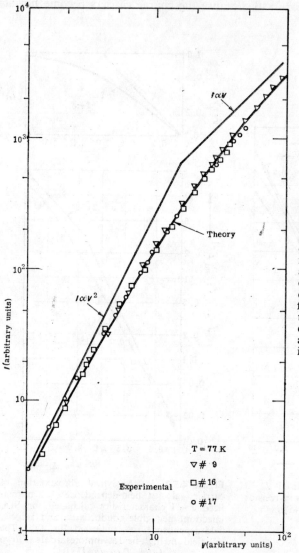

Figure 4.5.3. Comparison between the theoretical characteristic indicated in Figure 4.5.2. any experimental results measured for three silicon diodes, at 77 K. The SCLC current is carried by holes. Pulse measurements were necessary to avoid the diode heating. The comparison between theory and experiment yields the hole velocity field-dependence at 77 K, according to equation (4.5.12), where the critical field is $\mathscr{E}_c = 1.485$ kV cm^{-1} and the low-field mobility $v_{sat}/\mathscr{E}_c$ is 5520 cm^2 V^{-1} s^{-1}). (after Dascalu, [39]).

4.6 Transit-time Effects at Very High Frequencies

The square-law SCLC diode will be examined as an example. The entire problem is treated at length in a recent monograph [31].

4. SCLC Solid-state Devices

The non-stationary behaviour of an SCLC diode is described by the following equations

$$J = e\mu n \mathscr{E} + \varepsilon \frac{\partial \mathscr{E}}{\partial t} = J(t) \tag{4.6.1}$$

$$\frac{\partial \mathscr{E}}{\partial x} = \frac{e(n - N_D)}{\varepsilon} \tag{4.6.2}$$

($N_D = 0$ for the square-law diode)

$$V = \int_0^L \mathscr{E} \, dx = V(t), \tag{4.6.3}$$

where $n = n(x, t)$ and $\mathscr{E} = \mathscr{E}(x, t)$. The displacement current was added to the particle (convection) current in the first equation (Maxwell's continuity equation). The electrostatic potential difference V is defined by equation (4.6.3) provided that L is much shorter than the minimum wavelength of oscillation. Equations (4.6.1) and (4.6.2) determine the following non-linear equation with partial derivatives

$$\mathscr{E}\frac{\partial \mathscr{E}}{\partial x} + \frac{1}{\mu}\frac{\partial \mathscr{E}}{\partial t} + \frac{eN_D}{\varepsilon}\mathscr{E} = \frac{1}{\mu\varepsilon} J(t). \tag{4.6.4}$$

The general solution of this equation may be determined by using a boundary condition such that

$$\mathscr{E}(0, t) = \text{given function of time.} \tag{4.6.5}$$

A usual (although not entirely justified) assumption for SCLC injection is that the electric field at the injecting contact is always zero

$$\mathscr{E}(0, t) = 0 \tag{4.6.6}$$

that means that this contact behaves as an infinite reservoir of carriers at any current level.

In equation (4.6.4), $J(t)$ was assumed given. If, as usual, $V(t)$ is imposed and $J(t)$ has to be calculated the problem will become, in general, more difficult. A general procedure is to assume that for a given $V(t)$, $J(t)$ is a known function with unknown coefficients. These coefficients will be then determined from equations (4.6.3)—(4.6.5).

The small-signal problem is an example of such an approximative procedure. Consider a small alternating voltage superimposed on a steady-state diode bias V_0

$$V(t) = V_0 + \mathbf{V}_1 \exp i\omega t, \quad |\mathbf{V}_1| \ll V_0. \tag{4.6.7}$$

The second term in the right-hand is the alternating voltage (signal). $\mathbf{V}_1$ is a complex number (complex signal amplitude) and $i = \sqrt{-1}$ is the imaginary unit. Altogether $\mathbf{V}_1 \exp i\omega t$ is a rotating phasor, encountered in the electric engineering formalism.

The actual signal, a scalar function of time, is either the real or the imaginary part of $\underline{V}_1 \exp i\omega t$. The sum expressed by equation (4.6.7) is purely formal but this does not affect in any way the results because the two components of $V(t)$ will further separate each other.

The total current $J(t)$ should be, to a first approximation, of the same form

$$J(t) = J_0 + \underline{J}_1 \exp i\omega t, \quad |\underline{J}_1| \ll J_0, \tag{4.6.8}$$

where $\underline{J}_1$ will be determined by introducing $\mathscr{E}(x, t)$ from equation (4.6.4) into equation (4.6.3), where $V(t)$ is given by equation (4.6.7). This is essentially the procedure used by Wright [11] (see also [23]). However, a great simplification occurs if the local field will be assumed to be also equal to the stationary field plus a small sinusoidal perturbation [46]–[48]

$$\mathscr{E}(x, t) = \mathscr{E}_0(x) + \underline{\mathscr{E}}_1(x) \exp i\omega t, \quad |\underline{\mathscr{E}}_1| \ll \mathscr{E}_0. \tag{4.6.9}$$

By introducing $J(t)$ and $\mathscr{E}(x, t)$ in equation (4.6.4) and separating the small alternating terms from the non-alternating terms two equations will be obtained ($N_D = 0$ for simplicity):

$$J_0 = \varepsilon\mu \, \mathscr{E}_0 \frac{d\mathscr{E}_0}{dx}, \tag{4.6.10}$$

$$\underline{J}_1 = \varepsilon\mu \, \mathscr{E}_0 \frac{d\underline{\mathscr{E}}_1}{dx} + \varepsilon\mu \, \underline{\mathscr{E}}_1 \frac{d\mathscr{E}_0}{dx} + i\omega\varepsilon\underline{\mathscr{E}}_1 \tag{4.6.11}$$

exp $(i\omega t)$ was simplified from the second equation thus eliminating the time t. The product of two small quantities was neglected. Therefore, we obtain the linear differential equation (4.6.11) in $\underline{\mathscr{E}}_1 = \underline{\mathscr{E}}_1(x)$, where $\mathscr{E}_0 = \mathscr{E}_0(x)$ is given by equation (4.6.10) and is exactly the steady-state field distribution (Section 4.2). Again the substitution of T_x for x as an independent variable (Section 4.5) will be made. We have

$$dx = v_0(x) \, dT_x = \mu\mathscr{E}_0 dT_x \tag{4.6.12}$$

and equations (4.6.10)–(4.6.12) yield

$$\underline{J}_1 = \varepsilon \frac{d\underline{\mathscr{E}}_1}{dT_x} + \left(\frac{\varepsilon}{J_0 T_x} + i\omega\varepsilon\right) \underline{\mathscr{E}}_1 \tag{4.6.13}$$

and this is a first-order linear equation which can be integrated by standard methods, and by also using the boundary condition

$$x = 0, \; T_x = 0, \; \underline{\mathscr{E}}_1 = 0 \tag{4.6.14}$$

which is the result of introducing (4.6.9) in equation (4.6.6). The result of integration, $\underline{\mathscr{E}}_1 = \underline{\mathscr{E}}_1(T_x, \underline{J}_1)$ will be introduced in

$$\underline{V}_1 = \int_0^L \underline{\mathscr{E}}_1 \, dx = \int_0^{T_L} \mu \, \mathscr{E}_0 \underline{\mathscr{E}}_1 \, dT_x = \frac{\mu J_0}{\varepsilon} \int_0^{T_L} T_x \underline{\mathscr{E}}_1(T_x) \, dT_x \tag{4.6.15}$$

Because $\mathscr{E}_1$ is proportional to $\mathbf{J}_1$, integration of equation (4.6.15) will yield the ratio $\mathbf{V}_1/\mathbf{J}_1$ which is the diode small-signal impedance (a complex quantity). Equation (4.6.15) shows that Z will depend upon the total carrier transit-time T_L. This is the steady-state transit time (Table 4.2.1). In our calculations no difference was made between carriers with respect to their emission moment and all have approximately the same transit-time, T_L. This approximation can be used only in a small-signal theory [31].

The impedance is (for unit-area device)

$$Z = \frac{\mathbf{V}_1}{\mathbf{J}_1} = \frac{\mu J_0 T_L^3}{\varepsilon^2} \frac{1}{(i\theta)^3} \left[1 - i\theta + \frac{(i\theta)^2}{2} - \exp(-i\theta) \right], \quad (4.6.16)$$

where

$$\theta = \omega T_L \quad (4.6.17)$$

is the *transit-angle* (a normalized angular frequency). By normalizing the impedance to its low-frequency ($\theta \to 0$) value, which is the incremental resistance R_i (Table 4.2.1) one obtains [11], [23], [31], [45], [48]–[50]

$$\bar{Z} = \frac{Z}{R_i} = \frac{6}{\theta^2}\left(1 - \frac{\sin\theta}{\theta}\right) - i\frac{3}{\theta}\left[1 - \frac{2}{\theta^2}(1 - \cos\theta)\right]. \quad (4.6.18)$$

The diode series resistance is positive and frequency-dependent. The diode reactance is negative (capacitive) at all frequencies. The equivalent circuit may be represented as a conductance in parallel with a capacitance, both frequency- and bias-dependent. Figure 4.6.1 shows (solid lines) the normalized conductance and capacitance as a function of θ. Note that at $\theta = 2\pi$ the signal period is just equal to the carrier transit-time. The frequency characteristics depend upon the steady-state voltage through R_i and T_L. The low-frequency equivalent circuit may be found by assuming small θ in equation (4.6.18) [23], [46]

$$Z|_{\theta \to 0} \simeq R_i\left(1 - \frac{i\theta}{4}\right), \quad \theta = \omega T_L. \quad (4.6.19)$$

Because the diode admittance is

$$Y \simeq \frac{1}{R_i}\left(1 + \frac{i\theta}{4}\right) \quad (4.6.20)$$

and (Table 4.2.1) $R_i C_0 = T_L/3$

$$Y \simeq \frac{1}{R_i} + i\omega\left(\frac{3}{4}C_0\right), \quad C_0 = \frac{\varepsilon}{L}. \quad (4.6.21)$$

Therefore, the low frequency dynamic capacitance C_{dyn} shunting the incremental resistance is three-quarters of the geometrical capacitance $C_0 = \varepsilon/L$ (all capacitances are calculated per unit of electrode area). Note the fact that this value is considerably smaller than the quasi-stationary capacitance

$$C_{\text{st}} = \frac{\partial Q}{\partial V} = \varepsilon\frac{\partial \mathscr{E}(L)}{\partial V} = \varepsilon\frac{\partial}{\partial V}\left(\frac{3}{2}\frac{V}{L}\right) = \frac{3}{2}C_0. \quad (4.6.22)$$

The difference between C_{st} and C_{dyn} was explained [51] as an inductive effect associated to carrier inertia (to the finite time required for carrier transport).

The value of C_{dyn} was well confirmed by experiments [23], [34], [38], [39], [52]. The frequency characteristics were first measured by Shao and Wright [23] for a CdS diode, although somewhat blurred by high-frequency losses. Measurements

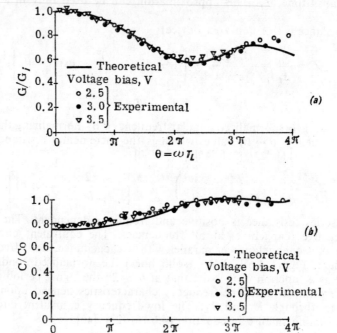

Figure 4.6.1. Normalized small-signal conductance (a) and capacitance (b) *versus* carrier transit angle θ. The solid line is the result of theoretical calculations of ideal SCLC square-law diodes (see equation (4.6.18)). The experimental data were obtained for several operating points of a square-law silicon diode. The devices used for measurements were specially constructed to provide very low (a few MHz) transit-time frequencies for a voltage bias in the square-law region ($p^+\nu p^+$ structures made from 110 μm thick, 40 kΩ cm silicon, the charge carriers are holes). (*after Dascalu*, [52]).

reported by Dascalu [52] for a square-law silicon diode at much lower frequencies (long transit-time due to large L) are indicated on Figure 4.6.1.

The striking difference between the SCLC solid-state diode and the planar vacuum diode (Table 4.2.1) is the fact that the latter exhibits a high frequency negative resistance whereas the former does not. The causes of this difference are somewhat complex [39], [53], [54] and require a detailed analysis (we refer the reader to chapter 7 of our book [31]).

Consider briefly the possibility of a high-frequency negative resistance in other SCLC solid-state diodes. The normalized frequency characteristics calculated for

$N_D \neq 0$ in equation (4.6.4) depend upon the parameter $\theta_r = \omega_r T_L$ used for steady-state calculations in Section 4.5. For large θ_r (low-current region on the $\Delta \to \infty$ curve in Figure 4.5.1) the frequency dependence of parallel conductance and capacitance is less important than indicated in Figure 4.6.1 and both remain constant (at $1/R_i$ and C_0, respectively) for $\theta_r \to \infty$, i.e. in the ohmic region [55]. The transit-time effects which explain frequency dependence such as in Figure 4.6.1, and may, in principle lead to a negative resistance, are annihilated by the dielectric relaxation of the injected charge ($\theta_r \gg 1$ means a dielectric relaxation time, ω_r^{-1}, much shorter than the carrier transit-time).

It seems that diffusion current in general has a detrimental effect upon a possible negative resistance effect [56].

Hot-carrier effects determine a more pronounced frequency dependence of device conductance, but the theory predicts a negative conductance only for negative-mobility devices [56]. This will be discussed later.

Trapping, surprisingly, has a favourable effect, although not in a direct way. Calculations [53], [57], [58] predicted a high-frequency resistance for a square-law 'dielectric' diode with high density 'slow' shallow traps (δ small, in Section 4.4). Deep trapping also favours, in a different way, a negative-resistance effect. This will be discussed when studying the punch-through diode, which is formally equivalent to an insulator SCLC diode with deep traps. It will be apparent later that the theory should introduce more appropriate boundary conditions than equation (4.6.6).

4.7 Noise in SCLC Diodes

Two broad categories of noise may be envisaged with respect to a one-dimensional solid-state diode region: injection noise and transport noise. The shot noise is a typical injection noise whereas the thermal noise belongs to the second category.

Robinson [59] studied both thermal and shot noise processes on the basis of the statistical mechanics and established a distinction between these two mechanisms of noise generation (sometimes confused in earlier literature). 'In a shot noise process the region in which the carriers interact with external fields is physically distinct from the region in which the statistical properties of these carriers are established. The carrier interacting with fields have no influence on, nor are influenced by processes in the same emitter region' [59]. In a thermal noise process, on the other hand, the 'interaction region is coincident with the generation region. During their interaction with external fields the carrier are in approximate thermal equilibrium with the lattice' [59].

The first account of noise in solid-state SCLC devices was given by Webb and Wright [60]. They applied Nyquist's theorem of thermal noise to the SCLC diode, as follows. The mean-square current fluctuation is given by

$$\overline{\Delta i^2} = 4k\,TG\,\Delta f, \qquad (4.7.1)$$

where Δf is the frequency band, such as the spectral intensity of the short-circuit noise current is

$$S_i(f) = 4k\,TG. \qquad (4.7.2)$$

Webb and Wright [60] assumed that the mobile carriers are in thermal equilibrium with the crystal lattice and T denotes the lattice temperature, and also postulated that the conductance G in equation (4.7.2) is the incremental conductance $G_i = dI/dV$. Van der Ziel has later shown [61] that due to the non-linear nature of the device the thermal noise hypothesis leads to a different formula. The open circuit thermal noise was calculated by evaluating a noise resistance R_n as the sum of noise resistances ΔR attributed to infinitesimal sections Δx of the one-dimensional diode, by assuming that the fluctuations in individual sections are independent

$$\overline{\Delta v^2} = 4kT\, R_n \Delta f = 4kT(\Sigma \Delta R)\, \Delta f = 4kT\, \frac{\Sigma \Delta V}{I}\, \Delta f = 4kT \left(\frac{V}{I}\right) \Delta f, \quad (4.7.3)$$

such that the spectral density of the open circuit noise voltage is

$$S_v(f) = 4kT \left(\frac{V}{I}\right) \quad (4.7.4)$$

and the spectral density of the short circuit noise current is

$$S_i(f) = 4kT \left(\frac{V}{I}\right) G_i^2 = 4kT \left(\frac{V}{I}\right)\left(\frac{dI}{dV}\right)^2. \quad (4.7.5)$$

For the square-law SCLC diode ($I \propto V^2$)

$$G_i = \frac{dI}{dV} = 2\cdot\frac{I}{V} \quad (4.7.6)$$

and equation (4.7.5) yields [61]

$$S_i(f) = 4kT \times 2G_i = 8\,kTG_i, \quad (4.7.7)$$

which differs from Webb and Wright's result (4.7.2) by a factor of two.

Equations (4.7.4) and (4.7.5) seem to be general because no reference was made to the structure and properties of the non-linear device. However, Nicolet [62] and van der Ziel [63] have later indicated that more detailed evaluation of thermal noise in non-linear solid-state devices invalidates equation (4.7.5), although it seems to be still valid for the square-law one-carrier SCLC diode (see below).

Another derivation of equation (4.7.7) was given by van der Ziel and van Vliet [64], on an entirely different basis. They started from the Langevin equation for electrons (positively charged)

$$J = e\mu_n n\mathscr{E} - eD_n \frac{\partial n}{\partial x} + H_1(x,t), \quad (4.7.8)$$

where $H_1(x,t)$ is the random source term describing the fluctuations. It was shown that no noise is associated with the drift component in equation (4.7.8), whereas the 'diffusion noise' in a section Δx can be attributed to carrier density fluctuation Δn_x in a section Δx such that

$$\overline{\Delta n_x^2} = \frac{4n(x)\,\Delta f \Delta x}{D_n}. \tag{4.7.9}$$

Because $\Delta j_x = -eD_n \Delta n_x/\Delta x$, the mean-square value of diffusion current at section x is

$$\overline{\Delta j_x^2} = \frac{4e^2 D_n n(x)\,\Delta f}{\Delta x} = 4kT \frac{e\mu_n n(x)}{\Delta x}\,\Delta f \tag{4.7.10}$$

(where the Einstein relation $eD_n = kT\mu_n$ was used), or

$$\overline{\Delta j_x^2} = \frac{4kT\,\Delta f}{\Delta R}, \tag{4.7.11}$$

where ΔR is the resistance of section Δx for a unit-area device. Therefore, equations (4.7.4)–(4.7.7) are again valid, because the diffusion noise source (uncorrelated) was written as a thermal noise source.

The Langevin method was also used to find the high-frequency thermal noise in an SCLC square-law diode. This was done by van der Ziel [65] and by Shumka [66] by using the thermal noise hypothesis for the conductivity associated with the drift term. By neglecting diffusion and including the displacement current, the Langevin equation becomes

$$J(t) = e\mu_n n(x,t)\,\mathscr{E}(x,t) + \varepsilon\frac{\partial \mathscr{E}(x,t)}{\partial t} + H_1(x,t). \tag{4.7.12}$$

All quantities in the above equation will be written as steady-state quantities plus small alternating current (a.c.) components. $H_1(x,t)$ has no steady-state part and its a.c. part is denoted by $\mathbf{H}_1(x,\omega)$. The calculations will be made as shown in Section 4.6 for the alternating current régime [31], [46], [48]. The thermal noise hypothesis

$$\langle \mathbf{H}_1(x_1,\omega)\mathbf{H}_1^*(x_2,\omega)\rangle = 4kTe\mu_n n_0(x_1)\,\delta(x_1 - x_2) \tag{4.7.13}$$

(where n_0 is the steady-state electron concentration and δ is the delta (Dirac) function) will be used. This equation indicates that the noise sources located at different distances x are uncorrelated.

The result is [65], [66]

$$S_v(\omega) = 4kT\,[2\,\mathrm{Re}\,Z(\omega)] = 4kTR_i\,\frac{3}{\theta^2}\left(1 - \frac{\sin\theta}{\theta}\right), \tag{4.7.14}$$

where $Z(\omega)$ is the small-signal diode impedance (4.6.18), R_i is the incremental resistance $R_i = 1/G_i$ and $\theta = \omega T_L$ is the transit angle. This formula predicts a considerable attenuation of the open-circuit noise above the transit-time frequency ($\theta > 2\pi$).

A large number of papers reported experimental results on noise in SCLC unipolar devices. The experimental proof of the validity of the above theoretical results was given recently by Golder, Nicolet and Shumka [67], [68]. They measured the noise of a silicon diode having the steady-state characteristic shown in Figure 4.7.1 [67]. The square-law region ($I \propto V^2$) is well described by the theoretical dependence (4.2.5). The results of noise measurements are reported in terms of the equiv-

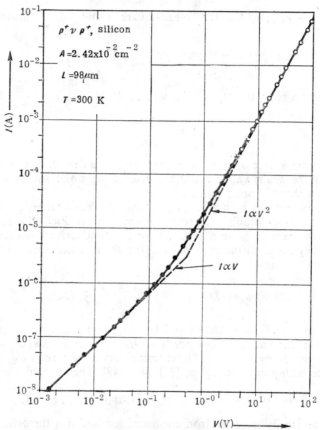

Figure 4.7.1. $I-V$ characteristic of a p^+vp^+ silicon sample used for noise measurements on SCL hole current. The solid line is the theoretical characteristic. Points marked by dots and by circles were measured with steady-state and pulsed current, respectively. The noise spectra shown in Figure 4.7.2 were measured for the operating points indicated by crosses. (*after Golder, Nicolet and Shumka,* [67]).

alent noise current I_{eq} whose shot noise $2eI_{eq}$ equals S_i, such as $I_{eq} = S_i/2e$. Figure 4.7.2 shows that the high frequency noise is indeed constant (white noise source) below the transit-time frequency (the low-frequency excess noise varies as $1/f$). Figure 4.7.3a indicates that I_{eq} is proportional to the incremental conductance G_i. For bias currents in the square-law region, equation (4.7.7) is quantitatively correct ($\pm 7\%$ error). Figure 4.7.3b shows that the more general equation (4.7.4)

SCLC Solid-state Devices

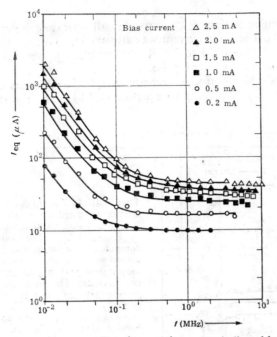

Figure 4.7.2. Noise spectra measured at six operating points indicated by crosses in Figure 4.7.1. The solid lines correspond to the least squares fit to the data. *(after Golder, Nicolet and Shumka, [67]).*

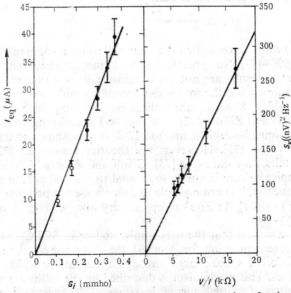

Figure 4.7.3. The white component measured at high frequencies of noise spectra shown in Figure 4.7.2, plotted against the incremental conductance G_i (a) and against V/I (b). The solid lines correspond to the theoretical dependence given by equation (4.7.7) and equation (4.7.4), respectively. *(after Golder, Nicolet and Shumka, [67]).*

is experimentally verified within ±6% error for all operating points from 0.2 to 2.5 mA (Figure 4.7.1). The noise current was also shown to increase almost linearly with the temperature T (from 133 to 300 K) [68] which is an obvious argument for the 'thermal' nature of the measured noise.

The same authors [68] reported a decrease of I_{eq} at frequencies comparable to the transit-time frequency, according to equation (4.7.14). These results are shown in Figure 4.7.4.

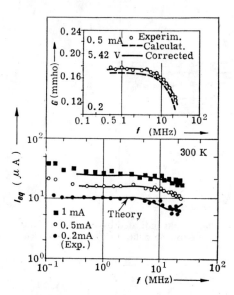

Figure 4.7.4. The high-frequency noise spectra measured for a few bias currents (the steady-state characteristic is indicated in Figure 4.7.1). The solid lines are computed theoretically ($I_{eq} \propto S_i \propto G(\omega)$), according to equation (4.7.14) ($G(\omega)$ is the small-signal high-frequency parallel conductance). The inset shows a comparison between $G(\omega)$ determined experimentally (circles) and theory (dashed curve) and also the theory corrected for a parallel ohmic conductance (solid line). (after Golder, Nicolet and Shumka, [68]).

A number of authors investigated the 'thermal' noise associated with hot carriers in SCLC diodes [69]—[73]. No agreement exists at this time between various theories and the experimental results are not yet conclusive. It was suggested, for example, that the temperature in the thermal noise formula should be the carrier effective temperature [69]—[71]. More detailed theories, including high-frequency effects, were given by Gisolf and Zijlstra [72] by using the Langevin method, and by Thornber [73] using the impedance-field method [74]. It was shown that the above two methods are equivalent [75]. However, the theories mentioned [72], [73] do not yield the same results. In contrast to [73], Gisolf and Zijlstra [72] have shown that the open circuit noise voltage is not proportional to the real part of the diode impedance and, consequently, it is not possible to define noise temperature T_n in a Nyquist type formula $S_v(f) = 4kT_n \, \text{Re} \, Z(\omega)$ except at very low ($\omega T_L \ll 1$) or at very high frequencies ($\omega T_L \gg 1$).

It is generally accepted that the mean noise source in SCLC solid-state devices is the thermal noise (except at low frequencies), the shot noise effect being suppressed (reduced) by the injected space charge, through a mechanism which is known from SCLC vacuum tubes. This mechanism is described by the following quotation from reference [76]. 'Bursts of emitted current in excess of the average depress the potential minimum, causing more electrons to be returned to the cathode, thereby reducing the amount of current passing the minimum. A small decrease in the emitted current has just the opposite effect. Every fluctuation in the emitted current has associated with it an almost-compensating fluctuation produced by the potential minimum. As a consequence, the noise current in the stream and that induced in the external circuit is only the small difference between the two fluctuations. The ratio of this residual noise to full-shot noise is called the shot noise reduction factor'. Van der Ziel [77] pointed out that the analogy between vacuum and solid-state SCLC devices is incomplete. In the vacuum diode the velocity distribution of emitted electrons is preserved during the transit, whereas in the solid-state diode the same distribution is destroyed by carrier collisions with the lattice. Therefore, vacuum diodes have convection current fluctuations plus velocity fluctuations at the potential minimum, whereas solid-state diodes can only exhibit convection current fluctuations [77]. The injected noise may be treated as a small fluctuation of the particle current at the cathode. If its mean-square value is denoted by $\overline{j_i^2}$ then the short circuit noise current is $\overline{j_i^2}$ times the shot noise reduction factor. This factor will be, however, equal to zero if idealized boundary conditions are assumed ($v(0) = 0$, $\mathscr{E}(0) = 0$, negligible diffusion). A total suppression of the injected noise will result. A more realistic boundary condition may be introduced by assuming a finite carrier concentration at the virtual cathode plane (defined by $\mathscr{E} = 0$). This implies that at the above plane the steady-state current is carried entirely by diffusion. However the diffusion current is neglected in calculating the small signal perturbations. This method was outlined in [31], by similarity to the shot noise treatment in SCLC vacuum devices [76]. For the square-law SCLC diode (by neglecting the height of the potential extremum at the virtual cathode and its distance from the actual cathode) one obtains the low-frequency shot noise reduction factor

$$\frac{\overline{J_1^2}_{\text{short circ.}}}{\overline{j_i^2}} = \left(\frac{J_0 T_L}{\mu e^2 p_0^2(0)}\right)^2 = \frac{9}{4}\left(\frac{\overline{n_0}}{n_0(0)}\right)^4, \qquad (4.7.15)$$

where $\overline{n_0}$ is the average value $L^{-1}\int_0^L n_0(x)dx$ of steady-state carrier concentration.

At transit-time frequencies, the shot noise effect decreases due to the transport phenomena, which reduce the coupling of the external circuit to the shot noise source at the cathode [31].

4.8 SCLC Triodes (Transistors)

Shockley [78] and Wright [79] suggested solid-state devices analogous to the SCLC vacuum triode. The problem of inserting a proper grid structure into the solid body seemed to be extremely difficult. An analog transistor which contains a burried grid was later fabricated by Zuleeg [80]. However, its operating principle differs considerably from that of the vacuum diode.

A 'true' solid-state triode was realized by Shumka [81]. This is a germanium device (Figure 4.8.1a), constructed by using mesa-etched isolation channels and an alloyed p^+ grid. The gate conductors are neither a source nor a sink for the injected electrons in the π region (about 10^{12} acceptors cm^{-3}). The gate provides an electrostatic control upon the carrier flow. The analogy to the vacuum tube triode is complete, because the output characteristics follow a $I \propto V^{3/2}$ law (Figure 4.8.1b). This is consistent with the assumption that $\mu \propto \mathscr{E}^{-1/2}$ at these field intensities. Clearly, Shumka's triode [81] is a laboratory device.

Among other SCLC solid-state triodes we quote here the SCL transistor with transversal injection suggested recently by Magdo [82].

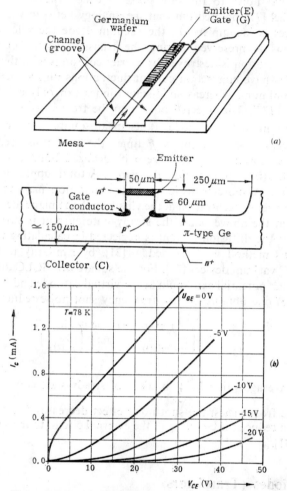

Figure 4.8.1. (a) Experimental germanium SCLC triode, (b) steady-state characteristics measured at 78 K. (after Shumka, [81]).

In all SCLC solid-state triodes mentioned above the mean current flows between grid regions. Other SCLC devices were suggested where the carrier flow passes through the control electrode (base). Structures using a metal base have been suggested [83]. Another proposal [84] was that of a heterojunction SCLC triode, later realized experimentally [85] as a CdS-silicon device. A similar structure, namely (intrinsic GaAs)−(p-type Ge)−(n-type Ge) (Figure 4.8.2) was discussed by Ladd and Feucht [86]. The emitter contact injects electrons into the intrinsic GaAs. The space-charge limited carrier flow (Fig. 4.8.2b) penetrates by diffusion through the high-conductivity p-layer, which acts as a control electrode (base), and reaches the collector by drift. The output (collector-emitter) characteristics are of the pentode-type, because the collector current is controlled by the base-emitter voltage and is almost independent on the collector bias.

The very high operating frequencies predicted [85], [87], [88] for the SCLC heterojunction triode will be probably deteriorated by the base series resistance [86] or by the slow carrier transport inside the base layer [89].

4.9 Conclusion

We have shown that a solid-state analogous to vacuum tube SCLC diodes and triodes can be constructed. Such devices would have, in principle, some advantages in the family of solid-state devices, namely low temperature dependence, good high-frequency response, and a parabolic steady-state characteristic. However, a practical version of the SCLC analog triode has not yet been realized. On the other hand, such a device would compete with the field-effect transistor which has been in production for many years.

The SCLC solid-state diode is impractical for power rectification (because of its high forward resistance). It may have advantages for low-power high-frequency applications such as detection, frequency conversion and analog multiplication (Section 4.2 and references [4]–[11]). A square-law characteristic can be obtained only for devices with a low doping-length product [31], [45]. This can be readily demonstrated as follows. The transition from ohmic conduction to SCLC square-law behaviour takes place around the crossover voltage V_c between the ohmic law $J = e\mu N\, V/L$ and the parabolic law (4.2.5). This voltage is

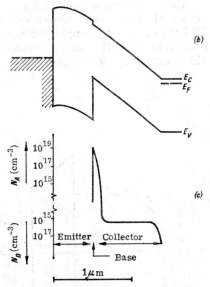

Figure 4.8.2. Energy band diagram at (a) thermal equilibrium, (b) for normal bias and (c) the doping profile for an SCLC heterojunction transistor. (after Ladd and Feucht, [86]).

$$V_c = \frac{8}{9}\frac{eNL^2}{\varepsilon} \approx \frac{eNL^2}{\varepsilon}, \qquad (4.9.1)$$

where N is the density of ionized impurities. On the other hand, the carrier mobility decreases considerably above a certain critical field $\mathscr{E}_c$, and we assume that the characteristic is modified by hot carrier effects above the bias $\mathscr{E}_c L$. This voltage must be very high as compared to V_c. Therefore the square-law region exists only if

$$NL \ll \frac{\varepsilon \mathscr{E}_c}{e}. \qquad (4.9.2)$$

For silicon, with $\varepsilon = 10^{-12}\,\text{F cm}^{-1}$ and $\mathscr{E}_c = 8\,\text{kV cm}^{-1}$, NL must be much lower than $5 \times 10^{10}\,\text{cm}^{-2}$. If the diode is 100 μm thick, the doping must be much lower than $5 \times 10^{12}\,\text{cm}^{-3}$, i.e. of the order of $1\ldots 2 \times 10^{11}\,\text{cm}^{-3}$, whereas the intrinsic concentration at room temperature is $1.45 \times 10^{10}\,\text{cm}^{-2}$. Therefore, silicon of very high resistivity (almost intrinsic) should be used.

Wide-gap semiconductors may be used to fabricate square-law diodes, but a high degree of crystalline perfection as well as techniques to provide almost ideal injecting contacts is required. However, SCL currents in insulators proved to be a useful tool in investigating lattice defects.

It is interesting to note that unipolar devices using injected space-charge, which do not have any counterpart into the vacuum-tube family does also exist. Figure 4.9.1 shows, as an example, the energy band diagram of the space-charge varactor [90]. The metal M1 injects electrons into the insulator layer 1, whereas the contact M2 is rectifying. The space charge injected into insulator 1 under forward bias

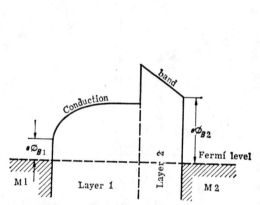

Figure 4.9.1. Energy band diagram for the space-charge varactor. (*after Howson, Owen and Wright,* [90]).

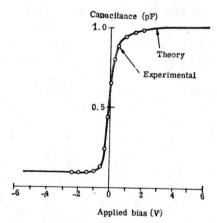

Figure 4.9.2. Capacitance *versus* voltage for an experimental space-charge varactor (silicon-silicon dioxide). (*after Howson, Owen and Wright,* [90]).

(M2 positive) accumulates at the interface and cannot be injected into insulator 2. The steady state current is zero and the device acts as a capacitor. For forward bias the device capacitance is practically equal to the capacitance of the insulator layer 2. Under reverse bias both insulator regions are free of injected mobile charge and the device capacitance is determined by the thickness and permittivity of both regions acting in series. Experimental devices were constructed by thermally growing silicon dioxide (layer 2) on epitaxial n silicon (layer 1) grown on an n^+ substrate (replacing M1). The capacitance-voltage dependence measured for such a varactor is shown in Figure 4.9.2. A capacity ratio as high as 9 to 1 with a relatively low transition range (2 V) was obtained [90]. Further experimental results together with an exact theoretical evaluation of device capacitance and loss resistance up to very high frequencies were later reported by Nigrin [91].

Although the analog SCLC devices proved to be, at least until now, of limited usefulness, the space-charge limitation of unipolar current is a very important concept and, in a certain sense, a fundamental phenomenon. Space-charge limited currents do indeed occur in many unipolar solid-state devices such as the transferred-electron amplifier, the junction-gate field-effect transistor, the metal-oxide-semiconductor (MOS) transistor (and MOS integrated circuits), and will be examined later in this book. These SCL currents were first observed in CdS light detectors around 1950 [3], and continue to occur in unexpected places, such as in chalcogenide-glass switches [92].

References

1. N. F. Mott and R. W. Gurney, 'Electronic processes in ionic crystals', Oxford, Clarendon Press, 1940.
2. G. T. Wright, 'Space-charge-limited solid-state devices', *Proc. IEEE*, **51**, 1642–1652 (1963).
3. M. A. Lampert and P. Mark, 'Current injection in solids', Academic Press, New York and London, 1970.
4. J. C. Cluley, 'Quarter-squares multiplier, high-speed analogue unit using square-law diodes', *Electron. Technol.*, **39**, 225–228 (1962).
5. J. H. Lepoff and A. M. Cowley, 'Improved intermodulation rejection in mixers', *IEEE Trans. Microwave Theory Techn.*, **MTT–14**, 618–623 (1966).
6. A. M. Cowley and H. O. Sorensen, 'Quantitative comparison of solid-state microwave detectors', *IEEE Trans. Microwave Theory Techn.*, **MTT–14**, 588–602 (1966).
7. J. G. Gardiner, 'Single-balanced modulators using square-law resistors (SCL diodes)', *Radio Electron. Eng.*, **37**, 305–314 (1969).
8. D. P. Howson and J. G. Gardiner, 'High-frequency mixers using square-law diodes', *Radio Electron. Eng.*, **36**, 311–316 (1968).
9. J. G. Gardiner, 'Cross-modulation and intermodulation distortions in the tuned square-law diode frequency converter', *Radio Electron. Eng.*, **37**, 353–363 (1969).
10. D. Dascalu, 'Square-law silicon punch-through diodes and their application as high-level parabolic detectors', *Rev. Roum. Phys.*, **17**, 401–404 (1972).
11. G. T. Wright, 'Transit time effects in the space-charge-limited silicon microwave diode', *Solid-St. Electron.*, **9**, 1–6 (1966).
12. W. Shockley and R. C. Prim, 'Space-charge limited emission in semiconductors', *Phys. Rev.*, **90**, 753–758 (1953).
13. G. T. Wright, 'Mechanisms of space-limited-current in solids', *Solid-St. Electron.*, **2**, 165–189 (1961).
14. E. I. Adirovichi, 'Electric field and currents in dielectrics', *Soviet Physics Solid-State*, **2**, 1282–1293 (1961); E. I. Adirovichi and L. A. Dubrovski, 'Dielectric electronics and the parabolic law of space-charge-limited currents' (in Russian), *USSR Sci. Acad. Reports*, **164**, 771 (1964).
15. N. Sinharay and B. Meltzer, 'Characteristics of insulator diodes determined by space-charge and diffusion', *Solid-St. Electron.*, **7**, 125–136 (1964).
16. D. J. Page, 'Some computed and measured characteristics of CdS space-charge-limited diodes', *Solid-St. Electron.*, **9**, 255–264 (1966).
17. A. Rose, 'Space-charge-limited currents in solids', *Phys. Rev.*, **97**, 1538–1544 (1955).
18. M. A. Lampert, 'Simplified theory of space-charge-limited currents in an insulator with traps', *Phys. Rev.*, **103**, 1648–1656 (1956).
19. M. A. Lampert, 'Injection currents in insulators', *Proc. IEEE*, **50**, 1781–1796 (1962).

20. M. A. LAMPERT. 'Volume-controlled current injection in insulators', *Rep. Progr. Phys.*, **27**, 329—367 (1964).
21. G. T. WRIGHT, 'Space-charge-limited current in insulating materials', *Nature (London)*, **182**, 1296—1297 (1958).
22. A. M. CONNING, A. A. KAYALI and G. T. WRIGHT, 'Space-charge-limited dielectric diodes', *Journ. IEE*, **5**, 595 (1959).
23. J. SHAO and G. T. WRIGHT, 'Characteristics of the space-charge-limited dielectric diode at very high frequencies', *Solid-St. Electron.*, **3**, 291—303 (1961).
24. M. A. LAMPERT and F. EDELMAN, 'Theory of one-carrier, space-charge-limited currents including diffusion and trapping', *J. Appl. Phys.*, **35**, 2971—2982 (1964).
25. W. G. TRODDEN, 'Space-charge-limited currents in cadmium sulphide with more than one discrete trapping level', *Brit. J. Appl. Phys.*, **18**, 401—404 (1967).
26. M. A. NICOLET, 'Unipolar space-charge-limited current in solids with nonuniform spacial distribution of shallow traps', *J. Appl. Phys.*, **37**, 4224—4235 (1966).
27. M. A. NICOLET, V. RODRIGUEZ and D. STOLFA, 'Unipolar interface-charge-limited current', *Surf. Sci.*, **10**, 146—164 (1968).
28. W. HWANG and K. C. KAO, 'A unified approach to the theory of current injection in solids with traps uniformly and non-uniformly distributed in space and in energy, and size effects in anthracene films', *Solid-St. Electron.*, **15**, 523—529 (1972).
29. E. HARNIK, 'Measurement of space-charge-limited current characteristics in high-resistivity materials', *J. Appl. Phys.*, **36**, 3850—3852 (1965).
30. A. MANY and G. RAKAVY, 'Theory of transient space-charge-limited currents in solids in the presence of trapping', *Phys. Rev.*, **126**, 1980—1988 (1962).
31. D. DASCALU, 'Transit-time effects in unipolar solid-state devices', Ed. Academiei, Bucureşti, Romania, and Abacus Press, Tunbridge Wells, Kent, 1974.
32. A. MANY, S. Z. WEISZ and M. SIMHONY, 'Space-charge-limited currents in iodine single crystals', *Phys. Rev.*, **126**, 1989—1995 (1962).
33. B. L. GREGORY and A. G. JORDAN, 'Experimental investigations of single injection in compensated silicon at low temperatures', *Phys. Rev.*, **134**, A1378—A1386 (1964); see also J. M. BROWN and A. G. JORDAN, 'Injection and transport of added carriers in silicon at liquid-helium temperatures', *J. Appl. Phys.*, **37**, 337—346 (1966).
34. U. BÜGET and G. T. WRIGHT, 'Space-charge-limited current in silicon', *Solid-St. Electron.*, **10**, 199—207 (1967).
35. A. VAN DER ZIEL, 'Normalized characteristic of $n-v-n$ devices', *Solid-St. Electron.*, **10**, 267—268 (1967).
36. B. L. GREGORY and A. G. JORDAN, 'Single-carrier injection in silicon at 76 and 300 K', *J. Appl. Phys.*, **35**, 3046—3047 (1964).
37. S. OKAZAKI and M. HIRAMATSU, 'Carrier density and mobility obtained from space-charge-limited current in *p*-type silicon', *Japan. J. Appl. Phys.*, **5**, 555—556 (1966); and 'Observations of space-charge-limited currents in *p*-type silicon', *Solid-St. Electron.*, **10**, 273—279 (1967).
38. O. J. MARSH and C. R. VISWANATHAN, 'Space-charge-limited current of holes in silicon and techniques for distinguishing double and single injection', *J. Appl. Phys.*, **38**, 3135—3144 (1967).
39. D. DASCALU, 'Space-charge-limited minority hole flow in very high resistivity silicon', *Rev. Roum. Phys.*, **17**, 675—708 (1972), and Ph. D. Thesis, Polytechnical Institute of Bucharest, 1970.
40. D. N. BOUGALIS and A. VAN DER ZIEL, 'Hot electron effects in single-injection silicon SCL diodes', *Solid-St. Electron.*, **14**, 265—272 (1971).
41. S. DENDA and M.-A. NICOLET, 'Pure-space-charge-limited electron current in silicon', *J. Appl. Phys.*, **37**, 2412—2424 (1966); V. RODRIGUEZ and M.-A. NICOLET, 'Drift velocity of electrons in silicon at high electric fields from 4.2 to 300 K', *J. Appl. Phys.*, **40**, 496—498 (1969).
42. V. RODRIGUEZ, H. RUEGG and M.-A. NICOLET, 'Measurement of the drift velocity of holes in silicon at high-field strengths', *IEEE Trans. Electron Dev.*, **ED—14**, 44—46 (1967).
43. T. E. SEIDEL and D. L. SCHARFETTER, 'Dependence of hole velocity upon electric field and hole density for *p*-type silicon', *J. Phys. Chem. Solids*, **28**, 2563—2574 (1967).

44. D. L. Scharfetter and T. E. Seidel, 'Analysis of the $I(V)$ characteristics of $p^+-n-\pi-p^+$ structures for the determination of hole velocity in silicon', *IEEE Trans. Electron Dev.*, **16**, 98–101 (1969).
45. D. Dascalu, 'Space-charge effects upon unipolar conduction in semiconductor regions', *J. Appl. Phys.*, **44**, 3609–3616 (1973).
46. M. Draganescu, 'Small-signal high frequency behaviour of the ideal dielectric diode' (in Russian), *Bull. Polytechn. Inst. Bucharest*, **27**, no. 5, 131–138 (1965).
47. D. E. McCumber and A. G. Chynoweth, 'Theory of negative-conductance amplification and of Gunn instabilities in 'two-valley' semiconductors', *IEEE Trans. Electron Dev.*, **ED–13**, 4–21 (1966).
48. A. Van Der Ziel and S. T. Hsu, 'High-frequency admittance of space-charge-limited solid-state diodes', *Proc. IEEE*, **54**, 1194 (1966).
49. W. Shockley, 'Negative resistance arising from transit-time in semiconductor diodes', *Bell Syst. Techn. J.*, **33**, 799–826 (1954), and 'High-frequency negative resistance device', U.S. Pat. 2794917, June 4, 1957.
50. H. Yoshimura, 'Space-charge-limited and emitter-current-limited injections in space charge region of semiconductors', *IEEE Trans. Electron Dev.*, **ED–11**, 412–422 (1964).
51. M. Draganescu and D. Dascalu, 'Detailed equivalent scheme for the dielectric diode at small signals', *Bull. Polytechn. Inst. Bucharest*, **27**, no. 5, 145–151 (1965).
52. D. Dascalu, 'Experimental evidence of transit-time effects in silicon punch-through diodes', *Electron. Lett.*, **5**, 196–197 (1969).
53. D. Dascalu, 'Small-signal theory of space-charge-limited diodes', *Int. J. Electron.*, **21**, 183–200 (1966).
54. D. Dascalu, 'Space-charge-waves and high frequency negative resistance of space-charge-limited diodes', *Int. J. Electron.*, **25**, 301–330 (1968).
55. D. Dascalu, M. Badila and N. Marin, 'Exact small signal theory of space-charge-limited majority carrier flow in semiconductors', *Rev. Roum. Phys.*, **15**, 1197–1199 (1970).
56. H. Kroemer, 'Detailed theory of the negative conductance of bulk negative mobility amplifiers, in the limit of zero ion density', *IEEE Trans. Electron Dev.*, **ED–14**, 476–492 (1967).
57. D. Dascalu, 'A high-frequency negative resistance in dielectric diodes with a high density of shallow traps', *Brit. J. Appl. Phys.*, **18**, 875–886 (1967).
58. D. Dascalu, 'Trapping and transit-time effects in high-frequency operation of space-charge-limited dielectric diodes. II. Frequency characteristics', *Solid-St. Electron.*, **11**, 391–400 (1968); also paper presented at 'Semiconductor Device Research Conference', Bad Nauheim (Germany), 17–21 April, 1967.
59. F. N. H. Robinson, 'Thermal noise, shot noise and statistical mechanics', *Int. J. Electron.*, **26**, 227–235 (1969).
60. P. W. Webb and G. T. Wright, 'The dielectric triode: a low-noise solid-state amplifier', *J. Brit. Inst. Radio Engrs.*, **23**, 111–112 (1962).
61. A. van der Ziel, 'Thermal noise in space-charge-limited diodes', *Solid-St. Electron.*, **9**, 899–900 (1966).
62. M.-A. Nicolet, 'Thermal noise in single and double injection devices', *Solid-St. Electron.*, **14**, 377–380 (1971).
63. A. van der Ziel, 'Nyquist's theorem for non-linear resistors', *Solid-St. Electron.*, **16**, 751–752 (1973).
64. A. van der Ziel and K. M. Van Vliet, 'H. F. thermal noise in space-charge limited solid state diodes – II', *Solid-St. Electron.*, **11**, 508–509 (1968).
65. A. van der Ziel, 'H. F. thermal noise in space-charge-limited solid-state diodes', *Solid-St. Electron.*, **9**, 1139–1140 (1966).
66. A. Shumka, 'Thermal noise in space-charge-limited solid-state diodes', *Solid-St. Electron.*, **13**, 751–754 (1970).
67. J. Golder, M.-A. Nicolet and A. Shumka, 'Thermal noise measurements on space-charge-limited hole current in silicon', *Solid-St. Electron.*, **16**, 581–585 (1973).
68. J. Golder, M.-A. Nicolet and A. Shumka, 'Noise of space-charge-limited current in solids is thermal', *Solid-St. Electron.*, **16**, 1151–1157 (1973).
69. D. N. Bougalis and A. van der Ziel, 'Hot electron effects in single injection silicon SCL diodes', *Solid-St. Electron.*, **14**, 265–272 (1971).
70. M.-A. Nicolet, H. R. Bilger and A. Shumka, 'Noise of hot holes in space-charge-limited germanium diodes', *Solid-St. Electron.*, **14**, 667–675 (1971).

71. A. ABDEL RAHMAN and A. VAN DER ZIEL, 'Hot electron effects in single-injection SCL diodes'. *Solid-St. Electron.*, **15**, 665–667 (1972).
72. A. G. GISOLF and R. J. J. ZIJLSTRA, 'Lattice interaction noise of hot carriers in single injection solid state diodes', *Solid-St. Electron.*, **16**, 571–580 (1973).
73. K. K. THORNBER. 'Some consequences of spatial correlation of noise calculations', *Solid-St. Electron.*, **17**, 95–97 (1974).
74. W. SHOCKLEY, J. A. COPELAND and R.P. JAMES, 'The impedance field method of noise calculation in active semiconductor devices', in 'Quantum theory of atoms, molecules and the solid state' (edit. P. O. Löwdin), pp. 537–563, Academic Press, New York (1966).
75. T. C. MCGILL, M.-A. NICOLET and K. K. THORNBER, 'Equivalence of the Langevin method and the impedance-field method of calculating noise in devices', *Solid-St. Electron.*, **17**, 107–108 (1974).
76. CH. K. BIRDSALL anf W. B. BRIDGES, 'Electron dynamics of diode regions', Academic Press, New York, 1966.
77. A. VAN DER ZIEL, 'Low frequency noise suppression in space-charge-limited solid state diodes', *Solid-St. Electron.*, **9**, 123–127 (1966).
78. W. SHOCKLEY, 'Transistor electronic imperfections, unipolar and analog transistors', *Proc. IRE*, **40**, 1289–1313 (1952).
79. G. T. WRIGHT, 'A proposed space-charge-limited dielectric triode', *J. Brit. Inst. Radio Engrs.*, **20**, 337–355 (1960).
80. R. ZULEEG, 'A silicon space-charge-limited triode and analog transistor', *Solid-St. Electron.*, **10**, 449–460 (1967), and 'Space-charge-limited current triode device', U.S. Patent, 3 409 812, Nov. 5, 1968.
81. A. SHUMKA, 'A germanium solid-state triode', *J. Appl. Phys.*, **40**, 438–439 (1969).
82. S. MAGDO, 'Theory and operation of SCL transistors with transverse injection', *IBM Journ. Res. Develop.*, **17**, 443–458 (1973).
83. M. M. ATALLA and R. W. SOSHEA, 'Hot-carrier triodes with thin film metal base', *Solid-St. Electron.*, **6**, 245–250 (1963).
84. G. T. WRIGHT, 'The space-charge-limited dielectric triode', *Solid-St. Electron.*, **5**, 117–126 (1962).
85. S. BROJDO, T. J. RILEY and G. T. WRIGHT, 'The heterojunction transistor and the space-charge-limited triode', *Brit. J. Appl. Phys.*, **16**, 133–136 (1965).
86. G. O. LADD, Jr. and D. L. FEUCHT, 'Performance potential of high frequency heterojunction transistors', *IEEE Trans. Electron. Dev.*, **ED–17**, 413–420 (1970).
87. S. BROJDO, 'Characteristics of the dielectric diode and triode at very high frequencies', *Solid-St. Electron.*, **6**, 611–629 (1963).
88. M. T. BRIAN, 'High frequency small-signal characteristics of the SCL heterojunction transistors', *Solid-St. Electron.*, **17**, 47–59 (1974).
89. M. T. BRIAN, personal communication (The University of Birmingham, England, 1969).
90. D. P. HOWSON, B. OWEN and G. T. WRIGHT, 'The space-charge varactor', *Solid-St. Electron.*, **8**, 913–921 (1965).
91. J. NIGRIN, 'Exact small-signal study on space-charge varactor', *Solid-St. Electron.*, **13**, 1267–1281 (1970).
92. J. ALLISON and V. R. DAWE, 'Interpretation of the preswitching behaviour of chalcogenide-glass switches in terms of a space-charge-injection mechanism', *Electron. Lett.*, **8**, 437–438 (1972).

Problems

4.1. Estimate the order of magnitude of the minimum transit-time which can be achieved in solid-state and vacuum SCL current diodes.

4.2. Show that the results for both devices presented in Table 4.2.1 can be obtained as particular cases of the same theory, when a generalized motion equation is used.

4.3. Consider an SCLC insulator diode with two trap levels. What kind of current-voltage characteristic may such a diode have? How can we obtain quantitative information about trapping levels from steady-state measurements?
Hint: See Figure 4.4.3.

4.4. Estimate the error which will be introduced by approximating the exact $J-V$ characteristic for SCLC régime obtained by introducing $\varDelta \to \infty$ in equations (4.5.8) and (4.5.9) (see Figure 4.5.1), if the device current is calculated by simply adding the ohmic current to the ideal SCL current (4.2.5) at any voltage.

4.5. Calculate the electric field $\mathscr{E} = \mathscr{E}(T_x)$ from equation (4.5.6) with equation (4.5.7) as boundary condition and discuss the physical situation inside the diode for various values of parameters θ_r and $\varDelta$, defined by equation (4.5.10).

4.6. By using equations (4.5.8) and (4.5.9) find the condition to be satisfied by N_D^+, N_D and L such as to provide a square-law region for the SCLC n^+nn^+ resistor.

4.7. Calculate the SCLC $J-V$ characteristic by assuming uniform doping and by approximating the carrier velocity by a constant (saturated velocity). Finally, discuss the validity of this result, with numerical data appropriate for silicon.

4.8. Find the small-signal impedance and the low frequency equivalent circuit for the SCLC semiconductor diode with the velocity-field dependence and the steady-state characteristic shown in Figure 4.5.2.

4.9. Discuss the possibility of constructing an SCLC semiconductor diode with very low transit-time when biased in the square-law region.
Hint: The diode should be constructed from a high resistivity semiconductor.

5

Punch-through Semiconductor Diodes. Microwave BARITT Diodes

The first device to be considered in this chapter is the floating base p^+np^+ transistor structure operating above the punch-through voltage, V_{PT}, but below avalanche breakdown. If the n-base doping is low (low V_{PT}) the hole current will be limited by the mobile space-charge. This is the low-field punch-through diode.

In a punch-through diode with the central region of moderate doping (V_{PT} of tens of volts) the mobile space-charge of minority carriers is negligible, whereas the carrier velocity is saturated because the electric field is quite high. The current density is injection-limited by the small barrier at the forward biased junction and increases steeply with the applied voltage, $V > V_{PT}$. The high-field punch-through diode exhibits a microwave negative resistance due to transit-time effects.

The operation of the third device is similar to that of the second one and it can be also used for microwave generation and amplification: this device is a metal-semiconductor-metal structure with Schottky-type contacts. When biased above the reach-through (or punch-through) voltage, it conducts a minority carrier flow carried by drift in the completely depleted semiconductor.

The current is thermionically emitted by the forward-biased Schottky contact. The term BARITT diode used for this device (and sometimes for its p^+np^+ counterpart) is the acronym for BARrier Injection Transit Time.

5.1 Space-charge-limited Current in a Punch-through p^+np^+ (or n^+pn^+) Structure

Shockley and Prim have shown theoretically [1] that space-charge-limited currents (SCLC) can occur in p^+np^+ (or n^+pn^+) structures biased above the punch-through voltage. Consider a p^+np^+ structure biased as a floating base pnp transistor (Figure 5.1.1). As the applied voltage increases, the depletion region at the collector (reverse-biased) junction widens and reduces the width of the neutral base. The current flowing in the external circuit is ($I_B = 0$)

$$I_C = I_E = \frac{I_{CBO}}{1 - \alpha_0} \qquad (5.1.1)$$

(where I_{CBO} is the saturation current of the collector junction) and increases somewhat because the amplification factor α_0 increases: the neutral base becomes narrower, recombination decreases and the transport factor increases. The current remains, however, very small [2].

As the applied voltage increases, at a certain reach-through voltage V_{RT} the two depletion regions eventually reach each other and the structure is 'punched-through'. Above the reach-through voltage, $V > V_{RT}$, the device behaviour changes completely. Most holes diffuse from the emitter over the potential barrier (Figure 5.1.1b) and are now injected *directly* into the collector depletion region, where they are accelerated by a high electric field towards the high-conductivity p^+ collector. This hole current increases rapidly with the decrease of the potential barrier and determines a steep increase of the current-voltage characteristic. The other components of the total current are indeed, negligible. The electrons are swept-out rapidly from the depleted region; thus only hole conduction

5. BARITT Diodes

is possible, and only from emitter towards collector, first by diffusion over a small barrier and then by drift. The floating-base transistor has become, therefore, a unipolar semiconductor diode. Because hole conduction takes place mainly in an n-type region, in that sense the device is a minority-

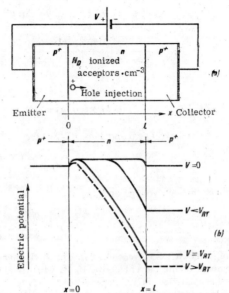

Figure 5.1.1. (a) Punch-through p^+np^+ structure biased as a floating-base transistor; (b) potential distribution inside the structure, the neutral base disappears at $V = V_{RT}$ (the reach-through voltage).

carrier diode, although in the absence of recombination or charge storage effects its internal behaviour and electronic properties are completely different from what we expect from a usual minority-carrier device (like a $p-n$ junction or a bipolar transistor).

The hole current injected over the potential barrier at an abrupt p^+n junction may be considered a thermionically emitted current. The emission process is somewhat similar to carrier emission into a semiconductor at a metal-semiconductor contact and is discussed in detail by Persky [3]. However, if the barrier becomes relatively small and the carrier concentration at the barrier is very high as compared to ionized impurity concentration such that the emitter acts as an ideal injecting contact, then the space-charge-limited current concept may be applied and leads to the following steady-state model (see Chapter 4)

$$J = ep v(\mathscr{E}) = ep\mu \mathscr{E} \qquad \text{(diffusion neglected, constant mobility)} \qquad (5.1.2)$$

$$\frac{d\mathscr{E}}{dx} = \frac{e(p + N_D)}{\varepsilon} \qquad \text{(space-charge created by mobile holes plus ionized donors)} \qquad (5.1.3)$$

$$V = \int_0^L \mathscr{E} dx \qquad = \text{applied bias} \qquad (5.1.4)$$

$$\mathscr{E}(0) = 0 \qquad \text{(SCLC boundary condition)} \qquad (5.1.5)$$

The top of the potential barrier defined by $\mathscr{E} = 0$ is located just at the emitter junction ($x = 0$) and its height is neglected as compared to the applied voltage V. These equations can be readily integrated by using the procedure indicated in Section 4.5. In fact, the result may be obtained directly by introducing in equations (4.5.8)—(4.5.10) $\Delta \to \infty$ (SCL injection) and by replacing N_D with $-N_D$

and θ_r with $-\theta'_r$, thus obtaining the parametric equations of the normalized $J-V$ characteristic

$$\frac{J}{J_{PT}} = \frac{32}{9} \frac{1}{\exp \theta'_r - \theta'_r - 1} \qquad (5.1.6)$$

$$\frac{V}{V_{PT}} = \frac{2[\theta'_r + 2(1 - \exp \theta'_r) + (\exp 2\theta'_r - 1)/2]}{[\exp \theta'_r - \theta'_r - 1]^2}, \qquad (5.1.7)$$

where

$$\theta'_r = \omega'_r T_L, \qquad \omega'_r = \frac{eN_D\mu_n}{\varepsilon} > 0. \qquad (5.1.8)$$

For $\theta'_r \to \infty$, we have $J \to 0$ and $V \to V_{PT}$, where

$$V_{PT} = \frac{eN_D L^2}{2\varepsilon} \qquad (5.1.9)$$

is the punch-through voltage and J_{PT} is

$$J_{PT} = \frac{9}{8} \varepsilon \mu_p \frac{V_{PT}^2}{L^3} \qquad (5.1.10)$$

(compare to the SCLC current (4.2.5)). The $J-V$ characteristic plotted in Figure 5.1.2 indicates a rapid increase of the current above V_{PT}*, whereas at very high current levels ($J \to \infty$, $T_L \to 0$, $\theta'_r \to 0$) the square-law SCLC dependence (4.2.5) is obtained. This characteristic was first derived by Shockley and Prim [1]. As shown above, the *entire* theoretical characteristic was obtained by assuming an ideal boundary condition corresponding to the SCLC régime. However the current limitation becomes effective (the slope of the $J-V$ characteristics becomes finite and decreases) as the current increases sufficiently to provide a large mobile injected charge. At very high current densities the density N_D of ionized donors becomes negligibly small and the current is limited by the mobile injected charge (square-law dependence, $J \propto V^2$).

The theoretical characteristic shown in Figure 5.1.2 is identical to the characteristic of SCL-current in an insulator with deep (completely filled) traps (equations (4.4.8)—(4.4.11) and the TRAP FILLING curve in Figure 4.4.2), where the current cannot flow until all trapping centres are not filled, at the applied voltage $V_{TFL} = \dfrac{eN_t L^2}{2\varepsilon}$ which is identical to V_{PT}, N_D being replaced by N_t.

The abrupt increase of current indicating the punch-through was indeed observed experimentally [2], [5], but the high-current square-law region was modified by hot carrier effects. Figure 5.1.3 shows a comparison between theory and experiment for $p^+ v p^+$ silicon structures at liquid nitrogen temperature. The dashed curve is the theoretical curve computed by including the hole-mobility field dependence (see equations (4.5.12)—(4.5.14) and Figure 4.5.2). The punch-through type behaviour was observed, in general, at low temperatures. At higher temperatures the conductivity of the high-resistivity central region and/or the injected carriers have an important effect. Figure 5.1.4 shows, for example, $J-V$ characteristics measured at various temperatures. The $T = 77$ K curve indicates a somewhat steep rise of characteristics around the theoretical punch-through voltage V_{PT}, whereas the room temperature characteristic resembles the $J-V$

---

* The punch-through voltage V_{PT} may be obtained by integrating the Poisson equation with zero hole density, from $x = 0$ to $x = L$. V_{PT} is larger than and, in many practical cases, almost equal to the reach-through voltage V_{RT} (if the emitter barrier height at punch-through may be neglected), see also reference [4].

5. BARITT Diodes

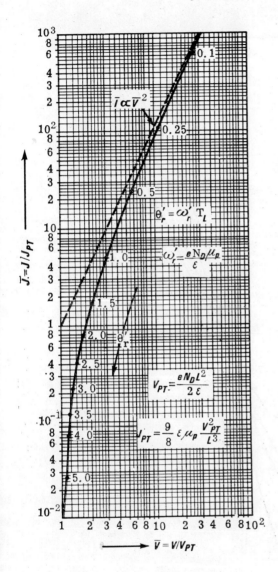

Figure 5.1.2. Normalized $J-V$ characteristic of a punch-through structure, according to the theory first given by Shockley and Prim [1].

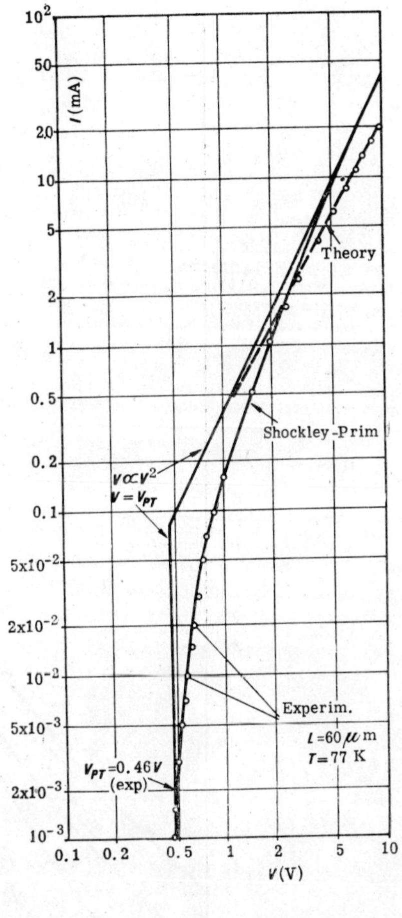

Figure 5.1.3. Comparison between the theoretical steady-state characteristic and experiment for a p^+np^+ silicon structure at liquid nitrogen temperature. The device is constructed from 38 600 Ωcm n-type silicon with alloyed Al contacts (p^+ regions). The measured hole current is indicated by circles, the Shockley-Prim theory (see Figure 5.1.2) is given by the solid line, and the high-field theoretical characteristic by the dashed curve (see Figure 4.5.2). *(after Dascalu, [5]).*

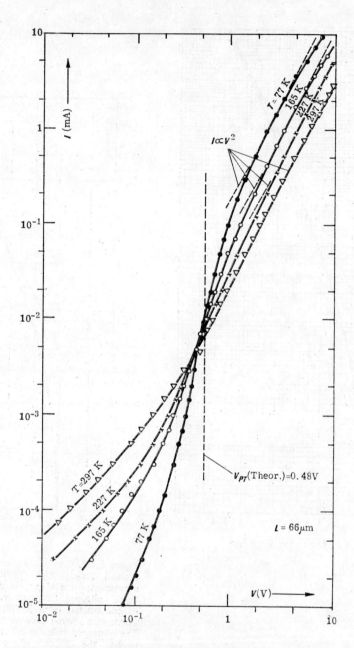

Figure 5.1.4. Temperature dependence of the steady-state characteristic of a p^+np^+ structure made from very high resistivity silicon. (*after Dascalu*, [5]).

characteristic of an SCLC resistor* (majority carrier SCL current). The linear low-current region on characteristics shown in Figure 5.1.5 does indeed indicate an ohmic-type behaviour. This behaviour could be attributed to the intrinsic conductivity of the central region. Note the fact that the

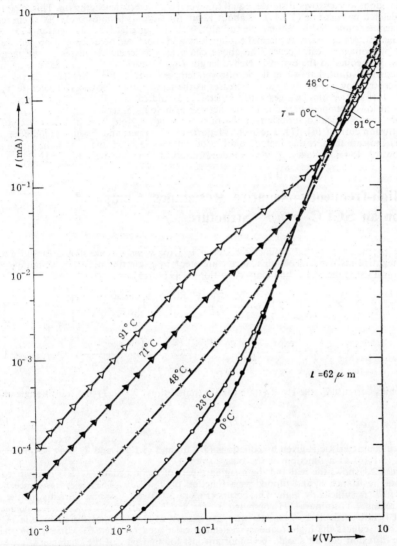

Figure 5.1.5. Temperature dependence of the $J-V$ characteristic for a high-resistivity silicon p^+np^+ structure. (*after Dascalu*, [5]).

intrinsic carrier density at 91 °C (Figure 5.1.5) is 9×10^{11} cm^{-3}, whereas the donor density in the central region is only 1.2×10^{11} cm^{-3}. It was shown [5] that the low-current conductivity varies with temperature (above 50 °C) approximately as the intrinsic conductivity does.

---

* See also Figure 4.7.1.

Below room temperature the intrinsic carrier concentration is negligible and the excess current in the punch-through region may be explained as follows [5]. The punch-through is defined by considering total depletion of barrier regions. However, the transition from neutral to totally depleted regions occurs over distances equal to several characteristic lengths L'_D. This characteristic length, defined by equation (2.3.17), is about 12 μm for n-type silicon used (38,500 Ω cm resistivity at room temperature) and is comparable to half the device length ($L/2$ is 25—40 μm). Therefore no distinct neutral region exists at thermal equilibrium and room temperature and no definite punch-through phenomenon can occur. The above characteristic length decreases as the temperature decreases, it is equal to the extrinsic Debye length $L_{DE} \simeq (\varepsilon kT/e^2 N_D)^{1/2}$ and becomes about 6 μm, for the resistivity quoted above, at liquid nitrogen temperature. On the other hand, carrier injection into the central low-resistivity region decreases as the temperature decreases. Therefore, the shape of the $J-V$ characteristics changes with temperature as indicated in Figure 5.1.4 (the SCL current increases at lower temperatures due to the increase of hole mobility).

A correct theoretical description of the punch-through region must include the contribution of the diffusion current [6], [7]. The results reported by Lampert and Edelman (computer calculations) [6] indicate the fact that including diffusion softens the slope and moves the punch-through region towards lower voltages. This was indeed observed experimentally [2], [5], [7].

5.2 High-frequency Negative Resistance Computed for an SCLC p^+np^+ Structure

Shockley [8] has shown that a punch-through p^+np^+ (or n^+pn^+) diode should exhibit a high-frequency negative resistance. The impedance computed* by neglecting diffusion, assuming constant mobility and using the SCLC boundary conditions is [9]—[11]

$$Z = \frac{J_0}{eN_D\varepsilon} T_L^2 \left[Q(i\theta - \theta'_r) + \frac{R(i\theta - \theta'_r) - R(-\theta'_r)}{i\theta} \right], \quad (5.2.1)$$

where

$$Q(\eta) = \frac{\exp(-\eta) + \eta - 1}{\eta^2}, \quad R(\eta) = \frac{1 - \exp(-\eta)}{\eta}, \quad (5.2.2)$$

J_0 is the bias current, T_L the steady-state carrier transit-time, $\theta = \omega T_L$ the transit angle and

$$\theta'_r = \omega'_r T_L, \quad \omega'_r = \frac{\varepsilon}{eN_D\mu_p}. \quad (5.2.3)$$

The device characteristic is given by equations (5.1.16) and (5.1.17) and T_L may be obtained from equation (5.1.16) as a function of the bias current.

Figure 5.2.1 shows the real and the imaginary part of the impedance (5.2.1) normalized to the incremental resistance R_i and plotted as a function of $\theta = \omega T_L$. The series resistance becomes negative if θ'_r is sufficiently high. This occurs (Figure 5.1.2) for low bias currents, in the punch-through region. Yoshimura [9] computed a very small negative resistance and concluded that the effect is not particularly useful. It was later shown [12], [13] that the decreasing of carrier mobility still reduces the high-frequency resistance calculated for the SCL current. On the other hand, the diffusion current should be important just for that part of the characteristic where the small-signal negative resistance was computed.

Although the negative resistance predicted by equation (5.2.1) is, by itself, less important at this time, we shall examine below the processes which explain this effect. Figure 5.2.1 indicates that the negative resistance occurs around the transit time frequency $f_T = 1/T_L$ ($\theta = 2\pi$). The transit-time effects can be examined in detail by using the method indicated by the author [11], [13], [14]. This is a quantitative analysis based upon the decomposition of the total alternating current.

---

* The method is outlined in Section 4.6.

5. BARITT Diodes

Equation (4.6.11) shows that the alternating current $\mathbf{J}_1$ may be expressed as a sum of three components:

$$\mathbf{J}_1 = ev_0\mathbf{p}_1 + ep_0\mathbf{v}_1 + i\omega\varepsilon\mathscr{E}_1. \tag{5.2.4}$$

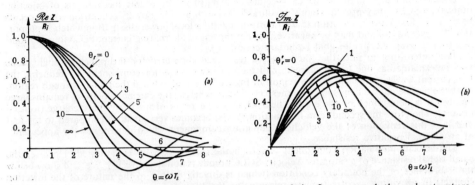

Figure 5.2.1. Theoretical small-signal frequency characteristic for a punch-through structure, where R_i is the incremental resistance, θ the transit angle, and θ'_r a parameter indicating the operating point (see Figure 5.1.2): (a) series resistance; (b) series reactance (capacitive). (*after Yoshimura,* [9]).

The first component is due to the charge-density modulation, the second due to velocity modulation and the third is the displacement current. The first two terms result from the alternating perturbation of the steady-state (particle) current $J = epv$. The first component corresponds to the modulation of the density of carriers moving with the steady-state velocity v_0. The second component is due to the velocity modulation of charge carriers injected by the steady-state current. It should be stressed that the second component is ohmic in nature (is proportional to $\mathscr{E}_1$) whereas the first is associated to the space charge, because

$$e\mathbf{p}_1 = \varepsilon \frac{d\mathscr{E}_1}{dx} \tag{5.2.5}$$

(obtained from the Poisson equation (5.1.3)) is non-zero only when the alternating field is non-uniform.

The negative resistance will be explained if the *internal* sources of alternating power generation are found. Only the first two components may dissipate or generate power. As far as the differential mobility $dv/d\mathscr{E}$ is positive, the velocity modulation component $ep_0\mathbf{v}_1 = ep_0\left(\dfrac{dv}{d\mathscr{E}}\right)\mathscr{E}_1$ dissipates power, because it is in-phase with the local electric field $\mathscr{E}_1$. On the other hand, the charge-density modulation component may either *dissipate or generate* power according to whether ep_1 is in-phase or in anti-phase with the alternating electric field $\mathscr{E}_1$. By integrating the Poisson equation (5.2.5) one obtains

$$\mathscr{E}_1(x) = \mathscr{E}_1(0) + \frac{1}{\varepsilon}\int_0^x e\mathbf{p}_1(x)\,dx. \tag{5.2.6}$$

For SCL injection $\mathscr{E}_1(0) = 0$ and $\mathscr{E}_1(x)$ is proportional to the total alternating-current charge contained between planes $x = 0$ and x. Due to the finite time required for charge propagation, $\mathscr{E}_1(x)$ and $\mathbf{p}_1(x)$ are not in-phase.

There are at least three important conclusions to the above discussion:

(a) the velocity-modulation component should be reduced to minimum (in positive-mobility diodes);

(b) the density-modulation component generates alternating-current power only in a certain part of the device*, this mechanism is frequency-dependent and becomes less efficient as the frequency increases [11] because the device is divided into a number of successive regions which generate and dissipate power, respectively;

(c) the boundary condition which specifies that $\mathscr{E}_1(0)$ is important: mechanisms of injection other than SCLC may provide more favourable conditions for power generation (this occurs because the first term in equation (5.2.6) corresponds to a local power dissipation which is proportional to $p_1(x)\,\mathscr{E}_1(0)$ and may be negative; in fact the entire alternating current behaviour is modified by changing the small-signal boundary condition).

It is interesting to note that the negative resistance computed for the punch-through diode may be explained as due to the supression of the velocity modulation component. By reducing the bias current we leave the square-law region in Figure 5.1.2 and enter in the punch-through region. Here the mobile charge density p_0 becomes very small, whereas the carrier velocity v_0 remains large because it is determined by the electric field which is the result of the fixed space-charge (ionized donors). Therefore the second term in equation (5.2.4) becomes very small as compared to the first and it is this reduction of the velocity modulation component which determines the appearance of an external negative resistance.

The velocity-modulation component is zero if the carrier velocity is saturated. However the small-signal resistance of a saturated-velocity SCLC† diode is non-negative [9]. We will show below the importance of the boundary condition (which is directly related to the nature of the injecting contact) in providing a correct phase relation for the "injected space-charge wave" [11], [13], [14] associated to the carrier-density modulation component, and thus providing an internal power generation. This observation [15] was essential in developing a suitable small-signal theory for experimental punch-through transit-time diodes.

5.3 Small-signal Behaviour of Punch-through BARITT Diodes

Microwave oscillations were recently obtained from punch-through silicon diodes [16]–[26]. Useful amounts of continuous microwave power (tens of mW) at oscillation frequencies of several GHz were achieved.

There are differences between the original structure suggested by Shockley [8] and these experimental devices. The central region of these diodes is an epitaxial layer of moderate resistivity (several ohm centimetres). Although the thickness of this region is low, the punch-through voltage is relatively high (due to higher impurity concentration), namely tens of volts. The average electric field inside the structure is, therefore, sufficiently high to provide carrier velocity saturation over most of the central region and short transit-times.

Other differences in comparison to the ideal model are: the *p-n* junctions are not abrupt and the mobile carrier concentration at the injecting contact, although high, cannot be assumed infinite and the current cannot be considered as fully space-charge limited, as will be discussed later.

Typical data for experimental devices are given below. Sultan and Wright [18] used *pnp* diodes made by boron diffusion into the *n*-type epitaxial layer grown on a *p*-type substrate. The epilayer was 6 μm thick with resistivity 4 Ω cm. The junction depth was 1 μm. The *p*-type surface was contacted with an evaporated gold

---

* For example, in SCLC diodes ($\mathscr{E}_1(0) = 0$), this component always dissipates power in the vicinity of the injecting contact, where the delay obtained by charge transport is insufficient.

† The carrier velocity cannot be saturated in the vicinity of the injecting contact, where the electric field intensity is very small. This region is assumed to be very thin.

layer. Circular contacts were defined by photolithography and mesa structures were obtained by chemical etch through the epitaxial layer. Devices were separated by scribing and mounted in varactor-type-packages*. Steady-state $J-V$ characteristics exhibiting a very sharp punch-through at about 23 V were obtained (these characteristics were symmetrical within 5% for either polarity of applied voltage) [18]. Figure 5.3.1 shows the bias dependence of CW (continuous-wave) microwave power, at several oscillating frequencies, for such a diode [18]. An optimum oscillation frequency and an optimum current density are apparent.

The experimental results indicate a small-signal negative resistance, therefore self-starting oscillator and small-signal amplifiers can be, in principle, constructed with these diodes. Because the SCL current model is no longer applicable, a new theory is necessary. A correct description should incorporate the diffusion current and the hot-carrier effects and the analysis requires a computer. We reproduce in Figure 5.3.2, after Sjölund [27] results of such a numerical analysis, in order to clarify the conduction processes in the vicinity of the injecting contact (forward-biased p-n junction). At low bias-current densities the electric field profile is practically linear and its slope is determined mainly by the doping density. This slope increases at higher current densities (200 A cm^{-2} in Figure 5.3.2a), thus indicating the effect of mobile injected charge.

In Figure 5.3.2b, the hole current is separated into its drift and diffusion components. The potential minimum is taken as the co-ordinate origin: here the diffusion current is maximum. To the left, inside the p^+ region, the hole current is constant†. The diffusion current is important in the vicinity of the potential maximum ($x = 0$) and negligible in the rest of the crystal.

The diode current may be considered to be controlled by the potential barrier at the forward-biased junction. At relatively low current densities the mobile charge is small and the magnitude of the barrier depends directly upon the applied bias, whereas the current strongly depends upon the barrier height. In that sense the current is *barrier-controlled*. This concept is especially useful in deriving the small-signal impedance, as shown below.

Wright and Sultan [25] suggested an approximate small-signal theory based upon the fact that diffusion is important only in the vicinity of the potential maximum and this maximum is situated very close to the emitting junction. They solved numerically the continuity equation and the Poisson equation for hole injection into a semi-infinite n-type semiconductor of given resistivity. At $x = 0$ the electric field is forced to be zero, the entire current is diffusion current and has a given value, J_0. The hole mobility μ_p is assumed constant because near the barrier the electric field is small. The steady-state current density may be written [25]

$$J_0 = e\mu_p p_0 \mathscr{E}_0 - eD_p \frac{dp_0}{dx} = -e\mu_p p_0 \frac{d\zeta_p}{dx}, \qquad (5.3.1)$$

---

* For other details on microwave diodes in mesa construction see also Chapters 6—8.

† The boundary condition for the iterative integration of the continuity equation and the Poisson equation is: the hole current inside the emitter is carried by drift and the hole density is constant and given by the semiconductor doping.

where ζ_p is the quasi-Fermi level for holes. Here, the quantity $-\mu_p d\zeta_p/dx$ represents the carrier velocity (not to be confused with the drift velocity). Wright and Sultan

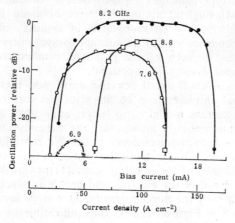

Figure 5.3.1. High-frequency continuous wave power generated by a silicon punch-through diode (see the text) *versus* the bias current, for several oscillation frequencies. (after Sultan and Wright, [18]).

[25] defined the *virtual source* as the plane $x = 0$ and the *injection velocity* v_{0s} as the carrier velocity at the source plane

$$v_{0s} = \frac{J_0}{ep_0(0)} = -\mu_p \frac{d\zeta_p}{dx}\bigg|_{x=0} \qquad (5.3.2)$$

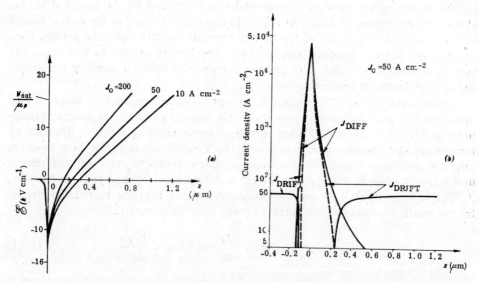

Figure 5.3.2. (a) Field distribution in the vicinity of the emitter junction, computed for a few bias-current densities; (b) drift and diffusion components of the total hole current for $J_0 = 50$ A cm^{-2}. These are the results of a numerical analysis of a silicon p^+np^+ structure with abrupt junctions (1.2 × $\times 10^{15}$ cm^{-3} doping and 7.9 μm thickness of the *n*-region). (after Sjölund, [27]).

5. BARITT Diodes

(note the fact that, by definition, the *drift* velocity at the source plane is zero). We reproduce in Figure 5.3.3 the injection velocity computed [25] for hole injection in *n*-type silicon at room temperature, and represented *versus* current density, with resistivity as parameter.

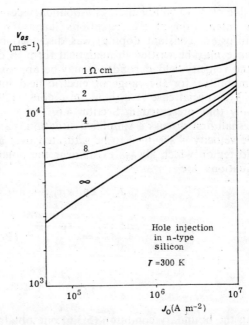

Figure 5.3.3. Injection velocity v_{0s} *versus* bias-current density for hole injection in *n*-type silicon, at 300 K, with silicon resistivity as parameter.(after Wright and Sultan, [25]).

The small-signal impedance will be calculated by considering only the drift component of the particle current. However, the effect of the steady-state diffusion current is introduced through the alternating current (a.c.) boundary condition written at the virtual source plane. Here the a.c. component of the particle current $\mathbf{j}_1 = \mathbf{j}_1(x)$ is

$$\mathbf{j}_1(0) = \sigma_1(0)\mathscr{E}_1(0), \tag{5.3.3}$$

where

$$\sigma_1(0) = e\mu_p p_0(0) = \frac{\mu_p J_0}{v_{0s}} \tag{5.3.4}$$

is the 'effective small-signal conductivity' [25] of the source plane. Note the fact that $\sigma_1(0)$ is the ohmic conductivity of the source plane and at this plane the entire particle current $\mathbf{j}_1(0)$ is due to velocity modulation. The charge-density modulation component is zero at $x = 0$ because this term is proportional to the steady-state drift velocity which is zero by definition*.

The total current may be written

$$\mathbf{J}_1 = \mathbf{j}_1(0) + i\omega\varepsilon\mathscr{E}_1(0) = [\sigma_1(0) + i\omega\varepsilon]\mathscr{E}_1(0) \tag{5.3.5}$$

---

* In the earlier theory of SCL injection (see Sections 4.6 and 5.2) the a.c. particle current density was assumed infinite [13] and the charge-density modulation at the injecting plane was considered finite. In the same theory $\sigma_1(0)$ is infinite (because $p_0(0)$ is infinite) and $\mathscr{E}_1(0)$ is zero.

and therefore

$$\underline{\mathcal{E}}_1(0) = \frac{\underline{J}_1}{\sigma_1(0) + i\omega\varepsilon}. \tag{5.3.6}$$

This value of $\underline{\mathcal{E}}_1(0)$ will be used as a boundary condition.

The integration of a.c. equations (arbitrary velocity-field dependence, negligible diffusion, constant doping) was discussed by Dascalu [13], [28]. The device impedance may be written in analytical form in certain special cases, for example when the velocity-field dependence may be approximated as linear (constant *differential* mobility) for the range of electric field intensities encountered at a given bias [13] (a special case is $v = \mu\mathcal{E}$, $\mu = $ const. [29]). The method of integration is essentially that presented in Section 4.6.

We shall first discuss a simplified situation, namely the 'high-field' diode, where the hole velocity is assumed to be saturated over almost the entire drift space. The low-field region which always exists near the potential maximum is neglected. The a.c. equations are

$$\underline{J}_1 = e\underline{p}_1(x)v_{\text{sat}} + i\omega\varepsilon\underline{\mathcal{E}}_1(x) \tag{5.3.7}$$

$$\frac{d\underline{\mathcal{E}}_1(x)}{dx} = \frac{e\underline{p}_1(x)}{\varepsilon} \quad \text{(Poisson)} \tag{5.3.8}$$

$$\underline{V}_1 = \int_0^L \underline{\mathcal{E}}_1(x)\,dx. \tag{5.3.9}$$

By using the boundary condition (5.3.6) one obtains

$$Z = \frac{\underline{V}_1}{\underline{J}_1} = \frac{L^2}{\varepsilon v_{\text{sat}}}\left[\frac{1}{i\theta} + \frac{\exp(-i\theta) - 1}{(i\theta)^2(1 + i\theta_s)}\right], \tag{5.3.10}$$

where

$$\theta = \omega\tau_L, \quad \theta_s = \omega\left[\frac{\varepsilon}{\sigma_1(0)}\right] \tag{5.3.11}$$

and θ_s was defined by Wright [15], [25] as the transit-angle of the source plane. The actual nature of the time constant $\varepsilon/\sigma_1(0)$ in equation (5.3.11) is, however, that of a *differential relaxation time* at the source plane [13]. The incremental (low-frequency) resistance is

$$R_i = Z(0) = \frac{L^2}{2\varepsilon v_{\text{sat}}}\left(\frac{2}{\sigma_1(0)}\frac{\varepsilon}{T_L} + 1\right). \tag{5.3.12}$$

For infinite source conductivity, $\sigma_1(0) \to \infty$, R_i is equal to the space-charge resistance (unit-area device)

$$R_i\big|_{\sigma_1(0) \to \infty} = R_{sc} = \frac{L^2}{2\varepsilon v_{\text{sat}}}, \tag{5.3.13}$$

whereas, in general $(\sigma_1(0) = \text{finite})$

$$R_i = R_{sc} + \frac{L}{\sigma_1(0)}. \qquad (5.3.14)$$

Because $\sigma_1(0)$ depends upon the bias current, equation (5.3.14) indicates that the non-linearity of the $J-V$ characteristic for a saturated-velocity diode, is associated to the finite a.c. conductivity of the virtual source. The above equations are valid not only for the punch-through p^+np^+ diode discussed here, but for any saturated-

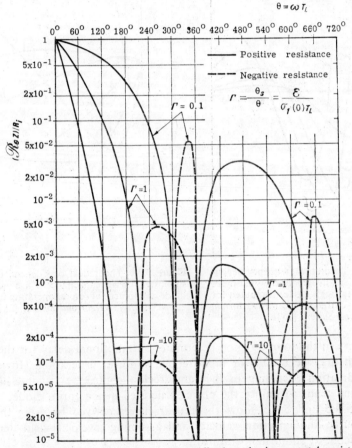

Figure 5.3.4. Series resistance normalized to the incremental resistance versus the transit angle for a barrier-injection saturated-velocity diode. The same results are valid for any uniform-velocity diode. $\sigma_1(0)$ is the differential conductivity at the injecting plane. (after Dascalu, [11]).

velocity diode, provided that we can introduce, according to equation (5.3.3), an a.c. conductivity of the source plane.

The real part of the impedance (5.3.10) is plotted in Figure 5.3.4 versus the transit angle, with $\theta_s/\theta = \varepsilon/\sigma_1(0) T_L$ as parameter. The device exhibits a transit-time negative resistance. Its maximum absolute value is, in general, very small in comparison to the incremental resistance. No negative resistance is obtained for $\sigma_1(0) \to \infty$ (SCL injection).

For the punch-through p^+np^+ diode the assumption of constant carrier velocity is a considerable simplification [25]. Wright and Sultan [25] outlined that the low

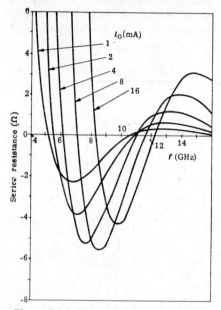

Figure 5.3.5. Theoretical frequency dependence of the small-signal series resistance of a punch-through diode, computed for several bias currents, I_0. (after Wright and Sultan, [25]).

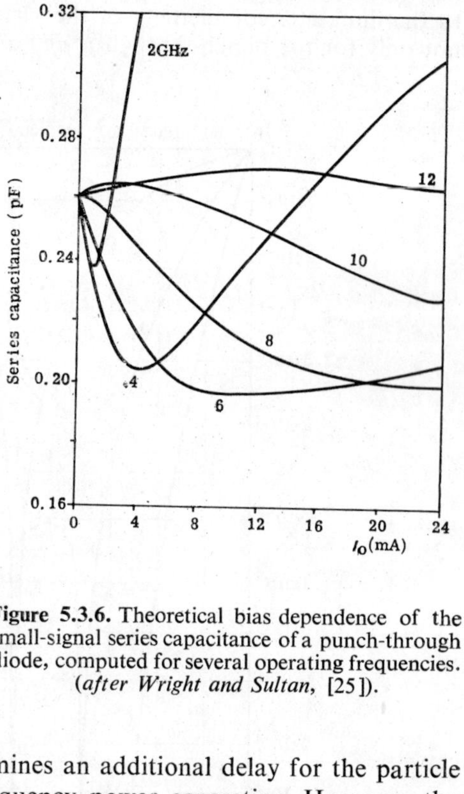

Figure 5.3.6. Theoretical bias dependence of the small-signal series capacitance of a punch-through diode, computed for several operating frequencies. (after Wright and Sultan, [25]).

carrier velocity in the source region determines an additional delay for the particle current which may be useful for high-frequency power generation. However, the velocity modulation occurring in the same region contributes to internal power dissipation and tends to reduce the power generated in the entire diode.

Figures 5.3.5 and 5.3.6 show theoretical curves derived by using the model of virtual source and injection velocity and assuming a velocity-field dependence (holes in silicon, at room temperature)

$$v = \frac{\mu_p \mathscr{E}}{1 + \mu_p \mathscr{E}/v_{\text{sat}}}, \quad (5.3.15)$$

where $\mu_p = 450 \text{ cm}^2 \text{ V}^{-1} \text{ s}^{-1}$ and $v_{\text{sat}} = 9.5 \times 10^6 \text{ cm s}^{-1}$. The p^+np^+ device has a 5 μm-thick, 10^{15} cm^{-3} doping density n-region and an active area of $1.25 \times 10^{-4} \text{ cm}^2$. The calculated punch-through threshold is 20 V [25].

The above data correspond to experimental devices (the experimental punch-through voltage varies between 20 and 21 V). Figures 5.3.7 and 5.3.8 indicate experimental results for small-signal series resistance and series capacitance. The agreement between experimental and theoretical results is very good.

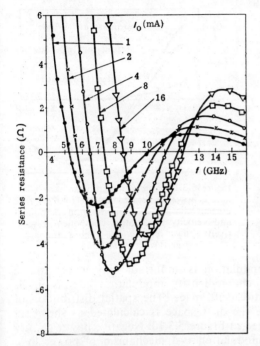

Figure 5.3.7. Experimental frequency dependence of the small-signal series resistance of a punch-through BARITT diode (the theoretical curves are shown in Figure 5.3.5). (after Wright and Sultan, [25]).

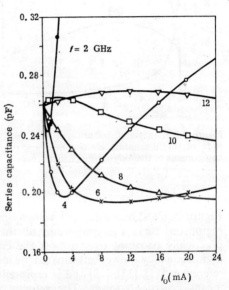

Figure 5.3.8. Experimental bias dependence of the small-signal series capacitance of a punch-through BARITT diode (the theoretical curves are shown in Figure 5.3.6). (after Wright and Sultan, [25]).

The negative-resistance behaviour may be qualitatively understood by using the concept of source conductivity $\sigma_1(0)$. At low bias currents $\sigma_1(0)$ is small and the phase lag between the injected particle current $\mathbf{j}_1(0)$ and the total current $\mathbf{J}_1$ is near to its maximum, which is $\pi/2$ (see equations (5.3.3) and (5.3.6)). As the bias current and $\sigma_1(0)$ increase, the above delay decreases and the negative-resistance domain moves to higher frequencies because a higher transit-time delay is necessary. On the other hand, the magnitude of the negative resistance decreases at higher bias currents because the source conductivity $\sigma_1(0)$ becomes very high and the phase lag of the injected current becomes small.

The details of device behaviour can be examined by using the method suggested by Dascalu [11], [13], [14], [28] and outlined in the preceding section. This method was used by Wright and Sultan [25] in an attempt to explain the microwave activity of their punch-through devices, and the result is shown in Figures 5.3.9 and 5.3.10. The computations were made for the device presented before, at a steady-state curren bias of 8 mA (64 A cm^{-2}) and a frequency of 8.5 GHz. These figures correspond to the maximum negative resistance computed for this particular structure.

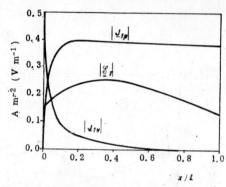

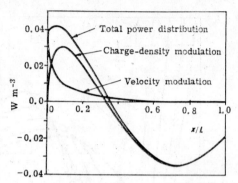

Figure 5.3.9. Spatial distribution of the amplitude of the alternating electric field and the components of the alternating particle current. (*after* Wright and Sultan, [25]).

Figure 5.3.10. Spatial variation of the alternating power generated by the total particle current, charge-density modulation component and velocity modulation component, respectively (power *generation* — below the axis). (*after* Wright and Sultan, [25]).

Figure 5.3.9 shows that the velocity-modulation is important only in the source region, whereas the amplitude of the charge-density modulation component is practically constant over almost the entire drift space. The spatial distribution of high-frequency power disssipated inside the drift space is calculated as shown in references [11], [13], [28] and is represented in Figure 5.3.10. Negative power density means *generated* power. The velocity modulation is a mechanism of power loss, whereas charge density-modulation generates power in about two-thirds of the interelectrode space, a region where the delay between $\mathbf{J}_{1p} = ev_0 \mathbf{p}_1$ and $\mathscr{E}_1$ is greater than $\pi/2$.

5.4 Noise Properties of Punch-through BARITT Diodes

There exist at least two sources of noise in BARITT diodes, the injection noise and the diffusion noise.

The *injection noise* is due to the random fluctuation of the injected particle current (see also Section 4.7). The effect of these fluctuations can be calculated by using the small-signal perturbation method, as follows. The total alternating current at the source plane ($x = 0$) is

$$\mathbf{J}_1 = \sigma_1(0)\,\mathscr{E}_1(0) + i\omega\varepsilon\mathscr{E}_1(0) + \mathbf{J}_i \qquad (5.4.1)$$

where $\mathbf{J}_i$ is the ω-frequency component of the injection noise, noise consisting in random fluctuations of the bias current, $\mathbf{J}_0$. This equation provides the boundary condition which replaces equation (5.3.6). Let us consider below the open-circuit noise for the saturated-velocity diode, first computed by Haus et al. [30]. From equations (5.3.7)–(5.3.9) and (5.4.1), with $\mathbf{J}_1 = 0$ (open-circuit), one obtains [11], [30]

$$\overline{V_1^2} = \left[\frac{L}{\sigma_1(0)}\right]^2 \frac{2(1-\cos\theta)}{\theta^2(1+\theta_s^2)} \overline{j_1^2}, \quad \theta = \omega T_L, \theta_s = \frac{\omega \varepsilon}{\sigma_1(0)} \quad (5.4.2)$$

If the injection noise is full shot noise, then (for a device of unit area)

$$\overline{j_1^2} = 2eJ_0 \Delta f \quad (5.4.3)$$

and the open-circuit noise voltage spectrum will be

$$S_v(\omega) = 4eJ_0 \left[\frac{L}{\sigma_1(0)}\right]^2 \left[\frac{1-\cos\theta}{\theta^2(1+\theta_s^2)}\right] \quad (5.4.4)$$

(note the fact that $\sigma_1(0)$ and θ_s also depend on J_0)*.

The frequency dependence of the open-circuit injection noise is indicated in Figure 5.4.1.

The *noise measure* defined as [27], [30],

$$M = \frac{2S_v^2(\omega)}{4kTR} \quad (5.4.5)$$

(where R is the small-signal series resistance) is sometimes used to appreciate the noise performance. The noise measure for the injection noise of a punch-through

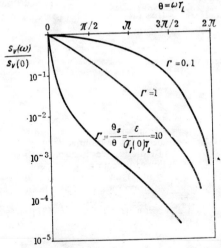

Figure 5.4.1. Frequency dependence of the open-circuit injection noise in a barrier-injection uniform-velocity diode ($\sigma_1(0)$ is the differential conductivity at the barrier). (*after* Dascalu, [11]).

diode was computed numerically by Sjölund [27] for a realistic model with field-dependent hole-velocity (see equation 5.3.15) and also including diffusion. Figure 5.4.2 shows the computed and measured small-signal resistance for three bias

---

* For SCL injection $\sigma_1(0) \to \infty$ and the injection is totally suppressed. If $\sigma_1(0)$ increases rapidly with J_0 then $S_v(\omega)$ may decrease with increasing bias current.

currents. Figure 5.4.3 shows (dashed curves) the noise measure, equation (5.4.5), computed for injection noise only. The noise is frequency-dependent due to transit-time effects. The noise measure decreases appreciably with increasing

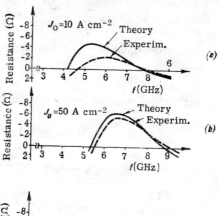

Figure 5.4.2. Frequency dependence of small-signal resistance of a punch-through BARITT diode. The solid lines are the results of a numerical analysis (including the diffusion current) for an abrupt junction p^+np^+ silicon structure at room temperature, with a $1.2 \times 10^{15} \text{cm}^{-3}$ doping and 7.9 μm thickness n-type junction. The dashed curves are experimental. (a–c) represent different bias current densities. The disagreement between theory and experiment at $J_0 = 200 \text{ A cm}^{-2}$ is probably due to the diode heating, which is not taken into account in calculations. (after Sjölund, [27]).

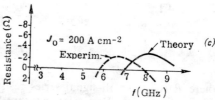

current density. This may be qualitatively explained as due to stronger space-charge reduction determined by higher space-charge densities (see also Figure 5.3.2 which is computed for the same diode, within the same model).

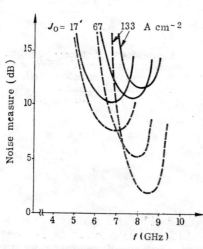

Figure 5.4.3. Theoretical frequency-dependence of the noise measure computed for a punch-through BARITT diode (see Figure 5.4.2) as follows: injection noise only (dashed lines) and injection noise plus diffusion noise (solid lines); the bias current density is taken as a parameter. (after Sjölund, [27]).

The *diffusion noise* [23], [27] is due to random velocity fluctuations of carriers within the drift space (see also the thermal, or diffusion noise discussed in Section 4.7). Sjölund [23], [27] computed the voltage spectrum for the diffusion noise by using the impedance-field method [31], [32] and the total (injection plus diffusion noise) is shown by the solid lines in Figure 5.4.3 (the diffusion constant was assumed independent of field and given by Einstein's relation). Due to the diffusion noise, the total noise measure increases with increasing current density. These results

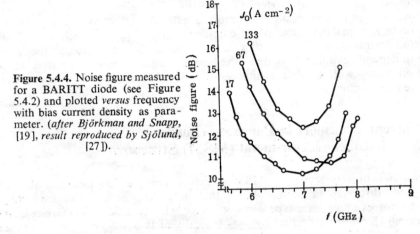

Figure 5.4.4. Noise figure measured for a BARITT diode (see Figure 5.4.2) and plotted *versus* frequency with bias current density as parameter. (*after Björkman and Snapp*, [19], *result reproduced by Sjölund*, [27]).

were confirmed by experiment as shown in Figure 5.4.4 (in this case the noise measure is almost identical to noise figure) [23], [27]. An optimisation carried out by Sjölund [23] indicates that noise measures significantly below 10 dB at 8 GHz cannot be expected. Lower noise measure may occur at lower frequencies. Both experimental and theoretical results indicate that the microwave punch-through BARITT diode is a low-noise device and may be used in low-power low-noise applications, for example in local oscillators.

5.5 Large-signal Behaviour of the Punch-through BARITT Diode

Approximate large-signal analysis for the punch-through transit-time diode were carried out by Rüegg [33], Sheorey, Lundström and Ash [34]. These authors considered a diode biased below punch-through and driven into conduction by large-amplitude voltage signals. The non-linearity of the punch-through characteristic and the sharp pulse approximation of the injection process are the salient features of these computations. The device described by such a model is not self-starting, although the experiment confirms the small-signal negative-resistance first outlined by Wright [15] and discussed here in Section 5.3.

The large-signal impedance for a device biased above punch-through was derived by Lacombe [35] and used to explain the nature of high-frequency power limitation (maximum power at a given frequency) [26]. However, the analysis is based upon the assumption of sinusoidal modulation of the barrier height, which cannot be justified *a priori*.

An improved punch-through diode is a $p^+np^-p^+$ structure [34], [36] with a narrow n region and a nearly trapezoidal electric field profile. The hole velocity is saturated in the almost entire conduction space (n and p^- regions), which decreases the power loss due to velocity modulation. On the other hand the applied voltage may increase well above the punch-through value, the maximum electric field still remaining below the critical value for carrier multiplication. This allows larger output powers.

Kawarada and Mizushima [36] reported extensive data on a computer simulation of large-signal behaviour of the complementary $n^+pn^-n^+$ punch-through diode by including diffusion and assuming that the carrier concentration at the injecting junction depends exponentially upon the barrier height. It was shown that the oscillation efficiency with a single tuning circuit is relatively low (10%) but can be improved by circuit techniques [36].

5.6 Current Transport in Punched-through Metal-semiconductor-metal (MSM) Structures

We consider in this section a MSM structure which differs from the SCL solid-state diode discussed in Section 4.3 in that both metallic contacts are rectifying. Consider for example an n-type semiconductor with two Schottky-barrier contacts (Figure 5.6.1). Such a structure consists basically of two Schottky diodes connected back to back [37]. No appreciable current will flow until the barrier forwardly biased decreases considerably to allow charge injection in the structure. This occurs only after the barrier at the reverse biased contact extends to the opposite contact such that the semiconductor is totally depleted of majority carriers. The current will by carried by holes. The situation is similar to that encountered in a p^+np^+ punch-through diode (section 5.1).

Consider, for simplicity, a symmetrical structure ($\phi_{n1} = \phi_{n2}$, $\phi_{p1} = \phi_{p2}$ in Figure 5.6.1). We denote by V_{RT} the reach-through voltage, when the two space-charge regions touch, and by V_{FB} the flat-band voltage, when the electric field is zero at the forward biased contact. In the p^+np^+ punch-through diode theory, V_{RT} (Section 3.1) was identified with the punch-through voltage V_{PT}. Here, the magnitude of the current will be determined by using the law of thermionic injection and we shall

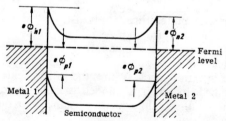

Figure 5.6.1. Energy-band diagram for MSM structure with Schottky contacts, at thermal equilibrium.

evaluate more precisely the barrier height. By neglecting the mobile charge density and assuming uniform doping, the electric field profile will be linear, as shown in Figure 5.6.2. The flat-band voltage (Figure 5.6.2c) can be readily calculated:

$$V_{FB} = \frac{eN_D L^2}{2\varepsilon}. \tag{5.6.1}$$

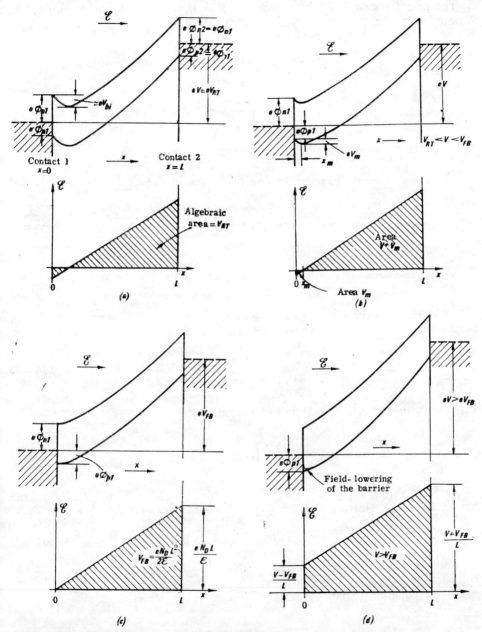

Figure 5.6.2. Energy-band diagrams and field profiles (neglecting mobile electric charge) for a symmetric MSM structure biased as indicated: (a) $V = V_{RT}$; (b) $V_{RT} < V < V_{FB}$; (c) $V = V_{FB}$; (d) $V > V_{FB}$ (V_{RT} is the reach-through voltage, V_{FB} the flat-band voltage; V_{bi} the built-in voltage of the metal-semiconductor junction).

The potential and field distribution for $V = V_{RT}$ is indicated in Figure 5.6.2a. A potential maximum of height V_m occurs at the distance $x = x_m$ from the forward biased contact ($x = 0$) when $V_{RT} < V < V_{FB}$ (Figure 5.6.2b). We have

$$x_m = \sqrt{\frac{2\varepsilon V_m}{eN_D}} \qquad (5.6.2)$$

$$L - x_m = \sqrt{\frac{2\varepsilon(V + V_m)}{eN_D}}. \qquad (5.6.3)$$

By eliminating x_m and L between the above three equations one obtains

$$V_{FB} = V + 2V_m + 2\sqrt{V_m(V + V_m)} \qquad (5.6.4)$$

or, approximately,

$$V_{FB} \approx V + \sqrt{2V_m V}, \quad V_m \ll V, \qquad (5.6.5)$$

such that

$$V_m = \frac{(V_{FB} - V)^2}{4V} \qquad (5.6.6)$$

and, because V_{FB} is very close to V_{RT}, such that V may be approximated by V_{FB} in the denominator of equation (5.6.6), we have

$$V_m \approx \frac{(V_{FB} - V)^2}{4V_{FB}} = \frac{V_{FB}}{4}\left(1 - \frac{V}{V_{FB}}\right)^2. \qquad (5.6.7)$$

From equations (5.6.1), (5.6.2) and (5.6.7) one obtains

$$x_m \simeq \frac{L}{2}\left(1 - \frac{V}{V_{FB}}\right). \qquad (5.6.8)$$

At reach-through $V = V_{RT}$, $V_m = V_{bi}$ (Figure 5.6.2a), i.e. the height of the barrier coincides with the built-in voltage *, such that

$$V_{RT} = V_{FB} - 2\sqrt{V_{FB}V_{bi}} \qquad (5.6.9)$$

For usual device parameters (see below) V_{FB} is quite large (tens of volts) as compared to V_{bi}, such that equation (5.6.8) is consistent with our assumption $V \simeq V_{FB}$ in equation (5.6.6).

---

* The forward drop on this metal-semiconductor junction is $(kT/e)\ln\left(1 + \dfrac{J}{J_{sat}}\right)$ and is very small compared to V_{bi} because J is comparable to J_{sat}. This drop cannot be longer neglected at higher current densities ($V > V_{RT}$): the potential barrier decreases with increasing V as shown by equation (5.6.6).

Voltages in excess of the flat-band voltage (Figure 5.6.2d) determine an electric field which is directed towards the semiconductor, at the forward biased (plus on metal for an n-type semiconductor) contact and its magnitude is

$$\mathscr{E}(0) = \frac{V - V_{FB}}{L}. \qquad (5.6.10)$$

The electric field at $x = L$ is directed towards the metal and has the value (Figure 5.6.2d) $\mathscr{E}(L) = \dfrac{V + V_{FB}}{L}$

If the avalanche breakdown occurs at the critical field $\mathscr{E}_{cr}$, then the breakdown voltage will be

$$V_B = \mathscr{E}_{cr} L - V_{FB} = \left(\mathscr{E}_{cr} - \frac{eN_D L}{2\varepsilon} \right) L. \qquad (5.6.11)$$

The flat-band voltage for a symmetrical silicon MSM structure is plotted in Figure 5.6.3 *versus* doping concentration, with device length as parameter. For a given doping the maximum V_{FB} is limited by the avalanche breakdown occurring at the reverse biased contact (which is equivalent to an abrupt asymmetric p-n junction, see Chapter 2).

We shall consider below the electron and hole transport when the contact 1 ($x = 0$ in Figure 5.6.1) is forward biased and the contact 2 ($x = L$) reverse biased, i.e. the contact 1 is made positive with respect to contact 2. When the contacts are

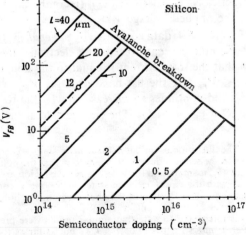

Figure 5.6.3. The flat-band voltage $V_{FB} = eN_D L^2 / 2\varepsilon$ for a symmetric silicon MSM structure, as a function of doping concentration and semiconductor thickness, L. The avalanche-breakdown limit ($V_{FB} < V_B$) is shown. (*after Sze et al.*, [37]).

inversely biased the phenomena are similar. Because the device has rectifying contacts we have to consider only the electron injection into the semiconductor at the contact 2 and hole injection into the semiconductor at the contact 1. The electron current is approximately equal to the thermionic emission current into the semiconductor (saturation current for the ideal thermionic characteristic of the Schottky barrier 2). This current increases slowly with increasing voltage due to the field-lowering of the barrier $e\phi_{n2}$ (image-force lowering and other contact phenomena, see Chapter 2).

The saturation current for thermionic emission of holes into the semiconductor should be much larger, because the barrier for holes, $e\phi_{p1}$, is, in general, considerably lower than the electron barrier. However, this saturation current is not attained as far as the electric field in semiconductor opposes hole injection. At voltages $V < V_{RT}$ the hole current is smaller than the electron current (see [37] and also the minority carrier injection at a Schottky barrier, Section 3.3). For $V_{RT} < V < V_{FB}$, the thermionic hole emission current over the barrier $e(\phi_{p1} + V_m)$ is *

$$J_{p1}(x_m) = A_p^* T^2 \exp[-e(\phi_{p1} + V_m)] \tag{5.6.12}$$

or, according to equation (5.6.7),

$$J_{p1}(x_m) = J_{p1\,sat} \exp\left[-\frac{e(V_{FB} - V)^2}{4kTV_{FB}}\right], \tag{5.6.13}$$

where

$$J_{p1\,sat} = A_p^* T^2 \exp\left(-\frac{e\phi_{p1}}{kT}\right) \tag{5.6.14}$$

is the thermionic emission saturation current for holes at contact 1 and A_p^* is an effective Richardson constant for hole emission (see Chapter 2). The hole current is practically equal to $J_{p1}(x_m)$ (recombination is negligible) and increases rapidly with the applied voltage $V > V_{RT}$.

Figure 5.6.4 (data computed by Sze et al. [37] for a practical silicon structure) shows that the hole current exceeds the electron current with several orders of magnitude such that the total current is equal to the hole current, equation (5.6.14), for $V_{RT} \leq V \leq V_{FB}$. At the flat-band voltage, V_{FB}, J attains the 'saturation' hole current $J_{p1\,sat}$, and continues to increase slowly due to the field-lowering of the metal-semiconductor barrier, $e\phi_{p1}$. The electric field at the contact is given by equation (5.6.10).

---

* The thermionic emission formula is simply corrected with the height of the electrostatic barrier in the semiconductor. It is assumed that the potential maximum is located very close to the contact (x_m is small) and the holes within the region $0 < x < x_m$ are maintained in intimate thermal contact with the empty electron states in the metal contact 1. Therefore the hole density has its thermal equilibrium value in the above region (the hole imref ζ_p is approximately constant at the Fermi level of metal contact 1) [38].

It is interesting to note that similar results were presented by Chu et al. [39] for a punched-through p^+np^+ structure. For $V > V_{RT}$ but as far as the hole density is negligible, the current is determined by thermionic emission over the potential barrier. A formula identical to equation (5.6.12) is obtained, except the fact that for the p^+np^+ structure $\phi_{p1} = 0$ [39]. The numerical value of the saturation current is therefore much higher than for a metal semiconductor contact. In fact, at higher current densities the space-charge-limitation mechanism occurs such that the thermionic saturation current is not attained. The theory was verified experimentally [39]. For very high resistivity punch-through diodes, the SCL current occurs at much lower current densities and the thermionic emission current is not observed (Section 5.1).

Assume, for example, that the field-lowering of the barrier is due to image-force only. According to the theory of Chapter 2, the current for $V > V_{FB}$ is

$$J \simeq J_p = J_p(0) = A_p^* T^2 \exp\left(-\frac{e\phi_{p1}}{kT}\right) \exp \frac{e}{kT} \sqrt{\frac{e}{4\pi\varepsilon} \frac{V - V_{FB}}{L}} . \quad (5.6.15)$$

The applied voltage has an uper limit, V_B, at the onset of avalanche breakdown (Figure 5.6.4).

Sze et al. [37] constructed MSM diodes using 11 Ω cm resistivity (4×10^{14} cm^{-3} doping), 12 μm thickness silicon wafers. Platinum of 500 Å thickness was sputtered onto both sides of the wafer and sintered to form a 1000 Å layer of PtSi. Chromium (300 Å) and gold (3000 Å) were then deposited by evaporation onto the PtSi contacts. The devices with areas of 2×10^{-3}, 5×10^{-4} and 1.25×10^{-4} cm^{-2} were separated by mesa etching. The barrier heights were determined from $I-V$ and $C-V$

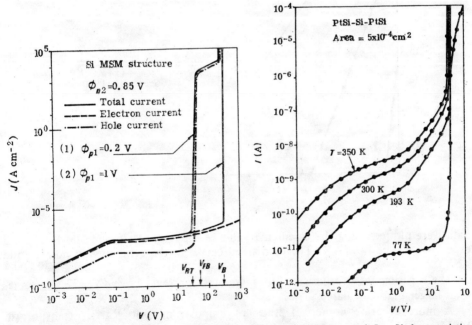

Figure 5.6.4. Theoretical current-voltage characteristics for an MSM silicon structure with $N_D = 4 \times 10^{14}$cm^{-3} and $L = 12\,\mu$m, at room temperature. The barrier heights are: case (1): $\phi_{n2} = 0.85$ V, $\phi_{p1} = 0.2$ V; case (2): $\phi_{n2} = 0.85$ V, $\phi_{p1} = 1.0$ V. (after Sze et al., [37]).

Figure 5.6.5. Measured $I - V$ characteristics of a Si MSM structure fabricated as shown in the text. (after Sze et al., [37]).

measurements on the control samples (with one ohmic contact) to be 0.85 ± 0.05 eV for electrons and 0.20 ± 0.05 eV for holes. Note the fact that the theoretical calculations in Figure 5.6.4 were made for $\phi_{n2} = 0.85$ eV and $\phi_{p1} = 0.2$ eV. Figure 5.6.5

shows $J - V$ characteristics at four temperatures. An abrupt increase of the current occurs around 30 V. The calculated V_{FB} is 46 V. The experimental value of V_{FB} at 77 K is only 41 V (Figure 5.6.5). The difference may be explained as due to a partial deionization of impurities [37]. It is interesting to note that the critical voltage at which the current rises rapidly has a negative temperature coefficient (Figure 5.6.5), whereas the avalanche breakdown voltage V_B has a positive temperature coefficient.

The theoretical formula (5.6.13) is compared to experiment in Figure 5.6.6. A plot of $\ln J$ versus $(V - V_{FB})^2$ should yield a straight line with slope $e/4kTV_{FB}$ and the intercept at $V = V_{FB}$ should give the hole saturation current. The experimental

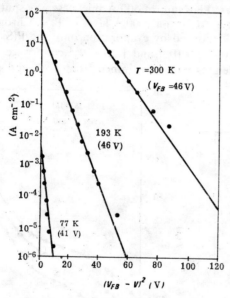

Figure 5.6.6. Experimental (points) and theoretical (solid lines) current plotted versus $(V_{FB} - V)^2$. (after Sze et al., [37]).

data were plotted using the calculated value of V_{FB} for 193 K and 300 K and the experimental one for 77 K. The theoretical curves fit the experimental points quite well.

The experiments also indicated an increase of current above V_{FB} (77 K), somewhat larger than calculated [37].

A more accurate numerical study of unipolar minority carrier transport in MSM structures was reported by El-Gabaly et al. [40]. The basis of their analysis is the current continuity equation for holes (PtSi—nSi—PtSi structure) including diffusion and field-dependent carrier mobility. The mobile space charge density is taken into account in the Poisson equation and more accurate boundary conditions at the hole-emitting contact are used *.

---

* The boundary hole concentration decreases above the thermal equilibrium value with increasing hole current injected in the semiconductor. The hole saturated current at the injected contact uses an effective Richardson constant corrected for reflexion probabilities at the interface.

The computed carrier distributions and electric field profiles for a diode with $L = 10$ μm, $N_D = 4.4 \times 10^{14}$ cm^{-3} (or 1.2×10^{15} cm^{-3}) at several bias densities are shown in Figures 5.6.7 and 5.6.8, respectively. Except in the immediate vicinity of

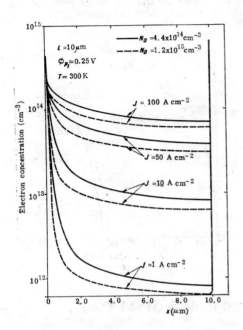

Figure 5.6.7. Mobile hole distribution computed at several current densities in a silicon MSM structure (unipolar conduction above the reach-through voltage). (after El-Gabaly et al., [40]).

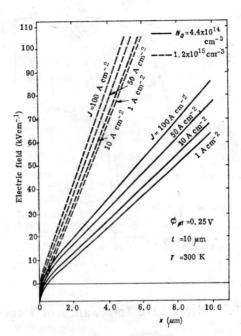

Figure 5.6.8. Field profiles corresponding to Figure 5.6.7. (after El-Gabaly et al., [40]).

the injecting contact ($x = 0$), the hole density is approximately constant and small compared to the ionized density, such that the electric field intensity varies almost linearly with x. A potential maximum occurs very close to the injecting contact ($\mathscr{E} = 0$ in Figure 5.6.8). Both the magnitude of this maximum and its distance from the contact decrease as the current density increases (a similar picture was obtained for the SCL semiconductor diode and for the punch-through p^+np^+ structure). The diffusion current is, of course, dominant in the neighbourhood of this maximum.

Figure 5.6.9 shows the computed $J - V$ characteristics for two device lengths, two doping levels and two temperatures. In Figure 5.6.9a the characteristics obtained by neglecting the mobile hole density are also shown. At low injection levels the current increase is exponential and the space-charge of mobile carriers is negligible. The hole charge is, however, important at practical current densities (tens of A cm^{-2}), below current saturation.

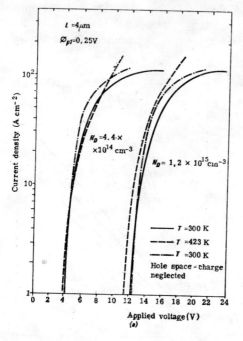
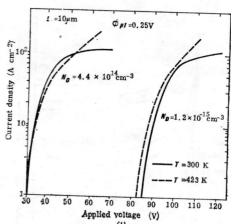

Figure 5.6.9. (a) Current-voltage characteristics for two different doping concentrations and at two different crystal temperatures ($L = 4\,\mu m$). Effect of ignoring space charge of injected holes is also shown. (b) Current-voltage characteristics for $L = 10\,\mu m$. (after El-Gabaly et al., [40]).

5.7 Microwave Impedance of Punched-through MSM Diodes

The simplest model of high-frequency diode operation assumes a saturated carrier velocity throughout the entire device. The small-signal impedance is then given by equation (5.3.10). The only parameter in this equation which depends upon the injection mechanism and upon the bias current is the small-signal source conductivity, $\sigma_1(0)$, defined by equation (5.3.3). Otherwise, the impedance is identical to that of the punched-through p^+np^+ diode.

The source conductivity, $\sigma_1(0)$, expresses the modulation of injected current by the electric field at the source plane. This parameter can be obtained from the steady-state equation (5.6.13). This expression was found by neglecting the mobile injected charge. We also neglect the contribution of this charge to the alternating electric field, $\underline{\mathscr{E}}_1$, and therefore $\underline{\mathscr{E}}_1$ is uniform * and [24]

$$\underline{V}_1 = L\underline{\mathscr{E}}_1. \qquad (5.7.1)$$

---

* The assumption of uniform alternating field implies, however, a uniform displacement current, $i\omega\varepsilon\underline{\mathscr{E}}_1$ and a uniform particle current, $\underline{j}_1 = \underline{J}_1 - i\omega\varepsilon\underline{\mathscr{E}}_1$, which is in contradiction with the calculation of the impedance, equation (5.3.10). It can be accepted that the particle current $\underline{j}_1$ is very small, and $\underline{\mathscr{E}}_1$ is *approximately* uniform. This requires that the reactive (capacitive) component of the impedance should be dominant and this is indeed so, because the absolute value of the device quality factor (Q) is quite high.

From equations (5.3.3), (5.6.13) and (5.7.1) one obtains [24], [30]

$$\sigma_1(0) = J_0 \frac{\varepsilon(V_{FB} - V_0)}{kTN_DL}. \tag{5.7.2}$$

The source conductivity first increases with the steady-state current J_0 and then decreases as the voltage bias V_0 approaches V_{FB}.

The normalized frequency characteristics of Figure 5.3.4 are also valid for the MSM diode. The incremental resistance (5.3.14) is

$$R_i = R_{sc} + \frac{L}{\sigma_1(0)} = \frac{L^2}{2\varepsilon v_{sat}} + \frac{2}{J_0} \frac{kT}{e} \frac{1}{1 - \dfrac{V_0}{V_{FB}}}. \tag{5.7.3}$$

Note that R_i becomes infinite as $V_0 \to V_{FB}$. This is consistent with the fact that the characteristic should saturate at $V_0 = V_{FB}$ (the field enhanced emission process which would determine an increase of J_0 when V_0 increases above V_{FB} is not contained in equation (5.6.13)). The parameter Γ (Figure 5.3.4) becomes

$$\Gamma = \frac{\varepsilon}{\sigma_1(0)T_L} = \frac{kTN_D v_{sat}}{J_0(V_{FB} - V)} \tag{5.7.4}$$

and has a minimum for a certain value of J_0. The above theory for the MSM structure was given by Haus et al. [30], by Coleman [38] and, in a somewhat modified form (including a correction for the low-field region near the source), by Chu and Sze [24]. The distance from the 'virtual source' at $x = x_m$ (see equation (5.6.8)) was simply subtracted from the total length [24], [38]. A more refined analytical theory,

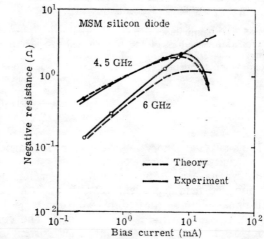

Figure 5.7.1. Comparison of experimental and theoretical small-signal negative resistance as a function of bias current for a silicon MSM structure ($N_D = 5 \times 10^{14} \text{cm}^{-3}$, $\phi_{p1} = 0.26$ V, $L = 12.5\,\mu\text{m}$, $A = 3 \times 10^{-4}\text{cm}^{-2}$). (after Chu and Sze, [24]).

due to Antognetti et al. [41], defines an effective diffusion velocity in the source region and is somewhat similar to the approach given by Wright and Sultan [25] (Section 5.3).

Microwave oscillations from an MSM diode were first reported by Coleman and Sze [42]. Figure 5.7.1 shows a comparison between theory and experiment [24] for an MSM diode with $L = 12.5$ μm, $N_D = 5 \times 10^{14}$ cm^{-3} and $\phi_{p1} = 0.26$ V hole injection barrier. The power is of the order of a few millimatts and the efficiency is above 1%. Other experimental results were communicated by a number of authors [17], [20], [38], [43]–[46].

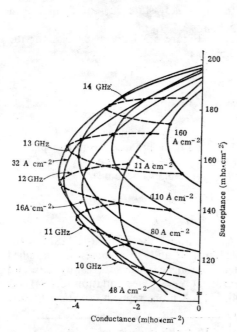 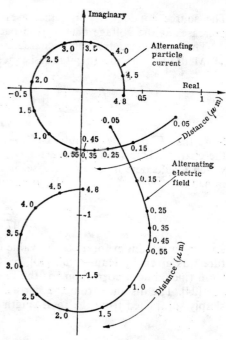

Figure 5.7.2. Computed small-signal admittance for various values of the bias current density ($N_D = 2.5 \times 10^{15}$ cm^{-3}, $L = 4.8$ μm, silicon, room temperature). (*after Matsumura,* [47]).

Figure 5.7.3. Phasor diagram of alternating electric field and alternating hole current as the position inside the diode changes (the distance from the hole emitting electrode is indicated in μm). The bias current and frequency are 80 A cm^{-2} and 12.8 GHz, respectively. The total alternating current is taken as reference (equal to +1). (*after Matsumura,* [47]).

The analytical theory cannot account for carrier diffusion and low drift velocity in the vicinity of the source contact. It was shown [41], [43], [44], [47] that this source region can have an important effect upon high-frequency device behaviour. The diffusion and field-dependent velocity were included in the computer analysis of small-signal régime reported by Matsumura [47]. The silicon device studied had $L = 4.8$ μm and $N_D = 2.5 \times 10^{15}$ cm^{-3}. The computed admittance is shown in Figure 5.7.2. The parameters are bias current density and frequency. Constant frequency (dashed) and constant current (solid) curves are shown. The maximum parallel conductance first increases and then decreases with increasing bias current, whereas the optimum frequency moves towards higher frequencies.

The phasor diagram of Figure 5.7.3 may contribute to the physical understanding of device operation [47]. The operating conditions for the above device are

$J_0 = 80$ A cm^{-2} and $f = 12.8$ GHz. The phasors of alternating field and alternating hole current are shown at each position in the device. The reference is the total current, equal to $+1$ (lying on the real axis, in the positive direction and having a unit magnitude). In the source region ($x < 0.55$ μm) holes travel predominantly by diffusion. Note the phase delay of about $\pi/2$ and the decay of the hole current phasor throughout this region. In the remainder of the semiconductor the charge carriers move approximately at their limiting velocity: this can be shown from the fact that the amplitude of the phasor remains approximately constant (Problem 5.7). For $x > 1.8$ μm the alternating particle current lags behind the alternating field by more than $\pi/2$; therefore microwave power is generated. However, a considerable amount of high-frequency power is absorbed in the source region. Matsumura [47] has also shown that N_D and L should be optimized in order to obtain maximum negative resistance and that the device resistance will be positive at any frequency if the semiconductor doping is too low or the device is too short ($N_D < 5 \times 10^{14}$ cm^{-3} and $L < 1.2$ μm).

The numerical analysis made by Stewart and Wakefield [48] confirms that a maximum of the negative resistance is attained for an optimum bias current density, J_0, of a few tens of A cm^{-2}, and that the negative resistance frequency range is reduced as J_0 increases. The theory is in complete agreement with the experimental results computed by Snapp and Weissglass [20]. The series negative resistance was computed as a function of temperature (300—500 K) and is shown in Figure 5.7.4.

The analysis made by Lee and Dalman [43] points out the contribution of the low-field source region to the negative resistance effect. If the processes in the source region are neglected and it is assumed that the hole injection is in phase with the raising and lowering of the barrier (as at the beginning of this Section), then the optimum drift angle through the diode is about $3\pi/2$ [24], [42]. However, an appreciable inductive delay of the particle current occurs in the low field source region, for example about 70° for an experimental diode studied by Lee and Dalman [43], whereas the drift transit angle is approximately $1.1\,\pi$.

It was suggested [11], [49], [50] that the MSM structure could operate as a microwave oscillator when biased in the saturation current régime (above the flat-band voltage, V_{FB}). By assuming saturated carrier velocity, the impedance is again given by equation (5.3.10) and the source conductivity has to be calculated by using the steady-state equation yielding the thermionic emission current as a function of the field at the barrier. Assuming a simple image-force lowering of the barrier (Chapter 2)

$$J_p = J_{p1\,\mathrm{sat}} \exp\left[\frac{e}{kT}\left\{\frac{e}{4\pi\varepsilon}\mathscr{E}(0)\right\}^{1/2}\right] \tag{5.7.5}$$

and the source conductivity is

$$\sigma_1(0) = \frac{eJ_0}{2kT}\left[\frac{e}{4\pi\varepsilon\mathscr{E}(0)}\right]^{1/2}. \tag{5.7.6}$$

The frequency characteristics for this mode of operation are similar, except the fact that the dependence upon bias current and device parameters is different. A small-signal calculation based upon the above model yields the results shown in Figure 5.7.5 for a 10 μm silicon diode with $J_{p_1} = 100$ A cm^{-2}. Figure 5.7.6 indicates the effects of the (relatively small) signal amplitude upon the series negative resistance. Note the fact that the negative resistance effect disappears at relatively low

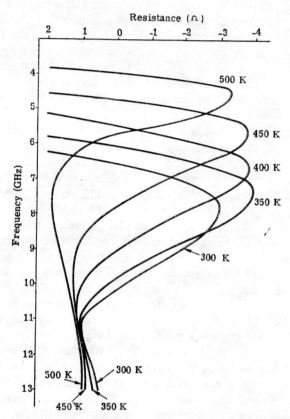

Figure 5.7.4. Small-signal negative resistance calculated for a Pt − n − p+ silicon device with 10^{15}cm^{-3} and 7.2 μm doping and thickness of n-region, 2×10^{-4}cm^{-2} active area, at various temperatures ($\phi_{p1} = 0.25$ V, $A_p^* = 32$ cm$^{-2} \cdot$K^{-1}, $\mu_p = 480$ cm^2 V^{-1}s^{-1}, $v_{p\,sat} = 8.6 \times 10^6$ cm s^{-1}). (after Stewart and Wakefield, [48]).

signal amplitudes, such that the device efficiency should be relatively low. At a given frequency (close to the optimum frequency which is slightly amplitude and bias dependent, according to Figures 5.7.5 and 5.7.6), the series resistance depends

5. BARITT Diodes

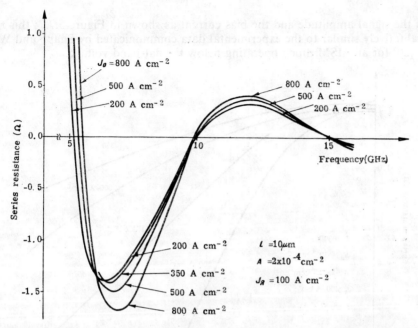

Figure 5.7.5. Small-signal series resistance computed for an MSM silicon structure operating in the 'saturation' current region (above the flat-band voltage), J_R is the hole thermionic emission current in the absence of the image-force lowering of the barrier (under flat-band conditions). *(after Dascalu and Brezeanu, [50]).*

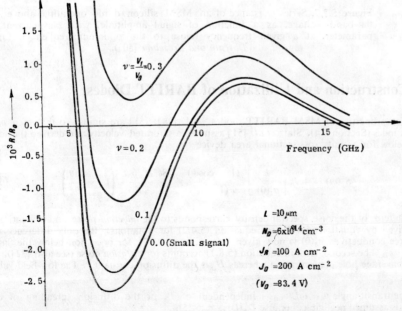

Figure 5.7.6. Series resistance of an MSM silicon diode operating above the flat-band voltage, as a function of frequency, with signal amplitude as parameter, at a given bias current. The normalizing resistance R_0 is equal to $L^2/\varepsilon v_{\text{sat}}$. *(after Dascalu and Brezeanu, [50]).*

upon the signal amplitude and the bias current as shown in Figure 5.7.7; this result is qualitatively similar to the experimental data communicated by Snapp and Weissglass [20] for an MSM diode operating below the flat-band voltage.

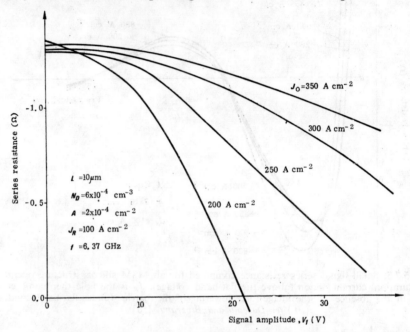

Figure 5.7.7. Series resistance of an MSM silicon diode operating above the flat-band voltage, as a function of signal amplitude, with bias current as parameter, at a given frequency (close to the optimum frequency). (*after* Dascalu and Brezeanu [50]).

5.8 Construction and Utilization of BARITT Diodes

The noise properties of MSM BARITT diodes [30], [38], [51] are similar to that of p^+np^+ (or n^+pn^+) diodes (Section 5.4). Statz *et al.* [51] assumed a saturated velocity and derived the formula given below for a unit cross-sectional area device

$$S_v(\omega) = 4eJ_0 \left[\frac{L}{\sigma_1(0)} \right]^2 \frac{(1-\cos\theta)}{\theta^2(1+\theta_S^2)} + \frac{8e^2 D_p \bar{p}_0 L}{\varepsilon^2 \omega^2} \left(1 - \frac{\sin\theta}{\theta} \right). \quad (5.8.1)$$

The first term of the noise spectral density corresponds to the *injection* noise and is identical with $S_v(\omega)$ given by equation (5.4.4) (see equation (5.4.2) for notations); the only difference is that the source conductivity $\sigma_1(0)$ is now given by equation (5.7.2), for operation below the flat-band voltage V_{FB}. The second term in equation (5.8.1) accounts for *diffusion* noise (see also [32]), where $\bar{p}_0$ is the average hole concentration whereas D_p is the diffusion constant *. The low-field value for

---

* The transit-angle $\theta = \omega L/v_{\text{sat}}$ is independent of D, or the diffusion spreading of carrier bunches is assumed negligible, because $D_p/L \ll v_{\text{sat}}$ [51].

D_p(15 cm²s⁻¹ for holes in silicon) was used. The total noise measure, equation (5.4.5), is plotted in Figure 5.8.1 *versus* the transit angle, θ, with the bias current, J_0, as parameter. The noise measure *versus* frequency exhibits a minimum due to the contribution of diffusion noise (arising from carrier velocity fluctuations in the drift region) at higher transit angles.

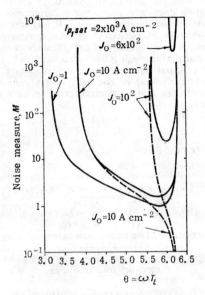

Figure 5.8.1. Noise measure M of an MSM diode as a function of transit angle θ for various current densities (below the saturation current corresponding to the flat-band voltage) for (a) shot noise only (dashed lines) and (b) shot noise and velocity fluctuation noise (solid lines), for silicon, with $\varepsilon = 12\,\varepsilon_0$, $v_{psat} = 10^7$ cm s⁻¹, and also $N_D = 4 \times 10^{14}$ cm⁻³, $J_{p1sat} = 2 \times 10^3$ A cm⁻², $f = 10$ GHz. (*after Statz et al.*, [51]).

The noise of BARITT diodes [16], [19], [27], [32], [38], [42], [43] is much below that of IMPATT diodes (see the next chapter) whereas the generated microwave power is considerably less. Its principal application should be in local oscillators [42]—[44], [46]. The fact that the semiconductor used is silicon makes it particularly attractive.

The BARITT diodes are usually fabricated in a mesa construction which is also used for IMPATT diodes and other microwave diodes. We describe below, the fabrication of an MSM diode, [44]. A ⟨111⟩ silicon wafer with a 23 μm, 4.2 Ωcm n-type epitaxial layer was the starting material. The substrate was mechanically and chemically polished to a thickness of only 6—8 μm. We note the recent advances of chemical thinning techniques (etching), as that described by Stoller *et al.* [52]. On both sides of the wafer palladium and gold were successively evaporated to cover the whole area. One side of the wafer was electroplated with gold (40 μm). Standard photoresist techniques were applied on the opposite side of the wafer to define circular gold contacts of 150 μm

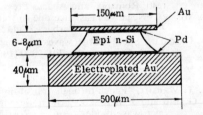

Figure 5.8.2. Schematic cross-section of a mesa Pd — Sin — Pd structure. (*after Freyer et al.*, [44]).

diameter. These contacts act as etching masks where the mesa structures (Figure 5.8.2) are etched out in a conventional mixture of several acids. The last step before individual diode separation and encapsulation was a heat treatment at 250 °C for 10 minutes in a hydrogen atmosphere, to form palladium silicide, which acts as a Schottky contact (Chapter 2). Note the fact that the entire semiconductor is active, thus reducing considerably the device series resistance. The electroplated gold reduces the device thermal resistance and will be soldered onto the case of the package.

An alternative is an $M-n-p^+$ construction where the active (drift) region is the n-type epitaxial layer. The normal operation is with the metal positively biased and the metal injecting minority holes in the depleted n-layer. The substrate should be as thin as practical, in order to reduce the series resistance. The structure will be mounted with the metal-junction down.

We note the attempt to construct a GaAs Schottky-barrier BARITT structure [53] and the interesting discussion about the occurrence of BARITT oscillations and avalanche transit-time oscillations (see the next chapter). It was concluded that there is a smooth transition between the reach-through characteristic and the avalanche breakdown characteristics, depending upon the semiconductor thickness and the efficiency of the injecting contact [53].

A punched-through p^+np^+ structure [18]–[20], [25] will be made by a p^+ diffusion in the n-type epitaxial layer grown on a p^+-substrate. The diffusion depth controls the drift length, L. The mesa technique is also used. An $n^+-p-\pi-n^+$ BARITT diode with low punch-through voltages (a few volts) was constructed recently [54].

References

1. W. SHOCKLEY and R. C. PRIM, 'Space-charge limited emission in semiconductors', *Phys. Rev.*, **90**, 753–758 (1953).
2. S. DENDA and M. A. NICOLET, 'Pure space-charge-limited current in silicon', *J. Appl. Phys.*, **37**, 2412–2424 (1966).
3. G. PERSKY, 'Thermionic saturation of diffusion currents in transistors', *Solid-St. Electron.*, **15**, 1345–1351 (1972).
4. G. T. WRIGHT, 'Depletion layer formation, space-charge injection and current-voltage characteristics for the silicon $p-n-p$ ($n-p-n$) structure', *Solid. St. Electron.*, **15**, 381–386 (1972).
5. D. DASCALU, 'Space-charge-limited minority hole flow in very high resistivity silicon', *Rev. Roum. Phys.*, **17**, 675–708 (1972).
6. M. A. LAMPERT and F. EDELMAN, 'Theory of one-carrier, space-charge-limited currents including diffusion and trapping', *J. Appl. Phys.*, **35**, 2971–2982 (1964).
7. R. STRATTON and E. L. JONES, 'Effect of carrier heating on the diffusion currents in space-charge-limited current flow', *J. Appl. Phys.*, **38**, 4596–4608 (1967).
8. W. SHOCKLEY, 'Negative resistance arising from transit-time in semiconductor diodes', *Bell. Syst. Techn. J.*, **33**, 799–826 (1954).
9. H. YOSHIMURA, 'Space-charge-limited and emitter-current-limited injections in space charge region of semiconductors', *IEEE Trans. Electron Dev.*, **ED–11**, 412–422 (1964).
10. C. K. BIRDSALL and W. B. BRIDGES, 'Electron dynamics in diode regions', Academic Press, New York, London, 1966.
11. D. DASCALU, 'Transit-time effects in unipolar solid-state devices', Editura Academiei, București, Romania, and Abacus Press, Tunbridge Wells, Kent, 1974.
12. T. MISAWA, 'Negative resistance due to transit-time in SCL current structure', *Solid-St. Dev. Res. Conf.*, Santa Barbara, California, June 1967.
13. D. DASCALU, 'Space-charge waves and high-frequency negative resistance of SCL diodes', *Int. J. Electron.*, **25**, 301–330 (1968).
14. D. DASCALU, 'Small-signal theory of space-charge-limited diodes', *Int. J. Electron.*, **21**, 183–200 (1966).
15. G. T. WRIGHT, 'Punch-through transit-time oscillator', *Electron. Lett.*, **4**, 543–544 (1968).
16. S. G. LIU and J. J. RISKO, 'Low noise punch-through $p-n-\nu-p$, $p-n-p$ and $p-n-$metal microwave diodes', *R.C.A. Rev.*, **32**, 636–644 (1971).
17. C. P. SNAPP and P. WEISSGLAS, 'Experimental comparison of silicon p^+-n-p^+ and $Cr-n-p^+$ transit-time oscillators', *Electron. Lett.*, **7**, 743–744 (1971).
18. N. B. SULTAN and G. T. WRIGHT, 'Punch-through oscillator – new microwave solid-state source', *Electron. Lett.*, **8**, 24–26 (1972).
19. G. BJÖRKMAN and C. P. SNAPP, 'Small-signal noise behaviour of companion p^+-n-p^+ and $p^+-n-\nu-p^+$ punch-through microwave diodes', *Electron. Lett.*, **8**, 501–503 (1972).

20. C. P. Snapp and P. Weissglas, 'On the microwave activity of punch-through injection transit-time structures', *IEEE Trans. Electron Dev.*, **ED–19**, 1109–1118 (1972).
21. S. Nagano, 'Negative conductance and d.c. voltage drop of a *pnp*-silicon BARITT diodes' (in Japanese), *Trans. IECE*, **55–B**, 481–482 (1972).
22. N. B. Sultan and G. T. Wright, 'Electronic tuning of the punch-through injection transit-time (PITT) microwave oscillator', *IEEE Trans. Microwave Theory Techn.*, **MTT–20**, 773–775 (1972).
23. A. Sjölund, 'Small-signal noise analysis of p^+-n-p^+ BARITT diodes', Electron Lett., **9**, 2–4 (1973).
24. J. L. Chu and S. M. Sze, 'Microwave oscillations in $p-n-p$ reach-through BARITT diodes', *Solid-St. Electron.*, **16**, 85–91 (1973).
25. G. T. Wright and N. B. Sultan, 'Small-signal design theory and experiment for the punch-through injection transit-time oscillator', *Solid-St. Electron.*, **16**, 535–544 (1973).
26. D. J. Delagebeaudeuf and J. Lacombe, 'Power limitation of punch-through injection transit-time oscillators', *Electron. Lett.*, **9**, 538–539 (1973).
27. A. Sjölund, 'Small-signal analysis of punch-through injection microwave devices', *Solid-St. Electron.*, **16**, 559–569 (1973).
28. D. Dascalu, 'Small-signal theory of unipolar injection currents in solids', *IEEE Trans. Electron Dev.*, **ED–19**, 1239–1251 (1972).
29. G. T. Wright, 'Small-signal characteristics of semiconductor punch-through injection and transit-time diodes', *Solid-St. Electron.*, **16**, 903–912 (1973).
30. H. A. Haus, H. Statz and R. A. Pucel, 'Noise measure of metal-semiconductor-metal Schottky-barrier microwave diodes', *Electron. Lett.*, **7**, 667–669 (1971).
31. W. Shockley, J. A. Copeland and R. P. James, 'The impedance field method of noise calculation in active semiconductor devices', in P. O. Lowdin (editor), 'Quantum theory of atoms, molecules and solid state', Academic Press, New York, London, 1966; pp. 537–563.
32. H. Johnson, 'A unified small-signal theory of uniform-carrier-velocity semiconductor transit-time diodes', *IEEE Trans. Electron Dev.*, **ED–19**, 1156–1166 (1972).
33. H. W. Rüegg, 'A proposed punch-through microwave negative-resistance diode', *IEEE Trans. Electron Dev.*, **ED–15**, 577–585 (1968).
34. U. B. Sheorey, I. Lundström and E. A. Ash, 'Analysis of punch-through-injection for a transit-time negative-resistance diode', *Int. J. Electron.*, **30**, 19–32 (1971).
35. J. Lacombe, 'Fonctionnement des diodes à injection thermoionique en mode de transit', *Rev. Techn. Thomson-CSF*, **4**, 467–480 (1972).
36. K. Kawarada and Y. Mizushima, 'Large-signal analysis on negative-resistance diode due to punch-through injection and transit-time effect', *Japanese J. Appl. Phys.*, **12**, 423–433 (1973); see also J.A.C. Stewart, 'BARITT-diode large-signal performance', *Electron. Lett.*, **10**, 193–194 (1974).
37. S. M. Sze, D. J. Coleman, Jr., and A. Loya, 'Current transport in metal-semiconductor-metal (MSM) structures', *Solid-St. Electron.*, **14**, 1209–1218 (1971).
38. D. J. Coleman, Jr., 'Transit-time oscillations in BARITT diodes', *J. Appl. Phys.*, **43**, 1812–1818 (1972).
39. J. L. Chu, G. Persky and S. M. Sze, 'Thermionic injection and space-charge-limited current in reach-through p^+-n-p^+ structures', *J. Appl. Phys.*, **43**, 3510–3515 (1972).
40. M. El-Gabaly, J. Nigrin and P. A. Goud, 'Stationary charge transport in metal-semiconductor-metal (MSM) structures', *J. Appl. Phys.*, **44**, 4672–4680 (1973).
41. P. Antognetti, A. Chiabrera and G. R. Bisio, 'Small-signal theory of transit-time diodes', *Solid-St. Electron.*, **16**, 345–350 (1973).
42. D. J. Coleman, Jr. and S. M. Sze, 'A low-noise metal-semiconductor-metal (MSM) microwave oscillator', *Bell Syst. Techn. J.*, **50**, 1695–1699 (1971).
43. C. A. Lee and G. C. Dalman, 'A low-noise Ku-band silicon diode oscillator', Paper at Internat. Electron Dev., Oct. 1971, Washington; 'Local oscillator noise in a silicon $Pt-n-p^+$ microwave diode source', *Electron. Lett.*, **7**, 565–566 (1971).
44. J. Freyer, M. Claassen and W. Harth, 'Fabrication of an epitaxial-silicon $Pd-n-Pd$ microwave generator', *Arch. Elektr. Übertr.*, **26**, 150–151 (1972).
45. H. Herbst and W. Harth, 'Frequency-modulation sensitivity and frequency-pushing factor of a $Pd-n-p^+$ punch-through microwave diode', *Electron. Lett.*, **8**, 358–359 (1972).

46. J. Helmke, H. Herbst, M. Claassen and W. Harth, 'F.M.-noise and bias-current fluctuations of a silicon $Pd-n-p^+$ microwave oscillator', *Electron. Lett.*, **8**, 158–159 (1972).
47. M. Matsumura, 'Small-signal admittance of BARITT diodes', *IEEE Trans. Electron Dev.*, **ED-19**, 1131–1133 (1972).
48. J. A. C. Stewart and J. Wakefield, 'Simulation of $Pt-n-p^+$ silicon punch-through device', *Electron. Lett.*, **8**, 378–379 (1972).
49. K. P. Weller, 'Small-signal theory of a transit-time-negative resistance device utilizing injection from a Schottky barrier', *R.C.A. Rev.*, **32**, 372–382 (1971).
50. D. Dascalu and G. Brezeanu, 'Large-signal-approach for a Schottky emission transit-time diode', in preparation.
51. H. Statz, R. A. Pucel and H. A. Haus, 'Velocity fluctuation noise in metal-semiconductor-metal diodes', *Proc. IEEE*, **60**, 644–645 (1972).
52. A. I. Stoller, R. F. Speers and S. Opresko, 'A new technique for etch thinning silicon wafers', *RCA Rev.*, **31**, 265–270 (1970).
53. H. T. Minden, 'Gallium arsenide dual Schottky barrier diodes', *Solid-St. Electron.*, **16**, 1185–1188 (1973).
54. D. Delagebeaudeuf, 'Low-voltage punch-through injection structure', *Electron. Lett.*, **10**, 166–167 (1974).

Problems

5.1. Derive the steady-state $J-V$ characteristic of a punch-through p^+-n-p^+ diode by assuming that the hole velocity is saturated.

Hint: By assuming SCL injection one obtains,

$$J = \frac{2\varepsilon v_{\text{sat}}}{L^2} V, \tag{P.5.1.1}$$

whereas under barrier-controlled injection conditions the characteristic should be non-linear (see equations (5.3.14)).

5.2. Calculate the low-frequency parallel capacitance of the saturated-velocity transit-time diode.
Hint: The small-signal impedance (5.3.10) may be written as a power series in the signal frequency.

5.3. Find the high-frequency series capacitance for the saturated-velocity transit-time diode and discuss its frequency and bias dependence. Compare to the results of more exact calculations, Figure 5.3.6.

5.4. Find the small-signal impedance of the punch-through diode with the steady-characteristic of Figure 5.1.2 by using the more refined boundary condition (5.3.6) (for $\sigma_1(0) \to \infty$ this impedance should be identical with that given by equation (5.2.1)), calculate the series resistance and discuss the possibility of high-frequency power-generation.

5.5. Show that the source a.c. conductivity, equation (5.7.2), has a maximum as the bias current increases and calculate the minimum value of $\Gamma = \varepsilon/\sigma_1(0)T_L$ for a silicon MSM structure with $L = 10\,\mu\text{m}$, $V_{FB} = 60$ V and an injection barrier $e\phi_{p1} = 0.29$ eV.

5.6. Explain, qualitatively, the frequency and bias dependence of the negative conductance computed for an MSM structure and reproduced in Figure 5.7.2.
Hint: Compare with the results of the elementary theory.

5.7. Show, quantitatively, that if the diffusion is neglected and the velocity modulation is very small, the phasor of the alternating particle current as a function of x will be constant in magnitude and its phase will be proportional to the transit time from $x = 0$ to x.

6

Impact-avalanche Transit-time (IMPATT) Diodes

6.1 Principle of Operation

The IMPact-Avalanche Transit-Time (IMPATT) diode is a source of power at microwave and millimetre-wave frequencies. Its active behaviour is due to a high-frequency negative resistance effect arising from both avalanche multiplication and carrier transit-time delay.

There exist several different IMPATT structures. Many of them consist of high-conductivity regions of opposite type separated by a depleted region. This structure is biased 'in reverse', as shown in Figure 6.1.1a. A large current may flow,

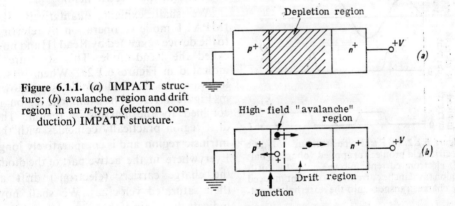

Figure 6.1.1. (a) IMPATT structure; (b) avalanche region and drift region in an n-type (electron conduction) IMPATT structure.

however at high applied voltages, where charge carriers are generated by impact-ionization inside the depleted region. Due to the fixed space-charge (ionized impurities) the electric field inside the depleted region is non-uniform and the impact-avalanche will occur only in a certain *avalanche region* where the electric field intensities are the highest. If, for example, the central region is n-type (lowly doped) then the avalanche will be localized in the vicinity of the p^+ region (at the $p-n$ junction). The holes and electrons generated in this avalanche region will drift in opposite directions: holes will be absorbed by the p^+ region and have little effect upon the total current flow, whereas the electrons will drift through the remaining portion of the depleted region, hence called the *drift region* (Figure 6.1.1b). Therefore, the IMPATT diode is *somewhat similar to a unipolar transit-time diode,* provided that the avalanche region is thin compared to the drift region and may be considered simply as an 'injection region' which supplies the flux of electrons: a pure unipolar electron current flows through most of the active region of the structure.

By analysing the punch-through transit-time diode, in the preceding chapter we have indicated that the small-signal power dissipation is due to carrier-velocity modulation and charge-density modulation. The carrier velocity modulation represents a power loss in positive-mobility semiconductors whereas the charge-density modulation may generate signal power if a proper phase relation between the electric field and the charge density is provided by charge transport and charge injection. The drift velocity is saturated in the drift region of almost all IMPATT structures (see below), the velocity modulation is practically zero and this mechanism of power loss is therefore eliminated*. The dynamic resistance of this saturated-velocity transit-time diode will be negative if the injection mechanism provides a proper phase relation between the injected particle current and the electric field.

We shall explain qualitatively the IMPATT mode of operation by referring to the device suggested by Read [1] and now called the Read diode. This structure is indicated in Figure 6.1.2a. When biased to avalanche the field profile is as shown in Figure 6.1.2b. The avalanche region is confined to the narrow n region. The drift region practically coincides with the intrinsic region and is comparatively long. Everywhere in the active part of the diode the charge carriers (electrons) drift at their saturated velocities. We shall now indicate the origin of the phase shift between the applied voltage signal and the alternating current, shift which determines the negative resistance behaviour. This phase shift has two components [2]. First, there is a phase delay caused by the avalanche process. The rate of generation of electron-hole pairs increases with both the electric field intensity and the density of carriers which are already present in the high-field region. Assume that a sinusoidal voltage signal (Figure 6.1.2c) is superimposed on a steady state voltage which is almost equal to the breakdown voltage V_B. As far as the total voltage is above V_B the carrier density in the avalanche region will grow rapidly with time. When the applied voltage falls below V_B, the carrier density starts to decrease. Because the carrier velocity is saturated, the particle current injected into the drift region is proportional to the

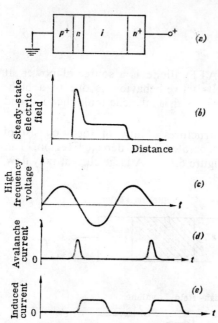

Figure 6.1.2. (a) Read-type structure; (b) static electric field profile; (c) applied voltage signal; (d) electron current injected into the drift region (avalanche current); (e) current induced by charge transport into the external circuit.

---

* The importance of carrier drift at saturated velocity for large-signal behaviour will be emphasized below.

charge density (this is a pure electron current) and has the form of a sharp pulse as indicated in Figure 6.1.2d. The injected particle current has a 90 degree phase lag with respect to the signal. This inductive delay is due to the inertia of the avalanche process and is frequency-independent.

An additional delay is frequency-dependent and provided by charge transport. The total current flowing through the drift region is the sum of the particle current and the displacement current:

$$J(t) = j(x, t) + \varepsilon \frac{\partial \mathscr{E}(x, t)}{\partial t}. \qquad (6.1.1)$$

By integrating this equation from $x = 0$ to $x = L$ one obtains

$$LJ(t) = \int_0^L j(x, t)\, dx + \varepsilon \frac{\partial}{\partial t} \int_0^L \mathscr{E}(x, t)\, dx \qquad (6.1.2)$$

or

$$J(t) = j_i(t) + C_0 \frac{dV}{dt}, \qquad (6.1.3)$$

where $j_i(t)$ is the so-called 'induced current' which is the spatial average of the particle current

$$j_i(t) = \frac{1}{L} \int_0^L j(x, t)\, dx \qquad (6.1.4)$$

and $C_0 = \varepsilon/L$ is the geometric capacitance of the drift space. The induced current is the only component which can generate power. Because the carrier velocity is saturated,

$$j_i(t) = \frac{v_{\text{sat}}}{L} \int_0^L en(x, t)\, dx = \frac{v_{\text{sat}} Q_n(t)}{L}, \qquad (6.1.5)$$

where $Q_n(t)$ is the total electron charge per unit area contained in the drift space at the time instant t. Due to charge injection in sharp pulses (Figure 6.1.2d) the induced current has approximately a rectangular form shown in Figure 6.1.2e; the pulse width is equal to the carrier transit-time $T_L = L/v_{\text{sat}}$. This transit-time should be, of course, shorter than the signal period. The fundamental of $j_i(t)$ is in antiphase with the sinusoidal voltage if the transit-time through the drift region T_L is half the signal period (this corresponds to a transit angle $\theta = \omega T_L = \pi$). The device does not behave as a pure negative resistance but as a complex impedance, due to the capacitive effect of the displacement current, and also as a non-linear element (current harmonics are present). The high-frequency power generated should be calculated by averaging the instantaneous power over a signal period. Such a calculation was done by assuming a rectangular current pulse starting at $\omega t = \pi$ and lasting for $\theta \leqslant \pi$. The generated h.f. power P was normalized to $V_1 I_0/\pi$ (where V_1 is the signal amplitude and I_0 is the average current) has a maximum for $\theta = 0.742\pi$, as indicated in Figure 6.1.3. The efficiency P/V_0I_0 depends upon θ and V_1/V_0 and is shown on the same Figure. For $\theta = \pi/2$ and $V_1/V_0 = 0.5$, the efficiency is above 30%, as first shown by Read [1].

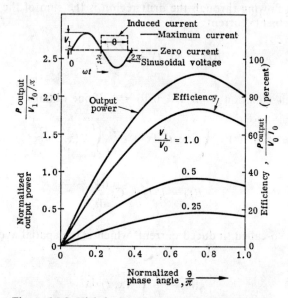

Figure 6.1.3. High-frequency generated power (normalized as shown) and efficiency computed for the waveforms indicated and represented versus the normalized transit-angle $\theta = \omega T_{\text{drift}}$ (T_{drift} is the carrier transit-time for the drift region), for several amplitudes V_1 of the sinusoidal voltage. The above results are applicable for a Read-type structure. V_0 and I_0 are bias voltage and bias current, respectively (*after Haddad, Greiling and Schroeder,* [2]).

If the average current is high the effect of mobile space-charge should be, however, taken into account. The sharp pulse of avalanche current (Figure 6.1.2d) injects a bunch of electrons which propagates through the drift region: this space charge decreases the electric field behind it and increases the field in front of it. The field in the avalanche region is therefore depressed and drops below the breakdown level before the applied voltage does. This determines a reduction of the phase lag between the avalanche current and the signal. If the bias current is increased the space-charge increases and the above phase lag will decrease. If the bias current is held constant, this phase lag increases with the amplitude of the high-frequency signal because the effect of injected space charge is less important [2].

The phase delay due to the avalanche process is slightly less than 90 degrees even at low bias currents, whereas the transit-time delay is proportional to the operating frequency. This determines a positive dynamic resistance at relatively low frequencies where the total delay is below 90 degrees. The diode resistance becomes negative above a certain cutoff frequency*. This cutoff frequency increases with increasing average current because of the space charge effect just explained.

The Read diode discussed above is convenient for an analytical description of phenomena and for their intuitive understanding. The efficiency predicted for such a structure is also high. There exist, however practical difficulties in constructing a device with such a doping profile with conventional fabrication techniques of impurity diffusion and epitaxial growth, although the ion implantation seems more promising. The majority of actual IMPATT diodes are asymmetrical pn junctions biased at breakdown, for example abrupt p^+nn^+ structures (n^+ being a high-conductivity substrate). We cannot actually separate in such a structure the relatively wide avalanche region from the drift region. Because the charge carriers generated by the avalanche process have different transit times, we expect a broad band and low negative resistance as compared to that of the Read structure. A more detailed analysis will indicate low-frequency and even static (zero-frequency) negative resistance. The complete description of IMPATT operation for a diode with a realistic doping profile does, however, require a computer simulation starting from basic differential equations describing current transport and charge conservation.

6.2 Static Analysis

The steady-state analysis of an IMPATT structure should yield the electric field distribution, the location of the avalanche region, the current and carrier distribution. A first-order calculation determines the electric field profile by neglecting the mobile space charge: this is valid at the onset of avalanche breakdown (very low currents). Figure 6.2.1 shows several IMPATT devices, their doping and internal field distribution. The $p^+n\nu n^+$ diode in Figure 6.2.1a is a slightly modified Read structure. An exact computation of the breakdown voltage V_B should use the field-dependent ionization rates, namely [3], [4]

$$\int_{\text{field region}} \langle \alpha \rangle \, dx = 1 \quad \text{defines } V = V_B \qquad (6.2.1)$$

where the ionization integrand $\langle \alpha \rangle$ is defined in Section 1.9. We recall the fact that if the electron and hole ionization rates, α_n and α_p, are equal then $\langle \alpha \rangle = \alpha_n = \alpha_p$. Because α_n and α_p are strongly field-dependent, $\langle \alpha \rangle$ gives a significant contribution to the ionization integral only into a high-field region called avalanche region and defined, for example (see also Figure 6.2.3), by

$$\int_{\text{avalanche region}} \langle \alpha \rangle \, dx = 0.95, \quad \text{at } V = V_B. \qquad (6.2.2)$$

---

* The avalanche (cutoff) frequency defined in the approximate small-signal theory has a somewhat different explanation as will be shown in due course.

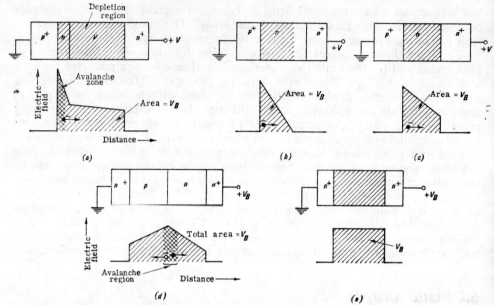

Figure 6.2.1. IMPATT structures and steady-state internal field distribution (all regions are uniformly doped): (a) modified Read diode; (b) non punched-through p^+nn^+ diode (n-type diode, electron conduction); (c) punched-through p^+nn^+ diode; (d) double-drift structure; (e) pin (uniform avalanche) diode.

An asymmetric abrupt pn junction may be used for IMPATT operation as a p^+nn^+ (or the complementary n^+pp^+) structure (Figure 6.2.1b). If the depletion length at breakdown is longer than the n-region thickness then the device will be punched-through at breakdown and the electric field profile is trapezoidal, as indicated in Figure 4.2.1c. A punched-through structure may be preferred because the series resistance introduced by the undepleted n region is eliminated and because the electric field in the drift region is everywhere above the critical value for velocity saturation. However, in a heavily punched-through structure the avalanche region becomes comparable or even thicker than the drift region. Moreover, such a device often operates in an anomalous mode which is basically different from the IMPATT mode.

Figure 6.2.2 shows the breakdown voltage, the maximum field at breakdown and the depletion-layer width at breakdown for asymmetric abrupt pn junctions. These data should be used only if the IMPATT p^+nn^+ (or n^+pp^+) structure is not punched-through. As an example, the actual V_B for a punched-through device is lower than that indicated by Figure 6.2.2.

Figure 6.2.1d shows a double-drift IMPATT structure. This is a symmetrical abrupt pn junction. The electron-hole pairs are generated in the central avalanche region which is shifted away from the metallurgical junction towards the contact which collects the more highly ionizing carriers (electrons in silicon). There are two drift regions: for electrons on the n-side and for holes on the p-side. This device acts essentially as two complementary single-drift IMPATT diodes.

Figure 6.2.1e shows a *pin* diode. The electric field is maximum and constant in the central intrinsic region and the avalanche region coincides with the drift region. Misawa [6] has shown that the dynamic resistance of such a diode is negative from zero frequency up to transit-time frequencies (see the next Section).

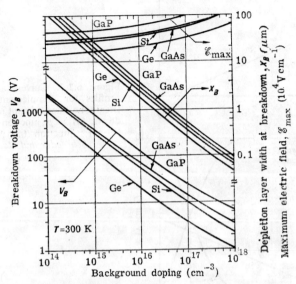

Figure 6.2.2. Breakdown voltage V_B, maximum electric field $\mathscr{E}_{max}$ and depletion-layer width at breakdown x_B, versus background impurity concentration for asymmetrical abrupt junctions, at room temperature. (*after Sze and Ryder*, [5]).

The electron ionization rate in silicon is much higher than the hole ionization rate. Therefore the properties of complementary structures are somewhat different. Figure 6.2.3 shows, for example, the current distribution and the electric field profile for silicon n^+p and p^+n abrupt junctions with the same background doping of 7×10^{15} cm^{-3}. The mobile space-charge is neglected. The electron and hole currents, J_n and J_p are normalized to the total steady-state current J_0 and computed by using equations (1.9.2) – (1.9.9) (the saturation current is neglected). The centre of the avalanche region, x_c, is defined by $J_p(x_c) = J_n(x_c)$, whereas its width is $x_a = |x_2 - x_1|$ where $J_p(x_2) - J_p(x_1) = 0.95 J_0$. The avalanche region is much more localized in the n^+p device (40% of the depletion width). The p^+n diode has a breakdown voltage higher by about 7%. Figure 6.2.4 shows the ratio of avalanche region width to the total depletion-layer width (non-punched-through structures) as a function of background doping for abrupt one-sided and symmetrical junctions. The complementary one-sided GaAs junctions show identical properties because $\alpha_n \approx \alpha_p$ [7]. The most advantageous is the silicon n^+p structure because the avalanche region is relatively thin. The background concentration is chosen to determine a suitable drift region length for a given operating frequency. Therefore, Figures 6.2.2 and 6.2.3 may be used for a first-order design of an IMPATT structure.

254   Bulk- and Injection-controlled Devices

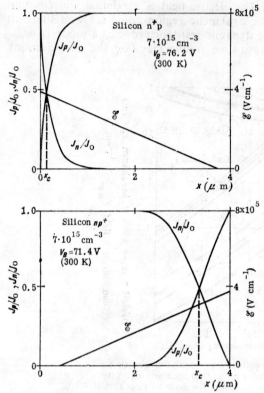

Figure 6.2.3. Current and electric field profiles for complementary Si asymmetrical (one-sided) abrupt junctions at breakdown: (a) n^+p junction (p-type diode); (b) np^+ junction (n-type diode). (*after Schroeder and Haddad*, [7]).

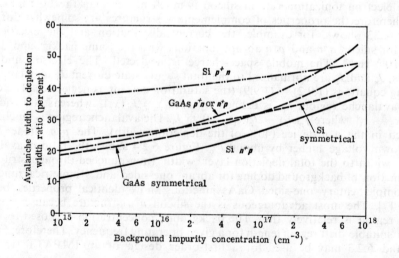

Figure 6.2.4. Ratio of avalanche region width to total depletion-layer width as a function of doping level (one-sided abrupt junctions). (*after Schroeder and Haddad*, [7]).

At high current-densities the current-voltage characteristics are influenced by the space charge of the mobile carriers generated by the impact avalanche process. Bowers [8] has made numerical calculations for silicon p^+nn^+ diodes with realistic

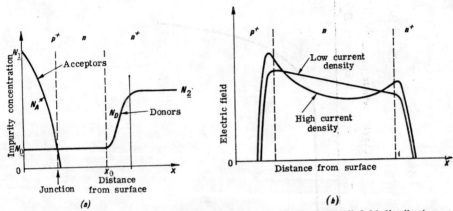

Figure 6.2.5. (a) Realistic doping profile for a p^+nn^+ IMPATT diode; (b) field distribution at low and, respectively, high current densities (schematic) (*after* Bowers, [8]).

doping profiles and showed that the incremental resistance may be either positive or negative. The shape of the doping profile (with diffused p^+n junction and out-diffusion from the low resistivity n^+ substrate) is shown in Figure 6.2.5a. The electric field profile for low current density (negligible mobile charge) for a punched-through p^+nn^+ device is shown qualitatively in Figure 6.2.5b. Also shown is a possible field distribution at high current densities, where the charge density of electrons and holes is comparable to the density of ionized donors in the n region. In the left part of the structure there exists an excess of holes and due to this positive charge the field gradient is larger than that determined by the doping of the n region. The electron charge towards the right part of the device also modifies the field gradient according to the Poisson equation. The electric field decreases in the middle of the structure but increases towards the ends and moves into the highly doped regions in order to maintain the larger current. The diode voltage is equal to the area under the $\mathscr{E} = \mathscr{E}(x)$ curve. This voltage may increase or decrease with the increasing current depending upon the diode parameters. In the latter case the incremental resistance is negative [8].

Figures 6.2.6 (a and b) show the $J-V$ characteristic and field profiles at three current levels computed for a non punched-through silicon p^+nn^+ diode [8]. As the current density increases the edge of the depletion layer moves slightly into the p^+ region and the maximum field increases in order to sustain the greater current. Near the right side of the depletion layer (Figure 6.2.6b) there is a rising of field due to an excess of electrons and therefore the depletion layer spreads farther into the n region. Because the total area under the field plot increases with an increase in current, the voltage across the diode increases and the incremental resistance is positive.

The situation is different for the p^+nn^+ diode whose $J-V$ characteristic and field profiles are plotted in Figure 6.2.6c and d [8]. The doping of the central n

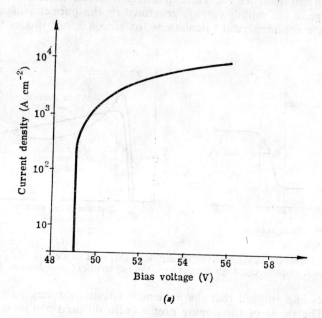

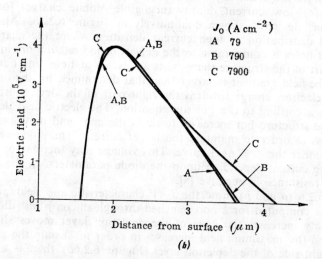

Figure 6.2.6. (a) $J-V$ characteristic and (b) internal field profile for a p^+nn^+ silicon diode with $N_0 = 1.8 \times 10^{16} \text{cm}^{-3}$, $N_1 = 2 \times 10^{19} \text{cm}^{-3}$, $N_2 = 5 \times 10^{17} \text{cm}^{-3}$, $A_0 = 1.17 \times 10^4 \text{cm}^{-1}$, $B_0 = 2.28 \times 10^4 \text{cm}^{-1}$, $x_0 = 8 \times 10^{-4}$ cm for the doping profile shown in Figure 6.2.5a, where
$$N_A = N_1 \,\text{erfc}\,(A_0 x),\ N_D = N_0 + N_2 \,\text{erfc}\,[B_0(x_0 - x)];$$

6. IMPATT Diodes

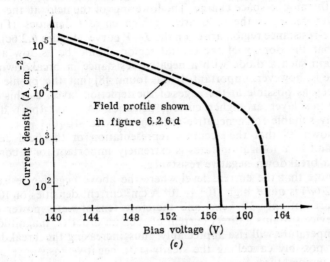

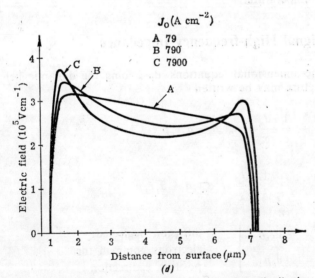

Figure 6.2.6. (c) $J - V$ characteristic and (d) field distribution for a diode with $N_0 = 10^{15} \text{cm}^{-3}$, $N_1 = 2 \times 10^{20} \text{cm}^{-3}$, $N_2 = 5 \times 10^{17} \text{cm}^{-3}$, $A_0 = 1.81 \times 10^4 \text{cm}^{-1}$, $B_0 = 2.28 \times 10^4 \text{cm}^{-1}$, $x_0 = 7.5 \times 10^{-4} \text{cm}$. Also shown (dashed) is the $J - V$ characteristic of another diode having $N_0 = 2 \times 10^{14} \text{cm}^{14}$, $N_1 = 2 \times 10^{20} \text{cm}^{-3}$, $N_2 = 5 \times 10^{17} \text{cm}^{-3}$, $A_0 = 1.81 \times 10^4$, $B_0 = 2.28 \times 10^4 \text{cm}^{-1}$, $x_0 = 7.5 \times 10^{-4} \text{cm}$. (after Bowers, [8]).

region is low and this diode is punched-through at breakdown. Because the field cannot penetrate appreciably into the n^+ region, the field profile is highly distorted by the mobile space charge. The lowering of the field at the centre of the depletion layer decreases the total area at high current densities (Figure 6.2.6d), and a negative-resistance region arises on the $J-V$ curve (Figure 6.2.6c). Bowers [8] has shown that the doping of the central region should be as low as possible in order to obtain such a diode with a negative-resistance in breakdown. The entire doping profile is, however, important. It was found [8] that the profile should be as abrupt or steep as possible at the edges of the depletion layer, as this prevents the spreading of this layer at higher current densities. A p^+in^+ (pin) diode should be particularly suitable for a negative-resistance device [4], [6].

It was shown [9] that the accurate representation of the field dependence of the electron and hole ionization rates is extremely important for a correct calculation of such a breakdown negative resistance.

We also note that the current level where the above negative resistance occurs [8] (Figure 6.2.6d) is quite high ($10^3 - 10^4$ A cm^2 current densities or $10^2 - 10^3$ mA for a diode area of only 10^{-4} cm^2 and therefore tens of watts power dissipation). Even if the device sustains a direct current of such an order of magnitude, the semiconductor temperature will rise appreciably, thus increasing the breakdown voltage [2] − [4] and possibly cancelling the steady-state negative resistance computed at the ambient temperature.

6.3 Small-signal High-frequency Impedance

The basic one-dimensional equations describing the dynamic behaviour of an IMPATT structure may be written

$$\frac{\partial n}{\partial t} = \frac{1}{e}\frac{\partial J_n}{\partial x} + \alpha_n v_n n + \alpha_p v_p p \tag{6.3.1}$$

$$\frac{\partial p}{\partial t} = -\frac{1}{e}\frac{\partial J_p}{\partial x} + \alpha_n v_n n + \alpha_p v_p p \tag{6.3.2}$$

$$\left.\begin{array}{l} J_n = -ev_n n \\ J_p = -ev_p p \end{array}\right\} \quad \begin{array}{l}\text{diffusion neglected,} \\ \text{electrons going in the} \\ \text{positive } x \text{ direction}\end{array} \tag{6.3.3}$$
$$\tag{6.3.4}$$

$$J = J_n + J_p + \varepsilon \frac{\partial \mathscr{E}}{\partial t} = \text{total current} = J(t) \tag{6.3.5}$$

$$\frac{\partial \mathscr{E}}{\partial x} = \frac{e}{\varepsilon}(N_D - N_A + p - n). \quad \begin{array}{l}\text{(Poisson} \\ \text{equation)}\end{array} \tag{6.3.6}$$

The generation rate in continuity equations (6.3.1) and (6.3.2) is assumed to be due only to impact multiplication. The ionization rates for electrons (α_n) and holes (α_p), as well as drift velocities (v_n, v_p) are field-dependent. The electron and hole currents are written by neglecting diffusion. The total electric current density J is independent of x^*.

Almost always the carrier velocities are assumed saturated (the electric fields in the avalanche region are at least an order of magnitude higher than critical field required for velocity saturation). The analytical computation has been developed by assuming $v_{n\,\text{sat}} = v_{p\,\text{sat}} = v_\text{sat}$ and $\alpha_n = \alpha_p = \alpha$. The latter condition is far from being satisfied for silicon and the results of analytical computations developed on such a basis could only illustrate some *qualitative* aspects of the IMPATT mode of operation.

Although the density of ionized impurities may be, in principle, position dependent, for analytical calculations, we have to refer to a specific doping profile. *The Read structure* (Figure 6.1.2) is chosen for simplicity. The active region of this structure divides itself into two distinct parts: the avalanche and drift regions which can be examined separately and the solutions will be matched at the boundary between them.

Drift region. Consider a one-dimensional drift region: a flux of electrons is injected at $x = 0$ and collected at $x = L$. The particle current injected at $x = 0$ is $j(0, t)$. If the electron velocity is saturated, then

$$j(x, t) = j\left(0, t - \frac{n}{v_\text{sat}}\right). \tag{6.3.7}$$

The steady-state current is

$$J_0 = J_n|_{x>0} = -en_0 v_\text{sat}, \tag{6.3.8}$$

J_n is conventionally negative, because it is directed towards negative x. If the boundaries of the drift region are fixed at $x = 0$ and $x = L$, then an increase ΔJ_0 of the current requires an increase $\Delta n_0 = J_0/ev_\text{sat}$ of the mobile charge density. By using the Poisson equation, the increase of the voltage drop on this drift region is $e\Delta n_0 L^2/2\varepsilon$. Therefore the incremental resistance of the drift space or the *space-charge resistance* (for a unit area device) is

$$R_\text{sc} = \frac{L^2}{2\varepsilon v_\text{sat}}. \tag{6.3.9}$$

The same space-charge resistance of a saturated-velocity drift space was calculated in Section 5.3 (equation 5.3.13). The high-frequency impedance of the drift space will be calculated as shown in Section 4.6 and the result (valid for large-signal operation as well) is

$$Z_\text{drift} = \left(\frac{\mathscr{E}_1(0)L}{\underline{J}_1} - \frac{1}{i\omega C_0}\right)\frac{1 - \exp(-i\theta)}{i\theta} + \frac{1}{i\omega C_0} \tag{6.3.10}$$

---

* $\partial J/\partial x = 0$ is obtained automatically by using equations (6.3.1), (6.3.2), (6.3.5) and (6.3.6). The numerical analysis discussed in the preceding Section was performed by using these equations with all time derivatives set equal to zero (steady-state).

where $\underline{\mathscr{E}}_1(0)$ and $\mathbf{J}_1$ are the alternating-current components of the electric field at $x = 0$ and the total current, respectively, $C_0 = \varepsilon/L$ is the unit-area geometrical capacitance of the drift space and $\theta = \omega T_L = \omega L v_{sat}^{-1}$ is the 'drift' transit-angle. The relation between $\underline{\mathscr{E}}_1(0)$ and $\mathbf{J}_1$ depends upon the injection mechanism*. Such a relation will be obtained below for avalanche injection.

Avalanche region. The dynamic behaviour of the avalanche region will be studied by assuming equal scattering limited velocities and equal ionization rates for electrons and holes. This is one of Read's original assumptions [1]. By adding and subtracting equations (6.3.1) and (6.3.2) one obtains respectively

$$\frac{1}{v_{sat}} \frac{\partial}{\partial t} j = \frac{\partial}{\partial x}(-J_n + J_p) + 2\alpha j \qquad (6.3.11)$$

$$\frac{1}{v_{sat}} \frac{\partial}{\partial t}(-J_n + J_p) = \frac{\partial j}{\partial x} \qquad (6.3.12)$$

and therefore

$$\frac{1}{v_{sat}} \frac{\partial^2 j}{\partial t^2} = v_{sat} \frac{\partial^2 j}{\partial x^2} + 2 \frac{\partial}{\partial t}(\alpha j), \qquad (6.3.13)$$

where $j = J_n + J_p = -e(p+n)v_{sat}$ is the total particle current. In this equation $\alpha = \alpha(\mathscr{E})$ and, at least formally, $\mathscr{E} = \mathscr{E}(j)$, such that (6.3.13) is a second-order non-linear equation with partial derivatives. This equation can be linearized for small-signal operation by assuming (Section 4.6)

$$\mathscr{E}(x, t) = \mathscr{E}_0(x, t) + \underline{\mathscr{E}}_1(x) \exp i\omega t, \quad |\underline{\mathscr{E}}_1| \ll \mathscr{E}_0 \qquad (6.3.14)$$

$$\alpha = \alpha(\mathscr{E}) = \alpha(\mathscr{E}_0) + \left.\frac{\partial \alpha}{\partial \mathscr{E}}\right|_{\mathscr{E}=\mathscr{E}_0} \times \underline{\mathscr{E}}_1 \exp i\omega t = \alpha_0 + \alpha_1 \underline{\mathscr{E}}_1 \exp i\omega t \qquad (6.3.15)$$

$$j(x, t) = j_0(x) + \mathbf{j}_1(x) \exp i\omega t = J_0 + \mathbf{j}_1(x) \exp i\omega t. \qquad (6.3.16)$$

One obtains

$$\frac{d^2 \underline{\mathscr{E}}_1}{dx^2} + \left[\frac{\omega^2}{v_{sat}^2} - \frac{2\alpha_1 J_0}{\varepsilon v_{sat}} + \frac{2i\omega \alpha_0}{v_{sat}}\right] \underline{\mathscr{E}}_1 = \left(\frac{2\alpha_0}{\varepsilon v_{sat}} - \frac{i\omega}{\varepsilon v_{sat}^2}\right) \mathbf{J}_1, \qquad (6.3.17)$$

which is a second-order linear equation with variable coefficients (α_0, α_1 depend on $\mathscr{E}_0$ and $\mathscr{E}_0 = \mathscr{E}_0(x)$). For a uniform avalanche region (like the intrinsic region in a

---

* $\underline{\mathscr{E}}_1(0) = 0$ in equation (6.3.10) corresponds to the SCL current injection. For barrier-controlled injection $\underline{\mathscr{E}}_1(0)$ is given, for example, by equation (5.3.6) and the diode impedance is (5.3.10).

pin diode) $\alpha_0 = \alpha_1 = $ constant* and equation (6.3.17) has constant coefficients; its general solution is

$$\mathscr{E}_1(x) = C_1 \exp i\lambda x + C_2 \exp(-i\lambda x) + \frac{\frac{2\alpha_0}{\varepsilon v_{sat}} - \frac{j\omega}{\varepsilon v^2_{sat}}}{\lambda^2} J_1 \qquad (6.3.18)$$

where

$$\lambda = \left(\frac{\omega^2}{v^2_{sat}} - \frac{2\alpha_1 J_0}{\varepsilon v_{sat}} + i\frac{2\alpha_0 \omega}{v_{sat}}\right)^{1/2} \qquad (6.3.19)$$

and the constants C_1 and C_2 will be determined by boundary conditions. The device impedance will be then calculated as $Z = \int_{-L_a}^{0} \mathscr{E}_1(x)\,dx/J_1$.

The above equations were used by Misawa [6] to calculate the impedance of a silicon pin diode. The device resistance was found to be negative at any frequency (we have indicated in the previous Section that the incremental resistance ($\omega \to 0$) should be negative) and has a maximum at a resonance frequency, as shown in Figure 6.3.1. At the same frequency the reactive part of impedance changes from inductive (lower frequencies) to capacitive. The physical understanding of the above negative resistance effect is facilitated by the following discussion. Consider a uniform avalanche region (Figure 6.3.2 a) and a non-uniformity in the middle of the structure creating an excess of both electrons and holes (Figure 6.3.2 b). The excess electrons and holes drift in opposite directions and their number increases due to avalanche multiplication. The electric field in the middle of this region decreases below the outside field (Figure 4.3.2 c). A significant decrease of the instantaneous voltage (equal to the area under the field profile) is possible, although the excess moving carriers induce an increasing current. This decrease occurs because the outside field does not have enough time to readjust to a higher value and preserve the breakdown condition $\int \langle \alpha \rangle \, dx = 1$ such as in the steady-state (see the explanation to Figure 6.2.5 b). Further, the field lowering decreases the ionization rate and produces a deficit of charge (Figure 6.3.2 d) which propagates and determines an increase of field (Figure 6.3.2 e) and then another cycle (Figure 6.3.2 b-e) starts. The propagation of electrons and holes is well described by the solution to equation (6.3.17), which contains two terms corresponding to two waves propagating in opposite directions (wave number λ defined by equation (6.3.19)). These waves move with an avalanche propagation velocity rather than the carrier drift velocity. Figure 6.3.1 does indeed show that the maximum of the negative resistance is not directly related to the drift transit-time L_a/v_{sat}. The resonance frequency increases with the increasing bias current $|J_0|$ because the avalanche wave grows faster when the number of carriers already present (in the absence of the signal) is larger (the theory indicates a proportionality to $|J_0|^{1/2}$).

---

* Because $\int_{-L_a}^{0} \alpha_0 \, dx = 1$, we have $\alpha_0 = 1/L_a$ ($L_a = $ length of the avalanche zone).

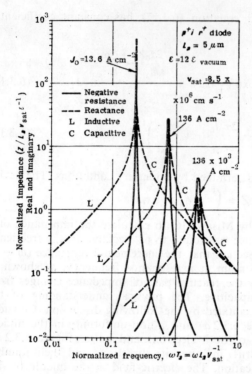

Figure 6.3.1. Normalized small-signal impedance *versus* normalized frequency for a p^+in^+ diode with $\varepsilon = 12\,\varepsilon_{\text{vacuum}}$, $v_{\text{sat}} = 8.5 \times 10^6$ cm s^{-1} (data representative for silicon) and active length $L_a = 5\,\mu$m (thickness of the intrinsic region). The impedance is normalized to $L_a^2 v_{\text{sat}} \varepsilon^{-1} = 2.77 \times 10^{-2}\,\Omega\,\text{cm}^2$, and the frequency is normalized to $v_{\text{sat}}/2\pi L_a = 2.71$ GHz. (*after Misawa, [6]*).

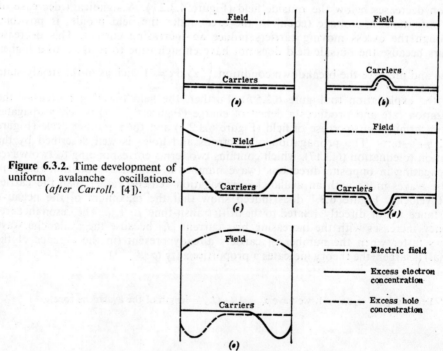

Figure 6.3.2. Time development of uniform avalanche oscillations. (*after Carroll, [4]*).

6. IMPATT Diodes

The above discussion shows that the avalanche region has associated with it a negative resistance effect due to the fields induced by moving charge carriers of opposite signs. However, to a first approximation, *the very thin avalanche region in a Read diode* has by itself only a reactive impedance and provides a correct phase for charge injection into the drift region, as shown below.

The qualitative description of the avalanche injection in the Read diode (Figure 6.1.2) corresponds to a relation of the form

$$\mathbf{j}_1(0) = \frac{\mathbf{V}_{1a}}{i\omega \mathscr{L}_a} \qquad (6.3.20)$$

where $\mathbf{V}_{1a}$ is the voltage drop on the avalanche region of length L_a and $\mathscr{L}_a$ is an equivalent inductance of the avalanche region. By assuming a uniform electric field $\mathscr{E}_1$ from $x = -L_a$ to $x = 0$ one obtains the ratio $\mathbf{j}_1(0)/\mathscr{E}_1(0)$ which is the boundary condition for the drift region and will be replaced in equation (6.3.10). The impedance of the avalanche region will be calculated by including the effect of the displacement current through the geometrical capacitance $C_a = \varepsilon/L_a$ (unit area device), in parallel to the above inductance, such that

$$Z_a = \frac{1}{i\omega C_a} \frac{1}{1 - \frac{\omega_a^2}{\omega^2}}, \qquad (6.3.21)$$

where

$$\omega_a = (\mathscr{L}_a C_a)^{-1/2}, \quad C_a = \frac{\varepsilon}{L_a} \qquad (6.3.22)$$

is the resonance frequency of the avalanche zone impedance, or, simply, *the avalanche frequency*.

The total diode impedance (unit area device) is

$$Z = Z_{\text{drift}} + Z_a = \frac{1}{i\omega C_a} \frac{1}{1 - \left(\frac{\omega_a}{\omega}\right)^2} + \frac{1}{i\omega C_0}\left[1 - \frac{1}{1-\left(\frac{\omega}{\omega_a}\right)^2} \frac{1 - \exp(-i\theta)}{i\theta}\right]. \qquad (6.3.23)$$

The series resistance may be written

$$R = \mathscr{R}eZ = \mathscr{R}eZ_{\text{drift}} = \frac{R_{sc}}{1 - \left(\frac{\omega}{\omega_a}\right)^2} \frac{1 - \cos\theta}{\theta^2/2}. \qquad (6.3.24)$$

The incremental resistance ($\omega \to 0$, $\theta \to 0$) is equal to the space-charge resistance R_{sc} of the drift region. However, R is positive below ω_a and *becomes negative for*

$\omega > \omega_a$. The Read diode exhibits, therefore, a transit-time negative resistance above the avalanche frequency. A detailed examination of equation (6.3.23) also shows that the diode reactance is inductive (positive) for $\omega < \omega_a$ and capacitive for $\omega > \omega_a$. In the above derivation the existence of the avalanche frequency is strictly related to the avalanche region. Its magnitude will be determined below by the following arguments. The displacement current in the avalanche region is assumed to be much larger than the particle current and therefore the alternating field is nearly uniform

$$\mathbf{J}_1 \simeq i\omega\varepsilon\mathscr{E}_1 \qquad (6.3.25)$$

$$\mathscr{E}_1(x) \simeq \text{const.} \qquad (6.3.26)$$

This assumption is reasonable at frequencies well above the avalanche frequency ω_a and this is the normal operation range for a Read diode. By using equations (6.3.25) and (6.3.26) in equation (6.3.17) one obtains

$$\left(\frac{\omega^2}{v_{sat}^2} - \frac{2\alpha_1 J_0}{\varepsilon v_{sat}}\right)\mathscr{E}_1 \approx -\frac{i\omega}{\varepsilon v_{sat}^2}\mathbf{J}_1 \qquad (6.3.27)$$

and

$$\mathbf{j}_1 = \mathbf{J}_1 - i\omega\varepsilon\mathscr{E}_1 \quad \text{may be written}$$

$$\mathbf{j}_1 = \frac{2\alpha_1 v_{sat} J_0 \mathscr{E}_1}{i\omega}, \qquad (6.3.28)$$

so that the avalanche inductance $\mathscr{L}_a$ defined by equation (6.3.20) becomes

$$\mathscr{L}_a = \frac{L_a}{2\alpha_1 J_0 v_{sat}} \qquad (6.3.29)$$

and the avalanche frequency (6.3.22) is

$$\omega_a = \left(\frac{2\alpha_1 v_{sat} J_0}{\varepsilon}\right)^{1/2} \propto J_0^{1/2} \qquad (6.3.30)$$

thus increasing with the bias current (J_0 is the absolute value) as shown elsewhere in this chapter.

A small-signal analysis of an arbitrary IMPATT structure should be performed numerically as shown by Misawa [6] and by Gummel and Blue [10]. The electron and hole velocity are assumed constant but not necessarily equal, whereas carrier ionization rates are unequal and field dependent. The static field and concentration distributions for a given bias current are used to solve the small-signal differential equation (which reduces to equation (6.3.17) for $v_n = v_p = v_{sat}$ and $\alpha_n = \alpha_p = \alpha$) with the aid of a digital computer. We reproduce over leaf some results obtained by Schroeder and Haddad [7] by using this method.

Figure 6.3.3 shows the frequency and bias dependence of the room-temperature small-signal admittance of two complementary abrupt-junction silicon diodes. Both diodes have a uniform background doping of 5×10^{15} cm^{-3} and an active

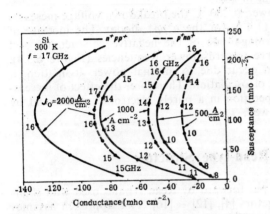

Figure 6.3.3. Small-signal admittance of complementary silicon abrupt-junction diodes, computed at room temperature (details in text). (*after Schroeder and Haddad,* [7]).

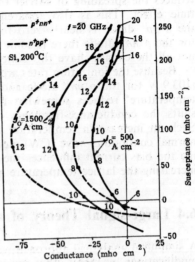

Figure 6.3.4. Small-signal admittance at 200 °C for the same silicon diodes as in Figure 6.3.3. The breakdown voltage is now 111.4 V for the n^+pp^+ diode and 103.9 V for the p^+nn^+ diode (both diodes are slightly punched-through at breakdown). (*after Schroeder and Haddad,* [7]).

width of 5 μm. The n^+pp^+ structure has $V_B = 97.4$ V and is slightly above punch-through at breakdown, whereas the p^+nn^+ diode has $V_B = 91.3$ V and is slightly below punch-through. The impedance of the same diodes at a higher temperature (200 °C) is shown in Figure 6.3.4. Both these figures show that there are two characteristic frequencies defined with respect to the diode admittance $Y = G + iB$, namely

a cutoff, frequency, ω_c
$$\begin{cases} G > 0 & \text{for } \omega < \omega_c; \\ G < 0 & \text{for } \omega > \omega_c; \end{cases} \qquad (6.3.31)$$

an avalanche resonant frequency, ω_a
$$\begin{cases} B < 0 & \text{for } \omega < \omega_a; \\ B > 0 & \text{for } \omega > \omega_a. \end{cases} \qquad (6.3.32)$$

Both these frequencies increase with increasing bias current. For the Read diode (simplified theory derived above) $\omega_c = \omega_a$.

Figures 6.3.3 and 6.3.4 also indicate that the negative conductance increases with the steady-state current. The optimum frequency (maximum absolute value of negative conductance) increases with increasing bias current. The admittances of the two complementary diodes are quite different. All the other parameters being the same, this difference is due to much higher ionization rate for electrons rather than for holes. The higher negative conductance of the n^+pp^+ diode is due to the narrower avalanche region as compared to the p^+nn^+ diode (see Figure 6.2.4), which reduces the spreading of carrier transit-times and determines a well-defined transit-time effect. This is also apparent from Figure 6.3.4, where the conductance of the n^+pp^+ diode is nearly zero for 18 GHz (unfavourable transit angle), whereas for the p^+nn^+ diode there is some negative conductance at all frequencies because not all the carriers have the same transit-angle.

Because the ionization rates are lower at 200 °C the breakdown voltage increases (103.9 V for the p^+nn^+ diode and 111.4 V for the n^+pp^+ diode). The increase of temperature* reduces somewhat the magnitude of the negative conductance and shifts the conductance-frequency curve towards lower frequencies (Figure 6.3.4). Such an increase in temperature can occur due to internal power dissipation in normal continuous-wave (CW) mode of oscillation. Therefore, the effect of increasing the bias current influences the operation in a purely electronic mode and by increasing the lattice temperature as well.

6.4 Large-signal Theory of a Read-type Structure

A simplified analytical approach was put forward by Read [1] and used with slight modifications by many other investigators [4], [12] — [16]. This approach assumes a very thin avalanche zone and yields a differential equation relating the avalanche current to the electric field in the avalanche region. The starting assumptions are: saturated and equal electron velocities

$$v_n = v_p = v_{sat} = \text{const.}, \qquad (6.4.1)$$

equal ionization rates

$$\alpha_n = \alpha_p = \alpha = \alpha(\mathscr{E}), \qquad (6.4.2)$$

and therefore one obtains equation (6.3.11) rewritten here for convenience

$$\frac{1}{v_{sat}} \frac{\partial j}{\partial t} = \frac{\partial}{\partial x}(-J_n + J_p) + 2\alpha j. \qquad (6.4.3)$$

This equation is solved approximately by assuming uniform total current [1]

$$j = J_n + J_p = j(t), \text{ independent of } x \qquad (6.4.3')$$

just as in the steady-state. This assumption is motivated by negligible transit-time effects in a *very thin avalanche region*, namely the transit-time through the avalanche region

$$T_a = \frac{L_a}{v_{sat}} \qquad (6.4.4)$$

---

* The effect of temperature upon small-signal impedance was also computed by Grierson [11] and compared with experimental data.

should be much longer than the signal period, or $T_a f \ll 1$, where f is the operating frequency. By integrating part by part equation (6.4.3) from $x = -L_a$ to $x = 0$, one obtains

$$T_a \frac{dj}{dt} = -(J_n - J_p)\Big|_{-L_a}^{0} + 2j \int_{-L_a}^{0} \alpha(\mathscr{E}) dx. \qquad (6.4.5)$$

The first term in the right-hand part of the above equation is evaluated as shown in Figure 6.4.1. The linear distribution of electron and hole currents in the avalanche

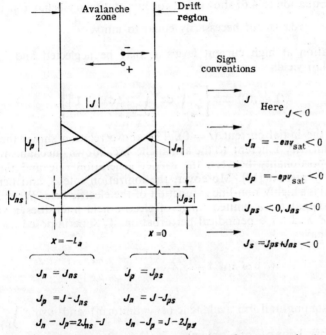

Figure 6.4.1. Notations and sign conventions for currents in the avalanche zone of an n-type IMPATT structure.

region is characteristic of equal ionization rates $\alpha_n = \alpha_p = \alpha$. J_{ns} is the electron component of the reverse saturation current density and is due to electrons thermally generated in the p^+ region, whereas J_{ps} is the hole component of the reverse saturation current, which is due to holes generated in both the space-region and the n^+ region. Here, j and $J_s = J_{ns} + J_{ps}$ are negative due to sign conventions (holes going in the negative x direction, see equations (6.3.3) and (6.3.4)). Equation (6.4.5) becomes*

$$\frac{T_a}{2} \frac{dj}{dt} = j\left(\int_{-L_a}^{0} \alpha dx - 1\right) + J_s, \quad \alpha = \alpha(\mathscr{E}). \qquad (6.4.6)$$

---

* j and J_s can be assumed both positive in equation (6.4.6) which is valid for absolute values of j and J_s. The same equation is valid for the complementary structure (with avalanche at an $n^+ p$ junction and holes going in the positive x direction).

For direct current conditions $(dj/dt = 0)$ we have

$$\frac{j}{J_s} = \left(1 - \int_{-L_a}^{0} \alpha dx\right)^{-1} \quad (6.4.7)$$

and for avalanche breakdown

$$j \to \infty \quad \text{and} \quad \int_{-L_a}^{0} \alpha dx = 1 \quad (6.4.8)$$

as expected. Equation (6.4.6) shows that for non-stationary behaviour the avalanche integral $-\int_{-L_a}^{0} \alpha dx$ is not necessarily equal to unity.

For operation at high current levels J_s may be neglected and integration of equation (6.4.6) yields

$$j(t) = j(0) \exp\left\{\frac{2}{T_a} \int_0^t d\tau \left(\int_{-L_a}^{0} \alpha dx - 1\right)\right\}, \quad (6.4.9)$$

where $j(0)$ is the initial current $(t = 0)$. The above relation shows the basic exponential non-linearity inherent to the avalanche breakdown mechanism and arising from the proportionality between the carrier generation rate and the number of carriers already present [13]. Moreover, the ionization rate α, and hence the ionization integral is a highly non-linear function of the electric field. We shall, however, assume at first that for a limited range of electric field intensities α varies almost linearly with $\mathscr{E}$ and for a periodical perturbation, $\mathscr{E}_1$ superimposed on the average value $\mathscr{E}_0$, we have

$$\alpha(\mathscr{E}) \simeq \alpha(\mathscr{E}_0) + \alpha_1 \mathscr{E}_1, \quad \alpha_1 = \left.\frac{\partial \alpha}{\partial \mathscr{E}}\right|_{\mathscr{E}=\mathscr{E}_0}. \quad (6.4.10)$$

By further assuming uniform avalanche ($\mathscr{E}$ = uniform) and using $\int_{-L_a}^{0} \alpha(\mathscr{E}_0) = 1$ (steady-state breakdown), one obtains

$$j(t) = j(0) \exp\frac{2\alpha_1}{T_a} \int_0^t V_{1a}(\tau) d\tau, \quad (6.4.11)$$

where $V_{1a}(t) = L_a \mathscr{E}_1(t)$ is the alternating voltage drop on the avalanche region. The current varies exponentially with the integral of the voltage. The reader may easily verify that if the signal amplitude is relatively small then a sinusoidal voltage will determine a sinusoidal alternating current j_1 such that

$$\omega \mathscr{L}_a \frac{dj_1(t)}{dt} = V_{1a}, \quad \mathscr{L}_a = \frac{T_a}{2\alpha_1 J_0}. \quad (6.4.12)$$

where $\mathscr{L}_a$ is just the avalanche inductance defined by equation (6.3.29), in the small-signal theory.

6. IMPATT Diodes

More generally, for a uniform avalanche approximation, the avalanche current is given by the first-order differential equation

$$\frac{T_a}{2} \frac{dj}{dt} = j[\alpha(\mathscr{E}) L_a - 1] + J_s \qquad (6.4.13)$$

where $\mathscr{E} = \mathscr{E}(t)$ is the electric field in the avalanche region. However, the assumption of a uniform particle current is questionable. It was shown that a correct quasi-static approximation (low frequencies) yields a time constant (called the intrinsic response time) $T_a/3$ instead of $T_a/2$ in equation (6.4.13) [4], [17]. Kuvås and Lee [18] derived the quasi-static approach for $\alpha_n \neq \alpha_p$, $v_n \neq v_p$ and even non-uniform avalanche. The correction to the above theory (which follows Read's approach) is due to the introduction of the effect of the mobile space charge upon the electric field. The general equation is still of the form (6.4.6) but with $T_a/2$ replaced by a time constant dependent upon $\mathscr{E}_a$ and with the ionization integral replaced by a more complicated function of ionization rates and $\mathscr{E}_a$, $\mathscr{E}_a$ being the instantaneous field at the boundary between the avalanche and the drift zone.

The same equation may be applied in principle for the avalanche region of any IMPATT configuration, although in practice it is difficult to define rigorously such an avalanche zone (for a p^+nn^+ diode for example).

Consider next the drift region. By using equations (6.1.3) and (6.3.7) one obtains

$$J(t) = \frac{1}{T_L} \int_{t-T_L}^{t} j(0, \tau) \, d\tau + C_0 \frac{dV_d(t)}{dt}, \qquad (6.4.14)$$

where $V_d(t)$ is the voltage drop on the drift region and $j(0, t)$ the particle current injected at t. The electric field $\mathscr{E} = \mathscr{E}_a$ entering in equation (6.4.6) or (6.4.13) also satisfies

$$J(t) = j(0, t) + \varepsilon \frac{d\mathscr{E}_a}{dt}. \qquad (6.4.15)$$

If the voltage drop on the avalanche region is negligible, $V(t) \simeq V_d(t)$, and equations (6.4.6), (6.4.14) and (6.4.15) form a system of integro-differential equations which describe the large-signal behaviour.

This system may be solved by using a computer and an iterative method [13], [14], [16]. Combined analytical and numerical solutions are also possible for a given periodic $V(t)$ and in particular for sinusoidal excitation [13].

Because the small-signal high frequency diode resistance is negative, self-oscillations are possible in a suitable circuit. If this circuit is sufficiently selective at a proper frequency, then the alternating voltage across the diode can be assumed sinusoidal. For any frequency and a given signal amplitude the total current can be derived and the device impedance for the fundamental frequency can be computed.

Results of such calculations are shown in Figure 6.4.2 [16]. The bias current is assumed constant for a given diode. The solid lines are contours of constant signal amplitude and the dashed lines are contours of constant signal frequency. Given

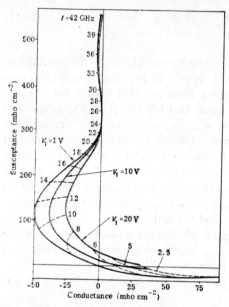
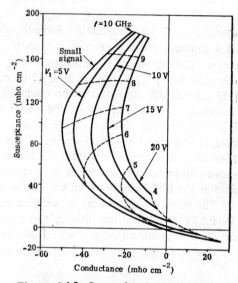

Figure 6.4.2. Large-signal admittance computed for sinusoidal voltage excitation (amplitude V_1). The Read-type IMPATT structure has a 1 μm avalanche zone and a 4 μm drift region. The bias current is $J_0 = 500$ A cm^{-2}. (*after Gupta and Lomax*, [16]).

Figure 6.4.3. Large-signal admittance of a Read-type diode (1.5 μm avalanche zone and 3.5 μm drift zone, 5×10^6 cm s^{-1} carrier velocity and $\alpha(\mathscr{E}) = 5.9 \times 10^6$ cm$^{-1} \times \exp[1.2 \times \times 10^6 (\text{V cm}^{-1})/\mathscr{E}]$ ionization rate), as a function of frequency and amplitude of sinusoidal voltage. The bias current is $J_0 = 500$ A cm^{-2}. (*after Blue*, [12]).

a single-frequency (infinite quality factor Q) circuit, the amplitude of permanent oscillations is determined by the circuit conductance which must be exactly compensated by the device negative conductance. For a real circuit, however, the oscillation frequency is not known *à priori* and the diode and circuit admittances must be compared at all frequencies to find the equilibrium oscillations.

Because an IMPATT structure exhibits broad-band negative resistance and large non-linearities, it may operate under multifrequency as well as single-frequency conditions. A multifrequency analysis can be performed within a reasonable computer time on a simplified model, such as that presented in this Section [12], [13]. It was shown that the negative conductance effect of a Read diode can be enhanced by having harmonics or subharmonics of the signal frequency present [2]. We illustrate this property by referring to the phase of the avalanche current with respect to the applied signal, following Blue [12]. Figure 6.4.3 shows the frequency and amplitude dependence of the device admittance operating with sinusoidal voltage. For lower frequencies the negative conductance first increases and then decreases with increasing signal amplitude. At 4 GHz the diode has a positive small-signal conductance but negative large-signal conductance. Figure 6.4.4 shows the magnitude and phase of the avalanche current injected in the drift region as a function of signal frequency and amplitude. It is clear that at 4 GHz the increased voltage

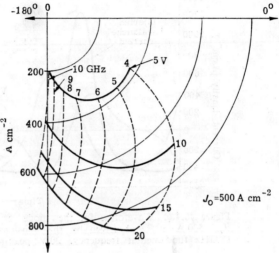

Figure 6.4.4. Magnitude and phase of the fundamental of avalanche current, as a function of frequency and voltage amplitude, for $J_0 = 500$ A cm^{-2}, the same diode as in Figure 6.4.3. (*after Blue*, [12]).

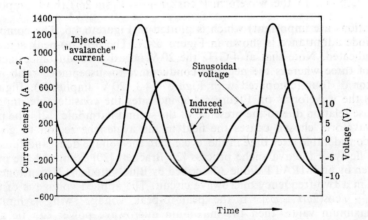

Figure 6.4.5. Waveforms for the diode of Figure 6.4.3. operating at 10 GHz (10 V signal amplitude) and $J_0 = 500$ A cm^{-2}. (*after Blue*, [12]).

amplitude improves (increases) the phase of the avalanche current with respect to the signal and therefore determines a (large-signal) negative resistance. Figure 6.4.5 also indicates the phase relation between the applied voltage and the extremely non-sinusoidal avalanche current.

The phase relation of the avalanche current can be also improved by a proper shaping of the voltage excitation. In an attempt to explain qualitatively the better experimental results obtained for an IMPATT diode in a microwave circuit resonant

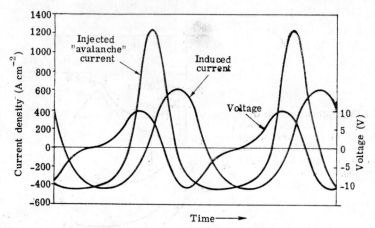

Figure 6.4.6. Waveforms for the diode of Figure 6.4.3 biased with $J_0 = 500$ A cm^{-2} and driven by the nonsinusoidal voltage signal shown (5 GHz fundamental frequency, 20 V peak-to-peak). (after Blue, [12]).

at two frequencies, the fundamental and its second harmonic [19], Blue [20] calculated the response to the waveform $V_1 \cos \omega t + \frac{1}{2} V_1 \sin 2\omega t$ (both amplitude and phase relations are important) which is plotted in Figure 6.4.6. The complex-plane plot of diode admittance is shown in Figure 6.4.7. The peak-to-peak signal magnitude is indicated. Note that at 4 GHz the 20 V negative conductance increases by a factor of three whereas the negative conductance-to-susceptance ratio increases* by a factor of four (compared with Figure 6.4.3, 20 V amplitude). Figure 6.4.8, as well as the waveforms on Figure 6.4.6, indicates the considerable improvement of the phase relation due to the presence of the second harmonic [12]. The particular voltage waveform chosen causes the field in the avalanche region to go through zero at a considerable later time in the cycle than the voltage does and thus increase the phase delay of the avalanche process [2]. Brackett [20] examined the maximum power given by an IMPATT diode (described by Blue's model [12] mentioned above) operating in a two-frequency microwave circuit. The applied voltage is of the form $V_1[\cos \omega t + \nu \cos(2\omega t + \varphi)]$. If the peak-to-peak voltage swing is limited to a certain maximum value, then the maximum microwave power can be generated at the transit-time frequency and with no harmonic interaction. An increase of the output power due to circuit tuning on the second harmonic can be, however, obtained at frequencies lower than the transit time frequency and for $\varphi \approx \pi$ as shown in Figure 6.4.9, at 4 GHz for diode with 6 GHz transit-time frequency. An optimum ratio of harmonic amplitudes is apparent.

---

* This increase reduces the effect of the diode series parasitic resistance R_s, because the total series resistance is $G(G^2 + B^2)^{-1} + R_s$ (where $G < 0$).

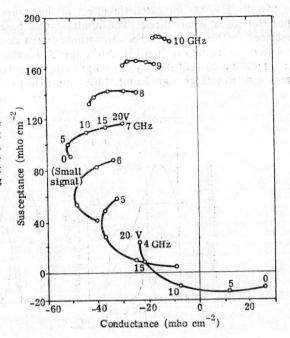

Figure 6.4.7. Complex-plane plot of diode admittance at fundamental frequency as a function of frequency and maximum alternating voltage, at $J_0 = 500$ A cm^{-2}. The alternating voltage is $V_1 \cos \omega t + \dfrac{1}{2} V_1 \times \sin 2\omega t$ (after Blue, [12]).

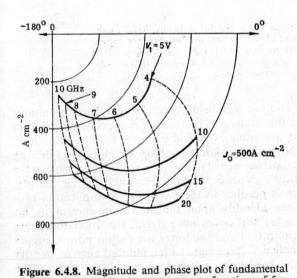

Figure 6.4.8. Magnitude and phase plot of fundamental component of avalanche current, as a function of frequency and alternating voltage amplitude, for $J_0 = 500$ A cm^{-2}. (after Blue, [12]).

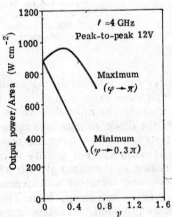

Figure 6.4.9. Output power for a 6 GHz germanium abrupt-junction IMPATT diode with $V_B = 50$ V and $J_0 = 340$ A cm^{-2}, with an applied signal $V_1[\cos\omega t + v\cos(2\omega + \varphi)]$. (after Brackett, [20]).

Misawa [14] has indicated the effect of the saturation current in the large signal operation of a Read diode. The analysis was based upon the Read-type model described in this Section. The saturation current density J_s is apparent in equation (6.4.6) or in equation (6.4.13). Although J_s is almost always much smaller than the operating current density, its magnitude is extremely important for large-signal operation. For large signal amplitudes the external voltage drops below V_B for a

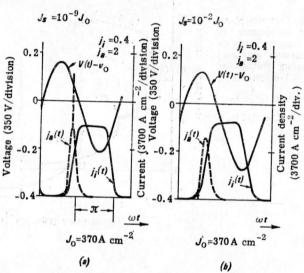

Figure 6.4.10. Waveforms computed for a 10 μm Read diode (1 μm avalanche region) at 5 GHz and for two different values of the saturation current density: (a) $J_s = 10^{-9} J_0 = 37 \times \times 10^{-8}$ A cm^{-2}; (b) $J_s = 10^{-2} J_0 = 3.7$ A cm^{-2}. (after Misawa, [14]).

large time interval during a cycle and the high-field avalanche zone becomes almost depleted of mobile carriers during this interval. The residual carriers originate from the avalanche multiplication in the preceding cycle and from the saturation current. As the diode voltage rises again above the breakdown voltage V_B, the carrier density starts to multiply from the residual carrier density. The time required to achieve the maximum current density is critically dependent upon the initial carrier density, because of the rapid growth which is peculiar to the avalanche multiplication process, and therefore will depend considerably upon the saturation current if this current is sufficiently high. The higher the saturation current, the lower the phase delay provided by avalanche multiplication and the lower the output power. Figure 6.4.10 shows the waveforms of applied voltage signal $V(t)$, induced current density $j_i(t)$ (see equation (6.1.4) and the avalanche injected current $j_a(t) = j(0, t)$ (the particle current density at the origin of the drift zone for a specified diode). When J_s is only $10^{-9} J_0$ (Figure 6.4.10 a), $j_a(t)$ has a phase delay below 90° with respect to $V(t)$. This delay decreases considerably as J_s is as high as 0.01 J_0 (1% of bias current), as shown in Figure 6.4.10 b. The microwave power-signal amplitude dependence for the same Read diode is plotted in Figure 6.4.11, with J_s/J_0 as parameter. Note that the output power decreases to 50% when J_s increases from $10^{-9}J_0$ to $10^{-2}J_0$*. The narrower the avalanche region, the larger the effect of the saturation current, as shown by the dotted curve on Figure 6.4.11.

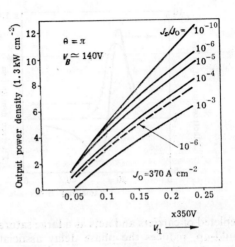

Figure 6.4.11. The output as a function of voltage amplitude for several values of the saturation current density. The diode bias current and frequency are as indicated in Figure 6.4.10. The dotted curve is computed for an avalanche region width of only 5% of the 10 μm depletion layer. (*after Misawa*, [14]).

There are other mechanisms than the thermally generated saturation current which can determine the initial carrier concentration necessary for the build-up of the avalanche process. Such a mechanism is minority carrier injection at an imperfect ohmic contact to the heavily doped region adjacent to the avalanche region [14], [21] or the tunnel current flowing at the *pn* junction (suggested by Read [1] and thought to be negligible after Misawa [22]). Another mechanism is minority carrier storage in the same heavily doped region and their subsequent reinjection into the avalanche region. It was shown by Misawa [23] that the storage effect can be introduced in the Read model as giving rise to an equivalent saturation current.

The minority carrier storage in a p^+nin^+ Read-type structure occurs as follows. The avalanche process creates an enormous electron concentration gradient at the p^+n interface due to the enormous electron density in the avalanche zone as compared to the very small electron density in the p^+ region. A net electron flow diffuses towards the p^+ region. During the very short signal period, these electrons cannot diffuse an appreciable distance into the neutral p^+ region and their recombination is insignificant. As long as the electron density at the p^+n junction still increases, electrons are continuously stored in the neutral p^+ region. As soon as the electron bunch has been formed and leaves the avalanche region drifting towards the n^+ contact, the electron density at the boundary of the avalanche zone decreases and the stored minority electron charge diffuses back from the p^+ region towards the

---

* These calculations are done for a fixed frequency corresponding to $\theta = \pi$. Part of the indicated decrease due to relatively high J_s, can be recovered by increasing the operating frequency and thus the transit-time delay [14].

avalanche zone. These processes are well illustrated by Figure 6.4.12, resulting from some approximate computations done by Misawa [23]*. The back-diffusion of stored minority carriers floods the avalanche region at the moment it should be

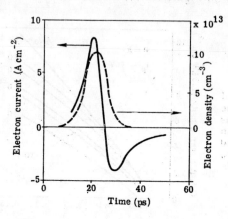

Figure 6.4.12. Computed electron density and electron current at the boundary between the avalanche region and the adjacent neutral region. The diffusion constant was taken equal to 26 cm^{2}s^{-1}. (after Misawa, [23]).

depleted of carriers and acts as a large saturation current which hastens the avalanche build-up, reduces the phase delay associated with the injection of the avalanche current and therefore reduces the negative resistance and the efficiency of high-frequency power generation [23]. Misawa [23] indicated that the electron current at the boundary between high field and neutral p^+ regions (Figure 6.4.12) influences the avalanche only when the avalanche zone is depleted. The Read theory may be assumed approximately valid by replacing the above electron current at the moment of maximum depletion (which takes place 1/4 to 1/2 cycle after the electron density of Figure 6.4.12 peaks) for the saturation current. The computed efficiency for a 10 GHz p^+nn^+ structure reduces from 26.6% to 12.7% due to an effective saturation current as high as 1/500 of the bias current [23]. It was concluded that, although it is theoretically possible to reduce this effect by a special doping profile (a p region is inserted between p^+ and n regions and thus 'isolates' the avalanche region), a more realistic experimental construction will be to replace the p^+n junction by a metal-semiconductor (Schottky) barrier, because the metal no longer stores minority electrons [23].

The Read-type large-signal models have the advantage of simplicity and are particularly useful for studies where we need a first-order description of the electronic processes inside the device, such as for amplifier operation or noise behaviour. This model permits an analytical formulation and reduces considerably the computer time required for a numerical analysis.

However, such approximate models have inherent limitations. Beside the difficulty of defining a narrow avalanche region in some practical structures, the Read

---

* The presence of minority carrier storage was first recognized in a computer simulation performed by using the program developed by Scharfetter and Gummel [24] but the accuracy of the solution is questionable because the thickness of the transition region from the high-field avalanche zone to the neutral p^+ region, where the above processes take place, is comparable to the length of one mesh interval [23].

model cannot account for unsaturated drift velocities and depletion layer modulation*. Both these effects are responsible for output power saturation at large signal levels. Note the fact that the output power computed by using the Read model and represented in Figure 6.4.11, does not show a definite tendency for saturation even at very large signal amplitudes. We recall the fact that the maximum output power and efficiency (32%) was predicted by Read [1] by arbitrarily limiting the signal amplitude to half of the steady-state voltage. An improved Read-type analysis [22] including non-equal ionization rates and an avalanche region of finite width but still assuming uniform particle current in the avalanche region predicts the effect of saturation of ionization rates at very high electric fields upon the Read diode efficiency. A constant efficiency was found for silicon diodes up to 100 GHz. The GaAs diodes efficiency is better at low microwave frequencies but decreases above 50 Hz and becomes comparable to that of silicon diodes beyond 100 GHz [22].

6.5 Computer Analysis of Large-signal IMPATT Mode

The large-signal results shown in the previous Section are based upon the assumption of constant particle current in the avalanche region and are valid, in principle, for a Read-type structure with a very narrow avalanche region. More practical structures like p^+nn^+ IMPATT's do not obey such a condition and an accurate approach is indeed necessary in order to optimize the device construction and operation for higher power generation and better efficiency.

A digital [7], [24]—[29] or an analog [30] computer should be used for such an analysis because the partial differential equations describing the device behaviour cannot be integrated analytically. These equations may be solved as they stand or the problem may be reduced to an iterative integration of an ordinary differential equation, as indicated by Greiling and Haddad [31]. Their approach, sketched below is a compromise between Read-type theories and a true computer simulation which will be described later. Greiling and Haddad [31] assume, as Read did [1] field-independent carrier velocities, equal ionization rates and given by an effective rate†, negligible diffusion and negligible recombination. The impact ionization may occur however, everywhere in the depletion layer. Equation (6.3.11) will be now integrated over the entire diode length. One obtains

$$\frac{dJ(t)}{dt} + A(t)J(t) = B(t), \qquad (6.5.1)$$

where $A(t)$, $B(t)$ depend upon the electric field, which in turn depends upon local density of the particle current

$$j(x, t) = J_n(x, t) + J_p(x, t) = J(t) - \varepsilon \frac{\partial \mathscr{E}(x, t)}{\partial t}. \qquad (6.5.2)$$

---

* Note the fact that at large applied signals the avalanche region extends appreciably into the drift region.

† The effective ionization rate gives the same steady-state breakdown voltage equal to that calculated with exact (unequal) electron and hole ionization rates.

The integral of the electric field is equal to the applied voltage which is assumed to be sinusoidal (all of the harmonics are shortened by the microwave circuit). The equation in $J(t)$ is first solved by assuming that the electric field is independent of $J(t)$. Then the effect of the space charge of the carriers at each point is introduced through the Poisson equation where the mobile charge density is approximately known from the preceding approximate computation. In the next iteration the total current is again calculated by using the new value found for the field, etc. The device admittance is derived by means of a Fourier analysis of $J(t)$. The iterative process is continued until the change in the admittance from one iteration to the next is less 1.5%.

Greiling and Haddad [31] reported results for silicon abrupt junction diodes with different dopings of the high resistivity region the doping profile shown in Figure 6.4.2. All diodes have a 5 μm active region which is punched-through at breakdown. Figure 6.5.1 shows for example the admittance of a silicon p^+nn^+ diode with 10^{15} cm^{-3} doping and a bias current of 500 A cm^{-2}. These results are to be compared with those shown in Figure 6.4.2 (note that the total active length and the bias current are the same). The dependence upon the signal frequency and amplitude is similar but the predicted negative conductance per unit area differs

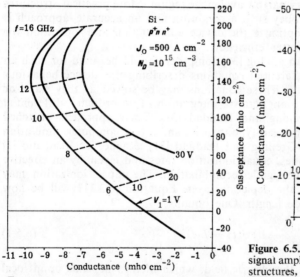

Figure 6.5.1. Admittance of a silicon p^+nn^+ diode (10^{15}cm^{-3} doping 5 μm active region) biased to $J_0 = 500$ A cm^{-2}. The diode is punched through at breakdown. (*after Greiling and Haddad*, [31]).

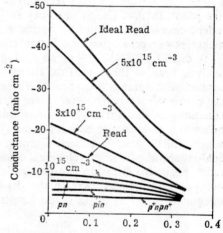

Figure 6.5.2. Conductance *versus* normalized signal amplitude for a number if IMPATT structures, all having a 5 μm active region and biased at $J_0 = 500$ A cm^{-2}. This conductance is computed at the optimum frequency (maximum absolute value of the small-signal negative conductance). (*after Greiling and Haddad*, [31]).

substantially. The difference may be explained mostly through the doping profile (and therefore the extent of the avalanche region) as indicated below.

Figure 6.5.2 shows the dependence of the parallel conductance upon the normalized signal amplitude at the optimum operating frequency (for maximum absolute

6. IMPATT Diodes

value of the small-signal negative conductance). The ideal Read diode results are reproduced after Blue [12] for a Read diode with a 1 μm avalanche region and 4 μm drift region. The Read diode curve in Figure 6.5.2 corresponds to a concentration of 10^{16} cm^{-3} in the n layer, whereas the doping of p and n regions in the p^+npn^+ and pn structures are both of the order of 10^{15}cm^{-3}.

Figure 6.5.3 shows the conduction current density distribution inside some of the above IMPATT structures at a phase angle equal to π, where the phase of the sinusoidal voltage is taken as reference (i.e. at the time instant when the voltage goes below its average value). In the *pin* diode, due to the large region where the charge carriers are generated, the carrier transit-angle varies greatly from one charge carrier to another, only a few carriers experience the same phase delay and the transit-time frequency-dependent negative resistance will be small. As the doping concentration increases (Figure 6.5.3) the avalanche region becomes narrower and the 'avalanche' current is generated in a sharper pulse [31] and the negative

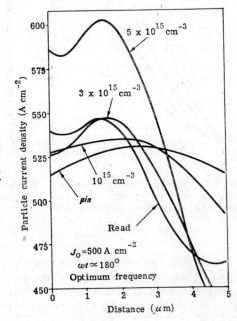

Figure 6.5.3. Conduction current density in the active region of structures indicated in Figure 6.5.2, at the moment the alternating voltage goes through zero and begins substrating from the bias voltage. (*after Greiling and Haddad,* [31]).

conductance at the optimum frequency is larger when all other parameters are the same.

As the signal amplitude increases the mobile space charge increases and the field in the avalanche region is flattened or smoothed out over a wider region. As shown above, a wider avalanche region determines a lower negative conductance at the optimum frequency. Figure 6.5.2 shows that at large signal amplitudes all diodes approach the behaviour of the *pin* diode. This is due to the above space-charge widening of the avalanche zone [31]. A similar phenomenon affects the diode efficiency, shown in Figure 6.5.4.

In the remainder of this section we shall describe the principle of computer simulation [24] and some results [7], [24], [26], [28], [29] illustrating details of device behaviour. Scharfetter and Gummel [24] used the following computational scheme. They included in equations the diffusion current, carrier generation and recombination through defects (Shockley-Read-Hall model), unequal ionization rates, field- and doping-dependent electron and hole mobilities. The doping profile may be arbitrary. The particular profile shown in Figure 6.5.5 was used for detailed calculations (this is a non-ideal silicon $p^+n\nu n^+$ Read structure) [24].

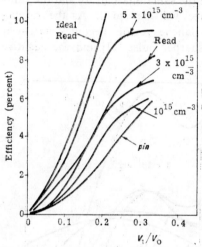

Figure 6.5.4. Efficiency *versus* normalized alternating voltage for the diodes of Figure 6.5.2, at optimum operating frequency and $J_0 = 500$ A cm^{-2}. (*after Greiling and Haddad*, [31]).

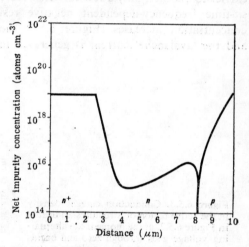

Figure 6.5.5. Doping profile of a diode used for computer simulation. (*after Scharfetter and Gummel*, [24]).

The calculations were performed for the diode biased by a direct-current generator and driven by a sinusoidal voltage generator, as shown in Figure 6.5.6 *a*. The coupling capacitor C serves to isolate the d.c. and a.c. portions of the circuit. In this way it was possible to control directly the frequency and amplitude of the signal and reduce the computer time. Essentially the same results were obtained with circuit shown in Figure 6.5.6 *b* (free-running oscillator) if the load inductance and conductance were properly chosen [24].

The structure was divided into a large number of cells and a standard difference approximations method was used to approximate the spatial derivatives in the Poisson equation and the continuity equations for each mesh point. Given the hole and electron distributions, at a certain initial moment, the electric field is determined by using the Poisson equation. From the particle currents and the net generation rates, the continuity equations yield the time derivatives of carrier

6. IMPATT Diodes

concentrations. Using these values the program computes the carrier concentrations an instant later (the time variable is also quantized) and then repeats the cycle with the new values for carrier concentrations in each point. For a given bias current and applied sinusoidal signal the process repeats until the difference between cycles practically disappear. The Fourier analysis for the limit cycle determines the

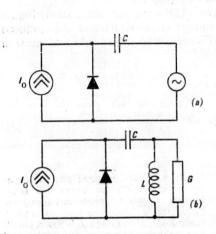

Figure 6.5.6. Diode in circuit: (a) voltage driven diode; (b) free-running oscillator.

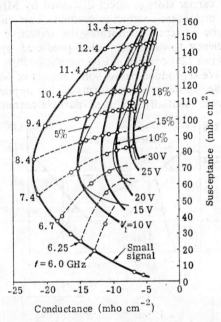

Figure 6.5.7. Diode admittance computed for a diode having the doping profile of Figure 6.5.5, biased at $J_0 = 200$ A cm^{-2} and driven by a sinusoidal voltage. The voltage amplitude and frequency is indicated. The efficiency (%) is also shown. (after Scharfetter and Gummel, [24]).

fundamental of the current and thus the diode impedance as a function of bias current, signal frequency and signal amplitude. The results of such numerical computations may be concentrated as shown in Figure 6.5.7. Note the constant frequency, constant amplitude and constant efficiency curves in the admittance plane. Qualitatively, the results are essentially the same as shown in Figures 6.4.2, 6.4.3, 6.4.7 and 6.5.1. Note the increasing efficiency with increasing voltage amplitude. The efficiency is lower than predicted by the simplified Read theory*. Scharfetter and Gummel [24] explained this result by the relatively large a.c. voltage drop on the avalanche region (neglected in Read's estimation of efficiency).

---

* For example, Figure 5.6.7 indicates 10% efficiency at 11.4 GHz and $V_1 = 30$ V signal amplitude. The average voltage is $V_0 = 87.7$ V [24]. For the corresponding V_1/V_0 ratio, Figure 6.1.3 predicts about 25% efficiency.

Misawa [26] has investigated an idealized $p^+n\nu n^+$ silicon structure, which is closer to the Read original proposal (the doping profile is abrupt in contrast to that shown in Figure 6.5.5). In addition, the p^+ region was replaced by a metal which forms a Schottky barrier contact to the n-type semiconductor. The avalanche multiplication occurs at this contact just as at an abrupt p^+n junction. However, the minority carrier storage effect discussed by Misawa [23] (Section 6.4) does not occur at the metal-semiconductor contact. This eliminates the premature build-up of avalanche current, ameliorates the inductive delay of this current and increases the efficiency towards the value predicted by Read [1]. The numerical data used by Misawa [26] correspond to an nSi — PtSi Schottky barrier.

We reproduce below some results of Misawa's [26] computer simulation. Figure 6.5.8a shows the waveforms of the applied voltage (sinusoidal) and the induced current (spatially averaged particle current). Figure 6.5.8b shows field and electron

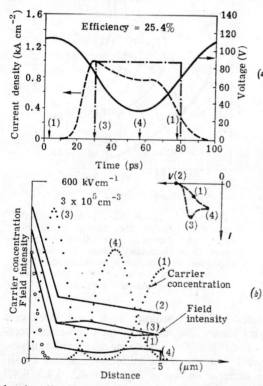

Figure 6.5.8. (*a*) Computed induced current waveform and (*b*) carrier and field distributions at several time instants during one cycle, for a silicon p^+(or metal)$n\nu n^+$ Read diode, at 10 GHz a $J_0 = 500$ A cm^{-2}. The donor densities are 2.1×10^{16} cm^{-3} and 9.5×10^{14} cm^{-3}, respectively. The active region length is 5 μm (n region width is slightly above 1 μm). The breakdown voltage at 300 K is 75.9 V. The saturation current is 1.5×10^{-7} A cm^{-2} (appropriate for a Schottky barrier). (*after Misawa*, [26]).

density distributions at several time instants during one cycle. The particle current waveform shown in Figure 6.5.8a departs considerably from the ideal rectangular shape corresponding to the simplified theory with 90° avalanche delay and a transit angle of π for the drift region which is also shown in Figure 6.5.9. The earlier start of the current is due to the current induced before the avalanche created electrons enter the drift region. A contribution also arises from holes created by the avalanche before they disappear into the metal. The dip in the current occurring when the signal has its lowest value (instant 4 in Figure 6.5.8) is due to the lowering of electron velocity below its saturated value. This lowering broadens the trailing edge of the electron bunch and makes the current cut-off less sharp. Actually, in the case shown in Figure 6.5.8 diffusion is primarily responsible for this slow falling of the current [26].

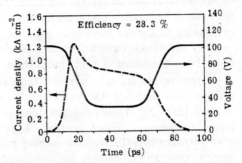

Figure 6.5.9. Waveforms of induced current and voltage computed for a 'rectangular' voltage excitation. (after Misawa, [26]).

The effects discussed above increase as the voltage amplitude increases and the efficiency is degraded. For the waveforms shown in Figure 6.5.8a the efficiency is 25.4% (only 12.7% would result if an ordinary p^+n junction is used instead of the Schottky-barrier contact). A slight improvement would result for a special voltage waveform, such as that shown in Figure 6.5.9 (the efficiency increases to 28.3%). This waveform avoids excessive high instantaneous voltages which may produce avalanche in a certain part of the drift region thus deteriorating the inductive avalanche phase relation. The lowest instantaneous voltage is not too low, such that the low-field drift effect is reduced. The waveform shown permits a larger amplitude at the fundamental frequency without increasing the peak-to-peak voltage [26].

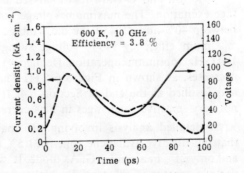

Figure 6.5.10. Waveforms of induced current and voltage computed at 600 K (sinusoidal-voltage 10 GHz). (after Misawa, [26]).

Misawa [26] also simulated the diode behaviour at 600 K (such a temperature increase may occur during the continuous wave (c.w.) operation at a high bias current). The waveforms indicated in Figure 6.5.8a are modified as shown in Figure 6.5.10*. The efficiency becomes as low as 3.8%. This is primarily due to the considerable increase of the saturation current of the Schottky barrier (7.7 A cm^{-2}). The result is the premature build-up of the avalanche current (a mechanism already discussed in the preceding section) and a reduction of the avalanche phase delay. This reduction

---

* The d.c. diode voltage increases because the ionization rates decrease with increasing temperature. The voltage swing is comparable in Figure 6.5.8 a and 6.5.10, respectively.

could be compensated by an increase in frequency which increases the drift phase delay. Conversely, at a lower frequency (8 GHz) the negative resistance disappears [26]. Other high-temperature effects are: the increase of carrier generation in the depletion layer (which further increases the number of carriers already present when the avalanche starts, thus accentuating the effect of the saturation current) and the decreasing of carrier mobilities together with the increase of the field required for velocity saturation (which increases the low-field effects at the same voltage swing, note the accentuated dip in the current waveform shown in Figure 6.5.10).

Extensive results of computer simulations on large-signal operation of silicon and gallium arsenide IMPATT diodes were reported recently by Grierson and O'Hara [28], [29]. Their program does not include carrier diffusion and generation-recombination on defect centres. The diodes analysed were *non-punch-through* p^+nn^+ structures with realistic doping profiles. Because diffusion is neglected, the electric conduction cannot be simulated outside the depletion layer (where the electric field is zero). This fact has two consequences. First, the time-varying resistance of the undepleted *n*-type region, whose width oscillates with the applied voltage, does not occur in this simulation. Secondly, the pulse of electrons reaching the limit of the depletion region may be left in the undepleted region where they stand as trapped. As the voltage increases again and the depletion region boundary penetrates through this bunch of 'trapped' carriers, they start to move again. This is an example of a non-physical situation arising in an imperfect (incomplete or approximate) computer simulation. It was shown that the effect of this anomalous trapping is small [28].

Grierson and O'Hara [28] analysed large-signal operation with sinusoidal voltage excitation. The maximum voltage amplitude was somewhat arbitrarily chosen between 40% and 45% of the d.c. voltage. It was found that for silicon diodes the maximum efficiency (43% voltage swing) remains approximately constant until 100 GHz (optimum operation frequency) whereas for GaAs diodes the efficiency decreases, as shown in Figure 6.5.11. This result was also found by Misawa [22] and justified at the end of Section 6.4. It was also found [7] that the efficiency is not very sensitive to changes in bias current (at the same voltage swing).

A detailed analysis involving the spatial field and carrier distribution shows that the effective avalanche zone is 31.5% of the depletion region of a silicon diode and only 21% in a similar GaAs diode. It was also found that the residual avalanche particle current is substantially higher in silicon p^+n diodes and results from the higher ionization coefficient for electrons. For both these reasons, the behaviour of the GaAs diode is close to the ideal behaviour of the Read diode and the efficiency is higher.

The high-frequency operation requires an increase in doping and therefore an increase of the breakdown field (see Chapter 2). This detail changes the behaviour of GaAs diodes because the slope of the ionization rate field-dependence decreases at higher voltages. Consequently, the ratio of the avalanche region width to the depletion layer width increases. The behaviour of a high-frequency (e.g. 100 GHz)

GaAs diode becomes therefore close to that of a silicon diode and the efficiency falls at higher frequencies as shown in Figure 6.5.11.

Whereas the experimental findings for GaAs diode efficiencies are as high as 19% in X band (quoted in [28]), the best efficiencies found for p^+n silicon diodes which were most extensively investigated, does not exceed 10%, which is consider-

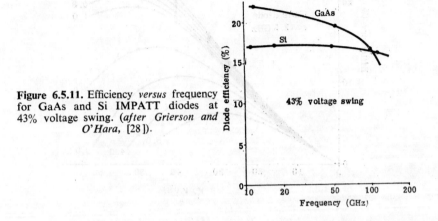

Figure 6.5.11. Efficiency *versus* frequency for GaAs and Si IMPATT diodes at 43% voltage swing. (*after Grierson and O'Hara*, [28]).

ably lower than the value 17% predicted by the computer simulation (Figure 6.5.11). The effect of the parasitic series resistance may explain the low efficiencies found for silicon diodes. The series resistance is due to the n^+ substrate and to the undepleted n region. Due to the much higher electron mobilities in GaAs, the silicon resistivity is higher at the same doping density. This fact may account for the low efficiencies achieved in silicon devices. Figure 6.5.12 shows the computed effect of a constant series resistance upon the efficiency of a GaAs and a silicon IMPATT diode. The peak of the diode efficiency occurs because the magnitude of the negative conductance decreases with increasing signal amplitude (see for example Figure 6.5.7) and thus at higher voltages the parasitic resistance has a larger effect. Therefore, at large signal amplitudes the output power experiences an anomalous decrease with the increasing signal amplitude.

The effect of series resistance can be reduced by letting the diode punch-through. Grierson and O'Hara [28] have found that the most favourable situation occurs if the diode is just punched-through at d.c. breakdown*. If the diode is heavily punched-through then the effect of the series resistance of the undepleted n layer will be reduced to zero. However, in this case the diode should be made much narrower and the diode susceptance will greatly increase (the depletion-layer capacitance increases). This effect decreases the magnitude of the diode series negative resistance. Therefore the diode area should be kept small in order to match the circuit load resistance, and a smaller area decreases the input (d.c.) power and, of course, the available microwave power. Another undesirable effect of lower negative

---

* The predicted efficiency is, however, too high in comparison with the experimental data. This is possibly due to the inability of the program to take into account the current flow from the n^+ substrate at lower instantaneous voltages [28]. This effect is discussed below.

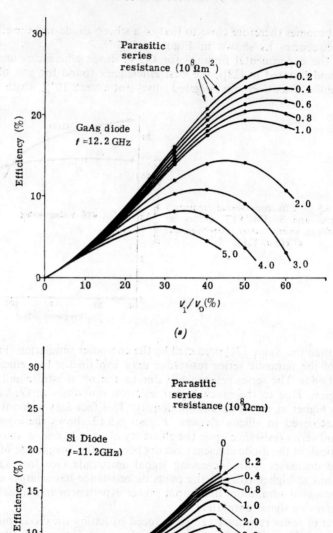

Figure 6.5.12. Effect of series resistance upon the voltage dependence of diode efficiency for: (a) a GaAs diode at 12.2 GHz, (b) a Si diode at 11.2 GHz. Both diodes are non punched-through p^+nn^+ structures (realistic doping profiles) biased at $J_0 = 10^3 \text{A cm}^{-2}$. (*after Grierson and O'Hara*, [28]).

resistance is the increasing influence of any series parasitic resistance*. On the other hand, Schroeder and Haddad [7] have also shown that for heavy punch-through the avalanche zone is wider, whereas the unsaturated velocities are important only at very large signals, these two phenomena having opposite effects upon diode efficiency. These authors found theoretically almost the same efficiency (10% for a silicon p^+nn^+ diode) with heavily punched-through and just punched-through structures. For n^+pp^+ diodes the efficiency should be as high as 15—16%, which is attributed to the narrower avalanche region [7].

Schroeder and Haddad [7] also evidenced (by using a modified Read-type computer analysis which takes into account diffusion inside the drift region) the phenomenon of carrier injection from substrate into the undepleted lowly-doped region. Figures 6.5.13—6.5.17 show the results computed for an abrupt junction p^+nn^+ silicon diode with 5×10^{15} cm^{-3} background impurity concentration and a 5 μm n-region. This diode is just punched-through at breakdown. Figure 6.5.13 shows the large-signal admittance for a 500 A cm^{-2} bias current density. At 60 V signal

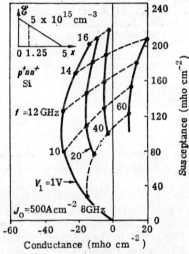

Figure 6.5.13. Large-signal admittance of a p^+nn^+ silicon diode. The effective avalanche width (small-signal operation) is 1.25 and the background doping is 5×10^{15}cm^{-3}. (*after Schroeder and Haddad,* [7]).

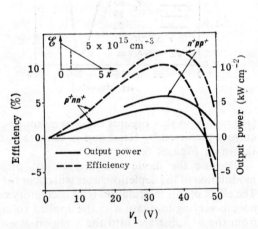

Figure 6.5.14. Output power and efficiency as a function of amplitude of the sinusoidal voltage for two complementary silicon diodes (details in text). (*after Schroeder and Haddad,* [7]).

amplitude the device conductance is positive. Therefore the high-frequency power and the efficiency will exhibit a maximum at a certain signal level. This is indeed shown in Figure 6.5.14, for $J_0 = 500$ A cm^{-2} and $f = 12$ GHz. The peak efficiency is 10.5% and the peak power is 4.5 kW cm^{-2}. Both occur for a signal amplitude $V_1 = 35$ V, which is 38% of the static breakdown voltage. For $V_1 > 40$ V the power

---

* The effect of such a parasitic resistance decreases with increasing bias current, because the magnitude of the negative resistance increases (almost linearly) with increasing current density.

and efficiency decrease rapidly and become zero at $V_1 = 47$ V. Also shown are the power and efficiency for the complementary n^+pp^+ diode, at $J_0 = 500$ A cm^{-2} and its optimum frequency, $f = 10$ GHz. The power saturation occurs at comparable signal levels for both diodes.

Figure 6.5.15 shows the waveforms of the induced current (spatially averaged particle current) at different signal amplitudes. For relatively low signal amplitudes

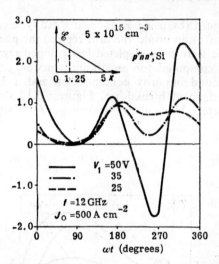

Figure 6.5.15. Induced current waveform (one period of the sinusoidal voltage) computed for a p^+nn^+ silicon diode ($f = 12$ GHz, $J_0 = 500$ A cm^{-2}) at several signal levels. (after Schroeder and Haddad, [7]).

($V_1 = 25$ V in this Figure) the waveform resembles somewhat the ideal rectangular current predicted by Read and is close to the results obtained by Misawa [26] and shown in Figures 6.5.8—6.5.10.

Because the diode analysed is just punched-through in the steady-state, the modulation of the depletion layer will occur for any signal level shown in Figure 6.5.15. The effect of this modulation is an oscillatory component of the induced current, component varying in-phase with the applied voltage. This is due to the electrons flowing from the n^+ substrate into the n region at $\omega t = 270°$, when the voltage is at minimum, and back into the substrate when the voltage rises again. This loss mechanism reduces the amount of power generated by the diode. This effect is particularly strong for $V_1 = 60$ V in Figure 6.5.15 and produces a negative induced current. The phenomenon is apparent from Figure 6.5.16, where the field profile, the electron density and electron current distributions are shown at four time instants during one cycle. The signal amplitude is $V_1 = 35$ V (maximum efficiency). The injection from the substrate first occurs in Figure 6.5.16 c, at $\omega t = 262°$. Here, the electric field at the nn^+ interface is negative and carries electrons from the substrate into the active region. The injected electron density almost neutralizes the ionized impurities (5×10^{15} cm^{-3} concentration). On the same figure we see how the two electron pulses drifting in the opposite directions (and injected from the avalanche region and the substrate respectively) meet each other [7]. Note that the diffusion current is important in the narrow region between 3.5—4.0 μm, where the electron current density does not vary as the electron density does. In Figure 6.5.16 the two charge 'pulses' already disappeared and the remaining electron density is extracted from the active region.

Figures 6.5.17 shows the dependence of high-frequency power and efficiency upon the bias current density, at $f = 12$ GHz and several values of V_1. At reason-

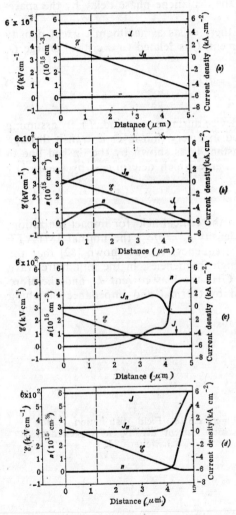

Figure 6.5.16. Current density, electric field and electron density distributions at four time moments during one high-frequency cycle (p^+nn^+ silicon diode discussed above with $f = 12$ GHz, $V_1 = 35$ V, $J_0 = 500$ A cm^{-2}): (a) $\omega t = 84°$ (the sinusoidal voltage is maximum at $\omega t = 90°$); (b) $\omega t = 173°$; (c) $\omega t = 262°$; (d) $\omega t = 351°$. *(after Schroeder and Haddad, [7]).*

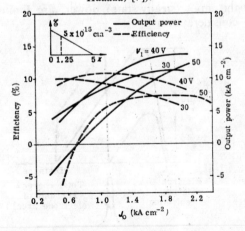

Figure 6.5.17. Output power and efficiency *versus* bias current density for the above p^+nn^+ silicon diode at $f = 12$ GHz. *(after Schroeder and Haddad, [7]).*

ably high signal levels and high bias currents the efficiency changes very little with the bias current. At higher current densities, the efficiency decreases due to the effect of the mobile space-charge upon the phase delay between current and voltage, as shown in Section 6.1. Of course, the efficiency should exhibit a maximum because at zero bias current the diode conductance is positive. Note the fact that the power saturation with increasing signal level will occur at higher voltages for higher bias current densities ($J_0 = 500$ A cm^{-2} and $V_1 = 50$ V yields a positive conductance, as already indicated in Figure 6.5.13 and 6.5.14). For any given V_1, there is an optimum J_0 for maximum efficiency. However, this maximum decreases at very large V_1, because as V_1 approaches V_B, the depletion layer modulation apparent in Figure 6.5.16 becomes more important and J_0 should be increased to overcome this loss. This increase leads to the degradation of the avalanche phase delay by the space-charge effect and the efficiency decreases.

It was shown by Scharfetter [32] that there exists a maximum current density J_c, for efficient operation. This critical current J_c is related to a critical charge Q_c

$$J_c = \frac{Q_c}{T} = fQ_c \qquad (6.5.3)$$

(Q_c is the total charge passing through the device during one period, T). By assuming that Q_c is sufficiently high and concentrated in a very thin layer in transit through the device, the field profile at a certain instant is as shown by the dashed line in Figure 6.5.18. The critical charge is that charge which determines, by Gauss law

$$Q_c = \varepsilon \mathscr{E}_c \qquad (6.5.4)$$

a maximum field inside the device equal to the critical field for impact ionization, $\mathscr{E}_c$. Therefore, above the critical current density $J_c = \varepsilon f \mathscr{E}_c$, the normal IMPATT operation is impossible. J_c increases with frequency. It was shown [32] that the above limitation is effective only at very high frequencies, in the millimetre wave region. For silicon diodes, below 80—100 GHz the bias current J_0, and therefore the input (steady-state) power, are limited by thermal and impedance matching considerations.

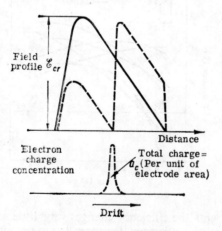

Figure 6.5.18. Field distribution at the critical current density, as suggested by Scharfetter, [32]).

The above concept of critical current density was revised by O'Hara and Grierson [29] by using the results of a computer simulation. Figure 6.5.19 shows the electric field profiles and carrier distributions at several time instants during the period of

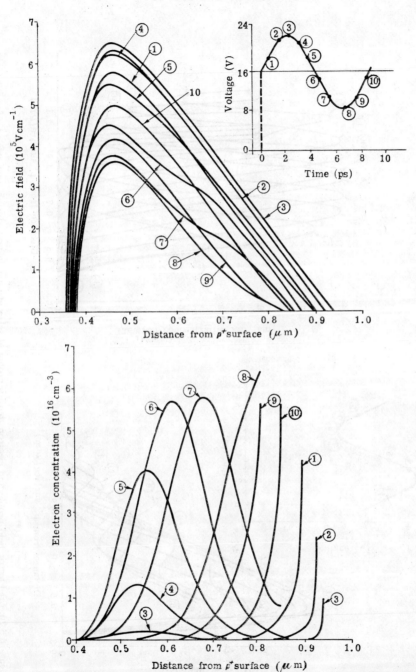

Figure 6.5.19. (a) Electric field profile and (b) electron concentration distribution at several time moments during one cycle of the sinusoidal voltage for a p^+nn^+ silicon diode (background doping 10^{17} cm^{-3}, $f = 113$ GHz, $J_0 = 2 \times 10^4$ A cm^{-2}). (*after O'Hara and Grierson,* [29]).

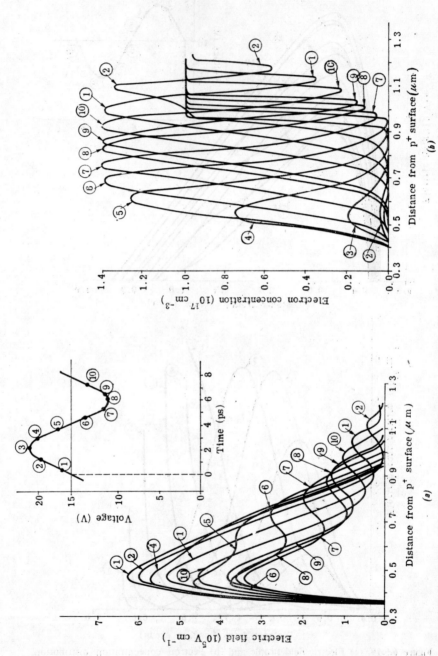

Figure 6.5.20. (a) Electric field profile and electron density profile at selected time instant during one period ($f = 128$ GHz, p^+nn^+ silicon diode, biased at $J_0 = 6 \times 10^4 \text{A cm}^{-2}$). (after O'Hara and Grierson, [29]).

the applied sinusoidal voltage, for a p^+nn^+ silicon diode with 10^{17} cm^{-3} background concentration, 113 GHz operating frequency and $V_1 = 0.44\, V_0$ ($V_0 = 15$ V) at a steady-state current $J_0 = 2 \times 10^4$ A cm^{-2}. Note the fact that the diode is never punched-through. Figure 6.5.20 shows results of simulation for the same diode at $J_0 = 6 \times 10^4$ A cm^{-2}, $f = 128$ GHz, $V_1 = 0.3\, V_0$ ($V_0 = 16.2$ V). The efficiency is only 5.3%, as compared to 16.4% for the operation described by Figure 6.5.20. The field profile exhibits indeed a secondary maximum (Figure 6.5.20 a) drifting through the device together with the charge pulse (Figure 6.5.20 b). The positive slope between the two field peaks indicate that the mobile electron density exceeds the ionized donor density. However, the electric field at the secondary peak does not exceed the critical field for the avalanche multiplication, the number of electrons generated here is negligible and the limitation mechanism described by Scharfetter [32] does not occur. The considerable reduction in efficiency is due to another mechanism. O'Hara and Grierson [29] have shown that the few holes generated at the secondary field peak reach the p^+n junction where they initiate a premature avalanching and degrade the inductive phase delay associated with the avalanche.

6.6 Intrinsic Noise of IMPATT Diodes

The IMPATT diodes operating in microwave amplifiers and oscillators are very noisy compared to other solid-state devices such as the transistors and Gunn diodes. On the other hand, the large noise generation makes the avalanche diode a useful noise source for microwave frequencies. Therefore, the noise problem deserves a special attention.

The physical sources of noise in an avalanche transit-time diode are [33],[34]:

(a) The avalanche noise associated with the random avalanche multiplication process; this avalanche multiplication starts from the saturation current which is by itself affected by the shot noise. Frequently, all fluctuations of the avalanche current injected into the drift region are assimilated to a kind of shot noise called the primary noise current.

(b) The diffusion noise or thermal noise associated to the drift region is probably negligible for most practical cases.

(c) The generation-recombination noise is small because the carrier transit-time through the space charge region (10^{-10}s) is much shorter than the lifetime of excess carriers (10^{-6}s).

(d) Flicker noise ($1/f$ noise) was found experimentally to be negligible.

(e) Low-frequency noise up converted to frequencies near the carrier due to the non-linearity of the diode which acts as a mixer. The low-frequency noise originates from the diode itself (both avalanche and thermal noise) and from bias current fluctuations generated by the external circuit.

The main sources of noise are (a) and (e). Only the avalanche noise will be considered here, because it can be assumed true 'intrinsic'.

Assume for the moment equal ionization rates $\alpha_n = \alpha_p = \alpha$. Deep into avalanche breakdown (very high multiplication ratio) the ionization integral $\int \alpha dx$ carried over the avalanche region is unity (see Section 1.9), i.e. the ionization probability is unity. Each transient of a pair of carriers will, therefore, produce *in average* one new carrier pair. However, this is only a probability because a certain pair may produce none, or two or more pairs in this transit and this fluctuation determines the noise. For simplicity the electric field is considered uniform in the avalanche region and practically constant during electron transit-time (very short) such that the probability of ionization by a given carrier during its transit time is constant. The mean time between successive collisions in a chain of ionizing events during an avalanche process is denoted by τ_x. It is assumed that the noise due to statistical variations of τ_x between ionizations in any given chain of ionizations is negligible. The noise occurs, as indicated above, because these chains of successive ionization start and stop in a random manner. Hines [35] used the above model for an infinitely thin avalanche region under short circuit conditions (constant voltage). He postulated that the noise events give step functions of current $\Delta i = ke/\tau_x$, where k may have various discrete values $-1, 0, +1, +2, +3$, etc. The Poisson probability distribution is used thus obtaining zero average value of Δi and e^2/τ_x^2 mean square value*. If $\alpha_n \neq \alpha_p$ the mean square value is corrected by an arbitrary factor, a^2 [35]. More rigorous computations use Langevin-type continuity equations (see Chapter 4) to derive the mean square value [37], [38]. The primary noise current of the short-circuited avalanche zone, calculated by Hines [36] is

$$\bar{I}_{no}^2(\omega) = \frac{2a^2 e I_0 \Delta f}{\omega^2 \tau_x^2} \tag{6.6.1}$$

where I_0 is the bias current. The circuit noise is calculated by using a Read-type small-signal model for the diode impedance. The theory predicts an infinite open-circuit noise voltage at ω_a, the avalanche frequency (6.3.30) (note the fact that the small signal series resistance (6.3.24) derived under the same assumptions also peaks to infinity for $\omega = \omega_a$). A maximum in the noise spectrum was indeed measured by Haitz and Voltmer [39]. More exact (numerical) computations were performed by Gummel and Blue [10] for a realistic structure with a wide avalanche region and also $\alpha_n \neq \alpha_p$ and same results are reproduced in Figure 6.6.1. The mean-square (open-circuit) noise voltage is constant at low frequencies

$$\bar{V}_n^2 = \frac{2e \Delta f}{\alpha_1^2 I_0} \left(1 + \frac{L + L_a}{L}\right)^2 \propto \frac{1}{I_0} \tag{6.6.2}$$

attains a maximum at a resonance frequency (which is proportional to $I_0^{1/2}$, compared to equation (6.3.30)) and then falls approximately as $1/\omega^4$ being almost proportional to I_0. The quasi-periodic frequency dependence is due to transit-time effects in the drift region. The space-charge smoothing occur at very high frequencies. However, the nega-

---

* Although the mean square value of these fluctuations happens to be the same as for the shot noise, the probability distributions in these two cases are totally different [36]. Hines [35], [36] shows that in an avalanche essentially all of the ionizing events are the direct results of previous ionizations and these fluctuations are, therefore, not independent and the noise cannot be, conceptually, regarded as a shot noise due to completely random events [36].

6. IMPATT Diodes

tive resistance also decreases with increasing frequency. The noise performances will be therefore described by the noise measure

$$M = \frac{\overline{V_n^2}}{4kT(-\mathcal{R}eZ)\,\Delta f}.\qquad (6.6.3)$$

Haus et al. [40] demonstrated that the optimum noise measure M_{opt} for a Read diode (equal ionization coefficients, very narrow avalanche region with uniform $\alpha_1 = \partial\alpha/\partial\mathcal{E}$) is obtained near 2π transit angles, and

$$M_{opt} = \frac{e}{2\alpha_1 kT}.\qquad (6.6.4)$$

However for α_1 large* and transit angles near 2π the negative series resistance becomes rather small and can be annihilated by a parasitic resistance.

An analytical theory of noise in small-signal operation of IMPATT diodes was suggested by Kuvås [41]. The Read model is used corrected to account for the transport delay in the avalanche region. The parameters are the ratio between the avalanche and drift zone widths (L_a/L), R_{sn}, the normalized series resistance R_S and the normalized bias current, $\overline{I_0}$. The normalized noise measure, M/M_{opt} is shown in Figure 6.6.2 *versus* the drift transit angle for zero series resistance. A wider avalanche zone (larger L_a/L) will improve the noise performances. The noise measure even decreases below M_{opt}, for large transit angles. However, there is little interest in operation above the transit-time frequency because of the small value of the negative resistance and the unavoidable series resistance. Figure 6.6.3 indicates the effect of the parasitic series resistance for the same normalized current as in Figure 6.6.2. Kuvås [41] also indicated that the minimum noise measure is reduced somewhat by increasing the bias current.

It was found experimentally that the minimum noise power is achieved under small-signal conditions. The noise is constant up to a certain signal level and then increases rapidly with the signal amplitude [33], [37], [38], [43]. Moreover, the large-signal noise is often different for diodes with similar small-signal behaviour. It is, therefore, evident, that a new noise mechanism sets in above a certain signal level. Hines [35] indicated that this excess noise is due to large percentage fluctuations of

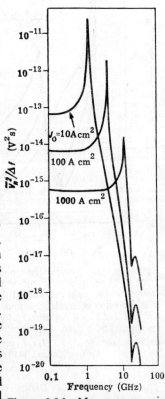

Figure 6.6.1. Mean-square noise voltage per bandwidth *versus* frequency, with the bias current as parameter, for a Si IMPATT diode with 10^{-4} cm^2 area, 5 μm total depletion width ($L+L_a$) and 1 μm avalanche region (L_a). (*after Gummel and Blue*, [10]).

---

* Both equations (6.6.2) and (6.6.4) show that higher $\alpha_1 = \partial\alpha/\partial\mathcal{E}$ give lower *small-signal* noise. Therefore a Ge diode should be less noisier than a silicon one. Equation (6.6.4) is strictly correct for equal ionization rates and for GaAs ($\alpha_1 = 0.419$ V^{-1}) yields $M = 47.7$ or 16.8 dB [40].

the current minimum in an oscillation cycle (initial current for the avalanche process, occurring approximately when the alternating voltage becomes positive). Sjölund [42] developed the idea of a large signal noise which originates from the shot noise corresponding to the above initial (minimum) current, I_{min}. He suggested a large-

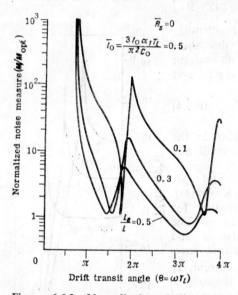

Figure 6.6.2. Normalized noise measure M/M_{opt} of an avalanche diode as a function of the transit angle over the drift region (θ), with the ratio of avalanche and drift region widths (L_a/L) as a parameter, for zero series resistance and a certain value of the normalized bias current $\bar{I}_0 = 3I_0\alpha_1 T_L/\pi^2$ $C_0 = 0.5$. (after Kuvås, [41]).

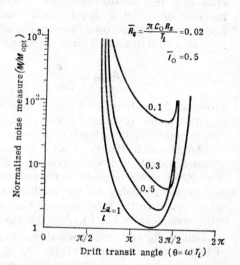

Figure 6.6.3. Normalized noise measure M/M_{opt} of an avalanche diode as a function of the transit angle over the drift region (θ), with the ratio of avalanche and drift region widths (L_a/L) as parameter, for the same value of the normalized bias current as in Figure 6.6.2 and a normalized series resistance $\bar{R}_S = \pi C_0 R_S/T_L = 0.02$. (after Kuvås, [41]).

signal noise inversely proportional to this minimum current, I_{min}, and found experimentally a proportionality to $I_{min}^{-1.1}$. A number of large-signal noise theories were developed recently [34], [36], [38], [42]–[45].

An analytical theory was given by Kuvås [37] for a Read diode described by the following equations

$$\frac{dj_a}{dt} + \frac{j_a}{M\tau_1} = \frac{J_s + j_{n0}}{\tau_1} \qquad (6.6.5)$$

$$J = j_a + \varepsilon \frac{d\mathscr{E}_{max}}{dt} \qquad (6.6.6)$$

where j_a is the avalanche current injected into the drift region, $\mathscr{E}_{max}$ is the instantaneous value of the electric field maximum, M is the instantaneous multiplication factor (equal to $[1 - \alpha(\mathscr{E})L_a]^{-1}$ in equation (6.4.13), τ_1 is the intrinsic response time ($T_a/2$ in equation (6.4.13) or, more precisely, $T_a/3$), J_s is the reverse saturation current density and j_{no} is a fluctuation corresponding to the primary noise current. The noise may be considered small compared to the large-signal components and treated as a perturbation. It was assumed that the noise is open-circuit terminated at all harmonics of the large-signal frequency. The noise power at the fundamental frequency was found to increase almost exponentially with the signal amplitude (at large signal levels) due to strong parametric interactions determined by the strong non-linear current response. This non-linearity is contained in equation (6.6.6), through $M\tau_1$ as a function of $\mathscr{E}$ (see also Section 6.4). The results of these calcula-

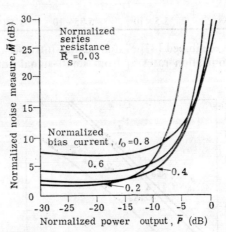

Figure 6.6.4. Normalized noise measure $\overline{M}$ as a function of normalized output power $\overline{P}$, with normalized bias current $\overline{I}_0$ as parameter, for a certain value of the normalized series resistance, $\overline{R}_s$ (the definition for $\overline{I}_0$ and $\overline{R}_s$ is different from those used for Figures 6.6.2 and 6.6.3). (after Kuvås, [37]).

tions were given in normalized form. The normalized noise measure $\overline{M}$ depends upon the normalized bias current $\overline{I}_0$, the normalized R.F. field and the normalized series resistance $\overline{R}_S$. The normalizing factors depend upon the particular diode structure. The generated microwave power can be written in a similar normalized form, $\overline{P}$. Therefore, general noise-power diagrams were obtained, as that shown in Figure 6.6.4. The noise at low power levels increases with the bias current because ω_a moves closer to the operating frequency (see Figure 6.6.1). The experimental results do indeed reproduce the shape of these curves, with a rapid increase at higher power levels. The optimum power-noise ratio will be therefore achieved at intermediate signal levels.

The analysis [37] also shows that the output power, efficiency and noise were proportional, respectively to τ_1^2, τ_1 and τ_1^{-1}. The diode will be therefore designed for long intrinsic response time of the avalanche region, thus wide avalanche regions and low saturated velocities are preferred. Low field dependence of ionization rates will give a larger increase in the output power than in the noise such that the power-noise ratio is improved [37]. Table 6.6.1 allows a comparison to be made between the semiconductor materials used for IMPATT diodes (see also Chapter 1).

Table 6.6.1
Ionization Rates and Drift Saturated Velocities [46] [47]

$$\alpha_n, \alpha_p = \alpha_0 \exp\left[-\left(\frac{b}{\mathscr{E}}\right)^m\right]$$

Semiconductor	m	α_n (cm^{-1})		α_p (cm^{-1})		$v_{n,\,sat}$	$v_{p,\,sat}$
		α_0 (cm^{-1})	b (V cm^{-1})	α_0 (cm^{-1})	b (V cm^{-1})	cm s^{-1}	cm s^{-1}
Si	1	3.8×10^6 2.40×10^6	1.75×10^6 1.60×10^6	2.25×10^7 1.80×10^7	3.26×10^6 3.20×10^6	1.1×10^7 1.05×10^7	1×10^7 [46] 7.5×10^6 [47]
Ge	1	1.55×10^7	1.56×10^6	1.0×10^7	1.28×10^6	6.0×10^6	6×10^6
GaAs	2	3.5×10^5	6.85×10^5	3.5×10^5	6.85×10^5	9.10×10^6	9×10^6

Goedbioed [47] evaluated the intrinsic response time τ_1 and the field derivative of ionization rates α_1 from small-signal admittance measurements and the results,

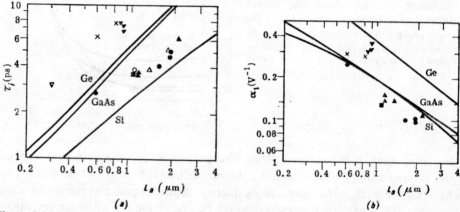

Figure 6.6.5. The intrinsic response time (a) and the field derivative α_1 of the average ionization coefficient (b) as functions of the length of the avalanche region, L_a. Solid lines are computed (the averaging for electrons and holes is not indicated here) using the data of Table 6.6.1. The experimental data obtained on various structures are indicated as follows: ▲ Si p^+n, ● Si^+n p, ■ n Si Schottky barrier, ▼ Ge n^+p, × $n-$ GaAs Schottky barrier, △ Si p^+n, ○ Si n^+p. (after Goedbloed, [47]).

shown in Figure 6.6.5, are close to theoretical predictions. Note the expected increase of τ_1 on the length of the avalanche region.

6.7 Experimental

The IMPATT diodes operate in avalanche breakdown, at high current densities. The high-frequency negative resistance is relatively small. Several problems of physical and technological interest should be solved, by device construction and design:

(a) Uniform breakdown in the semiconductor bulk.
(b) Appreciable heat dissipation. In many cases it is the dissipation capability which limits the output (R.F.) power.
(c) Very small series resistances.

We consider below the p^+nn^+ structure obtained by diffusing a p^+ region into the n-type epitaxial layer grown on the n^+ substrate. The thickness of the epi-layer and the diffusion depth are chosen so as to provide the desired drift length and therefore the required operating frequency

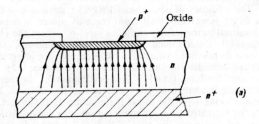
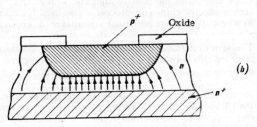

Figure 6.7.1. Field configuration in planar structures: (a) shallow junction and lateral (edge) avalanche breakdown; (b) deep junction and 'bulk' breakdown.

(see Figure 6.2.4). The background doping of the epi-layer is determined by using Figure 6.2.2. If the structure is just punched-through at breakdown then the depletion length will be equal to the thickness of the n-region (abrupt junctions). Otherwise a correction factor should be introduced. For an arbitrary doping profile the calculations are much more complicated. The outdiffusion from the substrate towards the n-region during epitaxial growth and diffusion should be taken into account.

The device can be constructed, in principle, in planar (Figure 6.7.1) or mesa technology (Figure 6.7.2) [48]. However, the mesa construction is largely preferred because of the difficulty to avoid

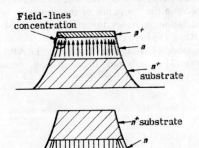

Figure 6.7.2. Field configuration in mesa structures: (a) negative bevel angle at the p^+n junction and 'surface' breakdown; (b) positive bevel angle at the p^+n junction and bulk breakdown.

'corner' breakdown occurring in planar devices (Figure 6.7.1 a). In mesa structure the junction is plane but the edge breakdown can still occur if the field lines concentrate as shown in Figure 6.7.2 a [49]. A proper bevel angle (between the surface and the junction plane) should be achieved such that the diameter decreases as going from the highly doped to the lowly doped side of the junction (Figure 6.7.2 b). Such an angle is considered positive. The importance of mesa profile in avoiding the surface breakdown was studied by using numerical methods [50], [51]. It was shown that for an improper (negative) bevel angle the maximum field occurs beneath the surface, either at the junction, or near the nn^+ interface [51]. The electric state of the surface is also important. Therefore the actual breakdown occurs at a voltage below that corresponding to bulk breakdown and is highly localized. Non-uniform breakdown can also occur due to the material non-uniformities or to non-uniform device temperature (the actual semiconductor temperature is higher in the centre of the structure and lower at the periphery).

The efficiency of commercial IMPATT devices is still relatively low (a few percent) and almost the entire power absorbed from the bias source is removed by heat dissipation. This dissipation is localized in the vicinity of the *pn* junction where the electric field is maximum. For efficient heat removal the diode should be mounted 'junction-down', as shown in Figure 6.7.2 to bring the junction as close as possible to the heat sink. Recently, the integral heat sink (or plated heat sink) technology reduced considerably the diode thermal resistance [52]. The technology is essentially the same as that used for mesa BARITT diodes (Chapter 5): the metallic contact on the junction side is electrochemically plated to a metal thickness of $50-200 \mu$m. The mesa structures are separated by chemical etching with the metal plate acting as a support. Then this plate is separated into pieces, one for each structure and mounted into the encapsulation. The plated electrode acts a micro-heat sink and eliminates the relative high and unreproducible thermal resistance of the direct welding of semiconductor chip. The process is also used for microwave transferred-electron devices (Chapter 8) where it has one particular advantage: it protects the semiconductor from damage during chip bonding.

The series resistance associated with the semiconductor substrate can be substantially reduced by thinning this substrate down to tens of micrometers. This thinning procedure (already mentioned in Chapter 6) is facilitated by the existence of the metal plate acting again as a mechanical support.

A number of improvements were suggested in device design and technology. *Schottky-barrier* IMPATT diodes [53]–[55] were constructed in an attempt to reduce carrier storage effects (see above) and lower the thermal resistance of the chip itself. Abrupt doping profiles obtained by ion implantation were used to produce a Read-type device [54], [56]. Four-layer structures have shown [57], [58] to exhibit better noise properties. Double-drift devices were constructed, especially for millimeter wave frequencies [59], [60]. Insulation of structures by proton bombardment was achieved [61] and yielded better results than mesa technique [61], etc. An important result of both theoretical and experimental research is that GaAs has definite advantages over Si when used as a material for IMPATT diodes. We note here that the high power and efficiency obtained from GaAs devices (for example 8 W, with 40% efficiency in X-band, 1973 result) can be, at least in part, explained by the negative-mobility of GaAs and its higher conductivity. The first property introduces an improvement of efficiency when operating at lower field intensities with non-saturated velocity [62]: the same effect degrades the efficiency in Si devices. The second property allows operation with non-punched-through devices because the resistance of the unswept layer is negligible: the mechanism of depletion width modulation increases the efficiency [63]. Efficiencies as high as 60%, in excess of that predicted by Read [1] were calculated [63].

Finally, we note that the so-called high-efficiency mode of avalanche diodes, or TRAPATT mode (TRApped Plasma Avalanche-Triggered Transit) [2]–[5] is not discussed here, because this operation mechanism is far from any unipolar-type mode of operation.

References

1. W. T. READ, Jr., 'A proposed high-frequency, negative-resistance diode', *Bell Syst. Techn. J.*, **37**, 401–446 (1958).
2. G. I. HADDAD, P. T. GREILING and W. E. SCHROEDER, 'Basic principles and properties of avalanche transit-time devices', *IEEE Trans. Microwave Theor. Techn.*, **MTT–18**, 752–772 (1970).

3. S. M. Sze, 'Physics of semiconductor devices', John Wiley, New York, 1969.
4. J. E. Carroll, 'Hot electron microwave generators', Edward Arnold Ltd., London, 1970.
5. S. M. Sze and R. M. Ryder, 'Microwave avalanche diodes', *Proc. IEEE*, **59**, 1140–1154 (1971).
6. T. Misawa, 'Negative resistance of pn junction under avalanche breakdown conditions', Part I and II, *IEEE Trans. Electron Dev.*, **ED–13**, 137–151 (1966).
7. W. E. Schroeder and G. I. Haddad, 'Non-linear properties of IMPATT devices', *Proc. IEEE*, **61**, 153–182 (1973).
8. H. C. Bowers, 'Space-charge-induced negative resistance in avalanche diodes' *IEEE Trans. Electron Dev.*, **ED–15**, 343–350 (1968).
9. D. L. Scharfetter, D. J. Bartelink and B. C. Deloach, Jr., 'Comments on static negative resistance in avalanching silicon p^+-i-n^+ junctions' *IEEE Trans. Electron Dev.*, **ED–16**, 970–972 (1969).
10. H. K. Gummel and J. L. Blue, 'A small-signal theory of avalanche noise in IMPATT diodes', *IEEE Trans. Electron Dev.*, **ED–14**, 569–580 (1967).
11. J. R. Grierson, 'Theoretical calculations on the effect of temperature on the operation of an IMPATT diode', *Electron. Lett.*, **8**, 258–259 (1972).
12. J. L. Blue, 'Approximate large-signal analysis of IMPATT oscillators', *Bell. Syst. Techn. J.*, **48**, 383–396 (1969).
13. K. Mouthaan, 'Non-linear characteristics and two-frequency operation of the avalanche transit-time oscillator', *Philips Res. Repts.*, **25**, 33–67 (1970).
14. T. Misawa, 'Saturation current and large-signal operation of a Read diode', *Solid-St. Electron.*, **13**, 1363–1368 (1970).
15. J. Nigrin, 'Analytical non-linear study on Read diode avalanche region', *Proc. IEEE*, **60**, 916–917 (1972).
16. M. S. Gupta and R. J. Lomax, 'A self-consistent large-signal analysis of a Read-type IMPATT diode oscillator', *IEEE Trans. Electron Dev.*, **ED–18**, 544–550 (1971).
17. J. J. Goedbloed, 'Noise in IMPATT-diode oscillators', Thesis, Technological Univ. Eindhoven, Philips Res. Repts. Suppl. 1973, No. 7.
18. R. Kuvås and C. A. Lee, 'Quasistatic approximation for semiconductor avalanches', *J. Appl. Phys.*, **41**, 1743–1755 (1970).
19. C. B. Swan, 'IMPATT performance improvement with second harmonic tuning', *Proc. IEEE*, **56**, 1616–1617 (1968); T. P. Lee and R. D. Standley, 'Frequency modulation of a millimeter wave IMPATT diode oscillator and related harmonic generation effects', *Bell Syst. Techn. J.*, **48**, 143–161 (1969).
20. C. A. Brackett, 'Peak a.c. voltage limitations in second-harmonically tuned IMPATT diodes', *IEEE Trans. Microwave Theor. Techn.*, **MTT–18**, 992–993 (1970).
21. D. R. Decker, C. N. Dunn and H. B. Frost, 'The effect of injecting contacts on avalanche diode performance', *IEEE Trans. Electron Dev.*, **ED–18**, 141–146 (1971).
22. T. Misawa, 'High frequency fall-off of IMPATT diode efficiency', *Solid-St. Electron.*, **15**, 457–465 (1972).
23. T. Misawa, 'Minority carrier storage and oscillation efficiency in Read diodes', *Solid-St. Electron.*, **13**, 1369–1374 (1970).
24. D. L. Scharfetter and H. K. Gummel, 'Large-signal analysis of a silicon Read diode oscillator', *IEEE Trans. Electron Dev.*, **ED–16**, 64–77 (1969).
25. A. L. Ward and B. J. Udelson, 'Computer calculations of avalanche-induced relaxation oscillations in silicon diodes', *IEEE Trans. Electron Dev.*, **ED–15**, 847–851 (1968); 'Computer comparison of n^+pp^+ and p^+nn^+ junction silicon diodes for IMPATT oscillators', *Electron. Lett.*, **7**, 723–724 (1971).
26. T. Misawa, 'Theoretical study of microwave oscillation efficiency in improved Read diodes', *Solid-St. Electron.*, **14**, 29–40 (1971).
27. N. Nakamura, H. Kodera and M. Migitaka, 'Computer study on GaAs Schottky barrier IMPATT diodes', *Solid-St. Electron.*, **16**, 663–667 (1973).

28. J. R. GRIERSON and S. O'HARA, 'A comparison of silicon and gallium arsenide large-signal IMPATT diode behaviour between 10 and 100 GHz', *Solid-St. Electron.*, **16**, 719–741 (1973).
29. S. O'HARA and J. R. GRIERSON, 'A study of the power handling ability of gallium arsenide and silicon, single and double drift IMPATT diodes', *Solid-St. Electron.*, **17**, 137–153 (1974).
30. R. KUVÅS and C. A. LEE, 'Non-linear analysis of multifrequency operation of Read diodes', *J. Appl. Phys.*, **41**, 1756–1767 (1970).
31. P. T. GREILING and G. I. HADDAD, 'Large-signal equivalent circuits of avalanche transit-time devices', *IEEE Trans. Microwave Theor. Techn.*, **MTT–18**, 842–853 (1970).
32. D. L. SCHARFETTER, 'Power-impedance-frequency limitations of IMPATT oscillators calculated from a scaling approximation', *IEEE Trans. Electron Dev.*, **ED–18**, 536–543 (1971).
33. M. S. GUPTA, 'Noise in avalanche transit-time devices', *Proc. IEEE*, **59**, 1674–1687 (1971).
34. J. J. GOEDBLOED, 'Noise in IMPATT-diode oscillators', Thesis, Technological University Eindhoven, Netherlands, November 1973, published in *Philips Res. Repts Suppl.*, No. 7, 1973.
35. M. E. HINES, 'Noise theory for the Read type avalanche diode', *IEEE Trans. Electron Dev.*, **ED–13**, 158–163 (1966).
36. M. E. HINES, 'Large signal noise, frequency conversion, and parametric instabilities in IMPATT diode networks', *Proc. IEEE*, **60**, 1534–1548 (1972).
37. R. L. KUVÅS, 'Noise in IMPATT diodes: Intrinsic properties', *IEEE Trans. Electron Dev.*, **ED–19**, 220–233 (1972).
38. G. CONVERT, 'Sur la théorie du bruit des diodes à avalanche', *Rev. Techn. Thomson–CSF*, **3**, 419–471 (1971).
39. R. H. HAITZ and F. W. VOLTMER, 'Noise of self-sustaining avalanche discharge in silicon: studies at microwave frequencies'; *J. Appl. Phys.*, **39**, 3379–3384 (1968).
40. H. A. HAUS, H. STATZ and R. A. PUCEL, 'Optimum noise measure of IMPATT diodes', *IEEE Trans. Microwave Theory Techn.*, **MTT–19**, 801–813 (1971).
41. R. L. KUVÅS, 'Small-signal noise measure of avalanche diodes', *Solid-St. Electron.*, **16**, 329–336 (1973).
42. A. S. SJÖLUND, 'Noise in IMPATT oscillator at large R. F. amplitudes', *Electron. Lett.*, **7**, 161–162 (1971); 'Noise at large RF amplitudes in IMPATT oscillators', communicated by the author.
43. A. SJÖLUND, 'Analysis of large-signal noise in Read oscillators', *Solid-St. Electron.*, **15**, 971–978 (1972).
44. F. DIAMAND, 'Noise in IMPATT-diode oscillators under large signal conditions', paper presented at the Europ. Microw. Conf., Montreux, Switzerland, 12–14 Sept. 1974.
45. J. J. GOEDBLOED, M. T. VLAARDINGERBROEK, 'Noise in IMPATT-diode oscillators at large signal levels', *IEEE Trans. Electron Dev.*, June 1974.
46. R. HULIN and J. J. GOEDBLOED, 'Influence of carrier diffusion on the intrinsic response time of semiconductor avalanches', *Appl. Phys. Lett.*, **21**, 69–71 (1972).
47. J. J. GOEDBLOED, 'Determination of the intrinsic response time of semiconductor avalanches from microwave measurements', *Solid-St. Electron.*, **15**, 635–647 (1972).
48. D. DE NOBEL and M. T. VLAARDINGERBROEK, 'IMPATT diodes', *Philips Techn. Rev.*, **32**, 328–344 (1971).
49. A. LEKHOLM and P. WEISSGLAS, 'Edge breakdown in mesa diodes', *IEEE Trans. Electron Dev.*, **ED–18**, 844–848 (1971).
50. A. BAKOWSKI and K. I. LUNDSTRÖM, 'Depletion layer characteristics at the surface of beveled high-voltage $p-n$ junctions', *IEEE Trans. Electron Dev.*, **ED–20**, 550–563 (1973).

51. I. Costea and D. Dascălu,'Numerical analysis of electric field inside mesa p^+-n-n^+ avalanche diodes', *Electron. Lett.*, **10**, 129–131 (1974).
52. R. A. Zettler and A. M. Cowley, 'Batch fabrication of integral heat-sink IMPATT diodes', *Electron. Lett.*, **5**, 693–694 (1969).
53. H.-C. Huang, P. A. Levine, A. R. Gobat and J. B. Klatskin, 'High-efficiency operation of GaAs Schottky-barrier IMPATT's', *Proc. IEEE*, **60**, 464–465 (1972).
54. T. Misawa, R. A. Moline and A. R. Tretola, '10 GHz Si Schottky-barrier IMPATT diode with hyperabrupt impurity distribution produced by ion implantation', *Solid-St. Electron.*, **15**, 189–193 (1972).
55. T. Watanabe, H. Kodera and M. Migitaka, 'GaAs 50 GHz Schottky barrier IMPATT diodes', *Electron Lett.*, **10**, 7–8 (1974).
56. J. J. Berentz, R. S. Ying and D. H. Lee, 'CW operation of ion-implanted GaAs Read-type IMPATT diodes', *Electron. Lett.*, **10**, 157–158 (1974).
57. S. Su and S. M. Sze, 'Design considerations of high-efficiency GaAs IMPATT diodes', *IEEE Trans. Electron Dev.*, **ED-20**, 541–543 (1973).
58. F. Diamand, 'Low-noise silicon IMPATT structure', *Electron. Lett.*, **9**, 406–407 (1973)
59. T. E. Seidel, R. E. Davis and D. E. Iglesias, Double-drift-region ion-implanted millimetre-wave IMPATT diodes, *Proc. IEEE*, **59**, 1222–1228 (1971).
60. W. C. Niehaus, T. E. Seidel and D. E. Iglesias, Double-drift IMPATT diodes near 100 GHz', *IEEE Trans. Electron Dev.*, **ED-20**, 765–771 (1973).
61. J. D. Speight, P. Leigh, N. McIntyre, I. G. Groves and S.O'Hara, 'High-efficiency proton-insulated GaAs IMPATT diodes', *Electron. Lett.*, **10**, 98–99 (1974).
62. B. Culshaw and R. A. Giblin, 'Effect of velocity/field characteristics on the operation of avalanche-diode oscillators', *Electron. Lett.*, **10**, (1974).
63. P. A. Blakey, B. Culshaw and R. A. Giblin, 'Efficiency enhancement in avalanche diodes by depletion-region-width modulation', *Electron. Lett.*, **10**, 435–436 (1974).

Problems

6.1. Consider the Read-type current waveforms shown in Figure 6.1.2. and show that the efficiency is given by

$$\eta = \frac{V_1}{V_0} \frac{\cos\theta_1 - \cos\theta_2}{\theta_2 - \theta_1} \qquad (P.6.1.1)$$

if the rectangular pulse of the external current occurs for $\theta_1 \leq \omega t \leq \theta_2$. Prove the results of Figure 6.1.3.

6.2. Discuss the practical way to compute the length of the avalanche region according to equation (6.2.2). Consider abrupt, linearly-graded and arbitrary profiles.

6.3. Consider the short-circuit and open circuit stability of a Read-type diode, with the small-signal impedance given by equation (6.3.23). Discuss the effect of a series inductance and resistance. Show the importance of bias current.

Hint: A Nyquist-type stability analysis is briefly presented in chapter 7.

6.4. Compute the phase velocity of the avalanche waves, equation (6.3.18), and discuss its dependence on the bias current.

6.5. Design a p^+nn^+ silicon diode for continuous wave oscillations at 10 GHz and 0.5 W output power, with a hypothetical efficiency of 4%, by using simplified results and diagrams presented in this chapter.

6.6. Show that the maximum steady-state power which can be dissipated in an IMPATT diode is approximately proportional to v_{sat} and $\mathscr{E}_c^2$.

6.7. Show that for a given impedance level and conversion efficiency, the generated power multiplied by the square of the operating frequency is approximately constant (Pf^2 rule).

6.8. Consider the operating limitations arising from thermal considerations: show that the maximum diode area is limited; the output power can be increased by series and parallel combination of devices.

6.9. Consider the particle current, equation (6.4.9), which depends upon a periodical voltage through the empirical law $\alpha = a \exp(-b/\mathscr{E})$, where $\mathscr{E}$ is uniform in the avalanche region. Show that the bias voltage depends upon the signal level (rectification effect): this rectification modifies the phase shift between current and voltage, thus reducing the output power.

7
Transferred-electron Diodes for Microwave Amplification

7.1 Introduction

One of the most exciting possibilities for microwave generation and amplification is to construct a negative-resistance diode by using a semiconductor material which exhibits a negative differential conductivity. Such a device would be, at least in principle, both simple and efficient. If the alternating component of the particle current density in a semiconductor is related to the local alternating field by

$$\underline{j}_1 = \sigma_1 \underline{\mathscr{E}}_1 \quad (7.1.1)$$

and σ_1 is real and negative, then a semiconductor resistor constructed from such a material will exhibit a broad band negative resistance.

Such a bulk negative dynamic conductivity was practically achieved by using the transferred-electron effect (Chapter 1). It was shown that up to very high frequencies (tens of GHz) the transferred-electron effect can be described by a carrier-velocity field-dependence exhibiting an N-shape. The negative slope of this velocity-field curve has the significance of a negative differential mobility (NDM): the (average) electron velocity decreases when the local electric field increases. The simplest transferred-electron device (TED) should be a semiconductor resistor like that shown in Figure 7.1.1. If the semiconductor conductivity is sufficiently

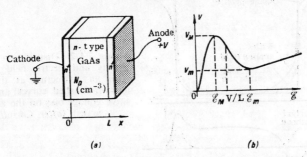

Figure 7.1.1. Semiconductor resistor using an NDM material.

high and the n-region is neutral, the steady-state $I-V$ characteristics will be the replica of the velocity-field characteristic. However, this is not really the case and we will see that even the simple n^+nn^+ structure of Figure 7.1.1 behaves in a complicated way, depending upon the actual semiconductor parameters and the external circuit.

Consider the electric charge distribution in the n^+nn^+ 'resistor' at thermal equilibrium (zero applied voltage). In the vicinity of the high-low n^+n junctions (Chapter 2) the semiconductor deviates from neutrality: electrons from the n^+ regions diffuse into the n region leaving uncompensated positively-charged donors. The extent of the space-charge region in the n-region is of the order of a few extrinsic Debye lengths:

$$L_{DE} = \left(\frac{kT\varepsilon}{e^2 N_D}\right)^{1/2} = \text{extrinsic Debye length (}n\text{-region)}, \qquad (7.1.2)$$

and in the n^+ region is much thinner. If the sample length, L, is much longer than L_{DE}, then the greatest part of the n region will be neutral at thermal equilibrium (Figure 7.1.2). An external voltage modifies the space-charge distribution as shown qualitatively in Figure 7.1.2 [1], [2].

The existence of a net space charge in the semiconductor bulk raises the question of the efficiency of the dielectric relaxation mechanism. The dielectric relaxation time, τ_r, is related to the dielectric permittivity ε and the electric conductivity σ by $\tau_r = \varepsilon/\sigma$. However, if the material behaviour is nonohmic, the dynamics of small perturbations from neutrality will be described by a differential relaxation time τ_d, related to a differential conductivity. In the case discussed here:

$$\tau_d = \frac{\varepsilon}{eN_D\mu_d}, \quad \mu_d = \frac{dv}{d\mathscr{E}}, \qquad (7.1.3)$$

where the differential mobility is negative, $\mu_d < 0$, if the electric field exceeds a certain threshold value, $\mathscr{E}_M$. A *negative* differential relaxation time means that a small space-charge (deviation from neutrality) in the semiconductor bulk will *grow*

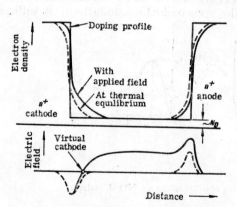

Figure 7.1.2. Distribution of electron density and electric field intensity in an n^+nn^+ structure at thermal equilibrium (dashed curves) and with positive voltage bias on the anode (solid curves). (*after* Carroll, [1]).

exponentially with the time constant $|\tau_d|$, instead of decay by the normal relaxation process! Because this space-charge is related to an excess or a deficit of mobile carriers (or both), it may move through the sample. If the transit-time of this disturbance is very short compared to $|\tau_d|$ ($\tau_d < 0$) it will not have time enough to develop and grow appreciably. The transit time is $T_L = L/v$ (v is the average velocity corresponding to an average field, $\mathcal{E}$), and the above condition requires

$$T_L \ll |\tau_d|, \quad \frac{L}{v} \ll \frac{\varepsilon}{eN_D|\mu_d|} \quad \text{or} \quad N_D L \ll \frac{\varepsilon v}{e|dv/d\mathcal{E}|}, \qquad (7.1.4)$$

i.e. *the doping-length product* of the sample should be sufficiently low in order to provide a stable field and charge configuration when a constant voltage bias is applied on the external electrodes.

However, when the relaxation time is long compared with the carrier transit time, the space-charge effects cannot be neglected (see, for example, Section 4.5) and therefore the steady-state $I-V$ characteristic does not follow the velocity-field characteristic. It will be shown that the $I-V$ characteristic does not, in general, exhibit a negative-resistance region. The high-frequency resistance can be, however, negative for a proper bias and operating frequency.

The carrier diffusion may have, under certain circumstances, an influence upon the dynamics of space-charge instabilities mentioned above and upon the establishment of a stable field profile as well. This influence is particularly important in high-conductivity samples.

Although the usual theory of transferred-electron devices assumes ideal ohmic contacts to the semiconductor bulk, it is a well-known laboratory reality that in many cases the actual contacts exhibit an anomalous, non-ohmic behaviour. The effect of such real contacts will also be discussed.

The present chapter refers to TED with an essentially stable internal field profile and a negative resistance which can be used for high-frequency (microwave) amplification.

The next chapter will be devoted in principal to an entirely different mode of behaviour, characterized by the cyclic propagation of large field disturbances through the device.

7.2 Steady-state Characteristic and Low-frequency Negative Resistance

The basic one-dimensional equations describing the unipolar (electron) conduction in a TED are: the continuity equation

$$J = env(\mathcal{E}) - e\frac{\partial}{\partial x}\{D(\mathcal{E})n\} + \varepsilon\frac{\partial \mathcal{E}}{\partial t}, \qquad (7.2.1)$$

and the Poisson equation*

$$\frac{\partial \mathcal{E}}{\partial x} = \frac{e}{\varepsilon}(n - N_D), \quad N_D = \text{effective donor density}. \qquad (7.2.2)$$

---

* All donors are assumed ionized and trapping is neglected. In general, N_D may be a function of x. We assume uniform doping, except when otherwise stated.

For algebraic convenience the sign conventions correspond to positively charged mobile particles (see also Chapter 4). The electron velocity, v, and electron diffusion constant, D, are assumed instantaneous functions of electric field intensity, $\mathscr{E}$. Both functions depend upon the particular-semiconductor considered and are determined experimentally or computed theoretically. We have shown in Chapter 1 that the above description of carrier transport, and particularly the introduction of the diffusion current and diffusion 'constant', are questionable. We retain, however, this 'classical' approach because it seems to be the only one which is analytically tractable.

By eliminating n through equations (7.2.1) and (7.2.2), one obtains a second-order non-linear differential equation with partial derivatives: $\mathscr{E} = \mathscr{E}(x, t)$ is the unknown function and $J = J(t)$ is assumed given. The particular solution for the problem considered is usually selected by imposing boundary conditions, for example, the electric field will be specified at the cathode, $x = 0$ and at the anode, $x = L$.

The voltage drop across the device is

$$V = \int_0^L \mathscr{E}(x, t)\, dx. \tag{7.2.3}$$

Formally, V will be calculated for a given device and a given $J(t)$. In many cases, however, V is impossed by the external circuit and $J(t)$ should be calculated.

In this section we consider the steady-state behaviour (the displacement current in equation (7.2.1) is zero). Let us assume that the diffusion current is negligible. A first-order differential equation will be obtained from equations (7.2.1) and (7.2.2) (J is considered a parameter), the particular solution $\mathscr{E} = \mathscr{E}(x; J)$ will be selected by imposing a proper boundary condition, namely the value of the cathode field, $\mathscr{E}_c$:

$$\mathscr{E}(0) = \mathscr{E}_c = \text{given, usually } \mathscr{E}_c = \mathscr{E}_c(J). \tag{7.2.4}$$

Then $V = \int_0^L \mathscr{E}(x; J)\, dx$ should yield the $J-V$ characteristic.

We prefer, however, a different method of integration. This is based upon the following property of the electric field. In a homogeneous and one-dimensional structure with negligible diffusion current, the electric field varies monotonously with x from the injecting $(x = 0)$ to the collecting contact $(x = L)$. In other words, (see the Poisson equation), there exists no neutral plane inside the structure $(0 < x < L)$ separating an accumulation and a depletion region. (Problem 7.1) [2]. Consequently, we shall integrate with respect to $\mathscr{E}$, instead of x. The change of variable is made by using

$$\frac{d\mathscr{E}}{dx} = \frac{e(n - N_D)}{\varepsilon} = \frac{J}{\varepsilon v(\mathscr{E})} - \frac{eN_D}{\varepsilon}. \tag{7.2.5}$$

Therefore equation (7.2.3) becomes ($\mathscr{E}(0) = \mathscr{E}_c$, $\mathscr{E}(L) = \mathscr{E}_a$):

$$V = \int_{\mathscr{E}_c}^{\mathscr{E}_a} \frac{\varepsilon \mathscr{E} v(\mathscr{E})\, d\mathscr{E}}{J - eN_D v(\mathscr{E})}. \tag{7.2.6}$$

Another independent integral equation will be found from $L = \int_0^L dx$:

$$L = \int_{\mathscr{E}_c}^{\mathscr{E}_a} \frac{\varepsilon v(\mathscr{E}) \, d\mathscr{E}}{J - eN_D v(\mathscr{E})}. \qquad (7.2.7)$$

The steady-state characteristic is given parametrically (and inexplicitly) by equations (7.2.4), (7.2.6) and (7.2.7) ($\mathscr{E}(L) = \mathscr{E}_a$ is a parameter). It can be easily shown (see Chapters 2 and 5) that in many cases the current injected in a unipolar structure can be written as a function of the cathode field, $J = J(\mathscr{E}_c)$. Hence occurs the dependence $\mathscr{E}_c = \mathscr{E}_c(J)$ postulated by equation (7.2.4). $J = J(\mathscr{E}_c)$ is the so-called *control characteristic*

$$j_c = j_c(\mathscr{E}) \text{ such that } j_c = J \text{ for } \mathscr{E} = \mathscr{E}_c, \qquad (7.2.8)$$

introduced by Kroemer [3], defining the injection properties of a contact (for example a Schottky-barrier metal-semiconductor contact) or an injection region in general (it may be a highly doped and very thin region introduced on purpose to enhance the field emission from a metal contact into the semiconductor, a damaged contact zone, Section 2.9, or the region between the virtual cathode and the actual cathode contact, chapters 4 and 5). We may also define a *neutral characteristic* [2] – [4]

$$j_n = (\mathscr{E}) = eN_D v(\mathscr{E}) \qquad (7.2.9)$$

and the steady-state equations (7.2.4), (7.2.6) and (7.2.7) can be written, respectively,

$$J = j_c(\mathscr{E}_c), \qquad (7.2.10)$$

$$V = \frac{\varepsilon}{eN_D} \int_{\mathscr{E}_c}^{\mathscr{E}_a} \frac{j_n(\mathscr{E}) \, \mathscr{E} \, d\mathscr{E}}{j_c(\mathscr{E}_c) - j_n(\mathscr{E})}, \qquad (7.2.11)$$

$$\frac{eN_D L}{\varepsilon} = \int_{\mathscr{E}_c}^{\mathscr{E}_a} \frac{j_n(\mathscr{E}) \, d\mathscr{E}}{j_c(\mathscr{E}_c) - j_n(\mathscr{E})}. \qquad (7.2.12)$$

The above results can be used either for a qualitative discussion of the field and charge configuration, or for an analytical or numerical computation of the steady-state $J-V$ characteristic [3], [4].

Let us consider a semiconductor sample constructed from a homogeneous negative mobility material, such as GaAs (see Chapter 1 for a detailed discussion of transport properties in GaAs). The sample length and doping are L and N_D, respectively. The neutral characteristic, $j_n(\mathscr{E}) = eN_D v(\mathscr{E})$, is shown in Figure 7.2.1. The structure has a 'cathode' at $x = 0$ and an 'anode' at $x = L$. The charge injection is provided by a contact at $x = 0$ or a certain injection region situated at $x \leqslant 0$. The injection is described by the control characteristic $J = j_c(\mathscr{E}_c)$ where J is the current density through the device and $\mathscr{E}_c$ the cathode field, the electric field intensity in semiconductor at $x = 0$. Figure 7.2.1 also shows an example of control characteristic j_c: this is a qualitative description of behaviour of a certain type of 'imperfect cathode', such as a Schottky contact or a heterojunction barrier. Due to the barrier lowering enhanced by the cathode field, the control characteristics increases sharply [3]. In Figure 7.2.1 j_c crosses j_n in the negative-slope region. The crossover

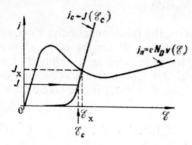

Figure 7.2.1. Neutral characteristic for an NDM semiconductor and control characteristic for a barrier-type contact, $\mathscr{E}_a$ is the anode field and $\mathscr{E}_c$ is the cathode field for the bias current shown (J) and a particular device length.

field and the corresponding current are denoted by $\mathscr{E}_x$ and J_x, respectively. It can be easily shown that for $J < J_x$ the entire semiconductor is depleted (because electron concentration at the cathode is below N_D and the electric field should vary monotonously), whereas for $J > J_x$ the entire semiconductor is accumulated. When the semiconductor is depleted, the electric field decreases with increasing x, starting from the maximum value $\mathscr{E}_c$ determined by the particular current density J (Figure 7.2.1). If the device is sufficiently long, the anode field will be well below threshold and the free carrier distribution looks like Figure 7.2.2. The cathode region, limited by the minimum of the carrier density, is situated in the negative-mobility field region and is potentially unstable (Section 7.1). Note the fact that the bias field V/L, which is the average value of the internal electric field (the voltage drop on the cathode-anode space is assumed equal to the external bias voltage), may be *above or below* the threshold field, $\mathscr{E}_M$. Assume now that the injected current increases. The cathode field also increases. The quantity under the integral in equation (7.2.12) decreases and, for a given device, $\mathscr{E}_a$ should increase and eventually rise above $\mathscr{E}_M$. The carrier distribution is modified as shown qualitatively in Figure 7.2.2. Note the fact that the imperfect (non-ohmic) cathode determines in this case:

(a) semiconductor depletion at low bias currents;

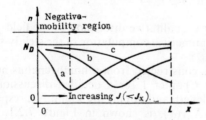

Figure 7.2.2. Carrier distribution inside an NDM sample with a non-ohmic injecting contact, for three values of the bias current (increasing from (a) to (c)).

(b) a negative-mobility region in the vicinity of the cathode, even if the bias field V/L is below its threshold value;

(c) the gradual extension of the negative-mobility zone in the entire semiconductor of length L, as the bias current increases.

7. Transferred-electron Diodes for Microwave Amplification

When J exceeds J_x (Figure 7.2.1) the semiconductor becomes accumulated, $\mathscr{E}_a$ increases above $\mathscr{E}_c$ and may attain high values (leading to avalanche breakdown) even at a moderate bias-current. This can be shown by using equation (7.2.12) where the quantity under the integral decreases rapidly as J increases above J_x, $\mathscr{E}_c$ remains almost constant and $\mathscr{E}_a$ should increase considerably to maintain the equality.

The possibility of a negative resistance on the static characteristic will now be discussed. The incremental (low-frequency) resistance can be obtained from equations (7.2.6) and (7.2.7) by differentiating with respect to $\mathscr{E}_c$, assuming $\mathscr{E}_a = \mathscr{E}_a(\mathscr{E}_c)$ and then eliminating $d\mathscr{E}_a/d\mathscr{E}_c$. The result is

$$\frac{dV}{dJ} = \frac{\varepsilon}{eN_D} \int_{\mathscr{E}_c}^{\mathscr{E}_a} \frac{(\mathscr{E}_a - \mathscr{E}) j_n(\mathscr{E}) \, d\mathscr{E}}{[j_c(\mathscr{E}_c) - j_n(\mathscr{E})]^2} + \frac{\varepsilon}{eN_D} \frac{1}{\left(\dfrac{dj_c}{d\mathscr{E}}\right)_{\mathscr{E}=\mathscr{E}_c}} \frac{\mathscr{E}_a - \mathscr{E}_c}{\dfrac{j_c(\mathscr{E}_c)}{j_n(\mathscr{E}_c)} - 1}. \quad (7.2.13)$$

We note in the expression of the second term in the right-hand side of equation (7.2.13), the differential cathode conductivity

$$\sigma_1(0) = \left(\frac{dj_c}{d\mathscr{E}}\right)_{\mathscr{E}=\mathscr{E}_c} \quad (7.2.14)$$

which is the slope of the control characteristic at a field corresponding to a given current. The same parameter was introduced by equation (5.3.3) as an effective small-signal conductivity of the source (cathode) plane. For an ohmic contact (space-charge limited (SCL) injection) this conductivity is infinite (see also Chapter 4) and the second term in equation (7.2.13) vanishes. Therefore, the first term of the incremental resistance, equation (7.2.13) may be considered as due to the bulk space-charge whereas the second one is due to the finite conductivity (or finite injection capability) of the cathode contact [4].

The bulk resistance

$$\left(\frac{dV}{dJ}\right)_{bulk} = \frac{\varepsilon}{eN_D} \int_{\mathscr{E}_c}^{\mathscr{E}_a} \frac{(\mathscr{E}_a - \mathscr{E}) j_n(\mathscr{E}) \, d\mathscr{E}}{[j_c(\mathscr{E}_c) - j_n(\mathscr{E})]^2} \quad (7.2.15)$$

is *always positive*, even if the differential mobility is negative ($dj_n/d\mathscr{E} < 0$)*. The cathode resistance is

$$\left(\frac{dV}{dJ}\right)_{cath} = \frac{\varepsilon}{eN_D} \frac{1}{\sigma_1(0)} \frac{\mathscr{E}_a - \mathscr{E}_c}{\dfrac{j_c(\mathscr{E}_c)}{j_n(\mathscr{E}_c)} - 1} \quad (7.2.16)$$

and has the sign of the cathode conductivity, $\sigma_1(0)$. The immediate results are:

(a) the incremental resistance is always positive under SCL injection conditions (Shockley's positive conductance theorem) [5] — [11]:

---

* When the semiconductor is accumulated $\mathscr{E}_c < \mathscr{E}_a$, $\mathscr{E} \leq \mathscr{E}_a$ and the integral is positive. The same situation takes place when the semiconductor is depleted ($\mathscr{E}_c > \mathscr{E}_a$, $\mathscr{E} \geq \mathscr{E}_a$).

(b) a *necessary* condition for a 'static' negative resistance is a negative value of the differential cathode conductivity (a negative slope of the control characteristic); this condition may be fulfilled if the emitting plane is situated in a medium of negative mobility as discussed by Kroemer [3], Dascalu [4], and Ryabinkin [12].

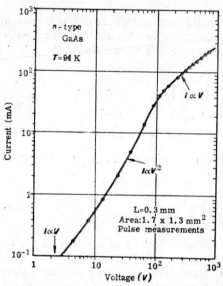

Figure 7.2.3. Experimental $I-V$ characteristic (pulse measurements) of an n-GaAs sample with ohmic contacts. (*after Yamashita*, [13]).

The physical explanation for Shockley's positive conductance theorem is that an increase in applied voltage will build up an internal space charge and increase the carrier density such that the decrease in carrier drift velocity is more than compensated and the current continues to increase with the increasing bias voltage [6], [7]. Figure 7.2.3 shows an experimental $I-V$ characteristic of SCL current in high resistivity n-type GaAs with ohmic contacts [13]. This characteristic shows a gradual transition from ohmic behaviour at low current densities to the square-law characteristic of the SCL current with field-independent carrier mobility and then a high-field linear region indicating the velocity saturation under SCL conditions.

Kroemer [8] generalized Shockley's theorem for arbitrary impurity distributions and geometries. However, computer calculations [14] and experiment [15] indicated a static negative resistance for particular geometrical configurations, e.g. a device with a steep expansion of the cross-section near the anode (a stationary high-field domain is accommodated in this region, see also below).

An important phenomenon so far neglected is carrier diffusion. Hauge [9] computed a static negative resistance in a GaAs sample with ohmic contacts by consi-

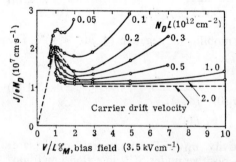

Figure 7.2.4. Calculated static $I-V$ dependence for $D' \approx \varepsilon v_0/eN_D$ between threshold and peak electric field and for several values of the doping-length product, $N_D L$. The assumed $v = v(\mathscr{E})$ relationship is also shown for comparison (dashed). (*after Hauge*, [9]).

dering a field-dependent diffusion coefficient $D = D(\mathscr{E})$, in equation (7.2.1). It was assumed that $D' = dD/d\mathscr{E}$ is positive near and above the threshold field for negative mobility. Hauge's results [9] are reproduced in Figure 7.2.4 and show that the negative resistance effect is more pronounced for higher doping-length products (carrier diffusion is increasingly important at higher doping levels). However, the actual $D = D(\mathscr{E})$ dependence is not particularly important from this point of view and high-conductivity samples may exhibit negative resistance even when D is constant [16]. It was shown that a stationary high-field domain (a dipole formed by an accumulation and a depletion layer, see also below) occurs near the anode contact [16]—[18]. A typical field and charge configuration near the anode is reproduced in Figure 7.2.5. Qualitatively, such an anode domain is maintained by a balance between carrier flow into the domain (drift in the forward direction, toward the anode) and carrier diffusion in the opposite direction. A sufficiently high background doping is necessary to provide the flow of electrons and sustain the anode domain. Outside the domain the semiconductor is neutral and the 'outside' electric field is below the threshold value for negative mobility. The behaviour of high-field domains will be studied in the next chapter and we shall see that an increase of the applied bias will be absorbed by the domain and the outside field will decrease thus reducing the external current (which is equal to the drift current outside the domain). A similar property is valid for a domain 'trapped' near the anode. This negative resistance effect is apparent on the computed steady-state characteristics in Figure 7.2.6. Below the threshold voltage the normalized $I-V$ characteristic follows the $v = v(\mathscr{E})$ characteristic of the semiconductor material. Above threshold, the current drops due to the occurrence of the stationary high-field anode domain. Once the domain is formed and 'trapped' near the anode contact, it is not immediately extinguished as the voltage decreases below the threshold value (hence the hysteresis

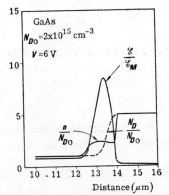

Figure 7.2.5. Distribution of the electric field and of the electron density for an anode-trapped domain. (*after Murayama and Ohmi*, [18]).

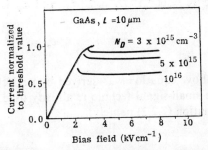

Figure 7.2.6. Normalized device current of an n^+nn^+ GaAs sample *versus* bias field for three donor densities. (*after Murayama and Ohmi*, [18]).

effect in Figure 7.2.5)*. The devices with static negative resistance are used for wideband amplification (Section 7.6) and as bistable switching elements (Section 8.5).

The steady-state $I-V$ characteristics of various TED's will be further discussed in Chapters 7 and 8.

---

* It was shown that the stable anode domain will not occur if the doping fluctuations are relatively large: the effect is less important for devices with higher doping-length product [18].

7.3 Small-signal Space-charge Waves

We shall consider below the behaviour of small perturbations of the electric field. Let us assume

$$n(x, t) = n_0 + n_1(x, t), \qquad (7.3.1)$$

$$\mathscr{E}(x, t) = \mathscr{E}_0 + \mathscr{E}_1(x, t), \qquad (7.3.2)$$

where $n_1(x, t)$ and $\mathscr{E}_1(x, t)$ are small space- and time-dependent perturbations superimposed on uniform steady-state carrier concentration and field, n_0 and $\mathscr{E}_0$, respectively. The semiconductor bulk is esentially neutral and $n_0 = N_D^*$. The electric field is so that *the differential mobility is negative* ($dv/d\mathscr{E} < 0$) and the dielectric relaxation frequency ω_D, defined by

$$\omega_D = -\frac{eN_D}{\varepsilon}\frac{dv}{d\mathscr{E}}\bigg|_{\mathscr{E}=\mathscr{E}_0} \qquad (7.3.3)$$

is positive. The small-signal (perturbation) part of the total current, equation (7.2.1), is [16]

$$J_1 = en_1v_0 - \varepsilon\omega_D\mathscr{E}_1 - eD_0\frac{\partial n_1}{\partial x} - ev_d n_1 + \varepsilon\frac{\partial \mathscr{E}_1}{\partial t}, \qquad (7.3.4)$$

where $v_0 = v(\mathscr{E}_0)$, $D_0 = D(\mathscr{E}_0)$ and v_d is a diffusion-induced velocity defined by

$$v_d = \frac{eN_D}{\varepsilon}\frac{dD}{d\mathscr{E}}\bigg|_{\mathscr{E}=\mathscr{E}_0} \qquad (7.3.5)$$

The above equation was derived by neglecting all second-order quantities according to the small-signal technique already used in Chapters 4–6. The small-signal form of equation (7.2.2) namely

$$\frac{\partial \mathscr{E}_1}{\partial x} = \frac{en_1}{\varepsilon} \qquad (7.3.6)$$

was also used. J_1 is spatially uniform, $\partial J_1/\partial x = 0$ and hence

$$\frac{\partial n_1}{\partial t} + (v_0 - v_d)\frac{\partial n_1}{\partial x} - \omega_D n_1 - D_0 \frac{\partial^2 n_1}{\partial x_2} = 0. \qquad (7.3.7)$$

---

* We have already seen (Section 7.2) that, depending upon the sample parameters, the nature of the contacts and the steady-state bias, the semiconductor bulk may depart more or less from neutrality. Note also that the electric field may be distorted by the time-dependent perturbation itself, if it grows large enough [7].

If the space-charge perturbation $n_1(x, t)$ is assumed to vary as $\exp i(\omega t - kx)$ — this is *the space-charge wave* — then equation (7.3.7) yields the dispersion relation

$$D_0 k^2 - i(v_0 - v_d) k - \omega_D + i\omega = 0, \qquad (7.3.8)$$

where k is the complex wave number [16]. The two solutions of equation (7.3.8), k_1 and k_2, correspond, respectively, to the forward wave and the reverse (backward) wave. The space-charge perturbation is the sum of these two waves

$$n_1(x, t) = n_1^{(1)} \exp(k_{1i}x) \cos(\omega t - k_{1r}x) + n_1^{(2)} \exp(k_{2i}x) \cos(\omega t - k_{2r}x), \qquad (7.3.9)$$

where the index r denotes the real part and i the imaginary part of the wavenumber ($k = k_r + ik_i$):

$$k_{1r}, k_{2r} = \pm \frac{1}{\lambda_D} \sqrt{1 - \beta^2 + \sqrt{(\beta^2 - 1)^2 + (\omega/\omega_D)^2}}, \qquad (7.3.10)$$

$$k_{1i}, k_{2i} = \frac{1}{\lambda_D} \left\{ \beta \mp \frac{1}{\sqrt{2}} \sqrt{\beta^2 - 1 + \sqrt{(\beta^2 - 1)^2 + (\omega/\omega_D)^2}} \right\}, \qquad (7.3.11)$$

where

$$\beta = \frac{v_0 - v_d}{2D_0} \lambda_D, \qquad (7.3.12)$$

and

$$\lambda_D = \sqrt{\frac{D_0}{\omega_D}} = \sqrt{\frac{\varepsilon D_0}{eN_D |dv/d\mathscr{E}|}}. \qquad (7.3.13)$$

At low fields, by replacing $dv/d\mathscr{E}$ and D_0 by their low-field values and using the Einstein relation, λ_D becomes equal to the extrinsic Debye length, equation (7.1.2). In general, however, λ_D is a bias-dependent parameter. For simplicity, λ_D is called below the Debye length. This is a characteristic length expressing the ability of charge to redistribute itself through the process of diffusion [1]. Its significance occurs from equation (7.3.7) when $\partial/\partial t \equiv 0$, $v_0 = 0$ (zero current-flow) $v_d = 0$ (constant D), $dv/d\mathscr{E} > 0$, $\omega_D = -D_0/\lambda_D^2$. This equation and its solutions become,

$$\frac{\partial^2 n_1}{\partial x^2} - \frac{n_1}{\lambda_D^2} = 0, \quad n_1 = \text{const} \exp\left(\pm \frac{x}{\lambda_D}\right), \quad \text{respectively.} \qquad (7.3.14)$$

We shall now discuss the wave-type solution (7.3.9) of equation (7.3.7) describing the evolution of a small space-charge perturbation in an essentially neutral *negative-mobility* semiconductor. The behaviour of the solution depends upon the value of the real parameter β. Four types of solution are distinguished, corresponding to four ranges of values taken by the parameter β. These ranges are determined by the following critical values: $\beta = 0, \pm 1$. Table 7.3.1 defines these modes of behaviour relative to parameters with physical significance, v_0, v_d, ω_D and D_0.

Table 7.3.1

Four types of Space-charge Wave Behaviour

Type	Parameter β	Physical parameters
I	$\beta > 1$	$v_0 > v_d$, $(v_0 - v_d)^2 - 4D_0\omega_D > 0$
II	$0 < \beta < 1$	$v_0 > v_d$, $(v_0 - v_d)^2 - 4D_0\omega_D < 0$
III	$-1 < \beta < 0$	$v_0 < v_d$, $(v_0 - v_d)^2 - 4D_0\omega_D < 0$
IV	$\beta < -1$	$v_0 - v_d$, $(v_0 - v_d)^2 - 4D_0\omega_D > 0$

We choose as essential parameters [16] the semiconductor doping, N_D, and the derivative of the diffusion coefficient with respect to the electric field, $D' = dD/d\mathscr{E}|_{\mathscr{E}=\mathscr{E}_0}$. The above four types of behaviour are described by four regions in an $N_D - D'$ semiplane, as shown in Figure 7.3.1 [16].

Following Ohmi and Hasuo [16], we consider only the *low-frequency region* defined by

$$\omega \ll \frac{|(v_0 - v_d)^2 - 4\omega_D D_0|}{4D_0}, \quad (v_0 - v_d)^2 \neq 4\omega_D D_0. \tag{7.3.15}$$

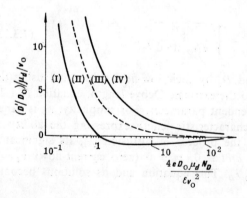

Figure 7.3.1. Diagram defining four types of space-charge wave behaviour (Table 7.3.1). The abscissa is proportional to the carrier doping density and the ordinate — to the derivative of the field-dependent diffusion coefficient. (*after Ohmi and Hasuo*, [16]).

The wave-number $k = k_r + ik_i$ will become

$$k_{1,2} = \frac{1}{\lambda_D}\left\{\pm \frac{\omega/\omega_D}{2\sqrt{\beta^2-1}} + i(\beta \mp \sqrt{\beta^2-1})\right\} \text{ for } \beta^2 > 1, \tag{7.3.16}$$

$$k_{1,2} = \frac{1}{\lambda_D}\left\{\pm \sqrt{1-\beta^2} + i\left(\beta \mp \frac{\omega/\omega_D}{2\sqrt{1-\beta^2}}\right)\right\} \text{ for } \beta^2 < 1. \tag{7.3.17}$$

Let us first assume, for simplicity, that $D = $ constant, $D' = 0$ and $v_d = 0$. Only I and II types of behaviour are possible, type I for $\beta > 1$ and type II for $0 < \beta < 1$. Because

$$\beta = \frac{v_0 \lambda_D}{2D} = \frac{v_0}{2\sqrt{\omega_D D}} = \frac{v_0}{2} \sqrt{\frac{\varepsilon}{eN_D/dv/d\mathscr{E}|D}} \qquad (7.3.18)$$

the first case (high β) corresponds to a low effect of diffusion as compared to that of the dielectric relaxation mechanism. For a given semiconductor this occurs for low semiconductor doping. The opposite is true for low β values. This distinction occurs because in our case the action of dielectric relaxation mechanism is opposite to that of the diffusion mechanism. Whereas in positive-mobility materials both dielectric relaxation and diffusion tend to disperse any charge bunches and to attenuate the space-charge wave, in negative-mobility materials the so-called dielectric relaxation is contrary to the normal relaxation and determines growing of space-charge perturbations (Section 7.1). In the limit case when $\beta \to \infty$ (negligible diffusion), equation (7.3.16) yields

$$k_{1r}, k_{2r} = \pm \frac{\omega}{v_0} \qquad (7.3.19)$$

and the corresponding phase velocity, v_p, is

$$(v_p)_{1,2} = \frac{\omega}{(k_2)_{1,2}} = \pm v_0 \qquad (7.3.20)$$

(the group velocities $d\omega/dk$ are here identical with the phase velocities). The forward wave moves in the positive x direction with the phase velocity v_{p1} equal to the steady-state drift velocity, v_0, whereas the reverse wave moves in the opposite direction with the same absolute velocity ($v_{p2} = -v_0$). The behaviour of these waves is, however, completely different. It can be easily shown that equation (7.3.16) yields for $\beta \to \infty$, $k_{1i} = \dfrac{1}{2\beta\lambda_D}$ and $k_{2i} = \dfrac{2\beta}{\lambda_D}$. A positive $k_{1i} = D_0/v_0\lambda_D^2 = \omega_D/v_0 > 0$ in equation (7.3.9) corresponds to a *growing forward wave* (due to the negative differential mobility), whereas a positive $k_{2i} = v_0/D_0 > 0$ corresponds to a decaying reverse wave. The characteristic length for the forward wave growth is $(k_{1i})^{-1} = v_0/\omega_D$ and is inversely proportional to the device doping, N_D. The total growth of the forward wave across a sample of length L is negligible if $L \ll v_0/\omega_D$ and hence the doping-length product $N_D L$ should be sufficiently low (Section 7.1). On the other hand, the characteristic length for the reverse wave decay is $(k_{2i})^{-1} = D_0/v_0 \ll \lambda_D (\beta \to 1$, diffusion unimportant).

When β is comparable to, but still higher than unity, the diffusion effects modify somewhat the above picture. For example the phase velocities are reduced as follows

$$v_{p1}, v_{p2} = \pm v_0 \sqrt{1 - \left(\frac{2D_0}{v_0 \lambda_D}\right)^2} \qquad (7.3.21)$$

(the higher the semiconductor doping, the lower the phase velocities).

The space-wave behaviour changes completely (type II) when $\beta < 1$ (this occurs in principle for strong diffusion effects, at high doping densities). Equation (7.3.17) should be used now. Note the fact that the phase velocity, $v_p = \omega/k_r$, will now be frequency-dependent but the wavelength = phase velocity × oscillation period will be frequency independent and finite at $\omega = 0$. In the limit of strong diffusion effect $(\beta \to 0)$ $k_r \to \pm 1/\lambda_D$ whereas $k_i \to \mp \omega/2\omega_D \lambda_D$. The phase velocity is

$$v_{p1}, v_{p2} = \pm \omega \lambda_D \quad (\beta \to 0) \quad (7.3.22)$$

and the space-charge wavelength, λ, is given by

$$\lambda = v_p \cdot \frac{2\pi}{\omega} = 2\pi \lambda_D \quad (\beta \to 0). \quad (7.3.23)$$

At low frequencies the wavelength is almost constant and equal to

$$\lambda = 2\pi \frac{2D_0}{\sqrt{4\omega_D D_0 - v_0^2}} \quad (0 < \beta < 1). \quad (7.3.24)$$

The frequency-independent wavelength suggests the possibility of wide band negative resistance and a negative static resistance ($\lambda =$ finite at $\omega = 0$). The upper frequency limit, $f_1 = v_0/2\pi\lambda_D$, where the growing wave disappears, is calculated from equations (7.3.10) and (7.3.11) and physically originates from the fact that a disturbance with the width less than the Debye length cannot build up even in semiconductors with negative differential mobility [10]. The oscillation wavelength, equation (7.3.24), decreases with increasing doping density.

We shall now consider the case when $D = D(\mathscr{E})$. When $D' = dD/d\mathscr{E}|_{\mathscr{E}=\mathscr{E}_0}$ is negative, v_d defined by equation (7.3.5) is negative such that the drift velocity v_0 in equations (7.3.18) — (7.3.24) should be replaced by $v_0 + |v_d|$. Note the fact that v_d is also doping dependent. When $D' > 0$, $v_d > 0$ and the phase velocity is $v_0 - v_d < v_0$. When the diffusion induced velocity, v_d, exceeds (at higher semiconductor dopings) the drift velocity, v_0, the situation changes completely. The space-charge disturbance will now travel in the opposite direction to the drift velocity of background electrons [16]. To explain this qualitatively consider an accumulation layer of electrons drifting from the left to the right and a positive $dD/d\mathscr{E}$. The diffusion coefficient on the right-hand side of the accumulation layer is larger than that of the left-hand side because the electric front is larger, so that the electrons in front diffuse faster than those at the back [16]. Consequently the velocity of the accumulation layer is reduced from v_0 to $v_0 - v_d$. When the doping density is high enough to change the sign of $v_0 - v_d$, the accumulation layer travels towards the cathode.

Types III and IV of behaviour (Table 7.3.1) correspond to $v_d > v_0$. Consider for example type IV of behaviour in the low-frequency range. Equation (7.3.16) will be used ($\beta < -1$) for the wave number k. The phase velocities of the forward and reverse waves, respectively, are

$$v_{p1}, v_{p2} = \pm \sqrt{(v_0 - v_d)^2 - 4\omega_D D_0}. \quad (7.3.25)$$

Table 7.3.2
Characteristic of Space-charge Wave and Device Performance [16]

Type of behaviour	Conditions	Growing space-charge wave			Device performance	
		Direction	Growing rate at $\omega = 0$	Characteristics	Small-signal impedance	Domain dynamics (see Chapter 8)
I	$v_0 - v_d > 0$ $(v_0 - v_d)^2 - 4\omega_D D_0 > 0$	forward	$\dfrac{v_0 - v_d - \sqrt{(v_0 - v_d)^2 - 4\omega_D D_0}}{2D_0}$	frequency-independent velocity $\sqrt{(v_0 - v_d)^2 - 4\omega D_0}$	static positive resistance and transit-time negative resistance	cyclically travelling domain from cathode to anode
II	$v_0 - v_d > 0$ $(v_0 - v_d)^2 - 4\omega_D D_0 < 0$	forward	$\dfrac{v_0 - v_d}{2D_0}$	frequency-independent wavelength $\dfrac{4\pi D_0}{\sqrt{4\omega_D D_0 - (v_0 - v_d)^2}}$	static negative resistance	anode trapped domain
III	$v_0 - v_d < 0$ $(v_0 - v_d)^2 - 4\omega_D D_0 < 0$	reverse	$\dfrac{v_d - v_0}{2D_0}$	frequency-independent wavelength $\dfrac{4\pi D_0}{\sqrt{4\omega_D D_0 - (v_d - v_0)^2}}$	static negative resistance	cathode trapped domain
IV	$v_0 - v_d < 0$ $(v_0 - v_d)^2 - 4\omega_D D_0 > 0$	reverse	$\dfrac{v_d - v_0 - \sqrt{(v_d - v_0)^2 - 4\omega_D D_0}}{2D_0}$	frequency-independent velocity $\sqrt{(v_d - v_0)^2 - 4\omega D_0}$	static positive resistance and transit-time negative resistance	cyclically traveling domain from anode to cathode

However, the forward wave decays as it travels, whereas the reverse wave builds up. Therefore the growing space-charge wave is that in the reverse direction and tends to determine a (large) field nonuniformity at the cathode end of the sample. Table 7.3.2, reproduced after Ohmi and Hasuo [16], summarizes the main properties of the four types of space-charge-wave behaviour.

Figures 7.3.2 shows the low-frequency phase and group velocity variation with the semiconductor doping, for a certain value of $D' = dD/d\mathscr{E} > 0$. Note the fact that, contrary to the results of many conventional treatments, the velocity of the space-charge wave is strongly doping-dependent through the diffusion effects discussed above. Figure 7.3.3 shows the frequency-independent wavelength *versus* normalized doping (regions II and III).

The above small-signal results describe the bulk properties of an NDM semiconductor. The device behaviour can be obtained only by specifying the contact properties. The small-signal impedance of a sample of length L with ohmic contacts will be calculated by using

$$\mathscr{E}_1(0, t) = 0, \quad \mathscr{E}_1(L, t) = 0, \quad \int_0^L \mathscr{E}_1(x, t) \, dx = V_1(t) \qquad (7.3.26)$$

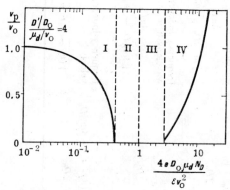
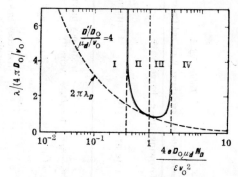

Figure 7.3.2. Normalized phase velocity as a function of sample doping, for a particular value of the derivative of the diffusion coefficient. Regions I—IV are defined in Table 7.3.1. (*after Ohmi and Hasuo*, [16]).

Figure 7.3.3. Wavelength of static space-charge wave as a function of doping density, for a particular value of the derivative of the diffusion coefficient. Dashed line shows the Debye length λ_D multiplied by 2. Regions I—IV are defined in Table 7.3.1. (*after Ohmi and Hasuo*, [16]).

and the alternating-current formalism introduced in Section 4.6. The impedance unit area device is [16]

$$Z = \frac{\lambda_D/\varepsilon\omega_D}{(1 - i\bar{\omega})\bar{L}} \left\{ -1 + \frac{2(\xi + i\eta)}{(1 - i\bar{\omega})\bar{L}} \times \right.$$

$$\left. \times \frac{\cosh \beta\bar{L} - \cos \xi\bar{L} \cos \eta\bar{L} - i \sinh \xi\bar{L} \sin \eta\bar{L}}{\sin \xi\bar{L} \cos \eta\bar{L} + i \cosh \xi\bar{L} \sinh \eta\bar{L}} \right\}, \qquad (7.3.27)$$

where

$$\bar{L} = L/\lambda_D, \quad \bar{\omega} = \omega/\omega_D, \tag{7.3.28}$$

$$\xi = \frac{1}{\sqrt{2}}\sqrt{\sqrt{(\beta^2-1)^2 + \bar{\omega}^2} + \beta^2 - 1}, \tag{7.3.29}$$

$$\eta = \frac{1}{\sqrt{2}}\sqrt{\sqrt{(\beta^2-1)^2 + \bar{\omega}^2} + 1 - \beta^2}. \tag{7.3.30}$$

There are two distinct types of behaviour of the device impedance. When calculated for regions I and IV (Figure 7.3.1), the resistance is positive at zero frequency but exhibits cyclical oscillations at higher frequencies, around the transit-time frequency and its harmonics (transit-time behaviour, see also Chapters 4—6). This is illustrated by Figure 7.3.4, where the frequency-dependence of the small-signal resistance is shown for a few dopings in region I and IV. Note the fact that the negative resistance tends to appear at lower frequencies in region I (Figure 7.3.4 a) as the donor density increases, while it occurs at higher frequencies with increasing N_D in region IV. This can be explained as follows. The device starts to exhibit

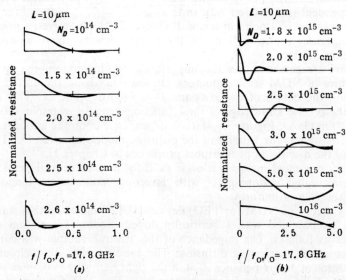

Figure 7.3.4. Frequency dependence of the small-signal resistance in region I (a) and in region II (b), calculated for $\mathscr{E}_0 = 3.5$ kVcm^{-1}, $D_0 = 687$ cm^{2}s^{-1}, $L = 10$ μm, $f_0 = 17.8$ GHz. (after Ohmi and Hasuo, [16]).

a negative resistance when the wavelength of the space-charge wave is reduced to about twice the diode length. However, the wavelength decreases with the increase of the doping density in region I because the phase velocity decreases (Figure 7.3.2), and increases with doping in region IV.

The second type of impedance behaviour, corresponding to regions II and III in Figure 7.3.1, is characterized by the possibility of a static negative resistance (SNR). This negative resistance occurs cyclically with the diode length as indicated by the diagram of Figure 7.3.5. The existence of this negative resistance is impossible for $L < \pi\lambda_D$. The occurrence of SNR at higher L depends upon the normalized

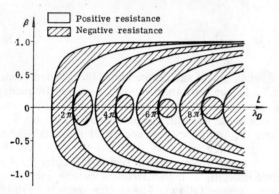

Figure 7.3.5. Diagram showing regions of static negative resistance (dashed area) in the normalized velocity, β, versus normalized diode length, L/λ_D. (after Ohmi and Hasuo, [16]).

effective velocity, β. The above results show that an SNR is possible even when D is field-independent (see Section 7.2): in this case $D' = 0$, $v_d = 0$ and the operation may still correspond to a point in region II (Figure 7.3.1). However, a positive $D' = dD/d\mathscr{E}$ decreases the lower limit of the doping density above which the SNR occurs (Figure 7.3.1).

The above treatment shows the importance of diffusion in the space-charge wave dynamics of NDM semiconductors. On the other hand the device behaviour depends upon the details of $v = v(\mathscr{E})$ and $D = D(\mathscr{E})$ dependence. This raises two questions: the precise knowledge of these functions (see for example reference [19] for $D = D(\mathscr{E})$ in GaAs, Figure 7.6.4), their dependence upon the lattice temperature and doping, as well as the validity of the diffusion model, equation (7.2.1), by itself (according to the discussion of transport properties in Chapter 1).

The theory presented in this section is valid for a quasi-neutral semiconductor. At lower semiconductor dopings or with 'imperfect' contacts the internal electric field is, however, non-uniform.

The early theory of NDM (or TED) devices [1], [20] indicated that stable amplification can be obtained with subcritically doped diodes. The notion of critical doping arises as follows. The impedance of the 'neutral' sample was computed as above by completely neglecting diffusion. The condition for short circuit stability requires no zeros of the impedance $Z(p)$ in the right half of the complex p plane (or the lower half of the complex frequency plane) [1], [20], [21]. By using data appropriate to GaAs one obtains [1]

$$N_D L > (N_D L)_{\text{crit}} = 5 \times 10^{11} \text{cm}^{-2}. \tag{7.3.31}$$

Subcritically doped diodes, $N_D L < (N_D L)_{\text{crit}}$, satisfy the short circuit stability criterion, which means that self-oscillations are not possible when the device is biased by a constant voltage. However, such a diode still exhibits a transit-time negative resistance and can be used for microwave amplification.

Equation (7.3.31) was derived by neglecting diffusion and assuming d.c. neutrality. Diffusion may be indeed less important in subcritically doped diodes. However, the space-charge effects cannot be neglected, especially in under SCL injection conditions (ohmic cathode).

7.4 Subcritically Doped Diode

We consider in this chapter the following model of a lowly-doped diode: one-dimensional geometry, uniform doping, negligible diffusion and ohmic injecting contact. Equations (7.2.6) and (7.2.7) will be used, where the cathode field is zero ($\mathscr{E}_c = 0$). The normalized steady-state characteristic is expressed parametrically by [2]

$$\bar{V} = \int_0^a \frac{\bar{\mathscr{E}}\bar{v}(\bar{\mathscr{E}}) \, d\bar{\mathscr{E}}}{\bar{J} - \nu \bar{v}(\bar{\mathscr{E}})}, \quad \int_0^a \frac{\bar{v}(\bar{\mathscr{E}}) \, d\bar{\mathscr{E}}}{\bar{J} - \nu \bar{v}(\bar{\mathscr{E}})} = 1, \quad (7.4.1)$$

where

$$\bar{V} = V/\mathscr{E}_M L, \quad \bar{J} = J/\varepsilon \mathscr{E}_M v_M L^{-1}, \quad (7.4.2)$$

$$\nu = \frac{e N_D L}{\varepsilon \mathscr{E}_M}, \quad a = \frac{\mathscr{E}(L)}{\mathscr{E}_M}, \quad (7.4.3)$$

$$\bar{v} = \bar{v}(\bar{\mathscr{E}}), \quad \bar{v} = \frac{v}{v_M}, \quad \bar{\mathscr{E}} = \frac{\mathscr{E}}{\mathscr{E}_M} \quad (7.4.4)$$

(v_M being the maximum (peak) velocity and $\mathscr{E}_M$ the threshold field). Note the fact that the normalized steady-state characteristic for a certain semiconductor material

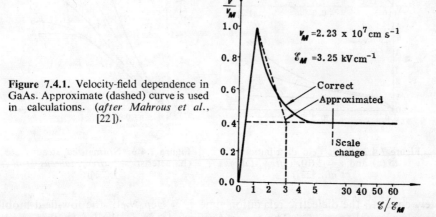

Figure 7.4.1. Velocity-field dependence in GaAs. Approximate (dashed) curve is used in calculations. (after Mahrous et al., [22]).

$v_M = 2.23 \times 10^7$ cm s^{-1}
$\mathscr{E}_M = 3.25$ kV cm^{-1}

($v = v(\mathscr{E})$ given) depends upon the single parameter ν, which for a given material is proportional to the doping-length product.

We illustrate the above theory with results obtained by Mahrous et al. [22]. The velocity-field dependence used in calculations is shown in Figure 7.4.1. The data used determine $\nu = 4.65 \times 10^{-11} N_D L$ ($N_D L$ in cm^{-2}). Figure 7.4.2 shows static field distribution inside the diode. For bias fields (V/L) below threshold ($\mathscr{E}_M$) the electric field is approximately uniform. The electric field is highly non-uniform in the anode region when V/L exceeds $\mathscr{E}_M$. The non-uniformities are less important for higher ν, as expected. When ν is relatively large, a large increase of the bias field does not substantially modify the field in the cathode region and hence the current changes very little. This is indeed shown by the normalized characteristics of Figure 7.4.3, which exhibit saturation for large ν. The experimental characteristics confirm this behaviour.

We shall now discuss the small-signal behaviour. The analytical procedure was presented in section 4.6 and exemplified in Chapters 4—6. The general small-signal theory is discussed in references [1], [2]. The diode impedance cannot be expressed analitically for any $v = v(\mathscr{E})$ relationship. We shall reproduce below results of numerical calculations [23]. Figure 7.4.4 shows the impedance diagram for different values of the two parameters: the normalized current density, $\bar{J_0}$, and the normalized doping-length product, ν. Note the fact that $\bar{J_0}/\nu = J_0/eN_D v_M = J/J_M$ where $J_M = eN_D v_M$ is the 'threshold' current calculated assuming neutrality. The impedance is normalized with respect to the low-current incremental resistance, R_{i0}. The angular frequency, ω, is normalized with respect to $1/\tau'_r$, where $\tau'_r = e\mathscr{E}_M/eN_D v_M$, approxi-

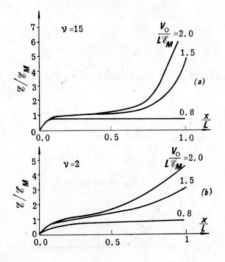
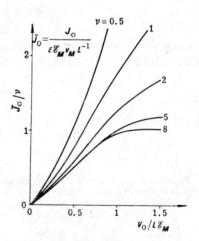

Figure 7.4.2. Static field distributions for $\nu = 15$ (a) and $\nu = 2$ (b). (after Mahrous et al., [22]).

Figure 7.4.3. Normalized steady-state characteristics. (after Mahrous et al., [22]).

mately equal to the dielectric relaxation time $\tau_r = \varepsilon/e\mu_n N_D$, if the low-field mobility μ_n is taken equal to $v_M/\mathscr{E}_M$. The $v = v(\mathscr{E})$ dependence is similar to that used by Mahrous et al. [22] (Figure 7.4.1). The impedance diagrams in Figures 7.4.4 a and b are calculated for $\nu = 3$. The high-frequency series resistance $\mathscr{R}_e Z$ becomes negative when the bias current is sufficiently high (Figure 7.4.4 *b*). A negative resistance occurs for lower bias currents if the diode has a higher doping-length product (the diode length is increased by maintaining a constant doping), as shown by Figures 7.4.4 *a* and *c*. The diagram in Figure 7.4.4 also exhibits a negative resistance for certain frequency ranges but indicates a qualitatively different behaviour: the $Z = Z(\omega)$ curve circles the origin.

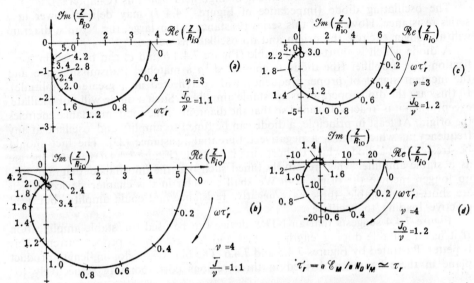

Figure 7.4.4. Nyquist diagram of the small-signal impedance computed for a GaAs diode (ohmic cathode, zero diffusion). (*after Holmstrom*, [23]).

It was shown [21], [23], [24] that the small-signal stability of a negative-resistance diode in circuit can be studied by using the Nyquist stability criterion known from the network theory. We only consider the short-circuit stability, i.e. the stability of a diode with a constant voltage source connected directly at its terminals. The diode will be short-circuit stable if the impedance diagram does not circle the origin. This is explained below by using the Nyquist criterion. The $Z = Z(\omega)$ diagram is plotted as a function of complex variable when ω follows a certain contour in the complex frequency plane, namely the real axis $\omega = -\infty$ to $\omega = +\infty$ and then an infinite semicircle in a clockwise direction. This infinite semicircle maps into an infinitesimal semicircle about the origin in the right half of the $Z(\omega)$ plane, since the diode behaves capacitively at very high frequencies (due to the displacement current, see also Chapter 4). $Z(\omega)$ has no poles because it can be calculated for any finite ω. The number of zeros minus poles in the lower half of the frequency plane is equal to the number of clockwise rotations around the origin. The impedance diagrams in Figures 7.4.4 *b* and *c* do not circle the origin, the impedance has no zeros in the lower half of the frequency plane and the diode is short-circuit stable. The complete impedance diagram corresponding to Figure 7.4.4 *d* circles the origin twice (only positive frequencies are shown) and the number of zeros in the lower-half of the frequency plane is 2. These are real-conjugate values of ω or complex-conjugate values of s (substituted for $i\omega$). The magnitude of the real part of these zeros is the frequency of oscillation that will result when the diode is short-circuited for the alternating signal. The imaginary part of these zeros corresponds to a negative resistance. In this situation small-signal oscillations will grow up in the circuit formed by the diode and its bias voltage: the diode is short-circuit *unstable**. The above growing high-frequency oscillation normally breaks in large-signal oscillations characterized by cyclic movement of high-field domains (Chapter 8).

The oscillating diode (impedance of Figure 7.4.4 d) may deliver power in a series resistance. However, if this series resistance is too large, the Nyquist diagram will no longer encircle the origin and no oscillations occur.

A diode which is short-circuit stable (Figure 7.4.4 b and c) can be used in a reflection-type amplifier (the diode is connected to a matched transmission line and the incident signal of proper frequency will be reflected at a greater amplitude). In this case the diode operates as a 'stable amplifier'. Such a diode will not oscillate, except when it is reactively tuned so that the diagram of the total impedance encircles the origin. At least in principle, a diode can be tuned to amplify and oscillate at any frequency at which it has a negative differential resistance [23]. The fact should be stressed that a diode which is short-circuit unstable (Figure 7.4.4 d) can still operate as a stable amplifier when reactively tuned such that the impedance diagram does no longer encircle the origin [24]. We shall consider in this chapter devices which are short-circuit stable: these have negative resistance and could amplify, but need reactive loading to oscillate.

Figure 7.4.4 suggests that an NDM device can be used for stable amplification if it has a suitable doping-length product and is biased by a proper current. This is better illustrated by Figures 7.4.5 and 7.4.6. The bias *versus* doping-length product plane in these figures is divided in three regions corresponding respectively to:

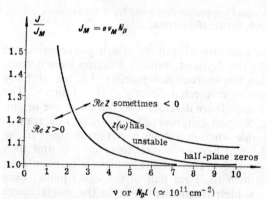

Figure 7.4.5. Threshold bias current density for negative resistance and for lower half-plane zeros of high-frequency impendace, as a function of normalized doping-length product. (*after Holmstrom*, [23]).

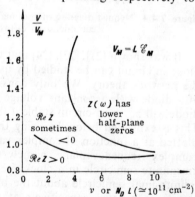

Figure 7.4.6. Threshold bias voltage for negative high-frequency resistance and for lower half-plane zeros of impedance, as a function of the normalized doping-length product. These voltage values correspond to the current values in Figure 7.4.5. (*after Holmstrom*, [23]).

---

* It can be shown that the diode is always open-circuit stable. However, a constant-current bias source is not used in practical circuits.

no negative resistance, short-circuit stability and short-circuit instability. Note the fact that for relatively high values of the doping-length product, stable amplification is possible for a certain, critically small, range of voltage bias. Measurements reported by Hakki and Knight [25] confirmed this behaviour (an 11% range of bias voltage for amplification was obtained for $N_D L \approx 7 \times 10^{10}$ cm^{-2}, and only 6.3% for $N_D L = 10^{11}$ cm^{-2}).

The impedance derived by McCumber and Chynoweth [20] by neglecting diffusion and assuming uniform steady-state field (d.c. neutrality) yields the following criterion for short-circuit stability

$$\alpha_R L < 2.09, \quad (7.4.5)$$

where

$$\alpha_R = \frac{eN_D |dv/d\mathscr{E}|}{v_0 \varepsilon} \quad (7.4.6)$$

is the space-charge growth constant due to dielectric 'relaxation' in a negative-mobility medium. Equations (7.4.5) and (7.4.6) calculated at threshold determine a critical doping-length product. The 5×10^{11} cm^{-2} value indicated by equation (7.3.31) is a usually accepted value obtained with relatively recent data for GaAs. However, stable amplification was obtained for devices with $N_D L$ considerably above this value.

Perlman et al. [26], [27] reported stable amplification by using n^+nn^+ diodes with a doping-length product in the range $1 - 2 \times 10^{12}$ cm^{-2}. Figure 7.4.7 shows the $I-V$ characteristic of a circuit loaded device with $N_D L \simeq 1.5 \times 10^{12}$ cm^{-2}. Both stable amplification and oscillations were observed and the corresponding bias ranges are shown [26]. Note the fact that stability was observed *beyond* an instability range, i.e. above a certain stabilization voltage V_A. Whereas the subcritically doped devices used in reflection-type amplifiers operate within a small bias

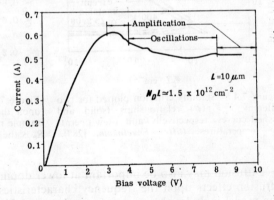

Figure 7.4.7. $I-V$ characteristic of a circuit loaded device. The decrease in current with increasing bias voltage is primarily due to device heating. (*after Perlam et al.*, [26]).

range just above the threshold voltage and are characterized by low power output and a narrow brandwidth, those 'supercritically doped' devices biased above V_A exhibit a stable negative conductance over a wide range of frequencies and power levels [26]. It was demonstrated experimentally that the stabilization voltage V_A

decreases with increasing temperature [26]. Pulse measurements have shown that the current drop with increasing bias voltage (Figure 7.4.7) is primarily due to device heating.

Engelmann [28] explained the above behaviour by using the same stability criterion, equation (7.4.5). It was shown that the critical doping-length product is, in fact, both field- and temperature-dependent. This is illustrated by Figure 7.4.8, which indicates a high-field stability range even at large $N_D L$ and a decrease of the minimum field for high-field stability with the increasing temperature. The experimental data indicate a transit-time behaviour of the negative conductance, in agreement with the space-charge wave theory (Section 7.3)*. At relatively large doping densities ($N_D = 1.5 \times 10^{15}$ cm^{-3} in Figure 7.4.7) the quasi-neutrality hypothesis used in deriving equation (7.4.5) is plausible. On the other hand, in the same doping range, diffusion effects should be important. Engelmann [28] shows that the diffusion will modify the stability criterion (by increasing the critical $N_D L$ product). Diffusion should reduce the stabilization voltage (especially at higher temperature) and broaden in frequency the negative conductance range.

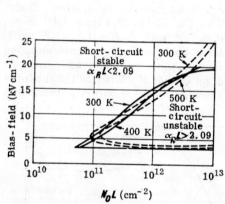

Figure 7.4.8. Stability criterion plotted for two different $v = v(\mathscr{E})$ relationships (solid and dashed curves, respectivelly) and a few operating temperatures. (after Engelmann, [28]).

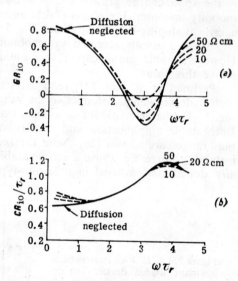

Figure 7.4.9. Effect of diffusion (dashed curves) upon the high-frequency characteristics with sample resistivity as parameter (n-type GaAs). R_0 is the low-field incremental resistance. (after Mahrous et al., [22]).

Diffusion may also be important at lower doping densities. Figure 7.4.9 indicates diffusion effects upon the frequency characteristics of a subcritically-doped device. The model includes the non-uniform electric field but still neglects diffusion in the

---

* If the more general criterion, equation (7.4.5), is used instead of equation (7.3.21), the high-bias stable amplification reported by Perlman et al. [26], [27] essentially corresponds to the operation of a subcritically-doped diode.

computation of the steady-state field. Diffusion is less important at higher sample resistivities. Kroemer [29] has, however, shown that the diffusion effects are appreciable even at zero doping (involving even the disappearance of the negative in the higher frequency ranges corresponding to the harmonics of the transit-time frequency). Certain theoretical [22], [30] and experimental data [25], [31] indicate that a minimum $N_D L$ is necessary in order to obtain a high-frequency negative resistance. Kroemer [29] demonstrated that the negative resistance should, theoretically, exist even in the limit of zero doping. The failure to obtain this, could be explained by trapping in high-resistivity samples or by the effect of parasitic series resistances [32].

7.5 Negative-resistance Diodes with Injection-limiting Cathodes

Figure 7.5.1a shows the computed steady-state field and mobile carrier concentration in a subcritically doped diode with an ohmic cathode (space-charge-limited injection). The electric field is below the threshold value for negative mobility, $\mathscr{E}_M$, in a wide cathode region. Figure 7.5.1b shows that the electric field can be raised above threshold in the entire cathode-anode space if a barrier-type cathode contact (like that discussed in Section 5.7) is used. The latter situation is preferable because a positive-mobility region determines ohmic losses instead of power generation.

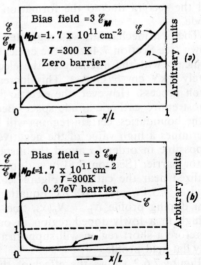

Figure 7.5.1. Computed static electric field profile and excess carrier concentration of a subcritically doped diode. Two different cathode boundary conditions are used: (a) SCL injection and (b) barrier-controlled injection, namely thermionic emission modified by image-force lowering of the barrier. (*after* Yu et al., [33]).

A more detailed analysis of the mechanism of alternating power generation inside the device [2], [33] — [36] takes into account both velocity modulation (ohmic local resistance) and density modulation (Section 5.3). The density-modulation component of the alternating particle current can generate high-frequency power, if a proper phase delay exists between the 'space-charge wave' associated with the density modulation and the local electric field. Such a delay is achieved by the transit-time effect and occurs in certain regions and at proper frequencies. In fact, the high-frequency transit-time negative resistance of the subcritically-doped diode with ohmic injecting contact is the result of the combined action of velocity-modulation (in negative-mobility regions) and density-modulation (due to transit-time effects).

Yu et al. [33] performed calculations and experiments for a subcritically-doped diode with a metal-semiconductor contact. The injection mechanism considered was the thermionic emission modified by the image-force lowering of the barrier (Section 5.7). It was experimentally found that the introduction of the cathode barrier improves the efficiency of power generation (an optimum value of the barrier exists [33]). However, the device behaviour is highly temperature-sensitive due to the thermionic emission injection mechanism.

Broad-band negative resistance can be achieved by using an injecting contact with very low and even zero differential conductivity, $\sigma_1(0)$ [2], [37] — [40]. In injection-controlled NDM devices ($\sigma_1(0) = 0$) the device resistance should be infinite at zero frequency and negative up to tens of GHz [2], [39], [40].

7.6 Stabilization Mechanisms in Supercritically-doped Transferred-electron Amplifiers

In this Section we shall briefly discuss two different stabilization mechanisms in supercritically doped amplifiers: the first due to the effect of a 'cathode-notch', and the second due to the formation of a stationary high-field domain near the anode.

The first mechanism was suggested by Charlton et al. [41], [42]. It was already shown (see Section 7.4) that the critical $N_D L$ product for instability increases monotonically with the intensity of the electric field (assumed uniform) above approximately 5 kV cm^{-1} (GaAs). This increase is considered responsible for the stabilization process. However, the electric field should be indeed sufficiently high in the almost entire cathode-anode space. Because ohmic contacts force zero electric field at the boundaries, certain regions will be subjected to a field above threshold which determines a high value of the negative mobility. These regions should be as short as possible in order to avoid the instability generated by them. A metal-semiconductor barrier (Section 7.5) or a doping reduction (a 'notch' in the doping profile) localized near the cathode (Figure 7.6.1 a) can determine an almost uniform field (Figures 7.5.1 b and 7.6.1 b). Equation (7.2.5), which is also valid for a non-uniform doping profile $N_D = N_D(x)$, does indeed show that the field gradient should be high in a lowly-doped region in order to compensate the decrease of $N_D(x)$*. Therefore a short region will support a large field gradient and will allow an essentially flat field profile.

Figure 7.6.2 shows the effect of the notch width on the stabilization voltage, V_A (defined in Section 7.4). The stabilization voltage decreases with the increase of the notch width (and also the notch depth). Increasing the temperature has a

---

* The same effect is achieved by reducing the velocity near the cathode and maintaining the doping uniform. The velocity reduction ('mobility notch') in the vicinity of the cathode may be due to a local damage of the crystal [42].

similar stabilizing effect (Section 7.4), confirmed by experiment [26], and the introduction of a doping slope (as indicated in Figure 7.6.1 a) as well*.

Figure 7.6.3 shows the computed frequency and bias dependence of the device negative conductance. A doping notch and a mobility notch have a similar effect

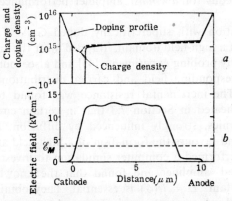

Figure 7.6.1. Doping (*a*) and electric field profile (*b*) of a transferred electron amplifier using cathode-notch stabilization. (*after Charlton and Hobson*, [42]).

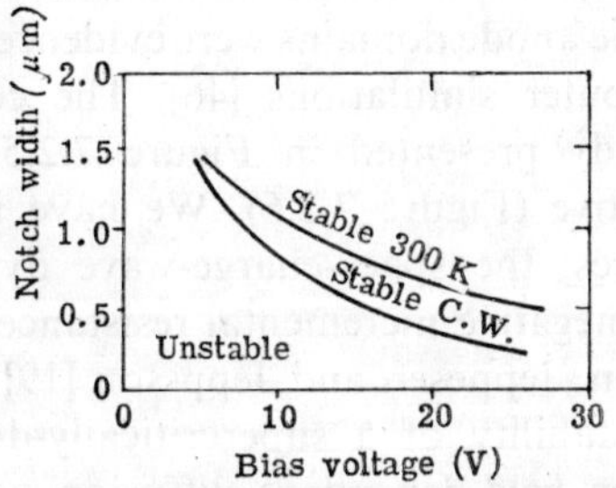

Figure 7.6.2. Relationship between the notch width and the stabilization voltage for a notch of 100 percent and a doping profile that slopes upwards from 10^{15} to 1.7×10^{15} cm^{-3} from cathode to anode. (*after Charlton and Hobson*, [42]).

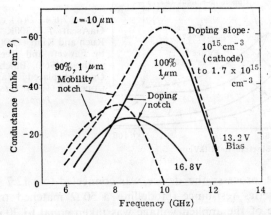

Figure 7.6.3. Computed frequency and bias dependence of the negative conductance for a device with a cathode doping notch (solid curves) or a cathode mobility notch (dashed curves). (*after Charlton and Hobson*, [42]).

---

* Talwar and Curtice [43], [44] studied both theoretically and experimentally the effect of doping and of the temperature gradient upon the behaviour of stabilized transferred-electron amplifiers.

(a reduction of negative conductance and a shift toward lower frequencies as the bias voltage increases). It should be stressed that such a notch also introduces a resistive effect and the device conductance may become positive [42]. The above calculated results were supported by experiment. We note the fact that a uniform field distribution should be advantageous for *low-noise* amplifier performance (see Section 7.8).

The experimental stable amplification with supercritically doped GaAs TED's was also explained by the formation of a high-field domain, 'trapped' near the anode. Stable anode domains were evidenced by probing experiments [45] and also found in computer simulations [46]. The corresponding field and carrier distribution was already presented in Figure 7.2.5. The incremental resistance was found to be negative (Figure 7.2.6). We have indicated in Section 7.3 that in certain circumstances, the space-charge-wave dynamics, strongly influenced by diffusion, leads to a negative incremental resistance and favours the formation of a high-field anode region. Jeppesen and Jeppsson [19] performed a computer simulation to investigate the stability of a supercritically-doped amplifier and found that the knowledge of the field-dependent diffusion constant, $D = D(\mathscr{E})$ is essential. They obtained* a stable behaviour only for the Copeland curve shown in Figure 7.6.4; note that in this particular case $dD/d\mathscr{E}$ is positive in a considerable range above the threshold field for negative mobility (see Sections 7.2 and 7.3)†. The computer simulated the

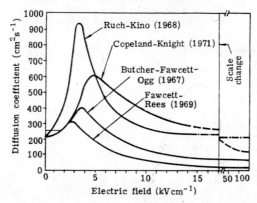

Figure 7.6.4. $D(\mathscr{E})$ characteristics for GaAs at $T = 300K$ as measured by Ruch and Kino (1968) and calculated by Fawcett and Rees (Monte Carlo method, 1969), Butcher, Fawcett and Ogg (displaced Maxwellian, 1967), Copeland (1971). (*after Jeppesen and Jeppsson,* [19]).

following transient process: the device was connected to a 12.7 V battery voltage through a 50 Ω series resistance (modelling a 50 Ω matched transmission line). The finite rise-time of the applied voltage was taken equal to 20 ps. Figure 7.6.5 a shows the transient diode voltage and current. The stable electric field and space-

---

* The calculations were performed for $L = 10\,\mu m$, $N_D = 1.5 \times 10^{15} cm^{-3}$, $T = 300$ K, 5 Ω low-field resistance, 12.7 V bias. The doping profile was flat and ohmic contacts were modelled by heavily doped n^+ regions.

† It was shown that a field-independent diffusion constant is also compatible with the existence of a high-field anode region [15], [16].

charge distributions achieved after 0.35 ns are indicated in Figure 7.6.5 b. The calculations were repeated for other doping levels and stability was observed down to 8×10^{14} cm^{-3}.

A peculiar characteristic of the above diffusion stabilized mode of operation is the fact that a minimum doping level is necessary [16] — [19], [47], [48]. Thim

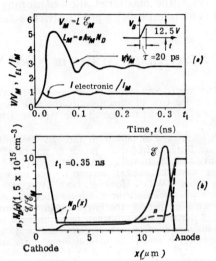

Figure 7.6.5. Results of device simulation indicating (a) device voltage and current variation in time, both normalized to their treshold values; (b) doping profile carrier and field distributions inside the device after the establishment of a stationary state. The device and circuit data are given in the text. (after Jeppesen and Jeppsson, [19]).

[17] obtained this doping density by using the following simple arguments. The anode accumulation layer consisting of mobile electrons does not disappear because it moves slowly into the anode and it has sufficient time to readjust its shape and reach a stable configuration due to the balance of two opposite processes: the space-charge growing due to the negative mobility existing in this high-field region (the anomalous dielectric 'relaxation') and the space-charge decay due to diffusion. The width of the layer is determined by diffusion and approximated by three times the Debye length λ_D, equation (7.3.13). The differential relaxation time $|\tau_d|$, equation (7.1.3), should be shorter than the transit-time $3\lambda_D/v_0$, such that the readjustment requires [17]:

$$\frac{\varepsilon}{eN_D|dv/d\mathscr{E}|} < \frac{\varepsilon}{v_0}\sqrt{\frac{eD_0}{eN_D|dv/d\mathscr{E}|}}, \qquad (7.6.1)$$

or

$$N_D > \frac{\varepsilon v_0^2}{2eD_0|\mu_d|}, \quad \mu_d = \frac{dv}{d\mathscr{E}}. \qquad (7.6.2)$$

The minimum doping for n-type GaAs is of the order of $5 - 10 \times 10^{14}$ cm^{-3}, depending upon the material parameters introduced in equation (9.6.2). Other authors [18], [19], [48] indicate an essentially similar stability criterion.

Another characteristic of diffusion-stabilized mode is the necessity of relatively uniform doping profile. If doping fluctuations exceed a certain critical value, stable anode accumulation layers are no longer possible [17]. These fluctuations determine a non-uniform bulk field and peaks occurring in the field profile may exceed threshold thus leading to instability. It was shown [18] that a notch in the doping profile near the cathode has a similar effect. The higher the semiconductor doping, the larger the fluctuation which still permits stability, because a higher donor density requires a lower bulk field to sustain the same current [18]. For usual doping levels (around 10^{15} cm^{-3}) the maximum allowed doping fluctuation is quite small (a few percent), and this condition is highly restrictive for device fabrication.

7.7 Space-charge Waves in NDM Semiconductor Layers

The negative-resistance diodes used in reflection amplifiers have a number of disadvantages such as: the lack of directionality (because they are two-terminal devices), the relative narrow bandwidth (due to the transit-time type negative resistance), the relative low power and efficiency.

A *distributed* amplification through the entire NDM semiconductor is, however, possible, because the high-frequency power is generated by a *bulk* effect. This leads to the concept of negative-mobility travelling-wave amplifier (TWA). The wave travelling in this device is the space-charge wave discussed in Section 7.3 and not the electromagnetic wave propagating in travelling-wave tubes (TWT)*. However, both TWA and TWT are based upon distributed amplification. In both cases the electron transit-time between cathode and anode can be many times longer than the oscillation period; the internal-gain mechanism provides an exponential increase of a wide band of signal as they propagate from the input toward the output. On the other hand the relative large distance existing between electrodes (long travelling path) determines reduced coupling (small capacitance) between output and input ports and allows very-high-frequency operation. It will be shown that the problem of heat removal from high-power devices (which is critical for semiconductors) is also less severe for TWA. The unidirectionality is provided by the direction of wave propagation (three- and four-electrode devices have been constructed). Many electronic functions can be realized by proper bias and excitation of device electrodes [51].

The solid-state TWA using an NDM semiconductor such as GaAs looks very promising, especially for microwave communications. However, this device raises a number of principal and, of course, technological problems. In order to achieve a high-gain TWA, the device should be long and the bias current should be large. These conditions make the $N_D L$ product large and determines instabilities. The oscillations can be suppressed in a number of ways, as discussed in this section.

Another important problem is the difficulty to achieve a sufficiently high (and uniform) electric field and provide the space-charge wave growth throughout the almost entire interelectrode space (Section 7.8).

The input and output coupling of the space-charge-wave to the external circuit can be realized in a number of ways: the coupling mechanism can either reduce or enhance the internal gain, as indicated in Section 7.9.

In this section we shall briefly discuss certain properties of space-charge waves propagating in a semiconductor slab of finite transversal dimensions. These waves are associated with the electron transport at velocity v_0 along the x direction. If the electric field $\mathscr{E}_x$ exceeds the threshold value, the growing electronic waves can be used for travelling wave amplification. It was shown above that a high doping-length $N_D L$ is required. Calculations and experiments show that the finite transversal dimension (thickness) of the semiconductor plate allows short-circuit stability up to quite high 'longitudinal' doping-length products.

The basic model was suggested by Kino and Robson [52]. The conduction takes place in a sheet of thickness $2a$ and infinite lateral extension. This sheet is bounded by a dielectric, such

---

* A true solid-state analog of the travelling-wave tube would utilize the interaction between drifting carriers in semiconductor and a slow-wave electromagnetic propagation circuit adjacent to the semiconductor surface [49]. Hines [50] suggested to replace the above circuit by a finely grained mosaic of insulated short metal stripes. The periodicity of these stripes determines a standing wave pattern and induce a negative resistance effect.

as in the symmetrical model shown in Figure 7.7.1. The outside medium can also be a semiconductor [53], [54] or a resistive material [55], [56], or the semiconductor can be embedded between two metal-backed dielectric sheets [57]. The basic semiconductor has an 'anisotropic' mobility: the differential mobility along the x axis is $\mu_x = (dv/d\mathscr{E})_0 < 0$ (field above threshold), whereas its transversal component μ_y is positive.

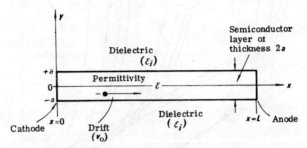

Figure 7.7.1. Semiconductor sheet in an infinite dielectric medium. The charge carriers drift in x direction with a velocity v_0.

Consider first the space-charge waves in the semiconductor plate bounded by two infinite dielectrics shown in Figure 7.7.1. The waves are treated as small perturbations: d.c. neutrality and negligible diffusion are assumed (see Section 7.3). The symmetric solution* for the potential perturbation is [52], [57]—[59]

$$\Phi_1(x, y, t) = \text{const} \times \cos(k_y y) \exp i(\omega t - k_x x), \tag{7.7.1}$$

where the wave numbers k_x and k_y are given by

$$k_y^2 = -k_x^2 \frac{\omega_x - \omega_{\text{eff}}}{\omega_y - \omega_{\text{eff}}}, \tag{7.7.2}$$

$$k_x = \frac{\omega}{v_0} - i\frac{\omega_{\text{eff}}}{v_0}, \tag{7.7.3}$$

$$\omega_{\text{eff}} = \frac{\varepsilon k_x \omega_x}{\varepsilon k_x + \varepsilon_i k_y \cot ak_y}, \tag{7.7.4}$$

$$\omega_x = \frac{eN_D \mu_x}{\varepsilon}, \quad \mu_x < 0, \tag{7.7.5}$$

$$\omega_y = \frac{eN_D \mu_y}{\varepsilon}, \quad \mu_y > 0. \tag{7.7.6}$$

The plate thickness, $2a$, occurs in the above relations, see equation (7.7.4), through the so-called Hahn boundary condition at the planes $y = \pm a$ (Fig. 7.7.1) (the continuity of potential and field, $\mathscr{E}_y$, is required). In the limit case of infinite transverse extension, with $a \to \infty$, one obtains an effective relaxation frequency $\omega_{\text{eff}} = \omega_x < 0$ and $k_y = 0$, such that the wave given by equation (7.7.1) becomes

$$\Phi_1(x, t) = \text{const} \times \exp\left(-\frac{\omega_x x}{v_0}\right) \exp i\left(\omega t - \omega \frac{x}{v_0}\right). \tag{7.7.7}$$

---

* There also exists an antisymmetric solution [57], [58].

This is a plane wave travelling with the v_0 phase velocity and having a positive growth rate, related to the negative relaxation frequency, $\omega_x < 0$. This solution was obtained in Section 7.3 in the limit case of negligible diffusion current.

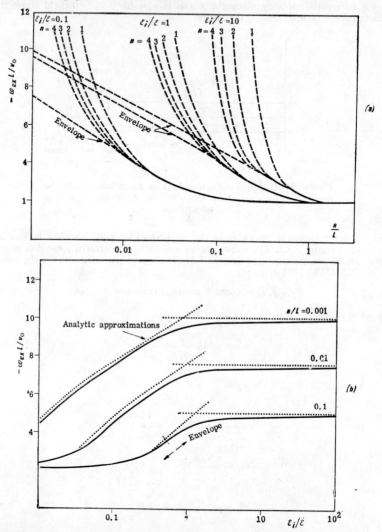

Figure 7.7.2. (a) Normalized critical $N_D L$ product *versus* sample thickness-to-length ratio for different orders of characteristic oscillations ($n = 1, 2, 3, 4$) and permittivity ratios $\varepsilon_i/\varepsilon$. Dashed curves are envelopes obtained with variable n. (b) Normalized critical N_D/L product *versus* degree of dielectric loading, with a/L as parameter. (*after Heinle and Engelmann*, [60]).

In the more general case, with finite a, there exist an infinite number of solutions, called lateral modes ($m = 0, m = 1, m = 2, \ldots\ldots$) and arising from the cotangent function, equation (7.7.4). The general solution corresponds to the superposition of waves determined by these solutions. In many cases only the fundamental mode ($m = 0$) is considered, because it exhibits the higher growth rate (this is valid for very small a/L as well as for $a/L \to \infty$, the results for intermediate a/L can be 'interpolated' by using the $m = 0$ solution [60].

The discussion of short-circuit stability is based upon the zeros of the sample impedance (Section 7.4). This impedance is calculated by imposing the boundary condition of zero field perturbation at $x = 0$ (ohmic cathode) (Section 7.3). There exist an infinite and denumerable pairs of complex-conjugate zeros, each of them corresponding to a characteristic oscillation. The lower oscillation frequency corresponds approximately to the transit-time frequency, and the higher ones to its harmonics. At finite a/L ratios the most stringent stability requirement is given by higher characteristic-oscillation orders ($n = 2, 3, \ldots$) rather than the fundamental one ($n = 1$). This is illustrated by Figure 7.7.2 a, which shows the (normalized) critical $N_D L$ product *versus* a/L, calculated for $\mu_y/\mu_x = -2$ and $\varepsilon = 0.1, 1$ and 10. The critical $N_D L$ is determined by the envelopes (dashed curves) obtained by considering n as a continuous variable. The critical $N_D L$ increases with decreasing a/L but much less than indicated when considering the fundamental oscillation only ($n = 1$). Figure 7.7.2 b shows the normalized critical doping-length product as a function of dielectrical loading (the ratio $\varepsilon_i/\varepsilon$). It can be seen that the admissible $N_D L$ product increases up to a finite maximum value when $\varepsilon_i/\varepsilon$ increases. The thinner the active layer, the higher the increase of $N_D L$ obtainable by dielectrical loading. Kino and Robson [52] indicated that the stability condition (derived by considering only the first-order oscillation, $n = 1$) is related to a critical doping-*thickness* product which is proportional to $\varepsilon_i/\varepsilon$. This led to the conclusion that a high-permittivity dielectric (such as barium titanate) should be used in order to suppress the instabilities. However, the exact theory including the higher-order instabilities [59]—[62] indicated that $(N_D L)_{crit}$ becomes almost independent of $\varepsilon_i/\varepsilon$ for $\varepsilon_i/\varepsilon \geqslant 1$ [60], [61] but still remains a function of a/L, even for very thin semiconductor plates [60], [62].

It was shown [63] that diffusion has an important effect upon the behaviour of relatively thin semiconductor plates, at small values of a/L. The longitudinal diffusion tends to suppress higher order oscillations and thus improves the stability. In particular, for $a/L \to 0$, $(N_D L)_{crit}$ becomes proportional to $\varepsilon_i/\varepsilon$, as shown in Figure 7.7.3.

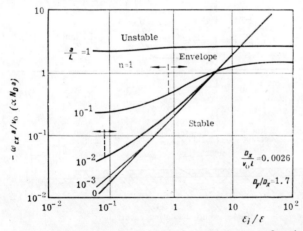

Figure 7.7.3. Normalized critical $N_D a$ product as a function of dielectric loading $\varepsilon_i/\varepsilon$, with a/L as parameter, for $D/v_0 L = 0.0026$ and $D_y/D_x = 1.7$ (D_x and D_y are the diffusion coefficients along the x and y directions, respectively). (*after Engelmann and Heinle*, [63]).

However, the theoretical results including diffusion were not confirmed by experiment. This occurs because certain assumptions of the theory are not met in practice. It was shown [64] that the current is confined near the centre of the plate when the semiconductor is depleted by a Schottky-barrier coupling contact (Section 7.9). Therefore a 'free surface' boundary condition at $y = \pm a$ should be used. This reduces the rôle of the diffusion current and increases the tendency for instability. Non-ohmic cathode contacts are also used in experimental devices: their influence upon stability seems to be small [65]. The non-uniformity of the electric field (the semiconductor layer is far from neutrality) should be also taken into account (Section 7.8).

The layer-type geometry ($a \ll L$) appears to be suitable for construction of travelling-wave amplifiers using NDM semiconductors (Section 7.9). It is interesting to note, however, that the lateral propagation of space-charge waves should be taken into account even in investigating the stability of a usual diode, where $a/L \gg 1$ [62], [66].

The stability of layered semiconductor structures, where coupled space-charge waves travel in parallel, in different semiconductors, was also investigated [53], [54]. Tesnzer [53] suggested an electronic control of the growth rate through the carrier drift velocities (the difference between these velocities in adjacent layers is a key parameter for the control of space-charge wave growth and decay). The physical and technological problems raised by such layered structures are, of course, more complex.

7.8 Field Profile in a GaAs Layer Biased above Threshold

The space-charge wave theory presented in the previous section assumes that the semiconductor layer is practically neutral and the longitudinal electric field is almost uniform and above the threshold value for negative mobility. However, there exist considerable difficulty in achieving a uniform field in epitaxial GaAs layers [51], [67], [68]. The basic structure is shown in Figure 7.8.1. Typically, a one-micron thick layer of n-type gallium arsenide of about 2×10^{15} donors per cm^3 is grown epitaxially on a semi-insulating substrate. Epitaxial n^+ contact-regions are also grown in grooves etched on the top surface [67]. The steady-state field profile along the surface exhibits

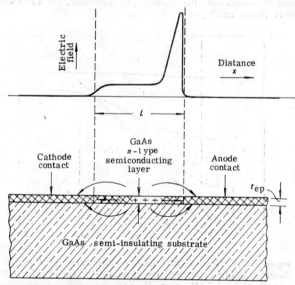

Figure 7.8.1. Thin-layer geometry with typical field profile. *(after Dean and Schwartz [67])*.

invariably a high-field region near the anode, as indicated in Figure 7.8.1. A similar high-field region occurs in a normal 'plane' diode, when biased above threshold (see Figure 7.4.2). The field profile falls steeply to zero at the anode end of the layer (Figure 7.8.1), which is a high-low n^+n junction.

Dean and Schwartz [67] suggested a simple one-dimensional analytical model which describes certain basic features of the practical structure (Figure 7.8.1) and compares well with more elaborate

7. Transferred-electron Diodes for Microwave Amplification

calculations. This model takes into account the two-dimensional field distribution upon the mobile space in the conducting layer, by simply introducing an effective dielectric constant

$$\varepsilon_{\text{eff}} = \varepsilon \frac{L}{t_{\text{ep}}} \tag{7.8.1}$$

in the one-dimensional Poisson equation

$$\frac{d\mathscr{E}_x}{dx} = \frac{e}{\varepsilon_{\text{eff}}}(n - N_D). \tag{7.8.2}$$

Here ε is an average value for the medium through which the field lines run. The ratio L/t_{ep} accounts for the fraction of electric field lines running outside the epitaxial layer. For large L/t_{ep}, ε approaches the dielectric constant for the passive dielectric outside the semiconductor (an average value should be used for an unsymmetrical structure). The one-dimensional surface current (diffusion neglected) is

$$J_s = ent_{\text{ep}}\, v(\mathscr{E}_x). \tag{7.8.3}$$

J_s is numerically equal to the current calculated per unit of device width (the device width is along the z axis, perpendicular to the plane of the section shown in Figure 7.8.1). The carrier drift velocity above the negative-mobility threshold is approximated by

$$v(\mathscr{E}_x) = \frac{v_M}{1 + \dfrac{\mu_M}{v_M}(\mathscr{E}_x - \mathscr{E}_M)}, \quad \mathscr{E}_x \geqslant \mathscr{E}_M. \tag{7.8.4}$$

By eliminating v and n between equations (7.8.1)–(7.8.3) and integrating with the 'boundary' condition $\mathscr{E}_x = \mathscr{E}_M$ at $x = x_M$, one obtains

$$\mathscr{E}_x - \mathscr{E}_M = \frac{v_M}{\mu_M}\left(1 - \frac{eN_D v_M t_{\text{ep}}}{J_s}\right)\left[\exp\left(\frac{\mu_M}{\varepsilon}\frac{J_s}{v_M^2}\frac{x'}{L}\right) - 1\right], \tag{7.8.5}$$

where $x' = x - x_M$. The above equation applies best in regions far from contacts where diffusion current is fairly negligible. Moreover, it is valid only in that region where $E_x \geqslant E_M$.

The field, equation (7.8.5), should be as uniform as possible. The quantity multiplying x'/L in the exponent should be as low as possible, for example less than unity. This puts an upper limit (dependent upon the semiconductor properties at a given temperature) on the surface current density, J_s. The coefficient $1 - eN_D v_M/J$ (where $J = J_s/t_{\text{ep}}$) may be positive or negative, corresponding, respectively, to accumulation and depletion, as shown in Table 7.8.1.

Table 7.8.1
Space-charge and Field Profiles

	Accumulation	Depletion
Current density, J	$J > J_M = eN_D v_M$ (high current)	$J < J_M = eN_D v_M$ (low current)
Coefficient $1 - eN_D v_M/J$	Positive	Negative
Space charge	Accumulation at $x = x_M$ and in the entire space $(0, L)$.	Depletion at $x = x_M$ and in the entire space $(0, L)$.
Electric field, $\mathscr{E}_x$	Increasing with x and determining a high-field anode region.	Decreasing with x and providing a better overall uniformity.
Cathode contact	Non-ohmic high-field cathode.	High-conductivity low-field cathode.

The critical value of the current density is just the 'threshold' value $J_M = eN_D v_M$ corresponding to neutrality. We have shown in Section 7.2 that the entire cathode-anode space should be either accumulated, or depleted (one-dimensional model with uniform doping and negligible diffusion). In the first case, $\mathscr{E}_x$ increases with x in the entire semiconductor layer (diffusion is still neglected). This situation corresponds to an ohmic cathode and leads to a high-field region near the anode. Both the cathode region ($0 < x < x_M$, with low electric field) and the anode region (where the carrier velocity is nearly saturated) introduce losses and reduce the efficiency of travelling-wave amplification. In the latter case, a non-ohmic cathode which supports a large field, determines the layer depletion at low current densities. This situation is illustrated by the current-field diagram shown in Figure 7.2.1.

Dean and Schwartz [67] have shown that in a correct two-dimensional model, the field spreading at the sharp edges of the contact determines an increase of the electric field in the vicinity of *both* contacts. Therefore, the more exact theories or numerical calculations indicate field profiles as those shown in Figure 7.8.2. The field minima situated on the anode side of the centre corresponds to a generally depleted condition [67], and this situation is characterized by better field

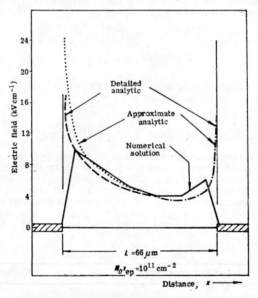

Figure 7.8.2. Results of analytic and numerical computations for the longitudinal electric-field profile in an thin layer biased above the threshold field ($L=66\mu$/m, $N_D t_{ep} = 10^{11}$cm^{-2}, $v_M = 2.2 \times 10^7$ cms^{-1} $\mathscr{E}_M = 3.67$ kV cm^{-1}, $\mu_M/\mu_n = 0.33$ μ_n being the low-field mobility, $\varepsilon_{\text{out}} = 10 \times \varepsilon_{\text{free space}}$, etc.). (*after Dean and Schwartz,* [67]).

uniformity. Figure 7.8.3 shows the importance of carrier density at the cathode (used as a boundary condition for numerical calculations). Carrier depletion at the cathode contact is favourable for field uniformity. A continuously increasing doping density toward the anode tends to raise the field in the centre and also improves the uniformity [67].

7. Transferred-electron Diodes for Microwave Amplification

Dean [68] suggested a special cathode composed of an electron-injecting n^+ contact complemented by an electron-blocking contact, which is a reverse-biased Schottky barrier. The Schottky contact occurs as an extension of the cathode (Figure 7.8.4), and depletes the cathode region operating like a gate of an MESFET. This depleting effect limits the electron injection and improves the field profile. Measured potential distributions in experimental devices are close to theoretical predictions (Figure 7.8.5). Similar results were obtained by Swartz et al. [69] by using a Schottky barrier contact strip parallel to the cathode.

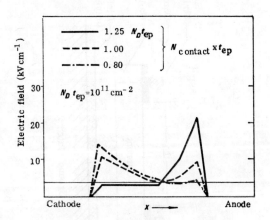

Figure 7.8.3. Field profiles with various carrier densities of free carriers available at the contacts (numerical solution for 50 V applied bias, and date indicated in Figure 7.8.2). (*after Dean and Schwartz,* [67]).

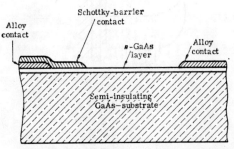

Figure 7.8.4. Two-terminal device with a Schottky-barrier extension at the cathode. (*after Dean,* [68]).

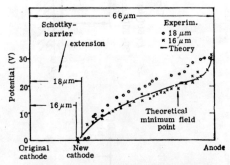

Figure 7.8.5. Measured potential profiles in two samples, compared with theoretical potential derived as shown in reference [67]). (*after Dean,* [68]).

7.9 GaAs Travelling Wave Amplifiers

Robson et al. [70] demonstrated a transferred-electron two-port amplifier constructed from a relatively thick, bulk-grown GaAs (Figure 7.9.1 a). The use of the travelling space-charge wave makes the device inherently unidirectional. The operation frequency was 1 GHz.

Kumabe and Kanbe [71] used the structure indicated in Figure 7.9.1 b. At the input, the high-frequency signal (applied between the cathode electrode and the ground plate) excites space-charge waves. The space-charge wave which flows into the anode vertically induces an output microwave

342 Bulk- and Injection-controlled Devices

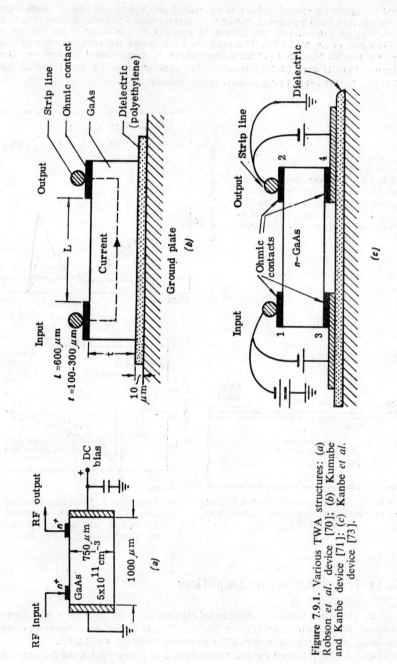

Figure 7.9.1. Various TWA structures: (a) Robson et al. device [70]; (b) Kumabe and Kanbe device [71]; (c) Kanbe et al. device [73].

signal. In this planar-type device the efficiency of coupling between the microwave field and the space-charge waves is low, because of the inactive regions occurring under the electrodes. Koyama et al. [72] used a $BaTiO_3$ slab as dielectric* to improve coupling and contribute to the stability by 'dielectric loading' (see Section 9.7). It was demonstrated that a similar effect is obtained by reducing the semiconductor thickness: the amplifier gain was increased up to 30 dB at 1.15 GHz.

A TWA with four ohmic contacts (Figure 7.9.1c) [73] has an increased coupling efficiency. In fact, this device is a combined structure involving two-terminal amplifiers (formed by 1 and 3, and 2 and 4 electrodes, respectively) and a travelling wave section, providing the undirectional

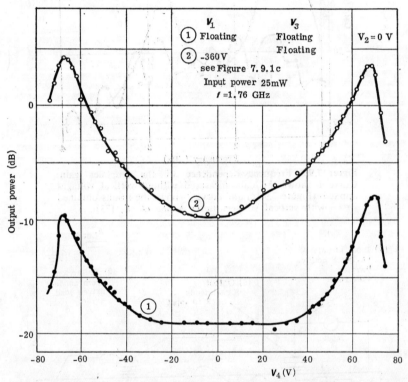

Figure 7.9.2. Curve 1 shows the amplification obtained with the output section of the device shown in Figure 7.9.1. c operating as a two-terminal amplifier. Curve 2 shows the voltage dependence of amplification when the travelling-wave section of the same device operates. The bias voltages are indicated in Figure 7.9.1c. (after Kanbe et al., [73]).

coupling between the input and output sections. Above the threshold voltage, the input and output sections exhibit negative resistance and provide amplification even when the travelling-wave section is not biased (lower curve in Figure 7.9.2). The output power, however, increases considerably when the travelling-wave section operates (upper curve in the same figure). The frequency

---

* Frey et al. [57] also used $BaTiO_3$ slabs bounding a 200 μm-thick GaAs crystal to improve coupling.

characteristics shown in Figure 7.9.3 are also highly relevant. Curve 1 shows the gain when electrodes 3 and 4 are floating: it exhibits a broad maximum at 2 GHz and small peaks with the period of 400 MHz. The former is related with the transit time in the coupling regions under

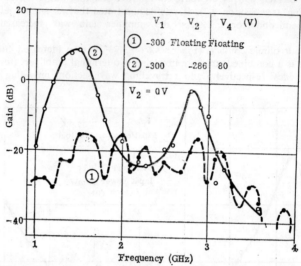

Figure 7.9.3. Frequency-dependence of the amplifier gain. Curve 1 shows the data measured without vertical voltages applied (Figure 7.9.1c) and curve 2 shows the results obtained with vertical voltages (*after Kanbe et al.*, [73]).

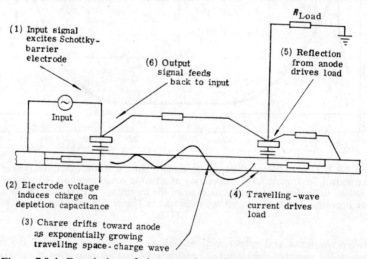

Figure 7.9.4. Description of the operating principle of the travelling-wave transistor shown in Figure 7.9.5. (*after Dean and Matarese*, [51]).

electrodes 1 and 4, and the latter is related with the transit time in the drift region. Curve 2 is obtained for normal bias. The gain shows two peaks at 1.4 and 2.8 GHz, which correspond to the fundamental and the second harmonic of the two 'vertical' amplifying sections [73].

7. Transferred-electron Diodes for Microwave Amplification

The TWA using an epitaxial GaAs layer should be preferred for a number of reasons. An epitaxially grown material exhibits a positive temperature coefficient of resistivity: on the other hand the direction of heat removal is perpendicular to the current flow and thermal resistance is reduced, therefore continuous wave (CW) operation is possible. Cathode-anode space can be reduced and control electrodes can still be placed between: the photolithographic techniques similar to those used in planar technology are used. The supplementary electrodes can be used either for correction of small fabrication variations, or for modulating the signal in different ways [51]. Finally, the thin-layer TWA exhibits better frequency performances: it operates at higher frequencies, possesses a broad band, etc. [51], [74], [75]. A two-terminal epitaxial-layer reflection amplifier can be also constructed [65]. The control of field profile and the coupling of the space-charge wave is provided by Schottky-barrier contacts [51], [64], [68], [69], [75].

Figure 7.9.4 describes the operation of the TWA reported by Dean and Matarese [51]. A top-view of the device is indicated in Figure 7.9.5. A voltage bias is applied between the two

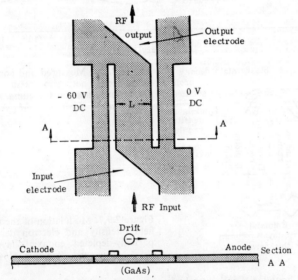

Figure 7.9.5. Cross-section and top view of the travelling-wave transistor. (*after Dean and Matarese*, [51]).

alloy contacts on the ends, with negative on the left. The bias applied between the input electrode (Schottky contact) and the alloy contact at the left controls the steady-state boundary condition at the cathode end of the travelling wave region (Section 7.8). The same bias controls the steady-state current (and provides a means for modulating the gain [51]). The bias on the output electrode can be used to modulate the phase shift. If the input bias is adjusted for oscillation, the output bias can be adjusted to modulate the frequency [51]. The device operation is extremely complex and will not be discussed here. Figure 7.9.6 shows the measured instantaneous net gain *versus* frequency. Recent theoretical results [64], providing a reasonable description of experimental data, are based upon a different approach of space-charge wave propagation. It was shown that in partially-depleted layers (biasing in depletion provides a better field uniformity) the electrons move to maintain charge neutrality and recede from the metallurgical surfaces of the *n*-type layer, as shown qualitatively in Figure 7.9.7. This corresponds to a 'free' transverse modulation of the electron density instead of the 'stiff-boundary' condition which assumes zero transverse electron current

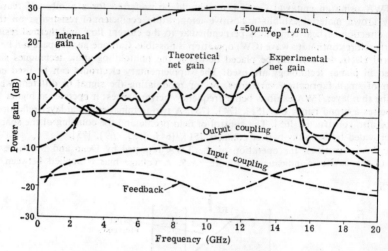

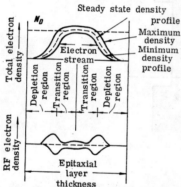

Figure 7.9.6. Measured and computed instantaneous net gain *versus* frequency for an experimental travelling wave transistor. (*after Dean and Matarese*, [51]).

Figure 7.9.7. Modulation of the transverse field intensity and electron density in a partially-depleted epitaxial layer. (*after Dean and Robinson*, [64]).

at the metallurgical boundaries and introduces a strong damping of the space-charge-wave growth by diffusion. The diffusion damping predicted by the new theory [64] is considerably weaker than that predicted by Engelmann and Heinle [63].

7.10 Noise in Transferred-electron Amplifiers (TEA)

The main source of noise in TEA is the thermal noise. The thermal noise at high frequencies can be evaluated by using the diffusion-impedance field method as for SCL diodes (Section 4.7) and BARITT diodes (Section 5.4).

Figure 7.10.1 shows the frequency-dependence of the noise figure calculated with homogeneous field profile, field-dependent diffusion and ohmic injecting contact [76]. Note the fact that inclusion of diffusion in calculations suppresses the transit-time oscillations of noise with increased frequency. At higher transit angles

($\theta > 3\pi$) the magnitude of the doping-length product is unimportant. Figure 7.10.2 indicates the effect of a non-ohmic cathode (with finite differential conductivity, $\sigma_1(0)$, see Section 7.5): the noise figure is reduced [76].

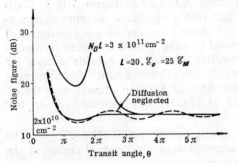

Figure 7.10.1. Noise figure *versus* transit angle (normalized frequency), illustrating the effect of the doping-length product. The cathode is ohmic. Diffusion is incorporated by using Copeland's data in Figure 7.6.4. The dashed curve is calculated by neglecting diffusion. (*after Källbäck*, [76]).

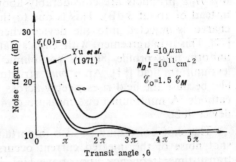

Figure 7.10.2. Noise figure as a function of transit time for three different cathode boundary conditions. (*after Källbäck*, [76]).

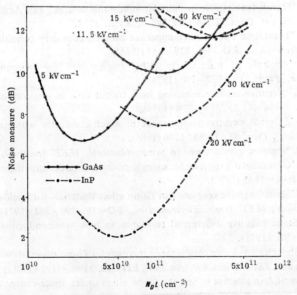

Figure 7.10.3. Variation of minimum noise measure M with $N_D L$ product for GaAs and InP at various bias fields. (*after Sitch and Robson*, [77]).

The minimum noise depends not only upon the doping-length product, but also upon the bias field and upon the details of $v = v(\mathscr{E})$ and $D = D(\mathscr{E})$ dependence. Figure 7.10.3 shows the minimum noise measure calculated for GaAs and InP

TEA's (uniform-field approximation) [77], with lower noise values for InP*. More exact numerical calculations confirm the results of Figure 7.10.2 for large $N_D L$ values. However, the low noise measures computed for InP devices at relatively low $N_D L$ products are considerably above those indicated in Figure 7.10.3 (7 dB instead of about 3 dB). This is due to the fact that a significant amount of space-charge is injected into the device when the doping density is low (Section 7.4) [77]. This is in agreement with Thim's statement that the noise is enhanced by a non-uniform field profile [80]. A non-ohmic injecting contact will provide a more or less uniform field [81]. An n^+pnn^+ structure [82] also exhibits a nearly flat field profile, because the field rises steeply across the depleted p region situated near the n^+ cathode. A minimum noise measure of 4 dB at 17 GHz was computed for an InP device [82].

The shot noise may be also important in injection-limited TED's [2], [39]. The shot noise of the injected current occurs attenuated in the external circuit at the transit-time frequency and above [36], [40].

References

1. J. E. CARROLL, 'Hot electron microwave generators', Edward Arnold Ltd., London, 1970.
2. D. DASCALU, 'Transit-time effects in unipolar solid-state devices', Ed. Academiei, Bucharest, and Abacus Press, Tunbridge Wells, Kent, 1974.
3. H. KROEMER, 'The Gunn effect under imperfect cathode boundary conditions', *IEEE Trans. Electron Dev.*, **ED–15**, 819–837 (1968).
4. D. DASCALU, 'Analysis of injection-controlled one-carrier flow in semiconductors', *Rev. Roum. Phys.*, **19**, 177–190 (1974).
5. W. SHOCKLEY, 'Negative resistance arising from transit time in semiconductor diodes', *Bell. Syst. Techn. J.*, **33**, 799–826 (1954).
6. H. KROEMER, 'External negative conductance of a semiconductor with negative differential mobility', *Proc. IEEE*, **53**, 1246 (1965).
7. H. KROEMER, 'Negative conductance in semiconductors', *IEEE Spectrum*, **5**, 47–56 (1968).
8. H. KROEMER, 'Generalized proof of Shockley's positive conductance theorem', *Proc. IEEE*, **58**, 1844–1845 (1970).
9. P. S. HAUGE, 'Static negative resistance in Gunn effect materials with field-dependent carrier diffusion', *IEEE Trans. Electron. Dev.*, **ED–18**, 390–391 (1971).
10. F. STERZER, 'Static negative differential resistance in bulk semiconductors', *RCA Rev.*, **32**, 497–502 (1971).
11. G. DÖHLER, 'Shockley's positive conductance theorem for Gunn materials with field-dependent diffusion', *IEEE Trans. Electron Dev.*, **ED–18**, 1190–1192 (1971).
12. YU. S. RYABINKIN, 'An attempt to find the Gunn effect under space-charge conditions', *Soviet Phys.-Semic.*, **2**, 977–978 (1969).

---

* The $v(\mathscr{E})$ and $D(\mathscr{E})$ dependences for InP are taken from Fawcett and Herbert, Figure 1.8.15 and reference [78] respectively. A detailed computer simulation for GaAs and InP TEA's shows that their gain and bandwidth are almost insensitive to the details of intervalley transfer [79]. However, the noise characteristics are substantially different.

13. A. YAMASHITA, 'Space-charge-limited currents and velocity-field characteristic in n-type GaAs", *Rev. Elec. Commun. Lab.*, **17**, 1089—1101 (1969).
14. K. TOMIZAWA, H. TATENO and S. KATAOKA, 'Computer analysis on the static negative resistance due to the geometrical effect of a GaAs bulk element', *IEEE Trans. Electron Dev.*, **ED—19**, 1299—1300 (1972).
15. H. TATENO and S. KATAOKA, 'Static negative resistance due to the geometrical effect of a GaAs bulk element', *Proc. IEEE*, **60**, 919—920 (1972).
16. T. OHMI and S. HASUO, 'Unified treatment of small-signal space-charge dynamics in field-effect devices', *IEEE Trans. Electron Dev.*, **ED—20**, 303—316 (1973).
17. H. W. THIM, 'Stability and switching in overcritically doped Gunn diodes', *Proc. IEEE*, **59**, 1285—1286 (1971).
18. K. MURAYAMA and T. OHMI, 'Static negative resistance in highly doped Gunn diodes and application to switching and amplification', *Jap. J. Appl. Phys.*, **12**, 1931—1940 (1973).
19. P. JEPPESEN and B. JEPPSSON, 'The influence of diffusion on the stability of the supercritical transferred electron amplifier', *Proc. IEEE*, **60**, 452—454 (1972).
20. D. E. MCCUMBER and A. G. CHYNOWETH, 'Theory of negative-conductance amplification and of Gunn instabilities in 'two-valley' semiconductors', *IEEE Trans. Electron Dev.*, **ED—13**, 4—21 (1966).
21. R. HOLMSTROM and H. DERFLER, 'Space-charge waves and stability of electron diodes', *IEEE Trans. Electron Dev.*, **ED—13**, 539—544 (1966).
22. S. MAHROUS, P. N. ROBSON and H. L. HARTNAGEL, ' The stability and reflection gain of sub-critically doped Gunn diodes', *Solid-St. Electron.*, **11**, 965—977 (1968).
23. R. HOLMSTRON, 'Small-signal behaviour of Gunn diodes', *IEEE Trans. Electron Dev.*, **ED—14**, 464—469 (1967).
24. S. MAHROUS and H. L. HARTNAGEL, 'Gunn-effect domain formation controlled by complex, load', *Brit. J. Appl. Phys.* (*J. Phys. D*), **2**, 1—5 (1969).
25. B. W. HAKKI and S. KNIGHT, 'Microwave phenomena in bulk GaAs', *IEEE Trans. Electron Dev.*, **ED—13**, 94—105 (1966).
26. B. S. PERLMAN, C. L. UPADHYAYULA and W. W. SIEKANOWICZ, 'Microwave properties and applications of negative conductance transferred electron devices', *Proc. IEEE*, **59**, 1229—1237 (1971).
27. B. S. PERLMAN, 'Microwave amplification using transferred-electron devices in prototype filter equalization networks', *RCA Rev.*, **32**, 3—23 (1971).
28. R. W.H. ENGELMANN, 'On 'supercritical' transferred-electron amplifiers', *Arch. Elektr. übertr.*, **26**, 357—359 (1972).
29. H. KROEMER, 'Detailed theory of the negative conductance of bulk negative mobility amplifiers, in the limit of zero ion density', *IEEE Trans. Electron Dev.*, **ED—14**, 476—492 (1967).
30. B. W. HAKKI, 'Amplification in two-valley semiconductors', *J. Appl. Phys.*, **38**, 808—818 (1967).
31. H. W. THIM, 'Temperature effects in bulk GaAs amplifiers', *IEEE Trans. Electron Dev.*, **14**, 59—62 (1967).
32. H. KROEMER, 'Effect of parasitic series resistance on the performance of bulk negative conductivity amplifiers', *Proc. IEEE*, **54**, 1980—1981 (1966).
33. S. P. YU, W. TANTRAPORN and J. D. YOUNG, 'Transit-time negative conductance in GaAs bulk-effect diodes', *IEEE Trans. Electron Dev.*, **ED—18**, 88—93 (1971).
34. D. DASCALU, 'Space-charge-waves and high-frequency negative resistance of space-charge-limited diodes', *Int. J. Electron.*, **25**, 301—330 (1968).

35. D. Dascalu, 'Transit-time effects in bulk negative-mobility amplifiers', *Electron Lett.*, **4**, 581—583 (1968).
36. D. Dascalu, 'Small-signal theory of unipolar injection currents in solids', *IEEE Trans. Electron Dev.*, **ED—19**, 1239—1251 (1972).
37. T. Hariu, S. Ono and Y. Shibata, 'Wideband performance of the injection-limited Gunn diode', *Electron. Lett.*, **6**, 666—667 (1970).
38. H. Zetsche, 'Stability criterion for Gunn diodes with injection-limiting cathodes', *IEEE Trans. Electron Dev.*, **ED—21**, 142—146 (1974).
39. M. M. Atalla and J. L. Moll, 'Emitter controlled negative resistance in GaAs', *Solid-St. Electron.*, **12**, 619—629 (1969).
40. D. Dascalu, 'Emitter-current limited injection in negative-mobility semiconductors in the limit of zero doping', *Electron. Lett.*, **8**, 185—186 (1972).
41. R. Charlton, K. R. Freeman and G. S. Hobson, 'Stabilization mechanism for 'supercritical' transferred-electron amplifiers', *Electron. Lett.*, **7**, 575—577 (1971).
42. R. Charlton and G. S. Hobson, 'The effect of cathode-notch doping profiles on supercritical transferred-electron amplifiers', *IEEE Trans. Electron Dev.*, **ED—20**, 812—817 (1973).
43. A. K. Talwar and W. R. Curtice. 'Effect of donor density and temperature on the performance of stabilized transferred-electron devices', *IEEE Trans. Electron Dev.*, **ED—20**, 544—550 (1973).
44. A. S. Talwar and W. R. Curtice, 'An experimental study of stabilized transferred-electron amplifiers', *IEEE Trans. Microw. Theor. Techn.*, **MTT—21**, 477—481 (1973).
45. H. W. Thim and S. Knight, 'Carrier generation and switching phenomena in n-GaAs devices', *Appl. Phys. Lett.*, **11**, 85—87 (1967).
46. M. P. Shaw, P. R. Solomon and H. L. Grubin, 'The influence of boundary conditions on current instabilities in GaAs', *IBM J. Res. Develop.*, **13**, 587—590 (1969).
47. J. Magarshack and A. Mircea, 'Stabilization and wide band amplification using overcritically doped transferred electron diodes', in Proc. Int. Conf. Microwave and Optical Generation and Amplif., pp.16.19—16.23 (1970).
48. P. Jeppsen and B. I. Jeppsson, 'A simple analysis of the stable field profile in the supercritical TEA', *IEEE Trans. Electron Dev.*, **ED—20**, 371—379 (1973).
49. L. Solymar and A. E. Ash, 'Some travelling-wave interactions in semiconductor theory and design considerations', *Int. J. Electron.*, **20**, 127—148 (1966).
50. M. E. Hines, 'Theory of space-harmonic travelling-wave interactions in semiconductors', *IEEE Trans. Electron Dev.*, **ED—16**, 88—97 (1969).
51. R. H. Dean and R. J. Matarese, 'The GaAs travelling-wave amplifier as a new kind of microwave transistor', *Proc. IEEE*, **60**, 1486—1502 (1972).
52. G. S. Kino and P. N. Robson, 'The effect of small transverse dimensions on the operation of Gunn devices', *Proc. IEEE*, **56**, 2056—2057 (1968).
53. J. L. Teszner, 'Etude bi-dimensionnelle des instabilités dans une lame mince d'arseniure de gallium en contact avec un autre semiconducteur à résistance positive', *Solid. St-Electron.*, **13**, 1471—1481 (1970).
54. P. Gueret, 'Stabilisation of Gunn oscillations in layered semiconductor structures', *Electron. Lett.*, **6**, (1970).
55. L. S. Metz and O. P. Gandhi, 'DC electric-field profile control in long epitaxial layers of GaAs', *Proc. IEEE*, **61**, 1048—1050 (1973).
56. L. S. Metz and O. P. Gandhi, 'GaAs carrier wave growth in the presence of a thin resistive surface layer', *IEEE Trans. Electron Dev.*, **ED—21**, 118—119 (1974).

57. W. Frey, R. W. H. Engelmann and B. G. Bosch, 'Unilateral travelling-wave amplification in gallium arsenide at microwave frequencies', *Arch. Elektronik Übertrag.*, **25**, 1−8 (1971).
58. R. W. H. Engelmann, 'Plane-wave approximation of carrier waves in semiconductor plates with nonisotropic mobility', *IEEE Trans. Electron Dev.*, **ED−18**, 587−591 (1971).
59. R. W. H. Engelmann, 'Raumladungswellen in Halbleiterplättchen', *Arch. Elektronik Übertrag.*, **25**, 357−361 (1971).
60. W. Heinle and R. W. H. Engelmann, 'Stability criterion for semiconductor plates with negative AC mobility', *Proc. IEEE*, **60**, 914−915 (1972).
61. K. R. Hofmann, 'Stability theory for thin Gunn diodes with dielectric surface loading', *Electron. Lett.*, **8**, 124−125 (1972).
62. W. Heinle, 'Some results of the two-dimensional space-charge wave theory for larger sample thickness', *Arch. Elektronik Übertrag.*, **25**, 598−600 (1971).
63. R. W. H. Engelmann and W. Heinle, 'Effect of diffusion on the small-signal stability of semiconductor plates with negative AC mobility', *Arch. Elektronik Übertrag.*, **28**, 66−70 (1974).
64. R. H. Dean and B. B. Robinson, 'Space-charge waves in partially depleted negative-mobility media', *IEEE Trans. Electron Dev.*, **ED−21**, 61−69 (1974).
65. R. H. Dean, 'Reflection amplification in thin layers of n-GaAs', *IEEE Trans. Electron Dev.*, **ED−19**, 1148−1156 (1972).
66. P. Gueret, 'Limits of validity of the l-dimensional approach in space-charge-wave and Gunn-effect theories', *Electron. Lett.*, **6**, 197−198 (1970).
67. R. H. Dean and P. M. Schwartz, 'Field profile in n-GaAs layer biased above transferred-electron threshold', *Solid-St. Electron.*, **15**, 417−429 (1972).
68. R. H. Dean, 'A practical technique for controlling field profile in thin layers of n-GaAs', *IEEE Trans. Electron Dev.*, **ED−19**, 1144−1148 (1972).
69. G. A. Swartz, A. Gonzalez and A. Dreeben, 'Electric-field profile and current control of a long epitaxial GaAs n layer', *Electron. Lett.*, 93−94 (1972).
70. P. N. Robson, G. S. Kino and B. Fay, 'Two-part microwave amplification in long samples of gallium arsenide', *IEEE Trans. Electron Dev.*, **ED−14**, 612−614 (1967).
71. K. Kumabe and H. Kanbe, 'Mechanism of coupling between space-charge waves and microwaves in a bulk GaAs travelling-wave amplifier', *Rev. Electrical Commun. Lab.*, **18**, 913−920 (1970).
72. J. Koyama, S. Ohara, K. Kawazura and K. Kumabe, 'Bulk GaAs travelling-wave amplifier', *Rev. Electrical Commun. Lab.*, **17**, 1102−1109 (1969).
73. H. Kanbe, K. Kumabe and R. Nii, 'High power GaAs travelling-wave amplifier', *Rev. Electrical Commun. Lab.*, **19**, 917−921 (1971).
74. R. H. Dean, A. B. Dreeben, J. F. Kaminski and A. Triano, 'Travelling-wave amplifier using thin epitaxial GaAs layer', *Electron Lett.*, **6**, 777 (1970).
75. H. Kanbe, N. Shimizu and K. Kumabe, 'Characteristics of a gallium-arsenide travelling-wave amplifier with Schottky-barrier contacts, *Electron Lett.*, **9**, 29−30 (1973).
76. B. Källbäck, 'Noise properties of the injection-limited Gunn diode', *Electron Lett.*, **8**, 476−477 (1972).
77. J. E. Sitch and P. N. Robson, 'Noise measure of GaAs and InP transferred electron amplifiers', paper at the 4th Europ. Microw. Conf., Montreux, 10−13 Sept., 1974.
78. C. Hammar and B. Vitner, 'Diffusion of hot electrons in n indium phosphide', *Electron Lett.*, **9**, 9−10 (1973).
79. H. Hillbrand, H. D. Rees and D. Jones, 'Theoretical characteristics of transferred-electron amplifiers', *Electron Lett.*, **10**, 87−89 (1974).

80. H. W. THIM, 'Noise reduction in bulk negative-resistance amplifiers', *Electron. Lett.*, **7**, 106—108 (1971).
81. B. KÄLLBÄCK, 'Noise performance of gallium-arsenide and indium-phosphide injection-limited diodes', *Electron Lett.*, **9**, 11—12 (1973).
82. J. E. SITCH, 'Computer modelling of low noise indium phosphide amplifiers', *Electron. Lett.*, **10**, 74—86 (1974).

Problems

7.1. Show that in a homogeneous and one-dimensional structure traversed by a unipolar current there is no internal neutral plane separating depletion and accumulation regions. Discuss the effect of diffusion current.
Hint: Assume the contrary and use the fact that v is a single-valued function of $\mathscr{E}$.

7.2. Derive the incremental resistance, equation (7.2.13).

7.3. Compute the $J-V$ characteristic of a negative-mobility sample with ohmic contacts by neglecting the fixed space-charge and diffusion. Assume a piecewise linear $v = v(\mathscr{E})$ characteristic. Show that the $J-V$ curve exibits an inflection point.

7.4. Show qualitatively, by using equations (7.4.1)—(7.4.3), that the electric field is highly non-uniform in the anode region of a negative-mobility sample biased above 'threshold' voltage $\mathscr{E}_M L$ (see, for example, Figure 7.4.2a).

7.5. Use the neutral characteristic and control characteristic concepts to prove qualitatively that a doping notch or mobility notch near the cathode will improve the field uniformity inside a negative-mobility diode (Section 7.6).

7.6. Derive the small-signal impedance of a neutral sample with ohmic contacts (diffusion neglected). Show that the real part of the impedance is negative when the differential mobility is negative and explain this fact by the propagation of the space-charge wave (special case of equation (7.3.9)) and its interaction with the alternating field.

7.7. Derive the impedance of a quasi-neutral sample biased in the negative-mobility region by neglecting diffusion and assuming a non-ohmic injecting contact. Discuss the effect of finite 'cathode' conductivity upon the device negative resistance.

7.8. Consider the conservation equation $dn_1/dt = -dn_2/dt = -n_1/\tau_{12} + n_2/\tau_{21}$ for electron population in the lower (1) and the upper valley (2), respectively. Assume $\tau_{21} = $ const and derive τ_{12} from the velocity-field relationship in GaAs. Show that the negative mobility effect will disappear at extremely high frequencies.

8
Transferred-electron Devices used as Oscillators and as Logic Elements

8.1 Introduction

We have considered in Chapter 7 the possibility of microwave amplification using the small-signal negative-resistance of a TED. In this chapter we discuss typical large-signal modes of operation of supercritical doped samples: the device is potentially unstable and large internal space-charge perturbations may occur. These large perturbations do indeed occur, as in the dipole-domain modes (Section 8.10), or their growth is limited by a proper external circuit, as in space-charge controlled modes (Section 8.11).

The Gunn diode will be discussed first. This is a structure characterized by a large doping-length product and therefore by large space-charge and field perturbations which propagate cyclically through the device giving rise to current oscillations. The perturbation usually takes the form of a high-field domain. This domain consists of an accumulation layer and an adjacent depletion layer and moves through the semiconductor together with the mobile carriers which form the accumulation layer. This domain forms in a high-field negative-mobility region of the devices, may reach a stable configuration during its transit and disappears when it reaches the anode. The behaviour of a Gunn diode may be quite diverse, depending upon the device geometry, semiconductor doping properties of electric contacts and also upon the external circuit [1]. A Gunn diode biased by a constant voltage generates a microwave oscillation corresponding to cycling occurrence and propagation of high-field domains. A similar device biased by a large oscillating microwave signal superimposed on a steady-state voltage may operate in the limited space-charge accumulation (LSA) mode. The formation of domains is inhibited by cycling falling of the bias field below the threshold value: the device exhibits a bulk negative resistance and allows high-efficiency generation and amplification.

If the high-field domain is trapped near the anode (Section 7.3 and 7.6), the TE diode has a negative incremental (static) resistance and can be used for bistable switching, as any negative-resistance element. Multi-terminal TED's can be used to perform a variety of logic operations [2].

We shall again use a very simple description of the transferred-electron effect based upon the static dependence of the average electron velocity upon the local electric field, $v = v(\mathscr{E})$. At the present state of the theory and knowledge of the device behaviour, this approach is satisfactory for most purposes.

This chapter is merely an introduction to the study of the above operation modes. Other recent books [1]—[3] should be consulted for details of device-circuit interaction, various logic devices, technological problems, etc. The analytical theory of high-field domains and of LSA mode is also presented in our book, reference [4].

8.2 Formation of a High-field Domain (HFD)

Consider a one-dimensional model, a diode consisting in a semiconductor slab of thickness L, sandwiched between two plane-parallel contacts. A steady-state bias voltage, V, is applied between these contacts. When the semiconductor doping

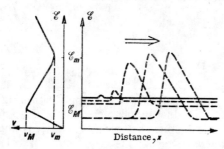

Figure 8.2.1. Growth and propagation of a high-field domain (HFD) in a negative-mobility semiconductor biased by a constant voltage, V. V/L is the average field or bias field, $\mathscr{E}_M$ is the threshold field, etc.

is sufficiently high and the cathode (negative) is an ohmic contact, the device behaves as a normal resistor. However, when the bias field $\mathscr{E}$ is increased above the threshold value for negative mobility, small space-charge non-uniformities may grow appreciably, as discussed below.

The basic equations are again

$$J(t) = env(\mathscr{E}) - e\frac{\partial}{\partial x}\{D(\mathscr{E})n\} + \varepsilon\frac{\partial \mathscr{E}}{\partial t}, \tag{8.2.1}$$

$$\frac{\partial \mathscr{E}}{\partial x} = \frac{e}{\varepsilon}(n - N_D) \tag{8.2.2}$$

(electrons treated as positively-charged particles). Figure 8.2.1 shows qualitatively the growth and displacement of a high-field perturbation, called high-field domain (HFD). Outside the local perturbation the field is uniform. The non-uniformity may be initiated, for example, by a notch in the doping profile: a local increase of the electric field is required to provide the current continuity through the lowly-doped notch. The field perturbation shown corresponds to a dipole formed by an accumulation layer at the upstream side and a depletion layer at the downstream side of the dipole. As far as the field intensity is above its threshold value, the moving dipole increases through the accumulation of new mobile carriers and extending of the depleted region; this is because the higher the field intensity the lower the velocity and the electrons pile up toward the trailing edge of the dipole.

Figure 8.2.1 shows that the external (outside) field, $\mathscr{E}_N$, decreases when the domain grows (the electric field is the same on both sides of the domain, Problem 8.1). When $\mathscr{E}_N$ falls below $\mathscr{E}_M$, the domain should tend toward a stable configuration (at the domain limits the mobility becomes positive) [5]—[12].

The current continuity equation, (8.2.1), yields

$$J(t) = eN_D v(\mathscr{E}_N) + \varepsilon\frac{\partial \mathscr{E}_N}{\partial t} = env(\mathscr{E}) - \varepsilon\frac{\partial}{\partial x}\{D(\mathscr{E})n\} + \varepsilon\frac{\partial \mathscr{E}}{\partial t} \tag{8.2.3}$$

and by using equation (8.2.2) one obtains

$$\frac{\partial}{\partial t}(\mathscr{E} - \mathscr{E}_N) = \frac{eN_D}{\varepsilon}[v(\mathscr{E}_N) - v(\mathscr{E})] + \frac{\partial}{\partial x}\{Dn\} - v(\mathscr{E})\frac{\partial \mathscr{E}}{\partial x}. \quad (8.2.4)$$

The above equation will be integrated with respect to x (at a given t), from a plane x_1, situated on the left of the domain (Figure 8.2.1) until a plane x_2 which remains always on the right of the domain. The result is

$$\frac{d}{dt}\int_{x_1}^{x_2}(\mathscr{E} - \mathscr{E}_N)dx = \frac{eN_D}{\varepsilon}\int_{x_1}^{x_2}[v(\mathscr{E}_N) - v(\mathscr{E})]dx, \quad (8.2.5)$$

or

$$\frac{\partial V_D}{\partial t} = \frac{eN_D}{\varepsilon}\int_{x_1}^{x_2}[v(\mathscr{E}_N) - v(\mathscr{E})]dx, \quad V_D = \int_{x_1}^{x_2}(\mathscr{E} - \mathscr{E}_N)dx, \quad (8.2.6)$$

where V_D is the *domain excess voltage*. The above equation describes the domain dynamics. It can be easily shown that *initially* the excess voltage V_D increases exponentially with a time constant numerically equal to the differential relaxation time, $\varepsilon/eN_D|dv/d\mathscr{E}|$.

8.3 Stable High-field Domains

Let us consider a stable non-uniformity moving toward the cathode with the uniform velocity v_D. If x' is the coordinate of a system which moves together with this non-uniformity, in the positive x direction, then

$$x' = x - v_D t \quad (8.3.1)$$

and the following changes will be made in equations (8.2.1) and (8.2.2)

$$\frac{\partial \mathscr{E}}{\partial x} \to \frac{d\mathscr{E}}{dx'}, \quad \frac{\partial(Dn)}{\partial x} \to \frac{d(Dn)}{dx'}, \quad \frac{\partial \mathscr{E}}{\partial t} = -v_D\frac{d\mathscr{E}}{dx'}, \quad (8.3.2)$$

such that

$$J = env(\mathscr{E}) - e\frac{d(Dn)}{dx'} - \varepsilon v_D\frac{d\mathscr{E}}{dx'}, \quad (8.3.3)$$

$$\frac{d\mathscr{E}}{dx'} = \frac{e}{\varepsilon}(n - N_D). \quad (8.3.4)$$

The spatial variable x' can be eliminated between the above two equations and the result is

$$J = env(\mathscr{E}) - \frac{e^2}{\varepsilon}(n - N_D)\frac{d(Dn)}{d\mathscr{E}} - ev_D(n - N_D). \quad (8.3.5)$$

On the other hand J can be written $J = eN_D v_N$, where $v_N = v(\mathscr{E}_N)$ is the 'external' velocity, and equation (8.3.5) becomes

$$\frac{e(n - N_D)}{\varepsilon} \frac{d(Dn)}{d\mathscr{E}} = n[v(\mathscr{E}) - v_D] - N_D(v_N - v_D). \tag{8.3.6}$$

The above equation is fundamental for the study of stable space-charge layers in NDM materials. There are four types of space-charge layers: high-field domains, low-field domains, accumulation layers and depletion layers [8], [12]. Equation (8.3.6) is quite general, the only limitation arises from the questionable superposition of drift and diffusion components of the particle current in equation (8.3.1) (see Chapter 1). Equation (8.3.6) should yield, at least in principle, $n = n(\mathscr{E})$ and, in conjunction with equation (8.3.4), the domain configuration, i.e. $n = n(x')$ and $\mathscr{E} = \mathscr{E}(x')$. However, equation (8.3.6), where $D = D(\mathscr{E})$ and $v(\mathscr{E})$ are assumed given (*static* relationships), is a non-linear differential equation and in most cases an analytical solution cannot be found.

We shall first consider a *field-independent diffusion coefficient*, $dD/d\mathscr{E} = 0$ [7]. In this case, equation (8.3.6) has the formal solution

$$\frac{n}{N_D} - \ln\frac{n}{N_D} - 1 = \frac{\varepsilon}{eN_D D} \int_{\mathscr{E}_N}^{\mathscr{E}} \left\{ [v(\mathscr{E}) - v_N] - \frac{N_D}{n}(v_N - v_D) \right\} d\mathscr{E}. \tag{8.3.7}$$

The above relation will be used to discuss the properties of stable dipole domains. High-field domains will be discussed first. The electric field increases with x' from $\mathscr{E}_N$ (outside the domain, on the upstream side), to the maximum value $\mathscr{E}_D$ at $x' = x'_D$ inside the domain, and then decreases and reaches again $\mathscr{E}_N$ (outside the domain, on the downstream side). Of course, $\mathscr{E}_N < \mathscr{E}_M < \mathscr{E}_D$, where $\mathscr{E}_M$ is the threshold field for negative mobility. We have

$$\left.\frac{d\mathscr{E}}{dx'}\right|_{x'=x'_D} = 0, \quad n\Big|_{x'=x'_D} = N_D. \tag{8.3.8}$$

At the 'top' of the domain ($x' = x'_D$) the displacement current is zero, see equation (8.3.2), the drift current is $eN_D v(\mathscr{E}_D)$ and the total current (uniform) should be equal to $eN_D v_N$ (its value outside the domain). However, $v(\mathscr{E}_D) \neq v_N$ because of the diffusion current existing at the top of the domain. On the other hand,

$$v_D = v_N \quad (D = \text{const.}). \tag{8.3.9}$$

The equality of the domain velocity, v_D, with the 'outside' velocity, v_N, can be demonstrated as follows. Let us perform the integral in equation (8.3.7) between $\mathscr{E}_N$ and $\mathscr{E} = \mathscr{E}_D$. Because $n = N_D$ for $\mathscr{E} = \mathscr{E}_D$, the left-hand side is zero and the integral should be also zero. Therefore we have

$$\int_{\mathscr{E}_N}^{\mathscr{E}_D} [v(\mathscr{E}) - v_N] \, d\mathscr{E} = \frac{eN_D^2 D(v_N - v_D)}{\varepsilon} \int_{\mathscr{E}_N}^{\mathscr{E}_D} \frac{d\mathscr{E}}{n(\mathscr{E})}. \tag{8.3.10}$$

The above relation should be valid if the integration is carried out both on the leading front of the domain and on the trailing edge. The integral on the left-hand side has the same values in both cases, but the right-hand side integral does not, because $n < N_D$ in the depletion layer and $n > N_D$ in the accumulation layer, as shown qualitatively in Figure 8.3.1. Therefore v_N should be equal to v_D, and

$$\int_{\mathcal{E}_N}^{\mathcal{E}_D} [v(\mathcal{E}) - v_D] \, d\mathcal{E} = \int_{\mathcal{E}_N}^{\mathcal{E}_D} [v(\mathcal{E}) - v_N] \, d\mathcal{E} = 0 \quad (D = \text{const.}) \quad (8.3.11)$$

The above equation defines the so-called 'equal areas rule', which is illustrated in Figure 8.3.2. When the external velocity $v_N = J/eN_D$ varies, the point having the

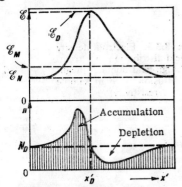
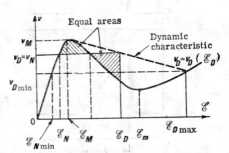

Figure 8.3.1. Electron density and field distribution inside a moving HFD.

Figure 8.3.2. The 'equal areas rule' for HFD's and the corresponding dynamic characteristic, $v_D = v_D(\mathcal{E}_D)$.

co-ordinates ($\mathcal{E}_D, v_D$) traces the *dynamic characteristic* in the $\mathcal{E} - v$ plane (Figure 8.3.2). This dynamic characteristic, $v_D = v_D(\mathcal{E}_D)$, is independent of N_D and the value of $D = \text{const}$. Note the fact that $v(\mathcal{E}_D) < v_D = v_N$ and the drift current of the top of the domain should be aided by the diffusion current, which is physically correct because at $x' = x_D'$ the electrons diffuse from the left (accumulation layer) toward the right (depletion layer, Figure 8.3.1).

The dynamic characteristic in Figure 8.3.2 determines the maximum possible domain field, $\mathcal{E}_{D\max}$. Note the fact that if $v = v(\mathcal{E})$ saturates at high field intensities $\mathcal{E}_{D\max} \to \infty$ and $\mathcal{E}_D$ may exceed the critical value for avalanche breakdown. This occurs if the device is sufficiently long to accomodate a wide domain with a high peak field, $\mathcal{E}_D$. The domain velocity may range between $v_{D\min}$ and v_M and this defines a range of current densities which can sustain a stable high-field domain. The field non-uniformity $\mathcal{E}_D - \mathcal{E}_N$, increases, as the current density J decreases below its threshold value $J_M = eN_D v_M$.

The exact shape of the domain is established by diffusion and depends upon N_D and D (field-independent). Equations (8.3.4) and (8.3.7) become, respectively,

$$x' = x_D' + \frac{\varepsilon}{e} \int_{\mathcal{E}_D}^{\mathcal{E}} \frac{d\mathcal{E}}{n(\mathcal{E}) - N_D}, \quad (8.3.12)$$

$$\frac{n}{N_D} - \ln \frac{n}{N_D} - 1 = \frac{\varepsilon}{eN_D D} \int_{\mathcal{E}_N}^{\mathcal{E}} [v(\mathcal{E}) - v_D] \, d\mathcal{E}, \quad (8.3.13)$$

where $v = v(\mathscr{E})$ is assumed known. Let us consider for example the case of negligible diffusion, $D \to 0$ (mathematically equivalent to $N_D \to 0$, high-resistivity material). The right-hand part of equation (8.3.13), tends to infinity, except when

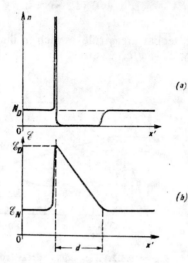

Figure 8.3.3. Electron density and field distribution in a 'triangular' high-field domain (negligible diffusion or high sample resistivity).

$\mathscr{E} = \mathscr{E}_N$ or $\mathscr{E} = \mathscr{E}_D$ and the integral is zero. $n = n(\mathscr{E})$ has two solutions, one for the accumulation layer ($n > N_D$) and the other for the depletion layer ($n < N_D$). Because the left-hand part of equation (8.3.13) should also tend to infinity when $D \to 0$, we have $n/N_D \to \infty$ for the accumulation layer and $n/N_D \to 0$ for the depletion layer. Therefore a triangular domain (Figure 8.3.3) is formed by an infinitely thin accumulation layer of infinite density and a completely depleted region of thickness

$$d \simeq \frac{\varepsilon}{eN_D}(\mathscr{E}_D - \mathscr{E}_N). \tag{8.3.14}$$

Note that in the depletion region the current continuity is provided by the displacement current.

We shall now discuss the case of *field-dependent diffusion coefficient*, $D = D(\mathscr{E})$. Equation (8.3.6) can be integrated as shown by Butcher et al. [10] to yield a relation analogous to equation (8.3.7). It was shown that [10], [12]

$$\frac{n}{N_D} - \ln\frac{n}{N_D} - 1 = \frac{\varepsilon}{eN_D}\int_{\mathscr{E}_N}^{\mathscr{E}}\frac{v(\mathscr{E}) - v_D}{D(\mathscr{E})}\,d\mathscr{E} + \frac{\varepsilon}{eN_D}(v_D - v_N) \times$$

$$\times \int_{\mathscr{E}_N}^{\mathscr{E}}\frac{N_D\,d\mathscr{E}}{n(\mathscr{E})D(\mathscr{E})} - \int_{\mathscr{E}_N}^{\mathscr{E}}\frac{D'(\mathscr{E})}{D(\mathscr{E})}\left[\frac{n(\mathscr{E})}{N_D} - 1\right]d\mathscr{E}. \tag{8.3.15}$$

The equal-areas rule is no longer valid and the domain velocity differs from the drift velocity in the neutral semiconductor, namely

$$v_D - v_N = \frac{\int_{\mathcal{E}_N}^{\mathcal{E}_D} \left[\left(\frac{n}{N_D}\right)_a - \left(\frac{n}{N_D}\right)_d\right] \frac{D'(\mathcal{E})}{D(\mathcal{E})} d\mathcal{E}}{\int_{\mathcal{E}_N}^{\mathcal{E}_D} \left[\left(\frac{N_D}{n}\right)_a - \left(\frac{N_D}{n}\right)_d\right] \frac{d\mathcal{E}}{D(\mathcal{E})}} \qquad (8.3.16)$$

where $D'(\mathcal{E}) = dD/d\mathcal{E}$ and the subscripts a and d refer, respectively, to the accumulation and depletion sides of the domain. Now let us assume that $D'(\mathcal{E})$ is positive in the field range considered. Because always $(n/N_D)_a > (n/N_D)_d$, the right-hand side of equation (8.3.16) is negative and the domain velocity is smaller than the external velocity. For fields close to threshold, $\mathcal{E}_N \lesssim \mathcal{E}_M \lesssim \mathcal{E}_D$ and at large doping densities, the domain velocity becomes negative, i.e. the domain will move in a direction opposite to that of carrier drift [12]. When $D'(\mathcal{E})$ is negative the domain velocity always exceeds the drift velocity outside the domain. It was shown that for bias fields sufficiently near the threshold field, $\mathcal{E}_N \lesssim \mathcal{E}_M \lesssim \mathcal{E}_D$, the domain velocity may be approximated by [10], [12]

$$v_D \simeq v_N - \frac{eN_D}{\varepsilon} D'(\mathcal{E}). \qquad (8.3.17)$$

The above equation should be compared with equations (7.3.5) and (7.3.8) which relate the phase velocity of the space-charge wave, v_p, to the drift velocity, v

$$v_p = v - v_d, \quad v_d = \frac{eN_D}{\varepsilon} D'(\mathcal{E}). \qquad (8.3.18)$$

Equations (8.3.17) and (8.3.18) are practically identical, which is indeed reasonable, because the domain velocity, equation (8.3.17) was derived in the small-signal limit ($\mathcal{E}_D - \mathcal{E}_N \ll \mathcal{E}_M$). We have shown that because of the field-dependent diffusion ($D'(\mathcal{E}) < 0$) the space-charge wave can propagate in a direction opposite to that of the carrier drift (Section 7.3). A computer simulation shows that for a large $D'(\mathcal{E})$ (positive) or for a large N_D, the high-field domain nucleated inside the cathode-anode space moves toward the cathode [13], [14] (see Section 8.7).

Butcher et al. [10] used equations (8.3.15) and (8.3.16), the velocity-field dependence shown in Figure 8.3.4, and the field-dependent diffusion coefficient shown in Figure 8.3.5 to compute data about stable high-field domains in GaAs. Figure 8.3.4 shows dynamic characteristics, $v_D = v_D(\mathcal{E}_D)$, which, this time, depend upon the doping density. Figure 8.3.6 shows the computed domain velocity, which exceeds the 'outside' drift velocity v_N, because $D'(\mathcal{E})$ is negative (Figure 8.3.5). Figure 8.3.7 shows the domain excess voltage, V_D, equation (8.2.6) as a function of the outside field, $\mathcal{E}_N$, with GaAs resistivity as parameter. The domain excess voltage

$$V_D = \int_0^L (\mathcal{E} - \mathcal{E}_N) dx \qquad (8.3.19)$$

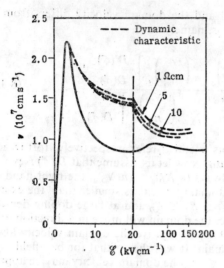

Figure 8.3.4. Velocity-field dependence for GaAs and dynamic $v_D = v_D(\mathcal{E}_D)$ characteristics, shown by dashed curves, computed for various resistivities. (*after Butcher et al.*, [10]).

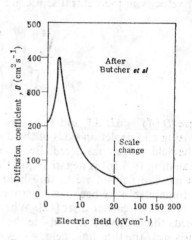

Figure 8.3.5. Field-dependent diffusion coefficient in GaAs. (*reproduced after Butcher et al.*, [10]).

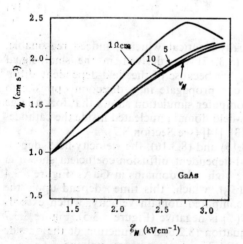

Figure 8.3.6. Domain velocity, v_D, and outside velocity, v_N, as functions of the outside electric field, $\mathcal{E}_N$. (*after Butcher et al.*, [10]).

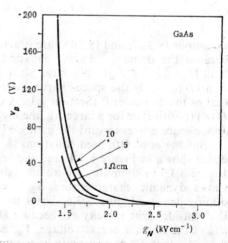

Figure 8.3.7. Domain excess voltage, V_D, versus outside (external) electric field, $\mathcal{E}_N$, for several resistivities of the GaAs sample. (*after Butcher et al.*, [10]).

is calculated by using equations (8.3.12) and (8.3.15), as well as Figures 8.3.4 and 8.3.5. Figure 8.3.8 indicates the maximum electron concentration inside the accumulation layer and the minimum electron concentration inside the depletion layer. For relatively high resistivities we have $n_{min} \ll N_D$ and $n_{max} \gg N_D$, in agreement

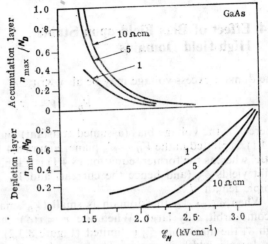

Figure 8.3.8. Maximum electron concentration inside the accumulation layer and minimum electron concentration inside the depletion layer as a function of the outside field, $\mathscr{E}_N$, for various GaAs resistivities. (after Butcher et al., [10]).

with the diffusionless theory (triangular domain, Figure 8.3.3). However, for a given semiconductor resistivity, the shape of the domain strongly depends upon the domain excess voltage, V_D, as shown by Figure 8.3.9, after Copeland [8]. For $\mathscr{E}_N = 2.2$ kV cm^{-1} and low V_D, the domain is nearly symmetrical (Figure 8.3.9a). For $\mathscr{E}_N = 1.45$ kV cm^{-1}, V_D increases by more than one order of magnitude, the

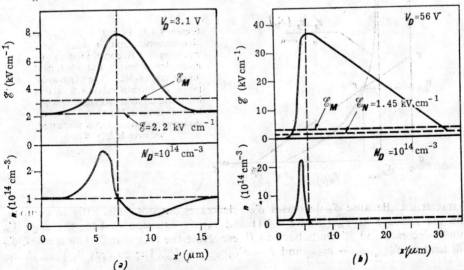

Figure 8.3.9. High-field domain configurations for $V_D = 3.1$ V (a) and $V_D = 56$ V (b) in a GaAs sample with $N_D = 10^{14}$ cm^{-3}. (after Copeland, [8]).

domain is nearly triangular and the depletion region is almost completely free of electrons (the electron density is below 0.001 N_D). The domain width increases, and the maximum electric field, $\mathscr{E}_D$, reaches quite high values. The behaviour of stable domains depends upon the external voltage bias as shown in the next section.

8.4 Effect of Bias Field upon Stable High-field Domains

The domain excess voltage may be also written

$$V_D = V - \mathscr{E}_N L, \qquad (8.4.1)$$

where V is the voltage bias (assumed constant) and L is the sample length. Equation (8.4.1) is plotted in the $V_D - \mathscr{E}_N$ plane, Figure 8.4.1 (the latter is a 'material characteristic' whereas the former, equation (8.4.1) is a kind of 'load line'*) [5]. The crossover point yields $\mathscr{E}_N$ (and hence the current) and V_D, for a given bias voltage, V, and sample length, L.

When $\mathscr{E}_N \to \mathscr{E}_M$ the domain is small (V_D small). When $\mathscr{E}_N$ decreases, V_D may become arbitrarily large. When the $v = v(\mathscr{E})$ characteristic does not saturate at high $\mathscr{E}$, the maximum $\mathscr{E}_D$ is limited (Figure 8.3.2), but V_D can still increase because the domain width increases. Therefore $V_D \to \infty$ for $\mathscr{E}_N \to \mathscr{E}_{N\min}$.

The crossover point, P, in Figure 8.4.1, corresponds to a stable domain. Let us assume that V_D increased accidentally and the point P rises in P' on the $V_D - \mathscr{E}_N$

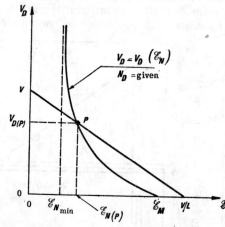

Figure 8.4.1. Typical $V_D = V_D(\mathscr{E}_N)$ characteristic and a load line, equation (8.4.1), corresponding to a given device length and a bias field above threshold. The crossover point P corresponds to a stable high-field domain propagating within the sample. When this domain disappears a new one is nucleated, because $V/L > \mathscr{E}_M$.

characteristic. Because $\mathscr{E}_N$ decreases, $\mathscr{E}_N L$ decreases but not sufficiently to compensate the increase in V_D (Figure 8.4.1) and satisfy equation (8.4.1): therefore V_D should decrease and P' tends back to P, etc. Note the fact that if V increases, $\mathscr{E}_N$ will tend to $\mathscr{E}_{N\min}$, $v_N \to v_{N\min}$ and $J \to eN_D v_{N\min} = J_{\text{sat}}$. The current flowing in the

---

* The domain is 'loaded' by the ohmic (low-field) sample resistance: the load line depends upon the device length.

8. Transferred-electron Devices used as Oscillators

external circuit during the domain propagation approaches a minimum value as the bias voltage is increased. The corresponding domain is called *saturated*.

The situation depicted in Figure 8.4.1 corresponds to an *astable-type behaviour*, characteristic for the so-called *Gunn oscillations* [1], [15], [16]. Let us assume that an increasing bias field V/L is applied to a GaAs sample with ohmic contacts. The steady-state current increases with V and follows the shape of $v = v(\mathscr{E})$ curve if the injected space-charge can be neglected (large $N_D L$). As far as V/L exceeds the threshold value $\mathscr{E}_M$, the NDM determines the growth of space-charge non-uniformities. We shall see later that in many cases an HFD is nucleated at the cathode and grows very fast (large $N_D L$) reaching a stable configuration. During the transit of this stable domain the current through the device is constant, and lower than the threshold value. When the domain reaches the anode, and a new domain forms, a current spike occurs in the external circuit. Because the excess voltage drop associated with the domain disappears, the field inside the crystal rises again above the threshold value and nucleates a new domain, etc. The experimental periodic waveform reproduced in Figure 8.4.2 corresponds to cyclical formation propagation and disappearance of high-field domains. By neglecting the time required for domain formation and disappearance, the frequency of current oscillations is equal to the transit-time frequency

$$f_T = \frac{1}{T_L} = \frac{v_D}{L} \tag{8.4.2}$$

where $T_L = L/v_D$ is the domain transit-time. For $L = 10 - 100$ μm and $v_D \approx$ $\approx 10^7$ cm s^{-1}, f_T has values between 1 and 10 GHz.

The situation shown in Figure 8.4.3 (two crossover points, P_1 and P_2) illustrates a *monostable-type behaviour*. When no high-field domain exists inside the crystal, $V_D \to 0$, the bias field is below the threshold value and the domain formation is impossible (at least as far as the internal field is uniform). Let us suppose that a

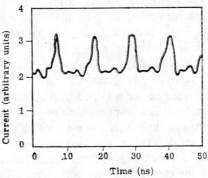

Figure 8.4.2. Gunn-type oscillations determined experimentally in a resistive circuit (constant voltage bias). The current peaks occur when a domain reaches the anode and disappears. The current variations during domain transit occur due to the sample non-uniformities (non-uniform doping or cross-sectional area). (*after Copeland*, [16]).

domain was nevertheless formed. There are two possible states corresponding to fully-formed HFD's propagating inside the crystal, P_1 and P_2. However, only the point P_2 describes a stable situation. The demonstration is the same as for the point P in Figure 8.4.1. The domain corresponding to the point P_1 is unstable with respect to perturbations. A small perturbation either leads to the disappearance of the (P_1) domain, or determines the transition into the (P_2) state.

An HFD can be nucleated by applying a short positive voltage pulse ΔV and thus increasing temporarily the bias field above the threshold value. This domain propagates through the crystal and disappears into the anode, thus giving rise to

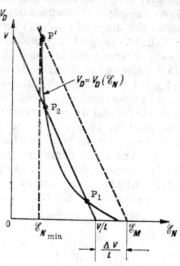

Figure 8.4.3. $V_D - \mathscr{E}_N$ plane indicating a monostable type of behaviour. Point P_2 corresponds to a stable HFD launched by a trigger pulse ΔV superimposed on the steady-state bias V, below the threshold value ($V < \mathscr{E}_M L$). Also shown is a load line for a minimum voltage, V_m, which can sustain the domains in a given device (for a certain L).

a negative current pulse: the pulse width is approximately equal to the domain transit-time. A new domain cannot appear until a new triggering pulse ΔV is applied. The main application of this type of operation is pulse regeneration. The pulse width cannot be changed in wide limits, because as the steady-state bias increases, the domain velocity tends to saturate.

If there is no crossover point in the $V_D - \mathscr{E}_N$ plane, the propagation of stable domains is impossible, even if such a domain is accidentally nucleated within the crystal. This is because the applied voltage cannot 'sustain' domains inside the device. We can also say that for a given bias, the crystal is too long and determines a too large 'ohmic' voltage drop. Note the fact that a series external resistance may have a similar effect (Problem 8.3).

When the device is biased by a constant current, $J = eN_D v_N = $ const, $\mathscr{E}_N = $ const and the 'load line' is parallel to the vertical axis: the crossover point with the $V_D - \mathscr{E}_N$ characteristic is always unstable. Indeed, it may happen that the current source yields exactly the current necessary to sustain the domain which moves toward the anode. However, when the bias current is slightly larger than these values, the electron density will increase inside the accumulation layer and decrease in the depletion layer thus increasing V_D and decreasing $\mathscr{E}_N$ and a stable situation cannot be achieved [5].

For similar reasons two stable domains cannot co-exist in a device. If one of them requires a larger current than the other, the excess voltage drop on the second one increases. As a result, the first one quickly disappears and all the excess voltage is absorbed by the second one requiring less current [5].

8.5 Quasi-stationary (QS) Characteristic and Bistable Switching

We consider below the current-voltage characteristic of a supercritically-doped sample (Gunn diode). During the domain propagation an increase of the domain voltage will determine (as shown above) a decrease of the current: this effect corresponds to a negative resistance of the quasi-stationary (QS) $I-V$ characteristic, which is the $I-V$ dependence during the domain propagation. This effect is due to the negative conductivity of the semiconductor region occupied by the domain in transit.

The QS characteristic will be obtained as follows. We shall plot first the low-voltage 'ohmic' region in the $v_N - V$ plane (Figure 8.5.1). Here, v_N depends upon $V = \mathscr{E}_N L$, exactly as $v = v(\mathscr{E})$. Above the threshold voltage, high-field domains are nucleated and travel toward the anode. The $v_D = v_D(\mathscr{E}_N)$ characteristic of Figure 8.4.1 will be transposed in Figure 8.5.1, by using $v_N = v_N(\mathscr{E}_N)$, according to the $v = v(\mathscr{E})$ relationship. Finally those two curves will be added, according to equation (8.4.1). The resulting curve has the shape of the QS characteristic, since $J = = eN_D v_N$.

The QS characteristic exhibits a *hysteresis*. When V increases gradually from zero, the domains occur above $\mathscr{E}_M L$, but when V decreases, they disappear below V_m which is considerably lower than $\mathscr{E}_M L$. The QS characteristic exhibits current saturation at higher voltages and this property is directly related with the shape of the $v_D = v_D(\mathscr{E}_N)$ curve and, therefore of the $v = v(\mathscr{E})$ curve. In the horizontal region of the QS characteristic the domains are 'saturated' (Section 8.4). Figure 8.5.2 shows the QS characteristic and the stationary (S), neutral characteristic, both in a normalized $v - \mathscr{E}$ plot. A particular load line is also indicated (the device is in series with a resistance and a constant voltage source). Crossover point 1 describes a stationary behaviour (stable state). Point 2 is unaccessible because of domain formation: during the domain propagation the QS characteristic is established.

Figure 8.5.1. Construction of QS characteristic.

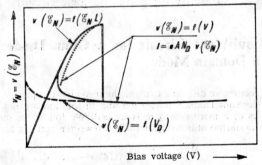

Bias voltage (V)

Point 4 represents a QS state on this QS characteristic: this state is unstable, it lasts only until the domain reaches the anode. Point 3 should represent in principle a stable state: it is situated in a positive-mobility region. Recent data on $v = v(\mathscr{E})$ characteristic in GaAs indicate velocity saturation in the high field range and exclude the existence of points like 3 in Figure 8.5.2. Even if such a point does exist on the static characteristic, it may be unstable because the carrier injection reduces the field near the cathode and allow domain formation. After the domain reaches the anode, the device returns to the 1 state. This behaviour corresponds to monostable-type operation (successive $1 \to 3 \to 4 \to 1$ transitions) induced by a positive triggering pulse which has a sufficient amplitude and duration: instead, a transition to 4

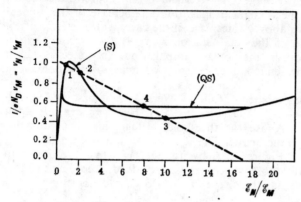

Figure 8.5.2. Stationary (S) and quasi-stationary (QS) characteristics in a normalized I (or v) versus $\mathscr{E}$ plot. The load line associated with a series load resistance and a battery is also shown.

(and back to 1) takes place (this type of monostable operation was already mentioned before, Figure 8.5.2). The use of Gunn diode as a switch (particularly for pulse discrimination by amplitude and duration) was theoretically discussed by Engelmann and Heinle [17], [18].

The bistable switching was obtained experimentally in supercritically doped GaAs sample [19] — [21]. For example a 41.5 per cent current drop with a 110 ps switching time was obtained in a 10 μm n^+nn^+ diode operating as a bistable switch into a 25 Ω resistive load. However, the high-field low-current state corresponds to non-uniform electric field, with a high-field domain 'trapped' near the anode [19], [21] (see also Sections 7.2, 7.3, 7.6 and 8.7). Short switching times and relatively large voltage signals, as compared with those given by tunnel diodes, make this device attractive for pulse and logic circuitry [17], [21].

8.6 Equivalent Circuit for a Gunn Diode Operating in Domain Mode

In this section we discuss a dynamic lumped circuit representation of a Gunn diode with a propagating domain. Domain nucleation and dissolution are not considered in this section: only the dynamics of a mature domain is studied. We follow the model suggested by Gunshor and Kak [22]. The startpoint is equation (8.2.6), rewritten here for convenience

$$\frac{dV_D}{dt} = \frac{eN_D}{\varepsilon}\int [v(\mathscr{E}_N) - v(\mathscr{E})]dx, \quad V_D = \int (\mathscr{E} - \mathscr{E}_N)dx. \qquad (8.6.1)$$

V_D is the domain excess voltage and the integrations are carried over the region including the domain. We shall first consider the diffusionless case $(D \to 0)$ [22]. The high-field domain is triangular (Figure 8.3.3) and equation (8.6.1) becomes

$$\frac{dV_D}{dt} = \int_{\mathscr{E}_N}^{\mathscr{E}_D} [v(\mathscr{E}_N) - v(\mathscr{E})]d\mathscr{E}. \qquad (8.6.2)$$

8. Transferred-electron Devices used as Oscillators

The differential capacitance of the triangular domain is

$$C_{D0} = \frac{eN_D A}{\mathscr{E}_D - \mathscr{E}_N} \qquad (D \to 0) \tag{8.6.3}$$

(see equation (8.3.14), where A is the area of the device cross section. Equations (8.6.2) and (8.6.3) yield

$$C_{D0}\frac{dV_D}{dt} = \frac{V - V_D}{R_0} - I_{DD}(\mathscr{E}_N, \mathscr{E}_D), \tag{8.6.4}$$

where $R_0 = L/AeN_D\mu_n$ is the low-field sample resistance, $v_N \simeq \mu_n \mathscr{E}_N$ (μ_n is the low-field mobility), $V = V_D + \mathscr{E}_N L$, and

$$I_{DD} = \frac{eN_D A}{\mathscr{E}_D - \mathscr{E}_N}\int_{\mathscr{E}_N}^{\mathscr{E}_D} v(\mathscr{E})d\mathscr{E}. \tag{8.6.5}$$

The total current through the device can be written

$$I = AeN_D\mu_n\mathscr{E}_N + A\varepsilon\frac{d\mathscr{E}_N}{dt}, \tag{8.6.6}$$

and becomes

$$I = \frac{V - V_D}{R_0} + C_0\frac{d}{dt}(V - V_D), \tag{8.6.7}$$

where $C_0 = \varepsilon A/L$ is the 'geometrical' device capacitance.

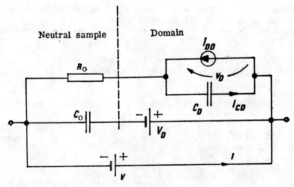

Figure 8.6.1. Equivalent circuit for dynamics of a mature domain, as derived from current continuity and Poisson equations. V_D is the domain excess voltage, I_{DD} is the current given by equation (8.6.5) and C_D is the domain capacitance (C_{D0} is its value calculated for zero diffusion), (after Gunshor and Kak, [22]).

Equations (8.6.4) and (8.6.7) define the equivalent circuit shown in Figure 8.6.1. Note the existence of the domain capacitance, $C_D(C_{D0})$, with the domain excess voltage across it, and of the voltage-controlled current generator I_{DD}, equation (8.6.5). Gunshor and Kak [22] appreciated that the dependence $I_{DD} = I_{DD}(V_D)$ may be approximated by the steady-state dependence $I = I(V_D)$ obtained from the $V_D = V_D(\mathcal{E})$ characteristic (Section 8.5), as indicated by Figure 8.6.2.

The diffusion effects can be formally included as follows [22]. Equation (8.6.1) yields

$$\frac{dV_D}{dt} = \frac{e\mu_n N_D x_d \mathcal{E}_N}{\varepsilon} - \frac{eN_D}{\varepsilon}\int_{x_1}^{x_2} v(\mathcal{E})dx, \quad (8.6.8)$$

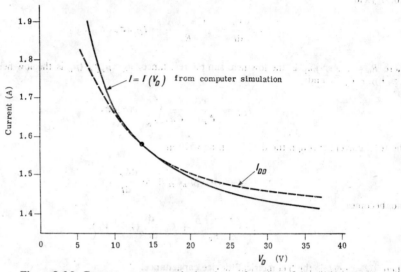

Figure 8.6.2. Current generator calculated as a function of domain excess voltage; the dashed curve shows I_{DD} computed by using the approximate equation (8.6.5), the solid line is $I = I(V_D)$ during domain propagation, obtained by computer simulation. (after Gunshor and Kak, [22]).

where x_1 and x_2 are the domain boundaries and the domain width $x_2 - x_1$ is approximately the depletion layer width, d. Equation (8.6.8) can be also written

$$\frac{dV_D}{dt} = \frac{1}{C_{D0}}\frac{V - V_D}{R_0} - \frac{eN_D}{\varepsilon}\int_{x_1}^{x_2} v(\mathcal{E})d\mathcal{E}, \quad C_{D0} = \frac{\varepsilon A}{x_d}. \quad (8.6.9)$$

The actual domain capacitance will now be $C_D \neq C_{D0}$. By rewriting equation (8.6.9) in the form given by equation (8.6.4), we obtain

$$C_D \frac{dV_D}{dt} = \frac{V - V_D}{R_0} - \hat{I}_{DD}, \quad (8.6.10)$$

$$\hat{I}_{DD} = \left(\frac{C_D}{C_{D0}} - 1\right) eAN_D\mu_n \mathcal{E}_N - \frac{eN_D C_D}{\varepsilon}\int_{x_1}^{x_2} v(\mathcal{E})dx. \quad (8.6.11)$$

We shall briefly discuss the utility of such an equivalent circuit. By defining the domain differential capacitance it is possible to relate the charge transferred to the circuit to the change in bias voltage. Kuru et al. [23] found that the charge transferred at the external device terminals, ΔQ_{ex}, in response to a step in bias voltage was less (approximately one-half) than the expected change in domain charge, ΔQ_D. It was later shown [22] that this is due to diffusion which broadens

8. Transferred-electron Devices used as Oscillators

the domain and reduces the capacitance to about one-half of its diffusion-free value. However, in general, $\Delta Q_{ex} \neq \Delta Q_D$, depending upon the nature of the transient experiment. Two cases are discussed below [24]. In the first case the bias is decreased to below the sustaining voltage such that the domain is quenched (Section 8.4). Figure 8.6.3a shows the excess domain voltage, V_D,

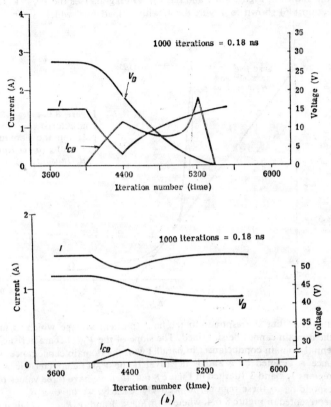

Figure 8.6.3. (a) Transient response to a bias step large enough to cause domain quenching (the initial value is 40 V, below threshold (70 V), and the final value is 23 V). The initial and final currents are the same. The voltage bias drops to the lower value in 0.1 ns. (b) Transient response to a small bias step (from 70 V to 65 V). The initial and final currents are almost the same. I_{CD} is the current charging the domain capacitance (Figure 8.6.1). (after Kak and Gunshor, [24]).

the external current, I (the transient voltage is so chosen that the final current is equal to the initial current), and the domain-charging (discharging) current I_{CD}. Figure 8.6.3b shows the second case the transient response to a small bias step. These data were obtained by computer simulation starting from the device basic equations. The applied bias has a finite fall time of 0.1 ns.

Figure 8.6.3 b indicates that the domain relaxes toward a new stable state after the applied voltage reaches its final value. A different process is illustrated by Figure 8.6.3a: V_D continues to decrease almost linearly whereas I_{CD} remains almost constant. Finally, a peak of the discharge current marks the irreversible process of domain dissolution (quenching). The charge transferred to the external circuit, ΔQ_{ex}, and the change in the domain charge, ΔQ_D, were calculated by using $I'(t)$ and $I_{CD}(t)$, respectively. It was shown that $\Delta Q_D \approx \Delta Q_{ex}$ up to about the middle of the current pulse $I_{CD}(t)$ in Figure 8.6.3a, and then ΔQ_{ex} saturates whereas ΔQ_D continues to increase to about twice ΔQ_{ex} at the end of the transient. The above behaviour can be understood by means of the equivalent circuit shown in Figure 8.6.1 and the $V_D - I$ diagram (see the $V_D = V_D(\mathscr{E}_N)$ dependence discussed in Section 8.6) shown in Figure 8.6.4, with a 'load line' which is tangent to the $V_D(I)$

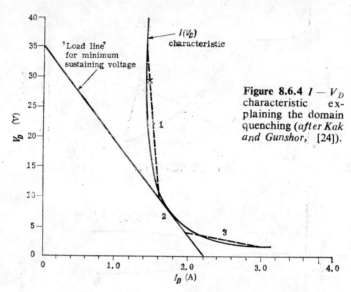

Figure 8.6.4 $I - V_D$ characteristic explaining the domain quenching (after Kak and Gunshor, [24]).

characteristic: this load line corresponds to the final transient voltage which is just at the limit beyond which the domain cannot be sustained. The slope of the $V_D(I)$ characteristic is related with a kind of differential domain conductance, in parallel with the domain capacitance, C_D. This non-linear conductance is denoted by g and is negative. If the $V_D(I)$ characteristic is approximated by three linear regions 1, 2 and 3 (dashed in Figure 8.6.4) then we have three values of g, namely g_1, g_2 and g_3, corresponding to these regions, respectively. Clearly, we have $|g_2 R_0| = 1$ for the particular situation represented in Figure 8.6.4, whereas $|g_1 R_0| < 1$ and $|g_3 R_0| > 1$. During the transient depicted in Figure 8.6.3a, V_D decreases and g becomes successively g_1, g_2 and g_3 (increases). In a simplified analysis the differential domain capacitance is assumed constant and the current through the geometrical capacitance C_0 (low-field device capacitance) is also neglected [24]. When the external voltage is constant, the domain capacitance discharges through an equivalent resistor formed by R_0 and $1/g$ in parallel. This equivalent resistor is positive when $|g|$ is small, namely $|gR_0| < 1$, and negative when $|g|$ is large, namely $|gR_0| > 1$. As far as the time constant is positive (region 1), the discharging current decreases; it remains constant in region 2 (zero-time-constant) and increases in region 3 where the time-constant becomes negative ($|g_3 R_0| > 1$). The domain excess-voltage decreases continuously and, finally, vanishes. When g changes continuously, only two distinct regions, separated by $|gR_0| = 1$ should be considered. Therefore the quenching process has two distinct phases, as indicated in Table 8.6.1.

The quenching of a mature dipole domain occurs in certain modes of operation (Sections 8.10 and 8.11). We also note that an equivalent circuit like that shown in Figure 8.6.1 can replace the device in a computer study of a microwave circuit. Such a lumped-circuit representation can be extended to multiple domains [25] or generalized to include domain formulation [25] and dissolution [26]. The computer time for simulation of device-circuit interaction is much reduced by using such a lumped-circuit approach.

Table 8.6.1

Phases of Domain Quenching

	First phase	Second phase
Defined by	$\|gR_0\| < 1$	$\|gR_0\| > 1$
Domain charge variation, ΔQ_D	$\Delta Q_D \approx \Delta Q_{ex}$	ΔQ_D sensibly higher than ΔQ_{ex}
Discharge current	Decreases	First increases (instability). Finally has to vanish.
Dominant discharge path	Initially the domain charge discharges essentially through R_0.	The discharge current circulates through the loop formed by C_D and the non-linear conductance g. Finally the domain and the associated loop disappears.
Dominant physical process	The rate of electron flow through the low-field region through the accumulation layer of the domain decreases, leading to a shrinkage of the domain as the discharge current flows in the external circuit	The collapse of the domain becomes uncontrollable by the external circuit (unreversible): the electrons from the accumulation layer are transferred directly to the depletion region [24].

8.7 High-field Domain Nucleation by a Doping Notch

W eshall qualitatively discuss domain nucleation by a doping notch. The formation of a mature domain when a few nucleation sites exist will then be discussed. The dynamics of high-field domains will be followed by using results of computer simulations.

Spatial fluctuations of doping concentration determine a non-uniform electric field and give rise to growing space-charge instabilities in a negative-mobility medium. Consider first an HFD nucleation at a doping notch somewhere in the semiconductor bulk. The notch width, Δx (one-dimensional model), satisfies $L_{DE} \ll \Delta x \ll L$, where L_{DE} is the extrinsic Debye length and L is the sample length. The local doping profile and electron distribution are shown in Figure 8.7.1a. When the bias field is below the threshold value, the electric field should be higher in the notch than anywhere in the bulk and thus provide the current continuity. If the applied bias is steeply raised above threshold, a high-field domain will be nucleated at the notch. The *rapid* variation is essential and this is explained below. In the final steady-state régime, continuity of the (drift) current requires a lower field (higher velocity) in the region with lower carrier concentration (Figure 8.7.1b). However, during the rapid transient the electric field is, momentarily, higher in the notch due to the displacement current. This displacement current is dominant in the first moment and therefore the field increase, $\Delta\mathscr{E}$, in a very short time interval, Δt, should be almost the same in the entire semiconductor (Figure 8.7.1b). If the voltage variation is slow, the displacement current will be negligible and will play no rôle in the transient.

A low-field domain can be nucleated in principle at a local increase of doping concentration. However the formation of a mature low-field domain is less probable or even impossible, as will be discussed later.

As the bias field rises above threshold, a number of high-field domains can be nucleated at different sites in the semiconductor sample. We have already shown (Section 8.4) that a single high-field domain can exist in the stable state. We shall now discuss the transient and see what happens between the nucleation of a few domains and the final state with a stable propagating domain.

Just after nucleation, the high-field domain retains the shape of the initial non-uniformity. As far as the space-charge perturbation is small (V_D small) and the velocity-field relationship can be assumed linear (constant negative mobility, μ_d), the domain excess voltage, V_D, increases exponentially with the time constant $\varepsilon/eN_D|\mu_d|$.

We shall derive a simple equation for the particular case of triangular domains (negligible diffusion). If x_1 and x_2 are the domain limits in equation (8.2.6), then the domain top will be situated at $x_D \simeq x_1$ and

$$V_D \simeq \int_{x_D}^{x_2} (\mathscr{E} - \mathscr{E}_N) dx \tag{8.7.1}$$

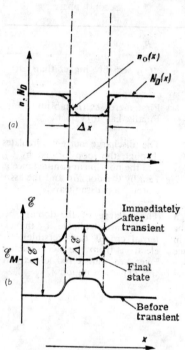

Figure 8.7.1. HFD nucleation by a doping notch: (a) local doping profile and steady-electron distribution; (b) electric field profile before and after the application of rapid increase in the bias voltage, the field profile which should be established after the transient is shown dashed. The notch width is $\Delta x \ll L$, whereas $\Delta x \gg L_{DE}$, the extrinsic Debye length.

(the voltage drop on the very thin accumulation layer was neglected). Then, because $n \simeq 0$ and $\partial\mathscr{E}/\partial x = -eN_D/\varepsilon$ in the depletion layer, equation (8.2.6) becomes

$$\frac{dV_{Dj}}{dt} \simeq \int_{\mathscr{E}_N}^{\mathscr{E}_{Dj}} [v_N - v(\mathscr{E})] d\mathscr{E}, \tag{8.7.2}$$

where $j = 1, 2, \ldots$ is the domain index. The field outside the domain is uniform and equal to $\mathscr{E}_N$ (Section 8.1), the 'outside' velocity, v_N, is also uniform. Let us consider, for example, the case of two domains nucleated simultaneously. When they detach from the initial nucleation site and grow enough to reach a quasi-triangular shape (low-conductivity samples, negligible diffusion), their dynamics is described by equation (8.7.2). As far as the integral in the right-hand part of equation (8.7.2) is positive, both domains continue to grow and absorb more voltage such that v_N decreases. The domain growth stops when the integral vanishes. This will happen first for the smaller domain (say domain 2, with $\mathscr{E}_{D2} < \mathscr{E}_{D1}$). Domain 2 will begin to decrease and finally will disappear, whereas domain 1 continues to grow, absorbing the voltage from the second one.

Let us assume that the difference between the two nucleation sites is small and the growing domains are nearly identical. When domain 2 begins to decrease, domain 1 still increases, but very *slowly* ($\mathrm{d}V_D/\mathrm{d}t$ in equation (8.7.2) is close to zero). Therefore, a long time will be necessary before the larger domain will absorb the voltage drop from the smaller one. During this time interval the domains continue to travel and one of them may reach the anode. Let us assume that this domain is the larger one. Then, when this domain disappears at the anode the bulk field $\mathscr{E}_N$ rises abruptly, $\mathrm{d}V_D/\mathrm{d}t$ in equation (8.7.2) changes its sign and the remaining domain begins to grow again. Because of this 'competition' between domains nucleated by centres of comparable size, the duration of one cycle of oscillation may be considerably longer than the domain transit-time.

Figure 8.7.2 obtained by computer simulation [25] illustrates the competition between two domains generated by two doping notches (20% of the nominal doping of 10^{15} cm^{-3}), one of 3 μm and other one of 5 μm (see Figure 8.7.2a) in a 100 μm long device. The device is initially in the ohmic state. Then the voltage bias is increased over threshold at a rate of 1 V ps^{-1} up to 42 V. Figure 8.7.2 b shows the nucleated domains. Figure 8.7.2c indicates that the larger domain increases at the expense of the smaller one. During this time the device current remains constant. Finally, with the extinction of the second domain, the device current seeks to a lower value (Figure 8.7.2d). It was shown [25] that in general the sum of the domain excess voltages remains approximately constant and therefore the circuit current is almost insensitive to domain extinctions taking place within the device. However, when one domain is absorbed at the anode, the device current is substantially modulated because the additional domains are unable to absorb quickly the voltage drop of the disappearing domain.

We shall present below part of the results of the computer simulation performed by Hasuo *et al.* [13], [14]. These authors identified five possible modes of high-field domains in an NDM sample with positive derivative of field-dependent diffusion coefficient above the threshold field, $\mathscr{E}_M$, namely:

$$D(\mathscr{E}) = D_\alpha(1 + \alpha \mathscr{E}/\mathscr{E}_M). \tag{8.7.3}$$

We recall the fact that the small-signal analysis of Ohmi and Hasuo [27] presented in Section 7.3 predicted four possible modes: a forward travelling domain, a stationary anode domain, a reverse travelling domain and a stationary cathode domain. The last two modes occur when $D'(\mathscr{E}) = \mathrm{d}D/\mathrm{d}\mathscr{E}$ is positive above $\mathscr{E}_M$ and therefore the diffusion-induced velocity (Section 7.3) is opposite to the drift velocity and can become greater than it. The above four modes can be obtained successively either by increasing D' or by increasing N_D (see Figure 7.3.1). This was effectively demonstrated by Hasuo *et al.* [13], [14] by a direct computer simulation. They considered a sample with a central doping notch and discovered the possibility of another mode, when the high-field domain is stationary inside the device, at the doping notch.

We discuss below the case when D is kept the same and the doping is varied to obtain these five modes [14]. The $v = v(\mathscr{E})$ curve used was that of GaAs. D_α

was taken equal to 200 cm²s⁻¹, whereas $\alpha = 1$ and $\mathscr{E}_M = 3.2$ kV cm⁻¹. The applied voltage was kept equal to 5 V after a rise time of 50 ps. Figure 8.7.3 indicates successive configurations of the field domain after the voltage was applied. The doping profile is also shown. Figure 8.7.4 shows the corresponding variation of the external current.

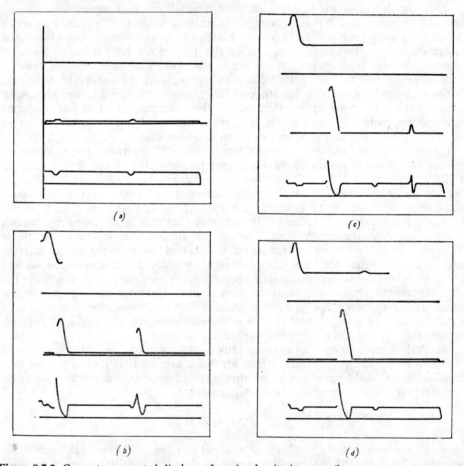

Figure 8.7.2. Computer generated displays of carrier density in space (lower curves), electric field in space (middle curves), and device current in time (upper curves on each figure $a-d$) for a 100 μm GaAs crystal with 5μm and 3μm notches in a nominal doping profile of 10¹⁵ cm⁻³.
(a) ohmic state, (b) two domains nucleated, (c) the second domain is nearly extinguished, (d) the disappearance of the second domain determines a slight reduction of the current (after Robrock, [25]).

Figures 8.7.3a and 8.7.4a are obtained for 10¹⁴ cm⁻³ nominal doping. A mature HFD does not form because the doping-length is too low. The steady-state profile established after about 100 ps is non-uniform. With a higher doping (4 × 10¹⁴ cm⁻³) an HFD forms at the doping notch and travels toward anode, where it disappears (Figure 8.7.3b). The process repeats cyclically (Figure 8.7.4b). Figure

8. *Transferred-electron Devices used as Oscillators*

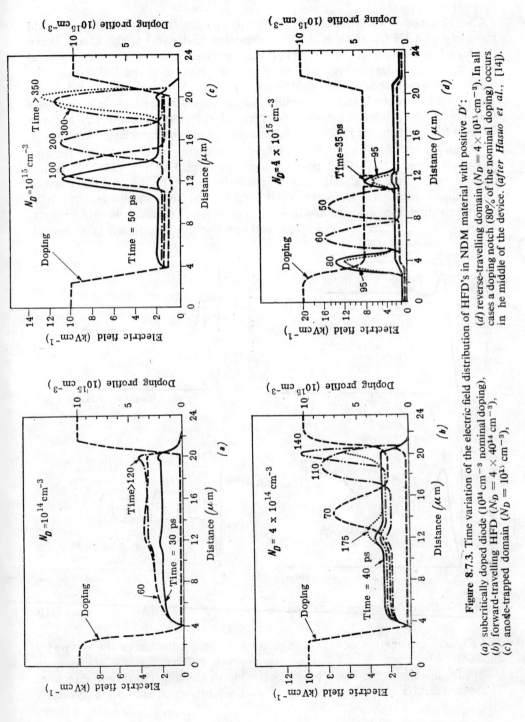

Figure 8.7.3. Time variation of the electric field distribution of HFD's in NDM material with positive D': (a) subcritically doped diode (10^{14} cm^{-3} nominal doping), (b) forward-travelling HFD ($N_D = 4 \times 40^{14} \text{ cm}^{-3}$), (c) anode-trapped domain ($N_D = 10^{15} \text{ cm}^{-3}$), (d) reverse-travelling domain ($N_D = 4 \times 10^{15} \text{ cm}^{-3}$). In all cases a doping notch (80% of the nominal doping) occurs in he middle of the device. (*after Hasuo et al.*, [14]).

8.7.3c, obtained at a still higher doping density (10^{15}cm^{-3}) shows the formation and travelling of a HFD, which is 'trapped' (remains stationary) at the anode. When the travelling domain adjusts its shape and becomes a stationary one, the total current increases slightly (Figure 8.7.4c) due to the increase of the field outside the domain. However, this 'outside' field is not high enough to nucleate a new domain. The possibility of anode-trapped domains was discussed before and we have shown that the device exhibits a negative incremental resistance (Section 7.2) and can be used for wideband amplification (Section 7.6) or bistable switching (Section 8.5).

When $N_D = 2 \times 10^{15}$ cm^{-3}, the high-field domain grows at the notch but remains stationary around this initial position (not shown in Figure 7.8.3). This occurs because the diffusion-induced velocity is almost equal to the average electron drift velocity (Section 7.3). When N_D is further increased, the domain travels in the reverse direction because the diffusion-induced velocity exceeds the drift velocity. At $N_D = 3.5 \times 10^{15}$ cm^{-3} a stationary cathode domain was found (not shown).

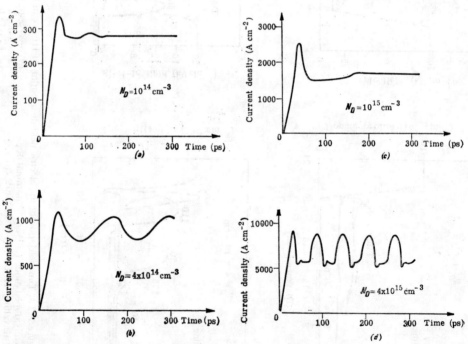

Figure 8.7.4. Current waveforms for the situations a—d in Figure 8.7.3. (*after Hasuo et al.*, [14]).

The physical explanation of this configuration is similar to that of the stationary anode domain, however, the current waveform is different because the domain is kept for a while at the notch before beginning to travel in the reverse direction. This phenomenon can also be seen on Figure 8.7.3d, which shows the cyclically reverse travelling-domain computed for $N_D = 4 \times 10^{15}\,\text{cm}^{-3}$. The external current (Figure 8.7.4d) peaks every time the domain reaches the cathode.

The data known at this time indicate a diffusion coefficient which decreases with the electric field above the threshold for NDM, in both GaAs and InP*. Therefore, only the two HFD modes illustrated in Figures 8.7.3b and c are physically realisable (and those two were indeed observed experimentally).

8.8 Effect of Electrode Boundary Conditions upon Behaviour of NDM Samples

The behaviour of real NDM devices cannot be understood without a knowledge of the nature of the contacts and an adequate description of their electrical properties. From a mathematical point of view, these contacts impose boundary conditions to the basic equations describing the semiconductor bulk. The cathode contact is by far the most important because it provides the charge carriers which sustain the charge transport.

We have discussed so far the behaviour of a domain-type perturbation in an NDM medium. The cyclic nucleation, propagation and disappearance of HFD's leading to Gunn oscillations was studied in Section 8.7 for a sample with ohmic contacts and doping non-uniformities. It was shown that the domain nucleation can take place at a doping notch. Such doping notches actually exist in practical devices. It was suggested to make internationally cathode notches in order to provide a reproducible behaviour of Gunn diodes or even to increase the efficiency of oscillations. We shall discuss this later. At this moment we assume that the semiconductor bulk is a perfect crystal (uniformly doped) and consider the rôle of the contacts. A few years after the Gunn effect was discovered [15] the behaviour of laboratory devices was extremely unreproducible and some times far from theoretical predictions [28], [29]. In many cases 'poor' contacts were blamed for the 'anomalous' behaviour of Gunn diodes. It is true that the electrical contacts to GaAs of InP samples are made in many different ways and the exact physical nature of these contacts is unknown. However, simplified electrical models were suggested [30]–[33] in order to explain the basic features of the 'anomalous' experimental behaviour.

Let us first discuss what we expect from an ideal crystal with ohmic contacts. If the doping-length product is slightly higher than its critical value (around $10^{12}\,\text{cm}^{-2}$ for GaAs), the space-charge non-uniformities take the form of *pure accumulation layers* nucleated at the cathode and propagating cyclically through the device [34]. The formation of an accumulation layer in the semiconductor, just near the cathode is somewhat natural, due to the charge injection provided by the ohmic contact (Figure 7.1.2). When the bias field increases above the threshold value for NDM, this primary accumulation layer grows. The electric field on the anode side of the layer is higher than that on the cathode side. As the field on the anode side increases

---

* Moreover, even the basic transport equations including diffusion are questionable, as shown in Chapter 1. However, the model discussed here and in Section 7.3 is still interesting because the domain velocity could be in principle vanished or reversed by diffusion-type forces and the possibility of the last three modes predicted by Hasuo *et al.* cannot be *a priori* ruled out for *any* negative-conductivity material.

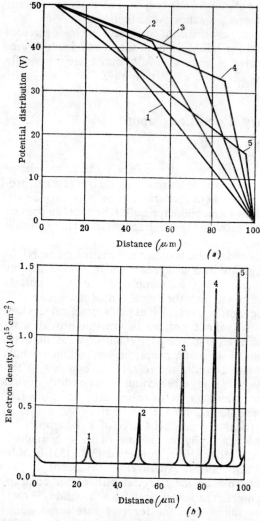

Figure 8.8.1. Dynamics of the pure accumulation mode in a 100 μm crystal, of 10^{14} cm^{-3} doping: (a) potential and (b) electron density distributions at intervals of 0.2 ns after voltage turn-on (constant voltage bias). (*after Kroemer*, [35]).

and the charge in the accumulation layer increases, the layer moves toward the anode whereas its upstream side falls into the positive mobility region and the layer 'detaches' from the cathode. The movement of such a pure accumulation layer is illustrated by Figure 8.8.1 [35]. Note the fact that the layer continues to grow as it travels toward the anode*. The 'outside' field and the total current change gradually. We recall the fact that Gunn oscillations are characterized by short current spikes separated by intervals when the current is almost constant. If the accumulation layer moves into the anode and disappears (which is not evident, a stationary anode accumulation layer could also occur†, see Section 8.7), then the cathode field rises above threshold and nucleates again an accumulation layer. Therefore current oscillations will be detected in a resistive circuit (constant bias voltage plus a small series resistance). The current waveform should be close to a sinusoid (for the reason shown above) and this improves the oscillation efficiency. Another feature of the pure accumulation mode is the fact that the oscillation frequency could be higher than the expected transit-time frequency, equation (8.4.2). This is because the accumulation layer moves faster than the electrons comprising it: the electron velocity inside the layer is higher than outside (due to the NDM effect) and

---

* It can be easily shown that a stable accumulation layer in transit is not compatible with a constant voltage bias (Section 8.9).

† The accumulation layer 'trapped' at the anode can be avoided by using a resonant circuit (the internal field is temporarily lowered by the external oscillation below threshold such that the accumulation layer collapses) [30], [36]. The device behaviour (called accumulation transit mode) is thus partly circuit-controlled (see below). We also note that a new accumulation layer can form before the preceding one disappears at the anode [34].

thus the electron gas ahead of the accumulation layer is compressed while at its end it is rarefied [34].

Kroemer [34] has shown that if the crystal is unhomogeneous, the doping fluctuations lead to the formation of 'secondary' space-charge layers, both of the accumulation and the depletion type, downstream to the 'primary' accumulation layer

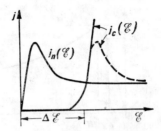

Figure 8.8.2. Control characteristic, $j_c(\mathscr{E})$, for a barrier-type cathode: the barrier is strongly reduced at intense electric fields (solid line). An embedded p-layer leads to essentially the same characteristic, except for the high field region (dashed line), where it coincides with the neutral characteristic, $j_n = eN_D v(\mathscr{E})$, but shifted to higher fields by $\Delta\mathscr{E}$. (after Kroemer, [30]).

formed at the cathode [35], [37]. If the $N_D L$ product is sufficiently high, the internal space-charge will reach a final configuration characterized by a stable high-field domain propagating toward the anode (only one pair accumulation layer — depletion layer will remain, see Sections 8.4 and 8.7). The $N_D L$ product was almost invariably high in early experiments because long devices were used to allow potential-probing experiments. For a long time the accumulation mode predicted by Kroemer [34] was not recognized as an experimental reality. However, more recent data and detailed calculations [36] support the existence of GaAs and InP devices operating in this mode.

In the rest of this section we shall discuss the effect of non-ohmic contacts, or in Kroemer's [30] formulation: imperfect cathode boundary conditions. We have already discussed this topic through several chapters of this book. We have indicated that an 'imperfect' (non-ohmic) cathode can be described by the so-called control characteristic, which relates the injected particle current to the cathode field, $j_c = = j_c(\mathscr{E}_c) = J$(Section 7.2). Only one type of control characteristic will be considered: that indicated in Figure 8.8.2. This characteristic almost coincides with the horizontal axis for $\mathscr{E} < \Delta\mathscr{E}$, and rises abruptly above this threshold, describing a barrier-type contact. This barrier type behaviour may occur due to a thin resistive layer containing a fixed space-charge (either trapped electrons, or a local excess of acceptors over the donor atoms)*. If the amount of fixed charge (Q per unit of electrode area) is sufficiently small, it would not seriously affect the low-field ohmicity of the crystal [30], but instead introduce a quasi-abrupt field discontinuity $\Delta\mathscr{E} = Q/\varepsilon$ across this thin layer. The control characteristic is the current density flowing through this charge sheet, as a function of the electric field on the anode side, $\mathscr{E}_c$. Let us consider the case of an embedded p-layer which is very thin compared to one mean free path [30]. When $\mathscr{E}_c < \Delta\mathscr{E}$, the field on the cathode side opposes to carrier flow; when $\mathscr{E}_c > \Delta\mathscr{E}$ the drift aids the diffusion current and becomes dominant, such that at higher $\mathscr{E}_c$ the control characteristic becomes identical with the neutral characteristic but shifted to higher fields by $\Delta\mathscr{E}$ (dashed line in Figure 8.8.2). When

---

* In the latter case the very thin p-type layer behaves like a 'punched-through base', particularly under applied bias conditions [30] (see also Section 5.1).

the contact barrier is reduced by the cathode field, the control characteristic rises continuously as shown by the solid line in Figure 8.8.2. However, apart from physical details (which are different from one case to another and, in fact, not precisely known), the only important thing for our discussion is the shape of $j_c(\mathscr{E}_c)$ up to the crossover point with the neutral characteristic, $j_n(\mathscr{E})$.

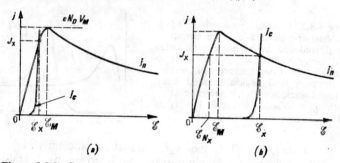

Figure 8.8.3. Control and neutral characteristics with a crossover point in the low-field positive-mobility region (a) and in the NDM region (b).

We consider two distinct cases: the first when the crossover point is situated in the low-field positive-mobility region of $j_n(\mathscr{E})$ (Figure 8.8.3 a), and the second, when the crossover point is situated in the negative mobility region. (Figure 8.8.3b). In the first case, the semiconductor is depleted at low current densities and becomes accumulated above $J_x = eN_D v(\mathscr{E}_x)$. If the doping length product is sufficiently large, the injected space charge is negligible and the electric field is almost uniform (except the cathode vicinity, where it is always close to $\mathscr{E}_x$, very steep control characteristic) and $J = eN_D v(\mathscr{E})$ where $\mathscr{E} \simeq V/L$. Therefore, space-charge instabilities are nucleated at the cathode for a current density approximately equal to the threshold (or peak) current $eN_D v_M$ and a bias voltage $V \simeq \mathscr{E}_M L$. The device behaviour is essentially that of a normal sample with ohmic contacts.

In the second case the crossover field $\mathscr{E}_x$ lies in the NDM region. The static analysis predicts that the crystal should be depleted for $J < J_x$ and accumulated for $J > J_x$. However, this static solution is not necessarily stable, because, at least in part, the bulk is in the negative-mobility region. The mathematical analysis does indeed show that the device becomes unstable when the current approaches the crossover current J_x. The internal space-charge and field distributions for $J < J_x$ depend not only upon the bias current, J, but also upon the crystal doping and length. Let us assume that the control characteristic is almost vertical and $\mathscr{E}_c \simeq$ const. For very low bias currents the Poisson equation yields the following value of the anode field

$$\mathscr{E}_a \simeq \mathscr{E}_c - eN_D L/\varepsilon. \tag{8.8.1}$$

If the doping-length product is low the field non-uniformity will be low ($\mathscr{E}_a \simeq \mathscr{E}_c$), the entire crystal will be in the negative-mobility region and the bias field will be above the theoretical 'threshold' bias $\mathscr{E}_M L$. When J increases the field becomes more uniform because the mobile charge neutralizes in part the ionized donor charge.

However, if N_DL is relatively high, the anode field will fall below the threshold field. When the current is allowed to flow, the space-charge non-uniformity takes the form of a depletion layer located near the cathode and the majority of the crystal remains neutral (large doping-length product). Figure 8.8.4 illustrates the field and space-charge distribution in both cases (high and low N_DL). Note the fact that the cathode depletion layer sustains a large field drop (the so-called cathode fall, first detected experimentally by potential probing in long samples [28]): the average field (which defines the bias field V/L) may be below the threshold value $\mathscr{E}_M$. We stress the fact that the threshold voltage for the onset of instabilities, V_{th}, does not coincide with $\mathscr{E}_M L$ because of the space-charge non-uniformities, and cannot be used to determine the threshold field for NDM.

We discuss below the steady-state characteristic and the instability in samples with the crossover field in the NDM region.

The voltage-current characteristic $J = J(V/L)$ can be represented in the $j - \mathscr{E}$ plane, Figure 8.8.3. Up to V_{th} this characteristic has a positive slope and is situated below the neutral characteristic. This can be qualitatively explained by the presence of the internal depletion region. When all other conditions are the same (equal N_D and $\mathscr{E}_c$), the longer diode sustains a higher voltage drop at the same current. Thus devices with larger N_DL product will exhibit a characteristic with a more pronounced tendency for saturation.

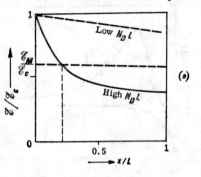
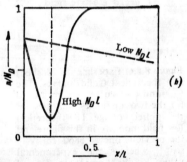

Figure 8.8.4. Field profile (*a*) and carrier distribution (*b*) for a certain cathode contact and bias current, when the sample has a large N_DL product (solid lines) and relatively low N_DL (dashed lines).

We consider now the effect of the cathode boundary condition for a certain sample (N_D and L specified). The characteristics labelled A, B_1, B_2 and C in Figure 8.8.5 and the corresponding non-stationary behaviour were obtained by computer simulation using the simplified boundary condition $\mathscr{E}_c = $ const. [31]. The characteristics B_1 and B_2 exhibit instabilities when the bias current reaches a certain critical value. This critical value (peak of curves B_1 and B_2 in Figure 8.8.5) is very close to the crossover current J_x (Figure 8.8.3b). We recall that according to the simplified theory neglecting diffusion (Section 7.2) the crystal becomes accumulated when J exceeds the crossover current. Note that threshold occurs when the bias field is below threshold for NDM and that the majority of the crystal bulk is in the positive-mobility region.

The characteristic C in Figure 8.8.5 exhibits a very pronounced saturation and no instability threshold. This occurs due to the extremely large cathode field (24 kV cm^{-1}) which introduces a very large cathode drop [31]. The carrier velocity is saturated and the negative differential mobility is very low. This explains intuitively the sample behaviour. The high-frequency resistance is negative and the device can maintain oscillations in a resonant circuit

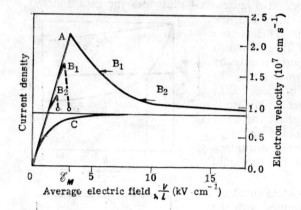

Figure 8.8.5. Neutral characteristic, $j_n(\mathscr{E}) = eN_D v(\mathscr{E})$ and numerically calculated $J - \mathscr{E}$ characteristics (where $\mathscr{E} = V/L$ is the bias field) for various cathode boundary fields: $\mathscr{E}_c = 0$ for curve A, $\mathscr{E}_c = 24$ kV cm^{-1} for curve C and indicated by arrows for curves B_1 and B_2. The sample is 100 μm long and has $N_D = 10^{15}$ cm^{-3}. The right- and left-hand ordinates are related by $J = eN_D v$. (after Shaw et al., [31]).

Figure 8.8.6. Experimental (dashed) and theoretical (solid) curves. The $(+)$ and $(-)$ curves are measured for the two opposite polarities of the applied voltage. The theoretical low field mobility in this figure and the previous one is 6860 cm^2 V^{-1} sec^{-1}, whereas the experimental value is 4000 cm^2 V^{-1} s^{-1} and hence the low field slopes differ. (after Shaw et al., [31]).

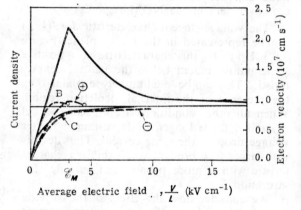

The above model is consistent with experimental data on various samples with different contact procedures (steady-state characteristics, measurements in a variety of circuit, potential-probing) [31]. Figure 8.8.6 compares experimental $I-V$ characteristics with theoretical curves ($\mathscr{E}_c$, unknown, is chosen for the best fit with experiment). The large asymmetry with respect to forward and reverse polarities is explained by the asymmetry (which is not necessarily large) of the fields at the contacts. Gunn-type oscillations for positive bias and broad-band negative resistance for negative bias were obtained.

A small-signal analysis of the device stability was performed by Kroemer [30] and by Grubin et al. [32], [33]. In both cases diffusion was neglected. Kroemer assumed a uniform electric field [30], whereas Grubin et al. [32], [33] introduced the field non-uniformity but assumed a vertical control characteristic (infinite value of the cathode differential conductivity, $\sigma_1(0) \to \infty$). These analyses confirm the high-frequency negative resistance of d.c. stable sample with positive incremental resistance (occurring for low $N_D L$ and high $\mathscr{E}_c$, saturated type $I-V$ characteristics), as well as short-circuit instability (Section 7.4) when the bias current approaches the crossover current in devices with high $N_D L$ and low $\mathscr{E}_c$ (but still in the NDM region).

The instabilities occurring in the latter type of devices are related to their ability to develop the space-charge perturbations (short negative differential relaxation time compared with the carrier transit-time). The nucleation of these space-charge perturbations can be easily explained by the static accumulation layer injected at the cathode when J exceeds the crossover current J_x. A high-field domain forms, detaches from the cathode and travels toward the anode. The domain formation reduces the 'outside' field $\mathscr{E}_N$ and thus reduces the current below J_x. Because of the specific cathode boundary condition, when J falls below J_x a new stationary depletion layer forms at the cathode during the domain propagation. We assume that the highfield domain reaches a stable configuration. Therefore

$$V = V_D + \mathscr{E}_N L + V_{CD}, \qquad (8.8.2)$$

where $V_D = V_D(J)$ is the domain excess voltage, $\mathscr{E}_N L = \mathscr{E}_N(J) L$ is the 'ohmic' voltage drop corresponding to the current J flowing in the external circuit (unit-area device), and V_{CD} is the excess voltage drop on the cathode depletion layer. Equation (8.8.2) is illustrated by Figure 8.8.7 which shows internal potential distributions at successive time moments during the propagation of the stable domain ($J = \text{const.}$).

We have shown in Section 8.4 that the lower the device current during domain propagation, J, the higher the domain excess voltage, V_D. Because J must be below J_x due to the boundary condition used, V_D should be above a certain minimum value necessary to sustain both the stable domain and the stationary cathode depletion layer. The applied voltage which nucleates the instability can be insufficient to sustain the propagation of a stable domain.

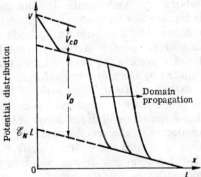

Figure 8.8.7. Mature domain propagation illustrated by internal potential distributions. A stationary depletion layer occurs ahead of the non-ohmic cathode and requires an additional potential drop, V_{CD}. (after Kroemer, [30]).

If the bias voltage cannot sustain a mature domain, the domain once nucleated will increase as it travels, but only up to a certain point. Then the domain will decrease ($dV_D/dt < 0$) and disappear. The process repeats cyclically* and determines oscillations with a frequency higher than the transit-time frequency. Moreover, this frequency will depend upon the circuit (bias and series resistance) which controls the travel path before the domain is quenched. Figure 8.8.8 illustrates the above consi-

---

* The cyclic occurrence of transit-time oscillations needs a special discussion when the domain disappears at the anode and not in the bulk [33].

derations with results of computer simulation (including field-dependent diffusion) [33]. Figure 8.8.8a indicates incomplete transit of the domain. Complete domain formation and transit followed by domain dissolution at the anode and a new

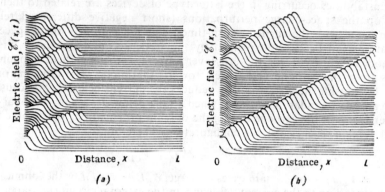

Figure 8.8.8. Space and time dependence of the electric field indicating: (a) incomplete transit and quenching of HFD's ($V = 1.5\,\mathscr{E}_M L$, $f_{osc} = 3.5\,\text{GHz}$), (b) complete transit of HFD's ($V = 1.8\,\mathscr{E}_M L$, $f_{osc} = 0.9\,\text{GHz}$). The cathode field is fixed at $4\mathscr{E}_M$. (after Grubin et al., reference [33]).

domain formation is obtained for a higher bias voltage, Figure 8.8.8b. Note in both cases the existence of a cathode fall. The propagation of a mature domain (Figure 8.8.8b) is accompanied by the formation of a stable stationary depletion layer at the cathode.

We summarize the two types of 'anomalous' behaviour discussed above and met in laboratory experiments. In one case Gunn-type oscillations are observed but with frequencies substantially higher than the transit-time frequency and circuit-dependent. The oscillation amplitude is also lower than expected because the oscillations break over at a bias current lower than the usual threshold current $eN_D v_M A$ and therefore the peak value of the current is reduced, whereas the 'valley' value is limited by the minimum value corresponding to saturated domains. We also note the existence of threshold voltage for oscillations which may be lower than $\mathscr{E}_M L$: the instabilities occur when the crystal bulk is biased in the positive-mobility range. Ocassionally, oscillations occur at large voltages compared to $\mathscr{E}_M L$. A distinct behaviour occurs in the second case: the $I-V$ curve exhibits a saturation, without the negative resistance and hysteresis effects characteristic of Gunn-type instabilities. Self-sustained oscillations do not occur in a resistive circuit but the device exhibits a high-frequency negative resistance and can sustain oscillations in a resonant circuit.

8.9 Stable Space-charge Layers Propagating in an NDM Semiconductor

We have discussed so far almost exclusively the behaviour of high-field domains (HFD). In Section 8.7 we mentioned the possibility of nucleation of a low-field domain (LFD) by a local increase in crystal doping. Other stable space-charge configuration are principially possible in an NDM medium and we discuss them *over leaf*.

An LFD can develop starting from a local decrease of electric field (Figure 8.9.1a) below the uniform value in the rest of the crystal. If this space-charge perturbation grows, the outside field increases (the voltage bias applied on a crystal of finite length is assumed constant). A stable configuration in transit will be reached only if the bulk is situated in a positive-mobility region.

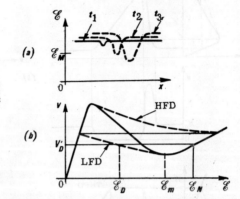

Figure 8.9.1. Low-field domain: (a) LFD formation, (b) dynamic characteristic for an LFD.

A stable LFD can exist only for an N-shaped $v - \mathscr{E}$ curve (Figure 8.9.1b). A theory of LFD's can be developed which is dual to that of HFD's. The minimum field inside the domain, $\mathscr{E}'_D$, is related to the domain velocity, v'_D, by the dynamic LFD characteristic shown in Figure 8.9.1b (the diffusion coefficient is assumed field-independent, and v'_D is equal to the 'outside' velocity v'_N, which corresponds to the 'outside' field $\mathscr{E}'_N$ on the high-field positive-mobility branch of the $v(\mathscr{E})$ characteristic). The external current $J = eN_D v'_N$ is lower than that corresponding to the propagation of an LFD. The external waveform for a cycling propagation of the LFD should exhibit negative spikes indicating the successive arrival of domain at the anode. As the domain disappears, the bulk field decreases below the 'valley' value $\mathscr{E}_m$ (Figure 8.9.1b) thus launching a new LFD.

The existence of stable LFD's will be impossible if the carrier velocity saturates at high field intensities, as in GaAs and InP at room temperature. However, even if a high-field raising branch of the $v(\mathscr{E})$ curve were to exist, the formation of LFD's would be difficult to find experimentally. This can be understood as follows. The occurrence of a bulk field perturbation like that indicated in Figure 8.9.1a has the same probability as the occurrence of a local increase in field (Figure 8.2.1). Perturbations of both kinds are determined by doping fluctuations. However, their evolution is different. When the applied bias field V/L rises above the threshold field $\mathscr{E}_M$, a high-field perturbation tends to increase whereas a low-field perturbation tends to disappear. Assume that V/L is only slightly above $\mathscr{E}_M$. When the LFD develops, the 'outside' field increases and drives the bulk further in the NDM region, whereas the internal field decreases and tends to introduce the perturbation in the positive-mobility region and therefore to reduce and dissolve it. In other words the disappearance of a small LFD can be described as follows. The field increasing above $\mathscr{E}_M$ reduces the electron velocity. Therefore the electrons tend to accumulate at the trailing edge of a travelling non-uniformity. This favours the formation and development of an HFD, whereas an LFD is inhibited [8]. An LFD can be formed by decreasing the field from very high values. Copeland [8] suggests that an increasing applied bias will first form an HFD domain which first saturates (Section 8.4) and then increases its width until it occupies the entire device and finally disappears determining a uniform electric field which biases the device in the high-field positive-mobility region of the $v(\mathscr{E})$ curve.

Figure 8.9.2 shows the carrier distribution and the field profile (related to the $v(\mathscr{E})$ characteristic) in various stable space-charge layers which may propagate in an NDM semiconductor. One indicated [9], [11] the possible existence of the 'trapezoidal' high-field domain (THFD) or 'flat-topped' domain, of Figure 8.9.2c and the trapezoidal low-field domain (TLFD) of Figure 8.9.2d. Such a trapezoidal domain would comprise a neutral layer between the separated accumulation and depletion layers. The current continuity requires that in a stable THFD the maximum field will determine a carrier velocity in the high-field positive-mobility branch which is equal to the domain

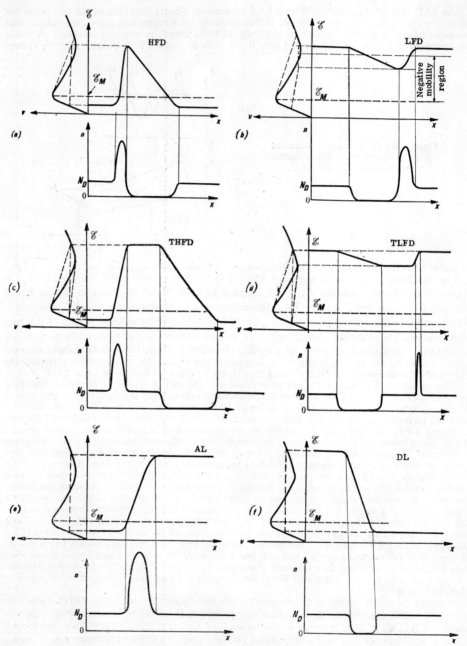

Figure 8.9.2. Electron distribution and field profile in stable space-charge layers propagating in an NDM semiconductor: (a) high-field domain (HFD), (b) low-field domain (LFD), (c) trapezoidal high-field domain (THFD), (d) trapezoidal low-field domain (TLFD), (e) accumulation layer, (f) depletion layer.

velocity ($D=$ const) and to the outside velocity v_N. In general, however, $v(\mathscr{E}_D)<v_D$ (Figure 8.3.2), except the case of saturated domains, when $v(\mathscr{E}_D)=v_D=v_{N\min}$ (Figure 8.9.2 c). As the voltage increases, the HFD has the usual (almost triangular) form up to the point where $\mathscr{E}_D$ reaches $\mathscr{E}_{D\max}$ ($v_{N\min}=v(\mathscr{E}_{D\max})$). Any further increase of the applied voltage will be absorbed by the 'flat-top' of the domain. The dynamics of THFD's was discussed by Heinle [11]. The existence of stable trapezoidal domains is again theoretically impossible if $v(\mathscr{E})$ saturates at large field intensities. This is because the crossover point between the 'dynamic' characteristic, $v_D=v_D(\mathscr{E}_D)$ and the static one, $v(\mathscr{E})$ is displaced toward infinity and always $v_D(\mathscr{E}_D) > v(\mathscr{E}_D)$.

All the four space-charge layers mentioned above, Figure 8.9.2a-d, are dipole domains which do not contain a net electric charge. Pure accumulation or depletion layers can also exist [8], [12]. In this case the uniform field $\mathscr{E}_{N1}$ on the cathode side is different from the field $\mathscr{E}_{N2}$ on the anode side (Figure 8.9.2e and f). We must have

$$J = eN_D v(\mathscr{E}_{N1}) = eN_D v(\mathscr{E}_{N2}), \quad v(\mathscr{E}_{N1}) = v(\mathscr{E}_{N2}) \qquad (8.9.1)$$

for a stable layer propagation. The existence of stable accumulation or depletion layers propagating* through the crystal is possible only if the velocity increases at high fields. On the other hand, the travel of stable accumulation or depletion layers cannot be sustained by a constant voltage bias. The voltage drop on the neutral regions on both sides of the layer

$$V_N = x_L \mathscr{E}_{N1} + (L - x_L)\mathscr{E}_{N2}, \qquad (8.9.2)$$

where $x_L = x_L(t)$ is the position of the layer (the layer width was neglected). Clearly $V_N = V_N(t)$ and the voltage drop on the layer, $V - V_N$, cannot be constant when $V =$ constant.

We have seen (Section 8.2) that a space-charge perturbation nucleated in the bulk is always a dipole. A pure accumulation layer can be launched by an ohmic-type cathode as discussed by Kroemer [34] (see also Section 8.8). Such layers grow as they travel toward the anode (the doping-length product is moderate) without reaching a stable configuration.

8.10 Gunn Diodes in Resonant Circuits. Domain Modes

We have shown that a Gunn diode placed in a series resistive circuit and biased above a certain threshold voltage will generate current oscillations. The high-frequency power obtained across the small series resistance is always low due to the small voltage swing. A practical Gunn oscillator is obtained by placing the diode in a resonant circuit: thus the power output and frequency stability is improved. However, a large voltage swing may occur across the diode during one oscillation cycle and can control the formation and extinction of domain. There are a few modes of oscillation called below resonant domain modes. We discuss them briefly. A detailed evaluation of their properties can be found in references [1], [38]—[41].

The diode parameters and diode model necessary for our discussion are: the diode doping (N_D), the diode length (L) and the diode $I-V$ characteristic presented in Section 8.5.

The essential feature distinguishing the resonant domain modes is the relation between the transit time of a mature domain, T_L, and the natural period of oscillation of the resonant circuit, T. The ratio between them is

$$\frac{T_L}{T} = \frac{Lf}{v_D} \approx \frac{Lf}{10^7}, \quad (L \text{ in cm and } f \text{ in s}^{-1}) \qquad (8.10.1)$$

---

* The effect of field-dependent diffusion upon the layer velocity was numerically computed by Copeland [8] and also discussed by Lampert [12].

if the domain velocity is approximated by the saturation velocity in GaAs (field-independent diffusion coefficient, saturated domains).

We have tacitly assumed that the domain formation and disappearance takes no time. The domain nucleation usually takes place at the cathode and therefore the full diode length, L, is available for domain transit, as assumed above. The domain is extinguished in two ways: either by collapsing at the anode or by quenching during its transit. In resonant domain modes this quenching occurs when the voltage across the diode falls below the value necessary to sustain the dipole domain. Anyway, the time scale for both domain formation and disappearance is given by the dielectric relaxation time, which is negative (domain growth) or positive (domain dissolution). An almost instantaneous domain growth requires a very short $\tau_d = \varepsilon/eN_D|dv/d\mathscr{E}|$, where $\mu_d = |dv/d\mathscr{E}|$ is the value of the negative differential mobility. It can be shown (Problem 8.4) that the mobile excess charge contained in the accumulation layer of a triangular domain (diffusion neglected) increases exponentially with the time constant $2\tau_d$ as far as the entire domain is in the negative-mobility region. The total time of growth depends, of course, upon the initial and final state of the domain. However, in all cases the growth time decreases when doping increases. For simplicity, we assume that the domain has enough time to form during one transit time *or* during one signals period if the doping concentration satisfies either

$$N_D L \gtrsim 10^{12} \text{ cm}^{-2} \text{ or } N_D/f \gtrsim 5 \times 10^4 \text{s cm}^{-3} \qquad (8.10.2)$$

(N_D in cm^{-3}, L in cm, f in s^{-1}, usual data for GaAs) [1]. For resonant domain modes both these criteria are required.

If the doping is somewhat higher than indicated by the above criteria the domain grows and collapses in a negligible time. Therefore a domain will be formed as soon as the diode voltage V increases above a certain threshold value, V_{th}, and will collapse as the voltage falls below the minimum voltage which can sustain the domain, V_m (see Sections 8.4 and 8.6). The dependence between the external current and the applied voltage can be established through the diode $I-V$ characteristic shown in Figure 8.10.1a (see Figure 8.5.2). This characteristic has two branches: the subthreshold branch and the stable domain branch. When the applied voltage increases and decreases back rapidly during one transit-time, the 'operating point' moves as shown by the arrows on Figure 8.10.1 a: the characteristic exhibits a hysteresis. We shall use the linearized characteristic of Figure 8.10.1b: in this case the ratio V_{th}/V_m is equal to the peak-current to the valley-current ratio. This ratio is directly related to the $v(\mathscr{E})$ dependence (Figure 8.5.1) and is a figure of merit of the device (the oscillator efficiency increases when the peak-to-valley current ratio increases).

We have so far ignored the capacitive currents through the device. Besides the low-field (geometrical) capacitance C_0, which can be easily included into the external circuit, the domain capacitance, which is voltage-dependent, should also be taken into account (Section 8.6). For simplicity we shall assume that all capacitive effects are concentrated in an averaged parallel capacitance, $\bar{C}$. Therefore the diode is replaced by a non-linear resistor having the hysteresis-type characteristics of Figure 8.10.1 b and the capacitance $\bar{C}$ which adds to the stray and circuit capacitances (Figure 8.10.2). The voltage appearing across the device is purely sinusoidal at the fundamental frequency of the resonant circuit.

8. Transferred-electron Devices used as Oscillators

We shall consider now three idealized modes of oscillation, referring to the circuit of Figure 8.10.2 and the characteristic of Figure 8.10.1 b.

(a) *Resonant transit-time mode.* The domain transit-time (T_L) is precisely equal to the period (T) of the sinusoidal voltage superimposed on a steady-state bias. The current waveform, Figure 8.10.3, is the same as when the device is placed in a resistive circuit. The current pulse occurs when the domain is extinguished at the

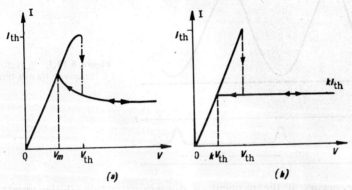

Figure 8.10.1. Exact (*a*) and approximate (*b*) $I - V$ characteristic of a Gunn diode. V_{th} is the threshold voltage for Gunn oscillations and V_m is the minimum external voltage which can sustain a mature dipole in transit.

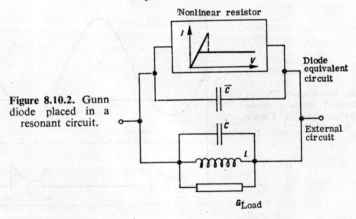

Figure 8.10.2. Gunn diode placed in a resonant circuit.

anode. The circuit has no direct effect upon pulse nucleation and extinction. Note, however, that the fundamental alternating component of the current is in antiphase with the voltage: this corresponds to a negative resistance effect which compensates losses in the circuit. In other words, the current pulses excite the voltage oscillations.

It is practically impossible to achieve a perfect syncronism by equalizing the natural transit-time frequency and the frequency of the resonant circuit. In practice the syncronism will be provided by the control of domain nucleation and domain extinction through the voltage signal, as discussed below.

(b) *The delayed domain mode*. In this mode the transit-time, T_L, is less than the signal period, T. The idealized waveforms are shown in Figure 8.10.4. The current pulse occurs during the interval the device is switched in its ohmic state (pre-threshold branch in Figure 8.10.1a, idealized linear in Figure 8.10.1b). The switching into

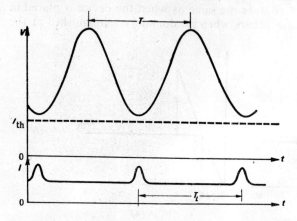

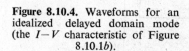

Figure 8.10.3. Waveforms for the resonant transit-time mode.

ohmic state occurs at the moment of domain extinction at the anode (assumed instantaneous, $t = T_L$ in Figure 8.10.4). At the moment $t = T$, when the voltage increases above threshold (V_{th}), the device is switched back into the stable-domain

Figure 8.10.4. Waveforms for an idealized delayed domain mode (the $I-V$ characteristic of Figure 8.10.1b).

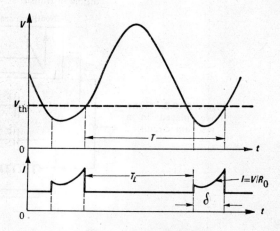

state: a domain is launched at the cathode and disappears at the anode at $t = T_L + T$, when a new oscillation cycle begins. Note that the domain is delayed from forming for a time $\delta = T - T_L$, $T_L < T$ in equation (8.10.1) and therefore $Lf < 10^7$ cm s^{-1}.

The current waveform can be optimized for output power by providing domain collapsing just as $V(t)$ falls below V_{th}. The efficiency is adversely affected by the existence of the saturation current (during domain propagation) which is k times the peak current: better efficiency are computed for lower k values [1].

The operating frequency is determined by the resonant circuit (including the device capacitance) and can be changed for a given transit-time frequency and a proper bias in an octave (ratio 1:2). A realistic analysis should take into account the finite time necessary for domain formation and disappearance: this time intervals are comparable to the transit time for X-band devices. This effect broadens the current pulse and increases the negative resistance [40]. An additional effect which has to be incorporated in calculations is the modulation of domain velocity due to the large-signal high frequency voltage [40].

(c) *Quenched-domain mode.* In this mode the voltage excursion extends below the minimum voltage necessary to sustain the domains, V_m. The domain is launched when $V(t)$ rises above V_{th} and quenched when $V(t)$ falls below V_m, as shown in Figure 8.10.5. An optimized current waveform is obtained when the minimum voltage is just equal to V_m [1]. The efficiencies obtained are lower than for the previous mode. Kroemer [35] pointed out that the anode region which is not travelled by the domain acts as a parasitic series resistance, except during that short portion of each cycle when a new domain is nucleated.

The oscillation period can be slightly longer than the transit-time, but as a rule is lower than it ($fL > 10^7$ cm s^{-1} from equation (8.10.1)); the operating frequency is circuit-controlled and can be several times the transit-time frequency.

We shall reproduce below several results of computer calculations for a Gunn diode operating in the quenched-domain mode. Khandelwal and Curtice [38] used an equivalent circuit similar to that presented in Section 8.6. During domain growth

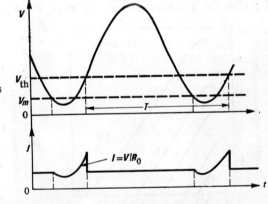

Figure 8.10.5. Idealized waveforms for quenched-domain mode.

and quenching the diode behaviour is modelled by different networks*. The current waveform is calculated by imposing a sinusoidal voltage on a steady-state bias (the harmonics are shorted by the resonant circuit). Figure 8.10.6 shows a typical com-

---

* The domain quenching network describes the domain disappearance after the external voltage drops *below* the domain-sustaining voltage, V_m: the variation is exponential with a time constant comparable with the low-field dielectric relaxation time. The domain-growth model is based upon simple equations discussed before. It was assumed that the domain excess voltage increases exponentially with a time constant $\tau_d = \varepsilon/eN_D\mu_d$, $\mu_d = |dv/d\mathscr{E}|$ until the current reaches a value equal to the stable domain device current. The corresponding time interval is the domain formation time which depends not only upon the particular device considered but also upon the time-dependent voltage $V(t)$.

puted waveform, indicating the domain travelling (region *A*), domain quenching (*B*), sub-threshold operation (*C*) and domain formation process (*D*). Figure 8.10.7 shows the computed admittance (related to the fundamental component of the current) for $f = 10$ GHz, as a function of the normalized alternating voltage, for

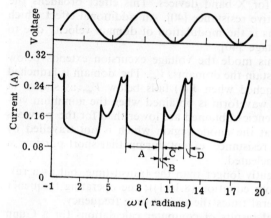

Figure 8.10.6. A typical current waveform of a Gunn diode operating in the quenched-domain mode, indicating domain transit (A), domain quenching (B), sub-threshold operation (C) and domain formation (D). The device data are: $N_D = 10^{15}$ cm^{-3}, $L \approx 16$ μm, $\mu_n = 5500$ cm^{2}V^{-1}s^{-1}, low field conductance $G_0 = 0.07143$ mho, low-field susceptance $B_0 = 0.0057$ mho, $V_{th} = 5.2$ V, current at threshold before the current drop ≈ 250 mA, $V_m = 4.47$ V, valley-to-peak velocity ratio $= 0.385$ (typical parameters for X-band $n^{++}-n-n^+$ epitaxial mesa structures). (*after Khandelwal and Curtice,* [38]).

different bias voltages (see the data shown in Figure 8.10.6). The conductance plot of Figure 8.10.7 a shows that at large signal amplitudes, V_1, the conductance becomes positive: this occurs because the instantaneous voltage $V(t)$ is in the positive-mobility region for a longer period. At small V_1, $V(t)$ does not go below V_m and the

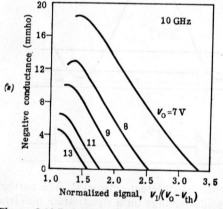
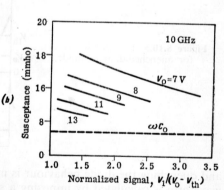

Figure 8.10.7 Calculated device conductance (*a*) and susceptance (*b*), versus normalized signal (V_1 is the amplitude of the sinusoidal voltage, V_0 is the steady-state component). The device parameters are indicated in Figure 8.10.6. (*after Khandelwal and Curtice,* [38]).

domain quenching does not occur. When V_0 is too large (above 15 V in this case) the quenched-domain mode does not occur because the domain becomes as wide as the device when the voltage is at maximum. Khandelwal and Curtice [38] conclude that the operation of GaAs X-band diodes is limited to bias voltages up to three times the threshold voltages.

Figure 10.8.7 b indicates that the device susceptance is two to four times the low-field susceptance (ωC_0): this is due to the domain capacitance. The device susceptance decreases with increasing V_0 and V_1 because the (average) domain excess voltage increases, the (average) domain width increases and the corresponding capacitance decreases (Section 8.6) [38].

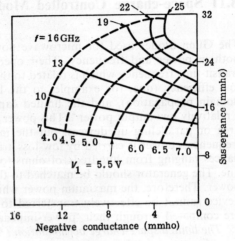

Figure 8.10.8. Calculated admittance for different frequencies and signal amplitudes at a bias voltage of 8 V. (after Khandelwal and Curtice, [38]).

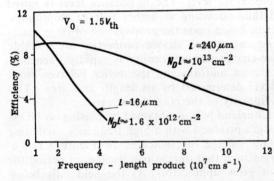

Figure 8.10.9. Efficiency as a function of frequency-length product for two devices of different length operating in the quenched-domain mode at $V_0 = 1.5\,V_{th}$. (after Khandelwal and Curtice, [38]).

Figure 8.10.8 shows the admittance plot calculated for various frequencies (f) and signal amplitudes (V_1) at a bias $V_0 = 8$ V. The device negative conductance is greatly reduced at high frequencies. A detailed examination of current waveforms [38] reveals that this occurs because the domain formation requires a large fraction of the period and because the capacitive current charging the domain increases considerably. Figure 8.10.9 shows the decrease of efficiency with increasing frequency-length product. According to equation (8.10.1), the maximum frequency of oscillation for the X-band device in Figure 8.10.9 ($L = 16\,\mu$m) is around four times the transit-time frequency (minimum 2% efficiency). The maximum frequency is, however, about twelve times its transit-time frequency for the longer device in

Figure 8.10.9 because the transit-time frequency is much longer and the domain formation and quenching time constants are comparatively short. The above model gives results consistent with experimental data.

8.11 Space-charge Controlled Modes

The Gunn diodes used for microwave power generation have limited performances both in power and frequency. Their operation frequency is comparable with the transit-time frequency which is related to the device active length. Both the relatively low efficiency (due, for example, to the large 'saturation' current flowing during domain propagation), and the limited capabilities of power dissipation determine a relatively low output power*. This power can be increased by increasing the device area or paralleling up devices, but the impedance level decreases. At microwave frequencies power is carried by low-loss transmission lines with characteristic impedances ranging from hundreds of ohms for waveguide to tens of ohms for coaxial line. The generator should be matched to the load in order to deliver the maximum power. Therefore, the maximum power which can be generated by a low-impedance device cannot be effectively transferred to the load. Hence the microwave devices are compared through their 'power-impedance product' [1].

The limited space-charge accumulation (LSA) mode of TED's removes the transit-time limitation and increases the impedance level. The impedance level is raised by increasing the device length and thus operating at higher voltage amplitudes. However, the operation can no longer be based upon the propagation of large space-charge non-uniformities. Instead, the circuit and device properties are suitably chosen so that any appreciable space-charge non-uniformity is rapidly quenched, so that the internal electric field is almost uniform and the device behaves much like a resistor with an impedance level determined by its length and area. This resistor is non-linear due to the non-linearity of the $v(\mathscr{E})$ characteristic.

The LSA mode first presented by Copeland [42], is obtained by biasing an NDM mobility (with supercritical doping-length product) with a high-frequency oscillating signal superimposed on a steady-state voltage, as indicated in Figure 8.11.1 a. The signal frequency is properly chosen such that during one oscillation period the space-charge non-uniformities cannot grow appreciably. As the field falls below the threshold field, the small space-charge accumulated during one period disappears by dielectric relaxation mechanism. During most of the period the crystal exhibits a negative differential mobility (Figure 8.11.1 b) and the device conductance is negative. Thus the device generates microwave power. The oscillator circuit consists in an NDM device (with a proper N_D and L) in series with a steady-state bias and a parallel resonant circuit tuned at a proper frequency.

We briefly discuss the LSA mode by assuming a uniform electric field

$$\mathscr{E}(x, t) = \mathscr{E}_b(t) = \mathscr{E}_{b0} + \mathscr{E}_{b1} \sin 2\pi ft. \tag{8.11.1}$$

---

* The output power decreases as the transit-time frequency increases because the shorter the device the smaller the voltage swing and thus the lower the a.c. power.

An electron moving at a velocity v in an electric field $\mathscr{E}$ absorbs the instantaneous power $e\mathscr{E}v$. The average power absorbed from the d.c. source is [42]

$$P_{dc} = \frac{e\mathscr{E}_{b0}}{T}\int_0^T v\,dt, \quad T = \frac{1}{f} \qquad (8.11.2)$$

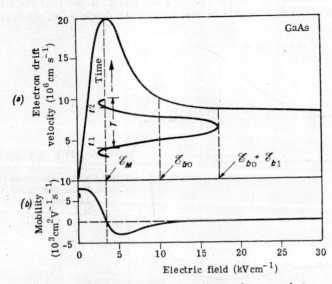

Figure 8.11.1. Field-dependence of average electron velocity (a) and differential mobility (b) for GaAs. A representative plot of bias field *versus* time is superimposed to illustrate the principle of the LSA mode. (*after Copeland,* [42]).

and the radio frequency power absorbed by the same electron is

$$P_{ac} = \frac{e\mathscr{E}_{b1}}{T}\int_0^T v\sin(2\pi ft)\,dt, \qquad (8.11.3)$$

where $v = v(\mathscr{E}) = v(t)$ according to Figure 8.11.1a. If $\mathscr{E}_{b0}$ and $\mathscr{E}_{b1}$ are properly chosen it is possible to obtain $P_{ac} < 0$, i.e. the electron delivers energy to the radio frequency field (or the d.c. energy is converted into a.c. energy). The conversion efficiency is

$$\eta_c = -P_{ac}/P_{dc}. \qquad (8.11.4)$$

Figure 8.11.2a shows $\eta_c = \eta_c(\mathscr{E}_{b1})$ for $\mathscr{E}_{b0} = 10\,\mathrm{kVcm^{-1}}$ in Figure 8.11.1. The ratio R_{ac}/R_0 is also shown (Figure 8.11.2 b) where R_0 is the low-field (ohmic) resistance

$$R_0 = \frac{L}{eN_D\mu_n A} \qquad (8.11.5)$$

and R_{ac} is an equivalent a.c. resistance

$$R_{ac} = \frac{(\mathscr{E}_{b1}L)^2}{2N_D LAP_{ac}}, \tag{8.11.6}$$

where P_{ac} is the average power corresponding to one electron and $N_D LA$ the total number of electrons in the sample [42]. Figure 8.11.2 shows that the effective load (numerically equal to R_{ac}) should be matched to the diode to get the maximum con-

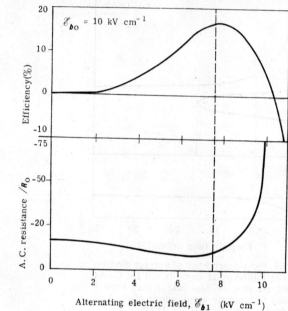

Figure 8.11.2. Efficiency (a) and dynamic resistance normalized to the low-field resistance (b) as functions of the alternating signal, $\mathscr{E}_{b1}$, for a d.c. bias field of 10 kV cm^{-1}, in GaAs. (after Copeland, [42]).

version efficiency. The maximum obtainable value depends upon $\mathscr{E}_{b0}$ and the exact shape of the $v(\mathscr{E})$ characteristic [43].

The assumption of uniform electric field will be valid only if the growth of the space-charge during one cycle is negligibly low and if this space-charge disappears completely during the positive-mobility interval (such that charge accumulation in successive cycles is impossible). A small-signal analysis yields the following results. The small excess of mobile electrons, $\Delta \rho_n$, in an accumulation layer propagating with these electrons satisfies [42]

$$\partial \Delta \rho_n / \partial t + (eN_D \mu_d / \varepsilon) \Delta \rho_n = 0 \tag{8.11.7}$$

and therefore

$$\Delta \rho_n(t) = \Delta \rho_n(t') \exp\left(-\frac{eN_D}{\varepsilon} \int_{t'}^{t} \mu_d dt\right), \tag{8.11.8}$$

where $\mu_d = dv/d\mathscr{E} = \mu_d(t)$. The space-charge growth during the interval (t_1, t_2) when $\mu_d < 0$ is measured by the growth factor

$$G_g = \exp(N_D/fh_g), \quad h_g = \left[-\frac{e}{\varepsilon T} \int_{t_1}^{t_2} \mu_d dt\right]^{-1}. \tag{8.11.9}$$

8. Transferred-electron Devices used as Oscillators

The corresponding decay factor during the interval $(t_2, t_1 + T)$ when $\mu_d > 0$

$$G_d = \exp(-N_D/fh_d), \quad h_d = \left[\frac{e}{\varepsilon T}\int_{t_2}^{t_1+T} \mu_d(t)\right]^{-1} \tag{8.11.10}$$

Both h_g and h_d are shown in Figure 8.11.3 as a function of the alternating field amplitude. The total growth factor $G_g G_d$ during one period should be below unity,

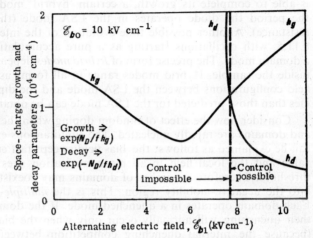

Figure 8.11.3. Space-charge control parameters h_g and h_d versus the alternating bias field, for $\mathscr{E}_{b0} = 10$ kV cm^{-1}, in GaAs. (after Copeland, [42]).

or $h_d < h_g$. Figure 8.11.3 shows that the space-charge control is possible only when $\mathscr{E}_{b1}$ exceeds a certain critical value. Copeland [42] also required a small variation of these space-charge non-uniformities by imposing arbitrarily $G_g < e^5$ and $G_d < e^{-6}$, which is equivalent to

$$6h_d < N_D/f < 5h_g \tag{8.11.11}$$

and for GaAs at room temperature

$$2 \times 10^4 < N_D/f < 2 \times 10^5 \text{ (cm}^{-3}\text{s)} \tag{8.11.12}$$

A low N_D/f slows down the rate at which the space charge can accumulate. However, N_D/f should not be too low because the rate of charge dispersion when the field is in the positive-mobility region also decreases. The range of permissible N_D/f values for a given $v(\mathscr{E})$ relationship also depends upon the bias field. The optimum ratio for maximum efficiency increases with the steady-state bias field [1]. A detailed examination of LSA mode requires a computer simulation [44]–[46].

There are considerable difficulties in achieving experimentally the theoretical predictions for the LSA mode. These difficulties originate both from the device and the external circuit. The efficiency is considerably deteriorated by the inhomogeneities of the active region (usually an epitaxial layer): these are doping non-uniformities and thermal gradients. The effect of non-uniform doping is well-known from computer simulations [37], [45], [47]. We reproduce after Taylor and Fawcett [45] the time development of internal field distribution in a GaAs sample subjected to a sinusoidal voltage, Figure 8.11.4. The effect of random doping fluctuations is the growth of dipolar space-charge layers simultaneously in several places (Figure 8.11.4 a) in contrast to the single (and small) accumulation layer characterizing the (theoretical) LSA mode. Figure 8.11.4 b shows the effect of a cathode notch: a high-field domain is nucleated and propagates across a certain distance until it is quenched. As long as the device voltage rises rapidly enough through the negative-mobility region, the domain growth is limited and the field everywhere in the active region still exceeds the threshold field. When the domain is able to complete its growth, a certain 'hybrid' mode results, when only part of the period the diode operates in the LSA mode (the bulk acting as a negative resistance). Another possible time-evolution of the internal field is shown in Figure 8.11.5, with oscillations starting as a pure accumulation mode and continuing as a domain mode. The precise form of *hybrid modes* depends upon the non-uniformities inside the sample. Hybrid modes range in all forms as a continuum of modes and field configurations between the LSA mode and the dipole domain. Higher efficiencies than those predicted for the LSA mode can be obtained [1].

Consider now the effect of random doping when the $N_D L$ product is high enough and domains are rapidly nucleated. The simultaneous existence of multiple domains can be explained as follows: the bias field sweeps fast enough through the threshold value and the local field in all mean nucleation sites rises simultaneously above threshold. Therefore a number of domains may coexist when the sample is driven into the negative-mobility region. This is the *multiple-domain mode* of oscillation. Each domain operates in a quenched mode. If the domains are comparable in size, their quenching will normally occur only when the bias field falls below threshold (because the internal quenching competition between domains, Section 8.7, is less effective; it was also assumed that the oscillation frequency is considerably above the transit-time frequency). The multiple-domain mode is somewhat similar to series operation of several devices. Higher output power is possible than with a device in a single-domain mode [1].

The non-uniform heating of the device affects the operation through the temperature-dependent $v(\mathscr{E})$ characteristic and reduces the efficiency of the LSA mode. It was shown through computer simulation that a considerable reduction in efficiency occurs when the cathode contact is made the heat sink; the temperature effect is smaller with anode connected to the heat sink [45].

The electron scattering rates limit the speed of response of the mean electron velocity v, to the applied field (chapter 1). Therefore at very high frequencies the static $v - \mathscr{E}$ dependence can no longer be used. Figure 8.11.6 shows dynamic $v-\mathscr{E}$ characteristics computed by using a two-valley model with electron population described by a Maxwellian statistics [48]. Note the apparent increase in threshold field as frequency is increased. The average negative mobility effect is reduced and the LSA-mode efficiency decreases appreciably above 40 GHz [48]. A more refined computation uses the displaced Maxwellian approach for electron distributions (Chapter 1), as Bosch and Thim [49] did, for example. Another approach was developed by Jones and Rees [36], [50], [51] who calculated the space- and time-dependent distribution function by an iterative method (see also Rees' results in Section 1.8). The last two methods have shown that the high-frequency LSA operation is modified by the finite time required for energy-relaxation processes. In other words, a delay arises in cooling of electrons returning from the satellite valleys back to the central valley [49]. Figure 8.11.7 illustrates the incomplete quenching

8. Transferred-electron Devices used as Oscillators

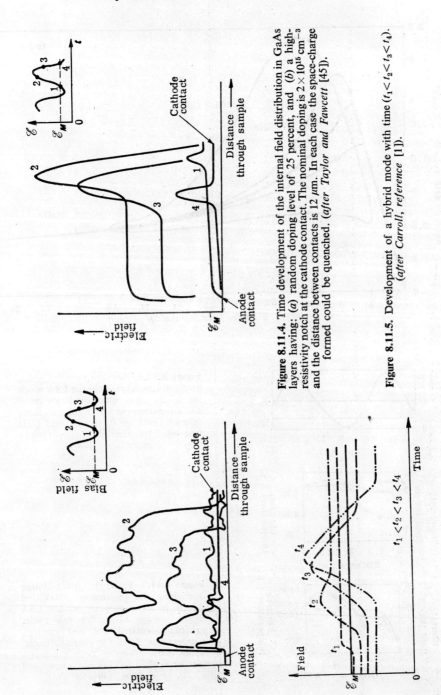

Figure 8.11.4. Time development of the internal field distribution in GaAs layers having: (a) random doping level of 25 percent, and (b) a high-resistivity notch at the cathode contact. The nominal doping is 2×10^{15} cm^{-3} and the distance between contacts is 12 μm. In each case the space-charge formed could be quenched. (*after Taylor and Fawcett* [45]).

Figure 8.11.5. Development of a hybrid mode with time ($t_1 < t_2 < t_3 < t_4$). (*after Carroll, reference* [1]).

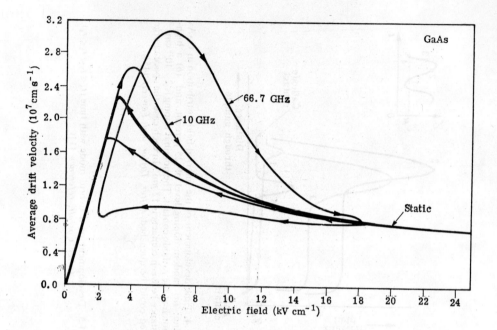

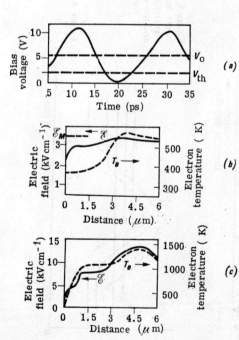

Figure 8.11.6. Average velocity *versus* field for a zero space charge LSA mode for a d.c. bias field of 10 kV cm^{-1} and a superimposed sinusoidal a.c. field of 8 kV cm^{-1} for 0, 10 and 66.7 GHz frequency, respectively. (*after Curtice and Purcell*, [48]).

Figure 8.11.7. Failure of space charge quenching due to electron relaxation effects in an improperly loaded LSA diode: voltage during first two cycles (*a*) and calculated field and electron temperature T_e in the central, lower, valley distributions at time instants 1 (*b*) and 2 (*c*). (*after Bosch and Thim*, [49]).

caused by this delay. It shows the first two cycles of an oscillation at 53 GHz that ceased after several cycles because of improper adjustment of the circuit parameters (it was shown that above 60 GHz stable oscillations cannot be obtained with any choice of circuit parameters). Figure 8.11.7 b shows that the electrons have not thermalized uniformly over the diode length by the time the voltage reached the threshold again (point 1 in Figure 8.11.7 a) but are still heated in the anode region, where large fields had been present before. The space-charge was not completely quenched in the same region and during the next cycle (Figure 8.11.7 c, point 2 in Figure 8.11.7 a) this space-charge perturbation grows in a dipole form coexisting with the space-charge layer nucleated at the cathode, in a hybrid mode. Thus the device does not operate in the LSA mode due to incomplete quenching arising from 'slow' relaxation of electrons heated by the electric field. The upper frequency limit of the LSA mode, according to Bosch and Thim [49], is around 60 GHz. Jones and Rees [50] indicated a considerably lower value of only 20 GHz. The experimental data reported by Barrera [52] confirm the optimum N_D/f value of about 10^5 s cm^{-3} found by Bosch and Thim [49]. The operation frequency of 65 GHz [52] is slightly above the limit predicted by theory [49] and could be associated with a certain hybrid mode. This mode occurs in short samples where the incompletely quenched space-charge drifts into the anode after the second or third period [49].

The space-charge growth control discussed above is dependent upon the time spent in different parts of the velocity-field characteristic. Therefore, the waveform of the applied field could be essential in determining the degree of space-charge control. It was experimentally observed that the best experimental efficiencies were normally obtained at N_D/f ratios well above the predicted optimal value of about 6×10^4 s cm^{-3} [42], whereas the voltage waveshape is non-sinusoidal [53]. This non-sinusoidal operation is known as the *LSA relaxation mode*. For example, Jeppesen and Jeppsson [53], [54] analyzed LSA diodes with N_D/f in the $1 - 5 \times 10^5$ s cm^{-3} range mounted in circuits which contain a short-circuited high impedance transmission line foreshortened to resonate the parallel capacitance of the

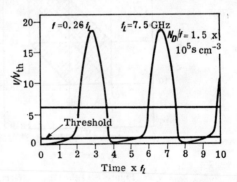

Figure 8.11.8. Voltage waveshape computed for a diode in the circuit shown. The diode circuit and parameters are $\mu_n = 8000$ cm^{2}V^{-1}s^{-1}, $R_0 = 5\Omega$, $N_D = 3 \times 10^{14}$ cm^{-3}, $C_0 = \varepsilon/R_0 N_D e \mu_n = 0.61$ pF, $v_M/v_m = 2.5$, $\mathscr{E}_m/\mathscr{E}_M = 6.0$, $R_L = 1$ kΩ, $L = 6.67$ nH, $f = 1.95$ GHz operating frequency, $V_0 = 6.1 \times V_{\text{th}}$ voltage bias. (after Jeppesen and Jeppsson, [53]).

diode, C_0. In a simplified model [46], [53] the diode is replaced by C_0 in parallel with a non-linear resistor whose characteristic is identical with $I = eN_D v\left(\dfrac{V}{L}\right)$ and the circuit is modelled by an inductance and a load conductance. A typical voltage waveshape is shown in Figure 8.11.8. A detailed simulation starting from the basic particle equations gives a qualitatively similar result. This waveshape, with a comparatively short interval spent in the negative-mobility region (between the 'peak' and 'valley' field), permits excellent control of space-charge and, in particular, reduces the sensitivity to doping gradients [53], [54]. The rapid variations in waveshape (voltage spikes) are due to the high N_D/f ratio, which determine a small effect

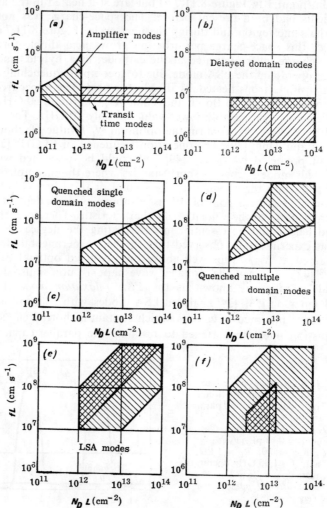

Figure 8.11.9. Mode charts. (*after Carroll*, [1]).

of the parallel capacitance (low values of the quality factor, Q) and therefore allows a rapid build-up of oscillations, preventing the domain formation [54]. Starting of the sinusoidal LSA mode is, indeed, a difficult problem. This is a typical large-signal mode and cannot operate until a large voltage swing occurs at the diode terminals. However, the small-signal gain mechanism in a supercritical doped device will lead to domain formation. If a domain is formed, it will tend to absorb the bias voltage in excess of threshold value and in a *long* sample the large domain will have a high peak field and avalanche breakdown is possible.

Figure 8.11.7 shows, after Caroll [1], the so-called mode-charts, indicating the value of fL and N_DL products for each mode (note that an $N_D/f = $ const curve in the $fL - N_DL$ plane is a straight line of slope 1). The delayed domain mode described in Section 8.10 is restricted to a quite narrow frequency range. However, in a circuit with suitable waveforms the domain formation can be delayed for more than half a cycle, though the efficiency may decrease considerably. The dark region in Figure 8.11.7 *b* indicates the more useful range. The 'area' occupied by quenched domain modes is justified in reference [1]. There are no distinct bounds to the specific areas. The distinct, limits indicated by different authors are derived within certain idealized theories, using particular values of the material constants. The operation mode in supercritical devices at frequencies above the transit-time frequency is decided by the circuit and/or by the device itself (material uniformity and contacts are important)*.

8.12 Various Phenomena in Practical Gunn Diodes

(a) *Impact Ionization in Gunn Diodes.* We have already shown that large internal fields can occur during dipole-mode operation of TED devices. When the bias field increases, both the domain excess voltage and the maximum electric field inside the domain increase and impact ionization may occur. If the maximum field is well above threshold for impact ionization, avalanche breakdown is possible. We shall briefly discuss below only the situation when the voltage is considerably below that necessary for avalanche breakdown and the degree of impact ionization is small. However, even such a low ionization level can change considerably the device behaviour. Heeks and Woode [56] have shown that excess holes produced by the ionization process are almost immediately trapped (lifetime below 1 ns) by traps corresponding to the oxygen level (0.64 eV below the conduction band). The excess electrons are localized at the hole trap sites, thus maintaining the neutrality. The excess electrons increase the apparent conductivity of the sample and thus the 'valley' current (Figure 8.10.3) flowing during domain propagation. The effect of electrons generated by impact

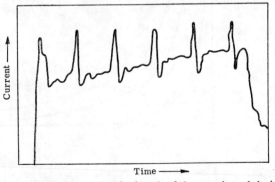

Figure 8.12.1. Current waveform measured for a Gunn diode at a bias level sufficiently high to cause impact ionization to occur in HFD's *(after Bohn and Herskowitz,* [57]).

ionization occurs in normal transit-time operation along the entire length of the sample and during each subsequent cycle such that the 'valley' current increases successively during several cycles as shown in Figure 8.12.1 [57]. When the current is sufficiently high (close to the peak value), the

* Gunn diodes in operation exhibit a negative-resistance bias effect: the average current decreases as the voltage bias increases: this may lead to low frequency oscillations [25] and impair stability.

sample current becomes erratic, thus indicating random domain formation: the transit-time oscillations are inhibited, as indicated in Figure 8.12.5 [57].

The threshold voltage for impact ionization, V_t, can be related to the threshold voltage for Gunn oscillations, V_{th}, as shown in Figure 8.12.3. This diagram indicates a maximum usable device

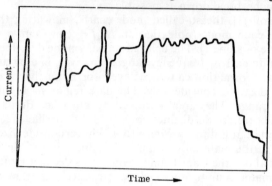

Figure 8.12.2. Current waveform indicating the inhibition of domain formation due to impact ionization. (*after Bohn and Herskowitz,* [57]).

ength for a given carrier concentration, for example when $N_D = 10^{15}\text{cm}^{-3}$, L cannot exceed 375 μm, since at this length impact ionization commences at the oscillation threshold.

(b) *Temperature dependence of Oscillation Frequency.* The variation of oscillation frequency of Gunn diodes with the ambient temperature is sometimes quite large (several MHz per K or even larger)

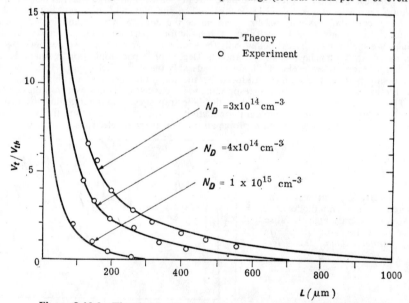

Figure 8.12.3. Threshold voltage for impact ionization V_{th} normalized to threshold voltage for Gunn oscillations as a function of device length. The ionization threshold is defined as the voltage which produces a 1-percent increase in the valley current over ten domain transits. (*after Bohn and Herskowitz,* [57]).

and limits device application in systems. The temperature sensitivity can be reduced either by circuit techniques (frequency control circuits or passive compensation techniques) [58] or by designing 'stable' devices.

The temperature sensitivity of frequency occurs mainly through the velocity-field dependence (see Chapter 1). In the pure transit-time mode f becomes $f(T)$ due to the temperature dependent transit-time: the domain velocity decreases slightly with increasing temperature and thus the transit-time frequency decreases (df/dT is negative). The peak-to-valley velocity ratio decreases

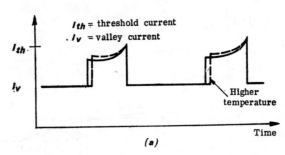

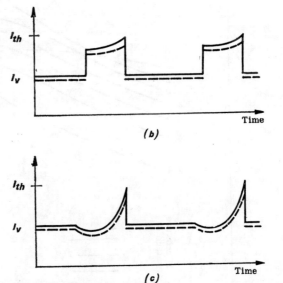

Figure 8.12.4. Current waveforms illustrating frequency-temperature effects: the dashed curve is for higher temperature. (a) Effect of temperature-dependent transit-time on the delayed domain mode. (b) Effect of temperature-dependent velocity-field characteristic on the delayed domain mode. (c) As (b) for quenched domain mode. (after Hobson, [59])

with increasing temperature and tends to increase f with increasing T, as qualitatively shown below [59]. Figure 8.12.4 illustrates the effect of increasing temperature upon the current waveforms: the solid lines are replaced by the dashed ones. The device generates oscillations in a resonant circuit. The large-signal negative resistance of the device should be such to compensate the equivalent positive resistances in circuit. By neglecting the temperature dependence of the device capacitance and the circuit elements, the device resistance should be invariant to the ambient-temperature variations. The oscillation frequency will vary as follows Figures 8.12.4 a and b show the effect of temperature upon the delayed domain mode (Figure 8.10.4). Figure 8.12.4 a illustrates the effect of increasing domain transit-time with increasing temperature. In order to maintain constant negative conductance, the voltage amplitude must decrease [59] such that the current is higher during the pulse (when the device behaviour is ohmic). Therefore, the quadrature component of the current is increased, because the 'centre' of the current pulse is shifted at a later time, the capacitive susceptance will be higher and the operation frequency should decrease [59]. The change of peak and valley velocities modifies the current waveform as shown in Figure 8.12.4 b (the transit-time is now assumed constant), decreases the quadrature component of the current and increases the frequency. Both mechanisms are operative in the delayed domain mode, the transit-time effect predominates at low f/f_T and the second effect at f/f_T close to unity. The transit-time effect is inexistent in the quenched-domain mode (Figure 8.12.4 c) and df/dT is positive.

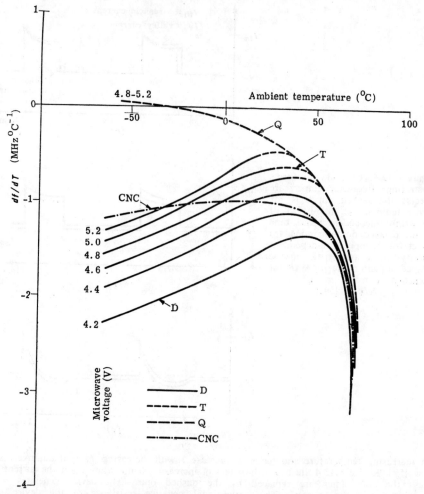

Figure 8.12.5. df/dT *versus* temperature for various amplitudes of the microwave voltage: D = delayed domain mode, Q = quenched-domain mode, T = transition region, CNC = constant negative conductance ($4 \times 10^{-3} \Omega^{-1}$). (*after Edridge,* [58]).

A quantitative analysis requires a computer simulation, even with a simple modelling of the device [58]–[59]. Figure 8.12.5 shows typical results computed for a GaAs Gunn diode [58]: df/dT is shown *versus* ambient temperature for a few amplitudes of the microwave voltage. The frequency variation for the quenched-domain mode is considerably smaller due to the elimination of transit-time dependence with temperature, the contribution to df/dT from other temperature dependent parameters almost cancel in a temperature range. The correct df/dT *versus* T curve also shown in Figure 8.12.5 was obtained by determining the relation between the microwave voltage and the ambient temperature when the negative conductance is held constant (see above). Such results were quite well verified by experiment [58]. The frequency variations in both modes increase considerably at high temperatures (Figure 8.12.5). The frequency decreases rapidly due to the effect of domain growth time: this time increases considerably at higher temperatures because the differential negative mobility decreases (Chapter 1).

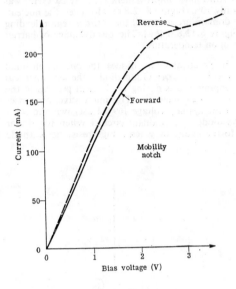

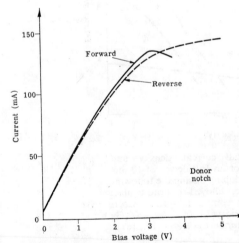

Figure 8.12.6. Typical asymmetrical experimental current-voltage characteristics of Gunn diodes explained by: (a) mobility notch, and (b) doping notch (*after* Gurney, [62]).

It was shown [59], [60] that by using n^+ contacts to GaAs instead of alloyed-metal contacts the frequency stability is considerably improved (100 kHz/°C or less). The disadvantage of the metal contacts is associated with the high-resistivity layer adjacent to the contact which acts as a temperature-dependent series resistance [60].

(c) *Contact Effects.* The lack of symmetry in $I-V$ characteristics of actual Gunn diodes was explained by contact effects [61]—[63]. Figure 8.12.6 shows two typical experimental current-voltage characteristics. The much more common type of characteristic is shown in Figure 8.12.6 *a* and is explained [62] by a cathode 'mobility notch', namely by the existence of a damaged cathode region with an increased number of scattering centres. This is due to the crystallographic damage occurring in the vicinity of the alloyed metal contact*. The second type of behaviour, Figure 8.12.6 *b*, is explained by the existence of a donor notch at the cathode. Computed characteristics agree qualitatively with the experimental ones [62]. Controlled damage of metal contacted diode *after* they were made and measured has determined a transition from type (b) (Figure 8.12.6) to type (a) of behaviour. Magnetoresistance measurements proved that the damage reduced the carrier mobility [63]. It was shown that the conductivity notch affects considerably the microwave behaviour of transferred electron diodes [65].

It was also shown that barrier-type cathode contacts to InP and GaAs Gunn diodes provide better efficiency than ohmic contacts [66]—[68]. Figure 8.12.7 shows for example $I-V$ characteristics and efficiency for the InP diodes with identical parameters, except the cathode which was made by different pocedures. The barrier-type contact reduces the current and allows a considerably higher efficiency. Figure 8.12.8 illustrates the effect of cathode boundary condition upon the space-charge dynamics inside a GaAs device (computer simulation). When the cathode is an n^+ graded contact (Figure 8.12.8 *a*), an accumulation layer is formed but at about 2 μm from the cathode and not just at the contact. This occurs because the low energy electrons injected from the n^+ contact need to gain from

---

* Crystallographic damage also occurs during bonding. The plated heat sink technology (also used for BARRIT and IMPATT diodes) provide damage-free bonding for TED's [64].

the field an energy equal to the intervalley separation energy: this requires a finite time (relaxation effects included, see also Section 8.11) and hence a finite propagation distance. The device efficiency is low (2% at 36 GHz) because the NDM region is confined to the anode. A low barrier contact (Figure 8.12.8 b) raises the cathode field to a value comparable with the threshold field and removes the above limitation: the accumulation layer is nucleated very close to the cathode. The efficiency is improved to 8.5% at 25 GHz. A still higher improvement (11% at 25 GHz) was obtained by raising the effective barrier height by only 10% (Figure 8.12.8 c). However, the nucleated instability is a dipole domain. After 12 ps the domain grows to fill the entire n region leading to a similar field configuration to that shown in Figure 8.12.8 b [66]. The disadvantage of barrier type contacts is the dependence of their properties upon temperature.

(d) *Non-uniform Doping and Geometrical Effects.* In practical Gunn diodes, the one-dimensional distribution of donors is not uniform. In epitaxial $n^{++} - n - n^+$ Gunn diodes the asymmetrical doping profiles occur due to outdiffusion or auto-doping effects during the growth process of the active layer. This determines an unsymmetrical $I-V$ characteristic. The microwave properties depend upon the doping profile and the polarity of the applied voltage [69]. Hasegawa and Suga [70] simulated the behaviour of an X-band GaAs diode in a resonant circuit. When the donor density, for example, increases from the cathode to the anode or when a high-resistance cathode

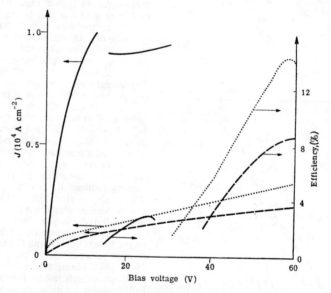

Figure 8.12.7. Experimental current density and efficiency *versus* voltage for an InP device of 10 mm length and 2×10^{15} cm^{-3} doping having the following cathode contacts: ohmic, alloyed-tin-bead contact (solid curves), pure silver contact alloyed at 425°C (dashed curves) and oxidized 80% Ag—20%Ga cathode (dotted curves). (*after Colliver et al.*, [67]).

region exists, the oscillation efficiency decreases. When a low-resistance region exists near the anode (for instance due to an autodoping effect) the efficiency is higher than for a uniform profile. The correspondent with experimental data is quite good [70]. The efficiency of Gunn diodes made from different epitaxial wafers now differ considerably due to the different doping profile [70].

Shah and Rabson [71], [72] have shown that the effect of doping and geometrical non-uniformities on domain dynamics are similar. Annular and spherical geometry devices have been shown to display a voltage-controlled oscillation frequency over a wide frequency range. Both doping

8. Transferred-electron Devices used as Oscillators

and geometry allow control of the current waveform and possible construction of built-in complex waveform generator at microwave frequencies. Proper choice of the device geometry and doping profile allow, at least in principle, a great flexibility in designing microwave Gunn devices.

(e) *Planar-type Structures.* A planar type structure (cathode and anode contacts on the same side of the active semiconductor layer, Figure 8.12.9 a) is attractive for a number of reasons [73]. Long devices can be constructed and low transit-time frequencies (of the order of 1 GHz) can be easily achieved: a similar operation in a sandwich structure will require about a 100 μm thick layer with severe heat dissipation problems. The heat flow path can be considerably reduced in

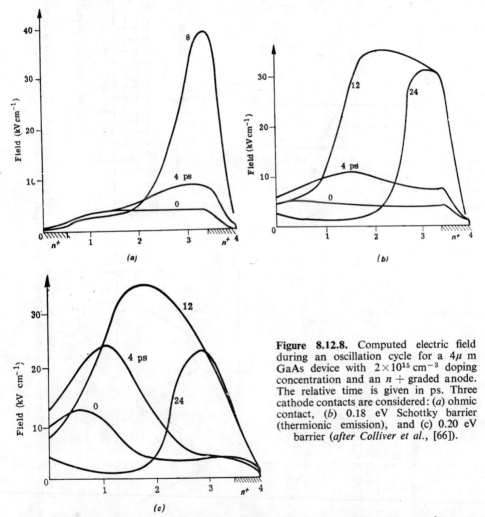

Figure 8.12.8. Computed electric field during an oscillation cycle for a 4μ m GaAs device with 2×10^{15} cm^{-3} doping concentration and an $n+$ graded anode. The relative time is given in ps. Three cathode contacts are considered: (a) ohmic contact, (b) 0.18 eV Schottky barrier (thermionic emission), and (c) 0.20 eV barrier (*after Colliver et al.,* [66]).

a planar (epitaxial layer) structure, as already mentioned in Chapter 7. The contacts made on one side of the material are an advantage for mass-production. A considerable freedom in device configuration (various geometries and more than two electrodes) and integrated circuit design (microstrip microwave circuits or logic circuits) is possible.

The planar GaAs structure shown in Figure 8.12.9 a was obtained by growing epitaxially a layer on a high-resistivity substrate and alloying tin (Sn) ohmic contacts with a nickel overlay.

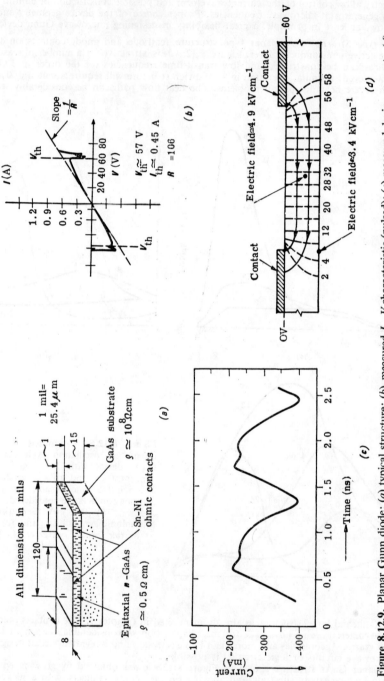

Figure 8.12.9. Planar Gunn diode: (*a*) typical structure; (*b*) measured $I-V$ characteristic (pulsed); (*c*) measured device current; (*d*) computed potential distribution (resistance-paper analog) for 60 V bias (coherent oscillations) (*after Dienst et al.*, [73]).

A typical measured $I-V$ characteristic is shown in Figure 8.12.9 b. The threshold voltage for Gunn-type oscillations (Figure 8.12.9 c) is about 57 V. The calculated potential distribution inside the coplanar device, Figure 8.12.9 d, shows that the electric field over most of the active layer is rather uniform (4.9 kV cm^{-1}) and appreciably higher than the threshold field for negative mobility.

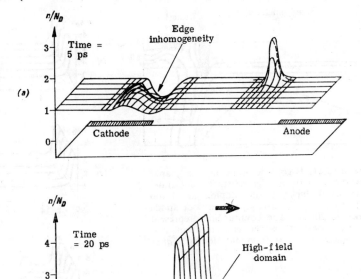

Figure 8.12.10. Two-dimensional high-field domain in a planar structure, results of computer simulation: (a) the domain is launched by the depletion due to the field inhomogeneity at the cathode edge: (b) stable shape of the domain (*after Reiser*, [74]).

When the bias voltage is below its threshold value (57 V), the electric field at the edges of the uniform field region falls below the threshold field, although the cathode field is above threshold. This parallel low-field path tends to determine very noisy oscillations or to supress them all [73].

The dynamics of space-charge perturbations in a planar NDM structure is a difficult three-dimensional problem. We quote below several pertinent results obtained by computer simulation. Figure 8.12.10 shows how the field inhomogeneity at the cathode edge launches a high-field domain [74]. Figure 8.12.11 gives an intuitive picture of domain penetration into the high-resistivity substrate, just before the domain enters into the anode.

The rapid lateral spreading of a nucleated dipole is indicated by Figure 8.12.12 [75]. Here the dipole domain is nucleated by a doping notch which is situated laterally in a sandwich-type structure and whose effect is apparent in Figure 8.12.12 a. The domain forms and propagates toward the anode (Figures 8.12.12 b-d). The lateral spreading velocity reaches values of the order of 10^8 cm s^{-1}: after the domain extended laterally its further growth and propagation is similar to what happens in a one-dimensional case. A similar analysis applied to a planar structure shows that a high-field domain is nucleated at the edge of the cathode, without any doping notch (see Figure 8.12.10 a), that the domain spreads laterally with a velocity-up to 10^8 cm s^{-1} until it fills the whole width and then grows and propagates similarly to the one-dimensional case [75].

(f) *InP Transferred-electron Devices*. It has been suggested by Hilsum and Rees [76] that InP is preferable as a TED material because the intervalley transfer mechanism involves three valleys and determines an N-shaped $v(\mathscr{E})$ characteristic, with a minimum followed by a region of positive slope and also a high effective diffusion coefficient. It was shown that accumulation layer

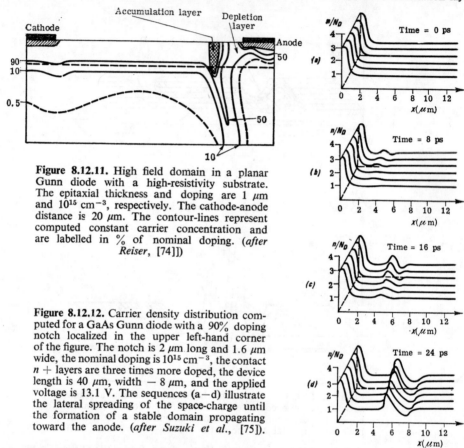

Figure 8.12.11. High field domain in a planar Gunn diode with a high-resistivity substrate. The epitaxial thickness and doping are 1 μm and 10^{15} cm^{-3}, respectively. The cathode-anode distance is 20 μm. The contour-lines represent computed constant carrier concentration and are labelled in % of nominal doping. (*after Reiser,* [74]])

Figure 8.12.12. Carrier density distribution computed for a GaAs Gunn diode with a 90% doping notch localized in the upper left-hand corner of the figure. The notch is 2 μm long and 1.6 μm wide, the nominal doping is 10^{15} cm^{-3}, the contact $n+$ layers are three times more doped, the device length is 40 μm, width — 8 μm, and the applied voltage is 13.1 V. The sequences (a–d) illustrate the lateral spreading of the space-charge until the formation of a stable domain propagating toward the anode. (*after Suzuki et al.,* [75]).

or LSA modes are favoured whereas domain modes are inhibited. Moreover, if domains do build up, no impact ionization will take place within them because a positive slope of the characteristic beyond the valley limits the field and produces 'flat-topped' domains, as indicated above in this chapter.

However, the high-field domains and impact ionization were experimentally observed in InP samples [77], [78] and this agrees with the saturating $v(\mathscr{E})$ characteristic determined recently for InP and presented in Chapter 1. The domain-mode operation shows certain peculiar instabilities [79], [80]. High-efficiency circuit-controlled oscillations (Section 8.11) were also obtained [81], [82]. It was shown [83] that although a higher peak to valley velocity ratio is available with InP, this does not necessarily lead to higher efficiency. In contacted devices, this high ratio is offset by the effects of the time required for the electrons to gain sufficient energy for transfer to upper valleys: this determines a 'dead zone' nearly the cathode which is 2—4 μm long in InP and only 1 μm in GaAs (Section 8.11). The latter effect could be overcome by a certain cathode injection control such that electrons enter the semiconductor already hot [83]. It was demonstrated that

non-ohmic type cathodes determine a better efficiency [66], [67], [83]. LSA-type operation in long devices is also possible but requires a high-quality thick epitaxial layer.
(g) *Noise.* Physical sources of noise in NDM devices were mentioned in Chapter 7. The noise in Gunn oscillators is discussed in references [1], [84]. A $1/f$ noise component related to surface effects may also be important [85].

8.13 Functional Devices

We have discussed so far the basic phenomena in NDM devices used for microwave generation and amplification and for bistable switching. Any negative-resistance device, the tunnel diode for example, can perform the above functions. The physical details and the parameters of actual device operation, for example the operation frequency or the response time, will be, of course, different. However, the NDM or *bulk-effect* devices are potentially capable of realizing a variety of complex electronic functions (both analogue and digital).

We shall discuss below domain-type operation of bulk-effect devices. It is possible to obtain a stable high-field domain propagating over a distance which is long compared to the width of the domain itself. The external current is related to the internal propagation of this domain and reflects instantaneously the change in the properties of the region of the conducting path traversed by the domain, as will be discussed in more detail below. Therefore, the control of the current waveshape can be distributed spatially throughout the whole drift path [86]. This is in contrast with the situation encountered in the majority of solid-state devices, where the device current is basically determined by the instantaneous potential at a narrow barrier.

Let us consider below that the applied bias is sufficiently high such that the domain velocity reaches a minimum, saturation value, $v_{D\text{min}} \simeq v_{\text{sat}}$ (see Section 8.4 for saturated domains). This property is valid irrespective of the uniformity or non-uniformity of the drift path. Let us consider a non-uniform drift path and denote by x_1 and x_2, respectively, the sections reached by the domain at the time moments t_1 and t_2. Therefore $I(t_1) = eA(x_1) N_D(x_1) v_{\text{sat}}$ and $I(t_2) = eA(x_2) N_D(x_2) v_{\text{sat}}$ and

$$\frac{I(t_1)}{I(t_2)} = \frac{A(x_1) N_D(x_1)}{A(x_2) N_D(x_2)}. \tag{8.13.1}$$

We have tacitly assumed that the domain has zero width. The current was computed as the external (neutral) current: the external drift velocity is equal to the domain velocity and approximately equal to the saturation velocity (Section 8.3). Both the cross-section of the conductive path, A, and the local doping, N_D, are assumed position dependent (see also references [86], [87]). Any variation in A and N_D will be reflected in the current waveshape. This is clearly illustrated by Figure 8.13.1: the current waveform is a replica of the device shape [88]. The above considerations are based upon the one-dimensional model of domain propagation. The theory should be modified to consider the phenomena occurring when the high-field domain passes a sudden increase or decrease of device cross-sectional area [87], [88]. The process of domain nucleation should be also reconsidered because the electric field is, in general, non-uniform in an arbitrary section of the drift path. A characteristic example is the slanted-cathode device [89], [90] shown in Figure 8.13.2 a. A resistive probe method was used to measure the internal potential distribution in such a GaAs sample, and the results are shown in Figure 8.13.2b and c, for the first and second periods of oscillation, respectively [90]. When the drive pulse voltage

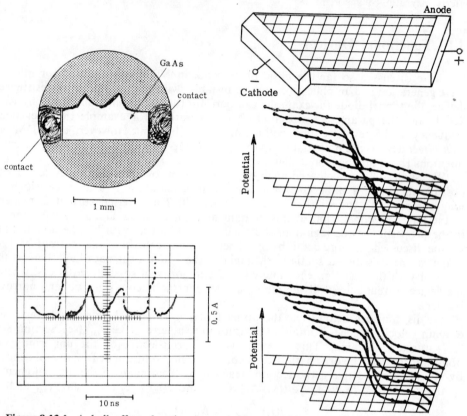

Figure 8.13.1. A bulk-effect function generator generates an M-shaped current waveform: (a) device shape, (b) current waveform. (after Shoji, [88]).

Figure 8.13.2. (a) Slanted-cathode device, (b) potential distribution in the device during the first period of oscillation, (c) the same during the second period. (after Shoji, [90]).

is well above threshold and the rise time is short, the first domain nucleates approximately parallel to the cathode (Figure 8.13.2 b). Without these conditions, during the second period of oscillation, the domain nucleates approximately parallel to the anode (Figure 8.13.2 c). A more detailed analysis shows that, in the second case, the domain nucleation begins *at the cathode edge* closest to the anode and then expands across the sample width [90]. The movement of an already formed domain can be described by a simplified theoretical model [90] which treats a small segment of a two-dimensional domain as a one-dimensional domain plus current sources and excess charges originating from the two-dimensional nature of the problem.

It was, therefore, shown that various current waveforms can be generated by modifying the conduction profile along the domain propagation path. Not only the variable doping and cross-sectional area can be used to vary the conduction profile, but the external radiation and additional electrodes, as well [86]. Dielectric and magnetic loading was also suggested [91].

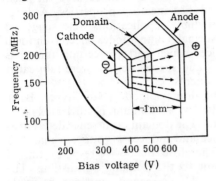

Figure 8.13.3. Voltage tunable Gunn oscillator using tapered bulk GaAs device. (*after Engelbrecht*, [92]).

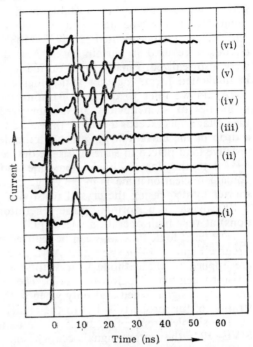

Figure 8.13.4. Traces of output current at different bias levels (increasing from (i) to (vi) for bulk-effect analogue to digital converter (the device profile is shown in the inset). (*after Sandbank*, [86]).

We shall consider now a device with a uniformly graded conductivity profile (for example the tappered bulk GaAs device shown in Figure 8.13.3 [92]). The distance traversed by the domain becomes proportional to the applied bias across the device. The transit-time oscillation frequency can be therefore controlled by the applied bias (Figure 8.13.3). A modified form of this voltage tunable oscillator can be used directly as an analog-to-digital converter [86]. The digital code can be implemented through a series of indentations along one edge of the epitaxial layer which form the device, as in the 4-level analog to digital converter sketched in Figure 8.13.4. The measured output current for a few bias levels is shown on the same figure. Trace (i) corresponds to the lowest bias level: no domain is launched. Trace (ii) shows a negative peak after the trigger pulse: a domain is launched but it is extinguished as it goes beyond the first low-conductivity region, etc. Finally, curve (vi) shows four separate increases in current as the domain travels through the four high-conductivity regions and is absorbed by the anode. Therefore the number of output pulses is directly related to the magnitude of the analog signal which biases the diode. The above device is one of the first Gunn-type functional devices [86]. Many others were suggested but their fabrication is difficult in the present state of technology. Logic devices and circuits have received a considerable attention: they are briefly examined in the next Section.

8.14 Gunn-effect Digital Circuits

We have already shown in this chapter that a Gunn diode can be used for pulse generation (astable or monostable mode) and has memory properties. A large number of Gunn-type devices and circuits for digital applications were suggested and, in part, experimented [2], [93]–[108]. Normal Gunn diodes in conventional circuits were used in certain cases. *New* devices which perform certain functions were also realized. Finally, both normal diodes and special devices were fabricated in integrated form, together with other components.

Again we refer to devices operating in the domain mode (Gunn-type). Basic properties of such devices are summarized below. A domain is nucleated after the voltage bias exceeds a definite threshold value. The domain occurrence determines a drop of the current flowing through the device and induces a voltage signal in a resistive load placed in series with the device. Therefore, the device response time is related to the domain nucleation time and not to the domain transit-time. The domain nucleation time is in the range of tens of pico-seconds and can be made shorter by decreasing the crystal resistivity. Hence, Gunn-effect devices are potentially capable of performing pulse generation and processing (including logic functions) at very high operation speeds [93].

The basic logic functions can be realized by using conventional diodes in resistive circuits. Figure 8.14.1a shows an AND circuit. The input signals (positive voltages) are applied to the cathode by resistors. The diode itself behaves as an ordinary resistor when biased below threshold. The diode is biased above threshold and gives and output signal when and only when both inputs have a signal [94]. The demonstration of this property is left to the reader.

The simple circuit of Figure 8.14.1 *b* realized the logic function called EXCLUSIVE-OR*: the output signal occurs when only one input has a signal. Assume that 1 input has signal whereas 2 input does not. The magnitude of minimum input

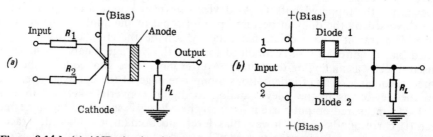

Figure 8.14.1. (a) AND circuit with a Gunn diode; (b) EXCLUSIVE-OR circuit with two Gunn diodes. (*after Hartnagel and Izadpanah*, [94]).

voltage determining domain nucleation is $V_{B1} = V_{th}(1 + R_L/R_0)$. When both inputs are connected together the minimum voltage necessary for domain nucleation is $V_{B2} = V_{th}(1 + 2R_L/R_0) > V_{B1}$. If $V_{B1} < V_B < V_{B2}$ the circuit will perform the required function†.

---
* This is a basic function in performing addition of binary numbers.
† It can also act as a comparator [93], [94].

The above logic function can also be performed by a single device: a Gunn effect device with two separate cathodes utilized as input terminals. Figure 8.14.2 shows computed potential distributions inside such a device in two cases: with

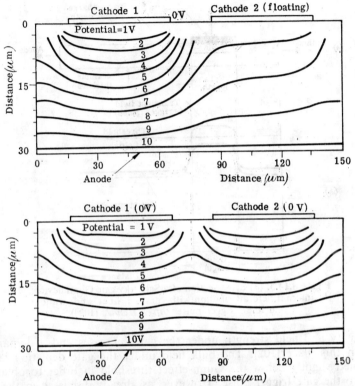

Figure 8.14.2. Computed potential distribution in a modified Gunn device with two small cathodes: (a) the right-hand cathode is floating; (b) both cathodes are at reference potential (1 Ω cm, GaAs). (*after Hartnagel and Izadpanah*, [94]).

bias applied to only one cathode and the second one floating (Figure 8.14.2 *a*) and with both cathodes connected together (Figure 8.14.2 *b*). In the first case a high-field region occurs in the vicinity of the biased cathode: the maximum field intensity reaches higher values than in the second case. Therefore for a proper input signal (acting as a voltage bias) it is possible to nucleate a domain, and thus determine an output pulse when and only when one cathode has an input signal.

Various bulk-effect devices performing directly logic functions can be imagined: these devices differ in geometrical configuration and in placement and nature of electrodes. Planar-type devices (with electrodes on the free surface of an epitaxial layer grown on a semi-insulating substrate) are advantageous for mass-production and integrated circuit fabrication [2], [96]—[101].

The suplementary electrodes are ohmic contacts, Schottky contacts [97], [102] or field effect electrodes (a metal plate separated by a dielectric from the semiconductor) [103]—[104]. Figure 8.14.3 shows for example a Schottky-gate Gunn device. The reverse-biased Schottky contact decreases the cross-section of the current-conducting channel. When a negative pulse is applied on this Schottky 'gate' the

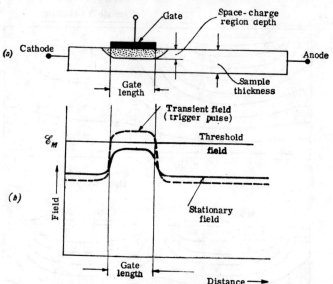

Figure 8.14.3. (a) Cross-section of a Schottky gate Gunn device; (b) internal field distribution with (dashed line) and without (solid line) trigger pulse. (*after Heime*, [102]).

current density and field strength under the gate increases and a domain can be nucleated in this place [102]. The bulk field under the gate depletion layer must be maintained slightly below the Gunn-effect threshold in order to obtain a high sensitivity to the gate trigger pulse. However, as already discussed in the previous chapter, the electric field in planar-type Gunn diodes is non-uniform. A high-field region was identified near the anode [106], [107]. Consequently, the average (bias) field should be well below the threshold field and then a quite high trigger pulse is necessary in order to launch a domain. In other words, the trigger sensitivity of the Schottky (or field-effect) gate is low. The situation can be improved for example by introducing a subsidiary Schottky anode above the region where the cross-sectional area expands [107], as shown in Figure 8.14.4.

High-speed low-power planar integrated circuits using Gunn-diodes, Schottky diodes and resistors was reported [96]—[98], [100]. The circuit elements are defined by photolithographic techniques on an epitaxial layer on an insulating (Cr-doped GaAs) substrate and separated by mesa-like etching. The device dimensions should be properly chosen. Lower cost and higher power capabilities require small-area and thin Gunn devices. A minimum length of the active channel is, however, required so that a domain launched at the cathode can build up. The epilayer thickness cannot be made arbitrarily small. Otherwise the energy contained in the stray field of a domain outside the device would be too large: the field inside the device would be too low to support domain formation. It was experimentally shown that the doping-thickness product should be higher than 10^{12}cm^{-2}; below this value the relative current drop due to domain formation decreases considerably and the Gunn-effect disappears at about 10^{11}cm^{-2} [96], [101], [108].

Figure 8.14.5a shows an integrated circuit with Schottky-gate Gunn devices in a series arrangement: a chain of four devices one of which is anode triggered

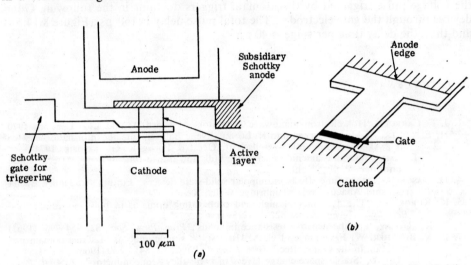

Figure 8.14.4. Schottky-gate coplanar Gunn devices: (a) normal structure; (b) with subsidiary Schottky anode. (*after Kurumada et al.*, [107]).

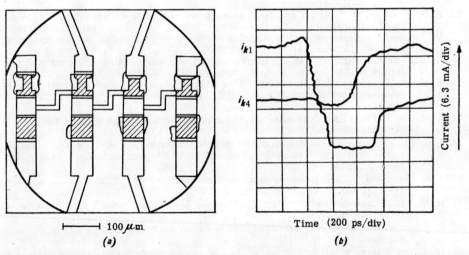

Figure 8.14.5. (a) Integrated GaAs circuit with four Schottky-gate Gunn devices, series arrangement. Epitaxial layer: $N_D = 1.3 \times 10^{16}$ cm^{-3}, $\mu_n = 6000$ cm^{2}V^{-1}s^{-1}, 3.5 μm thickness; Gunn device: 40 μm length, 18 μm width, 7 μm gate length, $R_0 \simeq 550\Omega$; load resistor: 6 μm length, 45 μm width, $\simeq 130\Omega$. Contact material: GaAs contact sheet; AuGe for alloyed contacts; Cr for Schottky contacts, (b) delay between the current pulses in the first (i_{k1}) and fourth (i_{k4}) Gunn device of the chain (6.3 mA/div versus 200 ps/div). (*after Mause et al.*, [96]).

and the following three are gate triggered [96]. The bias voltage supply is common, the load resistors are connected to the cathode terminals. The gate of a certain device is placed close to the cathode, see Figure 8.14.3, and connected to the anode of the next device. Figure 8.14.5 b gives the cathode currents i_{k1} and i_{k4} through the first and fourth Gunn device of the chain when a domain is launched in the first device: the voltage pulse triggered by this domain triggers domains in the following Gunn devices through the gate electrodes. The total pulse delay is 160 ps (Figure 8.14.5b) and thus the delay time per stage is 40 ps.

References

1. J. E. CARROLL, 'Hot electron microwave generators', Edward Arnold Ltd, London, 1970.
2. H. ELSCHNER, A. MÖSCHWITZER and K. LUNZE, 'Neue Bauelemente der Informationselektronik', Akademische Verlagsgesellschaft, Geest & Portig K.-G., Leipzig, 1974.
3. L. F. EASTMAN, 'Gallium arsenide microwave bulk and transit-time devices', Artech House, Tunbridge Wells, Kent, 1972.
4. D. DASCALU, 'Transit-time effects in unipolar solid-state devices', Editura Academiei, Bucharest, and Abacus Press, Tunbridge Wells, Kent, 1974.
5. K. KUROKAWA, 'The dynamics of high-field propagating domains in bulk semiconductors', Bell Syst. Techn. J., 46, 2235—2259 (1967).
6. B. K. RIDLEY, 'Specific negative resistance in solids', Proc. Phys. Soc., 82, 954—966 (1963).
7. P. N. BUTCHER, W. FAWCETT and C. A. HILSUM, 'A simple analysis of stable domain propagation in the Gunn effect', Brit. J. Appl. Phys., 17, 841—850 (1966).
8. J. A. COPELAND, 'Stable space-charge layers in two-valley semiconductors', J. Appl. Phys., 37, 3602—3609 (1966).
9. J. W. ALLEN, W. SHOCKLEY and G. L. PEARSON, 'Gunn domain dynamics', J. Appl. Phys., 37, 3191—3195 (1966).
10. P. N. BUTCHER, W. FAWCETT and N. R. OGG, 'Effect of field-dependent diffusion on stable domain propagation in the Gunn effect', Brit. J. Appl. Phys., 18, 755—759 (1967).
11. W. HEINLE, 'Principles of a phenomenological theory of Gunn-effect domain dynamics', Solid-St. Electron., 11, 583—598 (1968).
12. M. A. LAMPERT, 'Stable space-charge layers associated with bulk, negative differential mobility: further analytic results', J. Appl. Phys., 40, 335—340 (1969).
13. S. HASUO, T. OHMI and T. HORIMATSU, 'Five different modes of high-field domains due to field-dependent carrier diffusion', IEEE Trans. Electron Dev., ED—20, 476—481 (1973).
14. S. HASUO, T. OHMI and T. HORIMATSU, 'Influence of carrier concentration on the mode o high-field domains', IEEE Trans. Electron Dev., ED—20, 903—905 (1973).
15. J. B. GUNN, 'Microwave oscillations of current in III—V semiconductors', Solid-St. Commun., 1, 88—91 (1963); 'Instabilities of current in III—V semiconductors', IBM J. Res. Develop., 8, 141—159 (1964).
16. J. A. COPELAND, 'Bulk negative-resistance semiconductor devices', IEEE Spectrum, 4, No. 5, 71—77 (1967).
17. R. W. H. ENGELMANN and W. HEINLE, 'Proposed Gunn-effect switch', Electron. Lett., 4, 190—192 (1968).
18. R. ENGELMANN and W. HEINLE, 'Pulse discrimination by Gunn-effect switching', Solid-St. Electron., 14, 1—16 (1971).
19. H. THIM, 'Experimental verification of bistable switching with Gunn diodes', Electron. Lett., 7, 246—247 (1971).
20. M. HURTATO and F. J. ROSENBAUM, 'A CW Gunn diode bistable switching element', IEEE Trans. Electron Dev., ED—19, 1130—1131 (1972).
21. S. H. IZADPANAH, B. JEPPSSON, P. JEPPESEN and P. JONDRUP, 'A high current drop GaAs bistable switch', Proc. IEEE, 62, 1166—1167 (1974).

22. R. L. GUNSHOR and A. C. KAK, 'Lumped-circuit representation of Gunn diodes in domain mode', *IEEE Trans. Electron Dev.*, **ED−19**, 765−770 (1972).
23. I. KURU, P. N. ROBSON and K. S. KINO, 'Some measurements of the steady-state and transient characteristics of high-field dipole domains in GaAs', *IEEE Trans. Electron Dev.*, **ED−15**, 21−29 (1968).
24. A. C. KAK and R. L. GUNSHOR, 'The transient behaviour of high-field dipole domains in transferred electron devices', *IEEE Trans. Electron Dev.*, **ED−20**, 1−5 (1973).
25. R. B. ROBROCK, II, 'A lumped model for characterizing single and multiple domain propagation in bulk GaAs', *IEEE Trans. Electron Dev.*, **ED−17**, 93−102 (1970).
26. R. B. ROBROCK, II, 'Extension of the lumped bulk device model to incorporate the process of domain dissolution', *IEEE Trans. Electron Dev.*, **ED−17**, 103−107 (1970).
27. T. OHMI and S. HASUO, 'Unified treatment of small-signal space-charge dynamics in bulk-effect devices', *IEEE Trans. Electron Dev.*, **ED−20**, 303−316 (1973).
28. J. B. GUNN, 'Properties of a free, steadily travelling electrical domain in GaAs', *IBM J. Res. Develop.*, **10**, 300−309 (1966).
29. W. K. KENNEDY, 'Negative conductance in bulk gallium arsenide at high frequencies', thesis, Cornell Univ., Ithaca, New York, Sept. 1966 (quoted in reference [30]).
30. H. KROEMER, 'The Gunn effect under imperfect cathode boundary conditions', *IEEE Trans. Electron Dev.*, **ED−15**, 819−837 (1968).
31. M. P. SHAW, P. R. SOLOMON and H. L. GRUBIN, 'The influence of boundary conditions on current instabilities in GaAs', *IBM J. Res. Develop.*, **13**, 587−590 (1969).
32. H. L. GRUBIN, 'Exact solutions of the linearized equation for current flow in negative differential mobility elements', *IEEE Trans. Electron Dev.*, **ED−19**, 1294−1296 (1972).
33. H. L. GRUBIN, M. P. SHAW and P. R. SOLOMON, 'On the form and stability of electric field profiles within a negative differential mobility semiconductor', *IEEE Trans. Electron Dev.*, **ED−20**, 63−78 (1973).
34. H. KROEMER, 'Non-linear space-charge domain dynamics in a semiconductor with negative differential mobility', *IEEE Trans. Electron Dev.*, **ED−13**, 27−40 (1966).
35. H. KROEMER, 'Negative conductance in semiconductors', *IEEE Spectrum*, **5**, No. 1, 47−56 (1968).
36. D. JONES and H. D. REES, 'Accumulation transit mode in transferred-electron oscillators', *Electron. Lett.*, **8**, 567−568 (1972).
37. H. W. THIM, 'Computer study of bulk GaAs devices with random one-dimensional doping fluctuations', *J. Appl. Phys.*, **39**, 3897−3904 (1968).
38. D. D. KHANDELWAL and W. R. CURTICE, 'A study of the single-frequency quenched-domain mode Gunn-effect oscillator', *IEEE Trans. Microwav. Theory Techn.*, **MTT−18**, 178−187 (1970).
39. G. S. HOBSON, 'The external negative conductance of Gunn oscillators', *Solid-St. Electron.*, **12**, 711−717 (1969).
40. R. A. GOUGH and R. B. SMITH, 'The behaviour of a Gunn oscillator in the domain-delayed mode', *Int. J. Electron.*, **33**, 67−80 (1972).
41. P. R. SOLOMON, M. P. SHAW and H. L. GRUBIN, 'Analysis of bulk negative differential mobility element in a circuit containing reactive elements', *J. Appl. Phys.*, **43**, 159−172 (1972).
42. J. A. COPELAND, 'LSA oscillator-diode theory', *J. Appl. Phys.*, **38**, 3096−3101 (1967).
43. G. S. HOBSON, 'Velocity-field characteristics for LSA operation', *Solid-St. Electron.*, **15**, 1107−1112 (1972).
44. J. A. COPELAND, 'Theoretical study of a Gunn diode in a resonant circuit', *IEEE Trans. Electron Dev.*, **ED−14**, 55−58 (1967).
45. B. C. TAYLOR and W. FAWCETT, 'Detailed computer analysis of LSA operation in CW transferred electron devices', *IEEE Trans. Electron Dev.*, **ED−17**, 907−915 (1970).
46. W. O. CAMP, Jr., 'High-efficiency GaAs transferred electron device operation and circuit design', *IEEE Trans. Electron. Dev.*, **ED−18**, 1175−1184 (1971).
47. J. A. COPELAND, 'Doping uniformity and geometry of LSA oscillator diodes', *IEEE Trans. Electron Dev.*, **ED−14**, 497−500 (1967).
48. W. R. CURTICE and J. J. PURCELL, 'Analysis of the LSA mode including effects of space charge and intervalley transfer time', *IEEE Trans. Electron Dev.*, **ED−17**, 1048−1060 (1970).
49. R. BOSCH and H. W. THIM, 'Computer simulation of transferred electron devices using the displaced Maxwellian approach', *IEEE Trans. Electron Dev.*, **ED−21**, 16−25 (1974).

50. D. Jones and H. D. Rees, 'Electron relaxation effects in transferred-electron devices revealed by a new simulation method', *Electron Lett.*, **8**, 263—264 (1972).
51. D. Jones and H. D. Rees, 'Overlength modes of transferred-electron oscillators', *Electron Lett.*, **9**, 105—106 (1973).
52. J. S. Barrera, 'GaAs LSA V-band oscillators', *IEEE Trans. Electron Dev.*, **ED—18**, 866—872 (1971).
53. P. Jeppesen and B. I. Jeppsson, 'LSA relaxation oscillator principles', *IEEE Trans. Electron Dev.*, **ED—18**, 439—449 (1971).
54. B. I. Jeppsson and P. Jeppesen, 'LSA relaxation oscillations in a waveguide iris circuit', *IEEE Trans. Electron Dev.*, **ED—18**, 432—439 (1971).
55. W.-C. Tsai and F. J. Rosenbaum, 'Bias circuit oscillations in Gunn devices', *IEEE Trans. Electron Dev.*, **ED—16**, 196—202 (1969).
56. J. S. Heeks and A. D. Woode, 'Localized temporary increase in material conductivity following impact ionization in a Gunn-effect domain', *IEEE Trans. Electron Dev.*, **ED—14**, 512—517 (1967).
57. P. P. Bohn and G. J. Herskowitz, 'Impact ionization in bulk GaAs high field domain', *IEEE Trans. Electron Dev.*, **ED—19**, 14—21 (1972).
58. A. L. Edridge, 'Frequency stability of Gunn oscillators with variation of ambient temperature', *Solid-St. Electron.*, **15**, 1187—1196 (1972).
59. G. S. Hobson, 'Frequency-temperature relationships of X-band Gunn oscillators', *Solid-St. Electron.*, **15**, 431—441 (1972).
60. J. Bird, R. M. G. Bolton, A. L. Edridge, B. A. E. De Sa and G. S. Hobson, 'Gunn diodes with improved frequency-stability/temperature variations', *Electron. Lett.*, 299—301 (1971).
61. T. E. Hasty, R. Stratton and E. L. Jones, 'The effect of non-uniform conductivity on the behaviour of Gunn-effect samples', *J. Appl. Phys.*, **39**, 4623—4632 (1968).
62. W. S. C. Gurney, 'Contact effects in Gunn diodes', *Electron. Lett.*, **7**, 711—713 (1971).
63. W. S. C. Gurney and J. W. Orton, 'New techniques for the study of Gunn diode contacts', *Solid-St. Electron.*, **17**, 743—750 (1974).
64. R. E. Goldwasser and F. E. Rosztoczy, 'Gunn diodes for full band oscillators and amplifiers', paper presented at the Fourth European Microwave Conf., Montreux, Sept. 10—13, 1974.
65. B. W. Jervis and G. S. Hobson, 'Contact conductivity notch effects in indium phosphide microwave oscillators', *Solid-St. Electron.*, **16**, 945—950 (1973).
66. D. J. Colliver, K. W. Gray, D. Jones, H. D. Rees, G. Gibbons and P. M. White, 'Cathode contact effects in InP transferred electron oscillators', *Proc. of Fourth Int. Symp. on GaAs and related compounds*, Boulder, USA, 1972, paper 30, pp. 286—294.
67. D. J. Colliver, L. D. Irving, J. E. Pattison and H. D. Rees, 'High-efficiency InP transferred-electron oscillators', *Electron Lett.*, **10**, (1974).
68. M. P. Wase, B. W. Clark and R. F. B. Conlon, 'AgSn cathode contact in gallium-arsenide transferred-electron devices', *Electron. Lett.*, **9**, 189—190 (1973).
69. M. Suga and K. Sekido, 'Effects of doping profile upon electrical characteristics of Gunn diodes', *IEEE Trans. Electron Dev.*, **ED—17**, 275—281 (1970).
70. F. Hasegawa and M. Suga, 'Effects of doping profile on the conversion efficiency of a Gunn diode', *IEEE Trans. Electron Dev.*, **ED—19**, 26—37 (1972).
71. P. L. Shah and T. A. Rabson, 'Effect of various boundary conditions on the instabilities in transferred electron bulk oscillators', *J. Appl. Phys.*, **42**, 783—798 (1971).
72. P. L. Shah and T. A. Rabson, 'Combined doping and geometry effects on transferred-electron bulk instabilities', *IEEE Trans. Electron Dev.*, **ED—18**, 170—174 (1971).
73. J. F. Dienst, R. Dean, R. Enstrom and A. Kokkas, 'Coplanar-contact Gunn-effect devices', *RCA Rev.*, **28**, 585—594 (1967).
74. M. Reiser, 'Large-scale numerical simulation in semiconductor device modelling', *Computer Meth. Appl. Mech. Eng.*, **1**, 17—38 (1972).
75. N. Suzuki, H. Yanai and T. Ikoma, 'Simple analysis and computer simulation on lateral spreading of space charge in bulk GaAs', *IEEE Trans. Electron Dev.*, **ED—19**, 364—375 (1972).
76. C. Hilsum and H. D. Rees, 'Three-level oscillator: a new form of transferred electron device', *Electron. Lett.*, **6**, 277—278 (1970).
77. P. M. Boers, G. A. Ackett, D. H. Paxman and R. J. Tree, 'Observation of high-field domains in n-type indium phosphide', *Electron Lett.*, **7**, 1—2 (1971).

78. P. M. BOERS, 'Measurements on dipole domains in indium phosphide', *Phys. Lett.*, **34 A**, 329—330 (1971).
79. B. A. PREW, 'High-field-current instabilities in InP', *Electron Lett.*, **7**, 584—585 (1971).
80. D. H. PAXMAN, R. J. TREE and C. E. C. WOOD, 'Unstable domains in solution-grown epitaxial InP', *Electron. Lett.*, **8**, 241—243 (1972).
81. C. E. C. WOOD, R. J. TREE and D. H. PAXMAN, 'Solution-grown epitaxial InP for high-efficiency circuit-controlled microwave oscillators', *Electron. Lett.*, **8**, 171—172 (1972).
82. D. J. COLLIVER, K. W. GRAY and B. D. JOYCE, 'High-efficiency microwave generation in InP', *Electron. Lett.*, **8**, 11—13 (1972).
83. D. J. COLLIVER, 'Progress with InP transferred electron devices', paper presented at Cornell Conference No. 4, Microwave devices, circuits and applications, Cornell University, 1973.
84. H. A. HAUS, H. STATZ and R. A. PUCEL, 'Noise in Gunn oscillators', *IEEE Trans. Electron Dev.*, ED—20, 368—370 (1973).
85. P. KUHN, 'Noise in Gunn-oscillators depending on surface of Gunn diode', *Electron. Lett.*, **6**, 845—847 (1970).
86. C. P. SANDBANK, 'Synthesis of complex electronic function by solid state bulk effects', *Solid-St. Electron.*, **10**, 369—380 (1967).
87. M. SHOJI, 'Functional bulk semiconductor oscillators', *IEEE Trans. Electron Dev.*, ED—14, 535—546 (1967).
88. M. SHOJI, 'Dynamics of Gunn domains in functional bulk oscillators', Proc. Conf. Solid-St. Dev., Tokyo, 1969, *Suppl. J. Japan Soc. Appl. Phys.*, **39**, 12—18 (1970).
89. M. SHOJI, 'Theory of transverse extension of Gunn domains', *J. Appl. Phys.*, **41**, 774—778 (1970).
90. M. SHOJI, 'Two-dimensional Gunn-domain dynamics', *IEEE Trans. Electron Dev.*, ED—16, 748—758 (1969).
91. H. L. HARTNAGEL, 'Magnetic surface loading of Gunn oscillators and resulting new devices', *Solid-St. Electron.*, **13**, 931—936 (1970).
92. R. S. ENGELBRECHT, 'Solid-state bulk phenomena and their application to integrated electronics', *IEEE J. Solid-St. Circ.*, SC—3, 210—212 (1968).
93. H. L. HARTNAGEL, 'Theory of Gunn-effect logic', *Solid-St. Electron.*, **12**, 19—30 (1969).
94. H. L. HARTNAGEL and S. H. IZADPANAH, 'High-speed computer logic with Gunn-effect devices', *Radio Electron. Eng.*, **36**, 247—255 (1968).
95. S. H. IZADPANAH and H. L. HARTNAGEL, 'Gunn-effect pulse and logic devices', *Radio Electron. Eng.*, **39**, 329—339 (1970).
96. K. MAUSE, H. SALOW, A. SCHLACHETZKI, K. H. BACHEM and K. HEIME, 'Circuit integration with gate-controlled Gunn devices', 1972 *Proc. of Fourth Int. Symp. on GaAs and related compounds*, Boulder, U.S.A., 1972, paper 29, 275—285.
97. K. HEIME, 'Planare Gunn-Elemente mit Schottky-Steuerelektrode', Techn. Bericht Forschungsinst. Fernmeldetechn. Zentralamt (Deutsche Bundespost), Sept. 1971.
98. E. HESSE, K. MAUSE, H. SALOW and A. SCHLACHETZKI, 'GaAs Bauelemente für schnelle Digitalschaltungen', Techn. Bericht Forschungsinst. Fernmeldetechn. Zentralamt (Deutsche Bundespost), Oct. 1972.
99. A. SCHLACHETZKI and E. HESSE, 'Bauformen planaren Gunn-Elemente auf GaAs', Techn. Bericht Forschungsinst. Fernmeldetechn. Zentralamt (Deutsche Bundespost), Juni 1973.
100. K. MAUSE, 'Simple integrated circuit with Gunn devices', *Electron. Lett.*, **8**, 62—63 (1972).
101. A. SCHLACHETZI and E. HESSE, 'Current pulses in planar GaAs Gunn devices', *Solid-St. Electron.*, **17**, 633—635 (1974).
102. K. HEIME, 'Planar Schottky-gate Gunn devices', *Electron. Lett.*, **7**, 610—612 (1971).
103. N. HASHIZUME and S. KATAOKE, 'Control of high-field domain in GaAs by the field-effect and its application to functional devices', Third Conf. Solid-St. Dev., Tokyo, Sept. 1971.
104. N. HASHIZUME, M. KAWASHIMA and S. KATAOKA, 'Nucleation and control of departure of a high-field domain by a gate electrode', *Electron. Lett.*, **7**, 195—197 (1971).
105. K. TOMIZAWA, M. KAWASHIMA and S. KATAOKA, 'New logic functional device using tranverse spreading of a high-field domain in *n*-type GaAs', *Electron. Lett.*, **7**, 239—240 (1971).
106. K. KURUMADA, T. MIZUTANI and M. FUJIMOTO, 'High-field layers in planar Gunn diodes', *J. Phys. D: Appl. Phys.*, **7**, L49—L52 (1974).

107. K. Kurumada, T. Mizutani and M. Fujimoto, 'GaAs planar Gunn digital devices with subsidiary anode', *Electron. Lett.*, **10**, (1974).
108. A. Schlachetzki and K. Mause, 'Measurements of the influence of the *nd* product on the Gunn effect', *Electron. Lett.*, **8**, 640–642 (1972).

Problems

8.1. Show that when a high-field domain is nucleated in the bulk, the field intensities in the neutral regions on both sides of the domain should always remain equal.

Hint: Use the current continuity equation written on both sides of the domain. Electrostatic arguments can be also used.

8.2. Prove equation (8.3.15).

8.3. Consider the effect of a series resistance present in the bias circuit of a Gunn diode upon the domain nucleation and propagation (astable and monostable modes).

Hint: Use $V_D - \mathscr{E}_N$ diagrams (Figures 8.4.1 and 8.4.3).

8.4. Show that

$$\frac{dQ_D}{dt} = -\frac{Q_D}{2\tau_d}, \quad \tau_d = \frac{\varepsilon}{e\mu_d N_D} = \text{const.} < 0, \qquad (P.8.4.1)$$

where Q_D is the charge contained in the accumulation layer of a high-field domain and $\mu_d < 0$ is the negative differential mobility (see also [1], [4]).

8.5. Derive the magnitude of bias voltage and current for a GaAs Gunn diode with 10 GHz transit-time frequency and a dissipated continuous power below 4W (the device length, doping and cathode area should be indicated). Assume negligible diffusion and evaluate domain thickness, maximum electric field and growth time using the material parameters chosen above; discuss the significance of these results for the above device (*L* and *V* specified).

8.6. Use equation (8.7.2) to prove that a stable travelling high-field domain cannot exist in a Gunn diode biased by a constant current source.

8.7. Show that the triangular domain dynamics can be described by a kind of modified equal-areas rule [1], [4], [11] arising from

$$\int_{\mathscr{E}_N}^{\mathscr{E}_D} [v(\mathscr{E}) - v_D] d\mathscr{E} = 0. \qquad (P.8.7.1)$$

Then the graphical interpretation of the above equation on the $v(\mathscr{E})$ characteristics indicates that the domain grows when $v_N > v_D$ and discharges when $v_N < v_D$, etc.

8.8. Suggest a graphical method to discuss the propagation of high field domains in devices with a cathode depletion layer. Assume that the cathode drop V_{CD} is known as a function of J (or $\mathscr{E}_N$), as well as $V_D = V_D(J)$ (or $V_D = V_D(\mathscr{E}_N)$).

8.9. Explain the voltage tunability of the Gunn-diode oscillator shown in Figure 8.13.3, by using constant voltage contours in a $\mathscr{E}\text{-}x$ plane: the oscillation frequency is directly related to the distance travelled by the domain until V_D drops below the value necessary to sustain a domain [5].

8.10. Consider the Gunn device shown in Figure 8.14.3, where the Schottky gate is biased in reverse. The device is sensitive to gate triggering of the domain when the anode voltage for domain nucleation changes rapidly with the gate voltage. Discuss the dependence of the above 'sensitivity' upon the design data of this structure and the voltage bias [102].

9
Junction-gate Field-effect Transistor

9.1 Physical Principles

The junction-gate field-effect transistor (JGFET), proposed by Shockley [1], is based on a completely different operating principle than the junction (bipolar) transistor [1]–[4]. The JGFET is a majority-carrier (unipolar) device, essentially a semiconductor resistor whose cross-sectional area is modulated by the depletion region of a *pn* junction acting as a 'gate' for the conductive channel.

Many constructive forms are possible (discrete and integrated devices) [4]. Figure 9.1.1 shows a device made using the planar-epitaxial technology. An *n*-type layer is grown epitaxially on a heavily doped *p*-type substrate: the thickness and doping of this layer should be as uniform as possible. A *p*-type gate junction is achieved by diffusion. Note the fact that the diffusion front extends laterally and

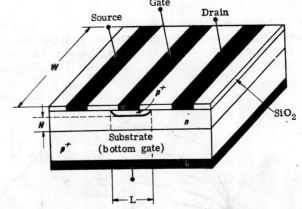

Figure 9.1.1. *n*-channe unction-gate field-effect transistor (JGFET) fabricated by planar-epitaxial methods.

increases the gate-junction area. The diffusion depth determines the channel cross-sectional area and should be precisely controlled*. Source and drain ohmic contacts provide the conductive path for the current flowing through the channel. Part of the epitaxial layer between the source and drain is not modulated by the gate depletion region and acts as a parasitic series resistance. There exist a source series resistance and a drain source resistance.

---

* A double diffusion technique (the channel and the gate are diffused successively) is also possible. However, in both cases the processes should be tightly controlled, thus making the fabrication difficult. Recent advances in ion implantation and Schottky barrier contacts ameliorates the performances and reproducibility of JGFET's (see below).

The device shown in Figure 9.1.1 is essentially symmetrical with respect to source and drain. In normal operation charge carriers move from source to drain. The JGFET of Figure 9.1.1 is an *n*-channel device (electrons flow from source to drain and the conventionally positive drain current flows from drain to source). The drain voltage with respect to source, V_D, is positive, the gate voltage relative to the source, V_G, is negative (the gate-channel *pn* junction is reversely biased). All sign conventions are correspondingly changed for a *p*-channel device. The p^+ substrate may be contacted and therefore acts as a second gate: the device becomes a field-effect tetrode. However, in most cases the top and bottom (substrate) gates are connected together and the transistor is symmetrical with respect to the longitudinal axis (plane) of the channel. A symmetrical transistor acts as two (identical) unsymmetrical transistors in parallel: therefore the computations done for a symmetrical structure do not essentially differ from those for an unsymmetrical one, as will be

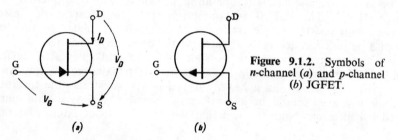

Figure 9.1.2. Symbols of *n*-channel (*a*) and *p*-channel (*b*) JGFET.

shown below. In certain cases the channel can be fabricated on an insulating substrate and the transistor is constructively unsymmetric.

The gate current, I_G, is the reverse current of a *pn* junction (a few nA or tens of nA) and can be neglected in most cases. Therefore, only the drain characteristics

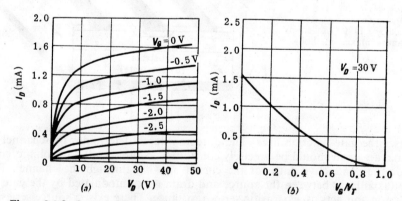

Figure 9.1.3. Output characteristics (*a*) and a transfer characteristic (*b*) for an *n*-channel JGFET (2 N 3066) at 25 °C.

$I_D = I_D(V_G, V_D)$ will be considered. Some experimental output characteristics (source grounded, Figure 9.1.2) are shown in Figure 9.1.3a and one representative transfer characteristic in Figure 9.1.3b.

As far as the drain voltage, V_D, is relatively small the device acts, between drain and source, as a resistor whose resistance is controlled by the gate voltage, V_G (linear region of the characteristic near the origin). As V_G becomes negative and its absolute value increases, the depletion regions of the gate-channel junctions become wider and reduce the effective (conductive) cross-sectional area of the channel (Figure 9.1.4a). Eventually, the conductive channel is totally "pinched-off" and the drain current becomes zero at a certain threshold, or pinch-off, voltage, V_T.

Consider a certain characteristic, for example $V_G = 0$. At a given V_D, the potential along the neutral channel changes from 0 to V_D. Thus both gate junctions become increasingly reverse biased as we proceed from the source toward the drain (Figure 9.1.4b). As the drain voltage increases, the average cross-sectional area for current flow is reduced. The channel resistance will increase. Figure 9.1.3a does indeed show that the channel resistance is non-linear.

As the drain voltage, V_D, is further increased, the two depletion regions touch near the drain, at the point X (the channel is pinched-off near the drain), as illustrated by Figure 9.1.4c ($V_D = V_{D\,sat}$). As V_D is increased above $V_{D\,sat}$ the depletion region near the drain will merely thicken and the point X will move slightly toward the source (Figure 9.1.4d) [3]. However, the potential drop between the source and the point X will still remain the same, $V_{D\,sat}$. The magnitude of the current flowing in the neutral channel will be determined by the above potential drop and will increase slightly with V_D ($> V_{D\,sat}$) because the effective channel length is somewhat reduced. The carrier flow injected into the depletion region at the point X is limited by the number of charge carriers arriving at this point. The continuity of the drain current through the depleted region is provided by the field existing in this region, as in a punched-through structure (Chapter 5).

Figure 9.1.3a shows a relative saturation of the drain current above a certain drain voltage $V_{D\,sat}$. The finite drain resistance beyond $V_{D\,sat}$ is associated with the reduction of the effective channel length. An idealized theory (Section 9.3) postulates a complete saturation with the onset of channel pinch-off near the drain (Figure 9.1.4c). The drain current vanishes when the gate voltage falls below a certain "threshold" value, which corresponds to the channel pinch-off *at the source end* of the channel.

At high drain voltages the drain current increases abruptly due to the onset of avalanche breakdown near the drain. This is due to the field determined by the drain-gate potential difference. The dependence of V_D at breakdown upon V_G is illustrated by Figure 9.1.5 (*p*-channel device).

Figure 9.1.6 illustrates the temperature dependence of the transfer characteristic of a JGFET. The change of device parameters with temperature is relatively small compared to that encountered in bipolar devices. The drain current at $V_G = 0$ increases with decreasing temperature due to the increase of carrier mobility. However, the converse is true at low current levels (V_G close to the threshold value), and this is due to the decrease of the width of the gate-to-channel barrier with increasing temperature (the absolute value of the threshold voltage increases with a temperature coefficient of the order of 2—3 mV°C^{-1} [4]). For a proper bias ($|V_G - -V_T| \approx 0.7$ V[4]) the change with temperature is minimized (see the crossover point of the characteristics in Figure 9.1.6).

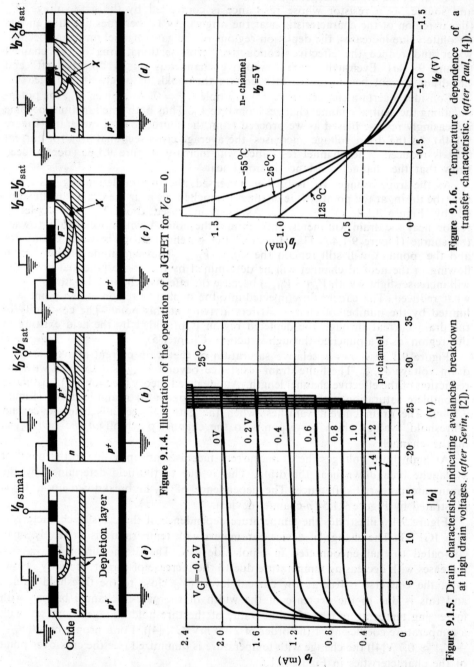

Figure 9.1.4. Illustration of the operation of a JGFET for $V_G = 0$.

Figure 9.1.5. Drain characteristics indicating avalanche breakdown at high drain voltages. (*after Sevin*, [2]).

Figure 9.1.6. Temperature dependence of a transfer characteristic. (*after Paul*, [4]).

The discrete JGFET has the advantages of high input impedance, low temperature and radiation effect. A version of this device, the Schottky-barrier gate FET, can be used in microwave circuits and fast integrated circuits (Section 9.7).

The elementary theory of JGFET is presented in Section 9.2—9.4. Certain results of a computer analysis of JGFET operation are given in Section 9.5. The alternating-current equivalent circuit and the device noise are discussed in Sections 9.6 and 9.8, respectively.

9.2 Ideal Field-effect Transistor: a Generalized Steady-state Model

Figure 9.2.1 shows an idealized model of a field-effect transistor (FET). The electrical conduction takes place inside a sheet-like channel of length L and width W. W, usually called the gate width is much larger than L, lateral effects (along the z axis) are neglected and thus the current flowing along the y direction is independent of z.

The electric potentials are as follows: the source end of the channel (Figure 9.2.1) is grounded, the drain has the potential V_D, the gate plane is equipotential at V_G and, finally, the channel potential is $V = V(y)$ (of course, $V(0) = 0$ and $V(L) = V_D$).

The total drain current, I_G, will be calculated as a function of the sheet conductivity of the channel per unit of gate area, σ_S. Consider electron conduction by drift from $y = 0$ to $y = L$. The sign

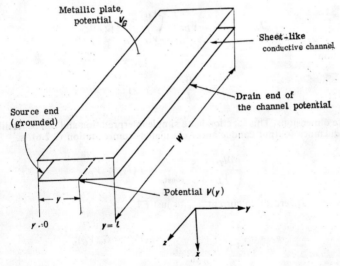

Figure 9.2.1. Idealized one-dimensional model of a field-effect transistor (FET).

conventions are: σ_S is considered positive and also the drain current I_D (which flows in the negative y direction). The longitudinal electric field carrying the electrons along the channel has the absolute value dV/dy and thus

$$I_D = \sigma_S W dV/dy. \qquad (9.2.1)$$

The mobile charge density inside the channel of a field-effect structure depends upon the configuration of the electric field: a two-dimensional electric field, $\mathscr{E} = \mathscr{E}(x, y)$ should be calculated. This is a difficult mathematical problem which is usually avoided by using Shockley's gradual approximation [1], [5]. The longitudinal component of the field, $\mathscr{E}_y$, is assumed to be very small as compared to the transversal component, $\mathscr{E}_x{}^*$. Therefore, the mobile charge at point y in the channel is primarily determined by the transversal field and this transversal field, $\mathscr{E}_x$, depends on the gate-to-channel potential difference $V_G - V(y)$. In a more general case, the conductivity $\sigma_S = \sigma_S(y)$ depends upon $V_G - V(y)$ (the carrier mobility may depend upon $\mathscr{E}_x$ as in a surface channel FET, Chapter 11), such that

$$\sigma_S = \sigma_S(y) \text{ through } \sigma_S = \sigma_S(V_G - V(y)). \tag{9.2.2}$$

In this generalized model we do not specify the physical mechanism by which the channel conductivity is modulated. The model is valid, at least approximately, for the junction-gate field effect transistor JGFET and for various types of insulated-gate field-effect transistors (IGFET's) [6].
Let us introduce

$$\chi = V_G - V(y) \tag{9.2.3}$$

in equation (9.2.2) and integrate equation (9.2.1) from

$$y = 0, \quad V(0) = 0, \quad \chi = V_G; \tag{9.2.4}$$

to

$$y = L, \quad V(L) = V_D, \quad \chi = V_G - V_D. \tag{9.2.5}$$

The result is

$$I_D = I_D(V_G, V_D) = \int_{V_G - V_D}^{V_G} F(\chi)\, d\chi, \tag{9.2.6}$$

when the function

$$F = \frac{W}{L} \sigma_S = F(\chi) \tag{9.2.7}$$

has conductance dimensions. This function has a simple interpretation and can be readily calculated. The drain conductance (output conductance) calculated from equation (9.2.6) is

$$g_d = \left. \frac{\partial I_D}{\partial V_D} \right|_{V_G = \text{const.}} = F(V_G - V_D) \tag{9.2.8}$$

and this conductance at zero drain current, g_{d0}, is just F†

$$g_{d0} = \left. \frac{\partial I_D}{\partial V_D} \right|_{\substack{V_G = \text{const.} \\ V_D \to 0}} = F(V_G) = G_0(V_G). \tag{9.2.9}$$

The steady-state characteristics $I_D = I_D(V_G, V_D)$ can now be calculated analytically, graphically or numerically when $G_0 = G_0(V_G)$ is known. A similar formula was derived by Shockley [7] for the JGFET, and later generalized for an arbitrary impurity distribution along the y direction [7].

---

* The validity of this assumption depends upon the particular FET structure considered, its geometrical and material parameters, the applied bias and usually fails in the vicinity of the drain, where the longitudinal field is stronger.
† Note the fact that, according to equation (9.2.6), $V_D = 0$ determines $I_D = 0$ (all $I_D - V_D$ characteristics pass through the origin).

9. Junction-gate Field-effect Transistor

The low current conductance G_0 increases with V_G above a certain *threshold voltage*, V_T. Below this voltage the conductance is zero and the current cannot flow. Figure 9.2.2 shows $G_0 = G_0(V_G)$ for a transistor normally open, $G_0(0) > 0$ (a depletion-type transistor), and for a transistor normally off (enhancement-type). The drain current can be calculated (Figure 9.2.3) as an area defined by

$$I_D = \int_{V_G-V_D}^{V_G} G_0(V_G)\, dV_G = I_D(V_G, V_D), \quad (9.2.10)$$

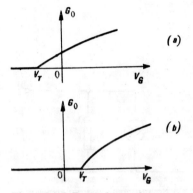

Figure 9.2.2. Dependence of channel conductance upon gate-to-source voltage for sp n-channel FET; depletion type $V_T < 0$ (a) and enhancement type, $V_T > 0$ (b).

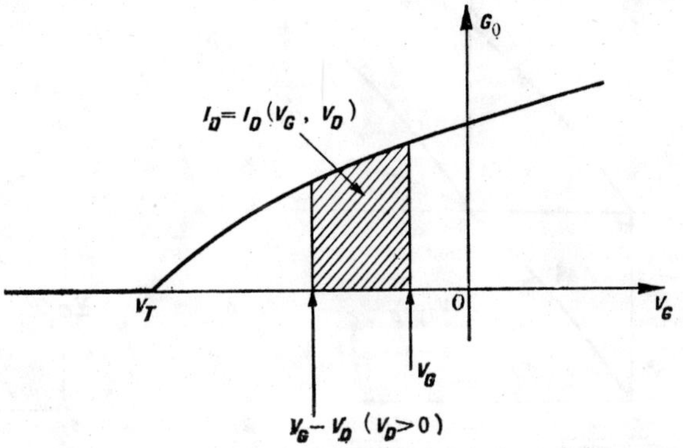

Figure 9.2.3. Illustrating the computation of I_D by using equation (9.2.10).

where V_G under the integral is regarded as an independent variable. Note the fact that for $V_D > V_{D\,\mathrm{sat}}$, where

$$V_{D\,\mathrm{sat}} = V_G - V_T, \quad (9.2.11)$$

the drain current, equation (9.2.10), remains constant at

$$I_D = I_{D\,\mathrm{sat}} = \int_{V_T}^{V_G} G_0(V_G)\, dV_G. \quad (9.2.12)$$

For $I_D = I_{D\,\mathrm{sat}}$ the potential difference between the gate and the channel becomes $V_G - V(L) = V_G - V_{D\,\mathrm{sat}} = V_T$ and thus the local channel conductance vanishes. However, near the saturation and above, the gradual approximation is no longer valid because the longitudinal field at the drain end of the channel is no longer negligible compared to the transversal field and the amount of mobile charge in the vicinity of the drain at a given moment is not directly related to $V_G - V(y)$, as it was assumed in equation 9.2.2. Despite this inconsistency we shall retain equations (9.2.10)–(9.2.12) for device operation. The physical mechanism responsible for current saturation should be clarified for each practical FET structure.

The ideal FET described by the above model has the following property: the entire family of steady-state characteristics

$$I_D = I_D(V_D) \text{ for } V_G = \text{const.} \quad (9.2.13)$$

can be obtained from a single characteristic $V_G = \text{const.}$, for example $V_G = 0$ (we assume $V_T < 0$). The proof is very simple. Equation (9.2.6) yields successively:

$$I_D(V_G, V_D) = \int_{V_G-V_D}^{0} F(\chi)\, d\chi - \int_{-V_G}^{0} F(\chi)\, d\chi = I_L(0, V_D - V_G) - I_D(0, -V_G) \quad (9.2.14)$$

and therefore $I_D(V_G, V_D)$ for a given V_D can be obtained from $I_D(0, V_D)$ characteristic. A graphical construction (suggested by Wedlock [8] for a junction-gate FET) is illustrated in Figure 9.2.4. The $I_D(V_G, V_D)$ characteristic is obtained from $I_D(0, V_D)$ by two successive translations along the horizontal and vertical axes, respectively. We stress the fact that the above demonstration is valid

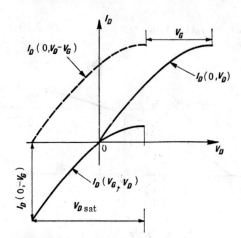

Figure 9.2.4. Graphical construction of a characteristic $I_D = I_D(V_G, V_D)$ for $V_G = $ const., by using $V_G = 0$ characteristic, according to equation (9.2.14). (*after Wedlock*, [8]).

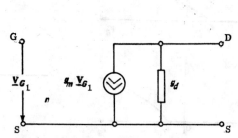

Figure 9.2.5. Small-signal low-frequency equivalent circuit of an idealized FET (common-source connection). V_{G1} is the phasor of the alternating component of the gate-to-source voltage.

for any type of FET, provided that: (a) the gradual approximation is justified, (b) the mobile charge in the channel can be written as shown by equation (9.2.2), (c) the unipolar current from source to drain is carried by drift (equation (9.2.1)).

The ideal FET discussed here has the small signal equivalent circuit of Figure 9.2.5 (common-source connection). The input conductance is zero (no gate current flows). The drain resistance is

$$g_d = \left. \frac{\partial I_D}{\partial V_D} \right|_{V_G = \text{const.}} = G_0(V_G - V_D). \tag{9.2.15}$$

The transconductance can be written

$$g_m = \left. \frac{\partial I_D}{\partial V_G} \right|_{V_D = \text{const.}} = G_0(V_G) - G_0(V_G - V_D). \tag{9.2.16}$$

Both formulas are valid below saturation drain voltage ($V_D < V_{D\,\text{sat}} = V_G - V_T$). The output (drain) conductance in saturation is zero

$$g_{d\,\text{sat}} = G_0(V_G - V_{D\,\text{sat}}) = G_0(V_T) = 0 \tag{9.2.17}$$

and the transconductance becomes

$$g_{m\,\text{sat}} = G_0(V_G) - G_0(V_G - V_{D\,\text{sat}}) = G_0(V_G) = g_{d0}. \tag{9.2.18}$$

Therefore, the transconductance in the saturation region is identical to the zero-current drain conductance at the same gate voltage (both of them do not depend upon the drain voltage).

9.3 Elementary Theory of Junction-gate Field-effect Transistor

Figure 9.3.1a shows schematically an unsymmetrical junction-gate FET. The channel is n-type (effective length L between source and drain, width W and total depth, from the metallurgical junction to the insulating substrate, H). When $V_D = 0$,

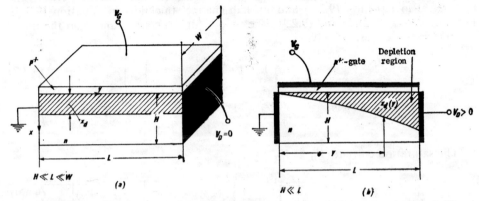

Figure 9.3.1. Unsymmetrical n-channel JGFET biased with $V_D = 0$ (a) and $V_D > 0$ (b).

the effective depth of the channel (neutral n region) is uniform and equal to $H - x_d$, where (Chapter 2)

$$x_d \simeq \left[\frac{2\varepsilon(V_{bi} - V_G)}{eN_D}\right]^{1/2} \qquad (9.3.1)$$

is the width of the depletion region, V_{bi} is the built-in potential of the pn junction and N_D is the channel doping. The p-side is assumed to be much heavily doped. The zero-current ($V_D = 0$) conductance of the channel is therefore

$$G_0(V_G) = G_N\{1 - [2\varepsilon(V_{bi} - V_G)/eN_D H^2]^{1/2}\}, \qquad (9.3.2)$$

where

$$G_N = \frac{e\mu_n N_D HW}{L} \qquad (9.3.3)$$

is the intrinsic channel conductance. According to the idealized theory of Section 9.2, the steady-state characteristics are obtained by integrating equation (9.2.10) and the result is

$$I_D = G_N\left\{V_D - \frac{2}{3}\frac{(V_D + V_{bi} - V_G)^{3/2} - (V_{bi} - V_G)^{3/2}}{(V_{bi} - V_T)^{1/2}}\right\}, \qquad (9.3.4)$$

where

$$V_T = V_{bi} - \frac{eN_D H^2}{2\varepsilon} \qquad (9.3.5)$$

is derived from $G_0(V_T) = 0$. Equation (9.3.4) is valid for

$$V_D < V_{D\,\text{sat}} = V_G - V_T. \tag{9.3.6}$$

The same result can be directly obtained by computing the current flowing through a channel of non-uniform depth (Figure 9.3.1b). If the structure is symmetrical (the channel is limited by two plane-parallel gates), the channel depth H in the above equations will be replaced by the total channel thickness (the distance between the two gates) in equation (9.3.3) and by half-channel thickness in equation (9.3.5). From equations (9.2.15) and (9.2.16) the output (drain) conductance and the transconductance are, respectively,

$$g_d = \frac{\partial I_D}{\partial V_D}\bigg|_{V_G=\text{const.}} = G_N\left[1 - \left(\frac{V_{bi} - V_G + V_D}{V_{bi} - V_T}\right)^{1/2}\right], \tag{9.3.7}$$

$$g_m = \frac{\partial I_D}{\partial V_G}\bigg|_{V_D=\text{const.}} = G_N\frac{(V_{bi} - V_G + V_D)^{1/2} - (V_{bi} - V_G)^{1/2}}{(V_{bi} - V_T)^{1/2}}. \tag{9.3.8}$$

The saturation current ($V_D \geqslant V_{D\,\text{sat}}$) is

$$I_{D\,\text{sat}} = G_N\left\{\frac{2}{3}\left[\left(\frac{V_{bi} - V_G}{V_{bi} - V_T}\right)^{1/2} - 1\right](V_{bi} - V_G) + \frac{1}{3}(V_G - V_T)\right\}. \tag{9.3.9}$$

Consider the simplified case $V_{bi} = 0$. The transfer characteristic for the saturation region can be written in normalized form:

$$\frac{I_{D\,\text{sat}}}{I_{DSS}} = 1 - 3\frac{V_G}{V_T} + 2\left(\frac{V_G}{V_T}\right)^{3/2}, \tag{9.3.10}$$

where I_{DSS} is the maximum current, at $V_G = 0$ (for $V_G > 0$ the gate-channel junction is forward biased), namely

$$I_{DSS} = -\frac{G_N V_T}{3} = \frac{WH^3}{6L}\frac{\mu_n e^2 N_D^2}{\varepsilon}. \tag{9.3.11}$$

Equation (9.3.10) is plotted in Figure 9.3.2 and compared to the empirical square-law relationship

$$I_{D\,\text{sat}} = I_{DSS}\left(1 - \frac{V_G}{V_T}\right)^2. \tag{9.3.12}$$

The differences are small. If the doping is non-uniform in a direction perpendicular to the junction gate the transfer characteristic $I_{D\,\text{sat}} = I_{D\,\text{sat}}(V_G)$ will be closer to the parabolic one [2], [4]. In practice the experimental characteristic is distorted by the presence of the source resistance (Problem 9.4). If the parasitic resistances were absent, the transconductance in saturation would be equal to the zero current drain resistance at the same gate voltage. This is not true when the source and drain resistances are taken into account, as illustrated by Figure 9.3.3.

9. Junction-gate Field-effect Transistor

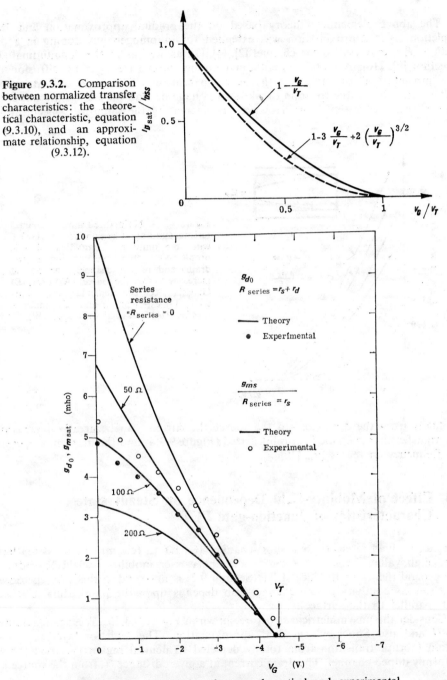

Figure 9.3.2. Comparison between normalized transfer characteristics: the theoretical characteristic, equation (9.3.10), and an approximate relationship, equation (9.3.12).

Figure 9.3.3. Comparison between theoretical and experimental low-current conductance, g_{d0}, and saturation transconductance, $g_{m\,sat}$. The calculated values are corrected for the series resistances as follows: $(g_{d0})_{\text{corrected}} = g_{d0}[1 + (r_s + r_d)g_{d0}]^{-1}$ and $(g_{m\,sat})_{\text{corrected}} = g_{m\,sat}[1 + r_s g_{m\,sat}]^{-1}$. (adapted from Grove, [3]).

436 Bulk- and Injection-controlled Devices

The above elementary theory based on the gradual approximation and the depletion layer approximation was extended for inhomogeneous doping in a direction, x, transversal to the channel [2], [4], [7], as well as in the longitudinal (y) direction [9], [10]. A certain longitudinal doping profile (realized by diffusion or ion implantation) can improve the device parameters. The pinch-off saturation voltage and the source to drain breakdown voltage depend only upon the impurity concentration near the drain. However, by increasing the impurity concentration

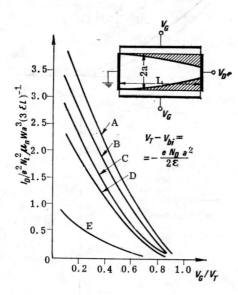

Figure 9.3.4. Normalized transfer characteristic (saturation region) for a JGFET with non-uniform longitudinal doping profile (N_L is the concentration at the drain end of the channel, the doping increases toward the source): (A) linear, (B) Gaussian, (C) complementary error function (D) exponential, and (E) uniform. (after DeMassa and Catalano, [10]).

gradually from the drain toward the source the current can be greatly increased. The transfer characteristic is also linearized (Figure 9.3.4) and the maximum operating frequency increases [9], [10].

9.4 Effect of Mobility-field Dependence on Steady-state Characteristics of Junction-gate FET

The electric field intensity in a short-channel JGFET (a few micrometers) reaches quite high values (tens of kV cm^{-1}) and the carrier mobility is field-dependent. The general theory of the ideal FET (Section 9.2) is not valid in this case, because the sheet conductivity of the channel also depends upon the longitudinal electric field (parallel to the surface).

Consider the unsymmetrical structure shown in Figure 9.3.1b. We again assume $L \gg H$ and use Shockley's gradual approximation. The depletion layer approximation (abrupt transition from totally depleted to neutral regions) is used for a uniformly doped channel. The drain current at a given distance, y, from the source is

$$I_D = eN_DW[H - x_d(y)]\,v(y), \tag{9.4.1}$$

where $v(y)$ is the electron velocity determined by the longitudinal electric field at the distance y, and

$$x_d(y) = \left[\frac{2\varepsilon(V(y) - V_G + V_{bi})}{eN_D}\right]^{1/2}. \qquad (9.4.2)$$

Many authors consider a velocity-field dependence of the form [11]–[13]

$$v = \frac{\mu_n \mathscr{E}_y}{1 + \mathscr{E}_y/\mathscr{E}_c}, \qquad (9.4.3)$$

where μ_n is the low-field mobility and $\mathscr{E}_c$ is a critical field.

Equations (9.4.1)–(9.4.3) yield [12]

$$\frac{I_D}{3I_p} = \frac{dv_0}{d\chi_0} \frac{1 - (v_0 + t_0)^{1/2}}{1 + Z_0 dv_0/d\chi_0}, \qquad (9.4.4)$$

where

$$v_0 = \frac{V(y)}{V_{bi} - V_T}, \quad t_0 = \frac{V_{bi} - V_G}{V_{bi} - V_T}, \quad \chi_0 = \frac{y}{L}, \qquad (9.4.5)$$

$$Z_0 = \frac{V_{bi} - V_T}{L\mathscr{E}_c}, \quad I_p = \frac{eN_D\mu_n HW}{3L}(V_{bi} - V_T) = I_{DSS} \qquad (9.4.6)$$

(see equations (9.3.3) and (9.3.5)). Equation (9.4.4) can be readily integrated by separating the variables. The boundary conditions are:

$$y = 0 \ (\chi_0 = 0) \to V(y) = 0 \ (v_0 = 0), \qquad (9.4.7)$$

$$y = L \ (\chi_0 = 1) \to V(y) = V_D \ (v_0 = v_{0D}), \qquad (9.4.8)$$

$$v_{0D} = \frac{V_D}{V_{bi} - V_T}. \qquad (9.4.9)$$

The result is

$$\frac{I_D}{I_p} = \frac{3v_{0D} - 2[(v_{0D} + t_0)^{3/2} - t_0^{3/2}]}{1 + Z_0 v_{0D}} \qquad (9.4.10)$$

and can also be written

$$\frac{I_D}{I_p} = \frac{3\dfrac{V_D}{V_{bi} - V_T} - 2\left[\left(\dfrac{V_D - V_G + V_{bi}}{V_{bi} - V_T}\right)^{3/2} - \left(\dfrac{V_{bi} - V_G}{V_{bi} - V_T}\right)^{3/2}\right]}{1 + V_D/\mathscr{E}_c L}. \qquad (9.4.11)$$

The above results (as well as those given in the previous section) are valid at most up to $V_D = V_{D\,\text{sat}}$. For $V_D > V_{D\,\text{sat}}$ we postulate

$$I_D = I_D(V_G, V_{D\,\text{sat}}) = I_{D\,\text{sat}} = \text{const}. \qquad (9.4.12)$$

$V_{D\,\text{sat}}$ is defined and computed as the value of V_D for which the drain conductance vanishes. Figure 9.4.1 shows that $V_{D\,\text{sat}} \ne V_G - V_T$ for the field-dependent mobility case. $Z_0 = 0$ corresponds to $\mu_n = \text{const}$. When Z_0 is comparable to unity, $V_{D\,\text{sat}}$ becomes considerably lower than $V_G - V_T$.

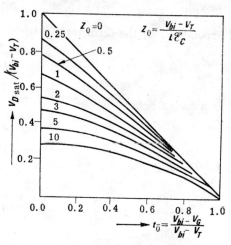

Figure 9.4.1. Effect of the mobility field-dependence on the drain saturation voltage. (*after Lehovec and Zuleeg*, [12])

The effect of mobility decrease on current-voltage characteristics is illustrated by Figure 9.4.2. Figure 9.4.3 shows the dependence of $g_{m\,\text{sat}}$, the transconductance in saturation, on the gate voltage for several values of the device parameter Z_0. For large Z_0 compared to unity $g_{m\,\text{sat}}$ changes only slightly with V_G: in this range the carrier velocity is almost constant.

The mechanism of current saturation in Shockley's model (constant mobility) is related to the channel pinch-off which occurs at the drain end of the channel, for $V_D = V_{D\,\text{sat}} = V_G - V_T$. The current saturation in the model with field-dependent mobility occurs before the pinch-off and a conductive channel of finite thickness conducts the saturation current in the drain region. When the carrier velocity is saturated at the value v_sat (equal to $\mu_n \mathscr{E}_c$, according to equation (9.4.3), the current continuity requires a neutral channel of constant thickness [12], [14], [15]

$$\delta_0 = \frac{I_{D\,\text{sat}}}{eN_D v_\text{sat} W}. \qquad (9.4.13)$$

Chiu and Ghosh [13] show, in contrast to the above assumption, that the channel continues to narrow toward the drain and saturated carrier velocity implies carrier accumulation near the drain.

Both mechanisms of current saturation discussed above (complete channel pinch-off and velocity saturation) were derived within the gradual channel approximation. Clearly, this approximation is inaccurate at large drain voltages, in the vicinity of the drain, where the electric field must be found from the two-dimensional Poisson equation including the effect of mobile charge carriers. Moreover, there exist practical structures which cannot be described, even approximately, by Shockley's model, such as structures with a short (narrow) gate compared to the channel

9. Junction-gate Field-effect Transistor

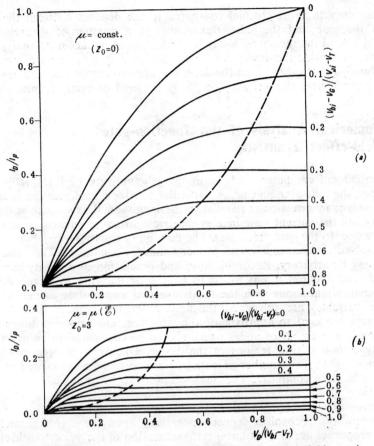

Figure 9.4.2. Effect of mobility reduction at intense electric fields on the output characteristics of the JGFET. (*after* Lehovec and Zuleeg, [12]).

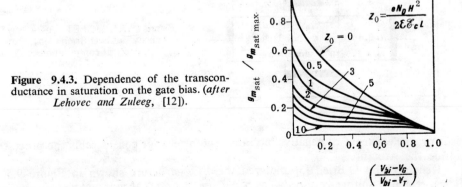

Figure 9.4.3. Dependence of the transconductance in saturation on the gate bias. (*after* Lehovec and Zuleeg, [12]).

or structures with short channel compared to the distance between the gates (or between the gate and the substrate). Although a number of theoretical papers [14]–[19] extend the theory by analytical calculations, an accurate analysis of physical processes inside the device can be performed only by using a computer. Such an exact analysis also shows that the depletion approximation is sometimes inacceptable: the conductive channel may be partly depleted or even accumulated.

9.5 Numerical Analysis of the Junction-gate Field-effect Transistor

Two dimensional computer analyses of the junction-gate FET [5], [20]–[25], are based upon the following scheme. A particular electrode configuration is considered in the xy plane (y direction is parallel to the longitudinal axis which is the axis of symmetry for the current flow in a symmetrical structure whereas x is normal to the above and to the gate junctions). The continuity equations and Poisson's equation are solved simultaneously with proper boundary conditions. The semiconductor doping may be arbitrary. Recombination and generation of minority carriers were taken into account but it was found to have little influence in silicon devices [5]. The differential equations with the local potential and mobile carrier densities as unknown variables are solved by the finite-difference method. Two-dimensional nodal arrays are used to approximate the structure. The principal difficulties arise due to the large number of nodes which should be considered in order to describe physically certain critical regions of the device. One array is necessary for each unknown variable. The number of data which have to be manipulated is too large even for a modern computer. The final solution can be approached by successive computations using arrays with smaller and smaller distances between nodes. The selection of boundary conditions is sometimes difficult.

The operation mechanism depends essentially upon the device geometry. The material properties are also important: the saturation of the velocity-field character-

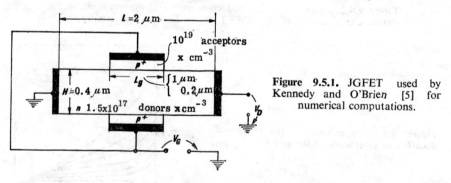

Figure 9.5.1. JGFET used by Kennedy and O'Brien [5] for numerical computations.

istic or the existence of negative mobility regions change considerably the processes inside the structure.

Kennedy and O'Brien [5] analyzed the FET structure shown in Figure 9.5.1. The n-region is uniformly doped. The p^+n gate junction is abrupt and unsymmetrical.

The dimensions of this structure are particularly small. The gate length L_g has several values, such that $2L_g/H = 5$ (long gate), $2L_g/H = 1$ (short gate) and $2L_g/H = 0.143$ (very short gate). Figure 9.5.2 shows the calculated potential distribution in two

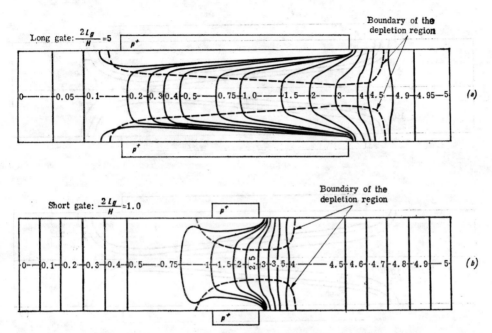

Figure 9.5.2. Calculated potential distribution (with field-dependent mobility) with $V_G = 0$ and $V_D = 5$ V. *(after Kennedy and O'Brien, [5])*.

structures. The boundary of the depletion region (considered at 50% depletion of mobile carriers) is also shown. Shockley's gradual approximation implies that in the neutral region the equipotential contours are perpendicular to the longitudinal axis of symmetry, whereas in the space-charge region these contours are parallel to the longitudinal axis. In an actual device the first requirement is approximately satisfied whereas the second one is not (Figure 9.5.2), especially for low $2L_g/H$ ratios.

Figure 9.5.3 shows constant carrier concentration contours (the percentage of carrier depletion is indicated) in a long-gate device, for three values of V_D. Constant mobility is assumed. The channel is never pinched-off completely when the source-drain electric current flows. A pinch-off point on the longitudinal axis is defined by the limit of the quasi-neutral channel (below 10% depletion). In a fashion consistent with Shockley's theory, Figure 9.5.3 shows that the pinch-off point moves toward the source with an increase of V_D. As this structure is biased more deeply into current saturation (which occurs for $V_D \simeq 3.6$ V) a change in V_D produces only a small relative change in the pinch-off point: this latter change is proportional to the variation in I_D [5].

If a field-dependent carrier mobility is assumed, electric current saturation will occur at a drain voltage below that required for pinch-off. For the structure consi-

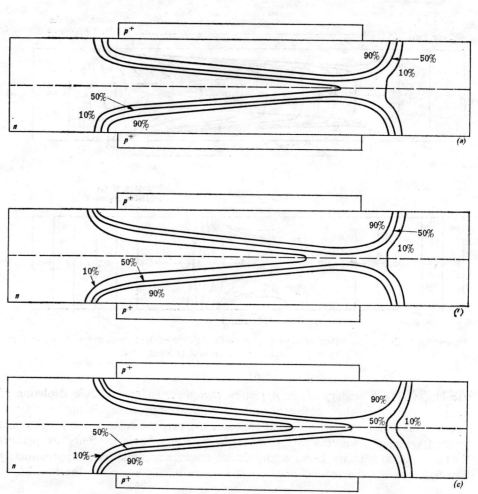

Figure 9.5.3. Calculated mobile electron distribution (constant mobility), for $V_G = 0$, $V_D = 4V$(a), $5V$(b), $7V$(c). The degree of electron depletion is shown in %. (*after Kennedy and O'Brien*, [5]).

dered in Figure 9.5.4, the current saturation occurs above to about 2.5 to 3.0 V ($V_G = 0$), whereas the voltage calculated for channel pinch-off is $|V_T| \simeq 3.6$ V. Figure 9.5.4 shows that the pinch-off point remains at a nearly fixed location regardless of the source-drain voltage and it is located close to the drain end of the gate junctions. It was shown [5] that a region of mobile carrier accumulation exists at the source side of the narrow conductive channel which provides the conduction path in the pinch-off region. An increase of V_D increases the density of these mobile carriers and thus the region of carrier depletion cannot extend toward the source. The mobile carrier accumulation is due to the necessity to maintain current conti-

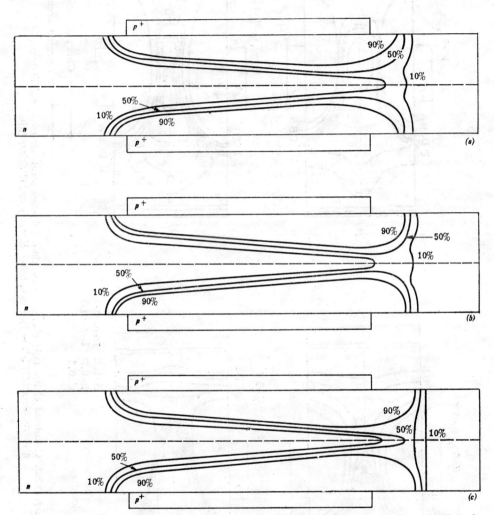

Figure 9.5.4. Calculated electrostatic charge distribution (field-dependent electron mobility), for $V_G = 0$, $V_D = 3V$ (a), $5V$ (b), $7V$ (c). The degree of mobile electron depletion is shown in %. (*after Kennedy and O'Brien*, [5]).

nuity in the presence of carrier velocity saturation. In addition, a region of carrier depletion is formed at the drain end of the channel to overcome the electrostatic charge in the accumulation region; thereby longitudinal electric fields of equal magnitude exist at the source and drain contacts [5].

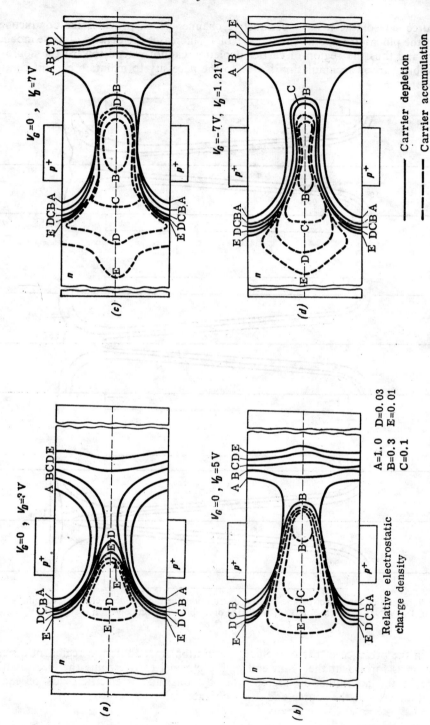

Figure 9.5.5. Calculated electrostatic charge distribution (field-dependent electron mobility) in a short-channel ($2L_g/H = 1$) device. (*after Kennedy and O'Brien*, [5]).

9. Junction-gate Field-effect Transistor

This carrier accumulation is most pronounced in structure with relative short gates, as illustrated by Figure 9.5.5*. Calculations for V_D between 1.0 and 3.0 V show that the onset of induced carrier accumulation is coincident with the first indications of current saturation. Figure 9.5.5 indicates that the pinch-off point moves toward the drain contact with an increase of V_D, corresponding to the expansion of the accumulation region. Here the 'pinch-off' point is simply the boundary between carrier depletion and carrier accumulation. The definition pinch-off is different from that introduced originally by Shockley [1]. Accumulated mobile carriers in the source-drain channel appear as an electrostatic barrier that limits the true channel pinch-off by the gate-junction space-charge layers.

The classical analysis of junction-gate FET's operation indicates that the conductive channel is completely 'pinched-off' by the space-charge layers of the gate junctions. When the transport of charge carriers is 'velocity-limited' the gate junctions constrict but do not pinch-off the source-drain conduction path and this constriction induces a region of carrier accumulation (Figure 9.5.5d). Because a large part of the applied V_D is supported by the double layer formed (the accumulation and depletion regions mentioned above) (Figure 9.5.5), this double layer cannot be eliminated in short gate devices by the application of a gate voltage. Instead, the channel is constricted and the drain current is reduced without a true pinch-off.

Kennedy and O'Brien [5] also calculated the drain characteristics. Figure 9.5.6 shows the effect of including the mobility-field dependence : the saturation is harder and occurs at lower drain voltages, the drain current decreases.

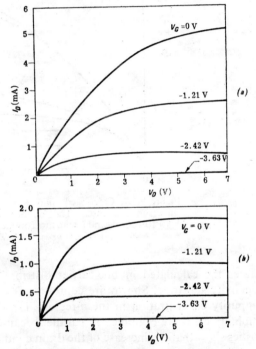

Figure 9.5.6. Electrical characteristics of the JGFET of Figure 9.5.1 (long gate, $2L_g/H = 5$): constant drift mobility (a) and field dependent drift mobility (b). (*after Kennedy and O'Brien*, [5]).

An important device parameter is the output (drain) resistance in saturation. Numerical computations reported by Kim and Yang [20], [21] and also analytical results of Grebene and Ghandhi [15], indicate that this conductance increases with decreasing $2L/H$ (or $2L_g/H$, L_g is equal or slightly longer than in their models).

---

* Kim and Yang [21] have also shown that in short-channel FET's carrier accumulation in the channel leads to a space-charge-limited current (régime) at higher drain voltages.

A substantial increase of saturation conductance can be expected in a junction-gate FET with a low-conductivity semiconductor substrate [24], [25]. The structure of Figure 9.5.1 is modified as shown in Figure 9.5.7 (inset). Figure 9.5.8 illus-

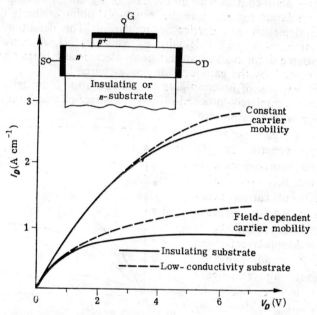

Figure 9.5.7. Calculated characteristics of the JGFET structure shown in the inset. The donnor density in the channel is 1.5×10^{17} cm^{-3}, and in the low conductivity substrate — 10^{12} cm^{-3}.
(*after* Kennedy and O'Brien, [25]).

trates the calculated mobile-carrier distribution within the device with low-conductivity substrate. Shown are contours of constant electron density related to the impurity concentration in the n-region. The constant mobility and field-dependent mobility cases are examined in parallel because the computed steady-state characteristics show that the increase of the drain conductance is related to both the substrate and the hot carrier effects. In Figure 9.5.8 a layer of excess mobile carriers lies essentially parallel to the metallurgical high-low (nn^-) junction. A detailed calculation shows that a significant part of the drain current arises from carrier transport through this layer of excess carriers on the lowly-doped side of the channel-substrate boundary.

In the presence of velocity saturation a process of carrier accumulation takes place in the pinch-off region (see above). Figure 9.5.8b shows that the accumulation region (forced by the gate-junction space-charge layer) expands in the low-conductivity substrate. The substrate provides this conduction through the accumulated drain region and increases the device conductance in saturation.

A similar numerical analysis for junction-gate FET's constructed from negative-mobility materials (GaAs and InP) was reported by Himsworth [22], [23]. The carrier accumulation in the channel was confirmed. Moreover, short-channel devices exhibit a negative-resistance drain characteristic, confirmed by experiment [26]. The absence of moving domains and Gunn oscillations is explained by the influence of the substrate [26] or of the high drain field [27].

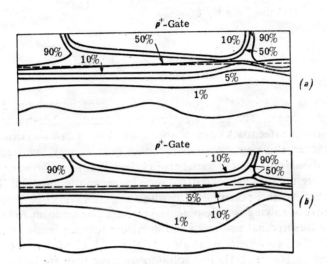

Figure 9.5.8. Calculated contours of constant majority-carrier density within the JGFET with low-conductivity substrate shown in Figure 9.5.7 with constant carrier mobility (*a*) and field-dependent carrier mobility (*b*). (*after Kennedy and O'Brien*, [25]).

A two-dimensional analysis of non-stationary behaviour of JGFET was reported by Reiser [28]. Another recent progress in JGFET analysis [29] consists in a two-dimensional particle model with a Monte-Carlo description of scattering events: one type of carrier for Si and two for GaAs devices (see Chapter 1). Such an analysis is necessary because the device dimensions are very small, comparable to the distance required for the establishment of the 'static' mobility condition (0.5 in GaAs) [29].

9.6 High-frequency Behaviour

We shall now briefly consider the high-frequency behaviour of the conventional junction-gate FET as described by Shockley's model presented in Section 9.3. The basic hypotheses are: gradual approximation, abrupt boundary between the depletion region and the neutral channel, negligible direct gate current [4], [30]—[32]. The case studied here is that of an unsymmetrical structure with a step (abrupt) gate-channel junction and uniformly-doped channel.

At low frequencies the alternating-current equivalent circuit is that in Figure 9.2.5 (small signal operation, common-source connection). At higher frequencies *reactive* effects should be taken into account. A *quasi-stationary* analysis introduces capacitances associated to the variation of the fixed charge in the channel, Q_f, namely

$$G_G = \left.\frac{\partial Q_f}{\partial V_G}\right|_{V_D=\text{const.}} = C_G(V_G, V_D) \tag{9.6.1}$$

and

$$C_D = \left.\frac{\partial Q_f}{\partial V_D}\right|_{V_G=\text{const.}} = C_D(V_G, V_D). \tag{9.6.2}$$

These capacitances are similar to junction-barrier capacitances and depend upon the steady-state voltages. The equivalent circuit in Figure 9.2.5 should be completed by a drain-to-gate (feedback) capacitance C_D and the gate-to-source capacitance $C_G - C_D$. However, this new circuit is incorrect even at very low frequencies. An exact analysis should include the capacitive current that flows between the channel and the gate: therefore the channel current is no longer continuous [30]. The problem of solving for the channel-potential $V(y)$ with the gate-channel current included is analogous to that of solving for the potential along a non-uniform RC transmission line, with both longitudinal resistance and shunt capacitance being voltage dependent. Such a lumped parameter equivalent of the FET divided in incremental elements was used by Hauser [31]. The solution for very high frequencies should be obtained by difficult iterrations. An approximation which uses an analog RC network with a few cells, was suggested by Das and Schmidt [32].

Figure 9.6.1 shows the equivalent circuit derived by Richer [30] for small-signal operation and low frequencies. The capacitances C_{ij} ($i, j = 1, 2$) are frequency

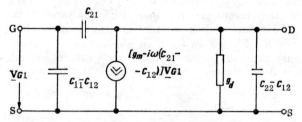

Figure 9.6.1. Small-signal equivalent circuit of a JGFET at relatively low frequencies.

independent and exhibit the bias dependence shown in Figure 9.6.2. C_{ij} are normalized to C_{gc}, which is gate-to-channel depletion capacitance with $V_G = V_T$ and $V_D = 0$ (complete pinch-off at zero drain current). C_{11} and C_{12} are identical with the quasi-stationary capacitances C_G and C_D. The negative output capacitance $C_{DS} = C_{22} - C_{12} < 0$ (Figure 9.6.2) and the delay introduced by $C_{21} - C_{12} > 0$

in the expression of the complex forward transadmittance are, however, imprevisible by intuition. Both effects are attributed to the inertia of charge carriers in transit.

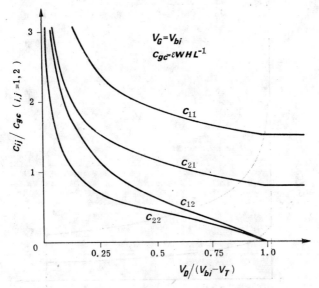

Figure 9.6.2. Bias dependence of low-frequency device capacitances C_{ij} (Figure 9.6.1) for the abrupt-junction unsymmetrical FET discussed in Section 9.3. C_{gc} is the depletion capacitance of the gate junction when $V_D = 0$ and $V_G = V_T$ (the channel is fully depleted). (*after* Richer, [30]).

The small-signal equivalent circuit in saturation takes the simple form of Figure 9.6.3 (note the fact that C_{12} and C_{22} become zero for $V_D > V_{D\,sat} = V_G - V_T$, $V_G = V_{bi}$, the output conductance is also zero: these results are valid at any gate

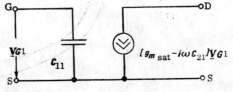

Figure 9.6.3. Small-signal low frequency equivalent circuit in the saturation region ($C_{12} = C_{22} = 0$, $g_d = 0$ in Figure 9.6.1).

bias). C_{12} and C_{21} depend upon the gate bias as shown in Figure 9.6.4. This dependence was experimentally verified as shown in Figure 9.6.5. The existence of C_{12} is therefore demonstrated and explains, in principle, the high-frequency fall-off of transconductance.

At high frequencies the transconductance $g_{m\,sat}$ will be replaced by a (forward) transadmittance

$$Y_{m\,sat} = g_{m\,sat}\left(1 - i\omega\,\frac{C_{21}}{g_{m\,sat}}\right) \simeq \frac{g_{m\,sat}}{1 + i\dfrac{\omega}{\omega_g}}, \qquad \omega \ll \omega_g = \frac{g_{m\,sat}}{C_{21}}. \qquad (9.6.3)$$

ω_g is defined over leaf as a cutoff frequency: however, the approximate expression for $Y_{m\,\text{sat}}$ is not valid for $\omega = \omega_g$. Moreover, the equivalent circuit of Figure 9.6.3 is valid only if [30]

$$\omega \ll \omega_0 = \frac{1}{\tau_0} = \frac{\mu_n |V_{bi} - V_T|}{L^2} = \frac{eN_D \mu_n H^2}{2\varepsilon L^2}. \tag{9.6.4}$$

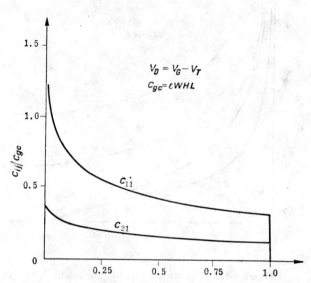

Figure 9.6.4. Saturation capacitances as a function of gate bias. (*after Richer*, [30]).

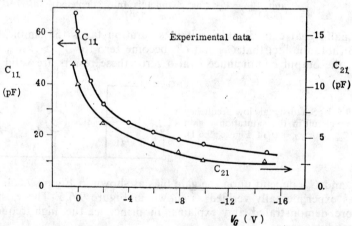

Figure 9.6.5. Experimental data for small-signal capacitances measured beyond pinch-off (in saturation). (*after Richer*, [30]).

The maximum operating frequency of an FET is sometimes evaluated [1], [3] at the ratio between the transconductance and the input capacitance (the input capacitance in common-source connection is defined with drain connected to source for the alternating signal, see Figure 9.6.1) in saturation (normal operating region for amplifiers), such that

$$\omega_{max} = \frac{g_{m\,sat}}{C_{11}}. \qquad (9.6.5)$$

$g_{m\,sat}$, C_{11} and ω_{max} are plotted in Figure 9.6.6 versus the relative gate bias and normalized to their values for $V_G = V_{bi}$, which is the limit for normal operation. Note the fact that in this limit case,

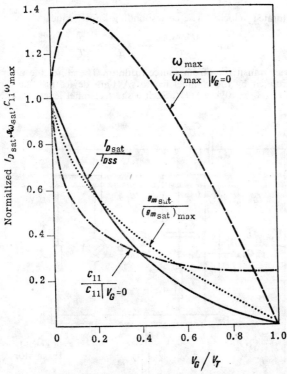

Figure 9.6.6. The drain current, transconductance, input capacitance and maximum frequency, all computed in saturation, normalized to their values at $V_G = 0$ and represented versus V_G/V_T. (see for example Das and Schmidt, [32]).

$g_{m\,sat}$ is equal to the channel intrinsic conductance G_N (see Section 9.3), whereas C_{11} is three times the gate-to-channel capacitance C_{gc}. It can be readily shown that

$$\omega_{max}\bigg|_{V_G=V_{bi}} = \frac{2}{3\tau_0}, \qquad (9.6.6)$$

whereas (Figure 9.6.4) at $V_G = V_{bi}$, $C_{21} = (2/5)C_{11}$ such that

$$\omega_y = \frac{5}{2}\omega_{max} = \frac{5}{3\tau_0}. \qquad (9.6.7)$$

Although there are different definitions for the cutoff frequency of the transconductance or maximum operating frequency [30]–[32], the device figure of merit for its high-frequency operation is always related to $1/\tau_0 = \omega_0$. Equation (9.6.4) shows that higher performances are obtained for higher channel conductivity and higher H/L ratios. High-mobility semiconductors and short-channel transistors (L/H as low as possible) are preferred. However, the shorter the channel length the higher the field intensity and the carrier mobility becomes field-dependent. The maximum attainable transconductance is

$$(g_{m\,sat})_{max} = \frac{2\varepsilon v_{sat} W}{H}, \qquad (9.6.8)$$

where v_{sat} is the saturated velocity. The corresponding limit frequency is

$$\omega_{max} = \frac{(g_{m\,sat})_{max}}{C_{gc}} = \frac{2v_t}{L} = \frac{2}{T_L}, \qquad (9.6.9)$$

where T_L is the carrier transit time from source to drain. Therefore, the ultimate limit of device performance is determined by the channel length. A 1 μm device has the theoretical frequency limit of about 32 GHz ($v_{sat} \simeq 10^7$ cm s^{-1})*. Such a short channel length is attainable by modern technology.

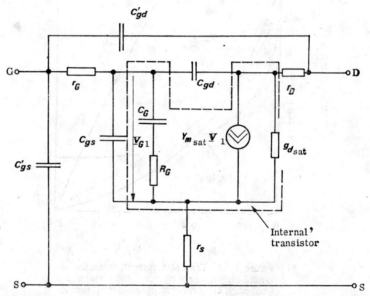

Figure 9.6.7. Equivalent small-signal network for a JGFET including the extrinsic elements corresponding to the semiconductor 'chip'. A symmetrical structure is considered: in an unsymmetrical one the parasitic resistances and capacitances associated to the substrate should be considered (see Figure 9.6.8). (after Das and Schmidt, [32]).

The high frequency performances of real devices is critically influenced by parasitic elements. Figure 9.6.7 shows a more complete equivalent circuit (grounded source, saturation region) including parasitic resistances and capacitances (series parasitic inductances and capacitances associated

---

* Ruch [33] has recently shown that for both silicon and GaAs, the *transient* carrier velocity may overshoot the saturation value at high fields (the electron distribution is first shifted in the momentum space and later energy relaxation becomes effective so that the distribution function spreads, and the drift velocity decreases, see Chapter 1) and this can partly explain very good experimental performance of GaAs FET's.

to the device terminals are neglected). The intrinsic circuit (inside the dashed lines) includes a finite drain conductance $g_{d\,sat}$ (always present in real devices) and a series gate resistance (which is a second-order effect derived from the exact computations [31]). The forward transconductance, $Y_{m\,sat}$, is complex. The parasitic source and drain resistances, r_s and r_d are associated with the channel regions which are not modulated by the gate. There is always a large, but finite series gate resistance, r_g. The gate-to-drain and gate-to-source capacitances, C_{gd} and C_{gs}, are internal parasitic capacitances. Direct capacitive coupling outside the semiconductor is possible in Schottky-gate FET structures (Section 9.7) and hence C'_{gd} and C'_{gs} in Figure 9.6.7. Figure 9.6.8 shows schematically

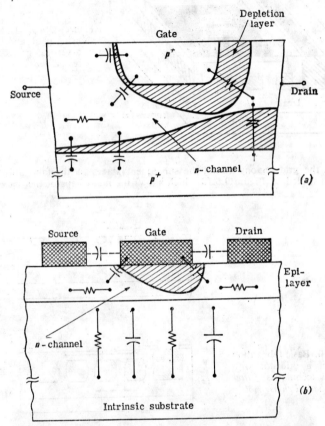

Figure 9.6.8. Schematic representation of the extrinsic elements of the equivalent circuit of (a) the conventional JGFET structure on heavily doped substrate and (b) the Schottky-gate FET structure on high resistivity substrate. (*after Das and Schmidt*, [32]).

the origin of parasitic elements in junction-gate and Schottky-gate FET structures [32]. There are these parasitic elements which effectively limit the very high frequency operation of the junction-gate FET.

The total (active plus parasitic) gate-drain and gate-source capacitances as well as the source and drain resistances are minimized in a vertical channel FET with an embbeded grid. Such a structure has also better power capabilities and is known as the gridistor [34] or the multichannel FET [35]. However, the gate series resistance, r_g, is quite large because the grid is formed from long grid fingers connected to an external gate wall as shown in Figure 9.6.9. Reducing r_g is essential for high frequency operation of this structure. One way to do this is by increasing as much as possible the impurity concentration in the grid bulk [34]. This involves the risk of impurity redis-

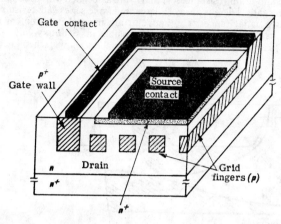

Figure 9.6.9. Isometric view showing plan and cross section of a vertical gridistor structure, (after Lecrosnier and Pelous, [36]).

tribution from the grid regions during the subsequent stages of the fabrication process [34], [35]. This is illustrated by Figure 9.6.10 which shows the main steps of the conventional planar

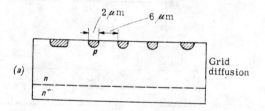

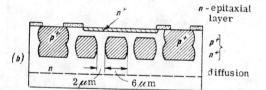

Figure 9.6.10. Basic fabrication sequences of a gridistor with conventional techniques. (after Lecrosnier and Pelous, [36]).

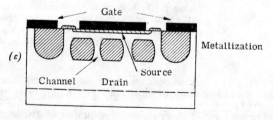

process [36]. First, p-type grid fingers are diffused in an n epitaxial layer. A second epitaxial layer is grown and thus the grid fingers are buried. Then, p^+ and n^+ regions are diffused to provide the gate wall and the source. Finally, the source, drain and gate are contacted by metal layers.

During the epitaxial growth and diffusions the vertical channels are narrowed whereas the parasitic capacitance increases as the effective gate area increases [36].

Ion implantation (see also Chapter 10) was successfully used to realize a high-conductivity grid with a sharp pattern. The n^+ source and the p^+ grid wall are diffused first (Figure 9.6.11). The boron implantation through a mask provides a p-type layer of 0.8 μm buried at 1 μm below

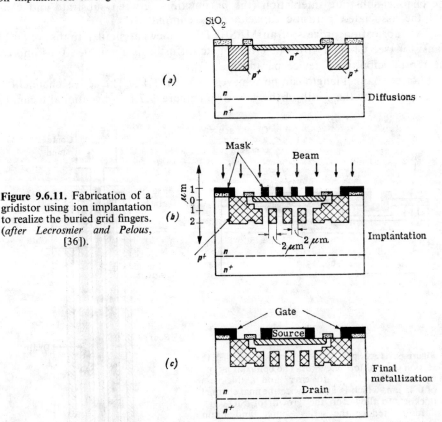

Figure 9.6.11. Fabrication of a gridistor using ion implantation to realize the buried grid fingers. *(after Lecrosnier and Pelous, [36]).*

the surface. There are no radiation-induced defects in the channel. The cross-section of the final structure is also shown. The experimental device has a 40 mA/V maximum transconductance ($V_G = 0$, $V_D = 15$ V), capacitances of a few pF and a maximum figure of merit $(g_{m\,sat})_{max}/2\pi(C_{gs}+C_{gd})$ of about 1 GHz. Due to the large active area, the same device is suitable for power applications [36].

9.7 Schottky-barrier Gate Field-effect Transistor

The *pn* junction gate in a FET can be replaced by a Schottky-barrier metal-semiconductor contact. The structure is also called the metal-semiconductor field-effect transistor (MESFET).

The operation of an MESFET is essentially the same as that of a junction-gate FET: the channel conductance is modulated by the depletion region of the Schottky barrier contact. The Schottky-barrier gate principle is particularly useful for those materials where good *pn* junctions are difficult to form, such as for GaAs. GaAs takes the advantage of its higher electron mobility and of its higher resistivity: the transistor substrate can be made semi-insulating ($\approx 10^8 \Omega$ cm) thus minimizing the channel-substrate interaction (the *pn* junction between substrate and channel and the associated parasitic capacitance is eliminated).

The fabrication process of an MESFET is somewhat similar to that of an MOS transistor (see Chapter 10). The critical gate diffusion is eliminated thus improving the reproducibility of device parameters [37].

Shorter channel length can be achieved in MESFET's. The active channel length is determined by the width of the gate strip (Figure 9.7.1, conventional technology)

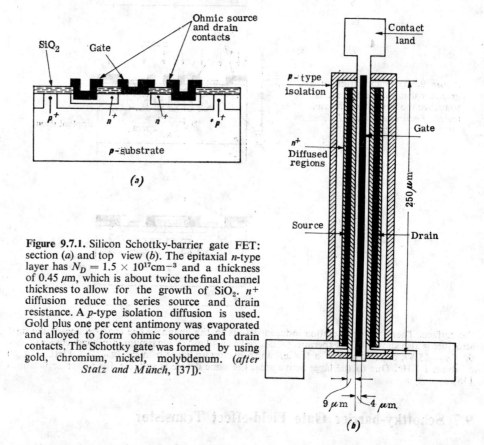

Figure 9.7.1. Silicon Schottky-barrier gate FET: section (*a*) and top view (*b*). The epitaxial *n*-type layer has $N_D = 1.5 \times 10^{17} \text{cm}^{-3}$ and a thickness of 0.45 μm, which is about twice the final channel thickness to allow for the growth of SiO_2. n^+ diffusion reduce the series source and drain resistance. A *p*-type isolation diffusion is used. Gold plus one per cent antimony was evaporated and alloyed to form ohmic source and drain contacts. The Schottky gate was formed by using gold, chromium, nickel, molybdenum. (*after Statz and Münch*, [37]).

and is usually of the order of 1 μm or a few μm. The channel is an epitaxial layer with a fraction of μm thickness. Higher transconductance require higher channel conductivity and higher W/L ratios. However, when L is small, the channel thickness, H, and doping, N_D, should be properly chosen in order to get a practical value for the pinch-off (or threshold) voltage, V_T. Figure 9.7.2 is useful for device design provided that the carrier mobility remains constant (silicon, 300 K, n-type channel). At a given value of $V_{bi} - V_T = eN_D L^2/2\varepsilon$, the transconductance (per unit length and width of the gate) can be increased by decreasing the channel thickness and increasing the doping level correspondingly [37]. $H = 0.2\,\mu\text{m}$ and $N_D \simeq$ $\simeq 10^{17}\,\text{cm}^{-3}$ yield a threshold voltage of about 3V ($V_{bi} = 0.8\,\text{V}$) and a maximum

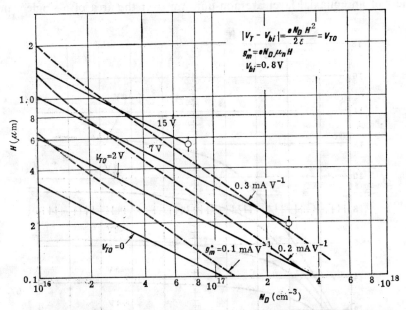

Figure 9.7.2. Diagram used for device design the circles indicate the avalanche breakdown limit for the Schottky contact. (*after Statz and Münch,* [37]).

transconductance of about 0.15 (W/L) mA/V and a few mA V^{-1} or tens of mA V^{-1} can be obtained.

Figure 9.7.3 shows experimental characteristics of a silicon device with $W = 250\,\mu\text{m}$, $L = 4\,\mu\text{m}$, $N_D = 1.5 \times 10^{17}\,\text{cm}^{-3}$, $e\mu_n N_D H = 200$ mmho, $V_{T\exp} = 6.3$ V ($V_{bi} = 0.8$ V), $r_s = r_d = 75\,\Omega$ (series source and drain resistances). The calculated maximum transconductance is $(g_{m\,\text{sat}})_{\max} = 4$ mA V^{-1} in fairly good agreement with the measured value of 3.3 mA V^{-1} (the field-dependent mobility may explain in part the difference). The total gate capacitance C is of the order of 1 pF and the figure of merit $g_m/2\pi C$ is comparable to 1 GHz.

An improved silicon MESFET [38] gives useful gain at a few GHz ($L = 1\,\mu\text{m}$, $W = 400\,\mu\text{m}$, reduced parasitics by a modified geometry, the epitaxial layer etched in the vicinity of the gate pads, etc.). The theoretical frequency performances are deteriorated by the source series resistance, r_s. A smaller resistance r_s could be obtained by reducing the gate-source spacing (electron beam exposure) or reducing the resistivity of this parasitic region by ion implantation [38].

A much higher maximum operating frequency can be achieved with a short channel GaAs MESFET [39], [40].

The ion-implanted channel MESFET's [40], [41] are characterized by greater reproducibility of device operating characteristics and higher yield. This is due to the reproducibility and uniformity of the ion-implantation doping process. The channel is created by ion implantation instead of epitaxial growth (the doping and thickness of an epitaxial layer are non-uniform over the area of a single wafer and

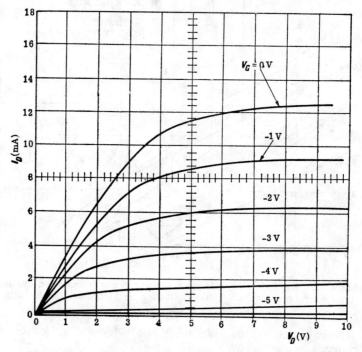

Figure 9.7.3. Experimental drain characteristic for a Schottky-barrier gate FET (gold on silicon). *(after Statz and Münch, [37])*.

differ substantially from wafer to wafer). An example of ion-implanted channel GaAs MESFET is shown in Figure 9.7.4 [49]. The 0.25 μm thick n-type layer for the channel is formed in the Cr-doped semi-insulating substrate by sulphur-ion implantation (a subsequent annealing at 800 °C activates electrically the implanted ions). The electron mobility is about 3000 cm^{2}V^{-1}s^{-1} (compared to the bulk value of about 4500 cm^2 V^{-1}s^{-1}). After the n-type layer was formed the fabrication process is similar to the conventional one, except for a chemical etching which defines the mesa channel for electrical isolation (Figure 9.7.4). The gate contact pad is located on the semi-insulating substrate: the gate capacitance and the leakage current are significantly lower than when pn junction isolates the gate from a conducting substrate. Gate to source capacitance are of the order of 1pF, and transconductances above 20 mA V^{-1}. f_{max} of 20 GHz is extrapolated for 2μm gate devices.

GaAs MESFET's find applications in microwave amplifiers [39], [42] and high speed integrated logic circuits [43]. GaAs MESFET's can be integrated along with

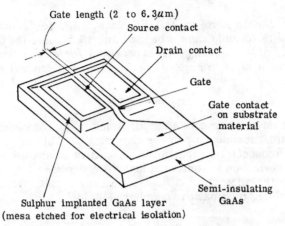

Figure 9.7.4. GaAs Schottky-barrier-gate FET with ion-implanted channel. (*after Hunsperger and Hirsch*, [40]).

Schottky diodes and resistors: they are formed in an *n*-type epitaxial layer and isolated by etching. Such a transistor exhibited a 15 ps internal delay in a large-signal switching test [43].

9.8 Noise of Junction-gate FET

There are essentially three basic sources of noise in FET's operation: the thermal noise, the shot noise, the generation-recombination noise. Figure 9.8.1 shows the noise equivalent circuit of a JGFET consisting of an ideal noise-free device and four noise generators. These are [44]:

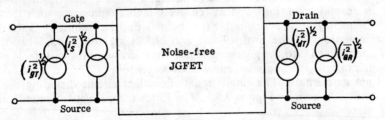

Figure 9.8.1. Network representation of noise sources in a JGFET.

(a) one output noise source (the device in common-source connection is represented by a two-port) due to the thermal noise of the conducting channel, $(\overline{i_{dT}^2})^{1/2}$;

(b) another output noise source associated to the generation-recombination effects, $(\overline{i_{GR}^2})^{1/2}$;

c) an input noise source representing the induced gate thermal noise, occurring at higher frequency due to the capacitive coupling between the gate and the channel, and denoted by $(\overline{i_{gT}^2})^{1/2}$;

d) a second input noise source related to the shot noise of the gate leakage current, $(\overline{i_g^2})^{1/2}$.

The *intrinsic thermal noise* in a JGFET arises due to fluctuations of channel resistance. Nyquist's formula cannot be used directly because the channel resistance is non-linear. A similar difficulty was experienced for diodes (see Chapter 4, for example). The drain noise computed for the 'classical' model of Section 9.3 is [4], [44], [45]

$$\overline{i_{dT}^2} = 4kT\Delta f g_m P \qquad (9.8.1)$$

where g_m is the device transconductance (common source connection) and P is a numerical factor depending upon the device bias*. This expression is valid at sufficiently low frequencies so that intrinsic reactive effects are still unimportant (Section 9.6). A correction at somewhat higher frequencies introduces new elements (capacitances) in the alternating current equivalent circuit (Section 9.6) and also the gate noise induced by capacitive coupling [44]–[46]:

$$\overline{i_{gT}^2} = 4kT\Delta f \frac{\omega^2 C_G}{g_m} Q \qquad (9.8.2)$$

In the above relation C_G is the input capacitance (effective gate-to-source capacitance) and Q another bias-dependent numerical factor. The noise currents given by equations (9.8.1) and (9.8.2) have the same physical origin and are statistically correlated.

Hot carrier effects upon the thermal noise sources should be accounted for in two ways. First, at high electric fields, when carrier mobility becomes field-dependent, the free-carrier temperature increases above the lattice and a correction should be applied to the thermal noise formula [46], [47]. On the other hand, at higher fields the field configuration inside the device is changed due to the velocity saturation, for example, and therefore the transconductance and the numerical factor in equation (9.8.1) are both affected [46].

The thermal noise sources are dominating the noise behaviour of a usual JGFET in normal conditions and at relatively high frequencies, such that the $1/f$ type low-frequency noise is negligible (see below) but the frequency is not too high such that equations of the form (9.8.1) and (9.8.2) are still valid. This frequency domain extends well into the gigahertz region for microwave transistors. We consider here as an example the analysis reported by Baechtold [46] for a GaAs MESFET with a 1 μm gate length. The small signal equivalent circuit including the thermal noise sources is shown in Figure 9.8.2. The following parasitic elements are taken into account:

(a) the source resistance, r_s, determined by the 1 μm gap between the source metallization and gate together with the resistance of the source contact;

(b) the gate metallization resistance, r_g;

---

* The applied external voltages are algebraically added to V_{bi} and normalized to the characteristic voltage $V_{bi} - V_T$ which is a physical constant of the device.

(c) the gate pad capacitance, C_L (formed by the pad — insulating oxide layer — epitaxial layer structure) and the associated series resistance, R_L, from the source to the epilayer under the pad;

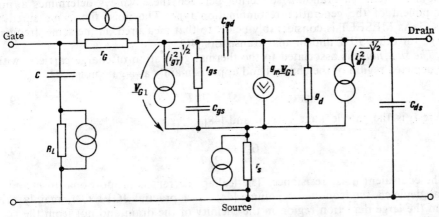

Figure 9.8.2. GaAs MESFET equivalent circuit. $R_L = 1000\ \Omega$, $r_g = 3\ \Omega$, $r_{gs} = 15\ \Omega$, $r_s = 35\ \Omega$, $g_d = 5 \times 10^{-4}$ mho, $C_L = 0.05$ pF, $C_{gs} = 0.2$ pF, $C_{dg} = 0.01$ pF, $C_{ds} = 0.05$ pF, $g_m = 0.016$ mho. (after Baechtold, [46]).

(d) a small source-to-drain capacitance, C_{ds}, between the drain metallization and the bottom of the transistor chip (connected to the source).

The thermal noise generators corresponding to the parasitic resistances, r_s, r_g and R_L are also indicated in Figure 9.8.2. The theoretical computations are based upon the circuit of Figure 9.8.2 where the intrinsic noise sources are evaluated

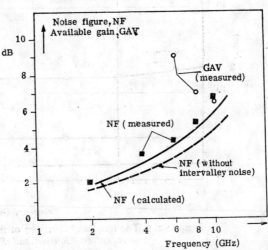

Figure 9.8.3. Optimum noise figure and gain as a function of frequency measured at $V_D = 4\text{V}$, $V_G = 0\text{V}$; —— calculated noise figure with intervalley scattering noise, – – – – calculated noise figure, without intervalley scattering noise; measured available power gain GAV with source impedance set for optimum figure. (after Baechtold, [46]).

including the hot carrier effects mentioned above. The channel is assumed relatively thin, with a short high-field region. The expected noise figure is shown by the dashed line in Figure 9.8.3, and is somewhat lower than the experimental value.

A better fit was obtained by including in the model the effect of intervalley scattering noise (solid line) [46]. It was shown in Chapter 1 that the high field conduction in GaAs is due to electrons situated in the central valley and in the satellite (upper) valleys as well. The random scattering between these valleys determines a noise component of the generation-recombination type. The noise figure measured for this GaAs MESFET is considerably below to that measured, in the same frequency range for other FET and bipolar transistors [46].

The shot noise is associated to the thermal generation of charge carriers within the depleted region of the channel and appears mainly as a gate noise:

$$\overline{i_S^2} = 2e\, i_g\, \Delta f = 4kT\, \Delta f \cdot G_{gs}, \qquad (9.8.3)$$

where i_g is the gate leakage current and [48]

$$G_{gs} = \frac{e}{2kT} i_g \qquad (9.8.4)$$

is the equivalent noise resistance. The leakage current is proportional to n_i and the total depleted semiconductor volume [44]. In a practial JGFET, i_g mainly arises from the large depletion region in the vicinity of the drain and not from the active channel (depleted) region [46]. Klaassen and Robinson [48] did indeed show that i_g depends linearly upon the gate-drain junction voltage (silicon transistors with

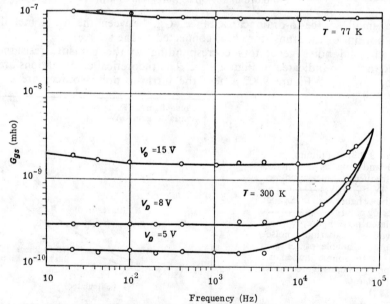

Figure 9.8.4. The frequency spectrum of the gate-leakage current shot noise. *(after Klaassen and Robinson, [48])*.

minor surface imperfections). Figure 9.8.4 shows the white noise spectrum of gate shot noise measured on a commercial device. Equation (9.8.4) was fairly well verified and i_g increased exponentially with temperature as expected.

The generation-recombination noise occurs as an 'excess' noise at relatively low-frequencies (below kHz or tens of kHz). At room temperature the generation noise in the gate-channel depletion region is the dominant source of low frequency noise. The corresponding noise is of the form [44], [48]—[50]:

$$\overline{i_{GR}^2} = \frac{k_1 \tau_1}{1+\omega^2 \tau_1^2}, \qquad (9.8.5)$$

where τ_1 is the trap time constant and K_1 is proportional to the trap density and also depends upon the device parameters and the applied bias. The time constant τ_1 increases rapidly when the temperature decreases whereas K_1 decreases monotonically with temperature [49]. It can be readily shown that when the noise is measured at a spot frequency a peak should be observed in the noise-temperature dependence. Two or more peaks were experimentally observed [49] thus indicating the presence of more than one trap level. Gold is known to act as a Shockley-Read-Hall (SRH) generation-recombination centre [44], [48], [49]. Other impurities or dislocations may have a similar effect [48], [49]. Two or more similar terms of the same form should be considered in equation (9.8.5). Such a model was used by Krishnan and Chen [50] to fit the experimental data (two generation levels were considered and a constant term, K_3, was added to account for the thermal noise component). An example is shown in Figure 9.8.5. The various curves were obtained for different irradiation levels (electron bombardment of the sample was performed, thus producing displacement of the lattice atoms and therefore traps or additional electron states in the forbidden gap). The results show that K_1 and K_2, which should be proportional to the trap density, increase with the irradiation dose. The time constants τ_1 and τ_2, are of the order of tenths of a millisecond and microseconds,

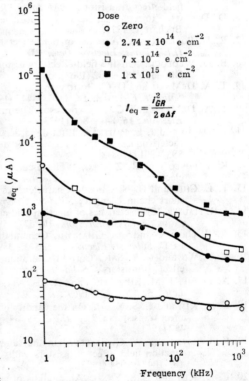

Figure 9.8.5. Frequency spectra of Texas Instruments 2N 2909 *p*-type channel JGFET at different doses of electron bombardment levels. (*after Krishnan and Chen*, [50]).

respectively. The first level was identified as due to SRH centres in the depletion region and the second one as due to SRH centres in the channel [50]. It should be stressed that the changes induced by irradiation in device conductance and transconductance are relatively small compared to the changes in the noise level. Although the gate leakage current increases very much with irradiation, this cannot account for the large change in the measured noise.

References

1. W. Shockley, 'A unipolar field-effect transistor', *Proc. IRE*, **40**, 1365–1376 (1952).
2. L. J. Sevin, 'Field-effect transistors', McGraw-Hill, New York, 1967.
3. A. S. Grove, 'Physics and technology of semiconductor devices', John Wiley, New York, 1967.
4. R. Paul, 'Feldeffekttransistoren', VEB Verlag Technik, Berlin, 1972.
5. D. P. Kennedy and R. R. O'Brien, 'Computer aided two-dimensional analysis of the junction field-effect transistor', *IBM Journ. Res. Dev.*, **14**, 95–116 (1970).
6. D. Dascalu, 'Injecția unipolară în dispozitive electronice semiconductoare', Ed. Academiei, Bucharest, 1972.
7. R. R. Bockemuehl, 'Analysis of field-effect transistors with arbitrary charge distribution', *IEEE Trans. Electron Dev.*, **ED–10**, 31–34 (1963).
8. B. D. Wedlock, 'On the field-effect transistor characteristics', *IEEE Trans. Electron Dev.*, **ED–15**, 181–182 (1968).
9. T. A. DeMassa and D. G. Goddard, 'Inhomogeneous channel resistivity field effect devices', *Solid-St. Electron.*, **14**, 1107–1112 (1971).
10. T. A. DeMassa and G. T. Catalano, 'The inhomogeneous channel FET: IGFET', *Solid-St. Electron.*, **16**, 847–851 (1973).
11. P. David and J. L. Pautrat, 'Effet de la dépendance mobilité-champ électrique sur les caractéristiques du transistor à effet de champ à jonctions', *Solid-St. Electron.*, **11**, 893–901 (1968).
12. K. Lehovec and R. Zuleeg, 'Voltage-current characteristics of GaAs J–FET in the hot electron range', *Solid-St. Electron.*, **13**, 1415–1426 (1970).
13. T. L. Chiu and H. N. Ghosh, 'Characteristics of the junction-gate field effect transistor with short channel length', *Solid-St. Electron.*, **14**, 1307–1317 (1971).
14. J. R. Hauser, 'Characteristics of junction field effect devices with small channel length-to-width ratios', *Solid-St. Electron.*, **10**, 577–587 (1967).
15. A. B. Grebene and S. K. Ghandhi, 'General theory for pinched operation of the junction-gate FET', *Solid-St. Electron.*, **12**, 573–589 (1969).
16. S. Y. Wu and C. T. Sah, 'Current saturation and drain conductance of junction-gate field-effect transistors', *Solid-St. Electron.*, **10**, 593–609 (1969).
17. A. Luque, J. M. Ruiz-Perez and R. Segovia, 'New analysis of FET saturation', *Electron. Lett.*, **5**, 420–422 (1969).
18. C.-K. Kim and E. S. Yang, 'On the validity of the gradual-channel approximation for field-effect transistors', *Proc. IEEE*, **58**, 841–842 (1970); see also *Proc. IEEE*, **59**, 431–432 (1971).
19. K. Lehovec and W. G. Seeley, 'On the validity of the gradual channel approximation for junction field effect transistors with drift velocity saturation', *Solid-St. Electron.*, **16**, 1047–1054 (1973).
20. C.-K. Kim and E. S. Yang, 'An analysis of current saturation mechanism of junction field-effect transistor', *IEEE Trans. Electron Dev.*, **ED–17**, 120–127 (1970).
21. C.-K. Kim and E. S. Yang, 'Carrier accumulation and space-charge-limited current flow in field-effect transistors', *Solid-St. Electron.*, **13**, 1577–1589 (1970).
22. B. Himsworth, 'A two-dimensional analysis of gallium arsenide junction field effect transistors with long and short channels', *Solid-St. Electron.*, **15**, 1353–1361 (1972).
23. B. Himsworth, 'A two-dimensional analysis of indium phosphide junction field effect transistors with long and short channels', **16**, 931–939 (1973).
24. M. Reiser, 'Two-dimensional analysis of substrate effects in junction FET's', *Electron. Lett.*, **6**, 493–494 (1970).

25. D. P. KENNEDY and R. R. O'BRIEN, 'Two-dimensional analysis of JFET structure containing a low-conductivity substrate', *Electron. Lett.*, **7**, 714—716 (1971).
26. T. W. TUCKER and L. YOUNG, 'GaAs negative conductance junction field effect transistor', *Solid-St. Electron.*, **17**, 31—34 (1974).
27. P. L. HOWER and N. G. BECHTEL, 'Current saturation and small-signal characteristics of GaAs field-effect transistors', *IEEE Trans. Electron Dev.*, **ED—20**, 213—220 (1973).
28. M. REISER, 'A two-dimensional numerical FET model for DC, AC and large-signal analysis', *IEEE Trans. Electron. Dev.*, **ED—20**, 35—45 (1973).
29. R. W. HOCKNEY, R. A. WARRINER and M. REISER, 'Two-dimensional particle models in semiconductor-device analysis', Report RCS23, Dept. Computer Science, University of Reading, Oct. 1974.
30. I. RICHER, 'The equivalent circuit of an arbitrarily doped field-effect transistor', *Solid-St. Electron.*, **8**, 381—393 (1965).
31. J. R. HAUSER, 'Small signal properties of field effect devices', *IEEE Trans. Electron Dev.*, **ED—12**, 605—618 (1965).
32. M. B. DAS and P. SCHMIDT, 'High-frequency limitations of abrupt junction FET's', *IEEE Trans. Electron Dev.*, **ED—20**, 779—792 (1973).
33. J. G. RUCH, 'Electron dynamics in short channel field-effect transistors', *IEEE Trans. Electron Dev.*, **ED—19**, 652—654 (1972).
34. S. TESZNER, 'Gridistor development for the microwave power region', *IEEE Trans. Electron Dev.*, **ED—19**, 355—364 (1972).
35. R. ZULEEG, 'Multi-channel field-effect transistor theory and experiment', *Solid-St. Electron.*, **10**, 559—576 (1967).
36. D. P. LECROSNIER and P. PELOUS, 'Ion-implanted FET for power applications', *IEEE Trans. Electron Dev.*, **ED—21**, 113—118 (1974).
37. H. STATZ and W. V. MÜNCH, 'Silicon and gallium arsenide field-effect transistors with Schottky-barrier gate', *Solid-St. Electron.*, **12**, 111—117 (1969).
38. W. BAECHTOLD and P. WOLF, 'An improved microwave silicon MESFET', *Solid-St. Electron.*, **14**, 783—790 (1971).
39. W. BAECHTOLD, 'X- and Ku-band amplifiers with GaAs Schottky-barrier field-effect transistor', *IEEE J. Solid-St. Circ.*, **SC—8**, 54—58 (1973); see also *Electron. Lett.*, **9**, 232—234 (1973).
40. R. G. HUNSPERGER and N. HIRSCH, 'GaAs field-effect transistors with ion-implanted channels', *Electron. Lett.*, **9**, 577—578 (1973).
41. R. A. MOLINE, W. C. GIBSON and L. D. HECK, 'An ion-implanted Schottky-barrier gate field-effect transistor', *IEEE Trans. Electron Dev.*, **ED—20**, 317—320 (1973).
42. C. A. LIECHTI and R. L. TILLMAN, 'Design and performance of microwave amplifiers with GaAs Schottky-gate field-effect transistors', *IEEE Trans. Microw. Theor. Techn.*, **MTT—22**, 510—517 (1974).
43. R. L. VAN TUYL and C. A. LIECHTI, 'High-speed integrated logic with GaAs MESFET's', *IEEE J. Solid-St. Circ.*, **SC—9**, 269—276 (1974).
44. M. B. DAS, 'FET noise sources and their effects on amplifier performance at low frequencies', *IEEE Trans. Electron. Dev.*, **ED—19**, 338—348 (1972).
45. A. VAN DER ZIEL, 'Gate noise in field effect transistors at moderately high frequencies', *Proc. IEEE*, **51**, 461—467 (1963).
46. W. BAECHTOLD, 'Noise behaviour of GaAs field-effect transistors with short gate lengths', *IEEE Trans. Electron Dev.*, **ED—19**, 674—680 (1972).
47. A. VAN DER ZIEL, 'Noise resistance of FET's in the hot electron regime', *Solid-St. Electron.*, **14**, 347—350 (1971).
48. F. M. KLAASSEN and J. R. ROBINSON, 'Anomalous noise behaviour of the junction-gate field-effect transistor at low temperatures', *IEEE Trans. Electron Dev.*, **ED—17**, 852—857 (1970).
49. J. W. HASLETT and E. J. M. KENDALL, 'Temperature dependence of low-frequency excess noise in junction-gate FET's', *IEEE Trans. Electron Dev.*, **ED—19**, 943—950 (1972).
50. I. N. KRISHNAN and T. M. CHEN, 'Effects of electron bombardment on the noise in junction gate field effect transistors', *Solid-St. Electron.*, **16**, 1233—1240 (1973).

Problems

9.1. Calculate the temperature coefficient of the threshold (pinch-off) voltage in a silicon JGFET.

9.2. Consider a JGFET with non-uniform channel doping. Show if the idealized theory of Section 9.2 may or may not be applied to this device (the gradual approximation and the abrupt depletion approximation are assumed to be valid).

Hint: Consider, separately, the cases when the doping is non-uniform along the x direction or along the y direction.

9.3. Can you modify the theory of Section 9.2 to describe a field-effect tetrode (with two independent gates on both sides of the channel)?

9.4. Find the effect of unmodulated source and drain series resistances upon the steady-state $I-V$ characteristics. Consider in particular the device transconductance, g_m, (see Figure 9.3.3) and $V_{D\,\text{sat}}$.

9.5. Calculate the mobile and fixed charges in a JGFET (the model of Section 9.3) as a function of applied voltages and derive the quasi-stationary capacitances.

9.6. Find the carrier transit-time in a JGFET and compare its minimum value to that of the dielectric relaxation time. Discuss numerical values for Si and GaAs devices.

9.7. The drain current, I_D, calculated with field-dependent carrier mobility can be obtained by simply replacing the low-field mobility in the expression of I_D computed with constant mobility by that value of $\mu = \mu(\mathscr{E})$ which is obtained when $\mathscr{E} = V_D/L$ (see for example equation (9.4.11)). Prove this result by using simple mathematical properties and discuss its range of validity (the transistor model, the theoretical approximation and the particular $\mu = \mu(\mathscr{E})$ relationship should be taken into account).

9.8. Discuss the relative importance of various noise sources as the temperature changes from 0 K up to large temperatures.

Part 3

Surface-controlled Devices

10

Metal-insulator-semiconductor Capacitor

10.1 Introduction

A very simple method of controlling the surface conductivity of a semiconductor is to apply an electric field normal to that surface. The electric field is due to the potential difference between the semiconductor bulk and a metallic plate parallel to the semiconductor surface. The applied field determines a redistribution of the mobile charge within a certain surface layer and, therefore, a modulation of the conductivity of this layer. The carrier mobilities also depend upon the surface field intensity [1].

The above-mentioned 'field-effect' is experimentally observed when the metallic electrode is very close to the semiconductor surface. A practical way to obtain this is to construct a metal-insulator-semiconductor (MIS) structure. The most known and widely used structure of this kind is the system metal-SiO_2-Si, where SiO_2 is thermally grown [2] and has a typical thickness between hundreds and thousands of angströms. The success of planar technology used for silicon transistors and integrated circuits is based upon the fact that the thermally grown silicon dioxide protects the semiconductor surface, particularly the vicinity of the planar *pn* junctions. Such a protection is necessary because the electrical properties of a free semiconductor surface exposed to the ambient are always unreproducible and unstable. Moreover, silicon dioxide has a compensating action upon the Tamm-Shockley surface states (i.e. the electronic states of a clean surface, Section 2.2). The same metal-SiO_2-Si system was used to construct surface-controlled electronic devices, hence called metal-oxide-semiconductor (MOS) devices. We shall often refer to the MOS system or MOS devices. Many properties discussed in connection to SiO_2-Si devices are representative for a metal-insulator-semiconductor (MIS) structure in general.

Let us consider the MOS structure shown in Figure 10.1.1a. The back contact to the semiconductor substrate is ohmic. The metal on top of the oxide is called gate and is biased as shown. No charge transport through the oxide is possible. We neglect the work function difference between the gate metal and the semiconductor and assume no oxide charge and no interface charge (trapped in the interface states). This is an *ideal* structure (Section 10.4) which has the band-energy diagram shown in Figure 10.1.1b when $V_G = 0$. Note that the bands of an ideal structure with zero gate potential are 'flat'.

A negative gate voltage, $V_G < 0$, will bend the bands up, as shown in Figure 10.1.2a [3]. The majority holes are attracted towards the semiconductor-oxide interface, thus forming an *accumulated* surface layer. The semiconductor bulk is neutral. On the other side of the oxide there is a negative electron charge on the gate, $Q_G < 0$, which compensates the hole charge in the semiconductor surface layer, $-Q_G > 0$. The readjustment of charge when V_G rises from $V_G = 0$ takes place through the external circuit (holes are injected into the semiconductor at the back contact) and is practically instantaneous (the dielectric relaxation time is very short, of the order o 10^{-12}s). Note the fact that the Fermi level in the semiconductor is

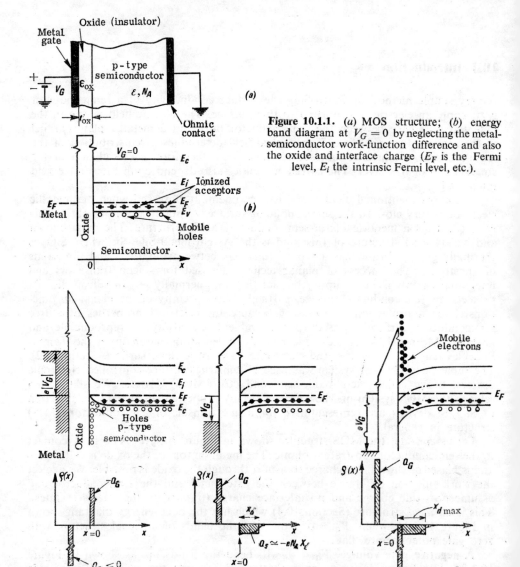

Figure 10.1.1. (*a*) MOS structure; (*b*) energy band diagram at $V_G = 0$ by neglecting the metal-semiconductor work-function difference and also the oxide and interface charge (E_F is the Fermi level, E_i the intrinsic Fermi level, etc.).

Figure 10.1.2. Effect of an external bias on the energy band diagram and charge distribution in an MOS structure: (*a*) $V_G < 0$, accumulated surface layer; (*b*) $V_G > 0$, depleted surface layer; (*c*) $V_G \gg 0$, *n*-type inversion layer at the semiconductor surface. (*after* Grove et al., [3]).

constant (thermal equilibrium, no electric current). The applied bias V_G is divided between the oxide and the surface layer.

A positive bias on the gate, $V_G > 0$ repells the holes from the oxide-semiconductor interface: the surface layer is *depleted* of mobile charge. The semiconductor charge density (per unit of surface area) Q_S (Figure 10.1.2b) is determined by the ionized donors, such that (per unit of surface area)

$$Q_S = -eN_A x_d, \qquad (10.1.1)$$

where x_d is the depletion layer width. In fact, the transition from the totally depleted surface layer to the neutral bulk is not abrupt, but gradual, as shown by the dashed line in Figure 10.1.2b. This transition takes place within a few Debye lengths (see Chapter 2 and also Section 10.2).

A higher gate voltage will bend down further the energy diagram thus moving the conduction band closer to the Fermi level. The surface layer will be *inverted*: minority electrons are pilled up at the surface and form an inversion layer (Figure 10.1.2c) which is separated from the neutral bulk by a depletion layer. The whole structure contains no excess charge such that the electron charge in the inversion layer, per unit of surface area, Q_n, is determined by

$$Q_S = Q_n - eN_A x_d = -Q_G. \qquad (10.1.2)$$

Let us assume that V_G changes from a positive value corresponding to Figure 10.1.2 b to the higher value corresponding to Figure 10.1.2c. This change has to determine an accumulation of electrons near the interface and a widening of the depletion region. Electrons have to be generated and accumulated at the surface and mobile holes have to be moved away from the edge of the neutral bulk thus widening the depletion region. However, this rearrangement cannot take place instantaneously. The time constants are the minority carrier generation lifetime, of the order of milliseconds to seconds, and the majority carrier relaxation time (dielectric relaxation time) which is very short (10^{-12}s is a typical value). The dynamic (alternating current) capacitance will depend upon the frequency of measurement. At very low frequencies the device behaviour is quasi-stationary. At higher frequencies the capacitance will be frequency dependent due to the minority carrier time constant.

The MOS capacitor can be used in principle for determination of semiconductor properties (minority carrier lifetime, doping), interface properties (interface state density, capture cross section) and oxide (insulator) properties (mobile or fixed charges within the oxide).

Another tool for surface studies is the gate-controlled diode structure shown in Figure 10.1.3 [2] and also discussed in this chapter.

The MOS transistor (Figure 10.1.4), widely used in the semiconductor integrated circuits consists of an MOS capacitor with two supplementary diffused regions, source and drain. When the surface under the gate is inverted, the conduction between the source and drain is possible through the high-conductivity surface sheet (inversion layer), which is isolated from the substrate (semiconductor bulk) by a continuous depletion region, as shown in Figure 10.1.4. Note the fact that the minority carriers in the conducting surface channel are provided by the source and the time constant for the rearrangement of the conductivity charge is the transit-time through the channel, which is very short, of the order of 10^{-9}s. The same device can be used as a tetrode if the source is not connected to the substrate. MOS transistors and tetrodes are discussed in the next chapter.

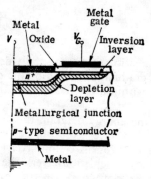

Figure 10.1.3. Gate-controlled diode structure.

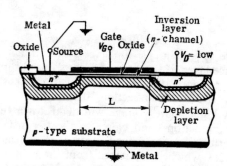

Figure 10.1.4. n-channel MOS transistor (MOST). The configuration of the surface channel and the depletion layer corresponds to a negligible drain-source voltage drop.

10.2 Electric Charge in the Semiconductor Surface Layer

Consider an MOS capacitor under steady-state bias. Provided that the electric conduction through the oxide (insulator) is impossible, the system is in thermal equilibrium. The electric charge contained in the surface layer can be calculated in terms of the surface field and the surface potential ϕ_s, measured with respect to the neutral semiconductor bulk. Such computations are usually done by assuming: (a) uniform semiconductor doping and fully ionized impurities; (b) nondegenerate statistics. The normalized electric potential $u = e\phi/kT$ is therefore given by equation (2.3.12), repeated here for convenience

$$\frac{d^2u}{dx^2} = \frac{1}{L_D^2}[\sinh(u-u_F) + \sinh u_F], \tag{10.2.1}$$

$$L_D = \left(\frac{kT\varepsilon}{2e^2 n_i}\right)^{1/2}, \quad u_F = \frac{e\phi_F}{kT} = \frac{(E_i)_{\text{bulk}} - E_F}{kT} = \operatorname{argsinh}\left(-\frac{N_D - N_A}{2n_i}\right) \tag{10.2.2}$$

The above equation can be integrated once, as follows

$$d\left(\frac{du}{dx}\right) = \frac{dx}{du}\frac{1}{L_D^2}[\sinh(u-u_F) + \sinh u_F]\,du, \tag{10.2.3}$$

$$\left(\frac{du}{dx}\right)^2 = \frac{2}{L_D^2}[u\sinh u_F + \cosh(u_F - u) - \cosh u_F]. \tag{10.2.4}$$

10. MIS Capacitor

Equation (10.2.4) satisfies the boundary conditions in the semiconductor bulk

$$y \to \infty: u \to 0 \quad (\phi = 0); \quad \frac{du}{dx} \to 0 \quad \text{(neutrality)}. \tag{10.2.5}$$

The solution $u = u(x)$ of equation (10.2.4) should also satisfy

$$y = 0: u = u_s = \frac{e\phi_s}{kT}; \quad \frac{du}{dx} = \frac{e}{kT}\frac{d\phi}{dx} = -\frac{e}{kT}\mathscr{E}_s, \tag{10.2.6}$$

$\mathscr{E}_s$ being the surface field, positive when directed towards the semiconductor (i.e. in the positive x direction).

The total charge inside the semiconductor per unit of surface area is, by the Gauss theorem:

$$Q_S = \varepsilon[\mathscr{E}(\infty) - \mathscr{E}(0)] = -\varepsilon\mathscr{E}(0) = -\varepsilon\mathscr{E}_s \tag{10.2.7}$$

and by also using equations (10.2.4) and (10.2.6), it can also be written [3]

$$Q_S = -2\frac{u_s}{|u_s|} e n_i L_D \{2[\cosh(u_s - u_F) - \cosh u_F + u_s \sinh u_F]\}^{1/2}, \tag{10.2.8}$$

where the factor $u_s/|u_s|$ was introduced to provide the correct sign of Q_s, as indicated by Table 10.2.1 (see also Table 2.3.1).

Table 10.2.1

Sign of Electric Charge in Semiconductor

	p-type semiconductor ($u_F > 0$)			n-type semiconductor ($u_F < 0$)		
	accumulation	depletion	inversion	accumulation	depletion	inversion
Surface potential, $\phi_s = \frac{kT}{e} u_s$	<0	>0	>0	>0	<0	<0
Total semiconductor charge, Q_S	>0	<0	<0	<0	>0	>0
Electron charge, Q_n	—	—	<0	<0	—	—
Hole charge, Q_p	>0	—	—	—	—	>0

The total semiconductor charge Q_S has three components: the mobile electron charge

$$Q_n = -e \int_0^\infty n\,dx, \tag{10.2.9}$$

the mobile hole charge

$$Q_p = e \int_0^\infty p\,dx \tag{10.2.10}$$

and the charge of ionized impurities. Q_n and Q_p can be calculated, in principle, by using $n = n_i \exp(u - u_F)$ and $p = n_i \exp(u_F - u)$ (see equations (2.3.5) and (2.3.6)) where $u = u(x)$ is given by equation (10.2.8). The solution $u = u(x)$ cannot be obtained, however, in analytical form.

We shall consider below the mobile charge in an inversion layer; for example the charge occurring at an inverted surface of a p-type semiconductor is

$$Q_n = - e \int_0^{x_i} n(x) \mathrm{d}x = - e \int_{u_s}^{u_F} \frac{n(u) \mathrm{d}u}{(\mathrm{d}u/\mathrm{d}x)}, \qquad (10.2.11)$$

where $\mathrm{d}u/\mathrm{d}x$ is given by equation (10.2.8). The charge Q_n was arbitrarily defined as the electron charge contained between the surface $x = 0$ and the plane $x = x_i$ where the semiconductor is intrinsic, such that $u(x_i) = u_F$. The electron charge contained in the surface space-charge layer on the right of $x = x_i$ (Figure 10.2.1) is anyway negligible. One obtains

$$Q_n = - \frac{u_s}{|u_s|} e n_i L_D \int_{u_F}^{u_s} \frac{\exp(u - u_F) \mathrm{d}u}{\{2[\cosh(u - u_F) - \cosh u_F + u \sinh u_F]\}^{1/2}} \qquad (10.2.12)$$

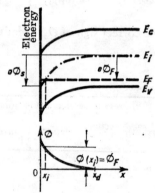

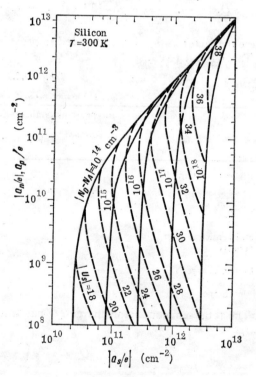

Figure 10.2.1. p-type semiconductor with surface inversion layer (between $x = 0$ and $x = x_i$) separated from the neutral bulk by a depletion layer ($x_i < x < x_d$).

Figure 10.2.2. The surface density of the mobile charge in the inversion layer as a function of the total semiconductor charge per unit of surface area, for a few doping concentrations (the surface potential in kT/e units is also shown). (*after* Grove et al., [3]).

(inversion layer charge only). The above integral can be performed only numerically. Such calculations were done for silicon at room temperature and the results are shown in Figure 10.2.2. The mobile charge in the inversion layer (p-type or n-type semiconductor) is plotted *versus* the total semiconductor charge (which is proportional to the surface field), the parameter being the effective doping concentration. The $|u_s|$ = constant curves are also shown.

Figure 10.2.2 shows that for a given semiconductor doping the mobile charge in the inversion layer increases abruptly with $|\mathscr{E}_s| \propto |Q_s|$ above a certain 'critical' surface field. For a very strongly inverted surface, at very high surface fields, the total net charge becomes indeed equal to the mobile charge in the inversion layer (see also Problem 10.1). This mobile charge occupies a very thin layer as compared to the depletion width (10...100 Å, [2]: see Problem 10.2) [4].

Numerical calculations also indicate the interesting fact that under very strong inversion conditions the total fixed charge remains constant. We can write in general

$$Q_S = Q_n + e(N_D - N_A)x_{de}, \qquad (10.2.13)$$

where the last term is the fixed space charge (per unit of surface area) and x_{de} is an effective width of the depletion region at the semiconductor surface. Figure 10.2.3 shows that $x_{de} \approx |Q_S|/e|N_D - N_A|$ below the surface field required for surface inversion (see Figure 10.2.2), whereas $x_{de} \simeq$ const. at high $|Q_S|$. Therefore, the depletion layer does not further penetrate into the semiconductor at very high surface fields.

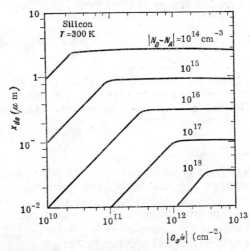

Figure 10.2.3. The effective depth of the depletion region as a function of the total charge density induced within the semiconductor. (*after* Grove et al., [3]).

This is due to the screening effect of the electron sheet at the inverted surface. Grove [3] explains the same situation by noting that under strong inversion even a very small increase of the band bending is sufficient to determine an enormous increase of the mobile charge in the inversion layer.

10.3 A Simplified Model of the Surface Layer

If the surface is depleted or inverted we can use the depletion approximation known from the *pn* junction and the Schottky barrier. Consider, first, that the surface of the *p*-type semiconductor is depleted. The semiconductor is assumed fully depleted from $x = 0$ to $x = x_d$ and neutral for $x > x_d$. The Poisson equation $d^2\phi/dx^2 = eN_A/\varepsilon$ yields a parabolic potential distribution

$$\phi = \phi_s(1 - x/x_d)^2, \quad \phi_s = eN_A x_d^2/2\varepsilon \tag{10.3.1}$$

which is fully determined by specifying the surface potential ϕ_s.

The condition for strong inversion is defined as follows. The electron concentration at surface, n_s, should exceed the ionized impurity concentration N_A^*. It can be readily shown that $n_s = p_{\text{bulk}}$ yields

$$\phi_{s(\text{inv})} = 2\phi_F. \tag{10.3.2}$$

At the onset of strong inversion, according to the discussion in the previous section, the depletion thickness should reach its maximum value

$$x_{d\max} = (2\varepsilon\phi_{s(\text{inv})}/eN_A)^{1/2} = [2\varepsilon(2\phi_F)/eN_A]^{1/2} \tag{10.3.3}$$

and the ion charge in the bulk remains approximately constant at the value

$$Q_B = -eN_A x_{d\max} \tag{10.3.4}$$

(per unit of surface area).

Grove [2] indicated that exact computer calculations prove satisfactorily the above simplified model[†]. In practice, this is a very useful simplification because the mobile electron charge can now be calculated simply as

$$Q_n = Q_S - Q_B, \tag{10.3.5}$$

where Q_B is given by equations (10.2.3) and (10.2.4) and Q_S is imposed by external conditions and calculated from electrostatic considerations, as shown below. No numerical calculations are necessary. The stronger the inversion, the better the above approximation. A more detailed calculation is necessary for weak inversion.

Note the similarity between equation (10.3.3) giving the depletion width for strong inversion and equation (2.7.6) giving the depletion width of an abrupt asymmetrical pn junction at thermal equilibrium ($\phi_{s(\text{inv})}$ stands for the built-in potential V_{bi}). There is an analogy between the physical situations described by these formulas, because the very thin inversion layer is separated from the p-type substrate by a depleted region, as in an n^+p junction. However, the n-type surface layer is electrically-induced and not formed by a metallurgical process as in an ordinary pn junction. Therefore the inverted surface layer can be considered a *field-induced* junction.

---

* In the true sense of the word, the surface is inverted when the electron concentration at surface exceeds the hole concentration, i.e. $n_s = p_s = n_i$.

† Hauser and Littlejohn [4] computed the carrier distribution within the inversion layer by neglecting the fixed space charge in this layer.

10.4 Ideal MIS (or MOS) Structure

The ideal MOS structure is defined by:
(a) the absence of any contact potential or work-function difference between metal and semiconductor;
(b) the absence of any charge within the oxide or at the oxide-semiconductor interface.

The energy band diagram is *flat*, as shown in Figure 10.1.1 for a *p*-type semiconductor. Any applied voltage on the gate with respect to the substrate will be divided between the oxide and the surface layer of semiconductor

$$V_G = V_{ox} + \phi_s. \qquad (10.4.1)$$

On the other hand, the total charge contained in this system is zero

$$Q_G + Q_S = 0 \qquad (10.4.2)$$

(Q_G is the electric charge on the gate and Q_S the total charge within the semiconductor, both measured per unit of surface area).

(a) MOS Capacitance

The differential capacitance of the MOS structure is

$$C = \frac{dQ_G}{dV_G} = -\frac{dQ_S}{dV_G} = -\frac{dQ_S}{dV_{ox} + d\phi_s} \qquad (10.4.3)$$

and may be written

$$C = \frac{1}{\dfrac{1}{C_{ox}} + \dfrac{1}{C_s}}, \qquad (10.4.4)$$

where

$$C_{ox} = -\frac{dQ_S}{dV_{ox}} \qquad (10.4.5)$$

is the oxide capacitance, and

$$C_s = -\frac{dQ_S}{d\phi_s} \qquad (10.4.6)$$

is the 'semiconductor' capacitance, associated with the surface layer. Because the electric field in the oxide, $\mathscr{E}_{ox}$, is constant and given by

$$\varepsilon_{ox}\mathscr{E}_{ox} = \varepsilon\mathscr{E}_s, \qquad (10.4.7)$$

where $\mathscr{E}_s$, the semiconductor surface field, is given by

$$-\varepsilon\mathscr{E}_s = Q_S \quad \text{(Gauss theorem)} \tag{10.4.8}$$

we have

$$C_{\text{ox}} = -\frac{dQ_S}{dV_{\text{ox}}} = \frac{d(\varepsilon_{\text{ox}}\mathscr{E}_{\text{ox}})}{d(t_{\text{ox}}\mathscr{E}_{\text{ox}})} = \frac{\varepsilon}{t_{\text{ox}}} \tag{10.4.9}$$

as expected.

The 'semiconductor' capacitance C_S, appearing in series to C_{ox}, is non-linear and much more difficult to be calculated. It can be, however, approximately evaluated on the basis of the simple model indicated in Section 10.3, as shown in Table 10.4.1.

Table 10.4.1

Semiconductor Surface-layer Capacitance

Surface layer	Surface charge	Capacitance, C_s
Accumulation	High mobile charge density (majority carriers)	Negligible
Depletion and weak inversion	Fixed space-charge (ionized impurities)	Depletion-layer capacitance (voltage dependent)
Strong inversion	High density mobile-charge (minority carriers) and ionized impurities in the depletion layer	Approximately equal to the depletion layer capacitance and constant (high frequency)

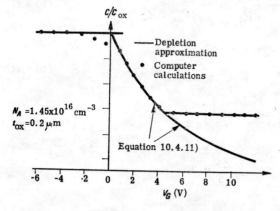

Figure 10.4.1. Computed $C - V$ characteristic of an MOS capacitor. (*after* Grove, [2]).

Figure 10.4.1 shows (solid line) the MOS capacitance, C, calculated on the basis of the above model. For $V_G > 0$ the surface is depleted or inverted. The depletion-layer capacitance is ε/x_d (per unit of surface area) where (see equation 10.3.1)

$$x_d = \left(\frac{2\varepsilon\phi_s}{eN_A}\right)^{1/2} \quad (p\text{-type semiconductor}). \tag{10.4.10}$$

It can be readily shown that the total capacitance C (C_{ox} in series with $C_S = \varepsilon/x_d$) is given by

$$C = C_{ox}\left[1 + \frac{2C_{ox}^2 V_G}{eN_A\varepsilon}\right]^{-1/2} \quad (10.4.11)$$

This formula is valid only for depletion or weak inversion at the semiconductor surface. Above a certain *turn-on voltage* V_T, the surface becomes strongly inverted. V_T can be found directly as

$$V_T = (V_{ox})_{\text{onset of strong inversion}} + \phi_{s(\text{inv})} \quad (10.4.12)$$

and because $V_{ox} = -Q_S/C_{ox} = -Q_B/C_{ox} = eN_A x_{d\max}/C_{ox}$, one obtains

$$V_T = -\frac{Q_B}{Q_{ox}} + 2\phi_F = \frac{[2eN_A\varepsilon(2\phi_F)]^{1/2}}{C_{ox}} + 2\phi_F. \quad (10.4.13)$$

For a given semiconductor, V_T and $C(V_T)$ depend upon the oxide capacitance and the semiconductor doping*.

If the differential capacitance is measured with a high-frequency signal then the mobile charge in the inversion layer cannot follow these rapid variations and the semiconductor capacitance is equal to the depletion capacitance. Therefore, the high-frequency capacitance in strong inversion is constant and equal to

$$C_{\min} = \frac{1}{\frac{1}{C_{ox}} + \frac{x_{d\max}}{\varepsilon}} = \left[\frac{t_{ox}}{\varepsilon_{ox}} + 2\left(\frac{\phi_F}{eN_A\varepsilon}\right)^{1/2}\right]^{-1}. \quad (10.4.14)$$

Figure 10.4.1 also shows exact results obtained by using a computer. Clearly, the depletion approximation yields qualitatively correct results, except near $V_G = 0$.

The 'semiconductor' capacitance (10.4.6) can be calculated by using equation (10.2.8). For a p-type semiconductor with inverted or depleted surface ($u_s > 0$), the result is†

$$C_S = -\frac{dQ_S}{du_s}\frac{e}{kT} = \frac{\varepsilon}{L_D}\frac{\sinh(u_s - u_F) + \sinh u_F}{\sqrt{2[\cosh(u_s - u_F) - \cosh u_F + u_s \sinh u_F]}} \quad (10.4.15)$$

The flat-band capacitance ($V_G = 0$ for an ideal MOS capacitor) is

$$C_{FB} = C_S\Big|_{u_s \to 0} = \lim_{u_s \to 0} C_S = \frac{\varepsilon}{L_D}(\cosh u_F)^{1/2}. \quad (10.4.16)$$

---

* The results derived here are equally valid for an n-type semiconductor with proper changes of sign and notations. For example N_D should be replaced for N_A and $|\phi_F|$ for ϕ_F in equation (10.4.13), etc.

† The same result is valid for $u_s < 0$ but the sign must be changed.

Figure 10.4.2 shows C_S normalized to $\varepsilon/\sqrt{2}\,L_D$ (about 305 pFcm^{-2} for silicon at 300 K), *versus* the normalized surface potential $u_s = e\phi_s/kT$, with u_F as parameter. The doping dependence of u_F for silicon at 300 K is shown in Figure 10.4.3. $C_S = C_S(u_s)$ shows a minimum for $u_s = u_{s\min}$, also plotted in Figure 10.4.3. C_S first decreases with increasing u_s because of the widening of the depletion layer and decreasing of the depletion-layer capacitance. For $u_s > u_{s\min}$, C_s increases rapidly with u_s due to the formation of the inversion layer (the mobile charge in the inver-

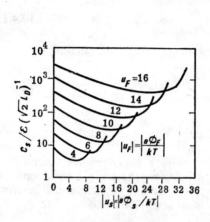

Figure 10.4.2. The MOS capacitance as a function of the surface potential (in kT/e units), for various u_F values. (*after Das*, [5]).

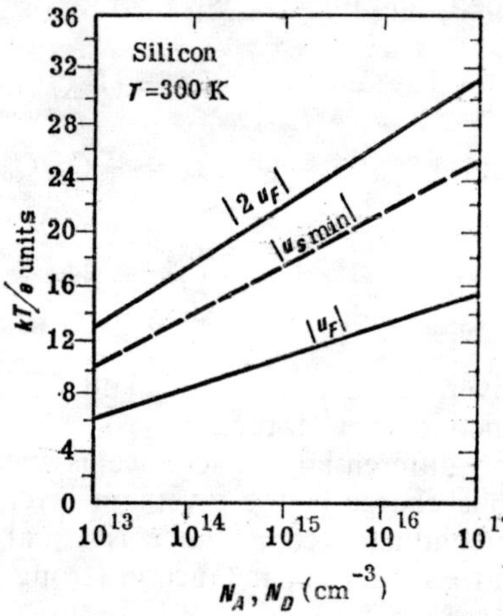

Figure 10.4.3. Dependence of u_F and $u_{s\min}$ on the semiconductor doping. (*after Das*, [5]).

sion layer depends exponentially upon the band bending and therefore upon u_s). Note the fact that $u_{s\min}$ is close to $2u_F$ (Figure 10.4.3), which is the surface potential for strong inversion (equation (10.3.2)). Of course, the capacitance plotted in Figure 10.4.2 is a low-frequency capacitance, calculated by assuming that the charge in the inversion layer corresponds to thermal equilibrium (the signal variation should be sufficiently slow to allow the re-establishment of this equilibrium in a negligibly short time).

Figure 10.4.4 shows a typical capacitance-voltage curve calculated for an ideal MOS capacitor (*p*-type substrate) [6]. The surface potential is indicated for several values of the capacitance. *Curve 1* is the low-frequency characteristic obtained by using equation (10.4.15) and also the relationship

$$V_G = \phi_S - \frac{Q_s}{C_{ox}}; \quad Q_S = Q_S(\phi_s). \tag{10.4.17}$$

Note the fact that for strong accumulation or strong inversion the total capacitance is practically equal to the oxide capacitance, C_{ox}. This occurs because the semiconductor capacitance C_S is very large and negligible in series with C_{ox}. The measurement frequency should be sufficiently low to allow a quick re-establishment of thermal equilibrium: in an ordinary MOS capacitor it is the generation-recombination mechanism which restores the equilibrium. This will be discussed later in more

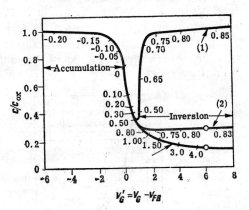

Figure 10.4.4. Capacitance-voltage curves computed for an ideal MOS capacitor: (a) low-frequency curve, (b) high-frequency curve, (c) deep depletion curve. The surface potential in volts is also indicated for each curve. (*after Zaininger and Heiman*, [6]).

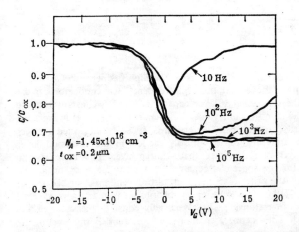

Figure 10.4.5. Experimental $C-V$ characteristics measured at various frequencies. (*after Grove*, [2]).

detail. Figure 10.4.5 illustrates the importance of the measurement frequency. When the structure is biased in accumulation (left-hand part of Figure 10.4.5) the frequency is unimportant because the majority carrier lifetime is extremely short, the equilibrium is established instantaneously (any practical measurement frequency is 'low'). Figure 10.4.5 shows that for the inversion layer, the frequencies above 10^3Hz can be considered 'high'. The high-frequency capacitance-voltage curve saturates at high-positive bias voltages, as already shown in Figure 10.4.1. Curve 2 in Figure 10.4.4 is also a high-frequency curve (the minority carriers in the inversion layer cannot follow the alternating signal used to measure the capacitance).

Curve 3 in Figure 10.4.4 is valid if the minority carriers do not follow both the alternating signal and the applied bias. The bias is usually applied as a voltage sweep. If this sweep is relatively fast, the minority carriers will not have time enough to accumulate at the surface during the interval the bias is high. An example of such a measurement is shown in Figure 10.4.6. Curve 3 is calculated with the depletion-layer capacitance as C_S at high positive V_G. This capacitance continues to decrease with increasing V_G (see equation (10.4.11) and the dashed curve in Figure 10.4.1) because the depletion layer continues to spread into the semiconductor (deep depletion régime). In fact, the capacitance-voltage curve is here analogous to that of a Schottky barrier.

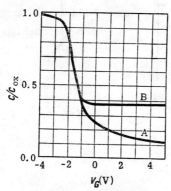

Figure 10.4.6. Experimental $C-V$ curves obtained by using a fast sweep (0.5ms/div, curve A) and a slow sweep (0.5 s/div. curve A). (*after* Zaininger and Heiman. [6]).

A recent refinement of the MOS capacitance theory refers to the high-frequency behaviour in strong inversion [7]—[10]. It was shown above that the high-frequency semiconductor capacitance in strong inversion is equal to the space-charge layer capacitance and is constant because the space-charge layer thickness reaches a maximum value, independent of bias. However, the effect of spatial redistribution of minority carriers in the inversion layer, so far neglected changes somewhat the above picture. The high concentration of minority carriers in the inversion layer will short circuit a thin (but finite) layer of the surface space-charge layer. This increases the total dielectric capacitance calculated for the space-charge layer. The error made by neglecting this redistribution is, however, small (of the order of 5 per cent, also depending upon the model used* to compute the capacitance [7]—[10]). The more exact theories also predict a very small minimum near the onset of strong surface inversion [7]—[9]. We briefly consider below the results communicated by Berman and Kerr [8]. They define a spatially-uniform quasi-Fermi level in the inversion region: there exists an internal equilibrium of the carriers in this region. This quasi-Fermi level changes in phase with the high-frequency signal and the total number of carriers in the inversion layer is held constant at a given bias. Then the total semiconductor capacitance is computed as a function of u_s and u_F following the approach presented above in this section (a similar numerical integration is necessary). A typical result is shown in Figure 10.4.7. The gradual saturation in the classical model is due to the extension of the space-charge depletion layer. This tendency is counteracted by the shunting effect of the inversion layer because this layer widens gradually with the increasing surface potential thus increasing the capacitance. The extension of the inversion

---

* Brews [11] examined recently the errors involved by different approaches in calculating the high-frequency MOS capacitance.

layer saturates slower than that of the depletion layer thus determining the shallow minimum of Figure 10.4.7.

This theory was compared with precision capacitance measurements [8]. In Figure 10.4.8 the theoretical curve with redistribution (1) is fitted to experimental points (a shift along the voltage axis is necessary to match the measurements at flat-band capacitance, see Section 10.5). Note the different shape of the 'classical' curves 2 and 3 (a slower saturation). The exact calculation of the theoretical $C-V$ characteristic in the absence of interface states and oxide charge may be useful in evaluating the properties of a real MOS structure (see below).

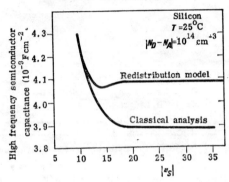

Figure 10.4.7. Calculated semiconductor capacitance *versus* normalized surface potential. The redistributional model predicts a minimum at a certain surface potential. (*after* Bermann and Kerr, [8]).

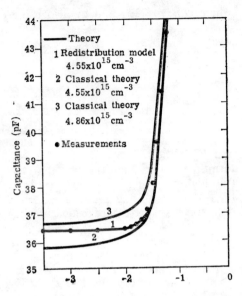

Figure 10.4.8. Comparison between theoretical and experimental capacitance-voltage data. (*after* Bermann and Kerr, [8]).

(b) Surface conductance

We shall calculate below the conductance presented by the inversion layer, which forms a conductive channel at the semiconductor surface, channel which is electrically isolated from the substrate (semiconductor bulk)*.

The local conductivity is

$$\sigma = \sigma(x) = en\mu_n \qquad (10.4.18)$$

For simplicity, the electron mobility μ_n is assumed constant (the effective surface mobility). Thus, only n depends upon the distance from the surface, x. We consider a uniform channel of length L and width W. The conductance of this channel is

$$g = \frac{W}{L} \int_0^{x_i} \sigma(x) dx = \frac{\mu_n W}{L} (-Q_n) > 0, \qquad (10.4.19)$$

---

* The conductance of an accumulated semiconductor surface can be calculated in a similar way.

Q_n being the electron charge in the inversion layer (depth x_i), as defined and calculated in Section 10.2. By using the simplified model introduced in Section 10.3, one obtains successively

$$g = \frac{\mu_n W}{L}[-Q_S + Q_B] = \frac{\mu_n W}{L}[C_{ox}(V_G - \phi_{s(inv)}) + Q_B], \quad (10.4.20)$$

where $Q_{s(inv)} = 2\phi_F$ and by also using equation (10.4.3) we have

$$g = \frac{\mu_n W}{L} C_{ox}(V_G - V_T), \quad V_T = 2\phi_F + V_B, \quad (10.4.21)$$

where

$$V_B = -\frac{Q_B}{C_{ox}} = \frac{[2eN_A \varepsilon(2\phi_F)]^{1/2}}{C_{ox}} \quad (p\text{-type substrate}) \quad (10.4.22)$$

and V_T is the turn-on voltage. The conductance (10.4.21) is finite only above this turn-on voltage. Note the fact that V_T increases with increasing substrate doping (the higher the impurity concentration, the higher the voltage required to deplete the surface and bring about the inversion). On the other hand, the turn-on voltage decreases with increasing oxide capacitance (decreasing oxide thickness for example).

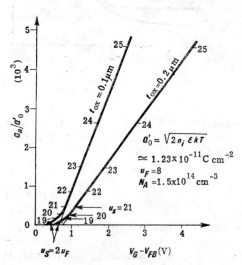

Figure 10.4.9. Dependence of mobile electron charge in the inversion layer upon the applied voltage ($V_{FB} = 0$ for an ideal MOS structure, see below) with oxide thickness as parameter. The normalized surface potential is indicated on each curve. (*after Das*, [5]).

However, the 'opening' of the inversion channel should be gradual and not abrupt as indicated above, because the transition from depletion to strong inversion is gradual. This is what the more exact calculations do indeed show. Figure 10.4.9 reproduces the voltage dependence of the electron charge in an inversion layer on a silicon surface at 300 K ($N_A = 1.5 \times 10^{14}\text{cm}^{-3}$). This dependence is calculated by using equations (10.2.8), (10.2.12) and (10.4.17). The values of the normalized surface potential are also shown. Note the fact that the linear $|Q_n| - V_G$ dependence occurs when the surface potential is a few kT/e above $\phi_{s(\text{inv})} = 2\phi_F$ (here $\phi_F = 8\,kT/e$), according to the simple theory of Section 10.2.3.

It is interesting to show that the 'tail' region of the Q_n versus V_G curve is related to the curvature of the high-frequency MOS capacitance-voltage curve [5]. We can approximate, under strong inversion conditions

$$\frac{d|Q_n|}{d(V_G - V_T)} = \frac{d|Q_n|}{dV_G} \simeq \frac{d|Q_S|}{dV_G} = C_{\text{ox}}\left(1 - \frac{d\phi_s}{dV_G}\right). \qquad (10.4.23)$$

On the other hand, by using successively equations (10.4.3), (10.4.5) and (10.4.1) one obtains

$$\frac{C}{C_{\text{ox}}} = \frac{dV_{\text{ox}}}{d(V_{\text{ox}} + \phi_s)} = \frac{d(V_G - \phi_s)}{dV_G} = 1 - \frac{d\phi_s}{dV_G} \qquad (10.4.24)$$

such that the derivative of $g = g(V_G)$ is proportional to $C = C(V_G)$. For example, the dependence $g = g(V_G)$ becomes linear when, under strong inversion conditions, the MOS capacitance approaches the oxide capacitance (see Figure 10.4.4). We stress the fact that the carrier mobility in the surface layer was assumed constant. However, at higher gate voltages, this 'surface' mobility becomes field-dependent and decreases with increasing V_G, as will be shown later.

10.5 Real MOS (MIS) Structures

There are a number of phenomena, so far neglected in the analysis of a metal-insulator-semiconductor structure and we list them below:
    (a) the existence of metal-semiconductor work-function difference;
    (b) the mobile electric charge trapped in surface states localized at the insulator-semiconductor interface;
    (c) the electric charge within the insulator layer;
    (d) non-uniform semiconductor doping profile;
    (e) degenerate statistics in the surface layer;
    (f) quantization of energy levels in the surface layer;
    (g) field-dependent effective surface mobility;
    (h) statistical distribution of electronic impurities in semiconductor.

We still consider a one-dimensional structure and assume that electric conduction through the relatively insulator layer is impossible.

In this Section we reconsider the theory by taking into account effects (a) through (c).

(a) Work-function Difference

Figure 10.5.1 shows the energy band diagram of an Al-SiO$_2$-Si structure at thermal equilibrium, with no applied bias. The surface of n-type silicon is accumulated

(Figure 10.5.1a) and the surface of p-type silicon is inverted (Figure 10.5.1b) [12]. The oxide and interface charge is, of course, assumed to be zero. The gate voltage which will bring the energy band diagram at flat-band condition is defined as the

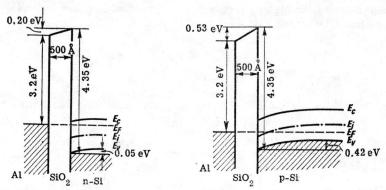

Figure 10.5.1. Band energy diagrams of the Al–SiO$_2$–Si system. The semiconductor doping is 10^{16} cm^{-3}, n-type (a) and p-type (b) and the oxide thickness is 500 Å. (after Deal and Snow, [12]).

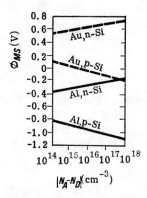

Figure 10.5.2. Barrier energy difference ϕ_{MS} for Al and Au gates on oxidized n- and p-type silicon, as a function of effective doping concentration. (after Deal and Snow, [12]).

flat-band voltage, V_{FB}, and, here, is just equal to the metal-semiconductor work-function difference divided by the electron charge

$$V_{FB} = \phi_{MS}. \tag{10.5.1}$$

The potential difference ϕ_{MS} may be either positive, or negative. Figure 10.5.2 indicates ϕ_{MS} computed for Au–Si and Al–Si, versus semiconductor doping.

The results obtained for the ideal MOS structure remain valid if V_G is replaced by an 'effective' gate voltage

$$V_G \rightarrow V'_G = V_G - V_{FB} = V_G - \phi_{MS}. \tag{10.5.2}$$

(b) Interface Charge

The interface charge is, usually, assumed to be an electronic charge localized in surface states and its surface density is denoted by Q_{SS} (conventionally positive).

Figure 10.5.3 indicates the energy band diagram and the charge distribution in an MOS system at thermal equilibrium, with zero gate voltage (the superscript zero indicates the value at $V_G = 0$). In the particular case depicted by Figure 10.5.3a,

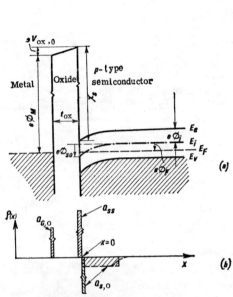

Figure 10.5.3. Effect of interface state charge and work function difference on the energy bands (a) and the charge distribution (b) in an MOS structure for zero gate voltage. (*after* Grove et al., [3]).

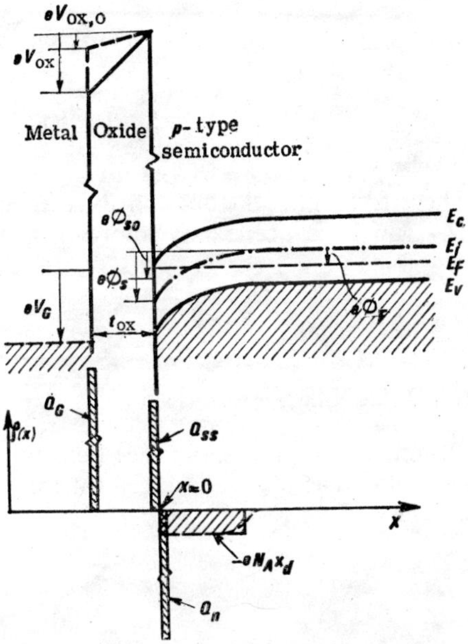

Figure 10.5.4. Energy bands (a) and charge distribution (b) in an MOS structure under applied gate bias. (*after* Grove et al., [3]).

the surface of the p-type semiconductor is inverted ($\phi_{s0} > 0$) due to both the work-function difference ($\phi_{MS} < 0$) and to the positive interface charge ($Q_{SS} > 0$). The total charge per unit of surface area shown in Figure 10.5.3b is, of course, zero. Note the fact that the work-function difference ϕ_{MS} can be written

$$\phi_{MS} = -\phi_{s0} - V_{ox,0}. \tag{10.5.3}$$

The external bias (see Figure 10.5.4 for $V_G > 0$) will be distributed across the oxide and the surface layer, such that

$$V_G = V_{ox} - V_{ox,0} + \phi_s - \phi_{s0} \tag{10.5.4}$$

or

$$V_G = V_{ox} + \phi_s + \phi_{MS}. \tag{10.5.5}$$

The total charge (Figure 10.5.4b) is zero (by Gauss theorem)

$$Q_G + Q_{SS} + Q_S = 0 \tag{10.5.6}$$

and

$$Q_G = \frac{V_{ox}}{C_{ox}}, \quad C_{ox} = \frac{\varepsilon_{ox}}{t_{ox}}. \tag{10.5.7}$$

By using equations (12.5.5) – (12.5.7) one obtains

$$V_G - \phi_{MS} + \frac{Q_{SS}}{C_{ox}} = \phi_S - \frac{Q_S}{C_{ox}} \tag{10.5.8}$$

The effect of work-function difference $e\phi_{MS}$ and of interface-state charge density Q_{SS} can be cancelled by applying a gate potential equal to the flat-band voltage

$$V_G \to V_{FB} = \phi_{MS} - \frac{Q_{SS}}{C_{ox}} \tag{10.5.9}$$

(c) Oxide Charge

Assume a charge with surface density Q' localized in oxide, at the distance x' from the metal. The charge induced into the semiconductor (with $\phi_{MS} = 0$ and $Q_{SS} = 0$) will be

$$Q_S = -\frac{x'}{t_{ox}} Q' \tag{10.5.10}$$

and the corresponding flat-band voltage

$$(V_{FB})_{\text{oxide charge only}} = \frac{Q_S}{C_{ox}} = -\frac{x'}{t_{ox}} \frac{Q'}{C_{ox}}. \tag{10.5.11}$$

If the oxide charge is spatially distributed with charge density $\rho' = \rho'(x')$, then, by using the superposition principle

$$(V_{FB})_{\text{oxide charge only}} = -\frac{1}{C_{ox}} \int_0^{t_{ox}} \frac{x'}{t_{ox}} \rho'(x') dx'. \tag{10.5.12}$$

The total flat-band voltage, including the effect of work-function difference, surface-state charge and distributed oxide charge is then

$$V_{FB} = \phi_{MS} - \frac{Q_{SS}}{C_{ox}} - \frac{1}{C_{ox}} \int_0^{t_{ox}} \frac{x'}{t_{ox}} \rho'(x') dx'. \tag{10.5.13}$$

10. MIS Capacitor

Clearly, the effect of electric charges localized either within the oxide or at interface may be included into an equivalent interface charge

$$(Q_{SS})_{\text{equiv}} = Q_{SS} + \int_0^{t_{\text{ox}}} \frac{x'}{t_{\text{ox}}} \rho'(x') dx' \qquad (10.5.14)$$

such that

$$V_{FB} = \phi_{MS} - \frac{(Q_{SS})_{\text{equiv}}}{C_{\text{ox}}}. \qquad (10.5.15)$$

We stress the fact that the above flat-band voltage is calculated by assuming that both the oxide charge and the surface state charge do not depend upon the applied gate voltage. We shall discuss below the origin of these charges for the SiO_2-Si system and it will be apparent that the above assumption is not always true.

If this assumption *is*, however, true, then all the results derived for the ideal MIS structure will remain valid, provided that V_G in Section 10.4 will be replaced by an 'effective' gate voltage

$$V_G \to V'_G = V_G - V_{FB} = V_G - \phi_{MS} + \frac{(Q_{SS})_{\text{equiv}}}{C_{\text{ox}}}. \qquad (10.5.16)$$

Both $C = C(V_G)$ and $g = g(V_G)$ curves will be translated towards the left by an amount algebraically equal to V_{FB}. This is shown in Figure 10.5.5. The fact that

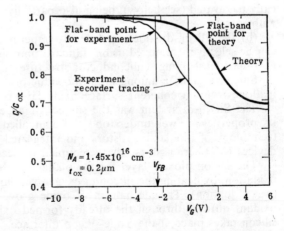

Figure 10.5.5. Combined effects of metal-semiconductor work function differences and charges within the oxide on the $C-V$ characteristics of MOS structures. (*after Grove*, [2]).

the experimental curve is indeed parallel to the theoretical one proves the assumption of field-independent oxide and interface charge. The flat-band voltage can be found graphically, as shown. If ϕ_{MS} and C_{ox} are known, $(Q_{SS})_{\text{equiv}}$ will be determined.

The actual turn-on (or threshold) voltage defined with respect to surface conductance (approximate strong inversion) will now be ($V_T - V_{FB} = 2\phi_F + V_B$, according to equation (10.4.21))

$$V_T = V_{FB} + 2\phi_F + V_B = \phi_{MS} - \frac{(Q_{SS})_{\text{equiv}}}{C_{ox}} + 2\phi_F - \frac{Q_B}{C_{ox}} \quad (10.5.17)$$

$$Q_B = -\frac{1}{C_{ox}}[2eN_A\varepsilon(2\phi_F)]^{1/2}, \quad C_{ox} = \frac{\varepsilon_{ox}}{t_{ox}}, \quad \phi_F = \phi_F(N_A). \quad (10.5.18)$$

The above formulas are valid for an *n-type inversion layer* formed at the surface of a *p*-type semiconductor. For a *p*-type inversion layer $-N_D$ should be replaced by N_A (ϕ_F is now negative) and the sign of Q_B must be changed (the fixed bulk charge is now positive). Note the fact that, for a given metal-insulator-semiconductor system, V_T depends upon semiconductor doping, insulator thickness the equivalent interface charge (per unit of gate surface) and upon the temperature (through ϕ_F and n_i, see Section 10.2).

10.6 The Silicon-dioxide Silicon System

The planar technology depends upon the properties of the SiO_2—Si system. This system is formed by thermal growth of a silicon dioxide layer on the silicon surface and has the following useful properties:

(a) the silicon surface covered by such an oxide is both chemically and electrically passivated (see below), being also partially protected from the influence of the ambient;

(b) the silicon dioxide is practically impermeable during impurity diffusion and can be selectively removed using standard photolithographic techniques, thus acting as a mask in defining semiconductor structures.

Silicon is unique among all semiconductors used for electronic devices, in that it alone forms a coherent surface oxide layer with the above properties [2], [13].

The SiO_2—Si system was the subject of an extensive research, the majority of its properties are well understood and controlled. This is the basis of silicon planar technology of bipolar transistors and integrated circuits and of surface-controlled devices (MOS transistors, charge-coupled devices, etc.).

The silicon dioxide layer for the planar or MOS process is obtained by thermal oxidation, by heating the silicon in oxygen vapours (dry oxidation) or water vapours (wet oxidation). The mechanism of oxide growing is discussed by Grove [2]. The oxidant diffuses through the already formed oxide layer and the interaction with silicon takes place at the oxide-silicon interface. The oxide thickness increases with the oxidation time, for example, as shown in Figure 10.6.1 [13]. A high-temperature process is faster but severe mechanical stresses may occur during cooling due to the low coefficient of thermal expansion of silicon dioxide. For similar reasons thick oxide films are avoided.

10. MIS Capacitor

A post-oxidation thermal treatment is necessary, as will be discussed below. The physical properties of the oxide layers depend upon this treatment and upon the layer thickness [2]. There exist practical limitations to the minimum oxide thickness obtainable, limitations arising from the initial non-uniformity of the oxidation

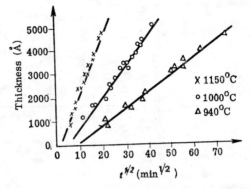

Figure 10.6.1. Thickness of SiO_2 film obtained by dry oxidation, as a function of the oxidation time, t, for three temperatures of the process. (*after Gregor*, [13]).

process and from the layer defects, such as pinholes occurring during oxide growth due to dust particles or micron-size structural defects occurring during thermal treatments.

Oxide and Interface Charge

(a) *Ionic charge within the oxide.* A major cause of concern with respect to the early MOS transistors was the drift of characteristics under applied bias at high temperatures. The instability corresponds to a drift of the flat-band voltage, V_{FB}. Figure

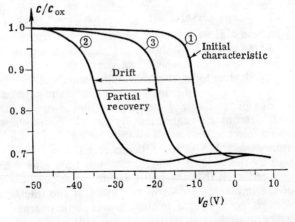

Figure 10.6.2. Shift in the $C-V$ characteristic due to the ionic contamination of the oxide. (*after Grove*, [2]).

10.6.2 shows the drift of the high-frequency $C-V$ characteristic after 30 minutes at 127 °C, with $V_G = +10$ V. The characteristic exhibits a partial recovery after 30 minutes heating at the same temperature, but with $V_G = 0$V. Figure 10.6.2

indicates the change of $(Q_{SS})_{\text{equiv}}$ in the expression of V_{FB} (the shape of $C = C(V)$ is not modified). It was shown that the drift and recovery are explained by the rearrangement of an ionic charge within the oxide. This is schematically explained by Figure 10.6.3. In a 'fresh' device, the ionic charge is located just near the metal interface (Figure 10.6.3a) and occurs due to ionic contamination during metallisation. No

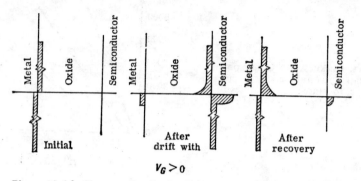

Figure 10.6.3. Charge redistribution in the MOS system at various stages of the drift process. (*after Grove*, [2]).

charge is induced in semiconductor. The application of a positive gate voltage exerts a repulsive force upon these positive ions. The ion mobility increases with increasing temperature. The time required for ionic transport is still quite large (minutes). The positive charge near the oxide-semiconductor interface (Figure 10.6.3b) reduces the flat-band voltage and produces the drift of $C(V)$ characteristic toward the left. The recovery is partial because not all charge returns just near the metal interface (Figure 10.6.3c).

The ionic contamination is mostly due to sodium [2]. The sensitivity of SiO_2 to contamination is greatly reduced by the formation of a phosphosilicate glass over the SiO_2. This glass both 'getters' sodium from the already deposited SiO_2 layer and acts as a barrier against further contamination [2]. The composition and thickness of the phosphosilicate glass layer should be controlled in order to avoid polarization effects which can occur in this layer [2]. In present-day surface-controlled devices ionic contamination is negligible.

It is possible, in principle, to control the surface potential by the electric charge of ions in the oxide. However, sodium is not suitable because its high mobility in SiO_2 determine instabilities. By use of ion implantation (see Section 10.8) it was found that the heavier alkali ions have a decreasing mobility in SiO_2 with increasing atomic weight. Sixt and Goetzberger [14] reported the control of the flat-band voltage by the dose of the Cs ions implanted into a bare silicon surface subsequently oxidized such that the entire Cs-doped region was converted into SiO_2. A fraction of the implanted ions will be found at the interface after oxidation. The flatband voltage increases due to the positive ionic charge. This voltage is stable during the temperature-bias stress (see above) [14].

(b) *Radiation-induced oxide space charge* [2], [15]. A possible effect of X-ray irradiation is the formation of electron-hole pairs and subsequent drift due to the electric field. A positive gate voltage will collect electrons and repel holes toward the oxide-semiconductor interface where they may be trapped. This produces a progressive accumulation of positive charge near the semiconductor interface and a modification of device characteristics. The radiation-induced oxide space charge can be annealed out by heating above 300 °C [2].

(c) *Surface-state charge or fixed oxide charge*. The so-called surface-state charge [2] is positive and located within 200 Å of the oxide-silicon interface. Its magnitude depends upon the oxidation and annealing conditions and upon the orientation of the silicon surface, as shown in Figure 10.6.4. Its density is practically unaffected

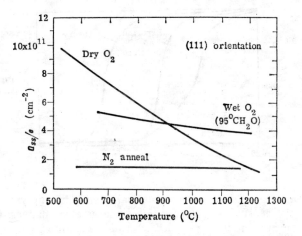

Figure 10.6.4. Dependence of the surface-state charge density on the ambient and temperature of the final heat treatment. (*after* Grove, [2]).

by the oxide thickness or by the type and impurity concentration in silicon and also independent of the applied bias. Careful experiments were made to distinguish between this charge and the oxide charge due to ionic contamination or radiation. The origin of the surface-state charge is in deviations from oxide stoichiometry. Grove [2] indicated the effect of excess silicon present in the oxide near the interface and waiting to react with oxidizing species that diffuse through the oxide. Recent experiments of Fowkes and Hess [16] indicate the effect of oxidation-reduction treatments and point out the effect of oxygen vacancies upon the fixed oxide charge and interface states.

Figure 10.6.4 shows that an inert-gas anneal reduces the density of surface-state charge to a value independent of the oxidation conditions. This value (of the order of 10^{11} cm^{-2}) is quite high, however, and its presence has important consequences upon the electrical behaviour of the silicon surface (see also below). Attempts to further reduce surface-state charge were also made by the following techniques [17]:

— introduction of gold, platinum or nickel after gate oxidation;
— the use of silicon wafers cut on the (100) plane (instead of normal (111) orientation);
— oxygen ion implantation near the silicon-oxide interface followed by a long heat treatment (450 °C) which results in an oxygen-silicon reaction and a decrease of the positive surface-state charge due to the silicon-reach layer at the interface [17].

(d) *Electric charge trapped in interface states.* We consider below the effect of energy levels located in the silicon band-gap at the surface. The Tamm-Shockley states caused by lattice interruption (Section 2.2) give for an atomic clean surface an enormous surface-state density of the order of 10^{15} cm^{-2}. The formation of the oxide layer, however, uses up most of the available chemical bonds such that the actual surface-state density at a clean SiO_2—Si interface is orders of magnitude smaller*.

The interface states are distributed in the band-gap. Those situated above the Fermi level are unfilled, whereas those below Fermi level are charged with electrons.

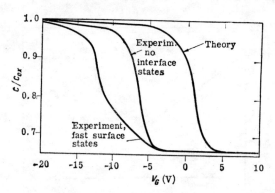

Figure 10.6.5. Effect of 'fast' surface states on the $C - V$ characteristic of an MOS structure. (*after Grove*, [2]).

By varying the surface potential (simply by applying a steady-state gate voltage V_G), the density of surface-trapped charge also changes. Therefore the theory developed in Section 10.5 is no longer valid because $(Q_{SS})_{\text{equiv}}$ depends upon V_G. The actual $C(V_G)$ curve is no longer parallel to the ideal theoretical characteristic. The negative charge trapped at the surface determines a positive displacement (in the positive V_G direction) of the MOS capacitance. This displacement increases with increasing V_G, such that the high-frequency $C(V_G)$ characteristic is distorted as shown in Figure 10.6.5.

The density and energy distribution of SiO_2—Si interface states were intensively studied (see Section 10.9). Initial investigations indicated a density of the order of 10^{11}—10^{12} cm^{-2}. A heat treatment at about 500 °C reduces this density below 5—10^{10} cm^{-2}[2]. This surface density is quite low, with no major effect upon the operation of planar devices, including some surface-controlled devices.

Effect of Oxide and Interface Charges†

The specific properties of the metal-insulator-semiconductor system are of great importance in determining the actual usefulness of MOS devices. For the SiO_2—Si system, it was shown that the effect of ionic contamination, ionizing radiation, fixed oxide charge and interface charge can be reduced and stabilized by adequate

---

* These states were known as 'fast' surfaces states [1], [2] because of their rapid interaction with the semiconductor.

† The mobile charge on the oxide surface not covered by metal determines an effect of fictitious mobile gate and is discussed by Grove [2].

thermal oxidation and post oxidation treatments. The fixed oxide charge remains, however, important. This positive charge attracts electrons toward the interface and *tends* to create an accumulation surface layer on an *n*-type semiconductor and inversion surface layer on a *p*-type semiconductor. It is this oxide charge, together with the work-function difference, which determines a definite asymmetry of MOS properties with respect to semiconductor type. We shall consider below the flat-band voltage, V_{FB}, and the turn-on voltage, V_T.

Figure 10.6.6 shows typical low-frequency $C = C(V)$ curves computed for the Al–SiO$_2$–Si system. Characteristic values of the normalized surface potential are

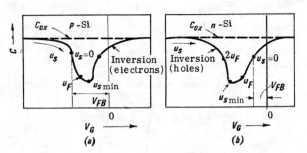

Figure 10.6.6. Typical $C - V$ plots of practical MOS structures with *p*- and *n*-type silicon substrates illustrating their respective flat-band positions. (*after Das*, [5]).

indicated ($u_s = 0$ for flat band, $u_s = u_F$ for surface (weak) inversion, $u_s = u_{smin}$ for minimum of semiconductor capacitance, see Figure 10.4.4 and $u_s = 2u_F$ for the onset of strong inversion). Consider, first, $V_{FB} = 0$ and compare the shape of $C = C(V_G)$ curves. The gate voltage for minimum capacitance will be positive for a *p*-type and negative for an *n*-type semiconductor. Assume, then, $Q_{SS} = 0$, such that $V_{FB} = \phi_{MS}$. Figure 10.5.2 indicates that $\phi_{MS} < 0$. Figure 10.5.1 shows the barrier energies and the usual state of surface for $V_G = 0$. Note that if Au is used as metal contact and $Q_{SS} = 0$, $V_{FB} > 0$ for an *n*-type semiconductor and $V_{FB} \approx 0$ for a *p*-type semiconductor. The use of Au instead of Al, is not, however, justified, because $Q_{SS} > 0$ will determine a negative shift of the $C(V)$ characteristic. Figure 10.6.6 indicates that for $V_G = 0$ the *p*-type semiconductor surface is inverted, whereas for the *n*-type substrate the surface is accumulated. This is of paramount importance for the MOS transistor: the *n*-type transistor (*n*-channel on *p*-type substrate) is already in a conduction state with no bias (normally conducting).

We shall examine below the turn-on voltage for the surface conductance $g = g(V_G)$. This voltage depends upon the oxide thickness and semiconductor doping. We reproduce below the results obtained by Kim [18]. He considered the turn-on voltage $V'_T = V_T - \phi_F$ corresponding to the onset of inversion ($u_s = u_F$) and not of the strong inversion ($u_s = 2u_F$)

$$V'_T = \phi_{MS} - \frac{Q_{SS}}{C_{ox}} + \phi_F - \frac{Q_B}{C_{ox}}, \tag{10.6.1}$$

where $Q_{SS} > 0$ is the fixed oxide charge and Q_B is the fixed bulk charge, negative for *p*-type substrate and positive for *n*-type substrate. Figure 10.6.7 shows the doping $-V'_T$ dependence for an n-type substrate, with Q_{SS}/e as parameter. The oxide thickness is $t_{ox} = 0.1\ \mu\text{m} = 1000\ \text{Å}$. Similar curves are represented in Figure 10.6.8 for a p-type substrate. The sign of V'_T is discussed in Table 10.6.1.

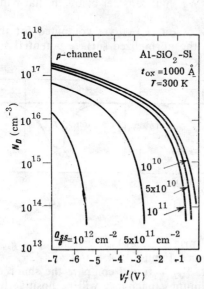

Figure 10.6.7. Voltage threshold for the onset of inversion as a function of doping density (n-type substrate) and interface state density. (*after Kim*, [18]).

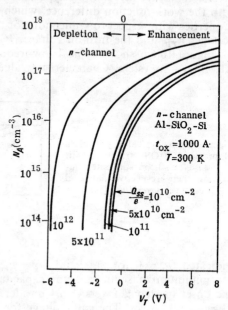

Figure 10.6.8. Voltage threshold for the onset of inversion as a function of doping density (p-type substrate) and interface state density. (*after Kim*, [18]).

Table 10.6.1

Turn-on Voltage V'_T

Substrate	Inversion channel	V'_T	Normal operation with inversion channel	State at $V_G = 0$
n-type	p-type	>0	$V_G < V'_T$	Inversion channel
		<0	$V_G < V'_T$	Depletion or accumulation
p-type	n-type	>0	$V_G > V'_T$	Depletion or accumulation
		<0	$V_G > V'_T$	Inversion channel

The above table shows that for normal operation of an inversion-channel MOS transistor (turned-off at $V_G = 0$), V'_T should be negative for n-type substrate and positive for p-type substrate.

Figure 10.6.7 shows that $V'_T < 0$ for n-type substrate, therefore p-channel MOS transistors can be readily fabricated. The substrate doping should be kept as low as possible ($N_D < 10^{15}$ cm^{-3}, for example) in order to make V'_T insensitive to accidental variations of resistivity (Q_{SS}/e is reproducible for a given technological process). On the other hand, n-type inversion channel MOS transistors are very difficult to fabricate and the majority of normal MOS transistors are p-channel type. This can be understood by examining Figure 10.6.8. The reproducibility of V'_T requires low doping, otherwise relatively small variations of N_A will determine a large dispersion of V'_T. For low N_A, $V'_T > 0$ or $V'_T \approx 0$ cannot be achieved except for very low oxide charge, which is practically difficult to obtain. The effect of oxide charge could be reduced by increasing the oxide capacitance per unit area, C_{ox}. However, the oxide thickness, t_{ox}, cannot be reduced below 1000 Å because of the increased number of deffects.

10.7 Surface Quantization and Surface Mobility in Inversion Layers

It was shown that an electric field perpendicular to the semiconductor surface can induce a surface inversion layer. This layer is separated from the semiconductor bulk by a depletion layer and acts as a conduction channel in surface-controlled devices, such as the MOS transistor. We have calculated from electrostatic considerations the total semiconductor charge per unit of surface area and have also found the minority carrier charge. The channel conductance was calculated by using the concept of effective field-independent mobility of minority carriers. An accurate calculation should, however, take into account *microscopic* processes into the very thin inversion layer [19].

a) Surface Quantization

The minority carriers in the inversion layer are bound in the very thin potential gap determined by the surface field. The motion of these carriers *in a direction perpendicular to the surface* should be, therefore, quantized, as first shown by Schrieffer [20]. Each energy level corresponding to the motion perpendicular to the surface is the bottom of a two-dimensional *sub-band* formed by a continuum of states (the motion parallel to the surface is not quantized). A correct solution to the quantization problem requires the exact potential distribution in the surface space-charge layer. This potential is determined from the Poisson equation. Note the fact that under strong inversion conditions the mobile carrier density cannot be neglected and this density depends upon the occupancy of the sub-bands mentioned above. It is, therefore, necessary to find a *self-consistent* solution to both Schrödinger's equation and Poisson's equation. This is usually done under the effective-mass approximation and with simplified boundary conditions at the semiconductor surface [21]–[24]. An analytical solution is available only in the electric quantum limit, when only the lowest sub-band is occupied [22], [24] (sufficiently low temperature).

Another approach to the quantization problem is the approximation of the potential well. Gnädinger and Talley [21] assumed a constant field within the semiconductor and an infinite potential barrier at the surface ('triangular' potential). Figure 10.7.1 shows the results of their calculations for silicon at room temperature. Figures 10.7.1a and b show the quantized energy levels and c and d the electron distribution (p-type substrate, 4×10^{15} cm^{-3} doping). The classical distribution is shown for comparison. The quantum-mechanical calculations bring out an important difference under strong-inversion conditions (Figure 10.7.1, b and d). It was later shown [25] that the surface barrier height should be taken finite. For a 3.2 eV barrier (SiO$_2$–Si system) the

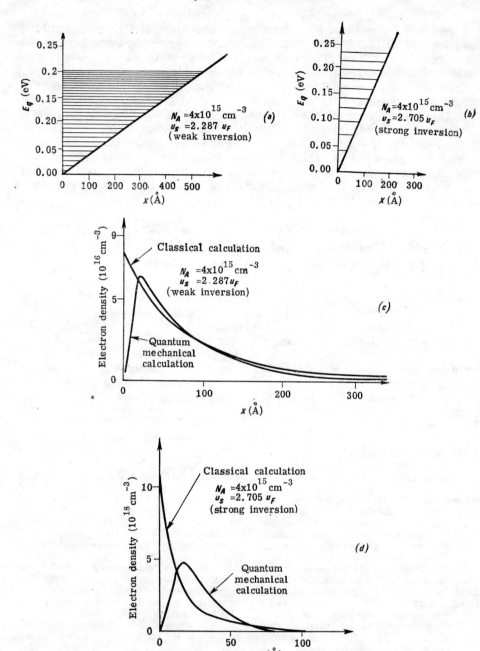

Figure 10.7.1. Quantized energy levels, E_q, in a linear potential well (a and b) and the electron density (c and d) as a function of distance from the interface. *(after Gnädinger and Talley, [21])*

electronic charge penetrates into the oxide a distance of 2 − 3 Å, and the charge distribution was found to be closer to the classical result, as indicated by Figure 10.7.2. The channel depth (defined as containing 99% of the mobile charge) is of a few hundreds of Å, decreasing with increasing surface potential to tens of Å.

The self-consistent approach to the surface quantization problem requires a numerical solution based on an elaborate computer program [22], [23]. The calculation starts with an initial estimate of the potential and then solves Schrödinger's equation and Poisson's equation successively until a sufficient accurate solution is obtained. The trial potential may be obtained, for example, by considering that all the carriers are on the lowest sub-band, a technique which is successful at low temperatures. The higher the temperature, the larger the number of sub-bands which should be handled by the program. Such calculations were performed by Stern [22] and by Pals [23] for p-type silicon (n-type inversion layer) at various temperatures and various substrate dopings. Stern [22]

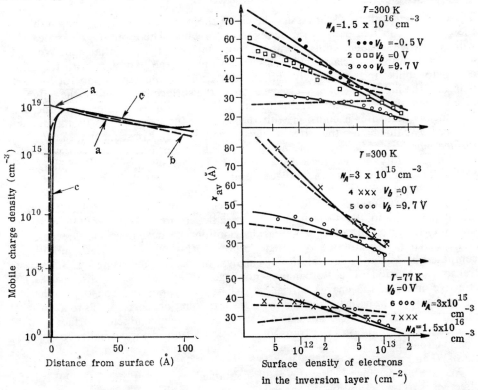

Figure 10.7.2. Charge density in an inversion layer: (a) classical distribution, (b) quantum-mechanical calculations with a finite barrier at the SiO_2−Si interface (3.2 eV), (c) quantum-mechanical calculations with infinite barrier. (after Laur and Jayadevaiah, [25]).

Figure 10.7.3. Average distance of electrons from the interface, x_{av}, versus the total number of electrons in the inversion layer of (100) silicon surface. Solid curves are calculated with quantization; dashed curves, $x_{av} + 15$ Å are calculated without quantization; circles, crosses and squares are measured points. (after Pals, [23]).

indicates that classical results are reasonably good for room temperature. However, quantum effects must be considered even at room temperature if the surface carrier density is very high.

We reproduce in Fig. 10.7.3 results obtained by Pals [23]. Solid lines were obtained by self-consistent quantum-mechanical calculations. Dashed curves are found without quantization.

Results of experimental measurements are also indicated. The average depth of the inversion layer was obtained by measuring the MOS capacitance of an MOS transistor with both source and drain grounded. The substrate is biased by V_b (for a.c. capacitance measurement is shortened to source and drain). The effect of doping and substrate bias can be understood by noting that an increase of the steepness of the potential well will decrease the average depth of the inversion layer and will increase the difference between the quantum-mechanical and classical calculations. The overall agreement with experiment is better for the theory based on surface-quantization. However, for room temperature and zero substrate bias the measurements do not make a significant difference between the two theories.

The surface quantization should also be taken into account for *accumulation* layers. These layers were less studied, perhaps because of theoretical and experimental difficulties*.

b) Surface Mobility

The effective depth of the inversion layer, of the order of 100 Å, is comparable to the carrier mean free path in the semiconductor bulk. Therefore the charge carrier will experience frequent collisions at the boundaries of the potential well. We shall assure, for simplicity, elastic collisions (reflexions) at the boundary with the semiconductor bulk and completely diffuse scattering at the semiconductor surface. Consider the movement of a charge carrier in the surface layer, under the influence of an external electric field parallel to the surface. A diffuse scattering to the surface means that after a collision with the surface, the carrier loses its momentum (the energy gained from the external electric field is lost). Consequently, *the effective surface mobility* (obtained from surface conductance measurements) is lower than the bulk mobility. We expect a decrease of the surface mobility with the increasing surface potential (absolute value) because of reduced depth of the inversion channel and the increased importance of surface scattering.

The effective surface mobility was first calculated by Schrieffer [20] by assuming a constant field in the surface layer, perpendicular to surface. A more accurate theory based upon the exact potential distribution was developed by Greene et al. [26] and later reformulated by other authors [27]–[29]. The surface mobility is computed by assuming the validity of classical mechanics and using the Boltzmann transport equation. The Boltzmann (nondegenerate) statistics is also used. The effective mass of the charge carrier is assumed isotropic and constant (spherical energy bands). The collision term in the transport equation corresponds to a constant relaxation time. Complete diffuse scattering at the semiconductor surface is used as a boundary condition. The transport equation is solved for a small perturbation (the external field parallel to the surface is assumed to be very small). Both inversion [27] and accumulation [28] layers were studied.

We shall reproduce below the results obtained by Pierret and Sah [29] for the effective surface mobility in inversion channels (silicon, room temperature)†. Figure 10.7.4a shows the electron surface mobility normalized to bulk mobility, versus $u_s - u_F$, with u_F as parameter (for an inverted p-type semiconductor, both u_F and u_s are positive). Figure 10.7.4b shows similar results for inversion layers in n-type silicon ($u_F < 0$, $u_s < 0$). For a typical MOS inversion-channel silicon

---

* Self-consistent calculations are more difficult to perform because of the greater spatial extent of the layer whereas the experiments are complicated by the fact that the transport properties of the surface layer cannot be separated from those of the bulk [21].

† More general plots are given in references [27], [28]. The surface mobility normalized to the bulk mobility depends upon u_s, u_F and the adimensional parameter $r = \lambda_x/L_{DE}$, λ_x being the unilateral mean free path, $\lambda_x = \mu_{BULK}(m^*kT/2\pi e^2)^{1/2}$ and L_{DE}, the extrinsic Debye length (2.3.18).

transistor, $|u_F|$ has a usual value between 9 and 15. The normalized surface potential u_s may vary between strong inversion and degeneracy such that

$$2u_F \lesssim u_s \lesssim u_F + 22 \quad \text{(electron conduction)} \tag{10.7.1}$$
$$u_F - 22 \lesssim u_s \lesssim 2u_F \quad \text{(hole conduction)}. \tag{10.7.2}$$

Figure 10.7.4 indicates that the surface mobility always decreases with increasing surface potential $|\phi_s| = |u_s|kT/e$. At relatively low surface fields (u_s close to $2u_F$), the effective mobility depends appreciably upon semiconductor doping (through u_F). When degeneracy is approached, this doping dependence becomes negligible (as might be expected from the 'metallic' behaviour of a degenerate semiconductor) [29].

Figure 10.7.5 shows a comparison between theory and experimental results obtained by measuring MOS transistors with p-type inversion channel. The dis-

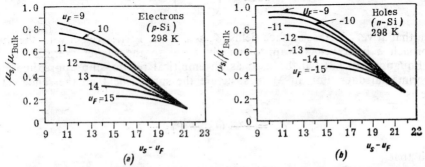

Figure 10.7.4. Effective surface mobilities computed for n-type (a) and p-type (b) inversion channels. (after Pierret and Sah, [29]).

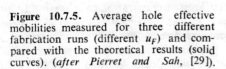

Figure 10.7.5. Average hole effective mobilities measured for three different fabrication runs (different u_F) and compared with the theoretical results (solid curves). (after Pierret and Sah, [29]).

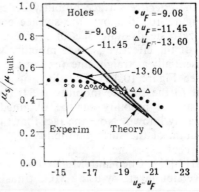

agreement obtained was explained by Pierret and Sah [29] by the improper modelling of the surface conditions and by experimental inaccuracies*. It is interesting to

---

* The same authors [29] calculated an anisotropic mobility which, for silicon, is even higher than the theoretical values shown in Figures 10.7.4 and 10.7.5.

note that the surface mobility is much less doping and surface-field dependent than predicted by theory and is roughly equal to half of its bulk value, in agreement with earlier measurements [29].*

For the purpose of device analysis the following analytical approximation is used

$$\mu_s = \mu_{\text{SURFACE}} = \mu_{\text{BULK}} \left(1 + \frac{\mathscr{E}_s}{\mathscr{E}_c}\right)^{-1} \tag{10.7.3}$$

where $\mathscr{E}_c$ is a certain critical field at which μ_s falls to half its bulk value. It was shown [5] that (10.7.3) is a reasonable approximation for Schrieffer's result [20] and

$$\mathscr{E}_c \mu_{\text{BULK}} \approx \left(\frac{3kT}{m^*}\right)^{1/2} = v_{th} \tag{10.7.4}$$

such that at very intense surface fields ($\mathscr{E}_s \gg \mathscr{E}_c$) the carrier velocity in the channel approach a limit velocity approximately equal to the carrier thermal velocity†.

Equation (10.7.3) will be used as an empirical relationship approximating the experimental data. For an MOS structure the surface field on the semiconductor side is

$$\mathscr{E}_s = \frac{C_{\text{ox}}}{\varepsilon}(V_G - V_{FB} - \phi_s) \tag{10.7.5}$$

such that

$$\mu_s = \mu_{\text{BULK}} \left(1 + \frac{V'_G - \phi_s}{\varepsilon \mathscr{E}_c / C_{\text{ox}}}\right)^{-1}, \quad V'_G = V_G - V_{FB} \tag{10.7.6}$$

and the *effective* surface mobility depends upon the gate voltage.

Figure 10.7.6 shows the surface conductance measured in an MOS transistor with zero steady-state bias of source and drain with respect to the semiconductor bulk. If the effective surface mobility were a constant, the $g = g(V_G)$ plot would be essentially a straight line (except the 'tail' near the extrapolated threshold voltage V_T, see Figure 10.4.9). The slope of $g = g(V_G)$ at higher values of $|V_G|$ is proportional to the surface mobility μ_s. It decreases due to the increasing surface field. The surface mobility can be obtained provided that the device parameters are accurately known. Note the fact that the gradual threshold introduces an apparent maximum of the surface mobility (g_{m0} in Figure 10.7.6).

Recent experimental data have shown a true maximum of μ_s at low values of V_G [31]–[33]. Chen and Muller [33] measured the average conductivity mobilities in inversion layers by using MOS transistors. The average mobility was determined

---

* The effective mobility in surface accumulation layers should be also lower than in the bulk and should decrease with increasing surface field. These 'majority' carrier surface mobilities were measured by Reddi [30] and are in qualitative agreement with theory [28].

† For silicon, at 300 K, with $m^* \approx 1.16\, m_0$ for electrons and $m^* \simeq 0.59\, m_0$ for holes ($m_0 = 9 \cdot 1 \cdot 10^{-28}$g is the free electron mass) these thermal velocities are $1 \cdot 1 \times 10^7$ cm s^{-1} for electrons and 1.51×10^7 cm s^{-1} for holes.

10. MIS Capacitor

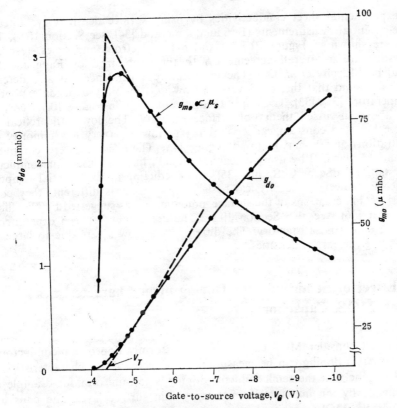

Figure 10.7.6. Experimental transconductance, g_{m0}, and drain conductance, g_{d0}, both measured at $V_D \to 0$ in a p-channel MOST. g_{d0} is just the channel conductance, equation (10.4.19). The slope of $g_{d0} = g_{d0}(V_G)$ is proportional to the surface mobility and to g_{m0}, and decreases at higher V_G, in good agreement with equation (10.7.6). (*after Das*, [5]).

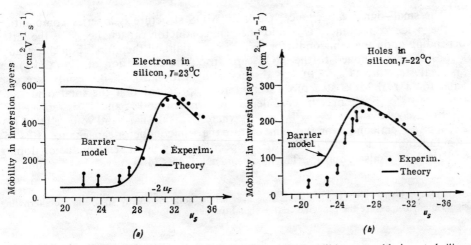

Figure 10.7.7. Experimental and theoretical results for carrier mobilities at weakly inverted silicon surfaces: (*a*) n-channel, (*b*) p-channel. (*after Chen and Muller*, [33]).

from experimental sheet conductivities by using surface densities evaluated from surface-potential measurements (Berglund's method [34], see Section 10.9). Typical results are shown in Figure 10.7.7 a and b for electrons and holes, respectively. The results are in overall agreement with those obtained by Fang and Fowler [31] and by Murphy et al. [32]. The temperature dependence was also determined and it was found that the low inversion (low $|u_s|$) mobility depends exponentially on temperature [31], [33]. The upper theoretical curves in Figure 10.7.7 correspond to the above discussed theory of surface mobility. The lower theoretical curves predict a low and constant carrier mobility at low $|u_s|$ (very low minority carrier concentration) and a maximum at higher $|u_s|$. These curves were computed by Chen and Muller [33] by using a 'barrier model', like that used for conduction in polycrystalline CdSe or CdS films [35], and predicting an exponential temperature dependence. However, the physical basis of this model is different: they consider [33] statistical fluctuations of the surface potential, u_s, as suggested by Nicollian and Goetzberger [36] (see also Section 10.9). The agreement between experiment and this crude model is satisfactory. The differences at low u_s are due to large errors in evaluating the carrier densities.

10.8 Effect of Semiconductor Doping upon Characteristics of MOS Capacitor

We shall first consider MOS capacitors on uniformly doped semiconductors and discuss how this doping can be measured by using the $C-V$ characteristics. Then we shall assume that the semiconductor doping is non-uniform, for example intentionally made by ion implantation in a very thin surface layer, and consider the effect upon the MOS capacitance curves and the turn-on voltage.

a) Using the MOS Capacitor in Determining the Semiconductor Doping

The small-signal capacitance, C, of an MOS structure is a series combination of the oxide capacitance, C_{ox}, and the semiconductor capacitance, C_S, the latter depending upon the semiconductor surface potential and therefore upon the steady-state bias, V_G, applied to the MOS capacitor. The calculation of $C = C(V_G)$ for an ideal structure ($V_{FB} = 0$) was discussed in Section 10.4. The results are also valid for a real MOS structure (Section 10.5), provided that the total oxide and interface-state charge may be considered independent of V_G, and V_G is replaced by an effective gate voltage $V'_G = V_G - V_{FB}$ where V_{FB} is the flat-band voltage (10.5.15). For a given semiconductor and temperature the semiconductor capacitance (10.4.15) is a function of u_s and u_F, the latter depending only upon semiconductor doping and the former also upon V_G. The *low frequency* $C = C(V_G)$ curve* has a minimum,

---

* The minority carriers in the inversion layer follow both the steady state and the alternating signal used to measure the capacitance (the signal frequency is low).

C_m, because $C_S = C_S(u_s)$ has a minimum (Figure 10.4.2). The ratio C_m/C_{ox} depends upon C_{ox} and the impurity concentration, is plotted in Figure 1.8.1 for silicon, at room temperature [6], and permits experimental determination of doping.

The *high-frequency* $C-V$ characteristic (see curve 2 in Figure 10.4.4) can be also used to find the effective doping. The minimum capacitance is denoted now

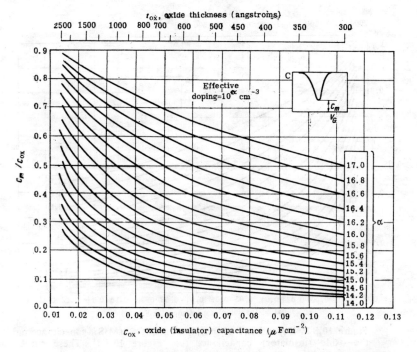

Figure 10.8.1. Dependence of normalized MOS (MIS) capacitance at room temperature on the oxide (insulator) capacitance, with $\alpha = \log_{10} |N_D - N_A|$ as parameter. For the case of SiO_2 films (3.82 dielectric constant), the upper abscissa with oxide thickness can be used. These curves are applicable when minority carriers follow both bias and the alternating signal (low-frequency $C-V$ characteristic). (*after Zaininger and Heiman*, [6]).

by C_{min} and is lower than C_m. The ratio C_{min}/C_{ox} is shown in Figure 10.8.2 *versus* C_{ox} with doping as parameter (see also equation (10.4.14), approximately calculated by using the simplified Grove model of Section 10.3). The high-frequency $C(V)$ characteristic also allows a rapid determination of semiconductor type: for a p-type semiconductor the capacitance decreases with increasing gate voltage, whereas for an n-type semiconductor the converse is true*.

If the minimum-capacitance technique is used to measure the resistivity of very thin epitaxial layers, then the surface depletion-layer will reach the substrate, the

---

* The semiconductor type is also apparent from the low-frequency characteristics, because these are asymmetric around C_m, as shown in Figure 10.4.4.

$C-V$ characteristic is distorted and the minimum capacitance is not directly related to the doping of the layer. In this case the *flat-band capacitance* method can be used. The flat-band ($u_s = 0$) semiconductor capacitance C_{FB} is given by equation (10.4.16). In Figure 10.8.3 the total capacitance at flat-band normalized to C_{ox} is shown *versus* C_{ox} with impurity concentration as parameter. However, C_{FB} cannot be identified

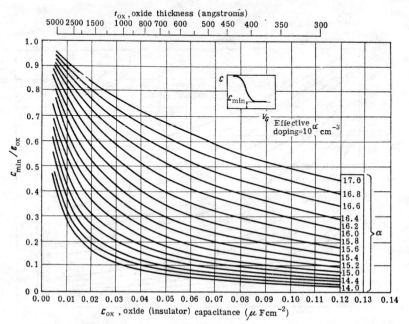

Figure 10.8.2. Dependence of normalized MOS (MIS) capacitance on the oxide (insulator) capacitance (see Figure 10.8.1). These curves are applicable when minority carrier follow bias but not the alternating signal (high-frequency $C - V$ characteristic). (*after Zaininger and Heiman,* [6]).

directly on the experimental $C-V$ curve, because V_{FB} is not known (it depends upon semiconductor capacitance and doping*). An approximate rule is the following. The voltage at which $C = 0.95\, C_{ox}$ (high-frequency curve) is offset one volt † from V_{FB}, provided that the surface-state density is low [6]. This result is independent of C_{ox} and doping if $C(V_{FB})$ is significantly lower than $0.95\, C_{ox}$. We see from Figure 10.8.3 that the rule is applicable for low impurity concentrations. This is the case of practical interest because it is the low doping domain where $x_{d\max}$ is large and the minimum-capacitance method cannot be used.

The slope of the region of the $C-V$ characteristic corresponding to the depletion régime also yields the semiconductor doping, provided that the interface-state effect is unimportant [6]. A fast ramp (see Figure 10.4.6) or an alternating voltage

---

* This dependence occurs through the metal-semiconductor work function difference Φ_{MS} (see **Figure** 10.5.2) and is unimportant for the minimum-capacitance method.

† This offset is positive for *n*-type and negative for *p*-type silicon. The latter case can be seen in Figure 10.4.4.

should be used as gate bias in order to avoid minority carrier accumulation at the interface and, therefore, the limitation of depletion layer depth to $x_{d\max}$: the measured capacitance is the depletion or deep depletion capacitance given by equation (10.4.11). Note that

$$\frac{dV_G}{d(C^{-2})} = \frac{eN_A\varepsilon}{2} \tag{10.8.1}$$

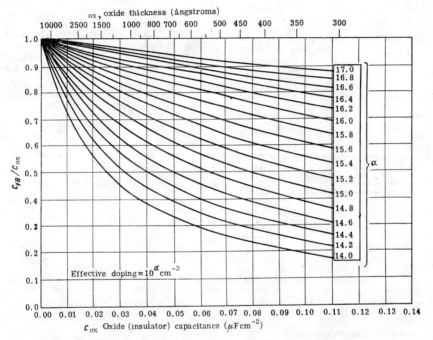

Figure 10.8.3. Dependence of normalized MOS (MIS) flat-band capacitance on the oxide (insulator) capacitance (see Figure 10.8.1). (*after Zaininger and Heiman*, [6]).

and the slope of the $d(C^{-2})/dV_G$ characteristic is directly related to the semiconductor doping, as shown in Figure 10.8.4.

If the doping is *non-uniform*, the $C^{-2} - V_G$ characteristic is no longer linear but the (variable) slope is still related to the doping. A similar situation occurs when the *pn* junction capacitance or Schottky barrier capacitance is used for determination of the impurity profile (see for example Section 3.6). Note the fact that here the depletion capacitance appears in series with the oxide capacitance and therefore the distance x_d in equation (3.6.4) which yields $N = N(x)$ needs a correction [37] (see Problem 10.7).

The major problem of semiconductor profiling with the MOS (MIS) capacitor is the effect of the interface states which can give a large and difficult to correct error. Fast signals should be used for capacitance measurements such as these

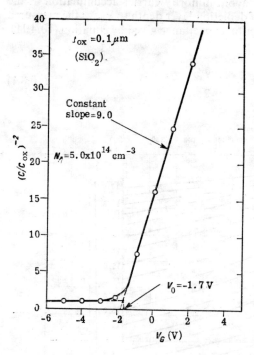

Figure 10.8.4. Evaluation of semiconductor doping from the $C - V$ data (deep depletion capacitance). The straight line follows the equation $(C_{ox}/C)^2 = A(V - V_{FB}) + 1$ where $A = 2C_{ox}^2/e\varepsilon \, |N_D - N_A|$ and $V_{FB} = \Phi_{MS} - Q_{SS}/C_{ox}$. (after Zaininger and Heiman, [6])

Figure 10.8.5. Doping density *versus* depth from Si−SiO$_2$ interface, calculated for an MIS capacitor (300 K temperature and 30 MHz fundamental frequency in the method described in reference [38]). The curve for an abrupt junction diode at zero bias is shown, indicating the range of x_d over which doping density can be accurately determined. Depth is measured from the edge of the abrupt doping discontinuity rather from the Si − SiO$_2$ interface. (after Nicollian et al., [38])

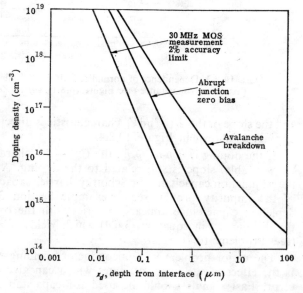

10. MIS Capacitor

interface levels cannot respond. The pulsed $C-V$ method and the second harmonic method, not discussed here, are convenient for doping profile measurements with minimal error due to interface states [38].

There are several advantages [38] of using the MIS capacitor for semiconductor profiling, instead of pn junctions or Schottky barriers: this capacitor can be easily fabricated on almost any semiconductor substrate, in a manner such that the doping profile is not disturbed (low-temperature deposition of the insulator), the method is not destructive and, finally, it enables accurate profile measurements to be made over a greater range of doping density and depth. This is illustrated by Figure 10.8.5, which shows the minimum depletion length to which N can be determined for an MIS capacitor and abrupt pn junction at zero bias. The minimum depth, for example, is a few Debye lengths, while in the abrupt junction diode is limited by the built-in depletion width. In both cases the maximum depth to which N can be determined is limited by avalanche breakdown.

b) Non-uniform Semiconductor Doping and $C-V$ Curves

Brotherton and Burton [39] calculated the MOS capacitance for a non-uniform semiconductor doping, namely (p-type substrate)

$$N_A = N_A(x) = N_{A\text{sub}}[1 + \alpha \exp(-\beta x)] \qquad (10.8.2)$$

If the depth of the high-conductivity layer is relatively large, such that it contains the entire surface space-charge region then the computed $C-V$ characteristics (Figure 10.8.6 a) will be similar to those expected from highly doped substrates. For example, both C_m/C_{ox} and the slope of the $C-V$ curve (for surface depletion) indicate that curve 2 corresponds to a higher semiconductor doping than curve 1. However, if the high-conductivity layer is very thin, a completely different picture will be obtained (Figure 10.8.6 b). The $C-V$ curves calculated for $\alpha = 10^2$ and 10^3 are almost identical in shape to that calculated for uniform substrate doping, but displaced towards higher voltages. Such a shift is normally interpreted as due to fixed interface charge or oxide charge (negative). Therefore abrupt changes of the doping profile in the immediate vicinity of the surface can introduce errors in estimating the interface state density (which would result here to be much higher than in reality).

The physical explanation of Figure 10.8.6 b is as follows [39]. A change in doping of two orders of magnitude takes place within a distance of a few hundred angstroms from the surface. Therefore, the surface space-charge region extends deep into the uniformly doped substrate and the minimum of the $C-V$ curve is about the same in all cases, because it is determined by $N_{A\text{sub}}$. On the other hand, a large proportion of the gate voltage drops across the highly-doped surface layer. This drop ΔV can be calculated approximately by evaluating the charge corresponding to the ionized impurities in excess, located just near the surface, ΔQ_A,

$$\Delta V \simeq \frac{\Delta Q_A}{C_{ox}} = \frac{e}{C_{ox}}\left[\int_0^x N_A \, dx - \int_0^x N_{A\text{sub}} \, dx\right] =$$

$$\qquad (10.8.3)$$

$$= \frac{eN_{A\text{sub}}}{C_{ox}}\int_0^x \alpha \exp(-\beta x) \, dx \simeq \frac{eN_{A\text{sub}}\alpha}{\beta C_{ox}}, \text{ for large } \beta x.$$

Note the fact that a uniform surface layer with an excess concentration $\alpha N_{A\text{sub}}$ and thickness $1/\beta$ has the same effect. Equation (10.8.3) yields a good approximation for the voltage shift indicated by the exact computation (Figure 10.8.6 b).

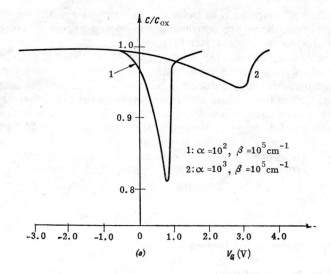

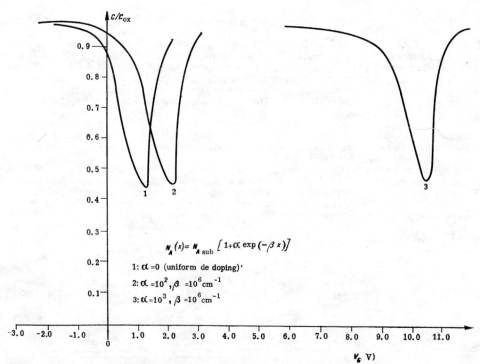

Figure 10.8.6. Theoretical $C - V$ curves for non-uniformly doped substrates ($V_{FB} = 0$, p-type substrate). (*after Brotherton and Burton*, [39]).

10. MIS Capacitor

The above considerations show that semiconductor profiling by the $C-V$ technique requires special care if the doping changes very much in a short distance from the surface because the exact form of the impurity distribution near the surface does not significantly affect the MOS capacitance*.

In the conventional MOS technology a non-uniform doping occurs due to the impurity redistribution near the surface during the thermal oxidation process used to grow the silicon dioxide layer. The modification of the $C-V$ curve due to this redistribution is minor (a shift of the order of 0.1V and a slight distortion), although it may result in an error in evaluating the surface-state charge [40]. The modification of the turn-on (threshold) voltage can be, however, important [41]–[46]. This is especially true for a p-type substrate, where the turn-on voltage is close to zero and highly doping dependent, as shown by Figure 10.6.8 (this substrate is used to fabricate n-channel MOS transistors and the threshold voltage should be positive to have a device turned-off at zero bias). Schottky [41] calculated boron redistribution during successive oxidation and heat treatments used in MOS transistor fabrication. The boron concentration in the p-type substrate results lowers near the surface (a total decrease of about 50% of the bulk doping occurs in a layer of one or a few micrometers) due to both the boron take up during the oxide growing and the boron diffusion in silicon [41]. The effect of this redistribution on the threshold voltage V_T', equation (10.6.1), can be approximately described by an apparent reduction of substrate doping and an apparent increase in oxide charge, thus reducing V_T'. The calculated reduction is about 1 V (1000 Å oxide) [41].

A non-uniform doping profile occurs during the ion implantation process, now frequently used to control the turn-on voltage.

c) Ion-implanted MOS Capacitors

The ion-implantation technology found an industrial application in fabrication of MOS devices and MOS integrated circuits [46], [47]. One possibility offered by this technology is the modification or adjustment of the threshold voltage by ion implantation in a very thin sheet at the semiconductor surface.

The implantation takes place through a very thin (1000 Å) oxide layer and the distribution of implanted ions is as shown qualitatively in Figure 10.8.7. The depth depends upon the energy used for acceleration (a few kV to hundreds of kV), and the concentration upon the dose (total number of ions per unit surface, of the order of 10^{12} ions cm^{-2}). The energy and the dose can be changed independently, leading to a great flexibility of the process, and both can be tightly controlled, thus providing the reproducibility. The process is self-masking because the ion implantation is effective only in regions covered by the very thin SiO$_2$ layer corresponding to the gate of MOS devices (the remainder of the wafer is covered by a thick oxide layer). The implantation is independent of the temperature. A subsequent heat treatment can recover the lattice defects produced by ion bombardment and also can produce a redistribution of the implanted ions. The gate metallization is completed, of course, after implantation.

The control of the threshold voltage is due, in principle, to the modification of fixed semiconductor charge (V_B term in equation (10.5.17)). An exact numerical calculation of potential distribution inside the semiconductor and the MOS capacitance was performed† by MacPherson [43] for boron ion implantation in n-type silicon. The distribution of implanted ions in silicon was assumed Gaussian [43], [46]

$$N_A(x) = \frac{N_\square}{\sqrt{2\pi}\,\Delta R_p} \exp\left[-\frac{1}{2}\left(\frac{x - R_p}{\Delta R_p}\right)^2\right], \tag{10.8.4}$$

where $N_\square$ (atoms/cm^2) is the implant dose, R_p is the effective implantation depth and ΔR_p the standard deviation. By modifying the dose $N_\square$ and keeping the ion energy constant, ΔR_p remains constant and R_p depends upon the location of the $x = 0$ plane (semiconductor surface). If the energy is properly chosen for a given oxide thickness, the distribution is centered at the SiO$_2$–Si interface

---

* If the surface region is highly doped and the depletion layer is confined to this region, the semiconductor depletion capacitance is very high and makes a negligible contribution to the total capacitance because it occurs in series with the much smaller oxide capacitance.

† The method of numerical integration of the Poisson equation is that indicated by Lindmayer [48] and also used by Panigrahi [40] and other authors. An alternative is indicated in reference [49].

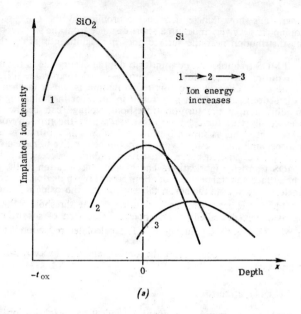

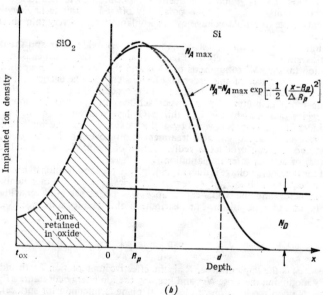

Figure 10.8.7. (a) Repartition of ions implanted through SiO_2 in a SiO_2-Si system, for different ion energies. The distribution of ions implanted in both SiO_2 and Si are Gaussian but with different parameters [46]. (b) In this figure the peak of the Gaussian describing ion density in the oxide is located in the semiconductor. The fraction of implanted ions stopped in the oxide can be readily calculated. If acceptors are implanted in an n-type semiconductor (as shown in this figure), a junction is formed at the depth d.

($R_p = 0$). Figure 10.8.8 a shows the calculated potential distributions in the semiconductor with implant dose as parameter, for a fixed surface potential. The impurity profiles ($R_p = 0$) are also shown. The charge implanted in silicon is one half of the total dose, $N_\square$. Figure 10.8.8 b displays potential distributions at a given dose, for various surface potentials. The flat-band condition for uniformly doped substrates has no equivalent for non-uniform profiles (when $u_s = 0$, the bands are not flat in the bulk, here a depletion layer of 0.2 µm exists). Also the condition $u_s = 2u_F$ for strong inversion is irrelevant because when $u_s = 2u_F$ the heavily implanted surfaces are far more strongly inverted than those not implanted [43]. The threshold voltage is now defined by $u_p = 2u_F$, where u_p is the peak potential (Figure 10.8.8 b). The calculated shift in threshold voltage due to boron implantation is positive and increases linearly with the dose $N_\square$, reaching about 4.3 V for $N_\square = 2 \times 10^{12}$ ions cm^{-2}. The measured shifts are slightly lower than predicted by theory and the difference may arise from the imperfect knowledge of the actual ion distribution in experimental devices [43].

The theoretical and experimental $C-V$ characteristics [43] are indicated in Figure 10.8.8 c and d. The shift of these characteristics towards positive voltages as the implant dose increases from zero to maximum is in complete agreement with the model of Brotherton and Burton [39] (see also Figure 10.8.6) and is explained by the fact that the penetration of boron ions in silicon is very small (compared to the total depletion width), of the order of hundreds of angstroms (Figure 10.8.8 a). If this distance is larger, the shape of the $C-V$ characteristics will be indeed modified [43] as predicted by Brotherton and Burton [39]. The qualitative explanation is similar (see above) in both cases although the situation is somewhat different: here the substrate is n-type and the surface is inverted to p-type, whereas before an n-type substrate with a highly-doped n^+ surface layer was considered [39].

Sigmon and Swanson [46] considered in detail the effect of the pn junction formed by ion implantation. In principle the inversion channel at the surface is built-in. However, the surface conduction depends upon the surface potential, as shown below. For simplicity, the Gaussian distribution of the p-type ions was assumed rectangular (depth d) and the substrate donor concentration-constant. The band-energy diagram in Figure 10.8.9 indicates the existence of a neutral p-type layer sandwiched between the surface depletion layer and the depletion layer of the pn junction. This neutral layer is a conductive channel separated from the bulk (and the surface). However, a larger positive gate bias will turn-off this channel by extending the surface depletion layer until it touches the junction depletion region. The $C-V$ curve will indicate this moment by a dip in capacitance. We shall consider below qualitatively the high frequency (small-signal) capacitance of such an implanted MOS capacitor, with a buried p-type layer beneath the gate oxide, in the n-type substrate. There are three capacitances in series: the oxide capacitance, C_{ox}, independent of voltage; the surface depletion layer capacitance, C_d, and, finally, the junction depletion capacitance, C_j. The total capacitance varies with the gate bias as shown in Figure 10.8.10 and the explanation will be given relative to the energy band diagrams shown in Figure 10.8.11 for various gate voltages.

For large negative gate voltages, the surface (p-type) is accumulated (Figure 10.8.11 a). The surface depletion layer capacitance tends to infinity ($C_d \to \infty$). The junction capacitance is determined by doping and by the built-in potential difference, $V_{bi} > 0$, and will remain at this constant value, C_{jd}, as far as its depletion is bounded by neutral regions until turn-off). Therefore the total capacitance $C = C_{ox} C_{jd}(C_{ox} + C_{jd})^{-1}$ is independent of voltage (Figure 10.8.10). Figure 10.8.11b shows the flat-band case. The flat-band voltage (Figure 10.8.11b) becomes:

$$V_{FB} = \Phi_{MS} - \frac{Q_{ss}}{C_{ox}} - V_{bi}. \qquad (10.8.5)$$

Increasing V_G above V_{FB} will decrease C because C_d becomes now finite and decreases with increasing surface depletion layer. This decrease will continue until the turn-off condition, when the neutral p-type layer disappears (Figure 10.8.11 c). For voltages above the turn-off the potential drop across the pn junction will decrease (Figure 10.8.11 d) and C_j will increase. The total capacitance of the structure will also increase. When the gate voltage is further increased the junction

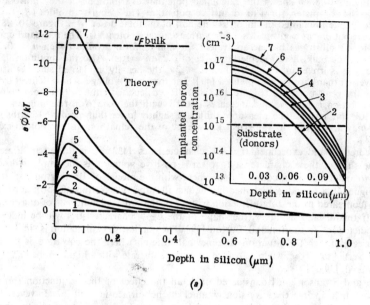

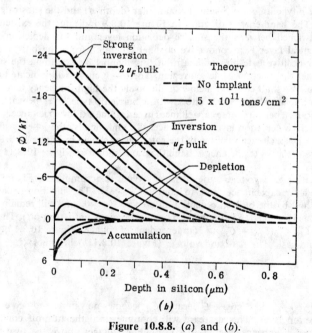

Figure 10.8.8. (a) and (b).

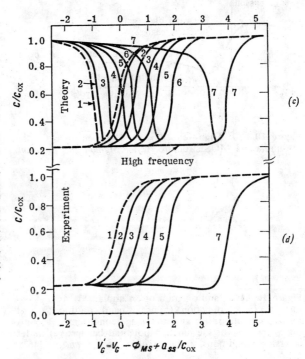

Figure 10.8.8. (a) Variation of potential with depth in silicon as a function of implanted boron dose, $N_\square$. (1) No implant; (2) 10^{11} ions cm^{-2}; (3) 3×10^{11}; (4) 5×10^{11}; (5) 7×10^{11}; (6) 1×10^{12}; (7) 2×10^{11}. Inset shows profiles assumed for the potentials displayed. The semiconductor is 3Ω cm, (100), n-type. $t_{ox} = 840$ Å. (after Mac Pherson, [43]) (b) Potential distributions at different surface potentials, for a given dose and with no implant [43]. (c) and (d) calculated and measured $C-V$ characteristics (measurement frequency was 1MHz). The implant energy was 26.6 keV (boron). The different doses are indicated in (a) [43]).

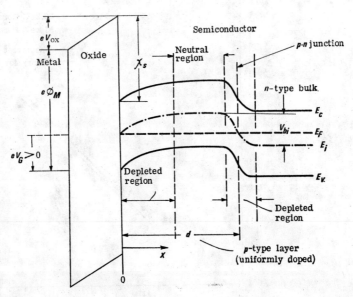

Figure 10.8.9. Band diagram of an MOS structure with an implanted p-layer in the n-type substrate (after Sigmon and Swanson, [46]).

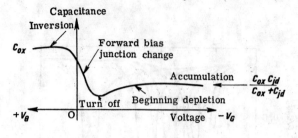

Figure 10.8.10. High-frequency $C-V$ dependence for an ion-implanted MOS structure. (*after Sigmon and Swanson*, [46]).

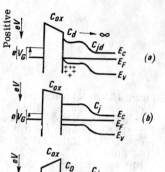

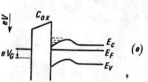

Figure 10.8.11. Band diagram of an implanted MOS structure: (*a*) in accumulation; (*b*) at flat-band condition; (*c*) at turn-on condition; (*d*) at gate voltage conditions determining a lowering of the junction potential barrier; (*e*) under inversion conditions. (*after Sigmon and Swanson*, [46]).

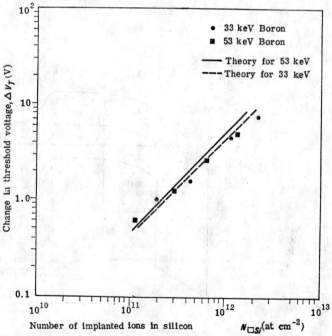

Figure 10.8.12. Measured shift in threshold voltage, ΔV_T, for MOS transistors, (*after Sigmon and Swanson*, [46]).

becomes forward biased and then the surface inverts to n-type (Figure 10.8.11 e). This shortens the depletion layer and junction capacitance to the substrate and the total capacitance reaches the oxide capacitance (Figure 10.8.10). The capacitance dip was indeed observed experimentally [42], [43], [46], see for example Figure 10.8.8 d.

If the n-type inversion layer occurs before the depletion region reaches the pn junction, the turn-off cannot be achieved by increasing the gate voltage. This occurs for heavier implants, because the depletion layer at inversion becomes narrower.

Figure 10.8.12 compares theory with experiment [46]. The agreement is fairly good for a wide range of implant doses and threshold voltage shifts (positive). The implanted junction depth (depending upon ion energy) is less important in determining ΔV_T than the dose. Above 2×10^{12} boron atoms per cm², the device cannot be turned-off with gate bias only, as predicted above.

10.9 Effect of Interface States

It was already shown (Section 10.6) that the mobile charge in the semiconductor surface layer can be trapped in states located at the semiconductor-oxide interface. The occupancy of these states depends upon the surface potential and therefore upon the applied bias. Not only is the ideal $C-V$ characteristic distorted by the interface states (Figure 10.6.5), but the static and dynamic characteristic of all solid-state devices containing the MOS structure are affected as well. A very low interface state concentration is essential for the MOS-technology. Declerck et al. [50] reported surface state densities as low as $1.0 - 1.5 \times 10^{10}$ cm^{-2}eV^{-1} for (111) substrates and a few 10^9 cm^{-2}eV^{-1} for (100) silicon substrates*. Modern planar technology can therefore provide nearly perfect Si—SiO$_2$ interfaces from the device point of view. However, the technological advances were achieved mainly by empirical means, without guidance from a well established physical model of the nature of interface states. In fact an exact experimental knowledge of the interface state parameters is relatively recent [50]—[52].

When the interface states are present the impedance of the MOS capacitor is no longer that of a pure capacitance: a loss mechanism involved in carrier capture and remission determines the occurrence of a conductance in parallel to the MOS capacitance. The impedance of the MOS structure was calculated by Berz [53], Lehovec and Slobodskoy [54], Forlani et al. [55]. An exact calculation of the MOS impedance including, in particular, the effect of interface states was reported recently by Temple and Shewchun [56]. This calculation is based upon a transmission line circuit model suggested by Sah et al. [57] which is the exact analog of the small signal transport equations in the semiconductor including explicitly phenomena as generation, recombination and trapping in interface states.

The interface-state parameters could be determined in principle by measuring the MOS capacitance, the MOS conductance or both. In practice, however the information contained in capacitance measurements is difficult to extract because the semiconductor capacitance C_s (affected by interface states) occurs in series with the oxide capacitance, C_{ox}. The oxide capacitance acts as a 'window' that permits only a partial examination of semiconductor capacitance [6]. C_{ox} should be as high as possible compared with C_s (thin oxide) and determined with great accuracy. On the other hand, the parallel conductance arises solely from the loss

---

* On the other hand the oxide charge density was reduced to 2×10^{11} cm^{-2} and $3 - 7 \times 10^{10}$ cm^{-2} for (111) and (100) silicon, respectively [50].

involved in the interface trapping and is a more direct measure of interface state properties [36]. This is illustrated by Figure 10.9.1 which shows capacitance and equivalent parallel conductance measured at two different frequencies and thus evidencing the interface-state effect. The largest capacitance spread is 14 per cent

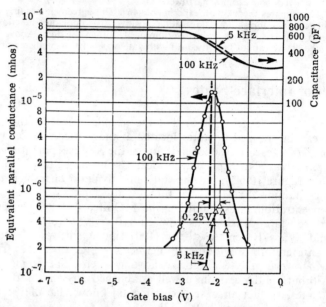

Figure 10.9.1. Capacitance and equivalent parallel conductance measured at 5 kHz and 100 kHz on p-type sample having an acceptor density of $2 \cdot 08 \times 10^{16}$ cm^{-3} and interface state density in the 10^{11} cm^{-2} eV^{-1} range. (*after Nicollian and Goetzberger*, [36]).

while the conductance is much more sensitive (the peak increases by over an order of magnitude) [36].

The so-called 'conductance technique' in determining the interface state properties was put forward by Nicollian and Goetzberger [36]. They examined the Si—SiO$_2$ interface and carefully demonstrated that the oxide loss is negligible, that the dominant loss is by capture and emission of carriers by interface states rather than states in the silicon space-charge region and that the loss is determined entirely by the majority-carrier time-constant in both the depletion and weak inversion regions (minority carrier density can respond to the applied signal by diffusion from the bulk and generation-recombination processes, which are too slow to follow a 50 Hz to 500 kHz signal [36], see also the next Section). For majority carriers the quasi-Fermi level and Fermi level are identical and therefore the alternating signal simply modulates the Fermi function defined in Chapter 1. It can be shown (Problem 10.10) that the effect of a single interface level can be described by a series RC network with capacitance $C_{ss} = e^2 N_{ss} f_0 (1 - f_0)/kT$ and time constant $\tau_{ss} = C_{ss} R_{ss} = f_0/c_n n_{s0}$, where N_{ss} is the density of states (cm^{-2}), c_n electron capture probability (cm^3 s^{-1}),

n_{s0} electron density established at the surface by the steady-state bias, and f_0 the Fermi function determined by the same bias. The equivalent parallel capacitance is

$$C_p = \frac{C_{ss}}{1 + \omega^2 \tau_{ss}^2} \qquad (10.9.1)$$

and the parallel conductance is G_p,

$$\frac{G_p}{\omega} = \frac{C_{ss} \omega \tau_{ss}}{1 + \omega^2 \tau_{ss}^2}. \qquad (10.9.2)$$

The total MOS impedance is indicated in Figure 10.9.2, where C_d is the depletion layer capacitance. The oxide capacitance C_{ox} is measured in the region of strong accumulation. The reactance of C_{ox} is subtracted from the total impedance. The parallel conductance of the resulting impedance (admittance) is given by equation (10.9.2). G_p/ω goes through maximum when $\omega \tau_{ss} = 1$, which gives τ_{ss} directly. The value of G_p/ω at the maximum is $C_{ss}/2$, thus yielding C_{ss}. Therefore the interface-state density and capture probability can be determined directly from experimental data on $G_p = G_p(\omega)$. Note that C_p occurs in parallel to the depletion layer capacitance which can be calculated by using the estimated doping (not accurately known near the silicon surface because of pile up or depletion of the dopand during the thermal growth of the oxide [36]). Therefore the conductance measurement is preferable.

However, a continuum of interface-state levels appears to be the characteristic of the Si–SiO$_2$ interface and equation (10.9.2) should be modified as follows [36], [51]

$$\frac{G_p}{\omega} = \frac{e \overline{N}_{ss}}{2 \omega \tau_{ss}} \ln(1 + \omega^2 \tau_{ss}^2), \qquad (10.9.3)$$

where $\overline{N}_{ss}$ is the density of states per cm^{-2} and per eV (assumed constant) and τ_{ss} is the corresponding lifetime $\tau_{ss} = c_n n_{s0}$, which may be written

$$\tau_{ss}^{-1} = v_{th} \sigma_n n_0 \exp|u_s| \qquad (10.9.4)$$

where σ_n is the capture cross section*, n_0 the bulk electron concentration and u_s the surface potential in kT/e units.

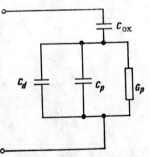

Figure 10.9.2. Equivalent circuit of the MOS structure including the elements (G_p, C_p) associated with the interface states.

The experimental curves yielding the total **parallel** conductance *versus* frequency are in general much broader than the theoretical curves obtained from the equivalent circuit of Figure 10.9.2, equations (10.9.3), (10.9.4), etc. This indicates a dispersion in the time constant. Nicollian and Goetzberger [36] suggested that this is due to fluctuations of the surface potential, u_s, such that the MOS capacitor can be considered as formed of many micro-capacitors, each with a definite value of

---

* Whereas $\overline{N}_{ss}$ varies slowly over an energy range of a few kT, the variation of σ_n with energy is much more important and a more general formula should be used [51].

u_s. The fluctuations arise because the built-in charges and charged interface states are randomly distributed in the plane of the interface [36]. The influence of the random distribution of ionized atoms in the space-charge region is shown to be less important [36]. To get the impedance of the MOS capacitor we have to multiply the parallel capacitance and conductance of the total admittance of a micro-capacitor (see Figure 10.9.2) with the relative probability of finding a normalized surface potential in the interval u_s to $u_s + du_s$ and then integrated, as shown by Deuling et al. [51]. The random distribution of charge is assumed to be Gaussian. The dispersion constant of this distribution is determined by comparing the theoretical curves with experiment [51]. This requires lengthy computations and even a computer program [50], [51]. It was shown that a direct fitting method using approximate analytical expressions is preferable [50]. A simplified version of these calculations was also suggested by Simonne [58]. Typical experimental results are shown in Figure 10.9.3 (dots) together with the best fit with the statistical model (solid lines). This comparison yields $\overline{N}_{ss}$, σ_n, the dispersion parameter and the average surface

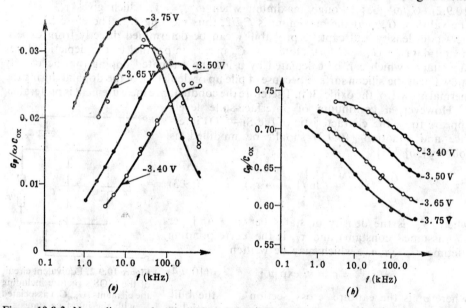

Figure 10.9.3. Normalized conductance (a) and capacitance (b) of an MOS capacitor as a function of frequency at a few gate voltages (n-type silicon sample 1000 Å thick SiO_2 thermally grown in dry oxygen at 1140 °C 30 min annealing in dry hydrogen at 350 °C). The points are experimental data (300 K) and the solid curves are the best fit obtained with the statistical model. (*after Deuling et al.*, [51]).

potential. We stress the fact that only those interface states lying within a few kT of the Fermi level contribute to the impedance. To detect states at different energies it is necessary either to change the bias (and thus move the conduction band edge at the surface and therefore the position of surface states relative to the Fermi level) or to modify the temperature (and therefore to move the Fermi level relative to the states). Deuling et al. [51] used both methods and obtained the interface state density and electron capture cross-section shown in Figures 10.9.4 and 10.9.5, respectively. The measured sample was epitaxial (111) silicon with 1200 Å oxide grown in dry oxygen and annealed for 30 min at 350 °C in a hydrogen atmosphere after deposition of the aluminium contact. Figure 10.9.4 shows that the interface state density does not explicitly depend on temperature. However, after warming to room temperature a new low temperature measurement gives slightly lower densities in the vicinity of the conduction band edge. This effect, observed many times [51], is still unexplained. The capture cross-section is again independent of temperature (Figure 10.9.5). No drift with temperature cycling was observed.

Recent measurements [50] indicate that only the conductance technique allows the measurement of densities as low as a few 10^9 cm^{-2}eV^{-1}. Declerck et al. [50] determined a uniform surface state density of a few 10^{-10} cm^{-2}eV^{-1} for a range of 0.25– –0.3 eV on both sides of the midgap (p-type and n-type (111) silicon substrates were used). This very low value of $\bar{N}_{ss}$ (compare with Figure 10.9.4) was obtained by improving the gate oxidation and the annealing process before metallization and by using a final low-temperature treatment [50].

The determination of interface-state parameters from capacitance measurements can be made by several methods [59]. The results are in general less accurate and the capture cross section cannot be determined. The only one of these methods which will be mentioned below is that originally suggested by Berglund [34] and named the low-frequency technique or the integration procedure. The method is based upon the determination of surface potential ϕ_s from a low-frequency differential capacitance measurement. According to equation (10.4.24)

$$\frac{d\phi_s}{dV_G} = 1 - \frac{C}{C_{ox}} \tag{10.9.5}$$

and the surface potential at any applied voltage can be determined by integrating the $1 - C/C_{ox}$ curve. Then the charge trapped in interface states is evaluated from the bias dependence of surface potential [34], [59], [60]. C is the *low-frequency* capacitance because all surface states must be in equilibrium at all times during the $C(V)$ measurement: they must follow both the steady-state bias and the alternating signal. Such measurements are difficult due to very long time constants involved in trapping kinetics. For example the minority carrier lifetime could be as high as 100 μs, the generation recombination mechanism in the inversion layer is very slow and test frequencies as low as 1 Hz should be used.

The Berglund method can be applied by using the low-frequency capacitance, C, determined by a quasi-static technique suggested by Kuhn [60] and others. The principle of the method is: a linear voltage ramp $V(t) = V_0 + \alpha t = V_G$ is applied to the capacitor and the charging current (displacement current) is measured. This current is

$$I(t) = \frac{dQ}{dt} = \frac{dQ}{dV_G}\frac{dV_G}{dt} = \alpha C(V_G), \tag{10.9.6}$$

where $C = \dfrac{dQ}{dV_G} = C(V_G)$ is the quasi-static differential capacitance (the variation of V_G has to be sufficiently slow), measured in farads. Typical voltage sweep

522   Surface-controlled Devices

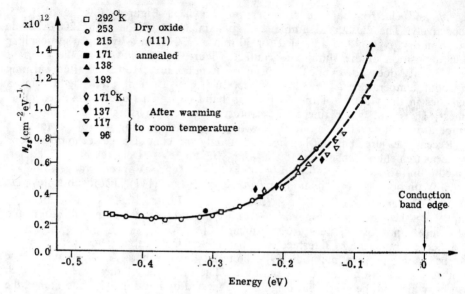

Figure 10.9.4. Surface state density (measured for the sample indicated in Figure 10.9.3) as a function of energy in the gap (relative to the conduction band edge). (*after Deuling et al.*, [51]).

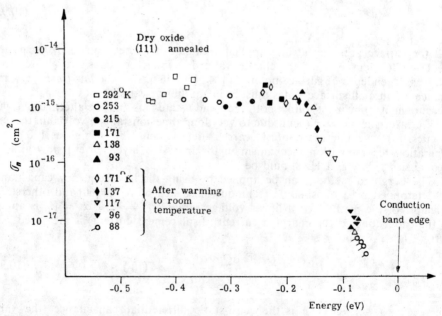

Figure 10.9.5. Electron capture cross-section of surface states as a function of energy. (*after Deuling et al.*, [51]).

rates should be low, in the range of 5 to 500 mV and give displacement current densities usually less than 10^{-8} A cm^{-2} [60]. This current is determined by measuring the voltage across a resistor R which together with the MOS capacitor and an operational amplifier forms an analog differentiator (see the inset in Figure 10.9.6).

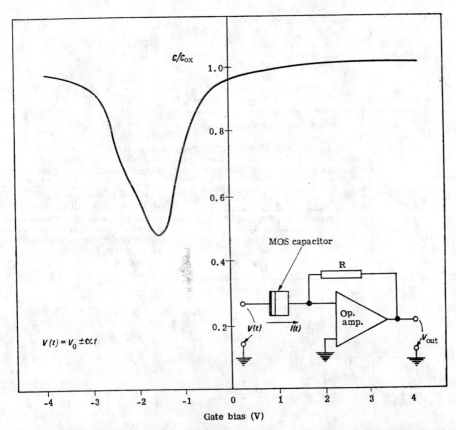

Figure 10.9.6. MOS $C - V$ curve obtained with the quasi-static technique (basic circuit shown in the inset) for an n-type silicon MOS capacitor ($N_D = 4.19 \times 10^{14}$ cm^{-3}, 800 Å thick oxide). (*after Kuhn*, [60]).

The simultaneous display of $V(t)$ and $V_{out} \simeq RI(t)$ on an $X - Y$ recorder will yield directly the $C(V_G)$ curve. This technique provides a very simple and rapid determination of low-frequency capacitance. The surface state density is obtained from the following considerations [60]. Figure 10.9.7 a shows the potential energy diagram for an MOS structure on an n-type substrate. Figure 10.9.7 b indicates the corresponding equivalent circuit [59], [60], where C_d, C_a and C_i are the depletion, accumulation and inversion capacitances, respectively and C_{ss} is the interface-state equivalent capacitance. The resistances represent: R_c and R_v — the capture and emission processes corresponding to interaction between the interface states and electrons and holes at the semiconductor surface; R_g — the depletion layer generation mechanism responsible for the formulation of the inversion layer. All these circuit elements are strongly dependent upon the surface potential. In the accumulation and depletion régimes the minority carrier concentration at the surface is

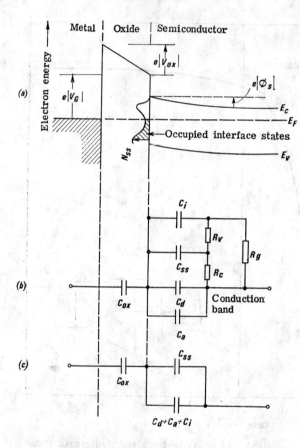

Figure 10.9.7. (a) Energy band diagram for an n-type MOS capacitor; (b) equivalent circuit for the MOS capacitor (all components are strongly dependent upon the surface potential); (c) reduced equivalent circuit under quasi-static conditions when thermal equilibrium is maintained. (*after Kuhn*, [60]).

very low and the interface states interact with the semiconductor surface only by capture and emission of electrons in the conduction band such that $R_v \gg R_c$ and the interface state time constant is $\tau_{ssc} = C_{ss} R_c{}^*$. In inversion, however, the inversion layer formation rate and the hole-capture rate from the valence band are important. The former process has the time constant $\tau_i \simeq C_i R_g$, of the order of one second for silicon [60]. If the voltage sweep is sufficiently slow and the inversion layer is maintained in thermal equilibrium the thermal generation term will effectively short the valence band to the conduction band and because in inversion $R_v \ll R_c$ the interface state relaxation constant will be $\tau_{ssv} = C_{ss} R_v$. Both τ_{ssc} and τ_{ssv} were

---

* This is the time constant introduced above in presenting the conductance technique. The same time constant is important in weak inversion.

determined by conductance measurements on n-type and p-type samples, respectively. These time constants depend exponentially upon the surface potential such that $\tau_{ssc} \ll \tau_{ssv}$ in depletion, whereas $\tau_{ssv} \ll \tau_{ssc}$ in inversion, and near the midgap both are of the order of 10^{-2}s. Therefore the longest interface-state time constant is of the order of 10^{-2}s. If the sweep rate is low enough to maintain the inversion layer in thermal equilibrium then the interface state will be also in thermal equilibrium and in the equivalent circuit of Figure 10.9.7 b, R_c, R_v and R_g may be replaced by short circuits. Therefore, the response to the slow ramp can be calculated by using the simplified circuit of Figure 10.9.7 c, where $C_S = C_i + C_d + C_a$ is the semiconductor capacitance, equation (10.4.6), of an ideal structure. The real MOS capacitance will include the interface-state capacitance

$$C_{ss} = C_{ss}(\phi_s) = e^2 \overline{N_{ss}}(\phi_s) = -\frac{dQ_{ss}}{d\phi_s} \qquad (10.9.7)$$

(the interface state charge Q_{ss} and capacitance C_{ss} are defined per unit area) such that (Figure 10.6.7 c)

$$\frac{1}{C(V_G)} = \frac{1}{C_{ox}} + \frac{1}{C_s(\phi_s) + C_{ss}(\phi_s)}. \qquad (10.9.8)$$

Note the fact that the oxide charge (assumed constant, fixed) influences the above relationship only through the dependence between ϕ_s and V_G. This dependence will be obtained by integrating equation (10.9.5) where $C = C(V_G)$ and C_{ox} are experimentally derived. Therefore the low frequency method (or the quasi-static technique) consists in the following steps:

(a) Ideal $C_S = C_S(\phi_s)$ curves are computed for the measured oxide capacitance, doping density and capacitor area.

(b) The experimental $1 - C(V_G)/C_{ox}$ curve is integrated to determine $\phi_s = \phi_s(V_G)$.

(c) The integration constant in $\phi_s = \phi_s(V_G)$ is found by comparing measured C/C_{ox} as a function of ϕ_s with the ideal $C = C(\phi_s)$ curve. These curves should coincide for strong accumulation and strong inversion, where the interface state capacitance effect is negligible. An example of such a comparison is shown in Figure 10.9.8.

(d) C_{ss} and $\overline{N}_{ss}$ are found by using equations (10.9.7) and (10.9.8). For the sample whose data are plotted in Figure 10.9.6 and 10.9.8, the final results are shown in Figure 10.9.9. Interface state distribution determined from conductance measurements is indicated for comparison*: two separate MOS capacitors are necessary, a p-type structure for the points below mid-gap, and an n-type structure above mid-gap. Only one sample is necessary to determine interface-state density over a large part of the energy gap if the quasi-static technique is used. Another advantage of this method is that the manipulation of experimental and ideal theoretical data can be carried out in a computer.

---

* A similar comparison was made by Declerck et al. [50].

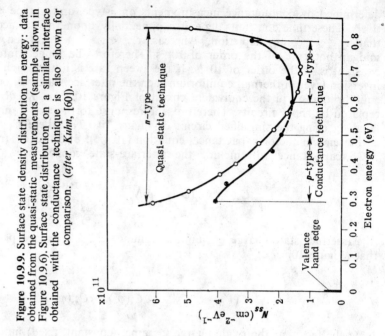

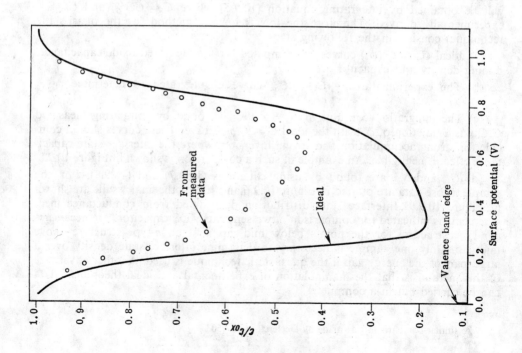

Figure 10.9.8. Measured and calculated capacitance as a function of surface potential, showing the expected agreement in accumulation and strong inversion. (*after Kuhn*, [60]).

Figure 10.9.9. Surface state density distribution in energy: data obtained from the quasi-static measurements (sample shown in Figure 10.9.6). Surface state distribution on a similar interface obtained with the conductance technique is also shown for comparison. (*after Kuhn*, [60]).

10.10 Determination of Minority Carrier Lifetime

The minority carrier lifetime is an important semiconductor parameter in device physics (see Section 1.11). It can be determined, in principle, from the transient response of the inversion layer formed under the gate of an MOS capacitor. However, care should be taken in defining and measuring the correct lifetime.

Consider for example the Shockley-Read-Hall recombination model for a single-level recombination centre. The recombination rate is given by [2], [59]:

$$U = \frac{pn - n_i^2}{\tau_{n0}(p + p_1) + \tau_{p0}(n + n_1)} \qquad (10.10.1)$$

where

$$\tau_{n0} = \sigma_n v_{th} N_t, \quad \tau_{p0} = \sigma_p v_{th} N_t \qquad (10.10.2)$$

and

$$p_1 = n_i \exp\frac{E_i - E_t}{kT}, \quad n_1 = n_i \exp\frac{E_t - E_i}{kT} \qquad (10.10.3)$$

(E_t, N_t, σ_n, σ_p are the energy level, the centre density, the electron capture cross section and the hole capture cross section, respectively, for the recombination level).

For low level injection of excess holes in neutral n-type semiconductor $n \gg p_1$, n_1 and the recombination rate U_p is approximately

$$U_p \simeq \frac{pn - n_i^2}{n\tau_{p0}} = \frac{p - p_{n0}}{\tau_{p0}} \qquad (10.10.4)$$

(compare to equation (1.11.6)). Therefore the *recombination lifetime* (for low level injection of excess minority carriers) is $\tau_{rec} = \tau_{p0}$. This is the so-called minority carrier lifetime usually encountered in device analysis.

Consider now a depletion region with negligible mobile carriers of both signs. A *bulk* generation rate

$$G_b = \frac{n_i}{\tau_g} \qquad (10.10.5)$$

can be defined, where τ_g is determined from equations (10.10.1), (10.10.3) and (10.10.5):

$$\tau_g = \tau_{p0} \exp\left(\frac{E_t - E_i}{kT}\right) + \tau_{n0} \exp\left(\frac{E_i - E_t}{kT}\right). \qquad (10.10.6)$$

The generation lifetime is the parameter determined by measuring the rate of inversion layer formation in an MOS capacitor. Note that in the 'symmetrical' case $\tau_{n0} = \tau_{p0}$, $E_t = E_i$ we have $\tau_g = 2\tau_{p0} = 2\tau_{n0}$.

However, the (thermal) bulk generation in the depletion region is only one possible mechanism for the formation of the inversion layer. Consider for example a p-type MOS capacitor biased by a large positive step of gate voltage at $t = 0$. A finite time is necessary for the inversion layer formation. Until then the surface is in the *deep-depletion* régime, which is, of course, a non-equilibrium state. Figure 10.10.1 illustrates the thermal generation in the bulk depletion region: the electron current creates the inversion layer at the surface, whereas part of the holes flowing towards the bulk neutralize the ionized impurities, thus reducing the width of the depletion region. Therefore, the width of the depletion relaxes as the inversion layer forms until the equilibrium is reached

[6]. Other mechanisms of inversion layer formation are illustrated on Figure 10.10.2 [61]. The thermal *surface* generation (process 2) takes place at a rate

$$G_s = n_i s_0, \tag{10.10.7}$$

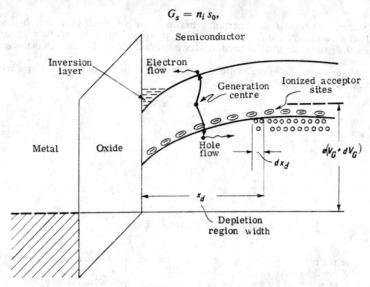

Figure 10.10.1. Relaxation of depletion region due to generation of electron-hole pairs in the depletion region of an MOS capacitor.

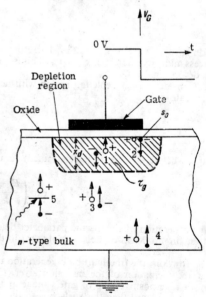

Figure 10.10.2. *n*-type MOS capacitor biased into deep depletion by a gate voltage step. The structure relaxes into its quasi-equilibrium inversion state as a result of a few mechanisms of electron-hole pair generation. Indicated are: (1) thermal bulk generation in the depletion region; (2) thermal surface generation in the depletion region at the Si—SiO$_2$ interface; (3) thermal bulk generation in the neutral semiconductor bulk; 4 thermal surface generation at the surface of the neutral bulk and (5) external generation (photons, energetic electrons etc.). *(after Schroeder,* [61]).

where s_0 is the surface generation velocity. We have to distinguish between the surface under the gate which gradually becomes inverted and the lateral depleted surface. Processes 1 and 2 are dominant for wide-gap semiconductors as silicon at room temperature and below. Thermal surface and bulk generation for the neutral semiconductor may be dominant for narrow-gap semiconductors (germanium) and at higher temperatures for wider bandgap materials [61]. The relaxation of the inversion layer for a deep-depleted surface can also occur by high-field effects at the semiconductor-oxide interface (avalanche generation, tunnelling, etc.) [62], processes which are not illustrated in Figure 10.10.2.

The method of pulsed MOS capacitor used to determine τ_g is discussed with reference to Figure 10.10.3. The differential capacitance is measured with a variable bias voltage: a bias pulse applied at $t = 10$ s in Figure 10.10.3 drives the capacitor from accumulation into deep depletion.

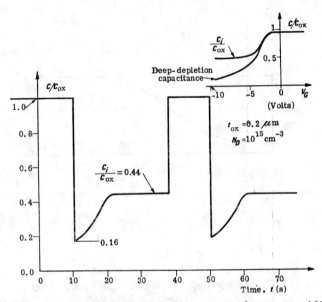

Figure 10.10.3. Experimental transient response for an n-type MOS capacitor driven in deep depletion by a gate pulse (the $C - V$ characteristic is shown in the inset). With an average capacitance of 0.3 C_{ox} during the transient lasting for 10 s, the minority carrier lifetime is approximately 8 μs (see the text), (after Zaininger and Heiman, [6]).

The quasi steady-state capacitance reached after about $T_0 = 10$ s. corresponds to the normal (high-frequency) inversion capacitance, C_i (see the inset in Figure 10.10.3). An approximate analysis uses the average normalized capacitance C_0/C_{ox} during transient, the average depletion layer width x_{d0} related to the above capacitance, and computes the transient current density flowing in the external circuit as $C_{ox} \Delta V/T_0$ where ΔV is the voltage step related to the flat-band voltage and also as $en_i x_{d0}/\tau_g = en_i x_{d0}/2\tau$, where τ is the minority carrier lifetime. For Figure 10.10.3 with $T_0 = 10$ s, $C_0/C_{ox} \simeq 0.3$, $\Delta V = 12$ V, τ is 8 μs (the oxide thickness is 2000 Å and $x_{d0} = 1.4 \mu$m) [6].

A more exact and complete information can be extracted from the pulsed MOS capacitor response by using the so-called Zerbst plott [63]—[65], indicated qualitatively in Figure 10.10.4 b. The (reciprocal) slope of the straight line portion is proportional to τ_g, whereas the intercept on the $-d(C_{ox}/C)^2/dt$ axis is proportional to the surface generation velocity, s_0 (for $C < C_1$ the response is mainly surface-generation dominated). A few comments about the surface effects are necessary. The total rate of the change of the inversion layer charge per unit of surface area is

$$\frac{dQ_{inv}}{dt} = \frac{en_i(x_d - x_f)}{\tau_g} + \frac{en_i(x_d - x_f)s_0 P}{A} + en_i s \frac{A_g}{A}, \qquad (10.10.8)$$

where $x_d = x_d(t)$ is the depletion layer width, x_f is the final depletion width, A_g is the area of the gate electrode, A is the area of the gate plus the lateral surface area of the depletion region extended beyond the gate, P is the peripheral length of the gate (the area of the depleted surface surrounding the gate is approximately $P(x_d - x_f)$), s is the surface generation velocity under the gate, $s = s(t)$. The first component of dQ_{inv}/dt is due to bulk generation, the second one due to the surface gener-

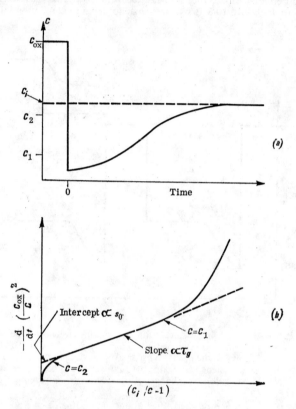

Figure 10.10.4. (a) Pulsed MOS capacitor $C-t$ response (b) A Zerbst plot of the pulsed response showing the linear portion. (after Schroder and Nathanson, [64]).

ation at the lateral portion of the space charge region and the third one due to surface generation under the gate. After an initial period of the transient the third components should be negligible because the surface is inverted. For small area devices the second term in equation (10.10.8) may be important and the linear region of the Zerbst plot indicates an 'effective' τ_g, which includes the effect of lateral surface generation [64], [65]. On the other hand, detailed measurements reported by Schroder and Guldberg [65] indicate that the surface generation may be important even when the surface is shielded by minority carriers (s_0 is reduced but does not vanish if the surface effects are very strong, such as for unannealed devices).

Other pulse methods are used, such as pulsing from inversion into depletion and thus the determination of recombination lifetime [66], etc.

The so-called linear-sweep MOS-C technique was suggested by Pierret [67]. The high-frequency capacitance is measured with the MOS structure biased by a linear gate voltage ramp ($dV_G/dt = \alpha$) which drives the device from inversion into even deeper inversion. The surface effects are t hereby reduced. The experimental $C-V$ curves for various sweep rates are indicated in Figure 10.10.5. During the initial period of the sweep the depletion region increases gradually thus decreasing the capacitance and increasing the minority carrier generation rate (only bulk generation in the depleted region is considered). As the time passes the generation rate within the semiconductor becomes sufficiently large to balance the rate at which charge is being added to the gate by the sweep. Therefore the depletion width and the high-frequency capacitance saturate (Figure 10.10.5). If carrier

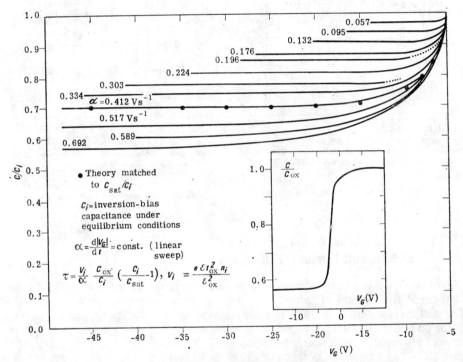

Figure 10.10.5. Experimental linear-sweep capacitance-voltage characteristics used to determine the minority carrier lifetime. (*after* Pierret, [67]).

generation at the $Si-SiO_2$ interface beneath the gate electrode cannot be ignored, the characteristic will never saturate and continue to slope downward because of the reduced surface generation associated with the filling of the surface levels. Non-negligible lateral effects and lateral surface generation will give rise to an increase in the $C-V$ characteristic [67], [68]. The convenient sweep rates lie from $0 \cdot 1$ V s^{-1} to 10 V s^{-1} and for a typical oxide thickness of 2000 Å the measured range of minority carrier lifetime, determined from the initial and saturation capacitances is $0 \cdot 1 \mu$s to 10μs. However, the practical range can be extended by changing the oxide thickness (see Problem 10.11).

The investigation of deep-depletion régime using the ramp-response method was reported by Bulucea [69]. The method is similar to that suggested by Kuhn [60] and presented at the end of the previous section. The ramp is, however, considerably faster and the current is correspondingly higher. A simpler arrangement can be used for displaying the $C-V$ characteristic on an oscilloscope: the voltage proportional to the current is obtained on a series resistor. Typical measured characteristics are represented in Figure 10.10.6. If the ramp is sufficiently fast, the oscilloscope will display the deep-depletion $C-V$ curve. For lower ramp slopes the minority carrier generation becomes important and the $C-V$ response modifies accordingly (Figure 10.10.6). A similar effect can be obtained by increasing the temperature because the (generation) carrier lifetime decreases (Figure 10.10.7). This measurement technique is particularly useful for investigation of the onset of surface avalanche breakdown in MOS structures biased in deep depletion.

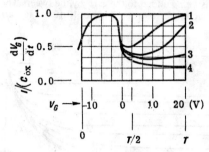
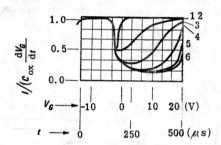

Figure 10.10.6. Ramp-response of an MOS capacitor ($N_D = 4 \times 10^{15}$ cm^{-3}, $C_{ox} = 32$ pF, $t_{ox} = 0.1$ μm, 10^{-3} cm² area, 2 ns minority carrier-generation lifetime) to four different ramps: 1: $T = 5$ ms; 2: $T = 500$ μ; 3: $T \times 50$ μs; 4: $T = 5$ μs (after Bulucea, [69]).

Figure 10.10.7. Ramp-response of an MOS capacitor (the same as in Figure 10.10.6) to a 500 μs ramp at six different temperatures: (1) 100°C; (2) 50°C; (3) 0°C; (4) -50°C; (5) -100°C; (6) -150°C. A soft breakdown takes place at $V_G = 20$ V for -150°C and 100°C which is not inhibited at the ramp speed used. (after Bulucea, [69]).

10.11 Surface Space-charge Region in Non-equilibrium Conditions

Figure 10.11.1 a shows the electric charge and energy diagram for an MOS structure at thermal equilibrium. Figure 10.11.1 b represents the non-equilibrium situation. The surface is inverted. A supplementary curvature of the energy bands occurs due to the potential difference between the semiconductor bulk (p-type) and the inversion layer. Such a potential difference appears in a gate-controlled diode structure (Figure 10.1.3) and in an MOS transistor (Figure 10.1.4) because of the potential drop along the surface. Consider for example the MOS transistor with a p-type substrate biased by a gate voltage V_G sufficiently high to invert all the surface along the channel, from source to drain. Assume a positive drain potential, V_D, and let $V = V(y)$ be the inversion layer potential at a distance y from the source along the surface channel (Figure 10.1.4). The potential drop $V(y)$ occurs on the field-induced junction formed by the inversion layer and the substrate and biases it in reverse. The depletion layer width depends upon the position along the channel, $x_d = x_d(y)$. The quasi-Fermi levels for electrons and holes on both sides of the field-induced junction will be separated by the bias $V = V(y)$, as for an ordinary pn junction. The source potential is zero and at the source end of the channel ($y = 0$) we have $V(0) = 0$. $V(y)$ reaches V_D at the drain end of the channel ($y = L$).

The total curvature of the energy bands is denoted by $\phi_s = \phi_s(y)$. A reasonable evaluation for $\phi_s(y)$ is

$$\phi_s(y) = \phi_{s0} + V(y), \quad (10.11.1)$$

where

$$\phi_{s0} = \phi_{s0}(V_G) \quad (10.11.2)$$

is the surface potential at thermal equilibrium for the same V_G. Note the fact that here $\phi_{s0} = \phi_s(0)$. Equation (10.11.1) is acceptable if we assume that the field along the surface is negligibly small compared to that perpendicular to surface: the electric charge is essentially determined by the gate potential V_G and therefore the position

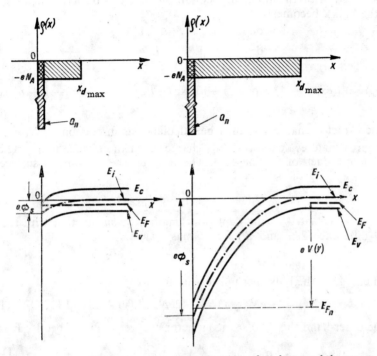

Figure 10.11.1. Charges in the semiconductor surface layer and the energy band diagram: (a) equilibrium case, (b) non-equilibrium.

of the electron quasi-Fermi level at surface with respect to the conduction band is the same as for $V(y) = 0$. This assumption is equivalent to the 'gradual approximation' in the classical theory of the field-effect transistor.

The electric charge in the semiconductor will be calculated as in the equilibrium case. By assuming non-degenerate statistics we have:

$$n = n_i \exp(u - u_{Fn}), \quad p = n_i \exp(u_{Fp} - u), \qquad (10.11.3)$$

where u_{Fn} and u_{Fp} are, respectively, the quasi-Fermi levels for electrons and holes in kT/e units. u_{Fp} is taken such as to coincide with the Fermi level in the bulk, u_F, whereas near the surface

$$u_{Fn} - u_{Fp} = \xi = \frac{eV(y)}{kT}. \qquad (10.11.4)$$

The Poisson equation can be written

$$\frac{d^2u}{dx^2} = \frac{1}{2L_D^2}[\exp(u-\xi-u_F) - \exp(u_F-u) + \exp u_F - \exp(-u_F)], \quad (10.11.5)$$

and, integrated with the boundary condition $du/dx \to 0$ for $u \to 0$ (zero electric field in the bulk), becomes

$$\frac{du}{dx} = \frac{u_s}{|u_s|}\frac{F(u,\xi,u_F)}{L_D}, \quad (10.11.6)$$

where

$$F(u,\xi,u_F) = \{[(\exp u - 1)\exp(-\xi) - u]\exp(-u_F) + [\exp(-u) - 1 + u]\exp u_F\}^{1/2}. \quad (10.11.7)$$

The semiconductor charge will then be calculated as in Section 10.2.

An approximate evaluation of this charge when the surface is inverted, closely follows the approach of Section 10.3. The total charge per unit of surface area at the distance y along the surface is

$$Q_s(y) = Q_n(y) - eN_A x_d(y), \quad (10.11.8)$$

where $x_d(y)$ is evaluated as the depletion length at a pn junction with the built-in potential difference $2\phi_F$ and a reverse bias $V(y)$:

$$x_d(y) = [2\varepsilon(V(y) + 2\phi_F)/eN_A]^{1/2}. \quad (10.11.9)$$

The fixed charge in the bulk (per unit of surface area) is

$$Q_B(y) = -eN_A x_d(y) = -[2eN_A\varepsilon(V(y) + 2\phi_F)]^{1/2}. \quad (10.11.10)$$

On the other hand the total semiconductor charge $Q_s(y)$ can be also found from

$$V_G - V_{FB} = \phi_s(y) - Q_s(y)/C_{ox}, \quad (10.11.11)$$

(see Section 10.5) where $\phi_s(y)$ is given by equation (10.11.1). The last equations determine $Q_n(y)$ as a function of $V(y)$ for a given V_G. $\phi_{s0} = \phi_{s0}(V_G)$ is assumed known from the thermal-equilibrium case.

The above results will be applied directly in the MOS transistor theory (Chapter 11).

10.12 Gate-controlled Diode Structure

The gate-controlled pn diode of Figure 10.1.3 is again represented in Figure 10.12.1a. We consider for analysis a portion of the circular gate, indicated by dashed lines in Figure 10.12.1a and shown idealized in Figure 10.12.1b [2]. The origin of the xy coordinates is placed at the surface end (under the gate) of the pn junction. The gate bias relative to the p substrate is so that the depletion region of the n^+p junction is continued by the depleted surface region: the surface is depleted or inverted. Consider the case when the n^+p junction is reverse-biased $V_J = V_R > 0$. The total band bending along the y direction in the semiconductor bulk is $V_{bi} + V_R$

(where V_{bi} is the built-in potential). The total band bending along the x direction is $e\phi_s = e\phi_s(y)$, as discussed in the previous section, equation (10.11.1). $V(y)$ is the electric potential of the surface relative to the neutral p-side of the junction (grounded). As far as the semiconductor surface is depleted but not inverted, to

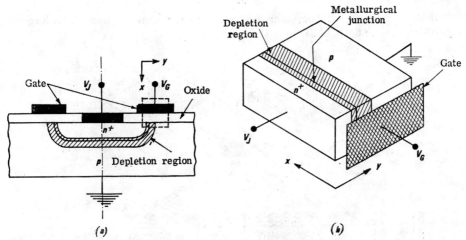

Figure 10.12.1. (*a*) The gate-controlled diode structure; (*b*) idealized representation of the region within the dashed frame in (*a*). (after Grove, [2]).

determine $V(y)$ one has to solve a problem of electrostatics to find the field (and potential) by taking into account fixed charges and applied external voltages. The electric potential map for a positive gate voltage is shown qualitatively on Figure 10.12.2*a*.

Figure 10.12.2*b* shows that a considerable band bending $e\phi_s(y)$ is necessary to invert the surface. Once the inversion layer is formed, it constitutes a high-conductivity sheet which has essentially the same potential as the n^+ region. Therefore $V(y) \simeq V_R$ in equation (10.11.1). On the other hand, at the onset of strong inversion in thermal equilibrium $\phi_{s0} \simeq 2\phi_F$. Therefore the surface potential in inversion is*

$$\phi_{s(\text{inv})} \simeq V_R + 2\phi_F. \qquad (10.12.1)$$

Note the fact that the turn-on (on threshold) voltage V_G for the onset for strong inversion will depend upon the reverse voltage applied to the n^+p junction, $V_G = V_G(V_R)$. We have (see equation (10.4.12)) for an ideal MOS structure

$$V_T = V_G \bigg|_{\substack{\text{onset of}\\ \text{strong inversion}}} = \phi_{s(\text{inv})} - \frac{Q_{s(\text{inv})}}{C_{\text{ox}}}, \qquad (10.12.2)$$

where

$$Q_{s(\text{inv})} = (Q_n + Q_B)_{\text{inv}} \simeq Q_{B(\text{inv})}. \qquad (10.12.3)$$

---

* The surface potential considered here is $\phi_s(y)$ for $y \to \infty$ or, practically, the surface potential a certain distance away from the n^+p junction, as illustrated by Figure 10.12.2b.

By also using equations (10.11.10) and (10.12.1) and replacing V_R by V_J (positive or negative) one obtains:

$$V_T \simeq 2\phi_F + V_J + \frac{1}{C_{ox}} [2eN_A\varepsilon(V_J + 2\phi_F)]^{1/2}, \qquad (10.12.4)$$

such that V_T increases steeper than linearly with the applied junction voltage, V_J (Figures 10.12.1 and 10.12.2).

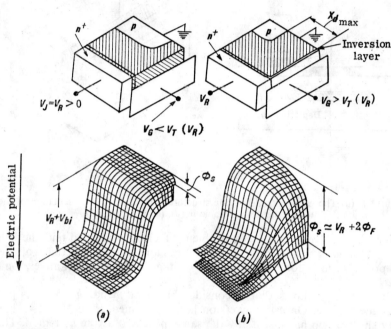

Figure 10.12.2. Reverse biased $p-n$ junction under the influence of surface fields (see Figure 10.12.1b): (a) surface depletion; (b) surface inversion. (*after* Grove, [2]).

Figure 10.12.3 shows the calculated low-frequency $C-V$ characteristics for a gate-controlled diode structure. The gate-to-substrate capacitance C is plotted *versus* the gate-to-substrate voltage, V_G, for several values of the junction voltage, V_J. Because the minority carriers can be rapidly supplied to the inversion layer by the external circuit (through the contacts of the *pn* junction), the low-frequency type characteristics will be practically observed up to very high frequencies. Until the onset of strong inversion the characteristic follows that corresponding to the depletion approximation. After the occurrence of strong inversion the capacitance increases rapidly towards the oxide capacitance.

For the real structure (see Section 10.5) the characteristics will be displaced along the voltage axis by the flat-band voltage.

The gate-controlled diode structure was used for surface studies [2]. We shall briefly mention below the effect of surface upon the reverse current and breakdown voltage of a *pn* junction. The gate-controlled diode has the advantage that the electric state of the surface can be controlled by the gate and thus the state of a free surface (without gate but with possible charges in the oxide, centers at the interface, etc.) can be simulated.

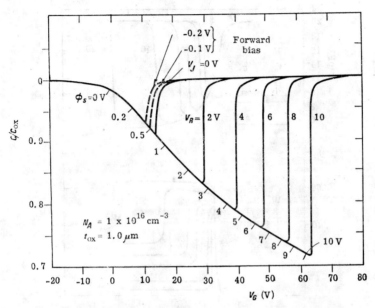

Figure 10.12.3. Theoretical low frequency gate-to-substrate capacitance-voltage characteristics of a gate-controlled structure under various applied junction bias conditions. Values of surface potential, ϕ_s, are also shown. *(after Grove [2])*.

Figure 10.12.4 shows the measured gate-to-substrate capacitance, C, and reverse current, I_R, of the *pn* junction *versus* the gate voltage, V_G, with the junction reverse voltage, V_R, as parameter. The room-temperature reverse current of a silicon *pn* junction is mainly due to generation of electron-hole pairs by the recombination-generation centres within the depletion region [2] (see also Section 10.10). Therefore the current increases with the extent of this depletion region. At a given V_R, I_R will be constant when the surface under the gate is accumulated, will increase gradually when the surface under the gate goes from accumulation to inversion (because the surface depletion region induced by the gate contributes to I_R) and then will remain constant as the surface is inverted. However a considerably higher (and almost constant) current is measured (Figure 10.12.4) when the surface is depleted (compare with the $C-V_G$ curves which clearly indicate the range of V_G when that depletion occurs). This supplementary current is due to the generation recombination centres at the oxide-silicon interface [2] and disappears when the surface is inverted because these centres are shielded from the bulk by the high-conductivity surface layer (however, see Section 10.10, this shielding is not always complete).

The avalanche-breakdown in a gate-controlled diode structure will be examined in the next section.

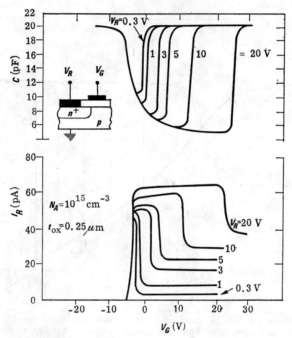

Figure 10.12.4. Reverse junction current and the gate-to-substrate capacitance as a function of gate voltage, with junction reverse voltage, V_R, as a parameter, measured for a gate-controlled diode structure. (*after Grove*, [2]).

10.13 Surface Breakdown and Avalanche Injection into the Oxide

Consider, for example, a *p*-type MOS capacitor biased by a positive gate voltage. When the semiconductor surface is depleted the highest field in the semiconductor occurs at the surface. If the field is sufficiently high, the avalanche breakdown will occur at the surface. However, in practice, avalanche breakdown in a silicon MOS capacitor occurs only in the deep-depletion régime. The ramp response method (Section 10.10) was used by Bulucea and Antognetti [70] to observe the onset of the surface avalanche breakdown in a small area MOS capacitor (Al—SiO$_2$—*n* type Si). The $C-V$ characteristic displayed by the oscilloscope follows the depletion curve until the onset of avalanche, when it rises to the oxide capacitance

(Figure 10.13.1). More abrupt rise of C at breakdown is obtained when the device is in inversion before the depleting voltage (the descendent ramp in Figure 10.13.1) is applied, because the effect of surface states is eliminated (or at least reduced); when the device is pulsed from quiescent accumulation (Figure 10.13.1b) these states generate electron-hole pairs and therefore trigger prematurely the avalanche process*. Oxide charging was observed and explained as the result of injection of hot electrons from the avalanche plasma into the oxide and subsequent trapping [70].

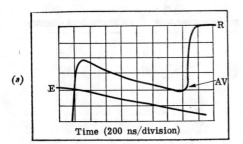 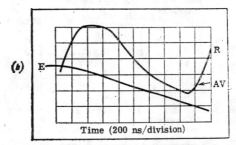

Figure 10.13.1. Ramp response of a miniature MOS capacitor pulsed into deep depletion from quiescent inversion (a) and from quiescent accumulation (b) by a voltage ramp of 10^7V s^{-1}. The voltage excitation is denoted by E, the displayed response (proportional to the device capacitance) is denoted by R. Both figures indicate the onset of avalanche multiplication (AV). (*after Bulucea and Antognetti,* [70]).

An alternative to obtain avalanche breakdown in an MOS capacitor is to apply to it a large alternating voltage [71], [72], such that for about half a cycle the device is driven into avalanche. Each time the surface is at avalanche an injection current flows through the oxide and its average value is recorded [72], [73]. The avalanche injection into the oxide can be accurately studied by using the gated *pn* diode (see the preceding section).

We shall first consider the surface avalanche breakdown in a gate-controlled diode and then discuss the avalanche injection into the oxide of such a structure. Figure 10.13.2 shows, after Grove [2], the depletion region at the edge of the diffused n^+ region, where the oxide is overlapped by the gate. A very large negative gate voltage determines a field-induced junction over *the n^+ region* (Figure 10.13.2a). Near the 'corner', the depletion region will be relatively narrow and therefore the electric field intensity will be high (the arrows indicate only the *direction* of the field), thus reducing the breakdown voltage of the gate-controlled diode below the normal 'bulk' value (the avalanche breakdown occurs at the surface). When the gate is at the same potential as the substrate, the field is still the highest near the corner region (Figure 10.13.2b) but it is lower than in the previous case, and the breakdown voltage is somewhat higher. When the gate is at the n^+-side potential ($V_G = V_R$), the breakdown will occur at a voltage approaching that of a plane junction (the same background doping) and somewhat higher than that of the same diffused junc-

---

* The device studied is a miniature avalanche MOS diode [70] with a very small area (of the order of 10^{-6} cm²) and therefore a very small volume and reduced triggering probability, such that it can withstand a transient voltage larger than the breakdown voltage without breakdown occurring.

tion without surface fields. Experimental data confirm the above qualitative considerations. Grove et al. [74] solved the two-dimensional Poisson equation and have found for an n^+p silicon structure ($N_A = 2.5 \times 10^{15} \text{cm}^{-3}$ and $t_{ox} = 1\ \mu\text{m}$) with $V_R = 100\ \text{V}$ and $V_G = 0 \ldots -100\ \text{V}$ that the corner field is always the highest. This maximum field will be approximately equal to the oxide field if the oxide is

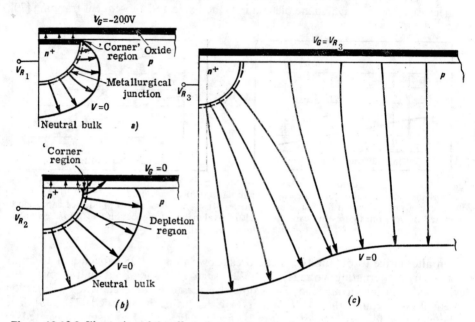

Figure 10.13.2. Illustration of the effect of gate voltage on the shape of the depletion region and the corresponding breakdown V_{R_1}, V_{R_2}, V_{R_3} (V_R is the potential difference between the n^+ region and the neutral p-type substrate). (after Grove et al., [74]).

relatively thin and the substrate impurity concentration is low (t_{ox} is much smaller than the depletion region width). If the critical field concept for avalanche breakdown is valid then in the above circumstances the breakdown voltage will change almost linearly with the gate voltage, and this dependence was indeed confirmed by experiment.

A more detailed investigation was reported recently by Bulucea et al. [75], [76]. The experimental breakdown voltage *versus* the gate voltage of quasi-perfect gate-controlled silicon p^+n diodes is shown in Figure 10.13.3 for three background resistivities. The two-dimensional field was numerically calculated for the *experimental* values of V_B and $V'_G = V_G - V_{FB}$ (points indicated by crosses on Figure 10.13.3). Figure 10.13.4 shows the distribution of the electric field at breakdown along the surface. The maximum field occurs just at the metallurgical junction and its magnitude changes slightly around $10^6\ \text{V cm}^{-1}$ (compared with the considerably lower value of $5 \times 10^5\ \text{V cm}^{-1}$ used previously by Grove et al. [74]).

The energy-band diagram along a field line near the corner of a gate-controlled p^+n silicon diode is drawn 'to scale' in Figure 10.13.5. The electron-hole pairs are generated in the vicinity of the p^+n junction. The electrons move (along the field line) towards the Si–SiO$_2$ interface. All electrons collected at the interface should move along this interface and constitute the reverse (breakdown) current of the p^+n junction. However, because the electrons accelerated by the field are highly energetic, a fraction of them can be thermionically emitted over the Si–SiO$_2$ potential

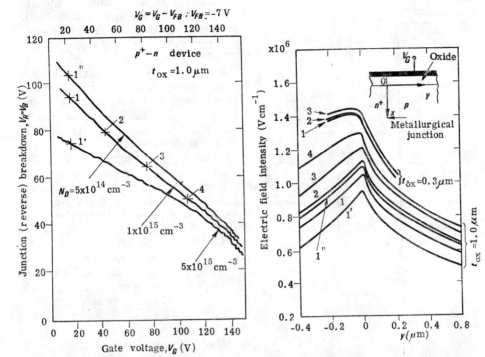

Figure 10.13.3. Junction breakdown voltage versus gate voltage for gate-controlled diodes with different background dopings. (after Bulucea et al., [76]).

Figure 10.13.4. Distribution of the electric field at breakdown, in silicon, along the Si–SiO$_2$ interface (results of two-dimensional computer calculations based on the measured breakdown voltage on experimental devices). (after Bulucea et al., [76]).

barrier into the oxide and (if trapping is unimportant) they are collected as a gate current. A detailed quantitative analysis of avalanche injection was given by Bulucea [75]. The same phenomenon was discussed recently by Pepper [77], Verwey and de Maagt [78], etc. The electron distribution in energy and the field-lowering of the barrier (Schottky effect) were taken into account.

Bulucea [75] has shown that the current flow along the surface can be considered one-dimensional and that the avalanche emission into the oxide is localized to a very narrow region of tens or hundreds of angstroms at the metallurgical junction. The injection coefficient I_G/I_R (where I_G is the gate current carried by electrons injected in the oxide and I_R the reverse current in breakdown) was computed and compares well with the experiment [75]. This coefficient is practically independent of I_R, V_G and substrate resistivity [76]. The calculation of injection coefficient is based upon the critical field concept, and the value of this field is accurately determined as shown over leaf (Figure 10.13.4).

Verwey and de Maagt [78] obtained recently high gate-current densities (up to 10 A cm^{-2}) in p^+n gate-controlled structures. It was shown that even at these high values, I_G can be considered as injection limited, although the space-charge-limitation mechanism in the oxide (accompanied by trapping) was expected (see Chapter 4) [77], [78].

A recent study of avalanche injection [72] (mentioned above) indicated that the electron current density injected from a (100) silicon surface is about one order of magnitude lower than that from a (111) surface (at the same bulk resistivity). On the other hand injection currents into silicon nitride (Si_3N_4) used as insulator were not observed although the energy barriers for electrons or holes are considerably lower ($\phi_n = 3.2$ eV, $\phi_p = 4.2$ eV for Si—SiO_2; $\phi_n = 2.05$ eV, $\phi_p = 1.95$ for Si—Si_3N_4). This was attributed to the possible trapping in the insulator* [72].

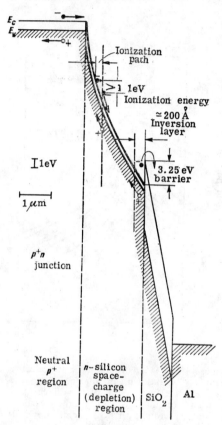

Figure 10.13.5. Energy-band diagram along a field line near the corner of a gate-controlled p^+n silicon diode showing avalanche multiplication in the depletion region and electron injection into oxide. (*after Bulucea*, [75]).

The importance of surface breakdown and avalanche injection is considerable. It was shown that electron avalanche injection into SiO_2 used for passivating the normal planar devices charges the oxide and creates new interface states [80] (a phenomenon which is still unexplained). The effects of these phenomena are the 'walk-out' of the breakdown characteristic of diodes and transistors (i.e. an increase of the surface-controlled breakdown voltage), a degradation in the low-current h_{FE} (h_{21e}) of bipolar transistors, changes in the threshold voltage and the transconductance of MOS transistors. The avalanche injection determines the occurrence of a gate current in MOS transistors and tetrodes [81]—[83].

---

* The measurement uses an MOS (MIS) capacitor and an alternating voltage (see above) [72]. The charge carriers are injected into the insulator when the field is large and are trapped; they are released and reinjected into semiconductor during the opposite alternance. The measured direct current is insignificant. A similar situation occurs for hole injection into SiO_2. The time-dependence of hole injection currents in SiO_2 was studied by Verwey [79] and the parameters of trapping centres were determined.

The avalanche injection can be used for memory devices (see Chapter 12). For example, a floating-gate MOS transistor can be used as a non-volatile memory element: the gate is charged by surface avalanche breakdown [84].

10.14 MOS (MIS) Varactor

This chapter presented the basic properties of the MOS capacitor and illustrated its usefulness as a tool in investigating the properties of the semiconductor bulk, insulator layer and also of the semiconductor-insulator interface. The MOS system

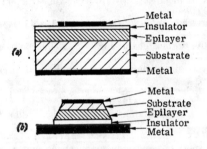

Figure 10.14.1. (a) Usual MIS structure, (b) MIS mesa device.

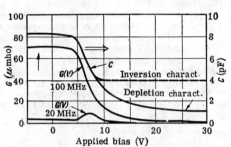

Figure 10.14.2. Experimental capacitance-voltage and conductance-voltage characteristics of an MIS mesa varactor. (after Stoev and Schumacher, [86]).

is the basis of surface-controlled devices (Section 10.12, Chapters 11 and 12). The MOS capacitor can also be used as an electronic device by itself: it is a non-linear capacitor [66], [85], [86]. A MIS (MOS) varactor constructed in mesa form (Figure 10.14.1b) was presented by Stoev and Schumacher [86]. The substrate is thinned (series resistance reduced) and the device is mounted with the top metal (on insulator) just on the heat sink, thus increasing the maximum power dissipation. The $C-V$ curve follows the depletion capacitance characteristic (Figure 10.14.2) at frequencies above 1 kHz (10 ms time necessary for inversion layer formation at the interface) [86]. The device is suitable for microwave applications (no recombination losses, etc.).

References

1. A. MANY, Y. GOLDSTEIN and N. B. GROVER, 'Semiconductor surfaces', John Wiley, New York, 1965.
2. A. S. GROVE, 'Physics and technology of semiconductor devices', John Wiley, New York, 1967.
3. A. S. GROVE, B. E. DEAL, E. H. SNOW and C. T. SAH, 'Investigation of thermally oxidized silicon surfaces using MOS structures', Solid-St. Electron., **8**, 145—163 (1965).
4. J. R. HAUSER and M. A. LITTLEJOHN, 'Approximations for accumulation and inversion space-charge layers in semiconductors', Solid-St. Electron., **11**, 667 (1968).
5. M. B. DAS, 'Physical limitations of MOS structures', Solid-St. Electron., **12**, 305—336 (19..

6. K. H. ZAININGER and F. P. HEIMAN, 'The C−V technique as an analytical tool', *Solid-St. Technol.*, **13**, May-June 1970.
7. M. J. McNUTT and C. T. SAH, 'High frequency space charge layer capacitance of strongly inverted semiconductor surfaces', *Solid-St. Electron.*, **17**, 377−385 (1974).
8. A. BERMAN and D. R. KERR, 'Inversion charge redistribution model, of the high-frequency MOS capacitance', *Solid-St. Electron.*, **17**, 735−742 (1974).
9. J. R. BREWS, 'An improved high-frequency MOS capacitance formula', *J. Appl. Phys.*, **45**, 1276−1279 (1974).
10. G. BACCARANI and M. SEVERI, 'On the accuracy of the theoretical high-frequency semiconductor capacitance for inverted MOS structures' *IEEE Trans. Electron Dev.*, **ED−21**, 122−125 (1974).
11. J. R. BREWS, 'Error analysis of high-frequency MOS capacitance calculations', *Solid-St. Electron.*, **17**, 447−456 (1974).
12. B. E. DEAL and E. H. SNOW, 'Barrier energies in metal-silicon dioxide-silicon structures', *J. Phys. Chem. Solids.*, **27**, 1873−1879 (1966).
13. L. V. GREGOR, 'Passivation of semiconductor surfaces', *Solid-St. Technol.*, **14**, 37−43, April 1971.
14. G. SIXT and A. GOETZBERGER, 'Control of positive surface charge in Si−SiO$_2$ interfaces by use of implanted Cs ions', *Appl. Phys. Lett.*, **19**, 478−479 (1971).
15. J. P. MITCHELL and D. K. WILSON, 'Surface effects of radiation on semiconductor devices', *Bell Syst. Techn. J.*, **46**, 1−80 (1967).
16. F. M. FOWKES and D. W. HESS, 'Control of fixed charge at Si−SiO$_2$ interface by oxidation-reduction treatments', *Appl. Phys. Lett.*, **22**, 377−379 (1973).
17. M. E. SPROUL and A. G. NASSIBIAN, 'Effect of O$^+$ implantation on silicon-silicon dioxide interface properties', *Solid-St. Electron.*, **17**, 577−582 (1974).
18. M. J. KIM, 'MOS-FET fabrication problems', *Solid-St. Electron.*, **12**, 557−571 (1969).
19. F. STERN, 'Surface quantization and surface transport in semiconductor inversion and accumulation layers', *J. Vac. Sc. Technol.*, **9**, 752−753 (1972).
20. J. R. SCHRIEFFER, 'Effective carrier mobility in surface charge layers', *Phys. Rev.*, **97**, 641−646 (1955).
21. A. P. GNÄDINGER and H. E. TALLEY, 'Quantum mechanical calculation of the carrier distribution and the thickness of the inversion layer of a MOS field-effect transistor', *Solid-St. Electron.*, **13**, 1301−1309 (1970).
22. F. STERN, 'Self-consistent results for *n*-type Si inversion layers', *Phys. Rev. B*, **5**, 4891−4899 (1972).
23. J. A. PALS, 'Experimental verification of the surface quantization of an *n*-type inversion layer of silicon at 300 and 77°K', *Phys. Rev. B*, **5**, 4208−4210 (1972).
24. J. A. PALS, 'Measurements of the surface quantization in silicon *n*- and *p*-type inversion layers at temperatures above 25°K', *Phys. Rev. B.*, **7**, 754−760 (1973).
25. J. LAUR and T. S. JAYADEVAIAH, 'Carrier concentration on the inversion layer of a MOS field effect transistor', *Solid-St. Electron.*, **16**, 644−646 (1973).
26. R. F. GREENE, D. R. FRANKL and J. ZEMEL, 'Surface transport in semiconductors', *Phys. Rev.*, **118**, 967−975 (1960).
27. N. B. GROVER, Y. GOLDSTEIN and A. MANY, 'Improved representation of calculated surface mobilities on semiconductors. I. Minority carriers', *J. Appl. Phys.*, **32**, 2538−2539 (1961).
28. Y. GOLDSTEIN, N. B. GROVER, A. MANY and R. F. GREENE, 'Improved representation of calculated surface mobilities in semiconductors. II. Majority carriers', *J. Appl. Phys.*, **32**, 2540−2541 (1961).
29. R. F. PIERRET and C. T. SAH, 'An MOS-oriented investigation of effective mobility theory', *Solid-St. Electron.*, **11**, 279−290 (1968).
30. V. G. K. REDDI, 'Majority carrier surface mobilities in thermally oxidized silicon', *IEEE Trans. Electron Dev.*, **ED−15**, 151−160 (1968).
31. F. F. FANG and A. B. FOWLER, 'Transport properties of electrons in inverted silicon surfaces', *Phys. Rev.*, **169**, 619−631 (1968).
32. N. ST. J. MURPHY, F. BERG and I. FLINN, 'Carrier mobility in silicon MOST's', *Solid-St. Electron.*, **12**, 775−786 (1969).
33. J. T. C. CHEN and R. S. MULLER, 'Carrier mobilities at weakly inverted silicon surfaces', *J. Appl. Phys.*, **45**, 828−834 (1974).

34. C. N. Berglund, 'Surface states at steam-grown silicon-silicon dioxide interface', *IEEE Trans. Electron Dev.*, ED–13, 701–705 (1966).
35. C. A. Neugebauer, 'Temperature dependence of the field-effect conductance in thin polycrystalline CdS films', *J. Appl. Phys.*, **39**, 3177–(1968).
36. E. H. Nicollian and A. Goetzberger, 'The Si–SiO$_2$ interface-electrical properties as determined by the MIS conductance technique', *Bell Syst. Techn. J.*, **46**, 1055–1133 (1967).
37. Y. Zohta, 'Rapid determination of semiconductor doping profiles in MOS structures', *Solid-St. Electron.*, **16**, 125–126 (1973).
38. E. H. Nicollian, M. H. Hanes and J. R. Brews, 'Using the MIS capacitor for doping profile measurements with minimal interface state error', *IEEE Trans. Electron Dev.*, ED–20, 380–389 (1973).
39. S. D. Brotherton and P. Burton, 'The influence of non-uniformly doped substrates on MOS C–V curves', *Solid-St. Electron.*, **13**, 1591–1595 (1970).
40. G. Panigrahi, 'Numerical calculation of low-frequency capacitance-voltage curves of MOS capacitors with non-constant doping profiles', *Electron. Lett.*, **9**, 43–44 (1973).
41. G. Schottky, 'Decrease of FET threshold voltage due to boron depletion during thermal oxidation', *Solid-St. Electron.*, **14**, 467–474 (1971).
42. M. Hswe, R. B. Palmer, M. L. Shopbell and C. C. Mai, 'Characteristics of p-channel MOS field effect transistors with ion-implanted channels', *Solid-St. Electron.*, **15**, 1237–1243 (1972).
43. M. R. MacPherson, 'Threshold shift calculations for ion implanted MOS devices', *Solid-St. Electron.*, **15**, 1319–1326 (1972).
44. G. Doucet and F. Van De Wiele, 'Threshold voltage of non-uniformly doped MOS structures' '*Solid-St. Electron.*, **16**, 417–423 (1973).
45. M. Kamoshida, 'Threshold voltage and 'gain' term β of ion-implanted enhancement-mode n-channel MOS transistors', *Appl. Phys. Lett.*, **22**, 404–405 (1973).
46. T. W. Sigmon and R. Swanson, 'MOS threshold shifting by ion implantation', *Solid-St. Electron.*, **16**, 1217–1232 (1973).
47. J. Bernard, 'Valorisation des technologies MOS par l'utilisation de l'implantation ionique', *L'Onde Electrique*, **54**, 15–22, Janvier 1974.
48. J. Lindmayer, 'Surface charge and surface potential in arbitrarily doped crystals', *Solid-St. Electron.*, **6**, 137–140 (1963).
49. H. el-Sissi and R. S. C. Cobbold, 'Numerical calculation of the ideal $C-V$ characteristics of nonuniformly doped MOS capacitors', *Electron. Lett.*, **9**, 594–596 (1973).
50. G. Declerck, R. Van Overstraeten and G. Broux, 'Measurement of low densities of surface states at the Si–SiO$_2$-interface', *Solid-St. Electron.*, **16**, 1451–1460 (1973).
51. H. Deuling, E. Klausmann and A. Goetzberger, 'Interface states in Si–SiO$_2$ interfaces', *Solid-St. Electron.*, **15**, 559–571 (1972).
52. J. A. Cooper, Jr., and R. J. Schwartz, 'Electrical characteristics of the SiO$_2$–Si interface near midgap and in weak inversion', *Solid-St. Electron.*, **17**, 641–654 (1974).
53. F. Berz, 'Variation with frequency of the transverse impedance of semiconductor surface layers', *J. Phys. Chem. Solids*, **23**, 1795–1815 (1962).
54. K. Lehovec and A. Slobodskoy, 'Impedance of semiconductor-insulator-metal capacitors', *Solid-St. Electron.*, **7**, 59–79 (1964).
55. F. Forlani, N. Minnaja and E. Pagiola, 'Equivalent circuit of a metal-insulator-semiconductor structure', *Solid-St. Electron.*, **10**, 9–20 (1967).
56. V. Temple and J. Shewchun, 'Exact frequency dependent complex admittance of the MOS diode including surface states, Shockley-Read-Hall (SRH) impurity effects and low temperature dopant impurity response', *Solid-St. Electron.*, **16**, 93–113 (1973).
57. C. T. Sah, R. F. Pierret and A. B. Tole, 'Exact analytical solution of high-frequency lossless MOS capacitance-voltage characteristics and valability of charge analysis', *Solid-St. Electron.*, **12**, 681–688 (1969).
58. J. J. Simonne, 'A method to extract interface state parameters from the MIS parallel conductance technique', *Solid-St. Electron.*, **16**, 121–124 (1973).
59. S. M. Sze, 'Physics of semiconductor devices', John Wiley, New York, 1969.
60. M. Kuhn, 'A quasi-static technique for MOS C–V and surface state measurements', *Solid-St. Electron.*, **13**, 873–885 (1970).
61. D. K. Schroeder, 'Bulk and optical generation parameters measured with the pulsed MOS capacitor', *IEEE Trans. Electron Dev.*, ED–19, 1018–1023 (1972).

62. H. Preier, 'Different mechanisms affecting the inversion layer transient response', *IEEE Trans. Electron Dev.*, ED−15, 990−997 (1968).
63. M. Zerbst, 'Relaxationseffekte an Halbleiter-Isolator-Grenzflächen', *Z. Angew. Phys.*, **22**, 30, May 1966.
64. D. K. Schroder and H. C. Nathanson, 'On the separation of bulk and surface components of lifetime using the pulsed MOS capacitor', *Solid-St. Electron.*, **13**, 577−582 (1970).
65. D. K. Schroeder and J. Guldberg, 'Interpretation of surface and bulk effects using the pulsed MIS capacitor', *Solid-St. Electron.*, **14**, 1285−1297 (1971).
66. J. Müller and B. Schiek, 'Transient responses of a pulsed MIS-capacitor', *Solid-St. Electron.* **13**, 1319−1332 (1970).
67. R. F. Pierret, 'A linear-sweep MOS-C technique for determining minority carrier lifetimes', *IEEE Trans. Electron Dev.*, ED−19, 869−873 (1972).
68. R. F. Pierret and D. W. Small, 'Effects of lateral surface generation on the MOS−C linear-sweep and C−t transient characteristics', *IEEE Trans. Electron Dev.*, ED−20, 457−458 (1973).
69. C. D. Bulucea, 'Investigation of deep-depletion régime of MOS structures using ramp-response method', *Electron. Lett.*, **6**, 479−481 (1970).
70. C. Bulucea and P. Antognetti, 'On the MOS structure in the avalanche régime', *Alta Frequenzy*, **39**, 734−740 (1970).
71. A. Goetzberger and E.M. Nicollian, 'MOS avalanche and tunnelling effects in silicon surfaces', *J. Appl. Phys.*, **38**, 4582 (1967).
72. H. C. Card, 'Factors affecting avalanche injection into insulating layers from a semiconductor surface', *Solid-St. Electron.*, **17**, 501−502 (1974).
73. E. M. Nicollian and C. N. Berglund, 'Avalanche injection of electrons into insulating SiO_2 using MOS structures', *J. Appl. Phys.*, **41**, 3052−3057 (1970).
74. A. S. Grove, O. Leistiko and W. W. Hooper, 'Effect of surface fields on the breakdown voltage of planar silicon $p-n$ junctions', *IEEE Trans. Electron Dev.*, ED−14, 157−162 (1967).
75. C. D. Bulucea, 'Avalanche injection in silicon planar semiconductor devices', Dr. eng. thesis, Polytechn. Inst. Bucharest (1973).
76. C. Bulucea, A. Rusu and C. Postolache, 'Surface breakdown in silicon planar junctions − a computer-aided experimental determination of the critical field', *Solid-St. Electron.*, **17**, 881−888 (1974).
77. M. Pepper, 'Electron injection into SiO_2 from an avalanching $p-n$ junction', *J. Phys.D: Appl. Phys.*, **6**, 2124−2130 (1973).
78. J. F. Verwey and B. J. De Maagt, 'Avalanche-injected electron currents in SiO_2 at high injection densities', *Solid-St. Electron.*, **17**, 963−971 (1974).
79. J. F. Verwey, 'Time dependence of injection currents in SiO_2', *Philips Res. Repts.*, **28**, 66−74 (1973).
80. J. F. Verwey, 'The introduction of charge in SiO_2 and the increase of interface states during breakdown of emitter-base junction of gated transistors', *Appl. Phys. Lett.*, **15**, 270−272 (1969).
81. R. G. Müller, 'Gate-enhanced vs channel-current induced breakdown for floating gate avalanche injection', *Solid-St. Electron.*, **17**, 503−505 (1974).
82. D. M. Erb, H. G. Dill and T. N. Toombs, 'Electron gate currents and threshold stability in the n-channel stacked gate MOS tetrode', *IEEE Trans. Electron Dev.*, ED−18, 105−109 (1971).
83. M. Pepper, 'Electron injection into SiO_2 in the n channel stacked gate MOS tetrode', *IEEE Trans. Electron Dev.*, ED−21, 174−175 (1974).
84. D. Frohman-Bentckowsky, 'Memory behaviour in a floating-gate avalanche-injection MOS (FAMOS) structure', *Appl. Phys. Lett.*, **18**, 332−334 (1971).
85. B. Schiek and J. Heimke, 'Untersuchungen an MIS-Varaktoren für Anwendungen als nichtlineare Reaktanzen', *Arch. Elek- übertr.*, **24**, 153−163 (1970).
86. I. Stoev and F. Schumacher, 'Preparation and RF properties of MIS mesa varactros', *Arch. Elek. übertr.*, **27**, 321−324 (1973).

Problems

10.1. Show that under very strong inversion conditions, $Q_n \to Q_s$ for a p-type semiconductor, the total charge may be writen

$$|Q_s| = 2en_s\lambda_{ns}. \qquad (P.10.1.1)$$

where $n_s = n_i \exp(n_s - u_F)$ is the surface electron density and $\lambda_{ns} = (\varepsilon kT/2e^2 n_s)^{1/2}$ is a certain characteristic length.

Hint: For large $|u_s|$ one obtains $Q_n/Q_s \to 1$ and

$$Q_s \simeq -2\frac{u_s}{|u_s|} en_i L_D \{2\cos h\,(u_s - u_F)\}^{1/2}.$$

10.2. Show that the characteristic length λ_{ns} defined in Problem 10.1 can be also written

$$\lambda_{ns} = \frac{kT}{e}\mathscr{E}_s^{-1} \qquad (P.10.2.1)$$

and compute its value at the onset of very strong inversion for the doping densities indicated in Figure 10.2.2.

10.3. Demonstrate that the higher the impurity concentration the better the depletion-type model described in Section 10.3, for depleted and inverted surface.

10.4. Show that the flat-band semiconductor capacitance, equation (10.4.16), can be written approximately C_S/L_{DE} (silicon room temperature). Give an interpretation of this result.

Hint: L_{DE} can also be written $(D_p\tau_r)^{1/2}$ where τ_r is dielectric relaxation time of the p-type semiconductor.

10.5. Assume that the metal electrode of an MOS capacitor is replaced by p-type (polycrystalline) silicon gate and compute the threshold voltage for an n-type silicon substrate of 10^{15} cm^{-3} doping. Consider <100> and <111> surfaces. The oxide thickness is 1000 Å.

10.6. Discuss the effect of temperature upon the $C-V$ characteristics determined at a particular frequency.

Hints: At elevated temperatures the generation is more active and the inversion layer builds up more rapidly. At liquid nitrogen temperature hysteresis phenomena are observed (automatic measurement).

10.7. Show that the impurity profile is given by

$$N(x) = -\frac{C^3}{e\varepsilon}\left(\frac{dC}{dV_G}\right)^{-1}, \quad x = \varepsilon\left(\frac{1}{C_{ox}} - \frac{1}{C}\right), \qquad (P.10.7.1)$$

where $C = C(V_G)$ is the MOS capacitance when the surface layer is depleted.

10.8. Evaluate the shift in threshold voltage determined by an implanted surface layer of uniform doping (Figure 10.8.9).

10.9. Consider the capacitance-voltage dependence of an ion implanted MOS structure (Figure 10.8.10). Calculate the flat band capacitance and the minimum capacitance when the implanted layer is uniformly doped (Figure 10.8.9).

10.10. Derive the (small-signal) interface-state admittance

$$y_{ss} = i\omega\frac{e^2}{kT}\frac{N_{ss}f_0(1-f_0)}{1 + i\omega f_0/c_n n_{s0}} \qquad (P.10.10.1)$$

by considering the capture and emission of majority carriers from a single-level interface state described by a Shockley-Read-Hall model.

Hint: The capture and emission rates are, respectively, $N_{ss}c_n[1 - f(t)]n_s(t)$ and $N_{ss}e_nf(t)$, where $f(t) = f_0 + \delta t$, etc.

10.11. Demonstrate the formula given for the minority carrier lifetime, Figure 10.10.5 determined from the linear sweep MOS$-C$ method. Evaluate the range of measurable τ for α ranging from 0.1 Vs^{-1} to 10 Vs^{-1} and t_{ox} from 500 Å to 3000 Å. Discuss the effect of temperature.

11

Metal-oxide-semiconductor (MOS) Transistor

11.1 Introduction

We have discussed in Chapter 9 the junction-gate field-effect transistor (JGFET). *The insulated-gate* field-effect transistor (IGFET) is an electronic device which may have various constructive forms [1]—[3]. In all cases the metallic gate is separated from the conductive channel by an insulating layer. This provides a very high input (gate) resistance ($10^{12}-10^{14}\ \Omega$), irrespective of the gate voltage polarity and even at elevated temperatures.

The only IGFET type discussed in this book is the MOS transistor [4]—[6]*. This is a surface-channel device. The conductive channel is an inversion surface layer formed between source and drain, which provides the conductive path for the source-drain current. The charge carriers are therefore minority carriers with respect to the semiconductor bulk. However, the physical principle of MOST operation is completely different from that of the minority-carrier transistor (normal *npn* or *pnp* device). The MOST (or MOSFET) is a unipolar device whereas the normal junction transistor is a bipolar one. The majority of surface-channel FET's are silicon devices and the silicon dioxide is used as insulating layer, hence the name metal-oxide-semiconductor transistor (MOST). Sometimes we refer to metal-insulator-semiconductor (MIS) transistor, where the insulator is not necessarily an oxide.

Figure 11.1.1 illustrates the device operation. When the drain potential is zero ($V_D = 0$), the surface potential and surface charge are controlled by the gate voltage. A gate voltage below a certain threshold voltage will change the surface conductivity to *p*-type (*n*-type substrate) (see Chapter 10). The surface inversion layer is separated from the semiconductor bulk by a depletion layer which continues with the depletion regions surrounding the source-to-substrate and the drain-to-substrate *pn* junctions (Figure 11.1.1 *a*). The same inversion layer communicates with the p^+ source and drain regions. With an appropriate potential applied on the drain, a hole current flows from source to drain through the very thin surface layer. However, the drain current produces a voltage drop along the channel. This drop is of such a polarity as to oppose the surface field produced by the gate bias. Figure 11.1.1 *b* shows that the channel becomes narrower toward the drain, whereas the depletion layer thickness increases in the same direction; near the drain this layer

---

* We only mention the depletion-type IGFET [7], [8] which is a majority carrier device similar to the junction-gate FET, except that the gate is separated from the partially depleted channel by an insulating layer. The JGFET-theory can be obtained as a limit case, when the insulator thickness tends to zero [3], [9].

11. MOS Transistor

merges with the depletion region of the drain junction, and this region is widened by the applied drain voltage. As the source-to-drain potential difference is further increased, the channel *pinches off* near the drain. The channel pinch-off occurs when the surface is no longer inverted. Although the mobile charge is repelled

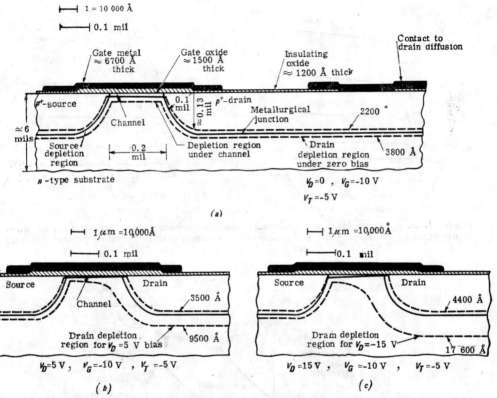

Figure 11.1.1. Scale drawing of a typical MOS structure in cross section: (*a*) $V_D = 0$; (*b*) $V_D = -5$ V; (*c*) $V_D = -15$ V. The source and substrate are both considered at ground potential. The channel is not drawn to scale: its typical thickness (depth in silicon) is of a few tens of angstroms. (*after Crawford*, [5]).

away from the surface, the current does not vanish. It is usually assumed that the drain current saturates at its pinch-off value. The mechanism of current conduction and the actual (finite) output resistance are discussed in Section 11.5. There is a certain similarity with the operation of a junction-gate FET (chapter 9). Note, for example, that as the source-to-drain potential difference continues to increase, the pinch-off point moves toward the source (Figure 11.1.1 *c*) and the effective length of the channel decreases.

Figure 11.1.2 *a* shows the drain (output) characteristics measured with both substrate and source grounded. The conventional signs of the transistor current and voltages are indicated on Figure 11.1.3, for both *n*-channel and *p*-channel MOST's. The drain current is zero at $V_G = 0$ (Figure 11.1.2 *a*) and the device operates in the *enhancement mode* (the threshold voltage is positive, the *p*-channel transistor is normally-off, i.e. turned off at $V_G = 0$). Figure 11.1.2 *b* exhibits current-voltage characteristics for a *p*-channel device which is normally open (a current flows at $V_G = 0$). Such a transistor can be turned off by increasing V_G above zero: this is a *depletion*

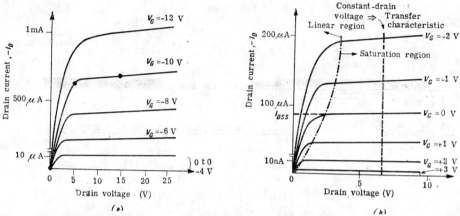

Figure 11.1.2. *p*-channel MOS drain characteristics: (*a*) enhancement mode device (the three points indicated on $V_G = -10$ V characteristic represent the three operating points of Figure 1.11.1); (*b*) depletion-mode device. (*after Crawford*, [5]).

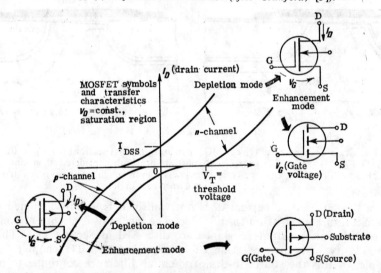

Figure 11.1.3. Transfer characteristics of *n*-channel and *p*-channel, enhancement-mode and depletion-mode transistors. Device symbols are also shown. In all cases the substrate is connected to the source.

mode device. This is better illustrated by Figure 11.1.3 which shows, qualitatively, the transfer characteristics of various MOST's (measured at $V_D = $ const. in the saturation region of the characteristics).

11. MOS Transistor

It was already shown (Chapter 10) that enhancement mode silicon transistors are normally *p*-type, because the positive oxide charge tends to invert the surface of *p*-type silicon with no gate voltage and therefore *n*-channel MOST are normally-open. The enhancement mode of operation is necessary for logic applications (inverter stages). Transistors operating in the depletion mode can be used as load devices in integrated circuits or as amplifiers. In the latter case *n*-channel devices should be preferred for their higher transconductances and higher figure of merit due to higher carrier mobility. However, for logic integrated circuits, complementary (both *n*-channel and *p*-channel) devices operating in the enhancement mode are highly attractive. The major application today of the MOST is in logic circuits [6]. With this in mind we discuss in this chapter certain problems specific to the integrated MOST, namely: threshold voltages and leakage drain currents, complementary transistors, short-channel devices, punch-through in structures with high substrate resistivities. Other parameter, which are important for amplifier applications (output resistance, high-frequency and noise behaviour), are also discussed.

11.2 Simplified Theory (Substrate Effect Neglected)

In this section we consider a simplified theory of the MOS transistor by neglecting the effect of fixed bulk charge contained in the depletion layer formed between the surface inversion layer and the neutral semiconductor substrate. The resulting simple equations are used for design purposes [5], [6]. On the other hand, the simple derivation which follows enables a direct comparison between the fundamental steady-state properties of MOSFET and JGFET transistors.

The generalized model of Section 9.2 is used. Within the gradual approximation, the channel conductibility is controlled by the transversal field, as indicated by equation (9.2.2), whereas the drain current is carried by drift due to the longitudinal electric field, equation (9.2.1). In an MOS transistor, equation (9.2.2) is valid provided that the bulk fixed charge is neglected and the inversion-layer mobility does not depend upon the longitudinal electric field. The absolute value of the surface conductivity is $\sigma_s = -\mu_n Q_n$ (*p*-type substrate) and the zero-current channel conductance, $G_0 = G_0(V_G)$ may be written*

$$G_0 = \frac{\mu_n W C_{ox}}{L}(V_G - V_T) = G_0(V_G). \qquad (11.2.1)$$

The steady-state characteristics will be found by integrating equation (9.2.6). For constant surface mobility, μ_n, the result is:

$$I_D = k[2(V_G - V_T)V_D - V_D^2], \quad k = \frac{\mu_n \varepsilon_{ox} W}{2 t_{ox} L}, \qquad (11.2.2)$$

---

* See equation (10.4.21), where the threshold voltage *for strong inversion*, V_T, is given by equation (10.5.17). See also Section 10.11 to demonstrate the validity of equation (9.2.2) for the MOSFET.

provided that
$$V_D < V_G - V_T = V_{D\text{sat}}. \tag{11.2.3}$$

When $V_D \geqslant V_{D\text{sat}}$, the drain current is saturated at
$$I_{D\text{sat}} = I_D\big|_{V_D = V_{D\text{sat}}} = k(V_G - V_T)^2. \tag{11.2.4}$$

The drain conductance below saturation ($V_D \leqslant V_{D\text{sat}}$) is given by equation (9.2.8):
$$g_d = G_0(V_G - V_D) = 2k(V_G - V_D - V_T). \tag{11.2.5}$$

The transconductance below saturation is given by equation (9.2.16):
$$g_m = 2kV_D \tag{11.2.6}$$

and the transconductance in saturation is equal to the zero-current drain conductance
$$g_{m\text{sat}} = 2k(V_G - V_T) = g_{d0}. \tag{11.2.7}$$

The device constant, k, may be written
$$k = \frac{\mu_n \varepsilon_{\text{ox}} W}{2 t_{\text{ox}} L} = k'\frac{W}{L}, \quad k' = \frac{\mu_n \varepsilon_{\text{ox}}}{2 t_{\text{ox}}}, \tag{11.2.8}$$

where k' (as well as V_T) are constants of the technological process (or process parameters)* and the ratio W/L depends upon the device (surface) geometry. The above equations are frequently used for MOSFET circuit design [5], [6].

The theoretical figure of merit, the so-called maximum operating frequency, $f_{\max}$, is defined by (see also Section 9.6)
$$f_{\max} = \frac{1}{2\pi} \frac{g_{ms}}{C_G}, \tag{11.2.9}$$

where
$$C_G = C_{\text{ox}} WL = \varepsilon_{\text{ox}} WL t_{\text{ox}}^{-1} \tag{11.2.10}$$

is the total gate capacitance. Therefore
$$f_{\max} = \frac{1}{2\pi} \frac{\mu_n V_{D\text{sat}}}{L^2} \tag{11.2.11}$$

---

* There are two basic versions of the p-channel MOS process (Al–SiO$_2$–Si) [6]: the high threshold (or standard) process with $V_T = -4$ V and $k' = 2 \times 10^{-6}$ A V^{-2} and the low threshold process, $V_T = -2$ V, $k' = 2 \times 10^{-6}$ A V^{-2}. The device transconductance is proportional to k' and can be increased using a high-permittivity insulator instead of SiO$_2$, for example silicon nitride or titanium dioxide (the relative permittivities are 4, 8 and about 80, respectively). The oxide thickness cannot be reduced below 500 Å due to the unavoidable faults (pinch-holes) present in such a thin film.

can be increased by reducing L, using an n-channel device (higher mobility) or operating at higher voltages. However, in a short-channel device at high drain voltages, $V_{D\,\text{sat}}$, the mobility decreases due to the very high longitudinal field thus limiting the available figure of merit (Problem 11.1)*. On the other hand, by increasing the gate voltage, $V_G = V_T + V_{D\,\text{sat}}$, the surface mobility decreases due to the increasing surface field. This effect was considered in Section 10.7. Equation (10.7.6) shows that the mobility decrease depends approximately upon the ratio $(V_G - V_T)/t_{ox}\mathscr{E}_c$, where $\mathscr{E}_c$ is a certain critical field. The carrier velocity approaches a limiting velocity which is of the order of bulk saturated velocity, v_{sat} [13]. The effect of the transversal (surface) field can be clearly distinguished from that of the longitudinal field by measuring $g_{d0}(V_D \to 0)$. Figure 10.7.6 shows a deviation of g_{d0} from the linear dependence due to the decrease of the surface mobility [13].

11.3 Theory of MOSFET Including Substrate Effect

The basic model of the MOS transistor [1], [4], [6] is introduced with reference to Figure 11.3.1. The fundamental assumptions are the gradual approximation and the depletion layer approximation (abrupt transition from fully depleted to

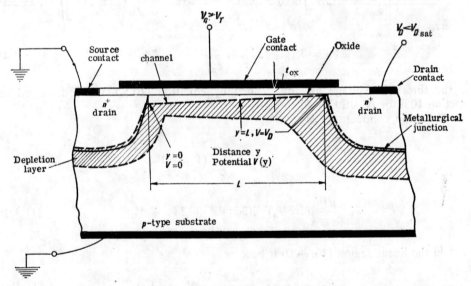

Figure 11.3.1. n-channel MOS transistor. The substrate is connected to the source.

neutral regions). The corresponding equations describing the charge balance in a section y (Figures 11.3.1) were derived in Section 10.11.

---

* The effect upon the device characteristics is discussed in references [10]–[12] (see also Chapter 9).

The steady-state drain current is the drift current determined by the mobile electron-charge density $Q_n(y)$ in the channel and by the longitudinal electric field of magnitude $dV(y)/dy$

$$I_D = -\mu_n Q_n(x) \, W \, dV/dy, \tag{11.3.1}$$

where W is the gate width (along the z axis, see Figure 11.3.1). By integrating equation (11.3.1) from $y = 0$ to $y = L$ one obtains

$$I_D = -\frac{\mu_n W}{L} \int_0^{V_D} Q_n \, dV, \tag{11.3.2}$$

where μ_n is the effective mobility in the inversion layer (assumed constant). Q_n can be expressed as $Q_n = Q_n(V)$ by using successively equations (10.11.8), (10.11.9), (10.11.11), (10.11.1), and the strong inversion approximation $\phi_{s0} \simeq 2\phi_F$ (Section 10.3). The final result is

$$I_D = \frac{\mu_n \varepsilon_{ox} W}{t_{ox} L} \left\{ \left[V_G - V_T - \frac{V_D}{2} \right] V_D - \right.$$

$$\left. - \frac{2}{3} \frac{t_{ox}}{\varepsilon_{ox}} (2\varepsilon e N_A)^{1/2} \left[(2\phi_F + V_D)^{3/2} - (2\phi_F)^{3/2} - \frac{3}{2} V_D (2\phi_F)^{1/2} \right] \right\}, \tag{11.3.3}$$

where

$$V_T = V_{FB} + 2\phi_F + \frac{t_{ox}}{\varepsilon_{ox}} (2eN_A \varepsilon \cdot 2\phi_F)^{1/2} \tag{11.3.4}$$

is the threshold voltage already defined by equation (10.5.17) and discussed in Section 10.6 for Al—SiO$_2$—Si transistors.

The drain conductance is

$$g_d = \frac{\partial I_D}{\partial V_D}\bigg|_{V_G = \text{const}} = \frac{\mu_n \varepsilon_{ox} W}{t_{ox} L} (V_G - V_T - V_D) +$$

$$+ \frac{\mu W}{L} (2\varepsilon e N_A)^{1/2} [(2\phi_F)^{1/2} - (2\phi_F + V_D)^{1/2}] \tag{11.3.5}$$

and in the linear region ($V_D \to 0$) it becomes

$$g_{d0} = g_d\big|_{V_D \to 0} = \frac{\mu_n \varepsilon_{ox} W}{t_{ox} L} (V_G - V_T). \tag{11.3.6}$$

The channel conductance at zero bias current does indeed vanish at the threshold voltage $V_G = V_T$. All equations are valid only for $V_G \geqslant V_T$. The transconductance can be written

$$g_m = \frac{\partial I_D}{\partial V_G}\bigg|_{V_D = \text{const}} = \frac{\mu_n \varepsilon_{ox} W}{t_{ox} L} V_D \tag{11.3.7}$$

11. MOS Transistor

and is independent of substrate resistivity. All the above formulas are valid for $V_D \leqslant V_{Dsat}$ where V_{Dsat} is defined by the onset of saturation (the drain conductance vanishes)

$$g_{dsat} = g_d(V_G, V_{Dsat}) = 0. \tag{11.3.8}$$

One obtains

$$V_{Dsat} = V_G - V_T + V_B + (V_B^2/4\phi_F) -$$
$$- V_B[1 + (V_B^2/16\,\phi_F^2) + (V_G - V_T + V_B)/2\phi_F]^{1/2} \tag{11.3.9}$$

and the drain current in saturation ($V_D \geqslant D_{Dsat}$) is

$$I_{Dsat} = \frac{\mu_n \varepsilon_{ox} W}{t_{ox} L} \{V_{Dsat}^2/2 + V_B V_{Dsat}[1 + (V_{Dsat}/2\phi_F)]^{1/2} -$$
$$- (4/3)\phi_F V_B[(1 + V_{Dsat}/2\phi_F)^{3/2} - 1]\}, \tag{11.3.10}$$

where

$$V_B = \frac{t_{ox}}{\varepsilon_{ox}} (2eN_A \varepsilon \times 2\phi_F)^{1/2}. \tag{11.3.11}$$

Figures 11.3.2 and 11.3.3 [14] show that both V_{Dsat} and I_{Dsat} decrease due to the finite resistivity of the semiconductor substrate. Equations (11.3.4) and (11.3.5)

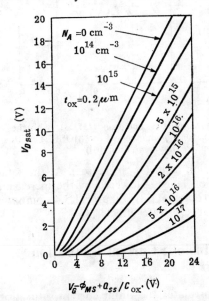

Figure 11.3.2. Decrease of V_{Dsat} due to the finite substrate resistivity. (*after Sah and Pao*, [14]).

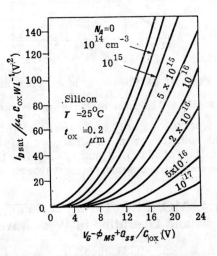

Figure 11.3.3. Effect of substrate resistivity on the transfer characteristic. (*after Sah and Pao*, [14]).

indicate that the drain conductance decreases and the threshold voltage increases with the decreasing substrate resistivity. The transconductance in saturation

$$g_{m\,sat} = \frac{\mu_n \varepsilon_{ox} W}{t_{ox} L} V_{D\,sat} \qquad (11.3.12)$$

decreases with the decreasing substrate resistivity (because $V_{D\,sat}$ decreases). All the above results indicate that the gate control of the mobile charge in the channel becomes less effective as the substrate conductivity increases. A higher V_G is necessary

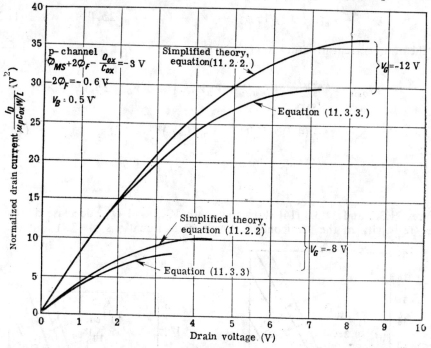

Figure 11.3.4. Comparison of the 'exact' drain characteristics, equation (11.3.3), with the results of the simplified theory, equation (11.2.2), for a p-channel silicon MOSFET with typical parameters. (reproduced *after Penney and Lau*, [6]).

to repel the majority holes and form an inversion layer (V_D increases). At the same bias voltages the mobile charge in the inversion layer is lowered by the increasing substrate conductivity and g_d decreases (see Problem 11.3). Figure 11.3.4 compares the computed characteristics with and without substrate effects.

11.4 MOSFET Theory Including Diffusion: Current Saturation and Operation in Weak Inversion

The classical MOST theory (Section 11.3) postulates the drain current saturation for $V_D > V_{D\,sat}$ where, by definition, the drain conductance vanishes at $V_D = V_{D\,sat}$. Apart from the 'mathematical' uncertainty, the physical situation is somewhat unclear because the inversion layer should disappear near the drain (channel pinch-off) at $V_D = V_{D\text{sat}}$. Pao and Sah [15] have shown that the current continuity in the drain region can be provided by carrier diffusion. This is physically reasonable because an enormous carrier concentration gradient occurs along the surface, toward the drain (the transition from the highly conductive inversion layer to the depleted drain region). From a mathematical point of view, by including the diffusion component in the current continuity equation [15] the saturation of the drain current is obtained automatically.

Other features of the Pao-Sah theory [15] are:

(a) The mobile charge in a given section, y, of the channel is accurately computed: the depletion approximation (abrupt edge of the depleted region) and the strong inversion approximation ($\phi_{\text{sinv}} \simeq 2\phi_F$, Section 10.3) are not used. Therefore, the weak inversion régime (low drain currents for gate voltages near threshold) can also be studied.

(b) The gradual approximation is still maintained. The mobile charge is determined by neglecting the field gradient along the surface (y direction). This assumption fails to be true at large V_D: the electric field in the drain region should be determined from a two-dimensional Poisson equation [15]. Therefore this theory does not accurately predict the drain characteristics in the saturation region. However, the computations are valid at low drain voltages and useful to predict leakage currents near threshold [16]–[19].

Consider an n-channel MOST (p-type substrate). The total channel current density (electron current, constant carrier mobility) is (Chapter 1)

$$J(x, y) \simeq e\mu_n n(x, y) \mathscr{E} + eD_n \frac{\partial n(x, y)}{\partial y}, \qquad (11.4.1)$$

where

$$\mathscr{E} \simeq \mathscr{E}_y = -\frac{dV}{dy} \qquad (11.4.2)$$

and $V(y)$ is the channel potential defined in Section 10.11. The current density (11.4.1) may be written [1], [15]

$$J(x, y) = -eD_n n(x, y) \nabla_y \xi, \qquad (11.4.3)$$

where u_{Fn} is the electron imref (in kT units) measured from the bulk Fermi level. The energy band diagram of an inverted surface in non-equilibrium conditions was represented in Figure 10.11.1 b. The corresponding potential diagram (in kT/e units) is shown in Figure 11.4.1. The drain current is

$$I_D = -eD_n W \int_0^{x_c} n(x, y) \frac{d\xi}{dx} dx \qquad (11.4.4)$$

(the electron conduction occurs until the depth $x_c = x_c(x)$ which will be defined below). By integrating from $x=0$ to $y=L$, taking into account equations (10.11.3)–(10.11.6) and (Figure 11.4.1)

$$\xi = u_{F_n}(y) - u_F = \xi(y), \qquad (11.4.5)$$

equation (11.4.4) becomes

$$I_D = \frac{eD_n n_i W L_D}{L} \int_0^{u_D} \int_{u_c}^{u_s} \frac{\exp(u - \xi - u_F)}{F(u, \xi, u_F)} \, du \, d\xi, \qquad (11.4.6)$$

where $u_D = eV_D/kT$ is the normalized drain voltage, and u_c is the normalized potential defining the depth x_c. Usually, x_c is defined as the potential where the electron concentration equals the intrinsic concentration, n_i. In this case $x_c = x_i = x_i(y)$,

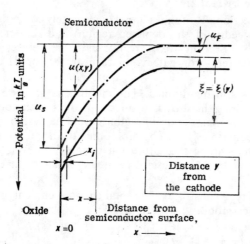

Figure 11.4.1. Potential diagram in kT/e units of an inverted surface layer on a p-type semiconductor, at the distance y, from the source of the MOSFET.

as shown in Figure 11.4.1. In this way the numerical evaluation of I_D requires less computer time [17]. However, for an accurate calculation of I_D in the weak inversion régime, u_c should take lower values [17] or even zero [16] ($u_c = 0$ corresponds to the neutral p-type substrate). The field function $F(u, \xi, u_F)$ is given by equation (10.11.7). The relation between the gate voltage, V_G, and the surface potential u_s, is (Section 10.11)

$$V_G = V_{FB} + \frac{kT}{e}\left[u_s + \frac{\varepsilon}{\varepsilon_{ox}} \frac{t_{ox}}{L_D} F(u_s, \xi, u_F)\right]. \qquad (11.4.7)$$

Figure 11.4.2 shows theoretical characteristics (dots) calculated by using the above equations. The experimental drain conductance in saturation is higher than expected. The discrepancy between theory and experiment is higher for devices with higher t_{ox}/L ratio, because the gradual approximation is far from being satisfied [15], [20]. The increasing substrate resistivity also deteriorates the validity of the gradual approximation. The device behaviour in saturation will be discussed later. In this section we consider the low-level current region.

For surface potentials satisfying $2 < u_s < 2u_F$, $F(u, \xi, u_F)$ can be approximated by

$$F(u, \xi, u_F) \simeq [(u - 1) \exp u_F]^{1/2} \qquad (11.4.8)$$

and the drain current, equation (11.4.6), becomes [17]

$$I_D = \frac{en_i D_n WL_D \exp u_s}{L(\exp 3u_F/2)(u_s - 1)^{1/2}} (1 - \exp - u_D) \qquad (11.4.9)$$

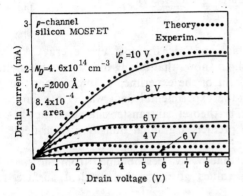

Figure 11.4.2. Comparison of theoretical and experimental characteristics of a p-channel MOSFET $V'_G = V_G - V_{FB}$ in the effective gate voltage (*after Pao and Sah*, [15]).

or

$$I_D = I_0(V_G)\left[1 - \exp\left(\frac{-eV_D}{kT}\right)\right], \qquad (11.4.10)$$

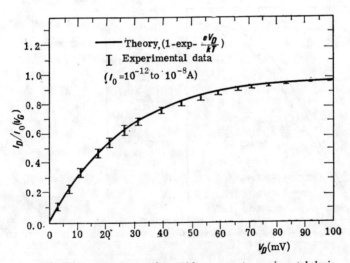

Figure 11.4.3. Comparison of low-current experimental drain characteristic with the theoretical curve, equation (11.4.10). (*after Barron*, [16]).

i.e., for a given gate voltage, the low-level current increases exponentially with the drain voltage. This dependence was verified experimentally, Figure 11.4.3, for a range of I_0 values from 1 pA to 0.1 µA.

Numerical calculations were performed [16], [17] by including the effect of surface states. In this case the flatband voltage, V_{FB}, in equation (11.4.7) will be replaced by

$$V_{FB} \to \phi_{MS} - \frac{Q_{ox}}{C_{ox}} + \frac{N_{SS}(u_S - \xi)}{C_{ox}} kT. \qquad (11.4.11)$$

The last term depends upon the surface state density (per unit energy) N_{SS} and upon the surface potential. Figure 11.4.4 *a* shows that taking into account the effect of surface states is essential for a good fit of the experimental low-current transfer characteristic. A very good match of the experimental data was also obtained at higher current levels (Figure 11.4.4 *b*). The theoretical transfer characteristic $I_D^{1/2} - V_G$ should be linear. The accurate theory predicts a gradual transition from depletion to strong inversion, a feature which is inexistent in the classical theory.

Van Overstraeten *et al.* [18] found that N_{SS} values obtained from the $I_D - V_G$ curves are in disagreement with the values given by independent surface-state

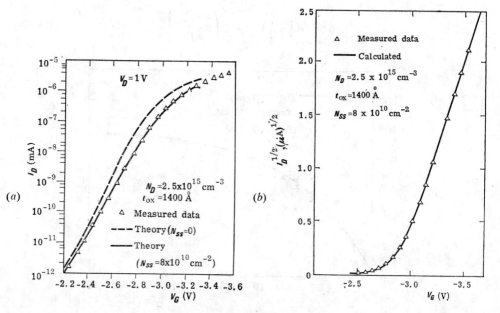

Figure 11.4.4. Comparison of numerically calculated transfer characteristic (including the interface state effect) with experimental data: (*a*) semilogarithmic plot indicating the exponential current-voltage dependence at low currents; (*b*) conventional $I_D^{1/2} - V_G$ plot. (after Barron, [16]).

measuring techniques (Section 10.9). It was shown that these discrepancies could be explained by the non-uniform surface potential caused by statistical fluctuations of the oxide charge, of the oxide thickness and of the bulk doping [18]. The oxide-charge fluctuations determine a probability $P(Q_{ox})$ to have an oxide charge Q_{ox} on an elementary area

$$P(Q_{ox}) = \frac{1}{(2\pi\sigma_q^2)^{1/2}} \exp - (Q_{ox} - \bar{Q}_{ox})^2/2\sigma_q^2 \qquad (11.4.12)$$

(a Gaussian distribution around the mean value $\bar{Q}_{ox}$ with a standard deviation σ_q in $C\text{cm}^{-2}$) [18]. The probability to find a surface potential u_s is also expressed by a Gaussian distribution around the mean value $\bar{u}_s$, with the standard deviation σ_s in kT/e units [18]. The equation (11.4.6) should be replaced by:

$$I_D = \frac{eD_n W n_i L_D}{L} \int_0^{u_D} \frac{1}{\sqrt{2\pi\sigma_q^2}} \int_{-\infty}^{+\infty} \exp - (Q_{ox} - \bar{Q}_{ox})^2/2\sigma_q^2 \times$$

$$\times \int_{u_F+\xi}^{u_S} \frac{\exp(u - \xi - u_F)}{F(u, \xi, u_F)} \, du \, dQ_{ox} \, d\xi. \qquad (11.4.13)$$

The oxide charge Q_{ox} is fluctuating around the mean value $\bar{Q}_{ox}$ as expressed by equation (11.4.12). Each value of Q_{ox} then determines another value of the surface potential u_s in the upper boundary of the integral over the potential u.

Figure 11.4.5 shows a comparison between theory with and without oxide charge fluctuations. The slope of the experimental $\ln I_D - V_D$ curve is smaller than that of the ideal curve and this yields a higher surface-state density than the real one. The discrepancy is explained by statistical fluctuations of the oxide charge. The differences indicated in Figure 11.4.5 for (100) silicon are smaller than for those obtained for (111) samples. Note the coincidence of all curves at high current levels: the fluctuations of the oxide charge are screened by the mobile charge [18].

The weak inversion region is important in certain applications, for example in complementary MOS integrated circuits (CMOS or COSMOS) [21]. Threshold voltages close to zero can be obtained and the weak inversion current has to be considered as a leakage current. In memory circuits this leakage reduces the refreshing time. In low-power CMOS circuits with static power dissipation in the nanowatt range the leakage currents of tens of nA and below are extremely important [19]. The drain-source leakage current also has a 'bulk' component due to the reverse-biased drain junction. This component is evidenced by the $V_{GS} \geqslant 2.2$ V characteristic in Figure 11.4.6. At higher $|V_{GS}|$ the surface leakage current (weakly inverted surface) predominates. This takes place above V_T' corresponding to the onset of surface inversion ($\phi_S = \phi_F$): this 'intrinsic' voltage was calculated for the Al—SiO$_2$—Si system and represented in Figures 10.6.7 and 10.6.8 (see reference [19] for Mo gate). Figure 11.4.6 shows that the drain leakage current increases above that of the reverse-biased drain junction, for gate voltages nearly 0.8 V less than the extrapolated threshold voltage, V_T.

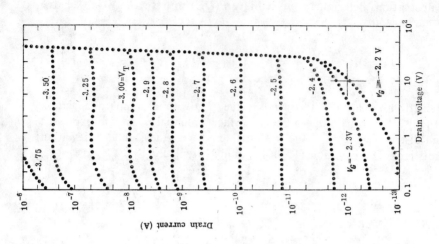

Figure 11.4.6. Log-log plot of $I_D - I_D$ characteristics ($I_D < 1\,\mu A$) for a p-channel MOS transistor (a 1×1 mil channel, 1 mil: 25.4μm, the long gate was used to give better $C-V$ characteristics). $V_T = -3V$ is the extrapolated threshold voltage, see for example Figure 10.7.6. Above $V_D = 20\,V$ the drain current increases rapidly due to the onset of punch-through or avalanch breakdown (Section 11.6). (after Gosney, [19]).

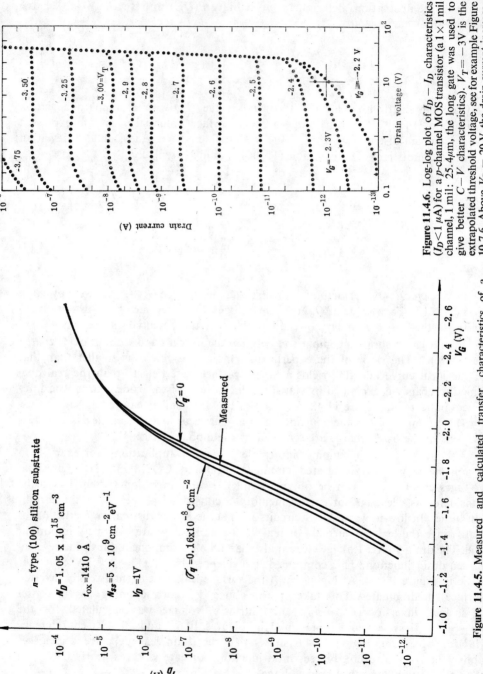

Figure 11.4.5. Measured and calculated transfer characteristics of a p-channel MOSFET made on (100) silicon substrate. (after von Overstraeten et al., [18]).

11.5 Analysis of MOSFET Saturation Region

The finite output conductance exhibited by MOS transistors operating in the saturation region of the drain characteristics was first attributed to the spreading of the depletion region near the drain [22]. When V_D increases, this depletion region extends toward the source and reduces the length of the channel (Figure 11.1.1). The extremity of the channel toward the drain is determined by $V(l) = V_{D\text{sat}}$, for $y > l$ the condition for surface inversion cannot be satisfied. The saturation current is approximately inversely proportional to the effective channel length, l, therefore

$$\frac{I_D}{I_{D\text{sat}}} \approx \frac{L}{l}, \qquad (11.5.1)$$

where $I_{D\text{sat}}$ is the current at the onset of 'saturation' (channel pinch-off near the drain). The drain conductance in saturation can be obtained by differentiating equation (11.5.1), where $l = l(V_D)$. The result is

$$g_{d\text{sat}} = \frac{\partial I_D}{\partial V_D}\bigg|_{V_G=\text{const}} = -I_{D\text{sat}} \frac{L}{l^2} \frac{\partial l}{\partial V_D} = \frac{I_{D\text{sat}}}{L(1 - l'/L)^2} \frac{\partial l'}{\partial V_D}, \qquad (11.5.2)$$

where

$$l' = L - l = l'(V_D) \qquad (11.5.3)$$

is the length of the drain region. The transconductance in saturation is

$$g_{m\text{sat}} = \frac{\partial I_D}{\partial V_G}\bigg|_{V_D=\text{const}} = \frac{L}{l} \frac{\partial I_{D\text{sat}}}{\partial V_G} - I_{D\text{sat}} \frac{L}{l^2} \frac{\partial L}{\partial V_G}. \qquad (11.5.4)$$

The output (drain) conductance will be calculated by assuming that the drain region is simply the depletion region of an abrupt pn junction reversely biased by $V_D - V_{D\text{sat}}$ (the built-in potential is neglected). Therefore

$$l' \approx \left[\frac{2\varepsilon(V_D - V_{D\text{sat}})}{eN_A}\right]^{1/2} \qquad (11.5.5)$$

and

$$g_{d\text{sat}} \approx \frac{I_{D\text{sat}}(2\varepsilon/eN_A)^{1/2}}{2L(1 - l'/L)^2(V_D - V_{D\text{sat}})^{1/2}}. \qquad (11.5.6)$$

The saturation conductance increases with the increasing substrate resistivity. The effect is stronger for short channel devices (small L).

The above results are not quantitatively verified by experiment. It was shown that the insulator thickness under the gate influences considerably the output conductance [23]. Devices with thin oxides, short channels and high-resistivity substrates depart appreciably from the above simplified model of drain conductance. The depth of the drain junction is also important [24]. All these results indicate that a more realistic description of the high-field drain region is necessary.

In an attempt to consider this complex situation, Frohman-Bentchkowsky and Grove [23] suggested the field distribution shown in Figure 11.5.1. The depletion region length is

$$l' \approx (V_D - V_{D\mathrm{sat}})/\mathscr{E}_T, \tag{11.5.7}$$

where $\mathscr{E}_T$ is an average field between the drain and the end of the inversion layer. Three contributions to $\mathscr{E}_T$ are considered: $\mathscr{E}_1$, due to the fixed space-charge in the reverse biased drain junction; $\mathscr{E}_2$, the fringing field due to the drain-to-gate potential drop; $\mathscr{E}_3$, the fringing field due to the potential difference $V_G' - V_{D\mathrm{sat}}$ (V_G' is the effective gate voltage $V_G + Q_{SS}/C_{ox}$). The above authors introduce two arbitrary and constant field-fringing factors corresponding to $\mathscr{E}_2$ and $\mathscr{E}_3$. The experimental results were readily matched by the above model. In particular, the saturation conductance increases with the oxide thickness since the field contribution due to the gate electrode is diminished thus leading to an increase of l'.

Schroeder and Muller [24] determined the field distribution in a MOSFET by a numerical solution of two-dimensional Poisson equation, and then evaluated the transconductance and the output conductance in saturation from the voltage depen-

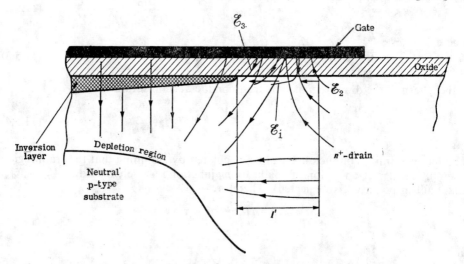

Figure 11.5.1. Electric field distribution in the drain region of an MOSFET operating in saturation. (*after Frohman-Bentchkowsky and Grove,* [23]).

dence of the effective channel-length, $l = l(V_G, V_D)$. The drain end of the channel ($y = l$) in this two-dimensional model was determined by the field reversal point. The reversal of the normal component of the surface field occurs at a drain voltage $V_D = V_{FR}$ given by $V_D + V_{bi} = V_G'$ and therefore $V(l) = V_G' - V_{bi}$, where V_{bi} is the built-in potential of the drain-substrate junction [24].

The Poisson equation is solved by a finite difference and relaxation methods: the two-dimensional device is filled with an array of points. A uniform array is unsuitable: there are more than two orders of magnitude difference between the largest and smallest characteristic dimensions (the channel length of a few μm and the inversion layer depth of a few hundreds of Å): the very large number of points may be prohibitive from the computing time and memory-size point of view [24]. Schroeder and Muller [24] first evaluated approximately the potential distribution in the entire device and then solved numerically the Poisson equation in a

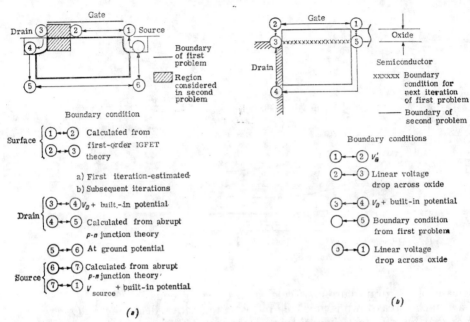

Figure 11.5.2. Boundary conditions for two problems solved alternately by successive iterations: (a) first boundary-value problem (entire device, oxide-semiconductor interface as a boundary); (b) second boundary value problem (a small drain region including oxide). (after Schroeder and Muller, [24]).

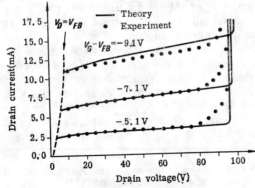

Figure 11.5.3. Drain characteristics in the saturation and breakdown regions (p-channel device with $N_D = 5 \times 10^{15}$ cm^{-3} (1 Ω cm), $L = 10$ μm, $t_{ox} = 0.12$ μm, drain diffusion depth $= d = 15$ μm. (after Schroeder and Muller, [24]).

certain region near the drain (insulator plus semiconductor). These two problems (Figure 11.5.2) were alternately solved until the calculated potential distributions match in the region common to both problems. Figures 11.5.3 and 11.5.4 indicate the comparison between the experimental points and the theoretical curves, derived from the above electrostatic calculations. The theoretical results are matched to the experimental ones at the field-reversal voltage, $V_D = V_{FR}$ (see above). The

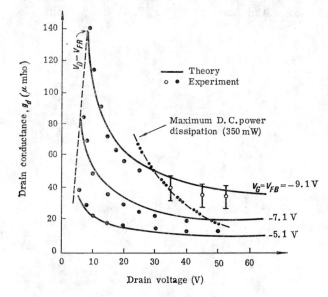

Figure 11.5.4. Output resistance in the saturation region for the same device as in Figure 11.5.3. (*after Schroeder and Muller,* [24]).

increase of the drain current is calculated by using equation (11.5.2)*. The onset of avalanche breakdown is predicted by using the critical field concept. Other results show that the effective channel-length decreases with the increasing of the drain-region depth. A sharper pinch-off and a smaller $g_{d\,sat}$ would be obtained in an ion-implanted device (shallow diffused drain).

Although the concept of channel-length modulation was shown to be useful in determining the electrical characteristics of MOSFET in saturation, the above electrostatic model did not clarify the conduction mechanism in the drain region. An exact analysis requires a numerical solution of coupled Poisson equation and current continuity equation written in two-dimensional form. Attempts to solve this problem were reported by Armstrong *et al.* [26]—[29], Vandorpe *et al.* [30]—[32], Kennedy [33], etc. The two-dimensional analysis is essential for the drain region but less important for the source, except for short channel devices made in low-resistivity substrates, when the punch-through may occur (see below). The boundary conditions are usually chosen from preliminary computations based upon simplifying assumptions. For example, the carrier flow was confined to an infinitely narrow strip

---

* Schroeder and Muller [24] have explicitly indicated that the widening of the drain-junction space-charge layer (equation (11.5.6), one-dimensional model, reference [22]) cannot correctly predict the drain resistance. Both the capacitive drain-gate coupling and the two-dimensional nature of the junction-channel interaction are important. The electrostatic action of the drain field upon the charge carriers in the channel (in low-conductivity substrates) results in a modulation of the drain current by the drain voltage, as discussed by Hofstein and Warfield [25].

near the surface [26] or the depletion approximation was used for the boundary between the space-charge region and the neutral substrate [28].

We consider below some results reported by Armstrong and Magowan [29]. They used the model shown in Figure 11.5.5. The two-dimensional Poisson (mobile

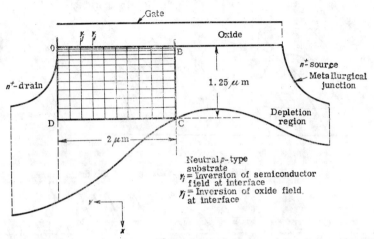

Figure 11.5.5. Model used by Armstrong and Magowan [29] for two-dimensional calculations.

electron density included) and continuity equations are solved in the domain OBCD. It was shown that a two-dimensional solution of the current flow equation is necessary only in the region between y_i and the drain junction, where y_i denotes the point where the transversal component of the electric field at the insulator-semiconductor interface, on the substrate side, changes the sign. From this point toward the drain the electric field forces the charge carriers to move away from the surface. The potential along the ODCB contour is obtained from the preliminary solution of the two-dimensional Poisson equation for the entire device, by neglecting the mobile charge. The mobile charge density at the semiconductor surface given by an approximate analytical solution. Other simplifying assumptions concerning the boundary conditions are made.

Figure 11.5.6 shows the mobile charge distribution in the drain region. Note the large mobile carrier density (also predicted by a semi-analytical model suggested by Popa [34]). Comparison of Figures 11.5.6 a and b shows the pronounced effect of the gate voltage upon the carrier flow magnitude and position. On the other hand the decrease of the carrier mobility has an insignificant effect (compare a and c), although the maximum electric field is sufficiently high to provide velocity saturation. The displacement of the conductive channel away from the surface is due to the progressive transverse carrier diffusion away from the oxide-semiconductor interface followed by space-charge limited drift through the bulk into the drain region [35]. In high-resistivity short-channel devices the conduction mechanism is similar to that described above for the drain region: the device is punched-through by the drain depletion region such that a space-charge-limited current flows through the substrate (see the next section).

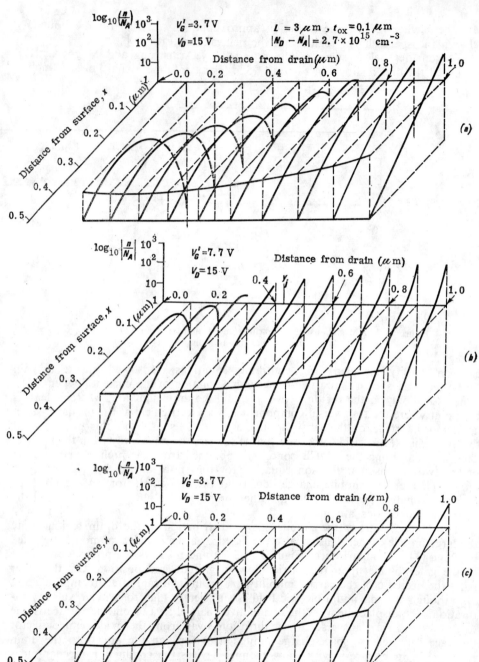

Figure 11.5.6. Mobile charge distribution in the drain (pinch-off) region of an n-channel MOST (Figure 11.5.5) with the following parameters $L = 3\,\mu\text{m}$, $t_{ox} = 0.1\,\mu\text{m}$, $|N_D - N_A| = 2.7 \times 10^{15}\text{cm}^{-3}$. (a) $V'_G = 3.7$ V, $V_D = 15$ V, constant-mobility analysis; (b) $V'_G = 7.7$ V, $V_D = 15$ V, constant mobility; (c) $V'_G = 3.7$ V, $V_D = 15$ V, field-dependent mobility analysis. (*after Armstrong and Magowan,* [29]).

11.6 Drain Voltage Limitations

Two specific phenomena which limit the maximum drain voltage of an MOST will be discussed: the punch-through mechanism and the avalanche (impact ionization) breakdown.

Figure 11.6.1 illustrates the effect of punch-through upon the drain characteristics of a p-channel MOST [36]. Note the large output conductance and the existence of a large current at large drain voltages (the 'triode' region) when the gate voltage has values in the cut-off region. In high-resistivity short-channel devices the drain depletion layer spreads out with increasing V_D and reaches the source thus changing the sign of the electric field along the edge of the source region and determining an injection of minority carriers in the semiconductor bulk (depleted). The substrate conduction between source and drain is by the space-charge-limited (SCL) flow mechanism (like in an n^+pn^+ structure, see Chapter 5). This conduction path appears in parallel to the normal channel formed by the inversion layer at the semiconductor-oxide interface. The triode-type characteristics are specific to the SCL current mechanism and clearly occur at large drain voltages when the gate voltage is below threshold and the surface conduction is non-existent. Above threshold the two mechanism operate in parallel and the normal pentode-type characteristics are distorted. The transition from pentode to triode-like characteristics was predicted by Neumark and Rittner [37] for thin-film insulated gate

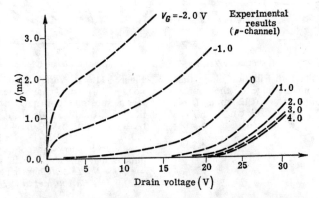

Figure 11.6.1. Effect of punch-through on output characteristics of a p-channel MOST. Reduction of gate voltage well below threshold does not eliminate punch-through current for large drain bias ($|V_D| > 20$ V). $L = 4.5$ μm, $N_D = 5.4 \times 10^{14}$ cm^{-3}, $t_{ox} = 0.14$ μm, diffusion depth for the source region, $d_S = 2.5$ μm. (after Bateman et al., [36]).

FET's when the channel length insulator thickness ratio decreases, and it was observed experimentally by Frohman-Bentchkowsky and Grove [23] for lowly doped short-channel MOST FET's. In both cases the triode-type characteristics are related to the fact that the mobile charge carrying the current is a sensitive function of the drain voltage*.

Bateman *et al.* calculated the punch-through voltage, V_{PT}, as the drain voltage which reverses the sign of the electric field at the source edge. They had done a purely electrostatic analysis based on a two-dimensional solution of the Poisson equation. The effective gate voltage, V_G', is set equal to zero. Figure 11.6.2 illustrates the onset of punch-through at the surface end of the source edge: as V_D is increased above V_{PT} this injection region spreads further down into the bulk. The dependence of V_{PT} upon the channel length, L, at various diffusion depths normalized to L and substrate doping concentrations is shown in Figure 11.6.3.

The punch-through mechanism is influenced by the gate bias. When the surface is accumulated, the punch-through should first occur at V_D higher than V_{PT}, and not at the surface but deep into the bulk. When the surface is strongly accumulated, the gate voltage can no longer control the punch-through current (see the

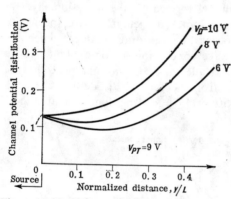
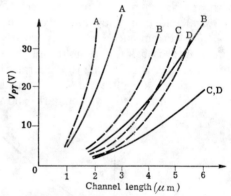

Figure 11.6.2. Voltage distribution along the channel region from the edge of the source diffusion. ($L = 3\mu m$, $t_{ox} = 0.12 \mu m$, $d_S/L = 0.2$, $|N_D - N_A| = 7 \times 10^{14} \text{cm}^{-3}$). (*after Bateman et al.*, [36]).

Figure 11.6.3. Variation of punch-through voltage V_{PT} with channel length L for different substrate doping concentrations and normalized diffusion depths d_s/L, respectively. (A) $6 \times 10^{15} \text{cm}^{-3}$; 0.2: (B) $1.4 \times 10^{15} \text{cm}^{-3}$, 0.35; (C) $7 \times 10^{14} \text{cm}^{-3}$, 0.2; (D): $7 \times 10^{14} \text{cm}^{-3}$, 0.35). Solid curves show for comparison the 'one-dimensional' punch-through voltage $V_{PT} = e|N_D - N_A|L^2/2\varepsilon$ (Chapter 5). (*adapted from Bateman et al.*, [36]).

right-hand part characteristics in Figure 11.6.1). However, a reverse bias applied to the substrate with respect to the source will tend to reduce the SCL flow in the depletion region, as demonstrated by Richman [39]. This property is important in

---

* A related but different mechanism occurs in the SCL insulated-gate surface-channel transistor, suggested and experimented by Wright [38]. This is an inversion-channel MOS transistor constructed in high-resistivity *n*-type silicon substrate. All lines of flux from the gate electrode terminate on mobile charge in the surface conducting channel. The characteristics are of the pentode type. The lowering of the substrate doping increases the device transconductance and reduces the parasitic capacitances (Section 11.8) but the output resistance in saturation also decreases (Section 11.5).

designing complementary MOS circuits in high resistivity substrates [40] where the parasitic SCL flow should be suppressed in order to achieve an effective cut-off [39]. When the surface layer is strongly inverted the bulk is screened from the gate, the SCL current is independent of the gate potential and flows in parallel with the normal surface-channel current.

Another basic process limiting the operating drain voltage is the avalanche breakdown, occurring in the vicinity of the drain-substrate junction. Analysis of the surface effects on $p-n$ junctions indicated that for bias conditions corresponding to saturation in an MOST, the avalanche breakdown occurs along the semiconductor-oxide interface, at the edge of the drain diffused region, and the breakdown junction voltage is well below the 'bulk' value corresponding to a diffused junction (Section 10.13). A similar type of surface breakdown occurs in MOS transistors biased below threshold. The 'sharp' breakdown at negative effective gate voltages is illustrated by Figure 11.6.4 a. Note the fact that breakdown voltage $V_{D(B)}$ decreases with decreasing gate voltage, which is consistent with the idea of surface breakdown (determined by the drain-gate potential difference). However, when the device is biased above threshold, the experimental characteristics exhibit a 'soft' breakdown, and the breakdown voltage decreases with the increasing gate voltage (Figure 11.6.4 a and b). The breakdown can be clearly distinguished from punch-through by the presence of the substrate current (drain and substrate currents will

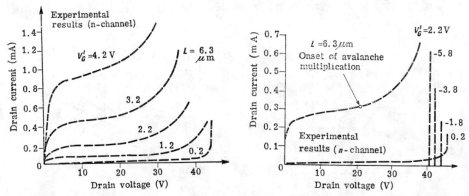

Figure 11.6.4. Output characteristics of an n-channel ($L = 6.3$ μm) silicon MOSFET indicating the effect of avalanche multiplication. (*after Bateman et al.*, [36]).

be measured separately). The increase in substrate current can only occur as a result of the electron-hole multiplication process which produce majority carriers in the drain depletion region and feed the substrate current [41], [42]*.

The modification of the breakdown characteristic when the gate voltage exceeds the threshold value is determined by the large current flowing inside the device. It was shown in Section 11.5 that the conductive channel departs from the surface in the vicinity of the drain (when the device is operated in saturation). It was suggest-

---

* An approximate theory was developed for the impact ionization current flowing in the substrate: its dependence upon V_D and V_G was predicted and verified experimentally.

ed [36] that the breakdown is now occurring in the bulk, in the high field region along the drain metallurgical junction where the charge density is maximum. Figure 11.6.5 indicates the measured $V_{D(B)} - V_G$ dependence for an n-channel device with relatively deep drain diffusion. The breakdown voltage for negative V'_G (below threshold) can be predicted by an electrostatic analysis, using the critical field concept. It was shown that the breakdown voltage increases with the channel length, with the oxide thickness, with the substrate doping and with the drain diffusion depth/channel length ratio. Figure 11.6.6 is representative for $V_{D(B)}$ dependence on device dimensions. At lower channel lengths the limitation of the drain voltage occurs due to punch-through. Punch-through is liable to occur for high resistivity, thick oxide, short channel devices with deep source and drain diffusion. The empirical factor of 1.7 multiplying the critical field in Figure 11.6.6 [36] is related to the distortion by the surface fields (Section 10.13) of the shape of the depletion region associated with the drain junction and by the presence of the conductive channel (Section 11.5).

A more accurate analysis of the high-field drain region was previously reported by Armstrong and Magowan [29]. Figure 11.6.7 shows the potential distribution indicating the depletion of the n^+ drain region near the oxide (Section 10.13), the point where the mobile electron density is maximum (see also Figure 11.5.6) and the point of maximum electric field. It can be seen that the higher the effective substrate doping, the lower the breakdown voltage. Low breakdown voltages are also likely to occur for devices with thin oxides. The computed drain characteristics

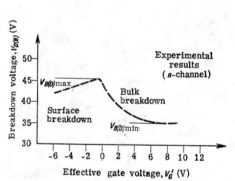

Figure 11.6.5. Measured breakdown voltage $V_{D(B)}$, as a function of the effective gate voltage V'_G, for an n-channel silicon MOST ($L = 6.3 \mu$m, $d_S = 2.5 \mu$m). Change-over point from surface breakdown to bulk breakdown occurs at $V'_G = 0$. (*after Bateman et al.*, [36]).

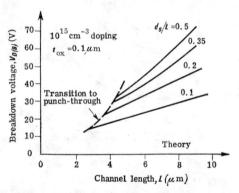

Figure 11.6.6. Theoretical values of breakdown voltage *versus* channel length for different normalized diffusion depths, d_S/L. The critical field is that used by Grove [4] multiplied with an empirical factor of 1.7. (*after Bateman et al.*, [36]).

indicate, however, a 'soft' breakdown characteristic and a breakdown voltage decreasing with increasing gate voltage, in agreement with the experimental data (Figure 11.6.4b). The injected mobile charge moves the position of the maximum field from the surface into the bulk (Figure 11.6.7) and reduces the breakdown voltage. The critical field concept for avalanche breakdown is questionable when the mobile carrier density is high. We also note that the *oxide* breakdown leading to device deterioration may be enhanced by the bulk electron-hole multiplication process (majority holes are directed toward the SiO_2—Si interface, Figure 11.6.7).

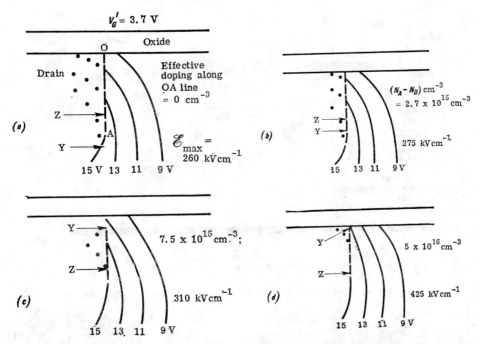

Figure 11.6.7. Potential distribution in the 'pinch-off' (drain) region of an *n*-channel MOST (two dimensional computer calculations). The effective substrate doping, the maximum field intensity and its location (point Y), as well as the position of maximum mobile charge density (Z) are indicated. (*after Armstrong and Magowan,* [29]).

11.7 Effect of Substrate Bias on Threshold Voltage

We have invariably assumed that the substrate contact is connected to the source. However, the substrate can be biased independently and acts as a second control electrode. In integrated circuits, the potential difference between the source and the substrate allows the control of MOST state (conducting or off).

The substrate action is explained with reference to Figure 11.7.1. When $V_{sub}=0$ the reverse bias of the channel-substrate field-induced junction is V and the built-in potential is approximately $2\phi_F$. A negative V_{sub} (*n*-channel transistor) determines a reverse bias of both source-substrate and drain-substrate metallurgical junctions. The same substrate bias increases the depletion width of the field-induced junction. V_{sub} simply adds algebraically to the built-in potential, because it introduces the same bias along the entire channel. Therefore $2\phi_F$ will be replaced by $2\phi_F - V_{sub}$

(where V_{sub} is negative) in the formula yielding the bulk fixed charge. This is a generalization of the theory of Section 11.3 based on the gradual approximation, and the simplified model of the surface layer given in Section 10.3.

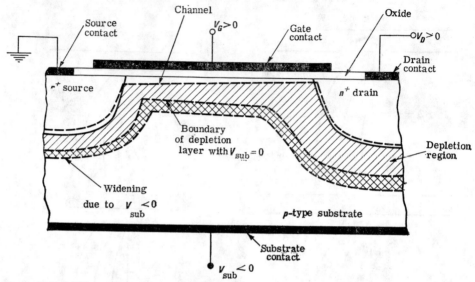

Figure 11.7.1. Effect of application of a substrate bias on the depleted region in a MOST structure.

In this Section we shall consider only the substrate-bias effect on the threshold voltage. Equation (11.3.4) should be modified as follows (n-channel device) [5], [43]

$$V_T = V_{FB} + 2\phi_F + \frac{1}{C_{ox}}[2eN_A\varepsilon(2\phi_F - V_{sub})]^{1/2} = V_T(V_{sub}). \quad (11.7.1)$$

Note the fact that the threshold voltage increases with increasing $|V_{sub}|$. In this way, the threshold voltage of an n-channel transistor may be easily increased above zero and this device will operate in the enhancement mode (normally-off). Figure 11.7.2 shows the computed V_T for an n-channel Al–SiO$_2$–Si transistor [43]. The effect of substrate bias is considerably higher for the thick-oxide device. In fact, the shift in V_T is inversely proportional to C_{ox}

$$\Delta V_T = K_1[(2\phi_F + |V_{sub}|)^{1/2} - (2\phi_F)^{1/2}], \quad K_1 = \frac{(2e\varepsilon N_A)^{1/2}}{C_{ox}} \quad (11.7.2)$$

For a p-channel MOSFET ΔV_T is negative:

$$\Delta V_T = -K_1[(2|\phi_F| + V_{sub})^{1/2} - (2|\phi_F|)^{1/2}], \quad K_1 = \frac{(2e\varepsilon N_D)^{1/2}}{C_{ox}} \quad (11.7.3)$$

and plotted in Figure 11.7.3.

11. MOS Transistor

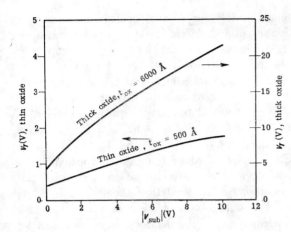

Figure 11.7.2. Theoretical threshold voltage, equation (11.7.1), *versus* substrate bias, plotted for 0.5×10^{11} cm^{-2} oxide plus interface charge, $N_A = 5.0 \times 10^{15}$ cm^{-3} and a $- 0.07$ V voltage shift in V_T due to boron depletion effect (Al—SiO$_2$—Si system, *n*-channel device). (*after Critchlow et al.*, [43]).

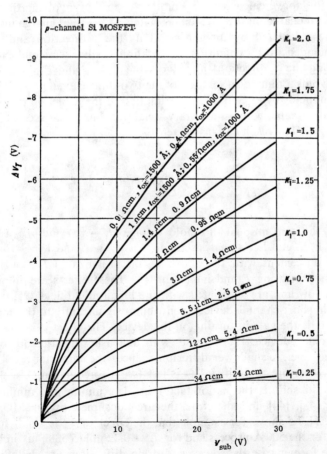

Figure 11.7.3. Threshold-voltage shift for a *p*-channel silicon MOST. (*after Crawford*, [5]).

Figure 11.7.4 shows that the measured V_T (n-channel devices) depends upon the channel length. Figure 11.7.5 also indicates that in short-channel transistors V_T depends upon the applied drain voltage. V_T is determined as an 'extrapolated' threshold voltage. It was previously shown (Section 11.4) that the device turn-on is gradual. Figure 11.7.6 exhibits experimental transfer characteristics for a 500 Å oxide n-channel device. The definition of a 'zero-current' threshold is imprecise. Therefore a parabola is fitted to the turn-on characteristic and extrapolated to 'zero' (see equation (11.2.4)). This can be conveniently done in a $I_D^{1/2}-V_G$ plot.

The dependence of V_T upon L (Figures 11.7.4 and 11.7.5) and V_D (Figure 11.7.5) cannot be explained by the classical analysis based on the gradual approximation. These effects occur for short-channel devices where the above approximation fails and are included by Lee [44] in an approximate analytical theory. He indicated that the usual definition of V_T involves charge conservation in every section, y, of the device (only the transversal component of the field, $\mathscr{E}_x$, is considered). Lee [44] replaces the above calculation by another definition involving the overall neutrality. The total charge on the gate metal plate and the total charge inside the semiconductor are calculated by integrating along the y direction and therefore V_D and L are included in V_T definition. Integration of the semiconductor charge in the bulk (along the x direction) introduces the source and drain junction depth, X_j. Lee [44] avoids a numerical solution of the two-dimensional potential distribution in the semiconductor by a regional approximation of the depletion regions (idealized in a rectangular shape, with an arbitrary weighting factor, ζ, accounting for their lateral extension). The result can be written

$$V_T = V_{T\infty} - \frac{\zeta}{C_{ox}L}[A_0 + A_1 X_j + A_2 X_j^2], \qquad (11.7.4)$$

where $V_{T\infty}$ is the conventional threshold voltage ($L \to \infty$) and A_0, A_1, A_2 depend upon N_A, V_D, V_{sub}, etc. Lee's analysis predicts a $V_T - L$ and $V_T - V_D$ dependence in qualitative agreement with the above results (Figures 11.7.4 and 11.7.5): V_T decreases with decreasing L and increasing V_D. These effects are important when L is relatively low and are associated with the longitudinal field, $\mathscr{E}_y$, which is relatively intense in short-channel devices, contributes significantly to the charge balance and therefore aids the gate in turning on the device. The effect is more pronounced for a thicker oxide device. V_T decreases with increasing X_j, the other parameters remaining the same. A shallower diffusion reduces the importance of lateral field extension and therefore the shift in V_T is less (see also equation (11.7.4)).

The threshold shift is the major factor in determining the minimum source-drain spacing [43], [44] and therefore the area occupied by an MOS integrated circuit. Figure 11.7.5 shows for example that the effect of V_D is small for $L > 0.15$ mil (3.31 μm) for the 500 Å process and for $L > 0.22$ mil (5.588 μm) in the 10,000 Å process (in the former case the source and drain diffusions are shallower).

11. MOS Transistor

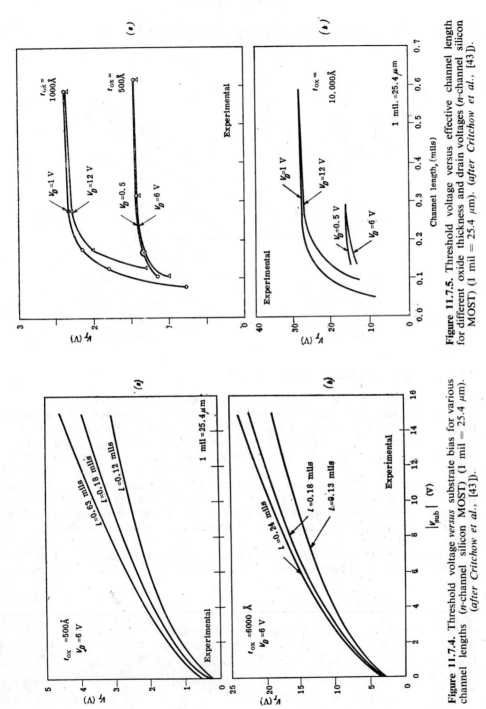

Figure 11.7.5. Threshold voltage versus effective channel length for different oxide thickness and drain voltages (n-channel silicon MOST) (1 mil = 25.4 μm). (*after Critchow et al.*, [43]).

Figure 11.7.4. Threshold voltage *versus* substrate bias for various channel lengths (n-channel silicon MOST) (1 mil = 25.4 μm). (*after Critchow et al.*, [43]).

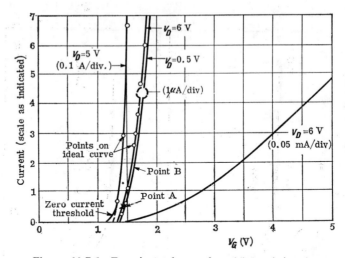

Figure 11.7.6. Experimental transfer characteristics for an n-channel device ($t_{ox} = 500$ Å). The current increases exponentially with V_G at very low values. The characteristic measured at higher current levels should be parabolic; a parabola is fitted to this characteristic and extrapolated to zero. The two points necessary to define the parabola are indicated by A and B. (after Critchow et al., [43]).

11.8 Dynamic Behaviour of a MOSFET

The dynamic parameters of a MOSFET are related to: the small-signal low-frequency equivalent circuit, the high-frequency small-signal characteristics, the large-signal equivalent circuit used for switching applications. The elements of the equivalent circuit are intrinsic and extrinsic. The intrinsic elements are inherent in the basic physical operation of the device and are associated with the gate-insulator-channel structure. For example, the intrinsic capacitances are associated with the charges stored on the gate electrode, in the channel and in the semiconductor bulk, just beneath the channel. The importance of extrinsic elements is strongly related to the particular construction: the transistor may be discrete or integrated, may be constructed in conventional technology (discussed up to now) or on an insulating substrate, etc.

We shall first consider the device (quasi-stationary) capacitances (compare to Section 9.6). Consider the device input capacitance (common source connection)

$$C_{gs} = \frac{\partial Q_{Gt}}{\partial V_G}\bigg|_{V_D = \text{const}}, \tag{11.8.1}$$

where Q_{Gt} is the total electric charge stored on the gate electrode (area LW, where the gate length was taken equal to the channel length, C_{gs} is an intrinsic capacitance). Let us assume that the oxide and interface charges are independent of V_G. Therefore

$$C_{gs} = \frac{\partial}{\partial V_G} W \int_0^L [-Q_s(y)]\,dy, \tag{11.8.2}$$

where $Q_s(y)$ is the semiconductor charge per unit of interface area at the distance y from source. From equations (10.11.1), (10.11.2) and (11.8.2) one obtains (within the gradual approximation)

$$C_{gs} = C_{gc}\left[1 - \frac{d\phi_{s0}}{dV_G} - \frac{1}{L}\frac{\partial}{\partial V_G}\int_0^L V(y)\,dy\right] \quad (11.8.3)$$

where $C_{gc} = WLC_{ox} = WL\varepsilon_{ox}/t_{ox}$ is the total oxide capacitance (or gate-to-channel capacitance).

Figure 11.8.1 shows the computed C_{gs}/C_{gc} dependence *versus* V_G for a given V_D [14]. Three different theoretical curves are indicated. All of them exhibit three regions from left to right: cut-off, saturation, and non-saturation (linear). In the central region, corresponding to saturation, C_{gs} is essentially independent of V_G. When $V_D = $ const. and V_G increases, the operating point moves into the linear region, whereas when V_G decreases it goes to cut-off. All curves merge with $C_{gs} = C_{gc}$ when V_G becomes very large compared to V_D. Curve 1 in Figure 11.8.1 corresponds to zero bulk charge and $\phi_{s0} = $ const, independent of V_G. It can be readily shown that in saturation $C_{gs} = (2/3)C_{gc}$, whereas in the linear region C_{gs} varies between C_{gc} and $(2/3)C_{gc}$. Equation (11.8.3) should be used, where $V(y)$ will be introduced as a function of V_G and V_D. Below threshold C_{gs} becomes infinite. This unphysical result (see also curve 2) is due to the neglect of $d\phi_{s0}/dV_G$: the device threshold is assumed abrupt (conduction occurs when ϕ_s equals $2\phi_F$). Curve 2 shows the effect of the bulk charge. Curve 3 is the exact one, obtained by using equation (11.8.3) where $\phi_{s0} = \phi_{s0}(V_G)$ and the fixed bulk charge are included in calculations. The cut-off region of curve 3 corresponds to the MOS capacitance

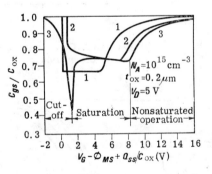

Figure 11.8.1. Dependence of the input capacitance (common-source connection) on the gate bias for an n-channel silicon MOST. Curves 1, 2, 3 are different theoretical results (see the text). (*after Sah and Pao,* [14]).

($I_D = 0$ and the semiconductor is at thermal equilibrium), namely to curve 1 in Figure 10.4.4, for surface accumulation and depletion (compare equation (11.8.3) where $V(y) \equiv 0$, with equation (10.4.28)).

In the linear region of the characteristics ($V_G \gg V_D$), $V(y)$ increases approximately linear with y and is almost independent of V_G. Therefore $C_{gs} \simeq C_{gc}[1 - d\phi_{s0}/dV_G]$ and, accordingly to equation (10.4.28), C_{gs} varies approximately as the low-frequency MOS capacitance when the surface is 'inverted'. In this case the structure (with the substrate connected to the source and with the drain at a low and *fixed* potential) behaves indeed as an MOS capacitor, except the small field distortion due to the potential drop along the channel. The source provides in a very short time the minority carriers necessary for the inversion layer formation. Therefore, even at high frequencies, the right-hand part of $C_{gs} = C_{gs}(V_G)$ curve is almost identical to the low-frequency $C-V$ characteristics of the gate-to-channel MOS capacitor biased in inversion. However, this part of the $C-V$ characteristic is displaced along the voltage axis and somewhat distorted by the application of $V_D \neq 0$. This is clearly illustrated by the theoretical curves (dots) reproduced in Figure 11.8.2. The theory is well fitted by experiment: note for example that the small dip in the $C(V_G)$ curve in the saturation region is indeed present in the experimental curve.

For small-signal circuit applications, the input capacitance of the device biased in saturation is a very important parameter. To a first approximation this capaci-

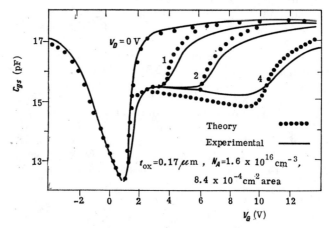

Figure 11.8.2. Gate input capacitance *versus* effective gate voltage for a few drain voltages: theory and experiment for an *n*-channel MOST. (*after Pao and Sah* [15]).

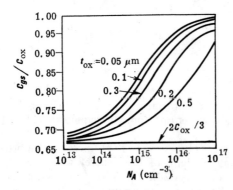

Figure 11.8.3. Gate input capacitance in saturation as a function of substrate doping for different oxide thickness (*n*-channel silicon MOST). (*after Sah and Pao*, [14]).

tance is almost independent of bias voltage and depends upon the substrate doping and oxide thickness as shown in Figure 11.8.3 [14].

The input gate capacitance C_G can be split into two capacitances C_{gs} and C_{gd} defined as shown below. The total gate charge can be written $Q_{Gt} = Q_{Gt}(V_{gs}, C_{ds})$ where V_{gs}, V_{gd} are the potential differences between gate and source, and gate and drain, respectively. This provides a symmetry with the gate as the reference point. However, the device cannot behave in a symmetrical way because of the fourth electrode: the substrate. However, when substrate effects are ignored (Section 11.2)

$$Q_{Gt} = \frac{2}{3} C_{gc} \frac{\Delta V_{gd}^3 - \Delta V_{gs}^3}{\Delta V_{gd}^2 - \Delta V_{gs}^2}, \qquad (11.8.4)$$

where $\Delta V_{gd} = V_{gd} - V_T$ and $\Delta V_{gs} = V_{gs} - V_T$. Then

$$C_{gs} = \frac{\partial Q_{Gt}}{\partial V_{gs}}, \qquad C_{gd} = \frac{\partial Q_{Gt}}{\partial V_{gd}} \qquad (11.8.4')$$

are given by [45]

$$C = \frac{2}{3} C_{gc}[1 - (1 + \delta)^{-2}], \qquad (11.8.5)$$

where $\delta = \Delta V_{gs}/\Delta V_{gd}$ for $C = C_{gc}$ and $\delta = \Delta V_{gd}/\Delta V_{gs}$ for $C = C_{gd}$. The reader can easily verify that at a given V_G, C_{gs} increases and C_{gd} decreases with increasing V_D. At $V_D = 0$ both capacitances are equal to $C_{gc}/2$. At $V_D = V_{Dsat} = V_G - V_T$, C_{gs} reaches $(2/3) C_{gc}$ whereas C_{gd} vanishes, such that C_G decreases from C_{gc} to $(2/3) C_{gc}$ when V_D increases (see above).

A more complete evaluation of bias dependence of intrinsic capacitances was reported by Armstrong and Magowan [46]. They used a one-dimensional approach

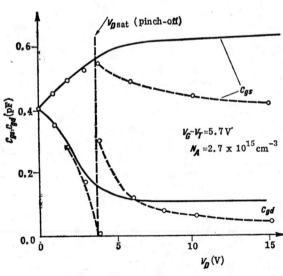

11.8.4. Variation of gate-source and gate-drain capacitance with drain bias (n-channel silicon device, $L = 3$ μm, $t_{ox} = 0.1$ μm, $C_{gc} = 0.816$ pF): dashed curves — one-dimensional analysis; solid curves — two dimensional analysis. (after Armstrong and Magowan, [46]).

based on the Frohman-Bentchkowsky and Grove model [23] (see Section 11.5) to find C_{gs} and C_{gd} in saturation (Figure 11.8.4). Note the discontinuity of C_{gd} at the onset of saturation. A two-dimensional computation removes this discontinuity and predicts a continuous increase in C_{gs}, instead of a peak at $V_D = V_{Dsat}$ (channel pinch-off near the drain). However, little error is made if C_{gs} is assumed saturated at the pinch-off value. When $V_D > 0$, $C_{gs} + C_{gd}$ decreases below C_{gc}^*.

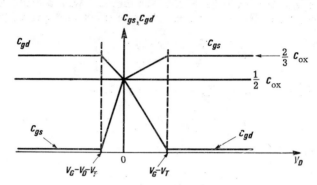

Figure 11.8.5. Approximate bias dependence of gate capacitances for an n-type MOSFET. (*after Meyer*, [45]).

The switching (transient) behaviour of the MOSFET is complicated by the fact that C_{gs} and C_{gd} are voltage-dependent (non linear). Sometimes a simple straight-line approximation is used, as shown in Figure 11.8.5 [45]. In many cases these capacitances are approximated by a constant average value, which should be an engineering approximation for the particular operation mode discussed [5], [45].

The high-frequency behaviour of an MOSFET was studied in a number of papers [47]–[52]. For a detailed discussion on the simple model of Section 11.2 the reader is referred to a recent monograph [53]. The analysis is similar to that performed for the JGFET (see also Section 9.6). The current continuity equation is written by including the displacement component, but only its transversal component (the longitudinal one is neglected, and this is a form of Shockley's gradual approximation). The solution of a non-linear differential equation with partial derivatives should be found. For the small-signal problem, the impedance can be expressed in terms of modified Bessel functions of complex argument. The result of theoretical calculations for a device biased in the saturation region is shown in Figure 11.8.6: the input admittance has a capacitive component, the forward transadmittance decreases with frequency, the output admittance and the feedback transadmittance are identically equal to zero (common-source connection). The characteristic frequency is $1/T_0$, where T_0 is a time constant related to the charging of the oxide capacitance through the channel resistance. For the saturation region, it happens to have T_0 almost equal to the carrier transit time, T_L, namely $T_0 = (3/4)T_L$. Figure 11.8.6 also indicates an experimental verification reported by Burns [48] for a silicon transistor grown on sapphire (parasitic elements of the equivalent circuit are negligible). An experimental device with a long time constant, T_0, was used, to facilitate the measurements.

A different approach was applied by Das [50] to the more complicated model including the substrate resistivity effects (Section 11.3). It was shown that the equations are analogous to those of a double RC transmission line having a uniformly distributed common resistance (the channel) but two separate capacitances distributed non-uniformly (the gate-to-channel and the substrate-to-channel capacitance). This transmission line was approximated by a few sections with lumped constants. An excellent agreement with experimental results (measured up to about 700 MHz) was also reported [50].

---

* A more general quasi-stationary analysis including the substrate as a fourth electrode was presented in reference [46].

Figure 11.8.7 illustrates the origin of a few extrinsic elements in the equivalent circuit. $C_{s\text{-sub}}$ and $C_{d\text{-sub}}$ are the barrier capacitances of the reverse biased source-substrate and drain-substrate junctions. They are voltage dependent according to $C = C_{j0}(1 + V_R/V_{bi})^n$ (see chapter 2), where V_R is the reverse bias of the junction and varies between $-1/2$ (abrupt step junction) and $-1/3$ (linearly graded junction). An engineering approximation for the usual operation range of MOS transistors yields $n = -1/2$ and $V_{bi} = 0.92$ V [6]. These are important stray (parasitic) capacitances.

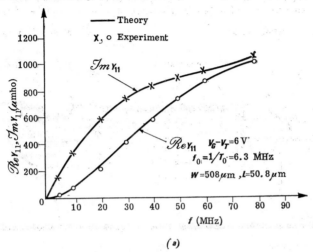

(a)

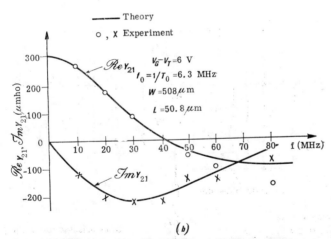

(b)

Figure 11.8.6. Theoretical (solid-lines) and experimental (circles and crosses) frequency characteristics of an ideal MOSFET (described by the theory of Section 11.2): (a) real and imaginary part of the input admittance, Y_{11}; (b) real and imaginary part of the forward transadmittance, Y_{21}. (*after Burns*, [49]).

They can be reduced by increasing the substrate resistivity but in this way the series resistances are enhanced.

C'_{gs} and C'_{gd} occur due to the overlap of the source and drain p^+ regions by the thin gate oxide (Figure 11.8.7). This overlap occurs, first, due to the lateral diffusion of the p-regions *under* the gate oxide, and, secondly, because the gate mask is intentionally overlapped onto the source and drain to provide the channel over the entire source-drain distance (mask misalignment is possible in the conventional process, auto-alignment of the gate is provided by a modern technology) [6]. These capacitances act as feedback elements and deteriorate the noise and amplifying performances.

The metallization pads for source, gate and drain combined with the oxide and the conductive substrate give parasitic MOS capacitors: C'_s, C'_g, C'_d, in series with resistances r'_s, r'_d, r'_g (Figure 11.8.7). The capacitance per unit of surface area is about an order of magnitude lower than the gate

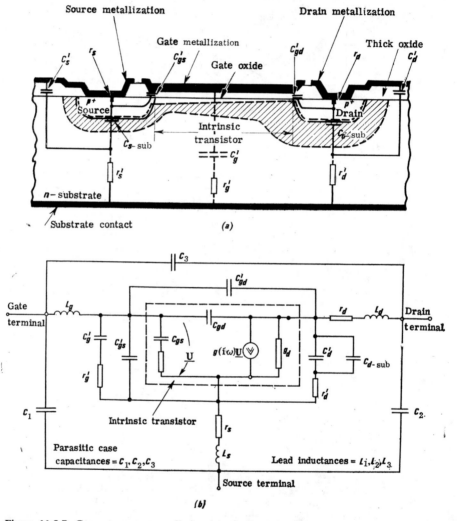

Figure 11.8.7. Common-source small-signal equivalent circuit of a MOSFET including parasitic elements.

oxide capacitance because of the thicker oxide isolating the interconnections from substrate. Series source and drain series resistances (usually small) are associated with the finite (sheet) resistance of p^+ regions and contact pads.

The total equivalent circuit of the semiconductor chip is similar to that of the JGFET (Section 9.6). Stray series inductances of the bending wires and parasitic case capacitances should complete this circuit. The parasitic elements in an integrated MOS circuit are discussed in reference [6].

Whereas the maximum operating frequency of the 'intrinsic' transistor can reach the gigahertz region, the operation speed of discrete and integrated MOST is considerably reduced by the parasitic elements. A definite improvement can be achieved by using a silicon film hetero-epitaxially grown on an insulating substrate.

11.9 Noise Sources in MOSFET's

The source of noise of the intrinsic MOS transistor are: the thermal noise of the channel, the $1/f$ type noise related to interface effects and the generation-recombination noise corresponding to centres in the depletion regions.

The thermal noise of the non-linear channel resistance may be written [2], [54], [55]

$$\overline{i_d^2} = 4kTg_{d0}\, \Delta f\, \frac{2}{3}\left[1 - \left(1 - \frac{I_D}{I_{D\text{sat}}}\right)^{3/2}\right]\frac{I_{D\text{sat}}}{I_D} \qquad (11.9.1)$$

where g_{d0} and $I_{D\text{sat}}$ are the low-current drain conductance and the saturation current, respectively, calculated for a given gate voltage. The simplified model of Section 11.2 was used. The drain noise decreases slowly with increasing current and remains constant in saturation.

The high-frequency thermal noise was computed by Shoji [55] by using a transmission line model of the channel (the steady-state characteristics are those given in Section 11.2, the alternating current behaviour corresponds to Figure 11.8.6). The induced gate noise $\overline{i_g^2}$ and the correlation noise $\overline{i_g^* i_d}$ (see also Section 9.8) were also derived. The analytical procedure is sketched in a recent book [53]. The noise is a complicated function of bias and frequency. The low-frequency components (series expansion for $\omega \to 0$) for the device biased in saturation are

$$\overline{i_d^2} = \frac{8}{3}kTg_{d0}, \quad \overline{i_g^2} = \frac{64}{135}\omega^2 kT \frac{C_{gc}^2}{g_{d0}}, \quad \overline{i_g^* i_d} = -\frac{4}{9}i\omega kTC_{gc}. \qquad (11.9.2)$$

The above results were generalized taking into account the finite substrate resistivity: the effect of substrate doping was found to be relatively small [56].

The low-frequency behaviour is dominated by the $1/f$ noise component associated with the insulator-semiconductor interface. It is clearly established that the wide $1/f$ spectrum extended down to very low frequencies originates from capture and emission of bulk minority carriers. A number of $1/f$ noise theories were suggested [57], [58]. For example Berz [59] put forward a theory based on the fluctuations in the occupancy of interface traps. He considered in particular the case when the traps are located in the oxide, near the interface, and the charge exchange between traps and semiconductor takes place by tunnelling (McWhorter's model, [60]). He found a $1/f$ spectrum when the oxide traps are distributed uniformly and a $1/f^\alpha$ spectrum ($1 < \alpha < 2$) where the trap density increases very rapidly with the distance from the interface. Similar models were developed by Christensson *et al.* [61], Hsu [62], Haslett and Trofimenkoff [58], etc. Experimental results were also reported [57], [62], [63]. A refined physical model was suggested by Fu and Sah

[57]: because a direct tunnelling from semiconductor conduction and valence bands to oxide traps is energetically improbable, an intermediate state is proposed. The mobile carriers are first trapped in interface states through a Shockley-Read-Hall process (see also Section 10.10) and then tunnel into or out of the oxide traps. This is illustrated by the energy band diagram of Figure 11.9.1. Consider a surface

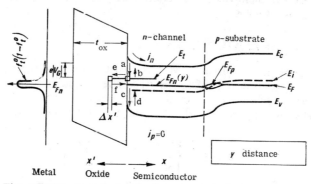

Figure 11.9.1. Energy band diagram and charge changing processes in the epitaxial n-channel MOS transistor (see the text). (*after Fu and Sah*, [57]).

state with energy level E_t which acts as a Shockley-Read-Hall recombination centre. It can capture an electron from the conduction band at the same position ($x = 0, y$) in the channel (process a). The electrons captured can be thermally excited to the conduction band (process b) or recombine with a hole captured (process c). A hole can be thermally excited from a surface state (process d). When an electron is trapped (or released) an electron will be injected from the n^+ source contact (or extracted) to maintain the same number of electrons: therefore, these processes will determine a net charge fluctuation at that point of the channel. This is not true for hole transitions because there is no hole current.

An electron captured by the interface states will tunnel into an oxide trap at the same energy level (process e) located at a certain distance from the interface and induce a charge $e(t_{ox} - x')/t_{ox}$ at ($x = 0, y$). The reverse process (f) will induce a charge $-e(t_{ox} - x')/t_{ox}$. The effect of all these fluctuations upon the drain current is obtained by integration along the channel length (y direction) and into the oxide, perpendicular to the interface (x' direction). The interface traps are distributed in energy in the entire band gap (Section 10.9). However, from occupancy probability considerations, only those traps situated at $E_{Fn}(y)$ will contribute to the noise (E_{Fn} is the quasi-Fermi level for electrons). When V_D is small the y dependence is negligible and only a narrow set of interface levels are involved in noise phenomena. The short-circuit noise current is

$$\overline{i_d^2} = 4\Delta f \frac{\mu_n^2 e^2 W}{L^3} V_D^2 \left(\frac{C_d}{C_{ox} + C_d} \right)^2 \times$$
$$\times \int_0^{t_{ox}} \left(\frac{t_{ox} - x'}{t_{ox}} \right)^2 \frac{N_T f_t^0 (1 - f_t^0) \tau_T}{1 + \omega^2 \tau_T^2} \, dx', \qquad (11.9.3)$$

where C_d is the depletion layer capacitance, f_t^0 is the equilibrium Fermi-Dirac occupation factor of the traps, N_T is the concentration of oxide traps and τ_T a time constant related to the interface state occupancy and tunnelling probabilities into the oxide traps. Figure 11.9.2 shows computed and experimental data. The parameters of

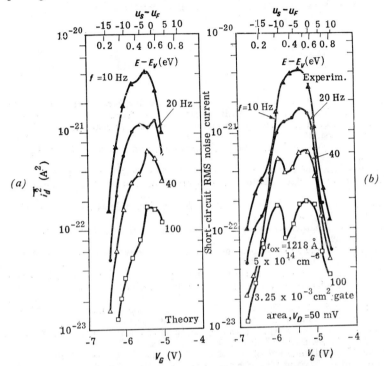

Figure 11.9.2. (*a*) Theoretical and (*b*) experimental noise power *versus* gate voltage, for different frequencies, illustrating the effect of oxide traps on the low-frequency noise. (*after Fu and Sah*, [57]).

oxide traps and interface states are obtained from $C-V$ measurements on the same sample. Other experimental data show a $1/f^\alpha$ dependence of the noise power, where α changes from 1 to about 2 as the surface changes from accumulation to strong inversion. It was proved that the noise power is proportional to the trap density. The noise spectra also depend upon the spatial distribution of the oxide traps. A $1/f^\alpha$ spectre with $\alpha > 1$ is obtained as N_T increases with increasing x' (as predicted by Berz [59]). It was suggested that the oxide traps arise from excess oxygen at the SiO_2-Si interface [57].

The generation-recombination (GR) noise is due to random emission of electrons and holes at the defect centres in the depletion region of the semiconductor. This component is usually negligibly low compared to the $1/f$ noise. Yau and Sah [64] verified their model of GR noise by introducing recombination centres (gold) in silicon MOST's used for experiments. The charge fluctuations in the impurity centres in the depletion region induce mobile charge fluctuation in the channel and also a charge fluctuation on the gate [64]. The frequency dependence of the equivalent gate noise resistance exhibits a plateau at low frequencies and a fall-off as $1/f^2$ at higher frequencies. On the other hand, the GR noise increases with V_D, whereas the thermal noise decreases with V_D and the $1/f$ noise is almost independent of V_D. An experimental verification was also reported [64].

References

1. S. M. Sze, 'Physics of semiconductor devices', John Wiley, New York, 1969.
2. R. Paul, 'Feldeffect-transistoren', VEB Verlag Technik, Berlin, 1972.
3. D. Dascalu, 'Injecția unipolară în dispozitive electronice semiconductoare', Ed. Academiei, Bucharest, 1972.
4. A. S. Grove, 'Physics and technology of semiconductor devices', John Wiley, New York, 1967.
5. R. H. Crawford, 'MOSFET in circuit design', McGraw-Hill, New York, 1967.
6. W. M. Penney and L. Lau (Editors), 'MOS integrated circuits', Van Nostrand, New York, 1972.
7. G. F. Neumark, 'Extension of the theory of thin-film transistors', *Solid-St. Electron.*, 7, 725−732 (1964).
8. J. S. T. Huang, 'Characteristics of a depletion-type IGFET', *IEEE Trans. Electron Dev.*, ED−20, 513−514 (1973).
9. J. E. Johnson, 'Physical processes in insulated-gate field effect transistors', *Solid-St. Electron.*, 7, 861−871 (1964).
10. G. T. Wright, 'Current/voltage characteristics, channel pinch-off and field dependence of carrier velocity in silicon insulated-gate field-effect transistors', *Electron. Lett.*, 6, 107−109 (1970).
11. A. Bar-Lev and S. Margalit, 'Changes of mobility along a depletion type MOS transistor channel', *Solid-St. Electron.*, 13, 1541−1546 (1970).
12. B. Hoeneisen and C. A. Mead, 'Current-voltage characteristics of small-size MOS transistors', *IEEE Trans. Electron. Dev.*, ED−19, 382−383 (1972).
13. M. B. Das, 'Physical limitations of MOS structures', *Solid-St. Electron.*, 12, 305−336 (1969).
14. C. T. Sah and H. C. Pao, 'The effects of fixed bulk charge on the characteristics of metal-oxide-semiconductor transistors', *IEEE Trans. Electron Dev.*, ED−13, 393−409 (1966).
15. H. C. Pao and C. T. Sah, 'Effects of diffusion current on characteristic of metal-oxide (insulator)-semiconductor transistors', *Solid-St. Electron.*, 9, 927−937 (1966).
16. M. B. Barron, 'Low level currents in insulated gate field effect transistors', *Solid-St. Electron.*, 15, 393−402 (1972).
17. R. J. Van Overstraeten, G. Declerck and G. L. Broux, 'Inadequacy of the classical theory of the MOS transistor operating in weak inversion', *IEEE Trans. Electron Dev.*, ED−20, 1150−1153 (1973).
18. R. J. Van Overstraeten, G. Declerck and G. L. Broux, 'The influence of surface potential fluctuations on the operation of the MOS transistor in weak inversion', *IEEE Trans. Electron Dev.*, ED−20, 1154−1158 (1973).
19. W. Milton Gosney, 'Subthreshold drain leakage currents in MOS field-effect transistors', *IEEE Trans. Electron Dev.*, ED−19, 213−219 (1972).
20. J. A. Geurst, 'Theory of insulated-gate field-effect transistors near and beyond pinch-off', *Solid-St. Electron.*, 9, 129−142 (1966).
21. R. M. Swanson and J. D. Meindl, 'Ion-implanted complementary MOS transistors in low-voltage circuits', *IEEE J. Solid-St. Circ.*, SC−7, 146−153 (1972).
22. V. G. K. Reddi and C. T. Sah, 'Source to drain resistance beyond pinch-off in MOS transistors', *IEEE Trans. Electron. Dev.*, ED−12, 139−141 (1965).
23. D. Frohman-Bentchkowsky and A. S. Grove, 'Conductance of MOS transistors in saturation', *IEEE Trans. Electron Dev.*, ED−16, 108−113 (1969).

24. J. E. SCHROEDER and R. S. MULLER, 'IGFET analysis through numerical solution of Poisson's equation', *IEEE Trans. Electron Dev.*, **ED—15**, 954—961 (1968).
25. S. R. HOFSTEIN and G. WARFIELD, 'Carrier mobility and current saturation in the MOS transistor', *IEEE Trans. Electron Dev.*, **ED—12**, 129—138 (1965).
26. G. A. ARMSTRONG, J. A. MAGOWAN and W. D. RYAN, 'Two-dimensional solution of the d.c. characteristics for the MOST', *Electron. Lett.*, **5**, 406—408 (1969).
27. G. A. ARMSTRONG and J. A. MAGOWAN, 'Modelling of short-channel MOS transistors', *Electron. Lett.*, **6**, 313—315 (1970).
28. J. A. MAGOWAN, W. D. RYAN and A. ARMSTRONG, 'Determination of Laplace-Poisson domain interface within semiconductor devices', *Proc. I.E.E.*, **117**, 921—926 (1970).
29. G. A. ARMSTRONG and J. A. MAGOWAN, 'The distribution of mobile carriers in the pinch-off region of an insulated-gate field-effect, transistor and its influence on device breakdown', *Solid-St. Electron.*, **14**, 723—733 (1971).
30. D. VANDORPE and N. H. XUONG, 'Mathematical 2-dimensional model of semiconductor devices', *Electron. Lett.*, **7**, 47—50 (1971).
31. D. VANDORPE, 'Etude bidimensionnelle du transistor MOS', *L'Onde Electr.*, **51**, 837—843 (1971).
32. D. VANDORPE, J. BOREL, G. MERCKEL and P. SAINTOT, 'An accurate two-dimensional numerical analysis of the MOS transistor', *Solid-St. Electron.*, **15**, 547—557 (1972).
33. D. P. KENNEDY, 'Steady-state mathematical theory for insulated-gate field-effect transistor', *IBM Journ. Res. Develop.*, **17**, 2 (1973).
34. A. POPA, 'An injection level dependent theory of the MOS transistor in saturation', *IEEE Trans. Electron Dev.*, **ED—19**, 774—781 (1972).
35. G. A. ARMSTRONG and J. A. MAGOWAN, 'Pinch-off in insulated-gate field effect transistors', *Solid-St. Electron.*, **14**, 760—763 (1971).
36. I. M. BATEMAN, G. A. ARMSTRONG and J. A. MAGOWAN, 'Drain voltage limitations of MOS transistors', *Solid-St. Electron.*, **17**, 539—550 (1974).
37. G. F. NEUMARK and E. S. RITTNER, 'Transition from pentode- to triode-like characteristics in field-effect transistors', *Solid-St. Electron.*, **10**, 299—304 (1967).
38. G. T. WRIGHT, 'Space-charge-limited insulated-gate surface-channel transistor', *Electron. Lett.*, **4**, 462—464 (1968).
39. P. RICHMAN, 'Modulation of space-charge-limited current flow in insulated-gate field-effect tetrodes', *IEEE Trans. Electron Dev.*, **ED—16**, 759—766 (1969).
40. P. RICHMAN, 'Complementary MOS field-effect transistors on high-resistivity silicon substrates', *Solid-St. Electron.*, **12**, 377—383 (1969).
41. H. MARTINOT and P. ROSSEL, 'Multiplication de porteurs dans la zone de pincement des transistors MOS', *Electron Lett.*, **7**, 118—120 (1971).
42. W. W. LATTIN and J. L. RUTLEDGE, 'Impact ionization current in MOS devices', *Solid-St. Electron.*, **16**, 1043—1046 (1973).
43. D. L. CRITCHLOW, R. H. DENNARD and S. E. SCHUSTER, 'Design and characteristics of *n*-channel insulated-gate field-effect transistors', *IBM Journ. Res. Develop.*, **17**, 430—443 (1973).
44. H. S. LEE, 'An analysis of the threshold voltage for short-channel IGFET's, *Solid-St. Electron.*, **16**, 1407—1417 (1973).
45. J. E. MEYER, 'MOS models and circuit simulation', *RCA Review*, **32**, 42—63 (1971).
46. G. A. ARMSTRONG and J. A. MAGOWAN, 'Derivation of simple expression for interelectrode capacitances of IGFET's as a function of bias condition', *Electron. Lett.*, **7**, 281—283 (1971).
47. A. van der ZIEL and J. W. ERO, 'Small-signal high-frequency of field-effect transistors', *IEEE Trans. Electron Dev.*, **ED—11**, 128—135 (1964).
48. J. R. BURNS, 'High-frequency characteristics of the insulated-gate field-effect transistor', *RCA Rev.*, **28**, 385—418 (1967).
49. J. R. BURNS, 'Large-signal transit-time effects in the MOS transistor', *RCA Rev.*, **30**, 15—35 (1969).
50. M. B. DAS, 'High frequency network properties of MOS transistors including the substrate resistivity effects', *IEEE Trans. Electron Dev.*, **ED—16**, 1049—1069 (1969).
51. E. M. CHERRY, 'Small-signal high-frequency response of the insulated-gate field-effect transistor', *IEEE Trans. Electron Dev.*, **ED—17**, 569—577 (1970).
52. P. ROSSEL, 'Methode de calcul des paramètres hautes fréquences des transistors MOS en régime de non pincement', *Electron. Lett.*, **8**, 614—616 (1972).

53. D. Dascalu, 'Transit-time effects in unipolar solid-state devices', Ed. Academiei, Bucharest, and Abacus Press, Kent, 1974.
54. C. T. Sah, S. Y. Yu and F. H. Hielscher, 'The effects of fixed bulk charge on the thermal noise in metal-oxide-semiconductor transistors', *IEEE Trans. Electron Dev.*, **ED−13**, 410−414 (1966).
55. M. Shoji, 'Analysis of high-frequency thermal noise of enhancement mode MOS field-effect transistors', *IEEE Trans. Electron Dev.*, **ED−13**, 520−524 (1966).
56. J. W. Haslett and F. N. Trofimenkoff, 'Gate noise in MOS FET's at moderately high frequencies', *Solid-St. Electron.*, **14**, 239−245 (1971).
57. H.-S. Fu and C.-T. Sah, 'Theory and experiments on surface $1/f$ noise', *IEEE Trans. Electron Dev.*, **ED−19**, 273−285 (1972).
58. J. W. Haslett and F. N. Trofimenkoff, 'Effects of the substrate on surface state noise in silicon MOS FET's', *Solid-St. Electron.*, **15**, 117−131 (1972).
59. F. Berz, 'Theory of low frequency noise in Si MOST's', *Solid-St. Electron.*, **13**, 631−647 (1970).
60. A. L. McWhorter, '$1/f$ noise and germanium surface properties', in R. H. Kingston (Editor) 'Semiconductor surface physics', Univ. of Pennsylvania Press, 1956; pp. 207−228.
61. S. Christensson, I. Lundström and C. Svensson, 'Low frequency noise in MOS transistors − I Theory', *Solid-St. Electron.*, **11**, 797−812 (1968).
62. S. T. Hsu, 'Surface state related $1/f$ noise in MOS transistors', *Solid-St. Electron.*, **13**, 1451−1459 (1970).
63. S. Christensson and I. Lundström, 'Low frequency noise in MOS transistors −II Experiment', *Solid-St Electron.*, **11**, 813−820 (1968).
64. L. D. Yau and C.-T. Sah, 'Theory and experiments of low-frequency generation-recombination noise in MOS transistors', *IEEE Trans. Electron Dev.*, **ED−19**, 170−177 (1969).

Problems

11.1. Derive the MOST figure of merit, f_{max} of equation (11.2.11), when the carrier velocity is saturated due to the intense longitudinal field in a short-channel device.

11.2. Derive the equation of the drain current with respect to V_S, V_G, V_D, source, gate and, respectively, drain potentials measured with respect to the substrate. Show that the source to drain current can be decomposed as follows:

$$I_D(V_S, V_G, V_D) = I_D(0, V_G, V_D) + I_D(V_S, V_G, 0) \qquad (P.11.2.1)$$

11.3. Show that the drain conductance can be written

$$g_d = -\frac{\mu_n \varepsilon_{ox} W}{t_{ox} L} Q_n(V_D) > 0, \qquad (P.11.3.1)$$

where $Q_n(V_D)$ is the mobile charge per unit of surface area at $y = L$.

11.4. Show that when the oxide thickness is negligibly small as compared with the depletion layer thickness at equilibrium $x_{d\,max,0}$ we have

$$V_{D\,sat} \simeq V_G - V_{FB} - 2\phi_F, \; t_{ox} \ll x_{d\,max,0} = \sqrt{\frac{2\varepsilon(2\Phi_F)}{eN_A}} \qquad (P.11.4.1)$$

11.5. Calculate the effect of temperature upon V_T and $g_{m\,sat}$.

11.6. Find the expression of the drain current when the surface mobility is field-dependent. Consider separately the effect of the longitudinal field and the transversal (surface) field.

11.7. Discuss the effect of interface states upon the behaviour of an $Al-SiO_2-Si$ MOS transistor.

11. MOS Transistor

11.8. The electrical breakdown ($8-10 \times 10^6$ V cm^{-1}) of the gate oxide occurs frequently due to accidental charging of the MOST's gate. Suggest a gate protection using a *pn* diode and discuss its practical realisation and usefulness.

11.9. Justify equation (11.4.3).

11.10. Show that the drain current equation (11.4.9) may also be written

$$I_D = -\frac{D_n W}{L} Q_n(0) \qquad \text{(P.11.10.1)}$$

when V_D is larger than a few kT/e. Q_n is the mobile electron charge at the source ($y=0$). Discussion.

11.11. Consider a *p*-channel MOS transistor operating as a tetrode: the substrate is the second controle electrode. Find the transconductances g_m and $g_{m\,\text{sub}}$ corresponding, respectively, to the gate and substrate action, and show that:

$$(g_m + g_{m\,\text{sub}})_{\text{sat}} = \frac{\mu_n \varepsilon_{\text{ox}} W}{t_{\text{ox}} L}(V_G - V_T). \qquad \text{(P.11.11.1)}$$

Compare with equations (11.2.7), (11.3.12) and discuss.

11.12. Show that the drain noise, equation (11.9.1),

$$\overline{i_d^2} = 4kT g_{d0} \Delta f (Q_{\text{tot}}/Q_{\text{tot, sat}})_{V_G = \text{given}}$$

where Q_{tot} is the total mobile charge in the channel.

12

Charge-coupled Devices and other MIS Structures used for Memory Applications

12.1 MOS (MIS) Memories

Memory circuits represent the major application area of MOS (MIS) devices. The elementary circuit is a memory cell which is able to store for a certain time interval information related to its electric state. A semiconductor memory cell should have the following characteristics: a small area and a simple technological process (resulting in high circuit density and high yields and thus determining a low cost), a light loading to the drive and sense circuitry (which 'write' and 'read' information), the ability to be written and read within a reasonable time interval, low power dissipation in retaining the information [1].

MOSFET devices are attractive for memory applications. They occupy a smaller area and require a simpler technological process (four masking steps and a single diffusion) than bipolar devices. Other favourable characteristics are: absence of minority carrier storage effects, conduction symmetry with respect to source and drain, practically zero gate current. Low power dissipation can be achieved by using high-resistance MOSFET's as loads (chapter 11) or complementary MOSFET circuits (both p- and n-channel devices in the same circuit, Chapters 10 and 11). MOSFET circuits have a relatively low operation speed: the long time delays are associated with the large parasitic capacitances and high impedance level of these devices.

The circuit concepts are far beyond the scope of this book. We note, however, that MOSFET's can be used in random access (read-write) memory cells (RAM),

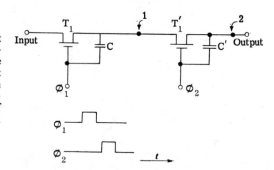

Figure 12.1.1. Bucket brigade shift register circuit and the corresponding clock signals, ϕ_1 and ϕ_2. The pulse ϕ_1 rises the potential of point 1 by capacitive coupling through C, and the input potential is transferred to node 1 at the end of this pulse, as shown for example in reference [1].

shift registers and read only memories (ROM's). The circuit principles are different from that of equivalent bipolar logic circuits and exploit the characteristics of MOSFET's. As an example we reproduce in Figure 12.1.1 a stage of a MOSFET shift register using the so-called 'bucket brigade' (BB) technique [1]. This is a simple (low-area) circuit which has no gain (the voltage attenuation at each transfer must be compensated after a number of stages). Note that the role of the capacitance C can be played by a parasitic gate-drain capacitance. The bucket-brigade operation will be mentioned later in this chapter.

ROM's with MOSFET devices can be obtained as follows. The information is written permanently during the fabrication process. The storage mechanism is simply the thickness of the gate oxides. We have seen that the threshold voltage, V_T, depends upon the oxide capacitance C_{ox} (V_T decreases when C_{ox} increases). Therefore, for a device with a normal thin oxide V_T is low. If the gate oxide is thick, V_T will be high. The response of the two devices to a signal pulse is different: the first device will conduct, whereas the high threshold device will not. The presence or absence of conduction can be detected as a logic signal. Since only a single device is required per bit of information the area occupied by such a memory is considerably reduced. The fact that the information can be written only through masks in the technological process, implies a considerable delay in fabrication. It is therefore desirable, at least for certain specialized circuits, to have means to write electrically the information to be stored.

In this chapter we shall discuss special MIS devices used for memory applications. All these are utilized in integrated circuits fabricated by planar techniques (with certain supplementary technological steps). These special devices belong to two general categories, as indicated in Table 12.1.1. In charge-transfer devices the information consists in the presence or absence of mobile charge packet held by the field effect under a certain electrode on the semiconductor surface. An array of such electrodes exists and the application of certain clock voltages determines the motion of the charge packet and the transfer of corresponding information. Only the charge-coupled devices are discussed somewhat in detail. In the second category, the so-called 'memory' transistors are special MIS devices where the information is stored due to the presence or absence of charge either in an insulating layer or in a floating 'gate'.

Table 12.1.1
Special MIS Devices for Memory Applications

	Charge-transfer devices	'Memory' transistors
Charge stored	Mobile carriers held by electric field in a surface layer	Charge trapped in an insulator or retained in an isolated (floating) gate
Time of storage	Very short (e.g. tens of milliseconds). The charge packets should circulate and the signal level should be restored (refreshed)	Very long (extrapolated for years): permanent or semipermanent
Operation	Charge packets transferred into an array of electrodes Volatile memory (information disappears when power supply is disconnected) Shift registers. RAM.	Charge injected into insulator or floating gate by a proper voltage signal. Nonvolatile. Sometimes can be erased. ROM.
Examples	Charge-coupled device (CCD) Bucket-brigade (BB) circuits Surface-charge transistors (SCT)	MNOS transistor MAOS transistor FAMOS transistor

12.2 Charge-coupled Semiconductor Devices

The charge-coupled device (CCD) concept was first described by Boyle and Smith [2] as follows. A mobile charge is stored in a potential well created by an electrode placed at the semiconductor surface. A variety of functions can be performed by having a means of injecting, transferring and detecting this charge. One method to transfer the charge by moving the potential wells is to form an array of MOS capacitors and excite properly the metallic gates of these capacitors. The charge is stored at the semiconductor surface and consists in minority carriers. This is a *surface* CCD [1]. Figure 12.2.1a shows the energy diagram of an MOS capacitor

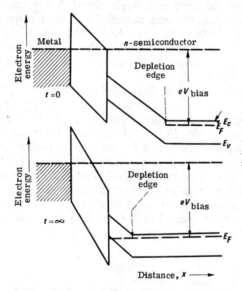

Figure 12.2.1. Energy diagram of an MOS (or MIS) structure: (a) biased in deep depletion, at $t = 0$, without charge at the interface; (b) at $t = \infty$, with charge stored at the interface. (*after Boyle and Smith*, [2]).

(on n-type semiconductor) biased in the deep depletion régime. If holes are introduced into the surface depletion layer, they will be collected at the semiconductor interface causing the interface potential to become more positive. The steady-state situation (reached when the valence band at the interface is approximately at the same energy as the Fermi level in the bulk) is shown in Figure 12.2.1b.

Figure 12.2.2 shows a three phase MOS CCD, formed by a linear array of MOS capacitors, with every third electrode connected to a common conductor. The quantities V_1, V_2, V_3 in Figure 12.2.2 are positive and the semiconductor is at zero potential. Initially electrodes 1, 4, 7 ... are biased by V_2 and all the others by $V_1 >$ $> V_T =$ threshold voltage for surface inversion (Chapter 10). The edge of the depletion region is schematically indicated by the dashed line. Assume, for example, that a positive mobile charge is placed under electrodes 1 and 7. Then a voltage $-V_3$ ($V_3 > V_2$) is applied to electrodes 2, 5, 8 ... and the charge will transfer to the right-hand potential minimum (Figure 12.2.2b). When the voltages are changed as shown in Figure 12.2.2c (the third phase) the transfer of charge by one position was completed. The structure shown in Figure 12.2.2 can be used as a shift register when a charge generator is added at one end (input) and a detector at the other end (output).

The charge generation can be provided by an electric signal through a forward biased *pn* junction or by surface avalanching, and also directly by radiation through

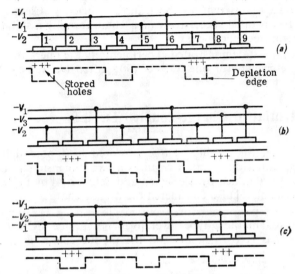

Figure 12.2.2. Illustrating the operation of a three-phase MOS charge coupled device (CCD). (*after Boyle and Smith*, [2]).

creation of electron-hole pairs. Detection may be accomplished by current sensing with a reverse biased *pn* junction or Schottky barrier or by utilizing the change of total MOS capacitance with charge [2].

Charge transfer in two dimensions along the semiconductor surface is possible. CCD's have the ability to perform logic operations. An imaging device can be obtained in principle by storing in a potential well the holes generated by the light image. A display device can be based on the inverse process consisting in reading the information (presence or absence of minority carriers) through the shift register action and then forward biasing the MOS structure to force the displacement of these carriers into the bulk where radiation recombination takes place [2].

Although the CCD principle is particularly simple, a number of conditions should be satisfied in order to provide a useful implentation of this principle. One essential requirement is that the storage time required for a pulsed MOS capacitor to reach its steady state (Figure 12.2.1*b*) should be as long as possible. The storage time is proportional to the equilibrium charge density and inversely proportional to the thermally generated current resulting from recombination in the depletion region and at the semiconductor-oxide interface. The charge 'signal' should be manipulated in a time shorter than the storage time (which may be as long as a few seconds) [2], [3].

The mobile charge stored in MOS-capacitor cells should be transferred as quickly as possible and with minimal loss. The time necessary for change transfer from one electrode to the next depends upon the mechanism of charge transport: this is accomplished by diffusion and drift and will be discussed below in more detail. The total charge transfer cannot be expected in the finite time between driving pulses and therefore the efficiency of transfer from one cell to the next will be below unity. Trapping in surface states reduces the transfer efficiency. Clearly, the number of cells which can be operated successively in a chain forming a shift register increases with increasing transfer efficiency.

Amelio et al. [3] first demonstrated experimentally the operation of silicon CCD's with about 98% efficiency for one transfer (90% after five transfers).

12.3 The Potential in Charge-coupled Devices

We have shown that the mobile charge is stored in a potential well. A correct description of potential distribution requires a two-dimensional analysis. We refer below to the results obtained by McKenna and Schryer [4] by neglecting the mobile charge density. (Chang [5] also considered the mobile charge).

The analysis was performed for a buried channel CCD (Figure 12.3.1). Here the charge is stored in the p-type layer [6]. The surface CCD is a special case, when the p-layer has zero thickness and the mobile charge is stored as minority

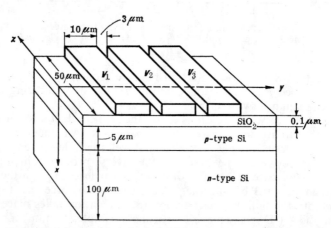

Figure 12.3.1. A buried channel CCD (a surface CCD is obtained when the thickness of the p-layer becomes zero). (after McKenna and Schryer, [4]).

holes in the surface inversion layer. The field is essentially two-dimensional, $\mathscr{E} = \mathscr{E}(x, y)$. The electrodes have zero thickness and the medium surrounding the electrodes has the same dielectric constant as the SiO_2. The n-type substrate is infinitely thick. The acceptor density is assumed to vary exponentially in the p-layer.

We refer the reader to the original paper [4] for a discussion of basic equations and boundary conditions. A simplified solution within depletion layer approximation was also discussed [4], [7]. Numerical results were reported for 10^{14} cm^{-3} doping of the substrate and 2×10^{15} cm^{-3} average doping density in the p-layer. Figure 12.3.2 shows the value of the potential at the potential minimum in the p-layer, for a three-phase burried layer CCD. The dashed curve, computed for no implanted surface charge in the gaps between electrodes (5 μm wide in this case), shows large potential wells *under these interelectrode gaps.* A CCD constructed with practically the same parameters did not work because of charge trapped in these wells [4].

The above difficulty can be eliminated in a number of ways. The solid curve in Figure 12.3.2 shows the effect of implating a uniform surface charge density of 0.8×10^{12} cm^{-2} in the gaps between electrodes. Figure 12.3.3 shows the 'channel'

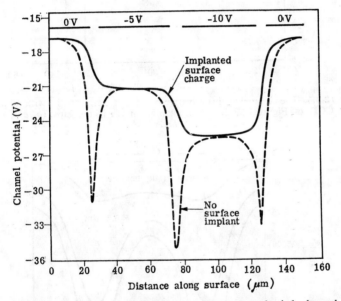

Figure 12.3.2. Channel potential in a three-phase buried channel CCD (see Figure 12.3.1) with 5 μm interelectrode gaps. The solid curve is computed for 0.8×10^{12} cm^{-2} × electronic charge surface charge density implanted in the gaps, the dashed curve — for no implanted surface charge. (*affter* McKenna and Schryer, [4]).

field computed at the potential minimum and plotted as a function of x. This field aids the charge transfer beneath the third electrode (at -10 V). Note that the field penetrates substantially under this electrode. Figures 12.3.4 and 12.3.5 show the effects of a double-level metallization layer. The upper metal level is continuous and biased by a certain voltage. Only two adjacent electrodes are considered. The potential wells decrease in magnitude when the dielectric between the two metallization levels becomes thinner (Figure 12.3.4). However, even with a

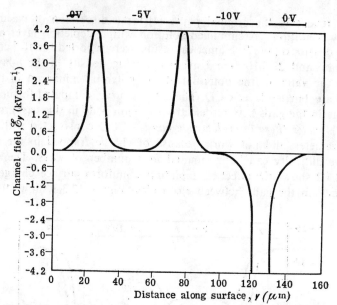

Figure 12.3.3. The channel field (derivative of the channel potential with minus sign) for the CCD of Figure 12.3.2, with implanted charge in the gaps. (*after McKenna and Schryer* [4]).

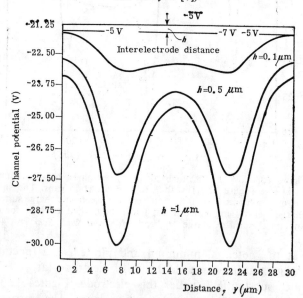

Figure 12.3.4. The channel potential in a buried channel CCD with double-level metallization. The lower level electrodes are 10 μm wide with 5 μm gaps. The upper level is a single electrode separated by a distance of 0.1 μm, 0.5 μm and 1 μm, as indicated. The other parameters are the same as in Figures 12.3.2 and 12.3.3. (*after McKenna and Schryer* [4]).

0.1 μm dielectric a slight potential well still exists between the plates. A higher potential difference between electrodes will also reduce (15 V for 0.5 μm dielectric thickness) the potential well between electrodes, as indicated in Figure 12.3.5. This effect increases when the distance between the two metallization levels decreases. The effect of the gap width (first metallization level) is very strong. The potential wells between electrodes are easily eliminated by reducing the gap width to, say,

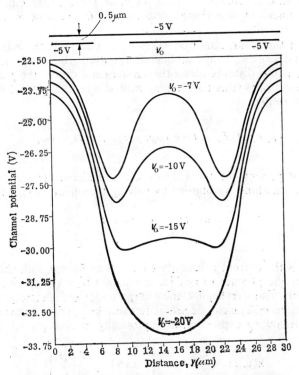

Figure 12.3.5. The channel potential in the CCD of Figure 12.3.4, with the separation of the metallization levels held fixed at 0.5 μm and the potential of the centre, lower-level electrode taking the values − 7, − 10, − 15 and − 20 V. (*after McKenna and Schryer*, [4]).

1 μm. However, such a small separation cannot be satisfactorily achieved in the conventional technology (photolithographic techniques). The first devices [3] had 3 μm metallization gaps, which are still very narrow for usual fabrication conditions. The second metal layer reduces somewhat the importance of the gap width but introduces a parasitic coupling between clock lines. Other constructions providing the charge transfer between adjacent gates were imagined and experimented (see, for example, [8]).

12.4 Charge Transfer Mechanisms in Charge-coupled Devices

The magnitude of the charge-transfer efficiency η (the percent of charge transferred from one stage to the next) is extremely important because the CCD shift registers have no mechanism for gain and for a given total transfer efficiency, the number of stages is limited by η for a single stage.

There are several mechanisms which degrade the transfer efficiency. The first is incomplete transfer of free charge because insufficient time is allowed for this transfer [9].

It was shown that charge motion takes place due to both drift and diffusion. The electric field arises due to the mobile charge itself (this is the self-induced field, $\mathscr{E}_{si}(y,t)$, where y is the distance along the surface) and due to the potentials applied to the external electrodes (this the fringing field, $\mathscr{E}_F(y, t)$. The total electron current density is therefore [9]:

$$j_n(y, t) = e\mu_n n(y, t)\,\mathscr{E}_s(y, t) + e\mu_n n(y, t)\,\mathscr{E}_F(y, t) + eD_n \frac{\partial n(y, t)}{\partial y}. \quad (12.4.1)$$

The three components of the current are considered separately. The charge transfer by diffusion alone is obtained by solving the diffusion equation

$$\frac{\partial n(y, t)}{\partial t} = D_n \frac{\partial^2 n(y, t)}{\partial y^2}, \quad (12.4.2)$$

which arises from the current continuity equation combined with the charge continuity equation in the absence of any recombination term (Chapter 1). Assume that at $t = 0$, when the transfer begins, the charge under the electrode of length L is uniform and equal to n_0. The charge profile remaining under the transferring elecrode approaches the following expression asymptotically in time [9]:

$$n(y, t) = \frac{4n_0}{\pi} \cos \frac{\pi y}{2L} \exp\left(-\frac{\pi^2 D_n t}{4L^2}\right) \quad (12.4.3)$$

and the total number of carriers remaining at time t is

$$N_{\text{tot}}(t) = \int_0^L n(y, t)\,dy = \frac{8}{\pi^2} N_{\text{tot}}(0) \exp\left(-\frac{\pi^2 D_n t}{4L^2}\right). \quad (12.4.4)$$

Aside from a small fraction of charge which decays very quickly, the decay of the total charge due to thermal diffusion is approximately exponential with a decay time constant $\tau_{\text{th}} \simeq L^2/2.5 D_n$ [9].

The magnitude of the self-induced fields directed along the Si–SiO$_2$ interface was first calculated by Engeler et al. [10] as shown over leaf. This field was taken equal to the gradient of the surface potential, ϕ_s. ϕ_s is calculated for the one-dimensional

MOS capacitor $C \simeq C_{ox}$ (the depletion capacitance in series with the oxide capacitance, C_{ox}, is assumed negligible) and assumed to be proportional to the signal carrier concentration

$$\phi_s(y, t) = \phi_{s0} - \frac{e}{C_{ox}} n(y, t), \tag{12.4.5}$$

such that

$$j_n(y, t) = e\mu_n n(y, t) \mathscr{E}_s(y, t) = \frac{e^2 \mu_n n(y, t)}{C_{ox}} \frac{\partial n(y, t)}{\partial y} \tag{12.4.6}$$

and the continuity equation yields

$$\frac{\partial n(y, t)}{\partial t} = \frac{e\mu_n}{2C_{ox}} \frac{\partial^2 n^2(y, t)}{\partial y^2}. \tag{12.4.7}$$

The numerical solution of equation (12.4.7) shows that after a very short time the decaying carrier profile maintains the same shape. This implies that $n(y, t)$ may be written $n(y, t) = h(y)g(t)$. Within this approximation equation (12.4.7) can be analytically solved and the result is [9]:

$$N_{tot}(t) = N_{tot}(0) \frac{t_0}{t + t_0}, \quad t_0 = \frac{L^2 C_{ox}}{1.57 e \mu_n n_0} \tag{12.4.8}$$

A more exact evaluation of the self-induced field was done by using the integral method which computes the resultant of image fields*. The last method allows to

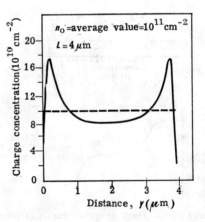

Figure 12.4.1. Cusping of charge stored in a square potential well due to self-induced fields (surface CCD). (after Carnes et al., [9]).

evaluate how the charge will rearrange itself within the potential well: Figure 12.4.1 indicates that the charge stored in a square potential well tends to accumulate at

---

* A line charge is considered at the $Si-SiO_2$ interface, perpendicular to the direction of transfer. The field created by this charge is computed by using the method of images and taking into account that the charge is 'reflected' by both the metallic gate (which acts as a ground plane) and the oxide-semiconductor interface (the semiconductor is treated here as a dielectric). The contribution of an infinite series of line charges is summed by integration and the total self-induced field is computed. When the charge is uniformly distributed along the gate length, L, the electric field can be written in analytic form [9].

the edges due to the mutual electrostatic repulsion. This is a stationary configuration. When the charge transfer is allowed, say, to the right in Figure 12.4.1, the self-induced field acts as a retarding field in the left half of the transferring region, thus hindering the transfer [9].

The fringing field along the Si—SiO$_2$ interface, $\mathscr{E}_F(y, t)$, arises due to the potential difference between adjacent electrodes. Analytical and numerical computations of this field were performed [9], [11], [12]. Again the *shape* of mobile charge distribution stabilizes and $n(y, t)$ can be expressed as a product solution. The time-dependent part decreases exponentially

$$n(y, t) = h(y) \exp(-t/\tau_F), \qquad (12.4.9)$$

where τ_F is a final decay constant of the remaining charge. A numerical solution for $h(y)$ and τ_F is necessary.

A correct description of charge transfer requires an account of all transfer mechanisms, and this can be done only by computer simulation [9], [13], [14]. Certain results are presented below, for the three-phase surface CCD shown in Figure 12.4.2a. The surface potential distribution is indicated in Figure 12.4.2b

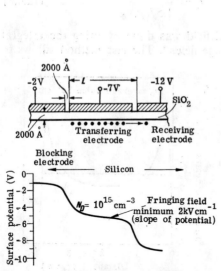

Figure 12.4.2. Three-phase surface CCD with 10^{15} cm^{-3} doping, studied by computer simulation. (*after Carnes et al.*, [9]).

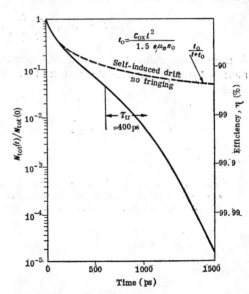

Figure 12.4.3. Remaining charge under the 4 μm long state (surface CCD with 10^{15} cm^{-3} substrate doping) normalized to the initial charge (at the beginning of the transfer) *versus* time. Dotted line indicates how the charge transfer would proceed in the absence of fringing fields. (*after Carnes et al.*, [9]).

and its slope yields the fringing field. Figure 12.4.3 shows the charge decay for a 4 μm-long gate, 10^{15} cm^{-3} substrate doping and a uniform 4.6×10^{11} cm^{-2} minority-carrier surface density (a uniform concentration is maintained by the balance between fringing fields and self-induced fields) [9]. A very short time interval (10 ns), diffusion is dominant. Then, for times less than 500 ps the charge transfer is dominated by self-induced drift. The dotted line shows the charge decay computed in the absence of any fringing field. The fringing field speeds appreciably the transfer. The computed transit time

$$\tau_{tr} = \frac{1}{\mu_n} \int_0^L \frac{dy}{\mathscr{E}_F(y)} \qquad (12.4.10)$$

beneath the electrode of length L is 400 ps. However, the charge transit takes more than 400 ps, partly because of the retarding effect of the self-induced fields on the left-hand side (see above). Figure 12.4.4 shows charge profiles for various times. After about 800 ps the self-induced field is negligible and the charge profile drifts to the right under the influence of the fringing field. Then, at about 1400 ps, the charge profile becomes stationary and decays exponentially in time, according to equation (12.4.9) where $\tau_F \simeq 100$ ps. Decreasing the substrate doping density increases the fringing field and decreases the time required for charge transfer. It was found that for closely spaced CCD (Figure 12.4.2 a) the time required to achieve $\eta = 99.99\%$ transfer efficiency is approximately $4\tau_{tr}$.

The two-dimensional computer simulation reported by Suzuki and Yanai [14] indicated certain details of the mobile charge transfer. The potential barrier which may occur in the interelectrode gap (see also Section 12.2) increases with the increas-

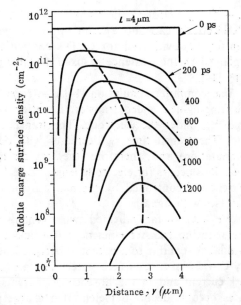

Figure 12.4.4. Charge profiles under the gate electrode for various times. Dotted line traces the charge peaks. (*after Carnes et al.*, [9]).

ing stored charge beneath the electrode. However, this potential barrier becomes smaller or even disappears when the donor density decreases or the gap width decreases (the increasing interelectrode potential difference has the same effect). This potential barrier has an important effect in that it limits the charge transfer:

Figure 12.4.5 illustrates how incomplete mobile charge transfer occurs when such a barrier does indeed exist. There exists a maximum value $Q_{s\,\mathrm{max}}$ of the stored charge (signal charge), Q_s, such that the potential barrier is still absent. When a charge $Q_s > Q_{s\,\mathrm{max}}$ is injected into the potential well and transferred, the excess charge

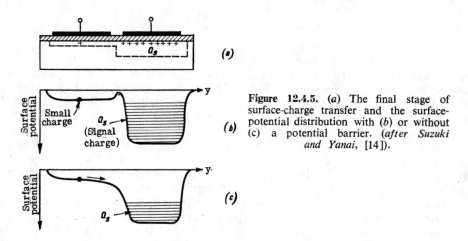

Figure 12.4.5. (a) The final stage of surface-charge transfer and the surface-potential distribution with (b) or without (c) a potential barrier. (after Suzuki and Yanai, [14]).

$Q_s - Q_{s\,\mathrm{max}}$ is lost in the transfer process. Numerical computations show that an interelectrode gap above 2 μm imposes a very low signal level*. However, good transfer was evidenced for CCD's constructed with 3 μm gaps [3], [15]. This was explained by the existence of fixed negative charges implanted in the interelectrode gaps [14].

The computer simulation [14] also reveals the importance of signal level upon the fringing-field aided transfer. Equation (12.4.10) yields the one-electron transit time, τ_{tr}, under the electrode. The transport is provided by drift in the fringing field. The transit time τ_{tr} decreases with decreasing signal level (from numerical computation) and, therefore, the transfer speed increases with decreasing signal level. This occurs for electrode length, $L < 5 - 10$ μm. For $L > 10$ μm the transfer efficiency is controlled by thermal diffusion [16] and the converse is true (speed increases with signal).

We have so far indicated two mechanisms determining charge loss and therefore reducing transfer efficiency: insufficient time allowed for charge transfer and residual charge due to a potential barrier. The third mechanism is trapping of minority carriers in fast surface (interface states). Consider a surface CCD with a low dark current and in the absence of charge signal over a long period: the interface states become deeply depleted. If a charge packet is then introduced and transferred successively from one electrode to another, a certain charge will be trapped under each electrode. Those carriers trapped by levels with short re-emission constants

---

* $Q_{s\,\mathrm{max}}$ is low compared to the charge which can be stored in the potential well.

compared to the transfer time will be re-emitted and transferred with the main packet. The carriers trapped by levels with longer time constants will form a residual charge which is transferred later and the transfer efficiency is degraded. The theory was developed by Tompsett [17] by using a Shockley-Read-Hall model. An approximate evaluation of transient behaviour with a single-level model within the charge-control approach was performed by Lee and Heller [18].

The interface-state effect can be reduced by using a background charge or *fat zeros* [17]. This background charge is transferred at all times between electrodes, in the absence of signal charge, and fills the interface states every cycle. The same fraction of this trapped charge will be then re-emitted into the succeeding charge packet each transfer cycle, such that each signal charge packet will lose as much net charge into the interface states as it gained from the previous packet. There are, however, several ways in which the transfer efficiency is still affected by interface states because the above compensation is imperfect: the edge effects (the effective area of charge storage depends upon the magnitude of the charge stored), the variable transfer time and trap occupancy (depending upon the signal level) [17]. The interface-state problem is eliminated in buried channel CCD's, where the charge packets are stored in the bulk.

12.5 Noise in Charge-transfer Devices

The primary operational limitations of charge transfer devices are due both to incomplete charge transfer and to the existence of random noise [19]. We shall briefly discuss below the basic noise sources present in charge-transfer devices [19]–[21].

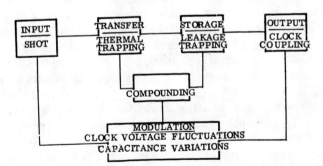

Figure 12.5.1. Sources of noise in a charge-transfer device.
(*after Thornber*, [19]).

Figure 12.5.1 shows schematically the physical sources of random noise. The input noise will be of shot noise nature if the charge packet is created by photon absorption (imaging applications) or if it is injected by an emission-limited mechanism.

The transfer noise is due to both thermal fluctuations and trapping fluctuations*. This transfer noise consist in fluctuations in the number of carriers in each charge packet. A charge packet reaching the output has been transferred typically 10^2 to 10^3 times. The noise in the output packet is an accumulation of noise acquired by this packet during each transfer. Note the fact that if at the end of a transfer phase a transfer noise of $+\Delta Q$ (charge fluctuation) has been added to the signal, by conservation of charge, a quantity of charge $-\Delta Q$ has been added to the charge left behind. These two contributions to the noise are correlated. The incomplete charge transfer discussed in Section 12.4 adds to the above correlation mechanism: one charge packet collects successively random varying charge quantities left behind by the preceding packets. This correlation leads to a suppression of transfer noise at low frequencies (Figure 12.5.2) compared to the clock frequency, f_0, [20]:

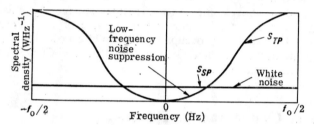

Figure 12.5.2. Noise spectral densities S_{TP} and S_{SP} plotted versus frequency f (f_0 is the clock frequency). (after Thornber, [20]).

here the averaging is done over long times and the fluctuations cancel each other (strong correlation between adjacent packets).

The storage noise arises from leakage-current fluctuations and trap-occupancy fluctuations. Clearly, the fluctuation of charge stored in one cell is independent of fluctuations of charge stored in other cells, and the corresponding noise is uncorrelated and has a 'white' (constant) spectrum. Figure 12.5.2 shows the noise spectral densities (of the short-circuit noise current) for transfer processes (S_{TP}) and storage processes (S_{SP}).

Figure 12.5.3 reproduces results of computations which include thermal noise, shot noise effects on the input signal and incomplete charge transfer. The signal-to-noise ratio is represented versus the number of transfers with the storage capacitance, C, as a parameter, for incomplete transfer (99.9% efficiency). The square of the signal charge is proportional to C^2, while the mean-square of the noise charge is proportional to C for both thermal ($\overline{q^2} \propto kTC$) and input shot noise [19], and therefore the signal-to-noise ratio decreases with decreasing C (smaller gate area).

---

* The thermal noise spectral density is calculated by the usual method. The interface (trapping) noise is found to be proportional to $\overline{N}_{ss}$ (number of interface states per unit area and per unit energy), namely $\overline{q^2} \propto e^2 kT N_{ss} A$ [20], [21]. Other noise contributions are not discussed here.

This will limit the practical size of CCD's. Figure 12.5.3 also shows that as the number of charge transfers increases, the thermal device nosie dominates the contribution of the input noise.

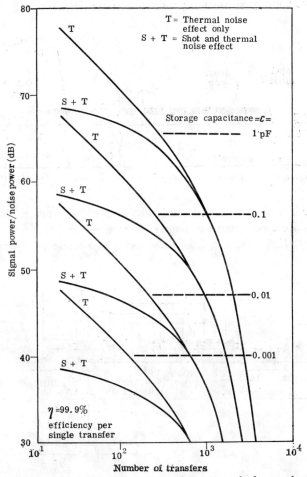

Figure 12.5.3. Signal-to-noise ratio computed for a charge-transfer device with 99.9% efficiency per single transfer, at various levels of storage capacitance, C. (*after Thornber*, [19]).

12.6 Various Charge-transfer Devices

Figure 12.6.1 shows a schematic cross section of a CCD 8-bit shift register [15]. This is a true *functional device* replacing an entire circuit. The input and output sections have the source-gate and gate-drain configuration of an MOSFET but there are 24 MOS gates (3 per bit) between these sections. Each of the transfer gates are 250 μm wide, 50 μm long and separated by 3 μm gaps. The electrodes are interconnected in triplets using two metallization leads on the oxide and one diffused cross-under. Thus the three-phase operation necessitates complex interconnections and reduces the conceptual advantages related to the simplicity of CCD principle [22].

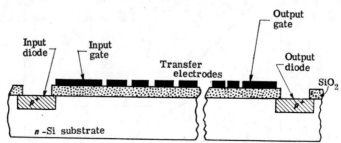

Figure 12.6.1. CCD shift register. (*after Tompsett et al.*, [15]).

Two-phase operation is also possible if the potential wells induced by the phase voltage pulses are unsymmetrical, namely deeper in the direction of the signal flow [12], [23]. Figure 12.6.2 shows a cross-section of a two-phase shift register constructed

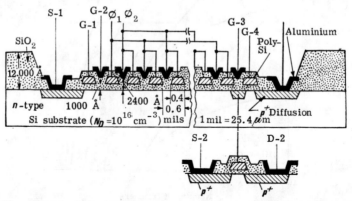

Figure 12.6.2. CCD 128-stage shift register. ϕ_1 and ϕ_2 are clock lines connected to silicon and metal gates. The input structure consists of source diffusion S−1 and input gates G−1 and G−2. The output can be detected as the current flow out of the drain diffusion D−1 or from the output MOS device S−2, D−2. (*after Kosonocky and Carnes*, [23]).

using silicon gates [23]. The technology is illustrated in Figure 12.6.3. The substrate is 1 to 0.5 μm *n*-type silicon. First a p^+ diffusion is made (*a*), followed by thermal growth of a 1000 Å-thick 'channel oxide', deposition of polycrystalline silicon film and definition of 'polysilicon gates' (*b*). The polysilicon gates are insulated by a thermally-grown oxide (*c*) and the channel oxide under the aluminium gates (*d*) is simultaneously grown to its final thickness. This construction has two supplementary advantages: first, the separation between gates can be made as small as the channel oxide (self-aligning gate construction), thus eliminating the interelectrod

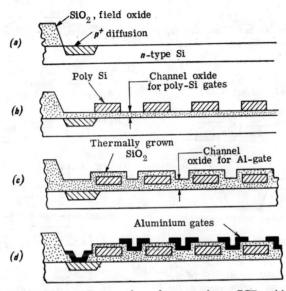

Figure 12.6.3. Construction of a two-phase CCD with polysilicon layers. (*after Kosonocky and Carnes*, [23]).

gap and accelerating the charge transfer*. Secondly, the channel oxide is always covered by metallization and thus is protected from the ambient.

The operation of the two-phase polysilicon gate shift register driven by the clock voltages $\phi_1(t)$ and $\phi_2(t)$ is illustrated by Figure 12.6.4. At time t_1 the signal charge is accumulated under the phase 2 gates and namely beneath the polysilicon gate which is situated closer to the silicon surface. In the second half-cycle (time t_2, Figure 12.6.4 c) the charge packet is transferred under the phase-1 gates. This mode of operation (Figure 12.6.4) is characterized by *complete charge transfer*. It can be shown that unidirectional charge flow can be obtained if only one phase voltage is changed and the second one is maintained constant (d.c. bias). Thus one-clock operation is possible. Furthermore, a true single phase operation can be achieved by replacing the d.c.-biased phase by a structure with fixed charge within the oxide, for example by ion implantation [23]. Charge storage in a metal-nitride-oxide-semiconductor (MNOS) structure can also be used [24].

Another mode of operation of the two-phase CCD is illustrated in Figure 12.6.5. The device parameters and clock-voltage levels are so that the barrier under the

---

* The alignment tolerances require only that the separation between the aluminium electrodes occurs over the polysilicon gates, and are by far less restrictive than in fabrication of the 3 μm gap CCD shift register in Figure 12.6.1.

aluminium gate is too high and the potential wells under the polysilicon gate can never be emptied. A background charge is maintained under the gate*. This is the *residual charge mode* or *bucket-brigade (BB) mode*, which is similar to the operation

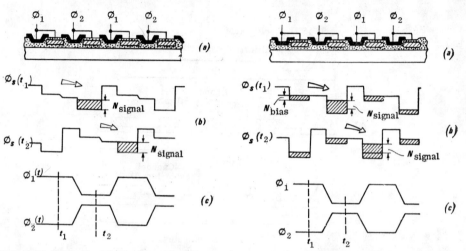

Figure 12.6.4. Complete charge-transfer operation of a two-phase CCD. The lower solid line in (b) and (c) shows the surface potential distribution in the absence of the stored charge: this charge is represented by the crossed area and the line above this area is the local surface potential in the presence of the above charge. (*after Kosonocky and Carnes,* [23]).

Figure 12.6.5. Bias-charge (bucket-brigade) mode of operation of a two-phase CCD. The bias charge, N_{BIAS} and the signal charge N_{SIGNAL} are indicated. (*after Kosonocky and Carnes,* [32]).

of the bucket-brigade MOSFET (IGFET) shift registers. In a MOSFET bucket brigade the bias charge regions (retaining the background charge) are replaced by floating diffused regions. The only difference is that the background charge density is enormous in the second case.

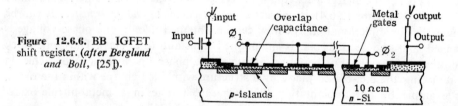

Figure 12.6.6. BB IGFET shift register. (*after Berglund and Boll,* [25]).

Figure 12.6.6 shows an IGFET bucket-brigade shift register. It consist of a series of *p*-type islands diffused into an *n*-type substrate and covered by a SiO_2

---

* Experimental work indicated a charge transfer loss of 5×10^{-5} per gate with background charge. Otherwise the charge loss is higher and determined by trapping in interface states (estimated to $N_{ss} = 1.2 \times 10^{11} \text{cm}^{-2} \text{eV}^{-1}$ for (111) substrates and $2.9 \times 10^{10} \text{cm}^{-2} \text{eV}^{-1}$ for (100) substrates) [23].

film (1000 Å). Note that each metal clock line overlaps a large part of one p island [25]. Note the similarity to the circuit shown in Figure 12.1.1. The operation can be understood as follows. We assume square-wave clock waveforms, oppositely phased. Let ϕ_1 be sufficiently negative and ϕ_2 positive. The odd numbered gates in Figure 12.6.6 induce a channel at the silicon surface and the p-type islands under the same gates are driven negative due to the large overlapping capacitance. The remaining p-islands are driven positive by ϕ_2, going positive. A number of MOSFET's are now in operation (for example the p-island under gate 2 acts as source, the 3 island (more negative) as drain and, for proper clock levels, holes flow through the channel under the gate 2. As a result, source potential drops and drain potential rises. At a certain moment the charge transfer ceases [25]. If the potentials of the clock lines are then reversed, the process repeats itself but the sources and drains reverse their roles. The channel under the odd gates are now conducting. A p-region which has accumulated a charge packet acting as a drain now becomes a source and injects this charge toward the next p region [26].

The technology of BB IGFET shift register is similar to normal MOS technology: there are no critical steps or restrictive photolithography tolerances. The silicon gate construction can also be used with several advantages associated with self-alignation and the availability of an upper metallization level for associated circuitry.

Berglund and Strain [22] studied comparatively both CCD and BB shift registers constructed with similar technology and found comparable performances. The operating mechanism, the physical and technological limitations are essentially the same, indeed. There are, nevertheless, differences. For example, the transfer efficiency in BB shift registers is limited not only by the insufficient time allowed for transfer or by surface states, but also by the existence of a finite IGFET drain conductance (the rate of transfer, i.e. the current, depends upon the drain voltage) [22] [25].

The surface-charge transistor (SCT) [27] is a charge-transfer device which has also 'active' properties: it realizes gain and performs control functions. The basic

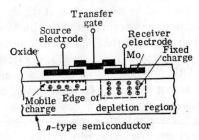

Figure 12.6.7. Cross section of a surface-charge transistor showing the surface depletion regions with and without mobile charge. (*after* Engeler *et al.*, [10]).

structure (Figure 12.6.7) has three overlapping electrodes which are separated from the semiconductor and from each other by thin insulating layers. Three adjacent surface depletion regions are formed beneath these electrodes, when appropriate voltages are applied to them. The first and third regions are charge-storage reservoirs and the second one controls the flow of charge between them. The transfer of minority carrier charge from the source (Figure 12.6.7) to the receiver electrode cannot occur unless the surface potential of the transfer region is below that of the source. The fringing fields are not essential and the source and receiver electrodes can be separated by any distance. We stress the fact that the amount of charge transferred is precisely controlled by the source-gate surface potential difference. If the transfer-gate capacitance is substantially smaller than the source capacitance, the amount of charge transferred is larger than the variation of gate charge which produced it: a gain in charge is provided by this structure [27]. SCT shift registers were constructed [28]. Only two phases are required to provide directionality of charge signal. The presence of upper (gate) electrodes eliminates the critical dependence upon both the interelectrode spacing and the ambient. We note the fact that the SCT principle can be used to construct a random-access memory system [29].

Finally, we note that CCD's provide a very promising approach to solid-state imaging systems [2], [30], [31]. The CCD cells are low-noise devices suitable for low-level applications [32]. Charge-coupled imagers can be obtained by using principially shift registers as lines of the sensor. The CCD array is exposed to light for a certain time (long compared to the total time necessary to transmit the entire image). This radiation generates minority carriers which are accumulated in potential wells. The stored charge pattern represents the integrated light intensity pattern and is then shifted by clock voltages and transmitted to an output detector and amplifier.

12.7 'Memory' Transistors

We briefly review a number of MIS-type 'memory' transistors used for permanent, non-volatile memories. The information is contained in the charge stored in a modified gate region and introduced here by an electric pulse. This charge modifies the threshold voltage such that when a 'read' pulse is applied the conduction state depends upon the presence or absence of the above charge.

The metal-nitride-oxide-semiconductor (MNOS) memory transistor is a normal MOS transistor, except the gate insulator which is formed by two layers: a thin SiO_2 film and grown on silicon surface and an SiN_4 (nitride) layer. The charge is stored at the SiO_3-SiN_4 interface, or inside the nitride. Several types of MNOS transistors have been described. Frohman-Bentchkowsky and Lenzlinger [33] presented a device with a relatively thick oxide (50–200 Å) and explained the charge storage by Fowler-Nordheim injection through the oxide, Poole-Frenkel current through the nitride and trapping inside the nitride. Ross and Wallwark [34] described thin oxide (15–25Å) devices with charge injection through direct tunnelling from the silicon into states located at the interface between the two insulators.

The energy band diagram of Si—SiO$_2$—Si$_3$N$_4$ is shown in Figure 12.7.1. Tunnelling of both electrons (Figure 12.7.2 a) and holes (Figure 12.7.2 b) is possible [35]. The injected charge remains trapped after the bias is removed. The discharge of trapped charge is possible by direct tunnelling back into the oxide, by Poole-

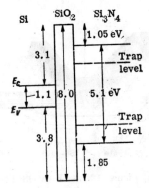

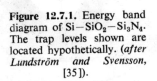

Figure 12.7.1. Energy band diagram of Si—SiO$_2$—Si$_3$N$_4$. The trap levels shown are located hypothetically. (after Lundström and Svensson, [35]).

Frenkel conduction through the nitride or by electron-hole recombination [36]. The discharge current limits the storage time and the number of readings. The information can be removed by exposure to ultra violet light or X-rays.

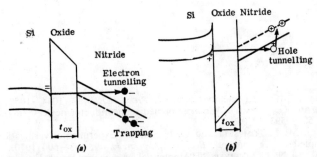

Figure 12.7.2. (a) Electron tunnelling into nitride at positive gate voltages; (b) hole tunnelling at negative gate voltage. (after Lundström and Svensson, [35]).

The metal-alumina-silicon-oxide (MAOS) memory transistor is based on a similar principle. A layer alumina layer (Al$_2$O$_3$) 700 Å-thick is chemically deposited on a 50 Å SiO$_2$ film. Experimental results indicated that the threshold voltage varies linearly for both negative and positive gate voltages and can be shifted *reversibly* by an appropriate electric pulse [37].

A transistor with electrically-alterable memory properties can be constructed by using a polysilicon floating gate situated between oxide and nitride [38]. The polysilicon gate is charged by injection of hot electrons due to avalanche breakdown at the substrate-drain *pn* junction, and discharged through the nitride when a large voltage is applied on the external gate (erasing operation).

The floating gate avalanche injection MOS (FAMOS) memory device represented in Figure 12.7.3 is essentially a *p*-channel MOS with a floating polysilicon gate [39]. The floating gate is charged by avalanche injection through the oxide: impact ionization and electron emission occurs from the depleted surface, in the

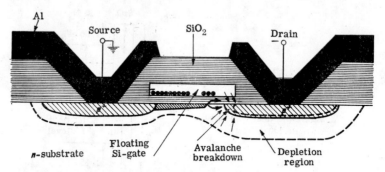

Figure 12.7.3. FAMOS biased for avalanche injection.
(*after Frohman-Bentchkowsky*, [39]).

p^+ drain region (Figure 12.7.3, see also Chapter 10). The charge-free situation is restored by ultraviolet and X-ray irradiation. The normal discharge current is very low, such that a shift of 1 V in threshold voltage should occur after minimum 10 years [39].

References

1. L. M. TERMAN, 'MOSFET memory circuits', *Proc. IEEE*, **59**, 1044–1058 (1971).
2. W. S. BOYLE and G. E. SMITH, 'Charge coupled semiconductor devices', *Bell. Syst. Techn. J.*, **49**, 587–593 (1970).
3. G. F. AMELIO, M. F. TOMPSETT and G. E. SMITH, 'Experimental verification of the charge coupled device concept', *Bell Syst. Techn. J.*, **49**, 593–600 (1970).
4. J. MCKENNA and N. L. SCHRYER, 'The potential in a charge-coupled device with no mobile minority carriers', *Bell Syst. Techn. J.*, **52**, 1765–1793 (1973).
5. W. H. CHANG, 'MIS array potential calculation', *Solid-St. Electron.*, **16**, 491–496 (1973).
6. R. H. WALDEN, R. H. KRAMBECK, R. J. STRAIN, J. MCKENNA, N. L. SCHRYER and G. E. SMITH, 'The burried channel charge-coupled device', *Bell Syst. Techn. J.*, **51**, 1635–1640 (1972).
7. J. MC. KENNA and N. L. SCHRYER, 'On the accuracy of the depletion layer approximation for charge coupled devices', *Bell Syst. Techn. J.*, **51**, 1471–1485 (1972).
8. R. H. KRAMBECK, R. J. STRAIN, G. E. SMITH and K. A. PICKAR, 'Conductively connected charge-coupled devices', *IEEE Trans. Electron. Dev.*, **ED-21**, 70–72 (1974).
9. J. E. CARNES, W. F. KOSONOCKY and E. G. RAMBERG, 'Free charge transfer in charge-coupled devices', *IEEE Trans. Electron Dev.*, **ED-19**, 798–808 (1972).
10. W. E. ENGELER, J. J. TIEMANN and R. D. BAERTSCH, 'Surface charge transport in silicon', *Appl. Phys. Lett.*, **17**, 469–472 (1970); see also 'Surface-charge transport in a multi-element charge-transfer structure', *J. Appl. Phys.*, **43**, 2277–2285 (1974).
11. J. E. CARNES, W. F. KOSONOCKY and E. G. RAMBERG, 'Drift-aiding fringing fields in charge-coupled devices', *IEEE Trans. Solid-St. Circ.*, **SC-6**, 322–326 (1971).

12. Charge-coupled Devices

12. W. F. KOSONOCKY and J. E. CARNES, 'Charge-coupled digital circuits', *IEEE Trans. Solid-St. Circ.*, **SC−6**, 314−322 (1971).
13. L. G. HELLER, W. H. CHANG and A. W. LO, 'A model of charge transfer in bucket brigade and charge-coupled devices', *IBM Journ. Res. Dev.*, **16**, 184−187 (1972).
14. N. SUZUKI and H. YANAI, 'Computer analysis of surface-charge transport between transfer electrodes in charge-coupled devices', *IEEE Trans. Electron Dev.*, **ED−21**, 73−83 (1974).
15. M. F. TOMPSETT, G. F. AMELIO and G. E. SMITH, 'Charge coupled 8-bit shift register', *Appl. Phys. Lett.*, **17**, 111−115 (1970).
16. R. J. STRAIN and N. L. SCHRYER, 'A non-linear diffusion analysis of charge-coupled-device transfer', *Bell Syst. Techn. J.*, **50**, 1721−1740 (1971).
17. M. F. TOMPSETT, 'The quantitative effects of interface states on the performance of charge-coupled devices', *IEEE Trans. Electron Dev.*, **ED−20**, 45−55 (1973).
18. H.-S. LEE and L. G. HELLER, 'Charge-control method of charged-coupled device transfer analysis', *IEEE Trans. Electron Dev.*, **ED−19**, 1270−1279 (1972).
19. K. K. THORNBER, 'Operational limitations of charge transfer devices', *Bell Syst. Techn. J.*, **52**, 1453−1482 (1973).
20. K. K. THORNBER, 'Theory of noise in charge-transfer devices', *Bell Syst. Techn. J.*, **53**, 1211−1262 (1974).
21. J. E. CARNES and W. F. KOSONOCKY, 'Noise sources in charge-coupled devices', *RCA Rev.*, **33**, 327−343 (1972).
22. C. N. BERGLUND and R. J. STRAIN, 'Fabrication and performance considerations of charge-transfer dynamic shift registers', *Bell Syst. Techn. J.*, **51**, 655−703 (1972).
23. W. F. KOSONOCKY and J. E. CARNES, 'Two-phase charge-coupled devices with overlapping polysilicon and aluminium gates', *RCA Rev.*, **34**, 164−202 (1973).
24. P. P. GELBERGER and C. A. T. SALAMA, 'A uniphase charge-coupled device', *Proc. IEEE*, **60**, 721−722 (1972).
25. C. N. BERGLUND and H. J. BOLL, 'Performance limitations of the IGFET bucket-brigade shift register', *IEEE Trans. Electron Dev.*, **ED−19**, 852−860 (1972).
26. K. K. THORNBER, 'Incomplete charge transfer in IGFET bucket-brigade shift registers', *IEEE Trans. Electron Dev.*, **ED−18**, 941−950 (1971).
27. W. E. ENGELER, J. J. TIEMANN and R. D. BAERTSCH, 'The surface-charge transistor', *IEEE Trans. Electron Dev.*, **ED−18**, 1125−1136 (1971).
28. J. J. TIEMANN, R. D. BAERTSCH, W. E. ENGELER and D. M. BROWN, 'A surface-charge shift register with digital refresh', *IEEE J. Solid-St. Circ.*, **SC−8**, 146−151 (1973).
29. W. E. ENGELER, J. J. TIEMANN and R. D. BAERTSCH, 'A surface-charge random-access memory system', *IEEE J. Solid-St. Circ.*, **SC−7**, 330−335 (1972).
30. G. F. AMELIO, W. J. BERTRAM, JR., and M. F. TOMPSETT, 'Charge-coupled imaging devices: design considerations', *IEEE Trans. Electron Dev.*, **ED−18**, 988−992 (1971); see also pages 992−996.
31. C. H. SEQUIN et al., 'A charge-coupled area image sensor and frame store', *IEEE Trans. Electron Dev.*, **ED−20**, 244−252 (1973).
32. J. E. CARNES and W. F. KOSONOCKY, 'Sensitivity and resolution of charge-coupled imagers at low light levels', *RCA Rev.*, **33**, 607−622 (1972).
33. D. FROHMAN-BENTCHKOWSKY and M. LENZLINGER, 'Charge transport and storage in metal-nitride-oxide-silicon (MNOS) structures', *J. Appl. Phys.*, **40**, 3307−3319 (1969).
34. E. C. ROSS and J. T. WALLMARK, 'Theory of the switching behaviour of MIS memory transistors', *RCA Rev.*, **30**, 366−381 (1969).
35. K. I. LUNDSTRÖM and C. M. SVENSSON, 'Properties of MNOS structures', *IEEE Trans. Electron Dev.*, **ED−19**, 826−836 (1972).
36. L. LUNDKVIST, I. LUNDSTRÖM and C. SVENSSON, 'Discharge of MNOS structures', *Solid-St. Electron.*, **16**, 811−823 (1973).
37. S. SATO and T. YAMAGUCHI, 'Study of charge storage behaviour in metal-alumina-silicon dioxide-silicon (MAOS) field effect transistor', *Solid-St. Electron.*, **17**, 367−375 (1974).
38. H. C. CARD and A. G. WORRALL, 'Electrically alterable avalanche-injection memory', *Electron. Lett.*, **9**, 14−15 (1973).
39. D. FROHMAN-BENTCHKOWSKY, 'FAMOS − a new semiconductor charge storage device', *Solid-St. Electron.*, **17**, 517−529 (1974).

Subject Index

Abrupt junction 117–119, 121, etc.
acceptor levels, 37, 34, 40, etc.
accumulation layer, 93, 95, 122, 154, 155, 353, 356, 358, 361, 371, 377, 378, 387, 470, 471, 478, 481, 495, 500
accumulation mode, 379
acoustic branch, 50, 51
acoustic vibration, 50
acoustic waves, 50
activation energy plot, 141, 142
analog-to-digital converter, 415
analog transistor, 199
anisotropic mobility, 501
anode, 169, 175, etc.
anode domain, 313, 332
astable behaviour, 363
asymmetrical junction, 118, 121
Auger process, 80
avalanche breakdown, 71, 72, 120, 132, 161, 208, 231, 252, 268, 269, 427, 428, 538, 539, 540, 566, 571, 572, 613
avalanche frequency, 251, 260, 263–265
avalanche generation, 529
avalanche inductance, 539, 542, 614
avalanche noise, 293
avalanche region, 247, 248, 251, 252, 254, 261, 263, 266, 269
average quantum velocity, 42

Band gap, 32, 36, 37, 92
band-to-band recombination, 80
BARITT (BARrier Injection Transit-Time) diode, 208, 216, 223, 226, 243, etc.
barrier height, 88–90, 92, 101, 105, 107, 112, 126

barrier lowering, 98, 100–103, 107
barrier-type contact, 379, 407
bevel angle, 299
bistable switching, 365, 366, 413
Bloch function, 30, 42, 48
Boltzmann equation, 44–46, 56–59, 63, 76, 79, 80, 135, 500
Bose-Einstein statistics, 52
boundary conditions, 377
Bravais lattice, 24, 25
Bragg reflection, 28, 32, 43, 49
breakdown voltage, 121, 152–154, 156–158, 253, 266, 427, 436
Brillouin zone, 28–33, 43, 45, 49, 50, 59, 62
bucket brigade, 592, 610
built-in voltage (built-in potential), 97, 117, 118, 122, 123, 128, 229, 230, 433, 476, 535, 573
bulk mobility, 500
bulk resistance, 311

Capture cross-section, 80, 519, 521
carrier accumulation, 438, 443, 445
carrier "cooling", 59, 75
carrier heating, 75
carrier inertia, 192
carrier injection, 86, 88, 108, 117
carrier mobility, 47, 65, 81, 427, 483, etc.
carrier temperature, 57, 58
carrier transit-time, *see* transit-time
carrier wave, 261
cathode, 169, 173, 175, 177, 377
cathode fall, 381
cathode notch, 330, 331, 398, 399 (*see also* doping notch)
cathode resistance, 311

channel, 425, etc.
channel-lenght modulation, 566
charge-coupled devices, 592, 594, etc.
charge-density modulation, 215, 219, 224, 329
charge-storage, 609
charge-transfer devices, 593, etc.
chemical bond, 27, 50
collection velocity, 136–139, 141, 149
collision cross-section, 48
collisions, 43–45, 47–49, 52, 55, 56, 77, 134, 500
compensated semiconductor, 41, 53
competition between domains, 372
complementary structures, 253, 254, 288
complementary (MOS) transistors, 551, 571, 592
computer simulation, 280, etc.
conductance, 191, etc.
conductance technique, 513, 525
conduction band, 32, 33, 37–40, 43, 44
conductivity, 46
conductivity anysotropy, 59, 60
conductivity effective mass, 52
contact resistance, 86, 114–116
continuity equation, 79–82
control characteristic, 309, 352, 379, 380
corner breakdown, 142
corner field, 152
Coulomb scattering, 53
covalent bond, 27, 37
critical doping-length product, 322, 336, 337
critical field for avalanche breakdown, 121, 168, 231, 253, 290, 357, 542, 566
critical field for field-dependent mobility, 202
crystal, 23
crystalline planes/directions, 24
cubic lattice, 24–26, 29, 31
current continuity equation, 171, 354
current density, 46, etc.
current saturation (FET), 427, 431, 438, 552, 554, 563
cuttoff frequency, 251, 265, 450, 452

Debye diffusion velocity, 138
Debye length, 96, 119, 122, 333, 471, 480
deep-depletion, 481, 508, 527–529, 531, 532, 539, 594
deep levels/traps, 165, 179

deep surface states, 104
defect scattering, 55
deformation potential, 50, 67
degenerate semiconductor, 40, 501
degenerate statistics, 39
delayed domain mode, 390, 405
density modulation, see charge-density modulation
density-of-state effective mass, 36
depletion approximation, 117–119, 122, 436, 440, 553, 557, 567
depletion capacitance, 165, 166, 168, etc.
depletion layer (depletion region), 93, 95, 97, 101, 106, 120, 134, 135, 165, 253, 254, 284, 354, 356, 371, 381, 387, 470, 478, 594
depletion layer modulation, 277, 288
depletion-type transistor, 431, 550
dielectric constant, 68, 77, etc.
dielectric loading, 336, 343
dielectric relaxation, 193, 306, 317, 327, 333
dielectric relaxation time, 185, 324, 470, 471
differential mobility, 62, 66, 68, 220, 314, etc.
differential relaxation time, 355, 383
diffraction condition, 28, 29
diffuse scattering, 500
diffusion coefficient (diffusion constant), 75, 312, 313, 318, 356, 358–360, 388
diffusion current, 75, 77–79, 81, 144, 173, 209, 214, 217, 235, 312, 328, 329, 337, 346, 357, 368, 377, 556, 600, etc.
diffusion-induced velocity, 314, 376
diffusion length, 82, 119, 144
diffusion noise, 195, 224, 226, 242, 293
diffusion velocity, 138, 139
dipole layer, 101, 107, 313
dislocations, 55
dispersion diagram, 50
dispersion relation, 315
displaced Maxwellian distribution, 57
displacement current, 184, 215, 264, 582, etc.
display device, 595
distributed amplification, 334
distribution function, 44, 57, 76, 77
domain capacitance, 367, 368, 370, 393
domain charge, 371, 424
domain dissolution, 384
domain excess voltage, 355, 359, 360, 362, 367, 368, 370, 372, 383, 393

Subject Index

domain modes, 387
domain quenching, 369, 370, 371, 388, 391, 392
domain transit-time, 363, 364, 372, 387–390, 405
domain velocity, 356, 359, 360
domain width, 362
donor levels, 37, 38, 40
doping-lenght product, 187, 307, 312, 322, 326, 334, 347, 348, 377, 380
doping notch, 330, 371, 372, 374, 377, 411
doping-thickness product, 337
Doppler shift, 50
double-drift structure (diode), 252, 300
double-injection, 184
drain, 425, etc.
drain conductance, 430 (resistance, 446), etc.
drain noise, 460
drain series resistance, 427, 456
drift angle, 239
drift current, 81, etc.
drift mobility, 54, 55, 64, 75, 77
drift of characteristics, 491
drift region, 247, 249, 250, 259, 269
drift velocity, 46, 49, 56–58, 60, 64, 65, 169, etc.
dyamond-type lattice, 26, 27, 31, 32
dynamic capacitance, 191
dynamic characteristic, 357, 360, 385

Edge effects, 152, 153, 155
effective density of states, 36, 39, 40
effective diffusion velocity, 137, 237
effective gate voltage, 487, 580
effective impurity density, 41
effective ionization rate, 72
effective mass, 42–44, 46, 52, 59, 110, 497, 500
effective mass for thermionic emission, 110
effective recombination velocity, 137
effective Richardson constant, 111, 138, 161, 235
effective (surface) mobility, 501, 502, 554
efficiency, 250, 285 etc.
Einstein relation, 75, 78, 136, 137, 195
electrochemical potential, 35
electrode-limited current, 170, 175
electron affinity, 88, 89, 104, 106, 124
electron density, 35, 43
electron distribution, 35

electronegativity, 90, 91
electron energy, 30, 31, 43, 44
electron irradiation, 147, 148
electron mass, 36, 48
electron momentum, 30, 42, 48, 76
electron-phonon scattering, 53
electron state density, 35, 40, 43
electrostatic screening, 101, 103, 106
electron temperature, 56, 62, 78, 460
electron traps, 104
electron wave, 30, 31
electron wave function, 43
elastic collisions, 45, 52, 77, 500
emission-limited current, 186
energy bands, 29, 31, 32, 43, 50, 106
energy conservation, 52, 78
energy diagram, 87
energy relaxation time, 56–58
energy surfaces, 32, 33
enhancement-type transistor, 431, 549, 574
equal areas rule, 357, 359, 424
equivalent circuit, 191, 366, 367, 448, 452, 587, 582, 584
equivalent noise current, 196
eutectic, 125
excess noise, 163, 164
external electric field, 48
external velocity, 356, 359, 360
extrinsic Debye lenght, 96, 97, 118, 122, 123, 128, 214, 306, 315

Fat zero, 605
feedback capacitance, 448
Fermi-Dirac integral, 39
Fermi-Dirac statistics, 35, 38, 44
Fermi level, 35, 38, 39, 41, 88, 90, 91, 105
feld-effect, 469
field-emission, 113–115, 132, 133
field-induced junction, 154, 476, 532, 539, 573
field penetration, 101, 104, 107
field plate, 153
finite-difference method, 440
flat-band capacitance, 479, 483, 547
flat-band voltage, 153, 228, 229, 232, 486, 488, 489, 491, 495, 504, 529, 537, 560
flat-topped domain, *see* trapezoidal domain

flicker noise, 293
floating-base transistor, 119, 208, 209
forbidden energy gap, 31, 32
Fowler-Nordheim injection, 612
frequency-lenght product, 393
fringing field, 602
functional device, 413, 415, 607

Gate, 425 etc., 469, etc.
gate-cotrolled diode, 471, 472, 532, 534, 537–539
gate-cotrolled Schottky diode, 153, 154, 164
gate-leakage current, 464
Gaussian distribution, 114, 115, 512, 513
Gauss theorem, 473, 477, 488
generation lifetime, 527
generation rate, 80
generation-recombination noise, 163, 293, 459, 463, 587
generation-recombination process, 79, 481
geometrical capacitance, 191, 367, 388
gradual approximation, 429, 432, 436, 441, 447, 466, 533, 551, 557, 574
gridistor, 453, 454
group velocity, 31, 42, 50
growth rate, 336
guard-ring diode, 156, 157, 161, 164
Gunn diode, 354, 365, 366, 387, 391, 403, 416
Gunn effect, 377
Gunn effect digital devices, 416
Gunn oscillations, 363, 378, 382, 384, 389, 447

Hall measurements, 55
Hahn boundary condition, 335
heat capacity, 52, 56
heat dissipation, 299, 300
heat transfer (conduction) 52, 56, 82
"heavy" electrons, 59
heterojunction, 123, 124, 128, 210
heterojunction triode, 200, 201
high-field domain, 312, 354, 355, 361–363, 365, 366, 377, 383
high-frequency negative resistance, 192, 193, 214, 217, 222, 239–241, 264, 321, 325, 330
high-level injection, 150

high-low junction, 122
holes, 43, 44
homojunctions, 122
hopping, 40
hot carriers, 49, 56, 57, 193, 210, 460
hybrid mode, 398, 399
hysteresis, 365, 384, 388

Ideality factor, 141
ionic bond, 27
ionic (oxide) charge, 491, 492
ionic contamination, 491, 493
ion implantation, 455, 458, 492, 493, 504, 511, 513, 516, 547, 609
ionization energy, 37
ionization integral, 72, 251
ionization rate (ionization coefficient), 71–73, 121, 251, 266, 298
image field, 601
image force-lowering, see Schottky effect
imaging device, 595
impact ionization, 71, 80, 403, 412
IMPATT (IMPact Avalanche Transit Time) diode, 247 etc.
impedance, 191, 214, 219, 236, 258, 337
impedance-field method, 198, 227, 346
impact cathode, 310
imperfect cathode boundary conditions, 379
impurity atoms, 37, 54, 55
impurity band, 40
impurity diffusion, 41
impurity ionization, 44
impurity profile, 166
impurity scattering, 53, 55, 63–65, 77
imref, see quasi-Fermi level
incomplete charge transfer, 604
incremental resistance, 91, 221
induced current, 249, 288
inelastic collisions, 53
infinite injection, 185, 189
injecting contacts, 86, 87, 124, 174
injection-limited current, see electrode limited
injection velocity, 218, 219, 222
input impedance, 429
insulated-gate field-effect transistor, 548, 569, 570
insulator, 32, 86, 177, 184

interelectrode gap, 597
interelectronic scattering, 55, 57, 77
interface charges, 104, 469, 470, 486, 487, 489, 490, 494
interface-state admitance, 543
interface state capacitance, 519
interface states, 105, 107, 471, 494, 506, 507, 509, 517—520, 523—525, 547, 585
interfacial layer, 87, 91, 101, 103—106
integral heat sink, 300
intervalley scattering noise, 461, 462
intrinsic carrier concentration, 36, 37, 214
intrinsic channel conductance, 433, 451
intrinsic circuit, 453
intrinsic Fermi level, 36
intrinsic noise, 293
intrinsic response time, 297, 298
intrinsic semiconductor, 38
isotropic scattering, 52, 61
inversion layer, 93, 95, 119, 154, 471, 474, 479, 480, 482, 484, 485, 496, 499, 500, 535, 548

Junction-barrier capacitance, 448
junction-gate field effect transistor, 448, etc.

Langevin equation (method), 194, 195, 198, 294
large-signal behaviour, 227, 241, 270, 276, 277,
lateral (domain) spreading, 411
lateral modes, 336
lattice, 23, 42
lattice constant, 27, 31, 42, 123
lattice defects (imperfections), 90
lattice mobility, 52, 53
lattice scattering, 55
lattice temperature, 56, 62, 63, 77, 79, 194
lattice vibration, 49, 50—52, 56
leakage current, 561, 606
"light" electrons, 59
limited space-charge accumulation (LSA), 354, 394
limit frequency, 450, 452
linear-sweep technique, 530
load line, 362, 364, 370
longitudinal modes, 51
low-field domain, 356, 372, 385
low-frequency excess noise, 196, 460, 585

Majority carrier(s), 133, 156, 481, 482
Maxwell continuity equation, 189
Maxwell equations, 81
Maxwellian distribution, 35, 56, 57, 59, 63, 64, 77, 108, 109, 398
mean free path, 48, 49, 110, 136, 500
mean free time, 47, 49
mean-square current, 193
memory applications, 592, 593
memory transistors, 593, 612
mesa structure, 160, 217, 243, 299, 543
metal-insulator contact, 178
metal overlap, 153, 164
metal-semiconductor contact, 86, 87, 92, 101, etc.
Miller indices, 24, 25
minority carrier injection, 122, 143, 146, 150, 157, 177, 275
minority carrier injection ratio, 145, 146, 149—151
minority carrier lifetime, 80, 151, 159, 471, 521, 527, 529, 532, 547
minority carriers, 80, 86, 234, 505—507, 548, 612
minority carrier storage, 275, 276, 282
mobility, 43, 47, 49, 52
mobility field-dependence, 186—188, 222, 436—439, 442, 445, 452, 466
mobility temperature-dependence, 52, 53
mode-chart, 402, 403
modulation effect, 146, 159
Monte-Carlo technique (method), 59, 63, 65—67, 447
monostable behaviour, 363, 366
momentum conservation, 52
momentum relaxation time, 45, 56, 57
MOS (metal-oxide-semiconductor) transistor, 549 etc.
multichannel FET, 453
multiple domain mode, 398
multiple domains, 370, 372
multiplication ratio, 72

Negative capacitance, 449
negative differential conductivity, 305
negative resistance, 62, 258, 305, 311—315, 322, 447

negative mobility, 62, 63, 65, 66, 68–70, 315, 355, 380, 388
neutral characteristic, 309, 352, 379, 380, 382
neutrality condition, 38
neutral impurities, 53
noise, 162, 193, 224, 242, 293, 413, 459, 585, 605
noise figure, 347, 461
noise measure, 225, 226, 243, 295, 297, 347
noise temperature, 162, 198
non-degenerate distribution, 35, 41, 56, 93, 117, 135
non-degenerate semiconductor, 39
non-ohmic contact, 86, 337, 339, 348, 352, 379, 383, 413
Nyquist formula, 162, 193, 198
Nyquist-type stability analysis, 303, 325, 326

Ohmic conduction, 180, 184, 187, 213, 312
ohmic contacts, 86, 114, 124, 125, 148, 156, 170, 307, 312, 320, 340, 343, 352, 354, 363, 377, 387, 417, 456
ohmic drop, 364, 383
one-sided junction, 118, 119
open-circuit noise, 194
optical branch, 50, 51
optical modes, 51, 53, 60
optical permittivity, 99, 102
optical scattering, 59, 63, 77
optimum frequency, 278
output characteristics, 426
output conductance (in saturation), 563, 564
outside field, 358, 360, 361
outside velocity, see external velocity
oxide breakdown, 573, 591
oxide capacitance, 477, 504, 527
oxide charge, 469, 470, 477, 486, 488, 489, 491, 493–495, 561
oxide passivation, 155

Particle conservation equation, 45
passivation, 490
permittivity, 63, etc.
phase shift (delay, lag), 248–250, 271, 272, 275, 279
phase velocity, 303, 317, 320, 326
phasor-doagram, 238

phonon(s), 48–53
phonon scattering, 52
phototransistor (heterojunction transistor), 124
pinched-off channel, 427, 438, 441, 445, 549, 557, 563, 569, 573
pinch-holes, 491, 552
pinch-off voltage, see threshold voltage (FET)
planar structure, 152, 160, 161, 299, 343, 409–411, 417, 418, 469, 490
plane wave, 30, 336
plated heat sink, see integral heat sink
Poisson equation, 77, 95, 96, 117, 171, etc.
polar scattering, 50, 63, 65
polysilicon gate, 608
Poole-Frenkel current (injection), 612
potential barrier, 87, 88, 108, 208, 604
potential energy, 30
potential probing, 379
potential well, 594, 596, etc.
power-impedance product, 394
primitive cell, 23, 25, 27, 50, 51
primitive vector, 23
pulse regeneration, 364
punched-through structure (diode), 119, 193, 208, 209, 211, 216, 221, 222, 226, 228, 236, 246, 248, 252, 278, 287, 299, 427, 566, 569
punch-through voltage, 208, 210, 214, 217, 570

Quantum state, 30
Quasi-Fermi level, 109, 135, 138, 141, 144, 149, 178, 179, 218, 482, 518, 532
quasi-Maxwellian distribution function, 57
quasi-stationary capacitance, 191, 448, 578
quasi-stationary characteristic, 365, 366
quasi-thermal equilibrium, 178
quenched-domain mode, 391, 392, 405

Radiation effects, 429, 455, 463, 464, 492, 493, 595, 614
ramp-response method (Bulucea), 531, 532, 538, 539
reach-through voltage, 208, 228, 229, 235
read-only memory, 593
Read-type structure (diode), 248, 250–252, 259, 263, 270, 279, 296, 303
reciprocal lattice, 27–29, 31, 50

recombination, 79, 80
recombination centers, 148, 537
recombination lifetime, 525
recombination velocity, 148
rectyfing contact, 157, 173
reflection amplifiers, 334, 345
relaxation frequency, 335
relaxation time, 45, 46, 48, 52, 56, 77
"reservoair-type" contact, 112, 170, 189
resistivity, 54, 55
resonance frequency, 261, 294
resonant transit-time mode, 389
reserse-travelling domain, 375
Richardson constant, 110, 111, 137, 168, 232

Saturated contact, 112
saturated domain, 363, 365, 384, 388, 413
saturated velocity, 61, 171, 207, 208, 221, 225, 236, 239, 240, 249, 259, 298, 438, 553
saturation current, 115, 135, 138, 139, 153, 170, 175, 232, 274, 293
saturation drain current, 434
saturation region, *see* current saturation
Schottky-barrier, 88, 91, 93, 99, 106−109, 111, 128, 152, 232, 300, 340, 341, 425, 507
Schottky-barrier gate, 453, 455
Schottky contact, 108, 109, 208, 228, 282, 417
Schottky diode, 108, 125, 131
Schottky effect, 98−104, 106, 107, 138, 153, 164, 232, 233, 239
Schottky-gate, 418, 419
Schrödinger equation, 30, 33, 497, 499
screening length, 104
self-induced field, 600
semiconductor, 32, 86, 177, etc.
semiconductor capacitance, 477, 479, 517
semiconductor charge (per unit surface), 471, 473, 476, 534
series resistance effect, 141, 158, 168, 285−187, 300, 425, 453
shallow levels (traps) 166, 179, 181
shift register, 608, 610, 611
Schockley-Read-Hall model, 280, 463, 527, 586
Schockley's positive conductance theorem, 311, 312
short circuit (in)stability, 322, 326, 327, 337

short circuit noise, 193
shot noise, 162, 163, 193, 198, 199, 225, 293, 384, 459, 460, 462, 605
slanted cathode device, 413, 414
silicide, 107, 125, 143, 157
silicon dioxide, 469, 490, 491
small perturbations, 335
small-signal problem, 189, 190, 260
soft breakdown, 571, 572
soft (reverse) characteristics, 152, 154, 159
solid-solid reaction, 125, 126
sound velocity, 50
source, 425, etc.
source conductivity, 223, 237, 240, 246
source plane, 218, 219
source (series) resistance, 434, 435, 456, 457, 466
space-charge controlled modes, 394
space-charge decay, 397
space-charge injection, 202, 236
space-charge growth, 396, 401
space-charge layer/region, 88, 95, 117, 122, 135, 532
space-charge-limited current, 169, 175, 177, 203, 208−210, 232, 246, 267, 269, 570
space-charge resistance, 220, 259
space-charge varactor, 202
space-charge waves, 314−316, 318, 329, 334, 335, 338, 341, 343, 345, 352
specific contact resistance, 115
spectral intensity, 193
spherical band approximation, 33, 45, 46, 52, 62
square-law characteristic, 171, 188, 189, 192, 196, 201, 207, 210, 312
stable amplifiers, 326, 327
stable domain, 364, etc.
stable space-charge layers, 384, 386
statistical distribution, 114
storage time, 149
strong accumulation, 525
strong inversion, 479, 480, 482, 485, 490, 495, 497, 501, 525, 536, 551, 554
sub-band, 497, 499
subcritical doping, 322, 323, 328−330
substrate effect, 452, 456, 553, 556, 558, 573, 575, 577
supercritically doped devices, 327, 330, 353, 365, 366

suppression of injection noise, 199
surface avalanche breakdown, 532, 538, 542, 571, 572, 595
surface charge, 478
surface-charge transistor, 611
surface conductance, 483, 495, 500, 502
surface conductivity, 469
surface depletion, 536
surface field, 473, 475, 478, 502
surface generation, 528, 530
surface generation velocity, 528, 530
surface inversion, 536
surface layer, 93, 95, 469, 470, 475
surface mobility, 500—503, 553
surface potential, 94, 472, 476, 480, 484, 494, 501, 519, 525, 533, 560
surface quantization, 497
surface states, 88, 90, 91, 104, 108, 493, 522, 526, 539, 560
surface trapping centers (levels), 87

Tamm-Shockley states, 90, 91, 469, 494
tapered device, 415
temperature-dependent ionization rates, 73, 74
temperature-dependent-velocity, 58, 60
tetrode (field effect) 426, 466, 591
thermal diffusion, 79
thermal equilibrium, 56, 57, 78--81, 87, 89, 93, 109, 118, 119, 487, 524, 525
thermal generation, 524, 525, 527
thermal noise, 163, 193—195, 198, 346, 459, 460, 585, 606
thermal velocity, 49, 50, 80, 502
thermal vibrations, 48
thermionic-diffusion theory, 135, 141, 145
thermionic emission, 109—112, 114, 123, 131, 134, 137—139, 145, 209, 231, 232, 239
Thomas-Fermi screening distance, 102
threshold current, 324, 340, 380
threshold field, 62, 65, 68, 310, 338, 339, 359, 362, 363, 372, 379, 384
threshold voltage (Gunn), 326, 352, 364, 365, 380, 381, 388, 389, 404

threshold voltage (FET), 427, 431, 457, 465, 496, 502, 511, 535, 542, 554, 561, 573, 574, 576, 593, 594; *see also* turn-on voltage
time-of-flight technique, 59
transconductance, 432, etc.
transfer characteristic, 426, 427
transfer efficiency, 596, 600
transferred-electron effect, 305, etc.
transit-angle, 191, 192, 214
transit-time, 99, 170, 172, 173, 183, 185, 191, 206, 214, 250, 307, 344, 383, 452, 472, 603
transit-time effects, 188, 192, 193, 222, 223, 247, 266, 334, 346, 363
transit-time frequency, 337, 348, 363, 378, 383, 394, 415
transmission line, 582
transport equations, 81
transversal modes, 51

Valence band, 32, 36—40, 43, 44
varactor (MOS), 543
velocity modulation, 215, 219, 224, 246, 248, 329
velocity saturation, 57, 60, 65, 216, 252, 365, 385, 444, 446, 447
virtual cathode (virtual source) 174, 175, 177, 218, 222, 237, 306

Wave function, 30, 107
wave number, 261, 315—317, 335
wave vector, 30, 44
weak inversion, 476, 479, 495, 504, 557, 558, 561
white noise component, 197, 462
Wigner-Seitz cell, 23, 28—30
work-function difference, 87, 88, 469, 477, 485, 487, 489

X-ray diffraction, 29

Zener breakdown, 121
Zerbst plot, 529, 530
zincblende-type lattice structure, 26, 27, 31, 32

Physical Properties of Ge, Si and GaAs at 300 K

Property	Ge	Si	GaAs
Atomic weight	72.6	28.09	144.63
Atoms (cm^{-3})	4.42×10^{22}	5.00×10^{22}	2.21×10^{22}
Lattice	Diamond	Diamond	Zincblende
Lattice constant (Å)	5.66	5.43	5.65
Density (gcm^{-3})	5.32	2.33	5.32
Band gap (eV)	0.67	1.12 ($=8$ for SiO$_2$)	1.43
Electron affinity (eV)	4	4.05	4.07
Effective density of states in conduction band, N_c(cm^{-3}) and valence band, N_v(cm^{-3})	1.04×10^{19} 6.0×10^{18}	2.8×10^{19} 1.04×10^{19}	4.7×10^{17} 7.0×10^{18}
Intrinsic concentration (cm^{-3})	2.33×10^{13}	1.6×10^{10}	1.3×10^{6}
Intrinsic mobility, electrons, holes μ_n, μ_p (cm^2 V^{-1} s^{-1})	3900 1900	1500 (1350) 600 (480)	8500 (8400) 400 (250)
Electron effective mass, m^*/m_0 (longitudinal; transversal)	1.58; 0.082	0.98; 0.19	0.068
Hole effective mass m^*/m_0 (light; heavy)	0.04; 03	0.16; 0.5	0.12; 0.15
Dielectric constant $\varepsilon/\varepsilon_0$ (relative permittivity)	16	11.8 (3.9 for SiO$_2$)	10.9 (12)
Breakdown field (V/μm)	$\simeq 10$	$\simeq 30$ ($\simeq 600$ for SiO$_2$)	~ 40
Melting point (°C)	937	1420 (~ 1700 for SiO$_2$)	1235
Thermal conductivity (W cm^{-1} °C^{-1})	0.64 (0.6)	1.45 (1.5) (0.014 for SiO$_2$)	0.35
Linear dilatation coefficient $\Delta L/L \Delta T$ (°C^{-1})	5.8×10^{-6}	2.6×10^{-6} (0.5×10^{-6} for SiO$_2$)	5.9×10^{-6}

Note: The above data are reproduced from R. Paul 'Halbeiterphysik', VEB Verlag Technik, Berlin, 1974; and A. S. Grove 'Physics and technology of semiconductor devices', J. Wiley, New York, 1967.